Fluorescent and Luminescent Probes for Biological Activity

BIOLOGICAL TECHNIQUES

A series of Practical Guides to New Methods in Modern Biology

Series Editor

DAVID B SATTELLE

Computer Analysis of Electrophysiological Signals
J Dempster (published November 1992)
Planar Lipid Bilayers
W Hanke and W-R Schlue (forthcoming)

CLASSIC TITLES IN THE SERIES

Microelectrode Methods for Intracellular Recording and Ionophoresis
RD Purves
Immunochemical Methods in Cell and Molecular Biology
RJ Mayer and JH Walker

BIOLOGICAL TECHNIQUES

Fluorescent and Luminescent Probes for Biological Activity

A Practical Guide to Technology for Quantitative Real-Time Analysis

Edited by

W.T. MASON

Department of Neurobiology
AFRC Institute of Animal Physiology and Genetics Research
Cambridge, UK

ACADEMIC PRESS
Harcourt Brace & Company, Publishers
London · San Diego · New York
Boston · Sydney · Tokyo · Toronto

ACADEMIC PRESS LIMITED
24–28 Oval Road
London NW1 7DX

United States Edition published by
ACADEMIC PRESS INC.
San Diego, CA 92101

This book is printed on acid-free paper

A catalogue record for this book is available from the British Library

ISBN 0–12–477829–1 (hardback)
ISBN 0–12–477830–5 (bench-top edition)

Typeset by J&L Composition Ltd, Filey, North Yorkshire
Printed and bound in Great Britain at The Bath Press, Avon

Series Preface

The rate at which a particular aspect of modern biology is advancing can be gauged, to a large extent, by the range of techniques that can be applied successfully to its central questions. When a novel technique first emerges, it is only accessible to those involved in its development. As the new method starts to become more widely appreciated, and therefore adopted by scientists with a diversity of backgrounds, there is a demand for a clear, concise, authoritative volume to disseminate the essential practical details.

Biological Techniques is a series of volumes aimed at introducing to a wide audience the latest advances in methodology. The pitfalls and problems of new techniques are given due consideration, as are those small but vital details that are not always explicit in the methods sections of journal papers. The books will be of value to advanced researchers and graduate students seeking to learn and apply new techniques, and will be useful to teachers of advanced undergraduate courses, especially those involving practical and/or project work.

When the series first began under the editorship of Dr John E Treherne and Dr Philip H Rubery, many of the titles were in fields such as physiological monitoring, immunology, biochemistry and ecology. In recent years, most biological laboratories have been invaded by computers and a wealth of new DNA technology. This is reflected in the titles that will appear as the series is relaunched, with volumes covering topics such as computer analysis of electrophysiological signals, planar lipid bilayers, optical probes in cell and molecular biology, gene expression, and *in situ* hybridization. Titles will nevertheless continue to appear in more established fields as technical developments are made.

As leading authorities in their chosen field, authors are often surprised on being approached to write about topics that to them are second nature. It is fortunate for the rest of us that they have been persuaded to do so. I am pleased to have this opportunity to thank all authors in the series for their contributions and their excellent co-operation.

DAVID B. SATTELLE SCD

Preface

The last decade has witnessed an extraordinary series of technical developments in chemistry, biology, computing, physical optics and microelectronics. The net result has been a range of important advances in the biological sciences which cut across virtually all disciplines, from basic research in cellular, neuro- and molecular biology through to the clinical sciences. Inevitably, these advances have also impacted on research and development in the pharmaceutical and agrochemical industries around the world.

At the heart of this revolution has been the development of optical probes for biological activity: probes able to be introduced into the living cell from whence they can act as reporters for cellular function. The sophistication of probe development has progressed to a level where thousands of such light-emitting molecules – fluorescent, bioluminescent and chemiluminescent – exist which can be accurately targeted to membranes, compartments and second messengers. These probes have in many cases displaced radioisotopes as standard research tools and provided outstanding improvements in speed and ease of detection.

Advances in computing, physical optics and microelectronics have been equally important because they have facilitated data capture and analysis under conditions where cell damage is minimized. The low light levels emitted by even the best optical probes are often much fainter than the human eye or fast photographic film can detect, yet as scientists we need to have information on the very fast, sub-second time scales on which most biological activity occurs. In this area, significant technical progress has permitted fast real-time imaging, rapid cell sorting and detection using flow cytometry and rapid three-dimensional image capture and reconstruction.

Such remarkable technical developments as real-time fluorescence ratio imaging and confocal microscopy have brought a new sophistication and power to the conventional light microscope, transforming it into a dynamic and essential tool able to provide a picture of life inside the single cell. However, it is important to realize that such developments would probably not have taken place were it not for the inspired activities of our scientific colleagues who have developed the optical probes and applied them to living cells.

This book in the Biological Techniques series has attempted to bring together some of the most important developments in this field under the cover of a single volume. It is intended to provide both new and existing scientists with a head-start into the practice of using optical probe technology, the problems, pitfalls and limitations likely to be encountered, the application to practical biological questions and the likely areas for future development. In line with this, the concept of the book is to focus on the practical hardware requirements and underlying principles of operation, the strategies leading to development of optical probes and finally their application to living material. Although this volume does not pretend to be exhaustive, it has attempted to highlight key areas of present and future development which have influenced and will continue to influence and enable scientific advance.

Perhaps the most important aspect of the science and technology discussed in this volume is that compared to even five years ago, the use and application of optical probe technology is now widely accessible to scientists around the world. In real terms, equipment costs have fallen while both ease-of-use and the range of important applications literature have increased. These factors are leading to an enormous growth in the scientific literature using optical reporters as routine tools.

It is without question that the next decade is likely to see major and continuing advances in many of the areas covered in this book, and we hope to be able to publish revised and expanded versions at regular intervals to enable the content herein to remain consistent with the dynamic state-of-the-art.

Finally, I must express my indebtedness to all of the authors found within this volume who have taken up much valuable time which they would probably have preferred to spend on original research, in order to develop the outstanding contributions you will find here. Their efforts, I hope you will agree, have produced a very special insight into an exciting and rapidly changing new dimension of the biological sciences, one which is gaining momentum all the time and challenging us to keep pace.

DR W.T. MASON

Contributors

S.R. Adams *Howard Hughes Medical Institute 0647, University of California San Diego, La Jolla, California 92093–0647, USA*

R. Aikens *Photometrics, 3440 East Britannia Drive, Ruscon, Arizona 85706, USA*

B.J. Bacskai *Howard Hughes Medical Institute 0647, University of California San Diego, La Jolla, California 92093–0647, USA*

G.R. Bright *Dept of Physiology & Biophysics, School of Medicine, Case Western Reserve University, Cleveland, Ohio 44106, USA*

A.K. Campbell *Dept of Medical Biochemistry, University of Wales College of Medicine, Heath Park, Cardiff CF4 4XN, UK*

M. Carew *Dept of Neurobiology, AFRC Institute of Animal Physiology, Babraham, Cambridge CB2 4AT, UK*

B. Chance *Joseph Research Foundation, Dept of Biochemistry & Biophysics, School of Medicine, University of Pennsylvania, Philadelphia, Pennsylvania 19104, USA*

L.B. Chen *Dana-Farber Cancer Institute, Harvard Medical School, 44 Binney Street, Boston, Massachusetts 02115, USA*

S.H. Cody *Dept of Physiology, University of Melbourne, Parkville, Victoria, Australia 3083*

L. Cohen *Dept of Cellular & Molecular Physiology, School of Medicine, 333 Cedar Street, PO Box 3333, Yale University, New Haven, Connecticut 06510–8026, USA*

I. Davison *Dept of Neurobiology, AFRC Institute of Animal Physiology, Babraham, Cambridge CB2 4AT, UK*

R. DeBasio *Center for Light Microscope Imaging & Biotechnology, Carnegie Mellon University, 4400 Fifth Avenue, Pittsburgh, Pennsylvania 15213, USA*

A.W. de Feijter *Meridian Instruments, Inc., 2310 Science Parkway, Okemos, Michigan 48864, USA*

J. Dempster *Dept of Physiology & Pharmacology, University of Strathclyde, Glasgow G1 1XW, UK*

P.N. Dubbin *Dept of Physiology, University of Melbourne, Parkville, Victoria, Australia 3083*

K. Florine-Casteel *Duke University Medical Center, Box 3712, M310 Davison Building, Durham, North Carolina 27710, USA*

M. Fricker *Dept of Plant Sciences, University of Oxford, South Parks Road, Oxford OX1 3RB, UK*

S. Gilroy *Department of Plant Biology, University of California, Berkeley, CA94720, USA*

J. Greve *Dept of Applied Physics, University of Twente, PO Box 217, 7500 AE Enschede, The Netherlands*

A. Gurney *Dept of Pharmacology, St Thomas's Hospital, Lambeth Palace Road, London SE1 7EH, UK*

K. Hahn *Department of Neuropharmacology, Scripps Research Institute, 10666 North Torrey Pines Road, La Jolla, CA 92037, USA*

R. Haugland *Molecular Probes, PO Box 22010, 4849 Pitchford Avenue, Eugene, Oregon, USA*

B. Herman *Laboratories for Cell Biology, Dept of Cell Biology & Anatomy, University of North Carolina School of Medicine, CB No. 7090, 108 Taylor Hall, Chapel Hill, North Carolina 27599–7090, USA*

H. Herweijer *Hoofdgroep, Gezondheidsonderzoek TNO, Instituut voor Toegepaste, Radiobiologie en Immunologie, Postbus 5815, 2260 HV Rijswijk, Lange Klelweg 151, Rijswijk, The Netherlands*

D. Hoekstra *Laboratory of Physiological Chemistry, University of Groningen, Bloemsingel 10, 9712 KZ, Groningen, The Netherlands*

M. Horton *Haemopoiesis Research Group, ICRF, St Bartholomew's Hospital, London, UK*

J. Hoyland *Dept of Neurobiology, AFRC Institute of Animal Physiology & Genetics Research, Babraham, Cambridge CB2 4AT, UK*

F.H. Kasten *Dept of Anatomy, Louisiana State University Medical Center, New Orleans, Louisiana 70119, USA*

J. Keij *Hoofdgroep, Gezondheidsonderzoek TNO, Instituut voor Toegepaste, Radiobiologie en Immunologie, Postbus 5815, 2260 HV Rijswijk, Lange Klelweg 151, Rijswijk, The Netherlands*

J.W. Kok *Laboratory of Physiological Chemistry, University of Groningen, Bloemsingel 10, 9712 KZ, Groningen, The Netherlands*

J. Kolega *Center for Light Microscope Imaging & Biotechnology, Carnegie Mellon University, 4400 Fifth Avenue, Pittsburgh, Pennsylvania 15213, USA*

I. Kurtz *Division of Nephrology, Dept of Medicine, UCLA School of Medicine, University of California at Los Angeles, Los Angeles, California 90024, USA*

G. Law *Dept of Neurobiology, AFRC Institute of Animal Physiology & Genetics Research, Babraham, Cambridge CB2 4AT, UK*

J.J. Lemasters *Laboratories for Cell Biology, Dept of Cell Biology & Anatomy, University of North Carolina School of Medicine, CB No 7090, 108 Taylor Hall, Chapel Hill, North Carolina 27599–7090, USA*

P.M. Lledo *Institut Alfred Fessard, CNRS, Gif-sur-Yvette, France*

L. Loew *Dept of Physiology, University of Connecticut Health Center, Farmington, Connecticut 06030, USA*

A. Lyons *Photonic Sciences, Old Station House, Robertsbridge, East Sussex, UK*

W.T. Mason *Dept of Neurobiology, AFRC Institute of Animal Physiology & Genetics Research, Babraham, Cambridge CB2 4AT, UK*

B.R. Masters *Dept of Anatomy and Cell Biology, USUHS, 4301 Jones Bridge Road, Bethesda, Maryland 20814–4799, USA*

T.J. Mitchison *Dept of Pharmacology, University of California, San Francisco, California 94143, USA*

J. Montibeller *Center for Light Microscope Imaging & Biotechnology, Carnegie Mellon University, 4400 Fifth Avenue, Pittsburgh, Pennsylvania 15213, USA*

J. Myers *Center for Light Microscope Imaging & Biotechnology, Carnegie Mellon University, 4400 Fifth Avenue, Pittsburgh, Pennsylvania 15213, USA*

W. O'Brien *Dept of Neurobiology, AFRC Institute of Animal Physiology & Genetics Research, Babraham, Cambridge CB2 4AT, UK*

C. Otto *Dept of Applied Physics, University of Twente, PO Box 217, 7500 AE Enschede, The Netherlands*

J.S. Ploem *Medical Faculty, University of Leiden, Wassenaarseweg 72, 2333 AL Leiden, The Netherlands*

P. Post *Center for Light Microscope Imaging & Biotechnology, Carnegie Mellon University, 4400 Fifth Avenue, Pittsburgh, Pennsylvania 15213, USA*

G.J. Puppels *Dept of Applied Physics, University of Twente, PO Box 217, 7500 AE Enschede, The Netherlands*

G.T. Relf *Regional Medical Physics Dept, Freeman Hospital, Newcastle upon Tyne NE7 7DN, UK*

G. Sala-Newby *Dept of Medical Biochemistry, University of Wales College of Medicine, Heath Park, Cardiff CF4 4XN, UK*

K. Sawin *Dept of Biochemistry and Biophysics, University of California, San Francisco, California 94143, USA*

G. Shankar *Haemopoiesis Research Group, ICRF, St Bartholomew's Hospital, London, UK*

C.J.R. Sheppard *Professor of Physical Optics, Dept of Physical Optics, University of Sydney, Sydney, New South Wales 2006, Australia*

E.R. Simons *Dept of Biochemistry, Boston University School of Medicine, 80 East Concord Street, Boston, Massachusetts 02118–2394, USA*

S.T. Smiley *Dana-Farber Cancer Institute, Harvard Medical School, 44 Binney Street, Boston, Massachusetts 02115, USA*

B. Somasundaram *Dept of Neurobiology, AFRC Institute of Animal Physiology, Babraham, Cambridge CB2 4AT, UK*

D.L. Taylor *Center for Light Microscope Imaging & Biotechnology, Carnegie Mellon University, 4400 Fifth Avenue, Pittsburgh, Pennsylvania 15213, USA*

S.S. Taylor *Department of Chemistry 0654, University of California San Diego, La Jolla, California 92093–0654, USA*

M. Terasaki *Building 36, Room 2A–29, National Institutes of Health, 9000 Rockville Pike, Bethesda, Maryland 20892, USA*

M. Tester *Dept of Botany, University of Adelaide, GPO Box 498, South Australia 5001*

J.A. Theriot *Dept of Biochemistry and Biophysics, University of California, San Francisco, California 94143, USA*

P. Tomkins *Photonic Sciences, Old Station House, Robertsbridge, East Sussex, UK*

R. Tregear *Dept of Neurobiology, AFRC Institute of Animal Physiology, Babraham, Cambridge CB2 4AT, UK*

J. Trosko *Dept of Pediatrics & Human Development, B240 Life Sciences, Michigan State University, East Lansing, Michigan 48824–1317*

R.Y. Tsien *Howard Hughes Medical Institute 0647, University of California San Diego, La Jolla, California 92093–0647, USA*

M. van Rooijen *Dept of Applied Physics, University of Twente, PO Box 217, 7500 AE Enschede, The Netherlands*

M. Wade *Meridian Instruments, Inc., 2310 Science Parkway, Okemos, Michigan 48864, USA*

D.A. Williams *Dept of Physiology, University of Melbourne, Parkville, Victoria, Australia 3083*

J-Y. Wu *Dept of Cellular & Molecular Physiology, School of Medicine, 333 Cedar Street, PO Box 3333, Yale University, New Haven, Connecticut 06510–8026, USA*

A. Zelenin *Deputy Director, Engelhardt Institute of Molecular Biology, Russian Academy of Sciences, Vavilov Street, Moscow 117984, Russia*

R. Zorec *Institute of Pathophysiology, University of Ljubljana, Ljubljana, Slovenia*

Contents

CHAPTER THIRTY-TWO
Video Imaging of Lipid Order   420
K. Florine-Casteel, J.L. Lemasters & B. Herman

Colour plates appear between pages 170 and 171

CHAPTER ONE

Fluorescence Microscopy

JOHAN S. PLOEM

Medical Faculty, University of Leiden, The Netherlands

1.1 INTRODUCTION

1.1.1 Applications of fluorescence microscopy

As a tool in microscopy, fluorescence provides a number of possibilities in addition to absorption methods. Fluorescence probes can, for instance, be selectively excited and detected in a complex mixture of molecular species. It is also possible to observe a very small number of fluorescent molecules – approximately 50 molecules can be detected in 1 μm^3 volume of a cell (Lansing Taylor et al., 1986). Furthermore, fluorescence microscopy offers excellent temporal resolution, since events that occur at a rate slower than about 10^{-8} s can be detected and measured with appropriate instrumentation. When confocal laser scanning is used in fluorescence microscopy, the theoretical limits of the spatial resolution (determined by the numerical aperture of the objective and the wavelength of the emitted fluorescence light) can be obtained in practice. In conventional microscopy, this is very difficult to obtain.

Immunofluorescence microscopy has been the most common application of fluorescence microscopy in cell biology (Coons et al., 1941). The possibility of detecting multiple regions, represented by specific antigens in the same cell, by selective binding of antibodies marked with fluorophores with different fluorescence colours is often used nowadays in in situ hybridization studies of, for example, DNA sequences in the interphase nucleus (Nederlof et al., 1990).

Fluorescence microscopy is also often used for the study of living cells (Kohen & Hirschberg, 1989). It is possible to measure, for example, the pH, free calcium and NAD(P)H concentration in the cytoplasm, as well as intercellular communications between cells. Flow cytometry as a specialized form of fluorescence microscopy (Melamed et al., 1990) permits the examination of biological surfaces when cells pass a beam of excitation light from a laser. A large number of cells can be analysed in a relatively short period of time by using several fluorescent probes in this technology.

1.1.2 The nature of fluorescence

Hot bodies that are self-luminous solely because of their high temperature are said to emit incandescence. All other forms of light emission are called luminescence. A system emitting luminescence is losing energy. Consequently, some form of energy must be applied from elsewhere and most kinds of luminescence are classified according to the source of this energy. One speaks, therefore, of electro-luminescence, radioluminescence, chemiluminescence, bioluminescence and photoluminescence. In the latter form of luminescence the energy is provided by the absorption of ultraviolet, visible or infrared light. Fluorescence is a type of luminescence in which light is emitted from molecules for a very short period of time, following the absorption of light. The emitted light is termed fluorescence if the delay between absorption and emission of photons is of the order of 10^{-8} s or less. Delayed fluorescence is the term used if the delay is about 10^{-6} s, while a delay of greater than about 10^{-6} s results in phosphorescence. All these phenomena can be seen in microscopy.

1.1.3 Fluorescent stains

Compounds exhibiting fluorescence are called fluorophores or fluorochromes. When a fluorophore absorbs light, energy is taken up for the excitation of electrons to higher energy states. The process of absorption is rapid and is immediately followed by a return to lower energy states, which can be accom-panied by emission of light. The spectral characte-ristics of a fluorochrome are related to the special electronic configurations of a molecule. Absorption and emission of light take place at different regions of the light spectrum (Fig. 1.1). According to Stokes's law the wavelength of emission is almost always longer than the wavelength of excitation. It is this shift in wavelength that makes the observation of the emitted light in a fluorescence microscope possible. The excitation light of shorter wavelengths is prevented from entering the eyepieces by using the appropriate dichromatic (dichroic) dividing mirrors (Ploem, 1967). It should be noted that the intensity of the emitted light is weaker than that of the excitation light, as the emitted energy is much smaller than the energy needed for excitation. For different fluorochromes this may vary and is known as the quantum efficiency of the fluorophore used.

Different fluorochromes are characterized by their absorption and emission spectra. The absorption or excitation spectrum is obtained by recording the relative fluorescence intensity at a certain wavelength when the specimen is excited with varying wave-lengths. The most intense fluorescence occurs when the specimen is irradiated with wavelengths close to the peak of the excitation curve. An example of an absorption and an emission spectrum is given in Fig. 1.1. Most excitation and emission curves overlap to a certain extent.

Decrease in fluorescence during irradiation with light is called fading. The degree of fading depends on the intensity of the excitation light, the degree of absorption by the fluorophore of the exciting light and the exposure time (Patzelt, 1972). Reduction in fluorescence intensity can also be due to modification in the excited states of the fluorophore. These physicochemical changes may be caused by the presence of other fluorophores, oxidizing agents, or salts of heavy metals. This phenomenon is called quenching. Prior to microscopy a decrease in the potential to fluoresce can also occur. Preparations are therefore best stored in the dark at 4°C. To reduce fading during microscopy, agents such as DABCO (1,4-diazobicyclo-2,2,2-octane), N-propylgallate and p-phenylenediamine should be added to the mounting medium (Gilot & Sedat, 1982; Johnson & Nogueira Araujo, 1981).

1.1.4 Specialized literature on fluorescence microscopy

A number of books have been published recently on (quantitative) fluorescence microscopy and its applications. A few interesting examples are the books by Rost (1991), Kohen and Hirschberg (1989), and Lansing Taylor et al. (1986). Also, specialized techniques of fluorescence microscopy such as laser scanning fluorescence microscopy and confocal laser scanning microscopy have found wide applications, and consequently have been included in most recent books dealing with microscopy.

1.2 MICROSCOPE DESIGN

A fluorescence microscope is designed to provide an optimal collection of the fluorescence signal from the specimen, while minimizing the background illumina-tion consisting of unwanted excitation light and autofluorescence. This requires rather sophisticated technology, since the specific fluorescence from the specimen can be several orders of magnitude weaker than the intensity of the exciting light. In the first place the fluorophore in the preparation must be excited with wavelengths as close as possible to the absorption peak of the fluorophore, assuming that the light source emits sufficiently in this wavelength region (Ploem,

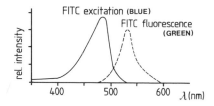

Figure 1.1 Excitation (absorption) and fluorescence (emission) spectra of fluorescein isothiocyanate (FITC).

1967). Secondly, the fluorescence emission collected by the optical system of the microscope must be maximized.

Strong excitation of the fluorophore with relatively efficient collection of the fluorescence is often not a good solution, since intense illumination may cause excessive fading of the fluorophore. Also, exciting light which does not correspond well with the excitation peak of the fluorochrome will often cause unnecessary autofluorescence of the tissue and optical parts, diminishing the image contrast. This contrast is determined by the ratio of the fluorescence emission of the specifically stained structures to the light observed in the background. For a good separation of exciting and fluorescence light the use of narrow-band excitation filters, which often have a relatively low transmission, is therefore necessary.

For easy visual observation, however, or photography with reasonably short exposure times, a sufficiently bright image is required. To that purpose a compromise between the intensity of the fluorescence and the level of background illumination must sometimes be accepted. If only a few fluorescent molecules are to be observed, not only the non-specific autofluorescence of tissue components, but also the level of autofluorescence of the glass components of the objective, immersion oil and the mounting medium can interfere with the observation of specific fluorescence. Laser scanning microscopy can provide a partial solution for these types of problems, as will be explained later in this chapter.

1.3 TYPES OF ILLUMINATION

A fluorescence microscope is a conventional compound microscope. There are two basic types of illumination for fluorescence microscopy (FM): transmitted illumination (Young, 1961; Nairn, 1976) and incident illumination (Ploem, 1967; Kraft, 1973). The illumination pathway of transmitted light illumination is shown in Fig. 1.2. A condenser focuses the excitation light onto a microscope field. The emitted fluorescence is collected by the objective and observed through the eyepieces. In this configuration it is essential that two different lenses are used: a condenser to focus the excitation light on the specimen and an objective to collect the emitted fluorescence light. For optimal observation of fluorescing images these two lenses, which have independent optical axes, must be perfectly aligned. This is not always easy to obtain and maintain in routine use. It should also be realized that focusing of the excitation rays by the condenser onto the specimen and focusing of the

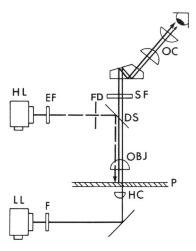

Figure 1.2 Schematic diagram of a microscope for fluorescence microscopy with transmitted ( ) and incident (– – –) illumination. LL/HL = light source; EF/F = excitation filter; FD = field diaphragm; DS = dichroic mirror; SF = barrier filter; HC = condenser; P = preparation; OBJ = objective; OC = ocular.

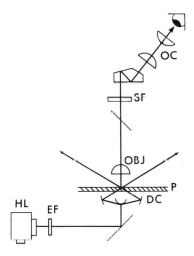

Figure 1.3 Schematic diagram of transmitted illumination with a dark-field condenser (DC). Other abbreviations as in Fig. 1.2.

objective for the observation of fluorescence are two different procedures.

For transmitted light illumination two types of condensers can be used. The excitation pathway either contains a bright-field condenser which allows all the exciting light to enter the objective or a dark-field condenser which illuminates the specimen with an oblique cone of light in such a way that no direct exciting light enters the objective (Fig. 1.3). The latter type of condenser, of course, facilitates separation of fluorescence from excitation light. Due to the fact that high-performance interference filters only became

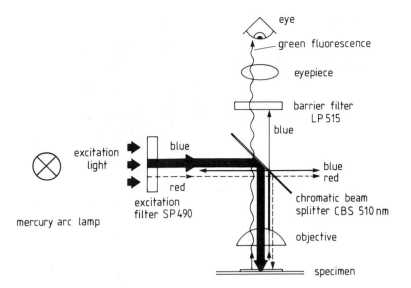

Figure 1.4 Light path in incident illumination. The vertical illuminator equipped with a chromatic beam-splitter has a high reflectance for blue excitation light and a high transmittance for green fluorescence light.

available after 1970, dark-field illumination using coloured glass filters was the best method to remove unwanted excitation light from the observing light path until 1970. In dark-field condensers part of the aperture is obscured to prevent the light from entering the objective, which must be used at a limited numerical aperture (NA) in order to avoid entrance of unwanted excitation light. Often a working aperture of less than about NA=0.7 is used, whereas good-quality objectives might have apertures of NA=1.4.

As mentioned above, originally, coloured glass filters of the Schott UG1, BG12, etc. type were used to select the excitation light. With these filters it was not possible to absorb all the excitation light with a barrier filter when bright-field illumination was used. Hence, the popularity of dark-field illumination, which did not put such high requirements on excitation and fluorescence filters. With modern high-performance interference filters it is now much easier to use full aperture transmitted bright-field illumination. Furthermore, dark-field condensers do not allow a combination of transmitted fluorescence with phase-contrast or differential interference contrast.

Incident or epi-illumination fluorescence microscopy is shown in Figs 1.2, 1.4 and 1.5. To focus the excitation light onto the specimen and to collect the emitted light from the fluorescing specimen only the objective lens is used. The advantage of epi-illumination is that the same lens system acts as objective and condenser. Focusing the objective onto the specimen results in proper alignment of the microscope, with the same alignment for excitation light and the observed fluorescence light. The illuminated field is the field of view.

To direct the excitation light onto the specimen, a

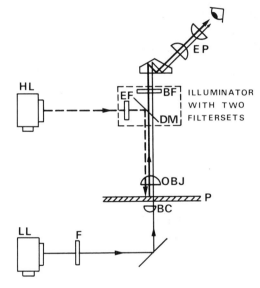

Figure 1.5 Excitation filters (EF), dichroic mirrors (DM) and barrier filters (BF) are mounted in one filter block which may contain up to four of such filter sets for different applications. HL/LL = light source; F = filter; BC = bright-field condenser; P = preparation; OBJ = objective; EP = eye pieces.

special type of mirror – a chromatic beam-splitter (CBS), also known as a dichroic mirror – is positioned above the objective. These mirrors have a special interference coating, which reflects light shorter than a certain wavelength and transmits light of longer wavelengths. Thus these mirrors effectively reflect the shorter wavelengths of the exciting light onto the specimen and transmit the longer wavelengths of the emitted fluorescence towards the eyepieces.

Also in incident illumination, a relatively small

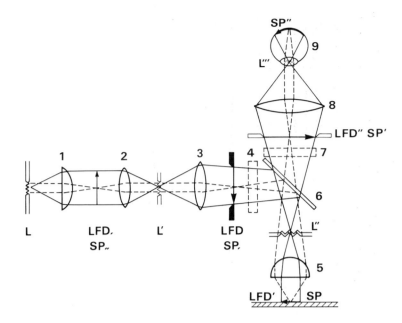

Figure 1.6 Images in Koehler illumination. ——— = imaging light path; – – – – – – = illuminating light path. L = light source; LFD = luminous field diaphragm; SP = specimen; 1 = collector; 2 and 3 = auxiliary lenses; 4 = excitation filter; 5 = objective/condenser in epi-illumination; 6 = chromatic beam-splitter; 7 = barrier filter; 8 = eyepiece; 9 = eye. L', L'', L''', LFD', LFD'', SP' and SP'' are forward images; LFD, , SP, and SP,, are backward images.

amount of exciting light may be reflected by the specimen or optical parts in the direction of the eyepieces. This unwanted excitation light is effectively deflected out of the observation light path by the same chromatic beam-splitter, blocking this light from reaching the eyepieces. In principle, the CBS acts thus as both excitation and barrier filter. In practice, an additional barrier filter is, however, still needed to eliminate any residual unwanted excitation light. Figure 1.4 gives an example of the use of a chromatic beam-splitter. Chromatic beam-splitters exist for separation of all regions of the light spectrum, from the UV (300 nm) to the far red (700 nm). They are mounted in units together with an excitation and barrier filter, especially selected for each separate wavelength range (Fig. 1.5). Units for UV, violet, blue, green and red excitation are provided by several microscope manufacturers.

Epi-illumination often makes use of a vertical illuminator allowing an illuminating light path according to the Koehler principle (Fig. 1.6). The image of the light source is focused onto an iris diaphragm which is conjugate with the entrance pupil or back aperture of the objective lens. This iris diaphragm therefore determines the illuminated aperture. Opening and closing of this aperture diaphragm results in an increase or decrease of the intensity of the illumination, without changing the size of the illuminated field. In addition, a field diaphragm is present which is brought into focus on the specimen plane. This diaphragm controls the size of the illuminated area of the

specimen without affecting the intensity of the illuminated area. Closing the field diaphragm as much as the specimen observed allows, generally increases the image contrast of the specimen due to the decrease in autofluorescence of optical parts and a further elimination of still remaining unwanted excitation light.

Moreover, epi-illumination permits an easy changeover or combination between fluorescence and transillumination microscopy, since the substage illumination remains available. Combinations of fluorescence with, for example, phase-contrast microscopy, differential interference contrast and polarization microscopy make it possible to compare the distribution of fluorescence in a specimen, while these transmitted light contrast methods give insight in the structure of the specimen.

In general, epi-illumination is also used in confocal fluorescence scanning microscopy, and in inverted microscopes used for the study of living cells (Ploem et al., 1978). In the latter instrument the epi-illuminator is mounted underneath the stage supporting the dishes or trays in which cells are grown or collected. Objectives for inverted microscopy should be selected such that they have a sufficient working distance to enable focusing on the cells on the plastic bottom of the tray. A disadvantage is that some plastics show a considerable autofluorescence when excited with short wavelengths. Preferably narrow-band long-wavelength blue or green excitation light should be used in combination with fluorochromes having absorption peaks in this wavelength area.

1.4 LIGHT SOURCES

Four major characteristics of light sources must be considered: (1) the spectral distribution of the emitted wavelengths; (2) the spectral density of the radiance of the arc or filament representing the radiant intensity per unit area; (3) the uniformity of the illumination in the microscope field; and (4) the stability of the light output over time and the spatial stability of the arc in high-pressure lamps.

The choice of light source is determined by the excitation spectrum of the fluorochrome and its quantum efficiency, the number of fluorochrome molecules that one wants to detect and the sensitivity of the detector used: human eye, film, photomultiplier or TV (CCD) camera. Halogen, mercury and xenon high-pressure arc lamps, and various laser light sources are available. Halogen and xenon lamps have more or less continuous emission spectra; mercury arc lamps have strong emission peaks, and laser light sources emit their energy in multiple lines. The choice of the light source depends also on the mode of illumination. With full field illumination halogen or arc lamps are suitable. In scanning illumination, as is mostly used in confocal fluorescence microscopy, multiple small spots in the specimen must be illuminated sequentially, and only an intense small light beam from a laser can provide sufficient photons to allow a relatively fast scanning of a microscope field. Laser light is coherent and can cause interference phenomena in the imaging of the microscope. With non-perfect excitation and barrier filter systems, leaking (unwanted) exciting laser light can cause interference images. Optical systems are therefore adapted to make laser light non-coherent for use in a microscope set-up.

Weak fluorochromes with low quantum efficiency (low Q) or low numbers of fluorochrome molecules require more excitation light for viewing than strong fluorochromes. Often strong light from an entire laser light source is concentrated on a small (0.5–1 μm) spot in the specimen, as in (confocal) laser scanning fluorescence microscopy.

Tungsten halogen (12 V, 50 and 100 W) lamps are suitable and inexpensive light sources for routine investigations, provided that the specimen emits fluorescence of sufficiently high intensity. These lamps can be used for both transmitted and incident light illumination, and can be switched on and off easily and frequently without damage to the lamp (Tomlinson, 1971).

A mercury lamp has peak emissions at, for example, 366, 405, 436, 546 and 578 nm, but also a strong background continuum. In the blue region, for instance, this continuum is still stronger than that given by a tungsten halogen lamp. If UV or violet, or green light is required, the mercury peaks at these wavelength ranges are preferred (Thomson & Hageage, 1975). Mercury lamps are available at 50, 100 and 200 W. It should be noted that the 100 W lamp has a smaller arc than the 50 and 200 W lamps. Ideally, the collecting lens of the lamphouse should provide an image of the arc onto the entrance pupil of the objective used for epi-illumination. It is clear that a collecting lens of fixed focal length cannot project different sizes of arcs in such a way that the entrance pupil of an objective is always homogeneously filled with an image of the arc for Koehler illumination. A zoom collecting lens should be constructed by the optical industry to solve this problem. Inhomogeneous illumination can thus not always be avoided. For homogeneous illumination of an entire microscopic field, which is desirable in fluorescence image analysis, the very large arc of a xenon 450 high-pressure lamp is sometimes used.

The mercury lamps have a limited lifetime (about 200 burning hours). They are mostly operated on AC current supply. The HBO 100 W can be operated on DC supply for increased stability in microfluorometry. The filter sets developed for fluorescence microscopy are mostly chosen in relation to the location of the major mercury emission peaks in the emission spectrum.

Xenon lamps emit a wide and flat spectrum of rather constant energy from UV to red, without strong peaks (Tomlinson, 1971). They are available as 75, 150 and 450 W with lifetimes of 400, 1200 and 2000 burning hours respectively. Xenon lamps should be handled with care, because even cold lamps are under relatively high pressure, and safety eye-glasses should be used during removal and replacement. The lamps are operated on DC current supply. Unfortunately the xenon 450 W DC operated lamp, which has a relatively long lifetime, needs a rather expensive power supply. For their use in microfluorometry the lamps should be burnt in, under conditions of low mechanical vibrations, e.g. during the night and with a voltage stabilizer to overcome large voltage fluctuations. This creates fewer and more stable burning points, resulting in greater stability of the arc.

Laser light sources emit strong lines which provide monochromatic radiation of very high energy. As such they provide, therefore, potential light sources for special purposes fluorescence microscopy applications that need such types of excitation light (Bergquist & Nilsson, 1975; Wick *et al.*, 1975). Lasers can provide continuous output of energy or operate in a pulsed mode. With the use of short pulses of excitation energy (1 μs to 1 ns), delayed fluorescence phenomena can be studied (Jovin & Vaz, 1989; Beverloo *et al.*, 1990). Lasers are also used in fluorescence scanning confocal microscopy (Wilke, 1983; see also Chapter 17).

Table 1.1 Lamps for fluorescence microscopy in order of intrinsic brilliancy of the arc.

Lamp	Mean luminous density (cd cm^{-2})	Wavelength region
Hg 100 W	170 000	Main peaks at
Hg 200 W	33 000	366, 405, 436, 546
Hg 50 W	30 000	and 578 nm
Xe 75 W	40 000	Continuum and
Xe 450 W	35 000	peaks > 800 nm
Xe 150 W	15 000	
Halogen		Little UV & violet emission; higher intensity towards longer wavelengths
Various lasers		Specific lines

Without aiming at confocal microscopy, it is possible to use laser scanning microscopy only for illumination of the field (Ploem, 1987). In this set-up, a vibrating mirror system is used to generate a meander of a few hundred thousand laser illuminated spots (0.5 μm) over the entire microscopic field in less than a second by using epi-illumination fluorescence microscopy and a photomultiplier for the recording of the fluorescence of each single spot. Since the energy of the entire laser output is concentrated on each 0.5 μm spot, an extremely high excitation energy is obtained. Since only a small pencil of light passes the objective lens at any moment for the illumination of one spot, the autofluorescence of glass in the objective contributing to the background light is low. It is especially low in relation to conventional microscopy, where the entire objective is filled with a massive excitation light beam needed to illuminate the entire microscope field simultaneously.

Modern fluorescence microscopy requires a range of light sources to meet the varying demands of the various applications. Very low irradiation may be required in combination with a very sensitive camera system, in order to avoid photo damage; extremely strong laser excitation may be wanted to kill living cells; and the wavelengths of the illumination will vary from deep UV (250 nm) to infrared. Since these types of illumination cannot be provided by a single light source, several lamp housings may be attached to one fluorescence microscope for an easy interchange of illumination.

1.4.1 Lamp housings

Correct alignment of the arc or high-pressure lamps is extremely important for the fluorescence yield. Therefore, the quality of a lamp housing can almost be judged by the stability of a correct alignment of the arc made in the factory, or by the efficiency of user accessible knobs for two directional arc alignment. It should be feasible to obtain a homogeneous illumination of the microscopic field and it should be possible to focus the lamp collector to project an image of the arc on the entrance pupil of the substage condenser with transmitted illumination or on the entrance pupil of the objective in epi-illumination.

Lamp housings usually have filter holders for inserting filters for infrared elimination and colour filters. Heat and infrared filters should be of the reflecting type rather than of the absorbing type, since these crack less frequently. These heat-reflecting filters should always be placed closer to the lamp than the coloured filters to prevent excessive infrared absorption by the latter.

1.5 FILTERS

Filters are very important components in the fluorescence microscope. Filter choice depends on the light source, and on the spectral characteristics, quantum efficiency and distance in wavelength between excitation and emission peak of the fluorochromes used. The main types of filters in fluorescence microscopy are: colour glass filters, interference filters, and a special type of interference filter placed at 45° to the light beam, known as dichroic mirrors or chromatic beam-splitters. (Fig. 1.4). Colour glass filters are mostly made by adding certain oxides of various heavy metals to the glass. Although to the naked eye a colour filter transmits only light from one colour, the transmission curve has in fact a fairly broad base. Thus while there will be a peak transmission of one colour, some light from the neighbouring regions of the spectrum will also be transmitted. The concentration of the added oxides and the thickness of the glass determine how much of the light is absorbed. The remaining light is transmitted. If the absorption extends into the infrared regions of the spectrum, it will cause a considerable heating and may lead to cracking if the filters are used in combination with a powerful light source such as high-pressure arc lamps. For this reason it is desirable to place a heat-reflecting filter between the colour filter and the lamp (e.g. Calflex filter from Schott). Colour filters transmit rather broad wavelength ranges and are therefore known as broad-band filters.

Interference filters consist of many layers of thin film with different refractive indices, sequentially deposited upon a flat glass surface. Interference filters transmit light of well-defined wavelengths resulting from the passage of light through layers of different refractive indices and from reflection by the surface of these layers. As the spectral characteristics of these filters depend on very precise maintenance of the gap between the semitransparent coatings, interference filters are made for very narrow tolerances and are accordingly much more expensive than glass filters. If the filters are tilted along the optical axis the spectral properties will change. Due to the construction of interference filters a shift towards shorter wavelengths occurs when the angle of incidence increases. Sometimes this shift is used in the fine-tuning of a filter to obtain a precise peak wavelength by introducing a small angle of the filter in relation to the optical path of the microscope.

A filter can be described according to its half bandwidth (HB) indicating the transmission width at 50% on either side of the transmission peak. The interference filters are defined into narrow-band and wide-band filters according to the wavelength band they transmit. Some interference filters do not have a symmetrical (bell-shaped) transmission curve but a sharp slope. When such a filter transmits light of longer wavelengths and blocks short wavelengths it is known as a long-pass (LP) filter. A filter which transmits short wavelengths and blocks long wavelengths is defined as a short-pass (SP) filter.

Recently, interference filters with very complex transmission characteristics have been developed for flow cytometry and fluorescence microscopy (Omega Optical Inc., USA) that enable the simultaneous excitation of two or three fluorochromes.

Filters can also be characterized by their position in the microscope (excitation or emission side). Consequently the terminology used by different manufacturers is quite confusing.

1.5.1 Excitation filters

Excitation filters are used to isolate a limited region of the light spectrum in correspondence with the absorption peak of the fluorochrome. In addition, almost all the light in the wavelength range of the fluorescence emission of the fluorochrome must be removed from the illumination light beam, since the barrier filters (used above the objective to block unwanted excitation light that otherwise would reach the eyepieces) are usually not perfect and will still transmit a very small amount of excitation light. Due to the fact that in many applications also very weakly fluorescing objects are to be observed, the amount of unwanted excitation light still passing the barrier filter must be minimized. The problem of eliminating unwanted excitation light from the observed microscope field becomes even more pronounced if the fluorochrome has excitation and fluorescent peaks which are rather close to each

other, like FITC. Filters with a high transmission close to the excitation peak of a fluorochrome and which also strongly block unwanted excitation light in the fluorescence wavelength range of such a fluorochrome are relatively difficult to manufacture and expensive.

The choice of an excitation filter must be made on the image contrast finally required for the intended application. Glass filters like the BG (blue glass) and UG (ultraviolet glass) filters are still in use. They have rather broad transmission characteristics. Interference filters are more selective. A disadvantage of these filters in the past was their low transmission value (30–60%). Modern technology has enabled the development of band-pass filters with high-transmission (90%) narrow-band characteristics and very good suppression of unwanted excitation light in the wavelength range of the expected fluorescence. Short-pass filters (SP) transmit shorter wavelengths and effectively block longer wavelengths (Rygaard & Olson, 1969; Ploem, 1971; Lea & Ward, 1974).

The recently developed filters for the simultaneous excitation of 2 or 3 fluorochromes should be used in combination with corresponding dichroic mirrors and barrier filters to allow observation of 2 or 3 fluorescence colours (Fig. 1.7).

1.5.2 Barrier filters

Barrier filters are used to block the unwanted excitation light in the wavelength range of the fluorescence emission. Mostly colour glasses are used with a high transmission for the longer wavelengths (90% or higher) and a very effective blocking of shorter wavelengths. Colour glass barrier filters absorbing short wavelength excitation light may fluoresce which may lead to a decrease in the image contrast. Barrier filters for some applications requiring an extremely dark background are therefore coated with an interference filter layer that will reflect most excitation light and prevent autofluorescence of the barrier filter.

In some applications not all the fluorescence light longer than a certain wavelength is wanted for observation, but only the fluorescence in a limited wavelength range (e.g. the narrow emission peak of FITC). This is achieved by adding an extra band or a short-pass interference filter to the barrier filter or by coating the colour glass barrier filter with an interference coating, selecting a narrow wavelength band. Such filter combinations can be defined as fluorescence selection filters.

Recently, barrier filters of the interference type have been manufactured which permit the observation of 2 or 3 fluorescence colours simultaneously (Fig. 1.7). Such filters have a complex transmittance curve with several wavelength bands of high transmission for

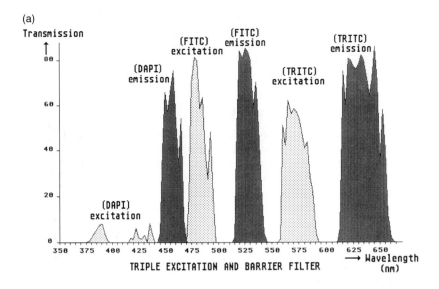

(a)

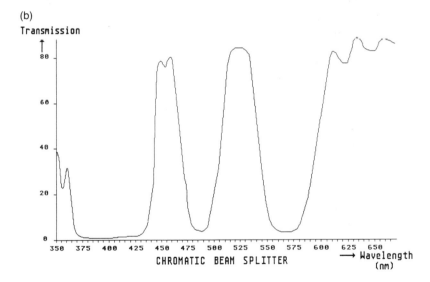

(b)

Figure 1.7 (a) Spectral characteristics of an excitation filter (lightly shaded area) and a barrier filter (darkly shaded area) that enable the simultaneous excitation of three fluorochromes (DAPI, FITC and TRITC) and the observation through the barrier filter of blue, green and red fluorescence (interference filters manufactured by Omega Optical Inc., USA). (b) Spectral characteristics of a chromatic beam-splitter (dichroic mirror) that must be used in combination with the excitation filter in (a) to simultaneously excite three fluorochromes in epi-illumination fluorescence microscopy (dichroic mirror manufactured by Omega Optical Inc., USA). (c) Spectral characteristics of an excitation filter (lightly shaded area), a chromatic beam-splitter (dichroic mirror) and a barrier filter (darkly shaded area) for the excitation of FITC and TRITC (interference filters manufactured by Omega Optical Inc., USA).

(c)

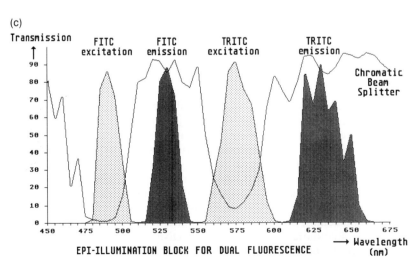

fluorescence and several wavelength regions for strong blocking of unwanted excitation light. Such filters must be used in combination with special excitation interference filters, exactly matching the transmission of the barrier filter.

1.5.3 Chromatic beam-splitters (CBS)

Chromatic beam-splitters (CBS), also known as dichroic mirrors, reflect light of wavelengths shorter than the specified wavelength and transmit light of longer wavelengths. They are placed at an angle of 45° to the optical axis and reflect excitation rays into the objective in epi-illumination, where the objective also serves as a condenser (Figs 1.2, 1.4 and 1.5). Recently, chromatic beam-splitters have been made for simultaneous fluorescence of 2 or 3 fluorochromes (Fig. 1.7) in epi-illumination. They should be used in combination with the appropriate excitation and barrier filters for dual or triple fluorescence excitation and observation.

1.5.4 Multi-wavelength epi-illuminators

Effective epi-illumination can only be achieved by combining a closely matched excitation filter, a chromatic beam-splitter and a barrier filter for each main fluorescence colour. They are usually mounted together in a filter block which can be inserted in an epi-illuminator (Fig. 1.5). Multi-wavelength vertical illuminators are available with sliding or revolving filter blocks permitting epi-illumination in several wavelength bands. Various filter combinations for different wavelengths are given in Table 1.2.

The newly developed combinations of an excitation filter, a chromatic beam-splitter and barrier filter, when mounted in one block, permit the excitation and observation of two fluorochromes with two different fluorescent colours (Fig. 1.7).

1.6 OBJECTIVES AND EYEPIECES

In epi-illumination the microscope objective also serves as a condenser. The obtained result therefore strongly depends on the choice of the objective. Not all objectives are suited for fluorescence microscopy. The glasses used for such objectives must show very little autofluorescence. This is especially important with very weak fluorescence signals. In testing an objective it is necessary to distinguish still remaining, unwanted excitation light, and autofluorescence of the mounting medium, the specimen and immersion oils from the autofluorescence of the objective itself. Objectives for fluorescence microscopy should have a

Table 1.2 Typical combinations of excitation filters, chromatic beam-splitters and barrier fluorescence emission filters (often combined in sets or blocks, which can be inserted in multi-wavelength epi-illuminators). Most filters, except the ones marked, can be obtained from all major fluorescence microscope manufacturers.

Excitation light	Excitation filter (nm)	CBS	Barrier filter	Fluorescence colours
UV (365 nm)	340–380 or 350–410	400 or 410	LP 430	Violet, blue-green, yellow, orange, red and infrared
Violet (405 nm)	350–460 or 420–490	455 or 460	LP 470	Blue, green yellow, orange-red, infrared
Blue (470 nm)	450–490 or 470–490	500 or 510	LP 520	Green, yellow-orange, red and infrared
Green (546 nm)	515–560 or 530–560	580	LP 580	Yellow, orange-red and infrared
Yellow (560 nm)	550–570[a]	595[a]	LP 635[a]	Orange, red and infrared
Orange (590 nm)	580–600[a]	620[a]	LP 660[a]	Red and infrared
Red (630 nm)	610–650[b]	660[b]	LP 670[b]	Infrared
UV, blue and green	Multiple transmittance and reflection bands[a,c]			UV, blue and green

[a] Glen Spectra Limited, UK.
[b] Chroma Technology Group, USA.
[c] Omega Optical Inc, USA.

relatively high numerical aperture in combination with a relatively low magnification. Examples of such objective lenses are oil-immersion objectives 10× with a numerical aperture (NA) of 0.45 and 40× with a NA of 1.30 (Leica, Germany). Also, dry objectives with relatively high NA, considering their magnification, are now manufactured. Water-immersion objectives made for fluorescence microscopy offer the advantage of avoiding the autofluorescence of some immersion oils and in addition permit the study of live cells by dipping the objective directly into the cultivating medium.

The fluorescence intensity obtained is proportional to the square power of the numerical aperture (NA) of both condenser and objective in transmitted and to the fourth power of the objective in epi-illumination. The brightness is inversely related to the magnification of the objective. Fluorescence microscopy thus preferably has to be carried out with objectives of high NA in combination with low-power eyepieces.

REFERENCES

Bergquist N.R. & Nilsson P. *Ann. NY Acad. Sci.* **254**, 157–162.

Beverloo H.B., Schadewijk A. van, Gelderen-Boele S. & Tanke H.J. (1990) *Cytometry* **11**, 784–792.

Coons A.H., Creech H.J. & Jones R.N. (1941) *Proc. Soc. Exp. Biol. Med.* **47**, 200–202.

Giloh H. & Sedat J.W. (1982) *Science* **217**, 1252–1255.

Johnson G.D. & Nogueira Araujo G.M. (1981) *J. Immunol. Methods* **43**, 349.

Jovin T.M. & Vaz W.L.C. (1989) *Methods Enzymol.* **172**, 471–513.

Kohen E. & Hirschberg J.G. (1989) *Cell Structure and Function by Microspectrofluorometry.* Academic Press, San Diego.

Kraft W. (1973) Fluorescence microscopy and instrument requirements. Leitz Technical Information 2, pp. 97–109.

Lansing Taylor D., Waggoner A.S., Murphy R.F., Lanni R. & Birge R.R. (1986) *Applications of Fluorescence in the Biomedical Sciences.* Alan Liss, New York.

Lea D.J. & Ward D.J. (1974) *J. Immunol. Methods* **5**, 213–215.

Melamed M.R., Lindmo T. & Mendelsohn M.L. (1990) *Flow Cytometry and Sorting.* John Wiley & Sons, New York.

Nairn R.C. (1976) *Fluorescent Protein Tracing.* E. & S. Livingstone, Edinburgh.

Nederlof P.M., Flier S. van der, Wiegant J., Raap A.K., Tanke H.J., Ploem J.S. & Ploeg M. van der (1990) *Cytometry* **11**, 126–131.

Patzelt W. (1972) *Leitz-Mitt. Wiss. u. Techn.* **V/7**, 226–228.

Ploem J.S. (1967) *Z. wiss. Mikrosk. u. mikrosk. Techn.* **68**, 129–142.

Ploem J.S. (1971) *Ann. NY Acad. Sci.* **177**, 414–429.

Ploem J.S. (1987) *Appl. Optics* **26**, 3226–3231.

Ploem J.S., Tanke H.J., Al I. & Deelder A.M. (1978) In *Immunofluorescence and Related Staining Techniques*, W. Knapp, K. Holubar & G. Wick (eds). Elsevier, Amsterdam, pp. 3–10.

Rost F.W.D. (1991) *Quantitative Fluorescence Microscopy.* Cambridge University Press, Cambridge.

Rygaard J. & Olson W. (1969) *Acta Pathol. Microbiol. Scand.* **76**, 146–148.

Thomson L.A. & Hageage G.J. (1975) *Appl. Microbiol.* **30**, 616–624.

Tomlinson A.H. (1971) *Proc. Microsc. Soc.* **7**, 27–37.

Wick G., Schauenstein K., Herzog F. & Steinbatz A. (1975) *Ann. NY Acad. Sci.* **254**, 172–174.

Wilke V. (1983) *Proc. of SPIE* **396**, 164–172.

Young M.R. (1961) *Q. J. Microsc. Sci.* **102**, 419–449.

CHAPTER TWO

Introduction to Fluorescent Probes: Properties, History and Applications

FREDERICK H. KASTEN

Department of Anatomy, Louisiana State University Medical Center, New Orleans, LA, USA

2.1 INTRODUCTION

The availability of sensitive and selective fluorescent probes for living cells has opened new horizons in cell biology. With the aid of the modern epifluorescence microscope and video intensification microscopy, in combination with fluorescent probes, fluorescent-labelled organelles and molecules can be visualized, measured, and the information stored. The fluorescence signal superimposed against a dark background permits sharper cytologic detail to be observed than with a comparably stained specimen in the ordinary light microscope. This enables cell organization and function to be analysed with a precision and clarity not previously possible (Rost, 1980; Willingham & Pastan, 1982; Sisken *et al.*, 1986; Spring & Smith, 1987; Taylor & Salmon, 1989). Single fluorescent microtubules have been detected (Sammak & Borisy, 1988). With the addition of the confocal principle applied to imaging in fluorescence and integration with computers, precise optical sectioning and analyses of living and fixed cells are possible. In the confocal system, the fluorescent contributions of out-of-focus areas are eliminated during laser scanning and the limits of resolution are extended. Enhanced imaging at high numerical apertures is realized. Also,

three-dimensional reconstructions and measurements are obtained based on the accumulated optical sections (Stelzer & Wijnaendts-van-Resandt, 1989; Wilson, 1990; Herman & Jacobson, 1990; Kohen *et al.*, 1991).

It is the purpose of the present chapter to summarize the nature of fluorescence, the properties of fluorescent probes, the historical developments leading from early use of fluorochromes to modern fluorescent probes, and to summarize their applications in living cells. Further details of probes employed for specific applications are given in chapters elsewhere in this volume. Additional information is to be found in other reviews (Kasten, 1967, 1981, 1983a, 1989; Waggoner, 1986; Wang & Taylor, 1989; Taylor & Wang, 1989; Haugland, 1992; Kapuscinski & Darzynkiewicz, 1990; Darzynkiewicz & Crissman, 1990; Herman & Jacobson, 1990).

2.2 NATURE OF FLUORESCENCE AND PROPERTIES OF FLUORESCENT PROBES

Fluorescence is a form of luminescence which occurs after photons of light are absorbed by a molecule known as a fluorophore, fluorochrome, or fluorescent probe at the ground electronic state. The molecule is raised to an excited state as a result of electron transfer to a higher energy orbit. This excess energy is dissipated when the electron returns to the original ground state, releasing a quantum of light. The time required for absorption is immediate, about 10^{-15} s, whereas the fluorescence lifetime is approximately 10^{-8} s. The fluorescence phenomenon was first described by Brewster in 1838. The term fluorescence was coined by Stokes in 1852. Phosphorescence is a type of luminescence that persists after the exciting light is turned off. It has a lifetime of several seconds or longer because the excited electron first stops at an intermediate triplet state before reaching the ground state. According to Stokes' law, the fluorescent light is of a longer wavelength than the absorbed light. The law was extended in 1875 by Lommel, who stated that the molecule must first absorb radiation in order to exhibit fluorescence.

The difference in energy levels associated with absorption and fluorescence characterizes the absorption and emission wavelength maxima. The absorption intensity or extinction coefficient, ε, reflects the probability of absorption. Fluorescein (FITC) has an extinction coefficient maximum of 75 000 cm^{-1} M^{-1}. Unusually high extinction coefficients are given by the algal-derived phycobiliproteins, which have multichromophore complexes. For instance, phycoerythrin has an extinction coefficient greater than 10^6 cm^{-1} M^{-1}.

The emission intensity relates directly to the quantum yield, ϕ, which is the ratio of quanta released to quanta absorbed. Fluorochromes have characteristic quantum yields of efficiencies that range from 0.1 to almost 1. For practical purposes, the quantum yield should be close to 0.4 or greater when the fluorochrome is bound to the cell structure or molecule. The fluorescence intensity of a probe is determined by the product of ε and ϕ.

The third important characteristic of a fluorochrome or fluorescent probe is the fluorescence lifetime or excited state lifetime, τ, which is the average time that a molecule remains in the excited state. Short fluorescence lifetimes permit the greatest sensitivity to be achieved since multiple excitations can be achieved if the molecule is quickly relaxed after a prior excitation event. Most fluorochromes have emission lifetimes on the order of nanoseconds. Fluorescein has a τ of about 4 ns. It has been pointed out that unusually long lifetimes can be valuable in high-sensitivity detection (Waggoner, 1986). In cases where scattered light and autofluorescence of short lifetimes create interference with the desired fluorescence signal, it is desirable to use long-lived fluorochromes in combination with an appropriate photomultiplier tube.

Another property of fluorescent probes that needs to be considered in selecting a suitable dye is the wavelength of maximum absorption or excitation. Vital probes of cell vitality (SITS), membranes (ANS, DPH, 'Long Name', NPN), and ions (fura-2, indo-1, quin-2, SBQ) all require excitation in the long-wave UV to produce fluorescence in the visible range. This requires suitable UV-emitting light sources with attending protection for personnel. Also, there may be interfering autofluorescence from native cytoplasmic flavins, flavoproteins and NADPH. In the case of fixed cells, these metabolites are unlikely to be a problem and blue-fluorescing DNA-binding probes, like DAPI and Hoechst 33258, are useful.

The photobleaching of some probes is a serious problem. This is commonly observed with fluorescein-labelled cells in the fluorescence microscope, especially during photographic exposures. It is not usually a problem in flow cytometry because the individual cells are in the laser beam only a short time. In the confocal microscope, laser photobleaching is reduced by cutting down the number of optical scans. However, when line-averaging is necessary to reduce background noise, bleaching can be observed with FITC and Nile red. Some chemical agents like phenylenediamine (Johnson *et al.*, 1982) and propyl gallate (Giloh and Sedet, 1982) in the glycerol mounting medium help to reduce fading, but this is not possible in studies of living cells. To counter this photobleaching effect, the light intensity may be reduced, sensitive video cameras can be used, and photographic film with high sensitivity can be employed to reduce exposure time. Colour film with an ASA rating of 3200 is available without pushing. High-sensitivity black-and-white film is also on the market.

Fluorescein substitutes have been sought and the Bodipy fluorochrome is now recommended (Haugland, 1990). The Bodipy fluorophore, boron dipyrromethene difluoride, is said to have high photostability (Wories *et al.*, 1985) and other desirable features. The absorption peak is similar to that of fluorescein (505 nm compared with 490 nm), the emission peak is at almost the same wavelength (520 nm compared with 519 nm), the extinction coefficients are almost identical (about 75 000 cm^{-1} M^{-1}) and the quantum yields are similar. In other ways, Bodipy overcomes certain deficiencies of fluorescein and seems too good to be true. Fluorescein is pH-sensitive in the physiological range, which limits its application in living cells. The emission curve of fluorescein exhibits a broadness on the long wavelength side, which causes some overlap with other dyes used in two-colour fluorescence. Also, fluorescein conjugates have negative charges, which limit their use in examining surface membranes and receptors. According to Haugland (1990), Bodipy offers advantages over fluorescein in addition to the improved photostability. Bodipy has a narrow emission spectrum with a large Stokes' shift and gives less overlap with certain red fluorochromes, like Texas Red. Bodipy is relatively lipophilic and can be bound to certain compounds for receptor studies that cannot be done with fluorescein.

For additional details on the nature of fluorescence and on the properties of fluorescent probes, the articles by Waggoner (1986) and Taylor and Salmon (1989) should be consulted.

2.3 HISTORICAL DEVELOPMENTS

2.3.1 Fluorescence microscope

The first fluorescence microscope was developed over 80 years ago by Heimstädt (1911) and Lehmann (1913) as an outgrowth of the UV microscope. The instrument used a high-powered arc lamp to generate UV light, a modified Wood's filter (nitrosodimethyl-aniline solution with copper sulphate) as a primary filter, a dark-field condenser, Uviol secondary filter, and quartz optics. The microscope was used to investigate the autofluorescence of bacteria, protozoa, plant and animal tissues, and bioorganic substances, such as albumin, elastin and keratin (Stübel, 1911; Tswett, 1912; Wasicky, 1913; Provazek, 1914). A history of these developments and subsequent technological advances are presented elsewhere (Kasten, 1983a, 1989).

2.3.2 Synthesis of coal-tar dyes and early uses

The first synthetic coal-tar dye, mauve or aniline purple (CI 50245),[*] was made accidentally by William Perkin in 1856 (cf. Perkin, 1906). This breakthrough was followed by feverish attempts on the part of many chemists to synthesize other dyes. Using oxidized aniline and the approaches suggested by Perkin, numerous dyes were produced, the first of which was magenta in 1859, also known as rosaniline or fuchsin (CI 42510). The great need for textile dyes with wide-ranging colours and resistance to bleaching by light stimulated further commercial interest. Between the time of Perkin's discovery and the invention of the fluorescence microscope, about 55 years, scores of new dyes became available. Among this group of dyes synthesized in the late nineteenth century were pararosaniline (CI 42500), methyl violet or gentian violet (CI 42535), crystal violet (CI 42555), methyl

Table 2.1 Common biological fluorochromes.

Fluorochrome	Year of synthesis	CI No.[a]
Acridine orange	1889	46005
Acridine red 3B	1891	45000
Acridine yellow	1889	46025
Acriflavine	1910	46000
Auramine O	1883	41000
Brilliant sulphoflavine	1927	56205
Calcein	1956	—
Chrysophosphine 2G	1922	46040
Congo red	1884	22120
Coriphosphine O	1900	46020
Eosin B	1875	45400
Eosin Y	1871	45380
Erythrosin B	1876	45430
Euchrysin	1922	46040
Flavophospine N	1887	46065
Fluorescein	1871	45350
Neutral red	1879	50040
Nile blue A	1888	51180
Oxytetracycline	1950	—
Pararosaniline	1878	42500
Phosphine 5G	1900	46035
Phosphine GN	1862	46045
Primulin	1887	49000
Proflavine	1910	—
Prontosil	1932	—
Pyronin Y (G)	1889	45005
Quinacrine (Atabrine)	1934	—
Rheonine	1894	46075
Rhodamine 3GO	1895	45210
Rhodamine 5G	1902	45105
Rhodamine 6G	1892	45160
Rhodamine B	1887	45170
Rhodamine G	1891	45150
Rhodamine S	1888	45050
Sulphorhodamine B	1906	45110
Thiazole yellow G	1893	19540
Thioflavine S	1888	49010

[a] The CI no. refers to the *Colour Index* no., a specific designation for the chemical structure of a dye as listed in the *Colour Index* (1971).

green (CI 42585), malachite green (CI 42000), brilliant green (CI 42040), safranin O (CI 50240), methylene blue (CI 52015), gallocyanin (CI 51030), and numerous azo dyes like Bismarck brown R (CI 21000). A small German firm known as Dr G. Grüblers Chemisches Laboratoriums first opened in 1880. Grübler tested and packaged the most desirable dyes for biologists and medical researchers.[*] This quality assurance of selected, high-quality German dyes was of great value to laboratory workers, who took advantage of the newly available dyes to stain

[*] CI stands for 'Colour Index' and the number following is that assigned in the 3rd edition (*Colour Index*, 1971).

[*] In 1897, the Grübler firm became known as Dr K. Hollborn & Söhne.

histological, haematological and bacteriological material and to develop new staining methods (Kasten, 1983b).

Other dyes produced during this period included xanthene and acridine derivatives, which were highly fluorescent. Some of the well-known xanthenes were pyronin Y (G) (CI 45005), rhodamine B (CI 45170), fluorescein (CI 45350), eosin Y (CI 45380), and erythrosin (CI 45430). Some of the early acridines were phosphine (CI 46045), acridine yellow (CI 46025), acridine orange (CI 46005), acriflavine (CI 46000), and coriphosphine O (CI 46020). A few fluorescent dyes were derived from other chemical groups, such as auramine O (diphenylmethane, CI 41000), Calcofluor white (stilbene, CI 40621), brilliant sulphoflavine (amino ketone, CI 56205), neutral red (azin, CI 50040), and pararosaniline (CI 42500). Table 2.1 lists some of the common fluorochromes and the year when each was synthesized. Common acridine dyes used in histology and histochemistry are described by Kasten (1973). General properties of dyes are given elsewhere (Harms, 1965; Lillie, 1977; Green, 1990).

In spite of the fact that many fluorescent dyes were available to microscopists by the beginning of this century, few were actually used. Histologists, cytologists and bacteriologists favoured strong-staining red, violet, blue and green dyes, which were largely non-fluorescent or only weakly fluorescent in solution. Pyronin Y and eosin Y, which are red dyes, were exceptional cases. Dyes of the acridine group, which usually stain cells yellow, were less commonly used. To illustrate with several examples, basic fuchsin, a mixture containing red-staining pararosaniline and rosaniline dyes, became an important nuclear stain in histology following its introduction by Waldeyer in 1863. It had a great impact in bacteriology, particularly in the Ziehl-Neelsen method for demonstrating acid-fast microorganisms like the tubercle bacillus. Gentian violet was introduced into microtechnique by Weigert and by Ehrlich in 1881 and 1882 and became an essential component of the Gram stain in 1884. Methylene blue was employed by Ehrlich as the first important vital stain in 1885; he demonstrated its affinity for nerve tissue. Ehrlich introduced many other dyes into the field, like his famous triacid mixture (methyl green, acid fuchsin, orange G).

The acridines and other fluorescent dyes were not used on fixed cells and tissues in fluorescence microscopy until the early 1930s, more than 20 years after the fluorescence microscope was developed. Krause's three-volume *Enzklopädie der Mikroskopischen Technik* (Krause, 1926a, b, 1927) did not mention any uses for acridine dyes and failed to include a discussion of the acridines, although other dye groups were included. The section on the fluorescence microscope in this otherwise valuable reference referred only to its application in detecting autofluorescence in tissue sections. Microscopists failed to appreciate the fact that fluorochromes could impart added sensitivity and clarity to stained tissues when viewed by fluorescence microscopy. It was mistakenly felt that microscopic observations by induced fluorescence via fluorochromes would introduce artifacts and misinterpretations.

Although fluorescent dyes were not utilized to stain fixed tissues and cells for many years, researchers made use of the dyes in other ways. Fluorescein was known to produce an intense yellow-green fluorescence in aqueous solution. Its sodium salt (uranin) produced a pale green fluorescence even when diluted 1 part to 16×10^6 parts of water (Fay, 1911). The dye was employed by Ehrlich (1882) to track the pathway of aqueous humour in the eye. In 1906, fluorescein was the first fluorescent dye to be used for tracing underground waters in the United States.

With the beginning of the First World War and the need to treat infected wounds, the efficacy of acriflavine as an antiseptic became established (Browning *et al.*, 1917). The dye was referred to as 'flavine' by British researchers. Proflavine, a close relative of acriflavine, was another useful antiseptic. Both of these dyes would later prove to be valuable fluorescent probes of nucleic acids. Many other diaminoacridine compounds were prepared for experimental and clinical trials, none of which proved to be superior to acriflavine (Browning, 1922; Albert, 1951). Tissue cultures were tested for their response to acriflavine and proflavine to determine cell toxicity levels and effects on bacterial-infected cultures (Mueller, 1918; Hata, 1932; Jacoby *et al.*, 1941). However, the culture system failed to aid in predicting the value of aminoacridine antiseptics as local chemotherapeutic agents (Browning, 1964).

2.3.3 First usage of fluorochromes in living cells

By the early 1900s, pharmacologists and experimental therapeuticists showed a great interest in the action of fluorescent dyes in sensitizing microorganisms to light. This dye-enhanced light inactivation became known as photodynamic inactivation. Research on this subject was stimulated by the appearance of an important volume by Tappeiner and Jodlbauer (1907). Acridine dyes were shown to be effective agents in treating trypanosomes (Werbitzki, 1909). Ehrlich's use of acriflavine for combating this protozoan in infected mice gave dramatic results (Ehrlich & Benda, 1913). The dye was referred to as 'trypaflavin' because of its influence on trypanosomes. Microscopists used the bright-field microscope to observe the binding of such dyes to microorganisms.

The protozoologist Provazek (1914) was apparently the first person to employ the fluorescence microscope to study dye binding to living cells. He added various fluorochromes and drugs (fluorescein, eosin, neutral red, quinine) to cultures of the ciliate *Colpidia* and viewed the induced fluorescence of the cells. He grasped the fundamental significance of this new experimental approach and stated that the object was:

> To introduce into the cell certain substances of different types, without regard as to whether they are stains or colorless drugs, on the assumption that they follow definite distribution laws and collect under certain circumstances in particular functional elements inside the cell so that they effectively illuminate the partial functions of the cell in the dark field of the fluorescence microscope.

2.3.4 Developments in vital and supravital fluorochroming*

The introduction of fluorochromes into fluorescence microscopy in 1914 marked a giant step forward in experimental cytology. The report by Provazek was the first to demonstrate vital fluorochroming. Previously, vital dyes, including fluorochromes, were tested on protozoa but observed only with the bright-field microscope. Supravital fluorochroming had its start in 1932 when Jancso, a Hungarian pharmacologist, injected several different fluorochromes into rodents previously infected with trypanosomes. Examination of blood smears in the fluorescence microscope revealed specific binding of the dyes to nuclei and basal bodies of the blood-borne trypanosomes. Acriflavine displayed especially strong binding to these structures. Because of the unusual interest in chemotherapy in the 1930s, attempts were made to locate fluorochromes that would bind to the malaria organism in infected animals and humans. Of the various fluorochromes tested, quinacrine (atabrine) was found by fluorescence microscopy to be selectively taken up by circulating plasmodia within 10 min after dye injection (Fischl & Singer, 1935;

---

* According to classical usage, vital fluorochroming or staining is the non-toxic staining of living cells or tissues in the organism. An example is the intracellular uptake of colloidal azo dyes like trypan blue by macrophages and Kupffer cells *in vitro*. The term intravital staining is sometimes used synonymously with vital staining. Supravital fluorochroming or staining means the addition of dyes to an *in vitro* solution containing cells previously removed from an organism. It may also refer to the staining of living cells within a recently killed animal. The stain binds to cytoplasmic organelles, like the binding of Janus green to mitochondria or of neutral red to cytoplasmic granules. Cell biologists today commonly apply the term vital stain or fluorochrome to dyes added to cultured cells.

Bock & Oesterlin, 1939; Patton & Metcalf, 1943). Acriflavine was added to fibroblast cultures and shown by fluorescence microscopy to inhibit cell division at the concentrations tested (Bucher, 1939). Additional details of the early investigations into vital fluorochroming are found in the volume by Drawert (1968).

From these studies, it became clear that certain aminoacridines bound preferentially to components in nuclei. Attempts to determine the mode of action of such binding led to the discovery that the dyes had an affinity for nucleic acids. This strong interaction was demonstrated with purified nucleic acids *in vitro* (DeBruyn *et al.*, 1953; Peacocke & Skerrett, 1956) and in fixed cells at a low pH range (Armstrong, 1956; Bertalanffy & Bickis, 1956; Schümmelfeder *et al.*, 1957). This is discussed further in Section 2.5.

2.3.5 *In vivo* fluorochroming*

After Ehrlich demonstrated that fluorescein could be used to follow the path of aqueous humour in the eye, the slit-lamp microscope was adapted by Thiel to observe the dye's fluorescence within the eye. The fluorescence microscope, which uses light transmitted through a condenser, could not be employed to examine opaque specimens from most living organs. In 1929, the fluorescence microscope was modified markedly by Philipp Ellinger, a pharmacologist at Heidelberg University, in collaboration with a young anatomist, August Hirt (Ellinger & Hirt, 1929a). Ellinger was interested to examine the microcirculation in the kidney of the exteriorized organ with the aid of fluorochromes previously injected into the animal. This new fluorescence microscope utilized vertical illumination that was directed into the microscope tube laterally and then passed through the objective to the specimen. The emitted fluorescence was transmitted back up the tube to the eye. Other essential components included appropriate filters and a water-immersion objective. The instrument was called an 'intravital microscope' and may be considered as the first epifluorescence microscope. Dilute fluorescein and acriflavine solutions were used to study the physiology of urine formation (Ellinger & Hirt, 1929b). During the Second World War, Hirt carried out unethical experiments on humans in which he hoped to examine human tissues *in vivo* with the intravital fluorescence microscope (Kasten, 1991).

The intravital fluorescence microscope attracted the attention of other researchers and a variety of fluorochromes were employed by them; brilliant

---

* *In vivo* or intravital fluorochroming refers to the non-toxic staining of living cells and tissues in the organism with observations by microscopy.

phosphine G (probably CI 46045), germanin S, primulin yellow (CI 49000), rheonin A (CI 46075), thiazole yellow (CI 19540), and thioflavine (probably CI 49010). During the 1930s, with this new technique, fluorescence observations were made of the microcirculation in living skin, liver, kidney, conjunctiva and adrenal gland (Singer, 1932; Franke & Sylla, 1933; Pick, 1935; Heuven, 1936; Grafflin, 1938; Schmidt-LaBaume & Jäger, 1939). For physiological studies, vital fluorochromes needed to be used under conditions of isotonicity, non-toxicity, and non-quenching. A review of these early investigations is given by Ellinger (1940), Price and Schwartz (1956), and Kasten (1983a). In recent years, intravital microscopy of the microcirculation has enjoyed renewed popularity as a result of access to modern epifluorescence instruments, scanning microfluorometry, sensitive charge-coupled video cameras, and time-frame generators for data evaluation (cf. Bollinger *et al.*, 1983; Witte, 1989). As an example, such instrumentation was utilized with acridine orange to quantify the hepatic microcirculation in rodents (Menger *et al.*, 1991).

2.4 APPLICATION OF FLUOROCHROMES IN HISTOLOGY AND MICROBIOLOGY

2.4.1 Histology

Until 1929, microscopic work with the fluorescence microscope was confined to observations of tissue and cellular autofluorescence (porphyrins, native cytoplasmic proteins, chlorophyll) and to the detection of living protozoa with fluorochromes. The first report of the use of a fluorochrome on fixed tissue sections was by the dermatologist Sigwald Bommer (1929). He employed a dilute solution of acriflavine on skin sections and observed a selective green-yellow fluorescence of cell nuclei. Bommer suggested the possibility of a fundamental cytochemical basis for the nuclear fluorescence he observed.* He raised the question: 'To what extent is it possible to establish definite affinities between certain stains and tissue constituents when used in a certain dilution and with a definite technique?'

The next milestone in this field was achieved in Vienna by the young pathologist Herwig Hamperl, in collaboration with Max Haitinger, an expert in fluorescence microscopy (Haitinger & Hamperl, 1933). They examined the staining properties of more

than 65 different fluorochromes on formalin-fixed frozen sections of animal tissues to see which ones would produce differential fluorescence. Out of this empirical survey, they recommended 35 fluorochromes worthy of use in normal and pathological histology. Hamperl (1934) extended the study to paraffin-embedded tissues and described the differential binding affinities of many fluorochromes and recommended staining methods. At this time, the word fluorochrome was created to mean fluorescent compounds which are bound selectively to individual tissue structures without disturbing the autofluorescence of other tissue elements (Haitinger, 1934). To distinguish these applied dyes from natural autofluorescing substances in cells, fluorochromes were also termed secondary fluorochromes and the process of staining as secondary fluorochroming. Autofluorescing substances like the porphyrins were called primary fluorochromes. In modern usage, the word fluorochrome has come to mean any fluorescent dye, regardless of its staining properties and effect on native autofluorescence.

The studies of Hamperl and Haitinger served as a foundation for later investigators, who applied the methods to their own fields and added modifications in staining protocols. It was clear that secondary fluorochroming produced brilliant colours in tissues and cells with striking contrast and sensitivity. The impact of Hamperl and Haitinger's work was felt first at the University of Vienna. Here, various investigators collaborated with Haitinger, who maintained a fluorescence microscopy facility and popularized the technique (see dedication volume to Haitinger by Bräutigam and Grabner, 1949). The initial histomorphological applications of the new fluorochroming methods were in the fields of botany (Haitinger & Linsbauer, 1933), the nervous system (Exner & Haitinger, 1936), pathology (Haitinger & Geiser, 1944; Eppinger, 1949), and cytology (Bukatsch, 1940).

When fluorochromes were used with plant material, some unusual colour combinations were produced; autofluorescence often persisted together with the induced fluorescence. For instance, when a section of wood was stained with Magdala red (CI 50375) and examined in the fluorescence microscope, there was revealed red-fluorescing cuticle, orange-red primary and secondary phloem, blue cork cambium, and intense red xylem (Haitinger & Linsbauer, 1935). With the use of certain fluorochromes, like coriphosphine O, two different colours were induced in the same cell, revealing orange-fluorescing cytoplasm and yellow-fluorescing nuclei (Bukatsch & Haitinger, 1940). This multicoloured fluorescence, known as metachromasia, was seen as well in mixed populations of live and dead cells when the mixture was treated

---

* Bommer was apparently unaware of the discovery of DNA in animal and plant cell nuclei by Feulgen and Rossenbeck in 1924 (Kasten, 1964) although Bommer and Feulgen were fellow faculty members at Justus-Liebig University in Giessen.

with acridine orange (Strugger, 1940). For instance, living epidermal plant cells fluoresced green at a pH of 5.7–8.0 and dead cells fluoresced red. Further details of Strugger's pioneering work with acridine orange will be discussed later.

2.4.2 Microbiology

Tubercle bacilli had been examined in the fluorescence microscope as early as 1917 by virtue of their autofluorescing properties (Kaiserling, 1917). The first report of the fluorochroming of microorganisms came 20 years later by Hagemann at the Hygienic Institute of the University of Cologne, who used berberine (CI 75160) to fluorochrome lepra bacilli (Hagemann, 1937a) and primulin (CI 49000) to visualize viruses (Hagemann, 1937b). Other fluorochromes that proved useful were thioflavine S (CI 49005, Hagemann, 1939; Levaditi & Reinie, 1939) and mordanted morin (CI 756609) and thioflavine (Hagemann, 1937c). Attempts were made to find a more sensitive staining method than the Ziehl-Neelsen procedure to detect tubercle bacilli. Eventually, auramine O (CI 41000) was found satisfactory in combination with an acid–alcohol treatment of the smears (Hagemann, 1938). Auramine O produced brilliant yellow fluorescence from stained organisms and the technique became popular in the United States through the work of Richards (1941). He examined the mechanism of staining and concluded that the specificity for the tuberculosis organism was due to dye binding to mycolic acid, the acid-fast component in the bacterial cell wall (Richards, 1955). With tuberculosis still a leading cause of death in the world among communicable diseases, laboratory detection still relies in part on the carbol–auramine O fluorescence method, often with a rhodamine or acridine orange counterstain. For reviews of the literature on fluorochroming of microorganisms, see Ellinger (1940), Strugger (1949), Duijn (1955), Price and Schwartz (1956), and Kasten (1983a, 1989).

2.5 INTRODUCTION OF ACRIDINE ORANGE INTO CELL PHYSIOLOGY, CYTOLOGY AND CYTOCHEMISTRY

Acridine orange (AO), a basic dye, was synthesized by Benda in 1889 and was produced by Badische Anilin & Soda Fabrik. Although AO was available from Dr K. Hollborn & Söhne, according to their catalogue of 1932, the dye was overlooked by Hamperl and Haitinger in their extensive survey of fluorochromes for possible value in fluorescence microscopy (Haitinger & Hamperl, 1933; Hamperl, 1934).

AO was first introduced as a fluorochrome into fluorescence microscopy, independently by Bukatsch and Haitinger (1940) and by Strugger (1940). Bukatsch and Haitinger found that AO was suitable as a vital fluorochrome in living plant cells, staining cell nuclei. Mitotic chromosomes were also fluorochromed (Bukatsch, 1940). These workers did not notice any unique fluorescent properties of AO.

Siegfried Strugger, a plant cell physiologist at the University of Münster, discovered the extraordinary ability of AO to fluorochrome live and dead cells in different colours (Strugger, 1940). This finding had its basis in a long series of papers published between 1931 and 1940 in which Strugger investigated the vital staining of cells with other dyes by bright-field microscopy. Most workers had not considered the influence of pH of the staining solution when examining fluorescence of tissues and cells; dyes were simply prepared in dilute solutions. However, Hercik (1939) reported that intravital staining of onion epidermal cells with fluorescein at pH 1.5 produced different fluorescence patterns, according to whether the cells were viable or not. As a cell physiologist, Strugger was acutely aware of the importance of pH in the binding between charged fluorochrome ions and intracellular constituents. He also recognized the influence that dye concentration might have on the presence of dissociated and undissociated forms of the dye in solution. His findings were made and reported during the war years. There were about 20 research papers published between 1940 and 1944, based on fluorescence microscopic investigations with AO and a few other fluorochromes. Immediately after the war, he was commissioned by the occupational authorities to summarize the German wartime research on cell physiology and protoplasm of plant cells (Strugger, 1946). He was recruited by the US government under Project Paperclip and did research in the United States. After returning to Germany, Strugger wrote two books, one of which dealt with fluorescence microscopy (Strugger, 1949). Strugger's seminal contributions led directly to the modern use of AO as a fluorescent probe for nucleic acids in fluorescence microscopy and flow cytometry. Because of the significance of Strugger's work, this will be covered in more detail here.

In his first and most important research work in fluorescence microscopy, Strugger systematically examined the uptake and storage of AO by living plant cells (Strugger, 1940). The fluorescence colour within cells was shown to depend in part on dye concentration, called the 'concentration effect'. Others later referred to this as the 'Strugger effect'. At low concentrations (1:5 000–1:100 000), the fluorescent colour was green whereas at high concentration (1:100), the colour was red. Intermediate dye concentrations produced a yellow colour. Also, Strugger showed that AO could discriminate between live and dead plant cells by fluorescence microscopy, according to the pH of the staining solution. At a pH of 5.7–8.0, living cells fluoresced green and dead cells appeared red. With dead cells, AO produced a red cytoplasm above pH 4.7. The nuclei were red at low pH and shifted to yellow-green, beginning at a pH of 6.8. Strugger's observations on bicolour fluorescence from populations of live and dead cells were confirmed (Bukatsch, 1941; Bucherer, 1943). Strugger and associates extended these vitality experiments to yeast, slime moulds, bacteria and sperm (Strugger, 1940/41; Strugger & Hilbrich, 1942; Strugger & Rosenberger, 1944). The AO method was important since previous techniques for distinguishing live and dead cells failed to give such striking and clear-cut results. Also, the observation of bicolour fluorescence with AO staining, a type of metachromasia, attracted attention to this phenomenon. Strugger pioneered in the use of fluorescent pH indicators in cell physiology (Strugger, 1941).

According to Strugger, the basis for bicolour fluorescence after vital fluorochroming depended on the relative binding of AO cations by cell proteins. In live cells, the concentration of bound dye cations was low, producing green fluorescence, whereas in damaged (yellow fluorescence) and dead (red fluorescence) cells, there would be a progressive increase in AO cation binding by proteins through electrostatic means. In viable cells, there would presumably be few accessible electronegative charges present on proteins. Following injury or cell death, Strugger postulated that a disturbance occurs in the submicroscopic protein scaffold, making accessible more negative charges. The copper-red fluorescence seen in dead cytoplasm would be the visible manifestation of the fine structural alterations. Strugger's AO method attracted the attention of Adolph Krebs, who systematically examined alpha particle radiation damage to cells with the aid of AO and fluorescence microscopy (Krebs, 1944). After the war, Strugger and Krebs worked together with Gierlach at Fort Knox, Kentucky, where additional radiation studies were done on onion cells using AO as a new experimental tool in radiation biology (Krebs & Gierlach, 1951; Strugger et al., 1953).

AO was also shown by Strugger to be favourable for determining the isoelectric point (IEP) of cellular proteins. In alcohol-fixed material, protein-containing structures emitted a green fluorescence below the IEP and a copper-red fluorescence above the IEP.

The interaction of AO with living cells was examined further by many workers. Strugger's interpretation of the differential fluorescence of living and dead cells has been questioned. The bicolour fluorescence was suggested to be related to cellular metabolic activity (Schümmelfeder, 1950), binding to DNA, mononucleotides in mitochondria, and polysaccharides (Austin & Bishop, 1959), lysosomes (Robbins & Marcus, 1963; Robbins et al., 1964), and nucleoprotein complexes (Wolf & Aronson, 1961). Dye solutions of AO exhibit metachromasia due most likely to the formation of species of dye monomers, dimers and polymers (Zanker, 1952; Steiner & Beers, 1961). AO–nucleic acid solutions form different complexes, including dye intercalation between base layers of DNA (green) and dye interaction with phosphate groups on nucleic acid surfaces, referred to as stacking (Loeser et al., 1960; Bradley, 1961; Steiner & Beers, 1961).

Binding of AO to fixed cells was studied in detail by Schümmelfeder (1948, 1956), who emphasized the importance of pH in the dynamics of AO binding to intracellular constituents. The application of the pH principle to determine the IEP of tissue proteins was verified (Schümmelfeder & Stock, 1956; Schümmelfeder, 1956). Under acidic staining conditions, the AO dye cation was shown by various workers to stain acid components, like the acidic mucopolysaccharides found in cartilage and mast cell granules, and nucleic acids of cells. Independently, three groups of investigators discovered that under controlled conditions of staining with AO, DNA of fixed interphase nuclei and chromosomes fluoresced yellow-green to green whereas regions rich in RNA (nucleolus, basophilic regions of cytoplasm) fluoresced orange to red (Armstrong, 1956; Bertalanffy & Bickis, 1956; Schümmelfeder et al., 1957). It was suggested that the colour differences were due to molecular size variations and configuration of the two nucleic acids (polymerization, denaturation) and not to intrinsic chemical differences between RNA and DNA (Schümmelfeder, 1958; Aldridge & Watson, 1963). An impressive microspectrofluorometric investigation of the mechanism of AO binding to purified and intracellular nucleic acids was done by Rigler (1966). He confirmed that the orderliness of the secondary structure of DNA (accessibility of DNA-phosphates) had a profound influence on AO binding.

It became apparent with AO fluorochroming that brilliant colour differences could be easily seen between cancer cells, with their hyperchromatic nuclei and high RNA content, and normal cells. The AO technique was incorporated into exfoliative cytology as a rapid screening test for cervical cancer and other malignancies (Bertalanffy & Bickis, 1956; Dart & Turner, 1959). AO proved to be a sensitive cytochemical fluorochrome for the detection and identification of nucleic acids in purified and viral-infected cells (Armstrong & Niven, 1957; Mayor, 1963). The organization of DNA in chromosomes was investigated using polarized fluorescence microscopy

Table 2.2 Fluorochromes used in biological microscopy.[a]

Fluorochrome	CI no.	Chemical group	Acidic or basic	Biological applications
Acid fuchsin	42685	Arylmethane	A	Counterstain, elastic fibres and other connective tissues supravital fluorochrome (plant cells), pH indicator
Acridine red 3B	45000	Xanthene	B	Histology
Acridine orange	46005	Acridine	B	Histology, cell viability, DNA intercalator, nucleic acids (fluorochromasia), cytodiagnosis, isoelectric point of proteins, bacteria, viruses, mast cells, supravital fluorochrome (plant and animal cells), protozoa, tumour localization, sperm, lysosomes, acid mucopolysaccharides, plant tissues, pH indicator, amyloid
Acridine yellow	46025	Acridine	B	Histology, insect tissues, cytodiagnosis, protozoa, viruses, tubercle bacilli, Schiff-type reagent
Acriflavine (trypaflavine, mixture of 3,6-diamino-10-methyl acridinium chloride and 3,5-acridinediamine or proflavine)	46000	Acridine	B	Histology, plant tissues, intercalating dye for nucleic acids (fluorochromasia), Schiff-type reagent, protozoa, viruses, vital stain, bacteriostatic agent, intravital fluorochrome, inhibits mitochondriogenesis
Alizarin red S	58005	Anthraquinone	A	Bone and bone growth
Alizarin cyanine BBS	58610	Anthraquinone	A	Histology
Aniline blue	42755	Arylmethane	A	Plant cell walls (callose), β-glucans, eosinophils, glycogen
Atabrine (see quinacrine dihydrochloride)				
Auramine O	41000	Arylmethane	B	Tubercle bacilli (acid-fast bacteria), bacterial counting, viability (plant tissue), blood cells, Schiff-type reagent; fluorescent complex with horse-liver alcohol dehydrogenase
Aurophosphine (see phosphine 5G)				
Basic fuchsin (see pararosaniline)				
Benzoflavine (see flavophosphine N)				
3,4-Benzopyrene				Lipids (suspect cancer agent)
Brilliant cresyl blue	51010	Oxazine	B	Lipids, vital staining of blood, reticulocytes in blood smears, protozoa, chromosomes
Berberine sulphate	75160	Natural plant alkaloid	B	Histology, wood tissue, vital fluorochrome (botany, protozoa), insect histology, bacteria, mitochondria, viruses, nucleic acids (nuclei), heparin, antibacterial and antimalarial agent, chromosome banding, mast-cell granules
Brilliant sulphoflavine	56205	Aminoketone	A	Protein stain, bacterial spores
Calcein (active part is DCAF, 3,6-hydroxy 24-bis-[N,N'-di-(carboxy methyl)-aminomethyl] fluoran (Fluorexone)	——	Xanthene	A	Bone growth, eosinophils
Calcein blue	——	Xanthene	A	Bone growth
Calcofluor white M2R (Cellufluor)	40622	Stilbene	A	Plant cell walls, β-glucans, microorganisms, vital fluorochrome, fungi in tissue sections
Chelidonium	——	Natural plant extract		Fat, nuclei
Chlortetracycline (Aureomycin)	——	Natural	A	Bone growth, antibiotic, membrane-bound Ca^{2+}
Chrysophosphine 2G	46040	Acridine	B	Nuclei, mast cells, wood, acid mucopoly-saccharides, Schiff-type reagent, amyloid
Congo red	22120	Disazo	A	Counterstain, amyloid, probe for conformation of nucleotide-binding enzymes

Table 2.2 Continued

Fluorochrome	CI no.	Chemical group	Acidic or basic	Biological applications
Coriphosphine O	46020	Acridine	B	Histology, nucleic acids (fluorochromasia), mast cells, haematology, wood, bacteria, plant cell walls, juxtaglomerular granules, fat, acid mucopolysaccharides, Schiff-type reagent, red fluorescence of diffuse neuroendocrine cells by 'masked basophilia'
Eosin B	45400	Xanthene	A	Counterstain, muscle, haemoglobin
Eosin Y	45380	Xanthene	A	Counterstain, muscle, plant tissue, immuno-fluorescence label, constituent of blood stains, histones
Erythrosin B	45430	Xanthene	A	Counterstain, supravital, dental disclosing agent for plaque, immunofluorescence label
Esculin (6,7-dihydroxycou-marin 6-glucoside)	——	Natural plant glucoside		Vital fluorochrome (protozoa), antimalarial agent
Ethyl eosin	45386	Xanthene	A	Counterstain
Euchrysine 2GNX	(see chrysophosphine 2G)			
Evans blue	23860	Disazo	A	Counterstain, fluoresces red when bound to protein, diagnostic aid (blood volume determinations), retrograde procedure for axonal branching with DAPI and primuline, teratogen, (suspect cancer agent)
Flavophosphine N (benzoflavine)	46065	Acridine	B	Histology, Schiff-type reagent, supravital fluorochrome (plant physiology), pH indicator
Fluorescein (uranin, Na salt)	45350	Xanthene	A	pH indicator, intravital fluorochrome (microcirculation), insect histology, viruses, diagnostic aid (ophthalmology and central nervous system tumours), immunofluorescence label, vital (plants), dental disclosing agent for plaque, circulation time
Geranine B	14930	Monoazo	A	Fat, cell nuclei, elastic fibres
Isamine blue (brilliant dianyl blue)	42700	Arylmethane	A	Connective tissue, nervous system, supravital
Magdala red	50375b	Azine	B	Fat, mucus, plant cell walls
Mercurochrome	——	Xanthene	A	Counterstain, leukocytes, protein-bound SH and S-S groups
Methylene blue	52015	Thiazine	B	Fat, histology, blood stain constituent, bacteriological stain, antidote to cyanide poisoning
Methyl green	42585	Arylmethane	B	Cell nuclei, polymerized nucleic acid (DNA), histology, gonococci, mast cells, (irritant)
Morin (3,5,7,2',4'-pentahydroxyflav-anol)	75600	Natural plant flavone	A	Cell nuclei (nucleic acids when complexed with aluminium ammonium sulphate), dye complexes with spirochaetes, trypanosomes, metal detection (Al)
Neutral acriflavine (see acriflavine)				
Neutral red	50040	Azine	B	Lipids, mast cells, bacteria, supravital fluorochrome, Schiff-type reagent, histology
Nile blue A	51180	Oxazine	B	Fat, differentiating melanins and lipofuscins, Schiff-type reagent
Nile red (Nile blue A oxazone)	51180	Oxazine	B	Lipid droplets, plant cell microsomes
Oxytetracycline	——	Natural	A	Bone growth, antibiotic
Pararosaniline (main component of fuchsin)	42500	Arylmethane	B	Cell nuclei, elastic tissues, Schiff reagent, bacilli, anti-schistosomal and topical anti-fungal agent, caries stain discloser (suspect cancer agent)
Phenosafranin	50200	Azine	B	Histology, Schiff-type reagent

Table 2.2 Continued

Fluorochrome	CI no.	Chemical group	Acidic or basic	Biological applications
Phloxine B	45405	Xanthene	A	Counterstain
Phosphine 5G (Aurophosphine)	46035	Acridine	B	Mast cells, acid mucopolysaccharides, Schiff-type reagent
Phosphine GN (phosphine 3R)	46045	Acridine	B	Lipids, cell nuclei, insect histology, nerve tissue, Schiff-type reagent
Primulin	49000	Thiazole	A	Intravital fluorochrome, plant cell walls, cell nuclei, protozoa, viruses, proteins
Procion red	18159	Monoazo	A	Bone growth
Procion yellow M4RS (MX-4R)	Reactive Orange 14	Monoazo	A	Vital fluorochroming of neurons and functional connections after introduction into cells by electrophoresis, label for newly forming bone
Proflavine (similar to acriflavine)				
Prontosil (sulphamido-chrysoidine)	——	Monoazo	B	Vital fluorochrome (insects and plants), connective tissue fibres
Pseudoisocyanin	——	Quinolin	B	Neurosecretion, cysteic acid groups in proteins
Pyronin Y (G)	45005	Xanthene	B	RNA preferentially (usually in combination with methyl green), single-stranded nucleic acids, bacteria, supravital fluorochrome (plant tissues), pH indicator, plasma cells
Quinacrine dihydrochloride (Atabrine)	——	Acridine	B	DNA, chromosome Q banding, vital fluorochrome (protozoa), nerve fibres, tumour localization, antimalarial and antihelminthic agent, (light-sensitive)
Quinine	——	Natural plant alkaloid	A	Vital fluorochrome (protozoa), insect histology, antimalarial agent, (light-sensitive)
Rheonin A	46075	Acridine	B	Histology, fungi, Schiff-type reagent
Rhodamine B	45170	Xanthene	Neut.	Vital fluorochrome (plant cell sap, mitochondria), bacteria, fat, viruses, metal detection, immunofluorescence label (suspect cancer agent)
Rhodamine G	45150	Xanthene	B	Histology, vital fluorochrome, viruses
Rhodamine 3GO	45210	Xanthene	B	Histology, Schiff-type reagent
Rhodamine 6G	45160	Xanthene	B	Vital fluorochrome (mitochondria), wood, (suspect cancer agent)
Rhodamine S	45050	Xanthene	B	Histology
Rhodanile blue (complex of Nile blue A and rhodamine B)	——	Oxazine-xanthene	B	Histological differentiation
Rhodindine (related to Magdala red)	50375a	Azine	B	Schiff-type reagent
Rhubarb	——	Natural plant extract		Histology
Rivanol	——	Acridine	B	Supravital (protozoa), bacteria, leukocytes, cell nuclei, Schiff-type reagent, mast cells
Rose bengal	45400	Xanthene	A	Fat, liver, bacteria, hepatic function determination
Safranin O	50240	Azine	B	Histology, nuclei and chromosomes, Schiff-type reagent, plant tissues, starch granules
Sanguinarine	——	Natural plant alkaloid		Insect tissues
Stilbene	40000	Stilbene	A	Proteins, insect tissues
Sulphorhodamine B (Lissamine rhodamine B 200)	45100	Xanthene	A	Immunofluorescence label
Tetracycline	——	Natural	A	Bone growth, mitochondria, cancer localization, antibiotic
Thiazin red R	14780	Monoazo	A	Fat, amyloid, proteins
Thiazole yellow G (Titan yellow)	19540	Monoazo	A	Histology, vital fluorochrome, Mg detection

Table 2.2 Continued

Fluorochrome	CI no.	Chemical group	Acidic or basic	Biological applications
Thioflavine S	49010	Thiazole	A	Histology (fluorochromasia), intravital fluorochrome (blood vessels), leukocytes, bacteria, amyloid, protozoa, myelin
Thioflavine T	49005	Thiazole	B	Histology, tubercle bacilli, phospholipids, mast cells, insect histology, amyloid
Titan yellow (see thiazole yellow G)				
TMPP (meso-tetra (4-N-methylpyridyl) porphine		Porphyrin	B	Chromatin (DNA)
TPPS (tetraphenylporphin sulphonate)		Porphyrin	A	Elastic fibres
Trypaflavine (see acriflavine)				
Trypan blue		Disazo	A	Dye exclusion test for cell vitality, teratogen, fluoresces red when bound to protein
Uranin (see fluorescein)				
Uvitex 2B		Stilbene	A	Fungi in tissue sections
Vasoflavine (see thioflavine S)				
Xylenol orange	——		A	Sites of calcification (bone growth)

[a] The word *fluorochrome* was coined by Haitinger in 1934 to denote fluorescent dyes used in biological staining to induce secondary fluorescence in tissues. Data listed above were derived from many sources. In addition to obtaining information from published research and review articles, other material was assembled from the *Colour Index* (1971), *Reichert's Fluorescence Microscopy with Fluorochromes* (1952), *Conn's Biological Stains* (1977), *Handbuch der Farbstoffe für die Mikroskopie* (Harms, 1965), and catalogues of the Aldrich Chemical Co., Eastman Kodak Co., Polysciences, Inc., and Sigma Chemical Co.

(MacInnes & Uretz, 1966). As a stain for DNA in chromosomes, AO ordinarily gives uniform fluorescence along the length of chromosome arms. However, the dye produces reverse banding (R-bands) when it is used after pretreatment with the antibiotics distamycin (AT-specific) or actinomycin D (GC-specific) or hot phosphate buffer (Comings, 1978; Gustashaw, 1991). Microfluorometry was employed to obtain quantitative information about the content of RNA and DNA in single cells and DNA molecular alterations (Rigler, 1966).

The metachromatic fluorochrome AO has proved to be a valuable nucleic acid probe in modern flow cytometry when conditions of dye binding are well-controlled. For reviews, see Melamed and Darzynkiewicz (1981) and Darzynkiewicz (1991). A review of the older AO literature is given by Kasten (1967). The chapter by Zelenin (Chapter 5) in the present volume gives additional details about AO as a fluorescent probe.

Hamperl and Haitinger, Strugger, and Schümmelfeder, dozens of fluorescent dyes have been tested for histological and histochemical specificity. Some have been used as vital fluorochromes. Table 2.2 gives a general listing of the dyes and their applications.

In modern cytochemistry and cell biology, fluorescent probes have been designed and employed as fluorescent tags. Some, like fluorescein, TRITC, phycoerythrin, and Bodipy, are covalently bound to isothiocyanates or linked to chlorotriazinyl derivatives and hydroxysuccinimido esters. They are water-soluble and have been attached to antibodies, lectins, hormones and other macromolecules. Other fluorescent probes are non-covalently linked to macromolecules, ions and organelles within cells. Some probes have been used to evaluate cell viability, pH and membrane potentials. A summary of many of these fluorescent probes is given in Table 2.3. Other chapters in this volume give further information about specific groups of probes.

2.6 GENERAL APPLICATIONS OF FLUORESCENT PROBES

Since the employment of fluorochromes in microscopy by the early pioneers, Bommer, Ellinger and Hirt,

ACKNOWLEDGEMENTS

It is a pleasure to recognize the contribution of Ms Paula Porter, who typed the manuscript. My research is supported in part by the Biological Stain Commission.

Table 2.3 Fluorescent probes used in modern cell and molecular biology.[a]

Name	Wavelength (nm) Excitation	Emission	Applications
Immunocytochemical fluorophores, conjugates and lectins			Tracers for various proteins, receptors, and mono- and polysaccharides
Allophycocyanin (AP)	620	660	
Allophycocyanin cross-linked (AP-XL)	650	660	
AMCA (7-amino-4-methylcoumarin-3-acetic acid	350	450	
Bodipy (borondipyromethene difluoride dye)	505	512	
CT-120 (coumarin 120 thiolactone)	345	410	
CT-339 (coumarin 330 thiolactone)	350	420	
Coumarin 138	365	460	
Dansyl chloride	340	578	
2,7'-Dichlorofluorescein	513	532	
4',5'-Dimethylfluorescein	510	535	
5-DTAF (5-(4,6-dichloro-triazinyl) aminofluorescein)	495	530	
Eosin-5-isothiocyanate	524	548	
Erythrosin-5-isothiocyanate	535	558	
FITC (fluorescein-5-isothiocyanate)	490	520	
Fluorescein anhydride (FA)	490	520	
3-HFT (3-hydroxyflavone thiolactone)	350	415	
Lucifer yellow CH	435	530	
N-Methylanthranyloyl	350	440	
NBD (nitrobenzoxadiazole)	468	520	
Phycocyanin (PC)	620	650	
Phycoerythrin B (PE-B)	545	576	
Phycoerythrin R (PE-R)	495, 545	578	
Princeton red anhydride (PRA)	490	580	
RITC (rhodamine B isothiocyanate)	570	595	
Sulphorhodamine B sulphonyl chloride (lissamine rhodamine B sulphonyl chloride)	570	590	
Sulphorhodamine 101 (sulphonyl chloride, Texas Red)	596	620	
TMRA (tetramethylrhodamine anhydride)	550	570	
TRITC (tetramethylrhodamine-5-isothiocyanate)	541	572	
XL-Allophycocyanin (XL-AP)	620	660	
XRITC (tetra-N-cyclopropyl-rhodamine isothiocyanate)	578	604	
Site-selective probes (vital)			
Acridine orange	490	590	Lysosomes, nuclei, fluorescent counterstain for retrogradely labelled neuronal tracers
Acridine orange-10-dodecyl bromide	493	520	Mitochondria
Alizarin complexone	580	645	Sites of calcification (bone growth)
AMA (3-amino-6-methocyacridine)	UV	Y,G	Lysosomes, nuclei (fluorochromasia)
Bis-ANS	385	500	Inhibitor of microtubule assembly and RNA polymerase
Bisbenzimide	360	530	Retrograde labelling of neuronal nuclei
Calcein	495	520	Sites of calcification (bone growth)
Calcein blue	375	435	Sites of calcification (bone growth)
Cascade Blue hydrazide	376, 389	423	Covalent labelling of microinjected cells
Colcemid-NBD	468	520	Tubulin polymerization detection
DACK (dansylalanyllysyl chloromethyl ketone)	350	450	Inhibitor of acrosin (mammalian sperm)

Table 2.3 Continued

Name	Wavelength (nm)		Applications
	Excitation	Emission	
DAPI (4',6-diamidino-2-phenylindole)	372	456	Tubulin polymerization detection without interfering with microtuble assembly, retrograde labelling of neurons
DASPEI (2-(4-dimethyl aminostyryl)-N-ethyl-pyridinium iodide)	429	557	Mitochondria (metabolic state affects fluorescence response)
DASPMI	429	557	Mitochondria (yellow), membranes (green), nucleus (red-orange)
DEQTC (1,3'-diethyl-4,2'-quinolylthiacyanine iodide)	502	529	Reticulocytes
5,7-DHT (5,7-dihydroxytryptamine)	UV-B	G	Injection induces fluorescence in distant populations of amacrine cells of retina
DiIC$_{18}$ (3) (1, 1'-dioctadecyl-3,3,3',3'-tetramethylindocarbocyanine perchlorate)	547	571	Long-term cell tracing *in vitro*
Dihydroethidium (*see* Nucleic acid probes and Cell viability probes)			
Dihydrorhodamine	UV (510)	320 (534)	Colourless but oxidizes inside cells to rhodamine 123 that vitally stains mitochondria
DiOC$_1$(3)	482	510	Reticulocytes
DiOC$_6$(3) (3,3'-dihexyloxacarbocyanine iodide)	478	496	Endoplasmic reticulum (also in fixed cells), mitochondria
DiOC$_7$(3) (3,3'-diheptyloxacarbocyanine iodide)	488	540	Penetration probe into spheroids or tumour cords, mitochondria of plant cells, distinguish cycling from non-cycling fibroblasts
DiSC$_1$(3) (3,3'-dimethyl-thiacarbocyanine iodide)	551	568	Endoplasmic reticulum
Dopamine	UV B	G	Injection induces fluorescence in distinct populations of amacrine cells of retina
DPPAO (lecithin analogue of acridine orange)	UV	G	Mitochondria (also in fixed cells)
Ethidium bromide	545	610	Useful Nissl fluorochrome after FITC labelling of nervous system
Evans blue	550	610	Retrograde labelling of neuronal cytoplasm
Fast blue (*trans*-1-(5-amidino-2-benzofuranyl)-2-(6-amidino-2-indolyl) ethylene dihydrochloride	360	410	Retrograde labelling of neuronal cytoplasm
FITC-dextran	490	520	Fluid phase pinocytosis, loading cells with macromolecules
FluoroBora 1 (3-(dansylamido) phenylboronic acid)	UV-B	Y-G	Lysosomes
FluoroBora 2 (3-(darpsylamidyl)-1-phenylboronic acid)	UV	B-W	Golgi apparatus
FluoroBora T-acriflavine (3-amino, 6-7'(7',8',8'-tri-)cyanoquinodimethane phenyl-boronic acid)	470	530, 590	Lipid- and water-soluble FluoroBora with trace of acriflavine penetrates cells and produces yellow-green chromatin and orange cytoplasm
Fluoro-Gold	323	408	Retrograde labelling of neuronal cytoplasm, fluoresces gold at neutral pH and blue at acid pH
Granular blue (2-(4-(4-amidinophenoxy) phenyl) indol-6-carbox-amidin-dihydrochloride)	375	410	Retrograde labelling of neuronal cytoplasm
Hexanoic ceramide-NBD (C$_6$-NBD-ceramide)	475	525	Golgi apparatus and lipid transport pathways in cytoplasm
Hoechst 33258 (bisbenzimide trihydrochloride)	365	465	Mycoplasma detection, chromosomal bands and interbands
Hydroethidine (*see* dihydroethidium under Nucleic acid probes and Cell vitality probes)			

Table 2.3 Continued

Name	Wavelength (nm)		Applications
	Excitation	Emission	
Lucifer yellow CH (LY)	430	535	Neurons and functional connections after microinjection into cells or fluid phase pinocytosis
Merocyanine 540	500	572	Mitochondria, binds to leukaemic cells, axons stain
MUA (4-methyl-umbelliferone derivatives)	340	430	Lysosomal enzymes, used to detect lysosomal storage diseases
NAO (10-N-nonyl-acridine orange chloride)	492	522	Mitochondria (also in fixed cells)
Nile red (Nile blue A oxazone)	450–500 515–560	530 605	Neutral lipids, cholesterol, phospholipids in cellular cytoplasmic droplets and lysosomes, foam cells and lipid-loaded macrophages, (excitation and emission spectra vary greatly according to hydrophobicity of environment)
NPN (N-phenyl-1-naphthylamine)	340	420	Detects early lymphocyte activation
Nuclear yellow (Hoechst S-769121)	360	460	Retrograde labelling of neuronal nuclei
Phallacidin-Bodipy	505	512	F-actin
Phallacidin-NBD	468	520	F-actin
Phalloidin-fluorescein	490	520	F-actin
Phalloidin-rhodamine	540	580	F-actin
Phallotoxin-phenylcoumarin	387	465	F-actin
Procion yellow M4RS (MX-4R)	488	530	Neurons and functional connections after electrophoretic injection into cells
p-Bis-(2-chloroethyl)-amino-benzilidene-cinnamonitrile fluorochromes (nitrogen mustard derivatives with stilbene-like structures)	360–400 450–480	520–550 550–590	Cell fluorochromes useful in combination with autofluorescing coenzymes
Pyronin Y	545	580	Mitochondria, arrests cells in G1 phase
Rhodamine 123	510	534	Mitochondria (increased accumulation and retention in mitochondria of carcinoma cells), also fixed cells
Rhodamine B hexyl ester, chloride	555	579	Endoplasmic reticulum
Rhodamine-dextran	570	595	Fluid phase pinocytosis, loading cells with macromolecules
Rhodamine 6G	530	590	Mitochondria
Texas Red-ovalbumin	596	620	Absorptive pinocytosis
Thiazole orange (TO)	509	533	Reticulocytes and malaria parasites (haematology and flow cytometry)
Thioflavin T	370	418	Amyloid plaque core protein (APCP), reticulocytes (flow cytometry)
'True Blue' (trans-1,2,-bis (5-amido-2-benzofuranyl) ethylene-dihydrochloride	373	404	Retrograde labelling of neuronal cytoplasm
Tubulin-DTAF	495	530	Microtubules
Tubulin-NBD	468	520	Microtubules
Xylenol orange	377	610	Sites of calcification (bone-growth)
Nucleic acid probes			
ACMA (9-amino-6-chloro-2-methoxyacridine)	430	474	DNA, chromosome Q-banding, AT-specific DNA
Acridine ethidium heterodimer	492 528	627 634	AT-rich DNA and total DNA according to excitation used
Acridine orange	490	530, 640	Distinguish and measure single-and double-stranded nucleic acids by fluorochromasia (intercalates into double-stranded nucleic acids, binds to phosphate groups), chromosome banding
Acriflavine-Feulgen	455	515	DNA
Adriamycin	480	555	Intercalates in GC-specific DNA, chromosome D banding, Y chromosome, antineoplastic anthracycline (mostly nuclear fluorescence)

Table 2.3 Continued

Name	Wavelength (nm)		Applications
	Excitation	*Emission*	
7-Amino-AMD (7-amino-actinomycin D)	555	655	Intercalates in GC-specific DNA
Auramine O-Feulgen	460	550	DNA
BAO-Feulgen (bisamino-) phenyloxadizole)	380	470	DNA
Bis-ANS	385	500	Inhibits RNA polymerase
Carminomycin	470	550	Antineoplastic anthracycline (cytoplasmic fluorescence)
Chromomycin A3	450	570	GC-specific DNA in presence of Mg^{2+}, antineoplastic antibiotic
DAMA (3-dimethylamino-6-methoxyacridine)	467	549, 617	RNA and DNA (fluorochromasia)
DAPI (4′,6-diamidino-2-phenylindole HCl)	372	456	AT-specific double-stranded DNA, chromosome Q banding, distinguish between yeast mitochondrial and nuclear DNA, viral and mycoplasma DNA infection in cells
Daunomycin (dauno-rubicin HCl)	475	550	DNA, chromosome D banding, Y chromosome, antineoplastic anthracycline (nuclear fluorescence)
Dihydroethidium (hydroethidine, reduced ethidium bromide)	370, 525	420, 605	DNA (hydroethidine is enzymatically oxidized in living cells to form ethidium bromide, which intercalates into nuclear chromatin (red), cytoplasm fluoresces blue-white in lipoidal pockets)
DiOC$_1$(3)	482	510	Nucleic acids
DIPI (4′,6-bis(2′-imidazol-inyl-4H,5H)-2 phenylindole)	355	450	AT-specific DNA, chromosome Q-banding
Ellipticine	450	525	RNA and DNA (intercalates into double-stranded nucleic acids)
Ethidium bromide (homidium bromide)	545	610	RNA and DNA (intercalates into double-stranded nucleic acids of fixed cells), viability assay
Ethidium monoazide	510	600	RNA and DNA (intercalates into double-stranded nucleic acids), viability assay
Hoechst 33258 (bisbenzimide trihydrochloride)	365	465	AT-specific DNA (vital and fixed cells), chromosome Q-banding, DNA synthesis quenching of fluorescence detects incorporation of 5-BrdU into DNA), viral and mycoplasma DNA infection in cells
Hoechst 33342 (bisbenzimidazole derivative)	355	465	AT-specific DNA (vital and fixed cells), may be effluxed rapidly from certain cells so as to prevent DNA binding
Homoethidium (*see* dihydroethidium)			
Hydroxystilbamidine	360	450, 600	AT-specific DNA (both peaks appear) and RNA (only 450nm emission seen)
Mithramycin (aureolic acid)	395	570	GC-selective DNA, antineoplastic antibiotic
Nogalomycin	480	560	DNA, antineoplastic anthracycline
Olivomycin	430	545	DNA, chromosome R banding, antineoplastic antibiotic
Oxazine 750 (0X750)	690	699	DNA (excitable by helium–neon laser)
Proflavine	455	515	Nucleic acids, AT-specific DNA
Propidium iodide	530	615	RNA and DNA (intercalates into double-stranded nucleic acids of fixed cells), viability assay
Pyronin Y	540	570	RNA (preferentially in presence of methyl green or Hoechst 33342)
Quinacrine dihydrochloride (atabrine)	436	525	AT-specific DNA, chromosome Q-banding, Y chromosome
Quinacrine mustard	385	525	AT-specific DNA, chromosome Q-banding

Table 2.3 Continued

Name	Wavelength (nm)		Applications
	Excitation	Emission	
Rhodamine 700(LD 700)	659	669	DNA (excitable by helium–neon laser)
Rhodamine 800(R 800)	700	715	DNA (excitable by helium-neon laser)
Rubidazone	480	560	Antineoplastic anthracycline (cytoplasmic fluorescence)
RuDIP (tris-(4,7-diphenyl-phenanthroline) ruthenium (III))	453 470	480 630	Metal complex to distinguish handedness of DNA helices
Thiazole orange (TO)	509	533	Nucleic acids (see site-selective probes)
Thioflavine T	422	487	Nucleic acids (see Site-selective probes)
TMPP (Meso-tetra (4-N-methylpyridyl) porphine)	436	655	DNA
4,5′,8-Trimethylpsoralen (trioxsalen)	338	420	Intercalates into double-stranded DNA and covalently adds to pyrimidines upon UV illumination
'True Blue' (trans-1,2-bis (5-amido-2-benzofuranyl) ethylene-dihydrochlorine)	373	404	AT-specific DNA
Protein probes and functional groups			
Acridine orange	490	530	Acidic groups of proteins after hot TCA extraction
Ammonium 7-fluoro-2-oxa-1,2-diazole-4-sulphonate	380	515	Thiol groups
ANS (8-anilino-1-naphthalene sulphonic acid)	385	485	Proteins (hydrophobic probe)
Anthracene-9-carboxaldehyde carbohydrazone	393	456	Aldehyde probe
Bis-ANS (1,1′-bi(4-anilino) naphthalene-5,5′-disulphonic acid, dipotassium salt)	385	500	Proteins (dimer of ANS, binds at multiple sites)
Brilliant sulphoflavine (BSF)	420	520	Histone proteins at pH 8.0 after DNA extraction, total proteins at pH 2.8
CPM (N-(4-(7-diethylamino-4-methylcoumarin-3-yl) maleimide	385–390	465	Thiol groups
DAB-ITC (4-N,N-dimethylamino-benzene-4′-isothiocyanate)	430		Amino acid probe
Dansyl chloride (5-dimethylamino-1-naphthalene-sulphonyl chloride)	335	500	Histone proteins and protamines
Eosin	522	551	Histones at pH 10.0 or higher
FITC (fluorescein-5-isothiocyanate)	490	520	Proteins
Fluoral-P (4-amino-3-pentene-2-one)	410	510	Aldehyde probe
Fluorescamine	390	460	Proteins at cell surface (primary amine binding creates fluorescent complex)
FDA (fluorescein diacetate)	490	520	Esterases
Fluorescein mercuric acetate	B	Y	Thiol groups of proteins (nuclear non-histone proteins after SH reduction to disulphides)
Formaldehyde-induced monoamine fluorophores	410–415	475, 525	Biogenic monamines (dopamine, noradrenaline, 5-hydroxytryptamine)
9-Hydrazine acridine	420	500	Aldehyde probe
'Long Name' (stilbene disulphonic acid derivative, optical brightener)	350	460	Proteins
Mercurochrome	B	Y	Thiol groups of proteins (nuclear nonhistone proteins after SH reduction to disulphides)
Mercury orange (1(4-chloro-mercury-phenyl-azo-2-naphthol))	450–490	600	Glutathione (non-aqueous solvents), thiol groups proteins (nuclear nonhistone proteins after SH reduction to disulphides)
MUA (4-methylumbelliferone acetate)	540	430	Esterases

Table 2.3 Continued

Name	Wavelength (nm)		Applications
	Excitation	Emission	
OPT (0-phthaldehyde)	UV	B	Polyamines at pH 7–9 (spermidine and spermine)
Primuline	365	455	Proteins
RITC (rhodamine B-isothiocyanate)	570	595	Proteins
Salicoyl hydrazine	320	400	Aldehyde probe
SBD-Cl (4-chloro-7-sulphobenzo-furan, ammonium salt)	380	510	Thiol groups
SITS (4-acetamido-4-isothiocyanatostilbenene-2, 2'-disulphonic acid, disodium salt)			
Sulphorhodamine 101	576	602	Proteins
XRITC (tetra-N-cyclopropyl-rhodamine isothiocyanate)	578	604	Proteins
Cell viability probes			
Acridine orange	490	530, 640	Vital fluorochrome (monomeric dye form is green in living cells, aggregated dye form is red in dead cells)
Calcein AM (acetoxymethyl ester)	495	520	Vital fluorochrome
Calcofluor white M2R (CFW)	UV	B	Stains non-viable animal cells, walls of live plant cells
CFDA (5(6)-carboxyfluorescein diacetate)	495	520	Cell vitality probe based on production of intracellular carboxyfluorescein by esterases, detects permeable channels between cells
Chrysophosphine 2G (euchrysine 2GNS)	435	515	Vital fluorochrome (similar to acridine orange)
Dihydroethidium (Hydroethidine, reduced ethidium bromide)	370, 535	420, 585	Nuclei fluoresce red in living cells due to production of ethidium bromide by enzymatic oxidation of dihydroethidine, cytoplasm blue to blue white
Dihydrorhodamine 123	320		Colourless but oxidizes inside cells to rhodamine 123 that vitally stains mitochondria
Ethidium monoazide	460	600	Fluorochromes dead cells, usable after cell fixation
FDA (fluorescein diacetate)	490	520	Cell vitality probe based on production of intracellular fluorescein by esterases, may be used in combination with propidium iodide
Fluorescein	490	520	Detection of fluorescence depolarization protein-bound dye) in living cells using fluorescence anisotropy measurements
Fluorescein digalactoside	490	520	Monitoring galactosidase gene activity by formation of intracellular fluorescein
FluoroBora T	470	590	Vital fluorochrome (*see* other FluoroBora derivatives under Site-selective probes and Membrane probes and receptors)
Hydroethidine (*see* dihydroethidium)			
Propidium iodide	470	615	Stains dead cells with membrane damage (used with green fluorescing vital dyes like fluorescein)
Pyronin Y	540	570	Vital fluorochrome (mitochondria)
Rhodamine 123	510	534	Vital fluorochrome (mitochondria), accumulates in carcinoma cells, sperm, distinguish cycling from non-cycling cells
SITS (4-acetamido-4-isothiocyanatostilbene-2,2'-disulphonic acid, disodium salt)	350	420	Vital fluorochrome
Vita blue dibutyrate-14 (VBDB-14)	524	570	Vital fluorochrome based on production of fluorescent derivative of fluorescein by esterases
Membrane probes and receptors (vital)			
Acridine orange	490	530, 640	Multilamellar liposomes
Acridine orange-10-dodecyl bromide	500	534, 568	Surfactant micelles (monomer-dimer spectral change)

Table 2.3 Continued

Name	Wavelength (nm) Excitation	Emission	Applications
9-Amino-acridyl propranolol	B	Y	β-Adrenergic receptors
ANS (8-anilino-1-naphthalene sulphonic acid)	385	485	Localizes at interfaces of hydrophilic and hydrophobic membrane regions
Anthroyl ouabain	362, 381	485	Cardiac glycoside receptors
α-Bungerotoxin-tetramethylrhodamine	518, 551	575	Cholinergic receptors
Dansyl lysine	340	515	Membranes with low cholesterol content, leukaemic cells
Dansyl phorbol acetate	341	496	Tumour promoter analogue binds to receptors
Dexamethasone rhodamine	542	566	Glucocorticoid receptors
DiIC$_{18}$ (3) (1,1′-dioctadecyl-3,3,3′,3′-tetramethylindocarbocyanine perchlorate)	549	568	Cationic lipophilic probe used to study fusion and lateral diffusion in membranes, retained for long periods within neurons
DPH (diphenylhexatriene)	351	430	Hydrophobic probe, fluorescence polarization probe of membrane fluidity
Fluorescamine	390	460	Cell surface proteins (non-fluorescent until bound to NH_2-groups)
FluoroBora P (3-(pyrenesulphamido-phenyl-boronic acid))	UV	W-Y,V	Hydrophilic areas (white-yellow), hydrophobic areas (violet)
'Long Name' (stilbene disulphonic acid derivative, optical brightener)	350	460	Intracellular membranes of granulocytes
Naloxone fluorescein	490	520	Opioid receptors
NPN (*N*-phenyl-1-naphthylamine)	340	420	Hydrophobic probe (non-fluorescent until bound to cell membranes or lipids)
1-Pyrene butyryl choline bromide	342	378–420	Synaptic localization
Rhodamine B octadecyl ester	560	590	Hydrophobic probe, membrane fusion assay
TMA-DPH	351	430	Outer plasma membrane, fluidity
Cell membrane potentials and pH (vital)			
ADB 1,4 diacetoxy-2,3-dicyano-benzene	351	450–476	Intracellular pH (emission peak shifts in alkaline state)
BCECF (2′7′biscarboxyethyl-5,6-carboxyfluorescein	500	530, 620	Intracellular pH
BCECF-AM (2′,7′bis-carboxy-ethyl-5,6-carboxyfluorescein tetraacetoxymethyl ester)	500	530, 620	Intracellular pH (transmitted into cells and enzymatically hydrolysed by esterases to BCECF)
Bis-oxonol (DiSBa-C$_2$ (3))	540	580	Membrane potentials
Carboxy SNAFL-2 (semi-naphthofluorescein)	485, 514 (acid) 547 (base)	546 (acid) 630 (base)	Monoexcitation–dual-emission pH indicator
Carboxy SNARF-1 (semi-naphthorhodofluor)	518, 548 (acid) 574 (base)	587 (acid) 630 (base)	Monoexcitation–dual-emission pH indicator
1,4-Diacetoxyphthalonitrile	350	420–440 500–580	Intracellular pH
DIDS (4,4′-diisothiocyano-2,2′-stilbenedisulphonic acid)	340	430	Anion transport inhibitor
DiOC$_7$(3) (3,3′-diheptyloxacarbo-cyanine iodide)	482	511	Membrane potentials
DiOC$_2$(3)	482	500	Membrane potentials
DiOC$_2$(5)	579	603	Membrane potentials
9-(*N*-Dodecyl)aminoacridine	430	475	pH gradients across membranes
Merocyanine 540	500	572	Membrane potentials, differentiation marker in cancer research
Oxonol V	609	645	Membrane potentials
Rhodamine 123	510	534	Mitochondrial and plasma membrane potentials
Vita blue dibutyrate-14 (VBDB-14)	609 (base) 524 (acid)	665 (base) 570 (acid)	Intracellular pH (dual fluorescence)

Table 2.3 Continued

Name	Wavelength (nm) Excitation	Emission	Applications
WW 781	603	635–645	Membrane potentials
Ionic probes (vital)			
Aequorin	———	469	Luminescent protein that emits light in presence of Ca^{2+}
9-Anthronyl choline iodide	320	420	Ca^{2+} binding to calmodulin
Calcein (active part is DCAF, 3,6-dihydroxy-24-bis-(N,N'-di (carbomethyl)-aminomethyl) fluoran (Fluorexone)	305, 490	520	Ca^{2+} and Mg^{2+}
CTC (7-chlortetracycline, Aureomycin)	345 400	430 520	Free Ca^{2+} near membranes
Fluo-3	506	526	Ca^{2+}
Fura-2	335, 362	505	Ca^{2+} (bound Ca^{2+} excites maximally at 335 nm while free dye excites at 362 nm ratio imaging)
Furaptra (mag-fur-2)	376 (low ion) 344 (high ion)	506 492	Mg^{2+} (microscopy and ratio imaging)
Indo-1	331	410	Ca^{2+}
Mag-indo-1	354 (low ion) 349 (high ion)	475 419	Mg^{2+} (flow cytometry)
PBF1	346 (low ion) 334 (high ion)	551 525	K^+
Quin-2	339	492	Ca^{2+}
Rhod-2	553	576	Ca^{2+}
SBFI	340/380	505	Na^+ microinjected into cells or by use of AM ester)
SPQ	344	450	Cl^-
TnC$_{DABZ}$ (troponin C dansyl-aziridinc)	340	514	Ca^{2+} binding to Ca^{2+}-specific regulatory sites of troponin C
TSQ	335	376	Zn^{2+} in presynaptic boutons

Information compiled from numerous sources, including published research articles and catalogues of companies that specialize in supplying fluorescent probes to immunologists and cell and molecular biologists. Many of the probes listed above are employed in flow cytometry. I have utilized the *Handbook of Fluorescent Probes and Research Chemicals* (Haugland, 1992) and published material supplied by Biomeda Corporation, Eastman Kodak, Sigma Chemical Co., and Polysciences, Inc. I am particularly grateful to Dr Paul Gallop (Massachusetts General Hospital), Dr Natalie S. Rudolph (Viomedics), Dr Richard P. Haugland (Molecular Probes), and Ms Larissa Korytko (Eastman Kodak Co.), who provided valuable information and gave generously of their time in discussing particular fluorescent probes. The spectral data shown represent published excitation and emission peaks. However, slight spectral shifts can be expected according to the solvent used, pH of the system, and whether or not the probe is bound to its substrate. In a few cases, colours are given (B = blue, G = green, UV = ultraviolet, W = white, Y = yellow). Trademarks are assigned for Bodipy (Molecular Probes), Cascade Blue (Molecular Probes), Cellufluor (Polysciences), FluoroBora (Childrens Hospital), Fluoro-Gold (Fluorochrome, Inc.), Hydroethidine (Prescott Labs.), Lissamine (Imperial Chemical Industries), and SNAFL, SNARF, and Texas Red (Molecular Probes).

REFERENCES

Albert A. (1951) *The Acridines, Their Preparation, Properties and Uses*. Arnold, London.

Aldridge W.G. & Watson M. (1963) *J. Histochem. Cytochem.* **11**, 773–781.

Armstrong J.A. (1956) *Exp. Cell Res.* **11**, 640–643.

Armstrong J.A. & Niven J.S.F. (1957) *Nature* **180**, 1335–1336.

Austin C.R. & Bishop M.W.H. (1959) *Exp. Cell Res.* **17**, 33–43.

Barch M.J. (ed.) (1991) *The ACT Cytogenetics Laboratory Manual* (2nd edition). Raven Press, New York.

Bertalanffy L.v. & Bickis I. (1956) *J. Histochem. Cytochem.* **4**, 481–493.

Bock E. & Oesterlin M. (1939) *Zbl. Bakteriol.* **143**, 306–318.

Bollinger A., Franzeck U.K. & Jäger K. (1983) *Prog. Appl. Microcirc.* **3**, 97–118.

Bommer S. (1929) *Acta Derm. Venereol.* **10**, 253–315.

Bradley D.F. (1961) *Trans. NY Acad. Sci.* **24**, 64–74.

Braütigam F. & Grabner A. (eds) (1949) *Beiträge zur Fluoreszenzmikroskopie.* Verlag Georg Fromme & Co., Vienna.

Browning C.H. (1922) *Nature* **109**, 750–751.

Browning C.H. (1964) In *Experimental Chemotherapy*, Vol. 2, Pt 1, *Chemotherapy of Bacterial Infections*, R.J. Schnitzer & Frank Hawking (eds). Academic Press, New York, pp. 2–36.

Browning C.H., Gulbransen R., Kennaway E.L. & Thorton L.H.D. (1917) *Br. Med. J.* **1**, 73–78.

Bucher O. (1939) *Z. Zellforsch.* **29**, 283–322.

Bucherer H. (1943) *Zbl. Bakteriol.* **106**, 81–88.

Bukatsch F. (1940) *Z. Gesamte Naturwissen.* p. 90.

Bukatsch F. (1941) *Zts. ges. Naturwiss.* **7**, 288 (cited Strugger, 1946, p. 4).

Bukatsch F. & Haitinger M. (1940) *Protoplasma* **34**, 515–523.

Colour Index (1971) (3rd edition). Society of Dyers and Colourists, Bradford, Yorkshire, UK.

Comings D.E. (1978) *Ann. Rev. Genet.* **12**, 25–46.

Dart L.H. Jr. & Turner T.R. (1959) *Lab. Invest.* **8**, 1513–1522.

Darzynkiewicz Z. (1979) In *Flow Cytometry and Sorting*, M.R. Melamed, P.F. Mullaney & M.L. Mendelsohn (eds). John Wiley, New York, pp. 283–316.

Darzynkiewicz Z. (1991) In *Flow Cytometry*, Z. Darzynkiewicz & H.A. Crissman (eds). Academic Press, San Diego, pp. 285–298.

Darzynkiewicz Z. & Crissman H.A. (eds) (1990) *Flow Cytometry*. Academic Press, San Diego.

DeBruyn P.P.H., Farr R.S., Banks H. & Morthland F.W. (1953) *Exp. Cell Res.* **4**, 174–180.

Drawert H. (1968) *Vitalfärbung und vitalfluorochromierung Pflanzlicher Zellen und Gewebe.* Protoplasmatologia Bd. 2. Teil D. Heft 3, Springer-Verlag, Vienna.

Duijn C. van Jr. (1955) *Microscope* **10**, 122–128.

Ehrlich P. (1882) *Dtsch. Med. Wschr.* **8**, 21–22, 36–37, 54–55.

Ehrlich P. & Benda L. (1913) *Ber. Dtsch. Chem. Gesel.* **46**, 1931–1951.

Ellinger P. (1940) *Biol. Rev. Cambridge Philos. Soc.* **15**, 323–350.

Ellinger P. & Hirt A. (1929a) *Z. Anat. Entwcklgeschl.* **90**, 701–802.

Ellinger P. & Hirt A. (1929b) *Arch. Exp. Pathol. Pharmakol.* **149**, 285–297.

Eppinger H. (1949) In *Beiträge zur Fluoreszensmikroskopie*, F. Braütigam & A. Grabner (eds). Verlag Georg Fromme & Co., Vienna, pp. 37–45.

Exner R. & Haitinger M. (1936) *Psychiatr. Neurol. Wschr.* **38**, 183–187.

Fay I.W. (1911) *Chemistry of the Coal-Tar Dyes*. D. Van Nostrand, New York.

Fischl V. & Singer E. (1935) *Z. Hyg. u. Infekt.* **116**, 348–355.

Franke F. & Sylla A. (1933) *Z. Exp. Med.* **89**, 141–158.

Giloh H. & Sedet J.W. (1982) *Science* **217**, 1252–1255.

Grafflin A.L. (1938) *J. Cell. Comp. Physiol.* **12**, 167–170.

Green F.J. (1990) *The Sigma-Aldrich Handbook of Stains, Dyes and Indicators*. Aldrich Chemical Co., Inc., Milwaukee.

Gustashaw K.M. (1991) In *The ACT Cytogenetics Laboratory Manual* (2nd edition), M.J. Barch (ed.). Raven Press, New York, pp. 205–269.

Hagemann P.K.H. (1937a) *Dtsch. Med. Wschr.* **63**, 514–518.

Hagemann P.K.H. (1937b) *Münch. Med. Wschr.* **84**, 761–765.

Hagemann P.K.H. (1937c) *Münch. Med. Wschr.* **20**, 761–765.

Hagemann P.K.H. (1938) *Münch. Med. Wschr.* **85**, 1066–1068.

Hagemann P.K.H. (1939) *Arch. Exp. Zellforsch.* **22**, 459–462.

Haitinger M. (1934) In *Abderhalden's Handbuch der Biologischen Arbeitsmethoden*, Abt. 11, Teil 3. Urban & Schwartzenberg, Berlin, pp. 3307–3337.

Haitinger M. & Geiser P. (1944) *Virch. Arch.* **312**, 116–137.

Haitinger M. & Hamperl H. (1933) *Z. Mikrosk. Anat. Forsch.* **33**, 193–221.

Haitinger M. & Linsbauer L. (1933) *Beih. Bot. Zbl.* **50**, 432–444.

Haitinger M. & Linsbauer L. (1935) *Bot. Zbl.* **53**, 387–397.

Hamperl H. (1934) *Virch. Pathol. Anat. Arch.* **292**, 1–51.

Harms, H. (1965) *Handbuch der Farbstoffe für die Mikroskopie*. Staufen Verlag, Kamp-Lintford.

Hata S. (1932) *Kitasato Arch. Exp. Med.* **9**, 1–71.

Haugland R.P. (1992) *Handbook of Fluorescent Probes and Research Chemicals*. Molecular Probes, Eugene, Oregon.

Haugland R.P. (1990) In *Optical Microscopy for Biology*, B. Herman & K. Jacobson (eds). Wiley-Liss, New York, pp. 143–157.

Heimstädt O. (1911) *Z. Wiss. Mikrosk.* **28**, 330–337.

Hercik F. (1939) *Protoplasma* **32**, 527–535.

Herman B. & Jacobson K. (1990) *Optical Microscopy for Biology*. Wiley-Liss, New York.

Heuven J.A. van (1936) *Ned. Tijdscher. Geneesk.* **80**, 1728–1732.

Jacoby F., Medawar P.G. & Willmer E.N. (1941) *Br. Med. J.* **2**, 149–153.

Jancśo N. von (1932) *Klin. Wschr.* **11**, 689.

Johnson G.D., Davidson R.S., McNamee K.C., Russel G., Goodwin D. & Holborow E.J. (1982) *J. Immunol. Meth.* **55**, 231–242.

Kaiserling C. (1917) *Z. Tuberc.* **27**, 156–161.

Kapuscinski J. & Darzynkiewicz Z. (1990) In *Flow Cytometry*, Z. Darzynkiewicz & H.A. Crissman (eds). Academic Press, San Diego, pp. 655–669.

Kasten F.H. (1964) In *100 Years of Histochemistry in Germany*, W. Sandritter & F.H. Kasten (eds) F.K. Schattauer-Verlag, Stuttgart, pp. 97–101.

Kasten F.H. (1967) *Int. Rev. Cytol.* **21**, 141–202.

Kasten F.H. (1973) In *Encyclopedia of Microscopy and Microtechnique*, P. Gray (ed.). Reinhold, New York, pp. 4–7.

Kasten F.H. (1981) In *Staining Procedures* (4th edition), G. Clark (ed.). Williams & Wilkins, Baltimore, pp. 39–103.

Kasten F.H. (1983a) In *History of Staining* (3rd edition), G. Clark & F.H. Kasten (eds). Williams & Wilkins, Baltimore pp. 147–185.

Kasten F.H. (1983b) In *History of Staining* (3rd edition), G. Clark & F.H. Kasten (eds). Williams & Wilkins, Baltimore, pp. 186–252.

Kasten F.H. (1989) In *Cell Structure and Function by Microspectrofluorometry*, E. Kohen & J.G. Hirschberg (eds). Academic Press, San Diego, pp. 3–50.

Kasten F.H. (1991) In *Historians and Archivists: Essays in Modern German History and Archival Policy*, G.O. Kent (ed.). George Mason Press, Fairfax, VA, pp. 173–208.

Kohen E., Kohen C., Hirschberg J.G., Santus R., Morlière P., Kasten F.H. & Ghadially F.N. (1991) In

Encyclopedia of Human Biology, Vol. 5, R. Dulbecco (ed.). Academic Press, San Diego, pp. 561–585.

Krause R. (1926a) *Enzyklopädie der Mikroskopischen Technik* (3rd edition), Vol. 1. Urban & Schwartzenberg, Berlin.

Krause R. (1926b) *Enzyklopädie der Mikroskopischen Technik* (3rd edition), Vol. 2. Urban & Schwartzenberg, Berlin.

Krause R. (1927) *Enzyklopädie der Mikroskopischen Technik* (3rd edition), Vol. 3. Urban & Schwartzenberg, Berlin.

Krebs A. (1944) *Strahlentherap.* **75**, 346 (cited in Strugger, 1946, p. 43).

Krebs A.T. & Gierlach Z.S. (1951) *Am. J. Roentgenol.* **65**, 93–97.

Lehmann H. (1913) *Z. Wiss. Mikrosk.* **30**, 417–470.

Levaditi R.O. & Reinie L. (1939) *C.R. Soc. Biol.* **131**, 916–919.

Lillie R.D. (1977) *H.J. Conn's Biological Stains* (9th edition). Williams & Wilkins, Baltimore.

Loeser C.N., West S.S. & Schoenberg M.D. (1960) *Anat. Rec.* **138**, 163–178.

MacInnes J.W. & Uretz R.B. (1966) *Science* **151**, 689–691.

Mayor H.D. (1963) *Int. Rev. Exp. Pathol.* **2**, 1–45.

Melamed M.R. & Darzynkiewicz Z. (1981) In *Histochemistry: The Widening Horizons of its Applications in the Biomedical Sciences*, P.J. Stoward & J.M. Polak (eds). John Wiley & Sons, Chichester, pp. 237–261.

Menger M.D., Marzi I. & Messmer K. (1991) *Eur. Surg. Res.* **23**, 158–169.

Mueller J.H. (1918) *J. Pathol. Bacteriol.* **22**, 308–318.

Patton R.L. & Metcalf R.L. (1943) *Science* **98**, 184.

Peacocke A.R. & Skerrett J.N.H. (1956) *Trans. Faraday Soc.* **52**, 261–279.

Perkin W.H. (1906) *Science* **24**, 488–493.

Pick J. (1935) *Z. Wiss. Mikrosk.* **51**, 338–351.

Price G.R. & Schwartz S. (1956) In *Physical Techniques in Biological Research*, Vol. 3, G. Oster & A.W. Pollister (eds). Academic Press, New York, pp. 91–148.

Provazek S. von. (1914) *Kleinwelt.* **6**, 37.

Reichert's Fluorescence Microscopy with Fluorochromes (1952) (2nd edition). Optische Werke C. Reichert, Wien.

Richards O.W. (1941) *Science* **93**, 190.

Richards O.W. (1955) In *Analytical Cytology*, R.C. Mellors (ed.). McGraw-Hill, New York, pp. 5/1–5/37.

Rigler R. Jr. (1966) *Acta Physiol. Scand.* **67** (Suppl. 267), 1–122.

Robbins E. & Marcus P.I. (1963) *J. Cell Biol.* **18**, 237–250.

Robbins E., Marcus P.I. & Ganatas, N.K. (1964) *J. Cell Biol.* **21**, 49–62.

Rost F.W.D. (1980) In *Histochemistry. Theoretical and Applied* (4th edition), Vol. 1, *Preparative and Optical Technology*, A.G.E. Pearse (ed.). Churchill Livingstone, Edinburgh, pp. 346–378.

Sammak P.J. & Borisy G.G. (1988) *Cell Motil. Cytoskel.* **10**, 237–245.

Schmidt-LaBaume F. & Jäger R. (1939) *Arch. Dermatol. Syphilas.* **179**, 531–542.

Schümmelfeder N. (1948) *Naturwissenschaften* **35**, 346.

Schümmelfeder N. (1950) *Virch. Arch.* **318**, 119–154.

Schümmelfeder N. (1956) *Z. Zellforsch.* **44**, 488–494.

Schümmelfeder N. (1958) *Acta Histochem.* Suppl. Bd. **1**, 148–151.

Schümmelfeder N. & Stock K.-F. (1956) *Z. Zellforsch.* **44**, 327–338.

Schümmelfeder N., Ebschner K.J. & Krogh E. (1957) *Naturwissenschaften* **44**, 467–468.

Singer E. (1932) *Science* **75**, 289.

Sisken J.E., Barrows G.H. & Grasch S.D. (1986) *J. Histochem. Cytochem.* **34**, 61–66.

Spring K.R. & Smith P.D. (1987) *J. Microsc.* **147**, 265–278.

Steiner R.F. & Beers R.F. Jr. (1961) *Polynucleotides.* Elsevier, Amsterdam.

Stelzer E.H.K. & Wijnaendts-van-Resandt R.W. (1989) In *Cell Structure and Function by Microspectrofluorometry*, E. Kohen & J.G. Hirschberg (eds). Academic Press, San Diego, pp. 131–143.

Strugger S. (1940) *Jena Z. Naturwiss.* **73**, 97–134.

Strugger S. (1940/41) *Z. Wiss. Mikrosk. u. Mikrosk. Tech.* **57**, 415–419.

Strugger S. (1941) *Flora* **35**, 101–134.

Strugger S. (1946) *Fiat Review of German Science* **52**, Biologie Teil 1, 1–50.

Strugger S. (1949) *Fluoreszenzmikroskopie und Mikrobiolologie.* Verlag M. & H. Schaper, Hannover.

Strugger S. & Hilbrich P. (1942) *Dtsch. tierärztl. Wschr.* **50**, 121–130.

Strugger S. & Rosenberger G. (1944) *Dtsch. tierärztl. Wschr.* (cited in Strugger, 1946, p. 5).

Strugger S. Krebs A.T. & Gierlach Z.S. (1953) *Am. J. Roentgenol.* **70**, 365–375.

Stübel H. (1911) *Pflügers Arch.* **142**, 1–14.

Tappeiner H. von & Jodlbauer A. (1907) *Die Sensibilisierende Wirkung Fluoreszierender Substanzen Gesammelte Untersuchungen über die Photodynamischer Erscheinung.* FCW Vogel Verlag, Leipzig.

Taylor D.L. & Salmon E.D. (1989) In *Fluorescence Microscopy of Living Cells in Culture Part A. Fluorescent Analogs, Labeling Cells, and Basic Microscopy*, Y.-I. Wang & D.L. Taylor (eds). Academic Press, San Diego, pp. 207–237.

Taylor D.L. & Wang Y-I. (eds) (1989) *Fluorescence Microscopy of Living Cells in Culture. Part B. Quantitative Fluorescence Microscopy – Imaging and Spectroscopy.* Academic Press, San Diego.

Tswett M. (1912) *Ber. Dtsch. Bot. Ges.* **29**, 744–746.

Venkataraman K. (1952) *The Chemistry of Synthetic Dyes*, Vol. 2. Academic Press, New York.

Waggoner A.S. (1986) In *Applications of Fluorescence in the Biomedical Sciences*, D.L. Taylor (ed.). Alan R. Liss, New York, pp. 3–28.

Wang Y-I. & Taylor D.L. (eds) (1989) *Fluorescence Microscopy of Living Cells in Culture. Part A. Fluorescent Analysis, Labeling Cells, and Basic Microscopy.* Academic Press, San Diego.

Wasicky R. (1913) *Pharm. Post.* **46**, 877–878.

Werbitzki F.W. (1909) *Zbl. Bakteriol.* **53**, 303–315.

Willingham M.C. & Pastan I.H. (1982) *Methods Enzymol.* **98**, 266–283.

Wilson T. (1990) *Confocal Microscopy.* Academic Press, London.

Witte S. (1989) *Res. Exp. Med.* **189**, 229–239.

Wolf M.K. & Aronson S.B. (1961) *J. Histochem. Cytochem.* **9**, 22–29.

Wories H.J., Kopec, J.H., Lodder G & Lugtenberg J. (1985) *Recl. Trav. Chim. Pays-Bas.* **104**, 288–291 (cited by Haugland, 1990).

Zanker V. (1952) *Z. Phys. Chem.* **199**, 225–258.

CHAPTER THREE

Intracellular Ion Indicators

R. HAUGLAND

Molecular Probes, Eugene, OR, USA

3.1 INTRODUCTION

During the last decade, the development of fluorescent indicators that selectively respond to biologically important ions has been very successful. These research tools have enabled bioscientists to measure slow or rapid flux of ions in single living cells essentially as it occurs. The availability of these indicators has stimulated the development of instrumentation used to quantitate the spatial and temporal changes in fluorescence. However, the indicators are not perfect and they have their limitations, particularly for quantitative measurement of ion concentrations. This chapter reviews the indicators currently used, their responses to ions, problems encountered and recent developments in indicator chemistry.

Fluorescent indicators have been described for most of the biologically important ions including hydrogen, calcium, magnesium, sodium, potassium and chloride. Certainly the most successful and widely used indicators have been for pH and intracellular calcium. During their development, several generations of probes have been designed to facilitate optical measurements and reduce experimental artifacts.

3.2 PROPERTIES OF INTRACELLULAR ION INDICATORS

Certain properties of ion indicators are expected of an optimal probe.

3.2.1 Stoichiometry

The stoichiometry of the indicator chemistry is a measure of the ratio of the number of indicator molecules and target ions. The best results are obtained when, on average, one indicator molecule interacts with one ion. This stoichiometry is common to most but not all of the indicators described in this chapter. For instance, the stoichiometry of calcium binding to aequorin is 3:1 (Blinks *et al.*, 1978) and to arsenazo III is 1:2 (Scarpa *et al.*, 1978).

3.2.2 Specificity

It is essential that the measured optical signal can be related to the ion that is to be probed. Competing equilibria that affect the measurement must be considered in the design of the probe and, where the effects cannot be eliminated, their binding and possible flux must be considered in the calculation of the indicator dissociation constants. For instance, intracellular magnesium ion concentrations are commonly near 1 mM, whereas the concentration of intracellular calcium ion is usually submicromolar. Therefore, calcium-sensitive indicators must have a high rejection of magnesium to respond to intracellular calcium. The acidity of the medium affects the response of virtually all of the physiological indicators. However, except for the pH indicator probes, this effect is usually low.

3.2.3 Dissociation constant

In most cases ion indicators have been used to estimate intracellular concentrations of the ions. Therefore, a fundamental property of the indicator must be an optical response that occurs at physiological concentrations of the specific ion. As is evident in several of the response curves in this chapter (see Figs 3.1, 3.2, 3.6 and 3.7, the maximum optical response occurs near the dissociation constant of the indicator, the ionic concentration at which half of the probe has bound ion and half does not. Thus, the dissociation constant should be matched to the expected intracellular ion concentration. Unfortunately, several conditions that may occur inside cells – including effects of pH, temperature, ionic strength and composition, viscosity and protein binding – may affect the dissociation constant in ways that are difficult to measure. Methods for calibration of the intracellular dissociation constant have been addressed and debated in several papers (Tsien, 1980; Grynkiewicz et al., 1985; Minta & Tsien, 1989).

The optical response of an ion indicator can typically be measured over about two orders of magnitude of ion concentrations, with its greatest sensitivity in a relatively narrow range near its dissociation constant (see, for example, Fig. 3.3). Matching the dissociation constant to the range of interest is, therefore, required. A major reason for the successful application of fluorescent calcium indicators is the exceptionally large increase in intracellular Ca^{2+} concentrations when cells are stimulated. The concentration of Ca^{2+} may change from a resting level as low as perhaps 10 nM to as high as 10 μM or even higher. On the other hand, the change in intracellular pH or Mg^{2+} concentration that occurs on stimulation is commonly less than a few tenths of a log unit. A calcium indicator such as fura-2, which has a dissociation constant of ~224 nM in 10 mM MOPS buffer at 37°C, is nearly saturated with Ca^{2+} above ~1 μM free Ca^{2+} (Grynkiewicz et al., 1985). Fortunately, it is possible to 'tune' the dissociation constant – and therefore the optical response – of most indicators by choosing appropriate chemical substituents and chelating or sensing moieties.

3.2.4 Optical response

Detecting the fluorescence of indicators is often the most sensitive technique for measuring ion concentrations in biological systems. However, other optical indicators, which respond with a change in absorption, or bioluminescent indicators (such as aequorin) can be used. Similar experimental criteria usually apply to their use. In any case, an optical property of the indicator must *change* when the indicator comes in contact with the target ion. In many cases this change is a shift in the absorption wavelengths. In fluorescence measurements, this is observed by a shift in the excitation wavelengths and a corresponding shift in the fluorescent emission wavelengths, or as an increase or decrease in the quantum yield or extinction coefficients of the indicator. With most indicators a combination of these changes is observed. Other changes in the optical properties of indicators can be used with success. For example, Lakowicz and colleagues used Calcium Green-1 – an indicator that does not undergo a spectral shift on ion binding – to determine calcium concentrations by measuring the decay time of the indicator fluorescent emission (Lakowicz et al., 1992).

The change in indicator fluorescence on ion binding typically can take several forms: (1) an increase or decrease in the quantum yield or the efficiency of fluorescence of the indicator with little change in either the absorbance or fluorescence spectra (e.g. fluo-3, Calcium Green, Magnesium Green and SPQ); (2) a shift of the absorbance, and therefore the fluorescence excitation spectrum, to shorter wavelengths with little shift in the emission maximum (e.g. quin-2, fura-2, Fura Red, mag-fura-2, SBFI, PBFI and the fluorescein- and NERF-pH indicators); or (3) a shift in both the absorption (excitation) and emission

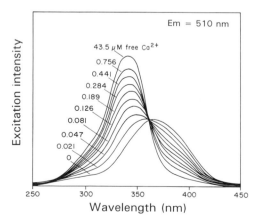

Figure 3.1 Excitation spectra of fura-2 in increasing concentrations of free calcium.

spectra to shorter wavelengths (e.g. indo-1, mag-indo-1 and the SNARF and SNAFL series of pH indicators).

A feature common to the spectral response of indicators that undergo an excitation or emission shift on ion binding is the presence of an 'isosbestic point' in the absorption, excitation and/or emission spectra, defined as the point at which the indicator fluorescence is insensitive to the characteristic being examined. Isosbestic points are common when two (and only two) species are in chemical equilibrium, such as the ion-free and ion-bound forms of an indicator. Examples of isosbestic points are seen in both the excitation spectra of fura-2 (Fig. 3.1) and the emission spectra of carboxy-SNARF (Fig. 3.2). Differences in the fluorescence quantum yields for the two species may result in the absorption and excitation isosbestic points being different. For instance, the isosbestic point of BCECF is at 465 nm in absorption measurements and at 439 nm in excitation measurements (Paradiso *et al.*, 1987). Absence of an isosbestic point in an indicator response curve may be evidence of additional equilibria or contamination by an ion-insensitive species. For indicators such as fura-2 that undergo a spectral shift on ion binding, when the fluorescence is excited at the isosbestic point, the intensity is directly proportional to the amount of dye in the sample and *independent* of the ion concentration. The intensity excited at any other wavelength depends both on the indicator concentration and the ion concentration. Because the intensities at any two wavelengths are expressed in arbitrary fluorescence units per mole of indicator, forming a ratio of the intensities removes the indicator concentration dependence. Thus the *ratio* of intensities measured at any two wavelengths depends on the ion concentration, but is *independent* of the indicator concentration. Although a ratio that is independent of dye concentration can be calculated for any indicator, the ratio for indicators such as fluo-3, Calcium Green and SPQ that do not undergo a spectral shift is also independent of the ion concentration, whereas for indicators such as fura-2 and SNARF, the ratio can be directly related to the ion concentration (Tsien, 1982; Whitaker *et al.*, 1991a). The excitation (or absorption) ratio for fluorescein and BCECF depends on the ion concentration, but the emission ratio does not, whereas both the excitation and emission ratios for the SNARF and SNAFL pH indicators can be related to the ion concentration (Fig. 3.3). Some indicators, such as indo-1 and mag-indo-1, are useful for both excitation and emission ratio measurements but are typically better suited for one of these methods – emission ratioing for indo-1 and mag-indo-1 – because the magnitude of the response is greater in that method.

The indicators that undergo a shift in the absorption

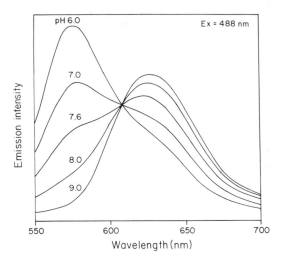

Figure 3.2 The pH-dependent emission spectra of carboxy-SNARF-1 when it is excited at 488 nm.

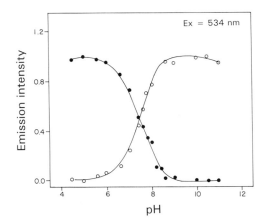

Figure 3.3 Titration curves constructed from carboxy-SNARF-1 emission ratios at 588 nm/604 nm (●) and 634 nm/604 nm (○) with excitation at 534 nm. The emission ratio is scaled to a maximum value of 1.0, representing the pure acid or base form of the indicator.

or emission wavelength upon ion binding have been widely recognized and extensively used. This has particular advantages for measurements in living cells. As long as the intensities are significantly above background, intensity ratioing can reduce problems in quantitation resulting from unequal loading or distribution of the probe, variations in cell thickness and susceptibility to indicator photobleaching and leakage from intact cells. Because they respond to calcium binding only by a change in intensity, it is more difficult to obtain quantitative information on ion concentrations using indicators such as fluo-3 and the new Calcium Green, Orange and Crimson. However, one can obtain excellent kinetic data using these indicators because they display large increases in quantum yield on calcium binding.

3.2.5 Other spectral properties

To be useful, it is essential that the fluorescence can be detected above any background due to the sample or other probes that may be present. The usual factors that affect fluorescence detection – high absorbance (i.e. high extinction coefficient at the excitation wavelength) and high quantum yield – also facilitate these measurements. It must be possible to excite the probe using illumination that is minimally absorbed by the cell or medium. While all of the probes described here can be excited beyond 330 nm, longer wavelength absorptions are sometimes preferred because reduced autofluorescence of the cells usually results. In some cases selection of the probe depends on the instrumentation available. This is particularly true of equipment that uses laser excitation including flow cytometers and confocal microscopes. Dual-emission indicators that can be excited at one of the principal wavelengths of the laser are preferred. The argon ion laser commonly used in these instruments has maximal output at 488 nm and 514 nm that is useful for the SNARF, SNAFL and NERF pH indicators. Some argon lasers also emit ultraviolet lines between 350 and 360 nm that can be used to excite indo-1 and mag-indo-1. Unfortunately, good dual-emission indicators for intracellular calcium, magnesium, sodium, potassium and chloride that can be excited by the argon laser have yet to be described; however, fluo-3, Magnesium Green and Calcium Green are optimally excited by this laser.

3.2.6 Indicator concentration

The indicator is used to measure the concentration of the ion with which it is in equilibrium. It is not intended to buffer the ion. Buffering of the ion will occur, however, if the indicator is present at high intracellular concentrations. The result will be the damping of high concentration transients of the ion, as has been observed with calcium ion binding by quin-2 (Tsien & Pozzan, 1989). To avoid these problems it is necessary to make the measurement using as low a concentration of the indicator as practical and to use sensitive detectors. But reducing the amount of indicator will also reduce the fluorescence signal so that indicators such as fura-2 – which has a much higher absorbance and higher fluorescence quantum yield than quin-2 or Fura Red – can be used at relatively low concentrations that do not appreciably buffer intracellular calcium.

3.2.7 Other indicator properties

High photostability of the probe is beneficial, particularly if the response of the probe cannot be ratioed. Reagents that are typically used to reduce dye bleaching, such as propylgallate, DABCO and *p*-phenylenediamine, obviously cannot be used with live cells. Low phototoxicity to cells is important and somewhat related to the fluorophore chosen. Brief excitation using low light levels and sensitive detectors is beneficial and usually preferred.

3.2.8 Cell loading

It is obviously essential to be able to get the probe inside the cell of which the ionic composition is to be measured. With the exception of the chloride indicator SPQ and its analogues, this has usually required conversion of the indicator to a form that will passively diffuse through the cell membrane and be converted to the indicator by intracellular esterase activity. Alternatively, the indicator may be loaded into the cell by artificial techniques that result in temporary membrane disruption (i.e. microinjection, scrape loading, fusion, electro- or chemical-permeabilization or by other means). For instance, plant cells have been loaded with the free acid form of calcium indicators by brief exposure to a relatively acidic culture medium (Bush & Jones, 1988a, b, 1990; Elliott & Petkoff, 1990). These membrane-disruptive techniques have been used with success that varies considerably with cell type. By far the most common method, however, has been the use of cell permeant esters – particularly acetoxymethyl (AM) esters – of indicator probes such as fura 2, Calcium Green and SBFI, simple acetate esters of some of the pH indicators or combinations of AM and acetate esters for some pH indicators.

Conversion of the relatively lipophilic AM esters back to the free indicator inside the cell requires the hydrolytic activity of intracellular enzymes that completely cleave the ester. Failure to remove even one of the AM esters from the calcium ion-chelating carboxylic acids of indicators such as fura-2 results in a highly fluorescent – but calcium-insensitive – probe. Presence of this species will result in an underestimation of intracellular calcium levels. Incomplete ester hydrolysis is best detected by lysing the cells into a medium and determining the resulting spectral response to the ion versus that of the parent indicator. Unfortunately, certain cells (in particular, plant cells) do not have sufficient intrinsic esterase activity. Furthermore, the relatively lipophilic AM esters can become trapped in membranes where they do not have access to esterases. It is important to use as little probe as practical because excess probe requires even greater enzymatic hydrolysis to achieve the correct product. The other by-products of AM ester hydrolysis – acetate, protons and formaldehyde – all have potential deleterious effects on living cells. However, except for cells that have been overloaded with AM esters, this is usually not a problem.

The lipophilic nature of AM esters gives them low solubility in aqueous medium. In most cases they have been added to the culture medium from stock solutions in dimethylsulphoxide (DMSO) to give a final concentration of about 1–10 μM. Passive uptake through the membrane and intracellular hydrolysis results in a final intracellular indicator concentration of 25–100 μM or even higher. Labelling of some cells can be improved by addition of a non-ionic dispersant such as Pluronic F-127. Incubation at room temperature to 37°C for 10 min to 1 h is usually sufficient for intracellular hydrolysis for most AM esters, although a much longer incubation has been recommended for SBFI-AM (Negulescu & Machen, 1990). Adherent cells on a slide usually require use of less reagent than do cell suspensions.

3.2.9   Other artifacts

A number of artifacts have been reported that are common to most of the indicators. Among the most significant problems are compartmentalization, protein binding and sometimes leakage or secretion. Although it is usually desired to measure the ion concentration in the cytoplasm and the cellular esterase activity responsible for AM ester hydrolysis is usually concentrated in the cytoplasm, it is common for most of the indicators to redistribute into organelles, particularly into the nucleus. Since this is a time-dependent phenomenon, spectral measurements should be commenced as soon as indicator loading is sufficiently complete. Loading at room temperature rather than at 37°C has reportedly reduced compartmentalization (Di Virgilio et al., 1990).

Protein binding of indicators has been reported to be a common property (Blatter & Weir, 1990). Although it is probable that this has an effect on both the dissociation constant of the indicator and its spectral properties, few people have attempted to accommodate this artifact in the calculations of free ion concentration. For cells that can be loaded by relatively invasive techniques, the best means for avoiding both protein binding and compartmentalization may be to use indicator conjugates of polar polymers such as dextrans. In the case of BCECF-dextran conjugate, it was shown that the conjugation prevented binding to intracellular proteins (Bright et al., 1989). So far only dextran conjugates of fura, indo, Calcium Green, BCECF, SNARF and SPQ have been described (Biwersi et al., 1992). Relatively lipophilic indicators that have low intrinsic quantum yields, such as Fura Red and the near-membrane calcium indicators described by Etter and colleagues (1992) may bind to cell membranes with resulting fluorescence enhancement and spectral shifts.

Once hydrolysed intracellularly to the free indicator, most calcium- and magnesium-ion indicators are well retained in viable cells because of their high ionic charges. Certain cells, such as macrophages, however, actively secrete the indicator (Steinberg et al., 1987; Di Virgilio et al., 1988). Secretion can sometimes be blocked by the addition of drugs such as probenecid. On the other hand, pH indicators typically leak from cells; their retention times are primarily determined by the number of ionic charges on the indicator. BCECF, with 4–5 negative charges at physiological pH is the best retained pH indicator that is not polymer-conjugated, whereas fluorescein, with only 1–2 negative charges, usually leaks from cells within minutes. Passive leakage is significantly retarded by cooling the cells or enhanced by heating. The chloride indicator, SPQ, has been reported to leak from rabbit proximal convoluted tubules (PCT) cells within minutes of loading (Krapf et al., 1988). Because ratio measurements are relatively independent of indicator concentration, the actual amount of dye retained in the cell is not particularly important, as long as sufficient dye is present to obtain adequate signal. However, for dyes that cannot be ratioed – fluo-3, for example – loss of dye can result in a considerable underestimation (or overestimation) of the ion concentration. For imaging experiments, where cells typically are continuously perfused with fresh medium, the dye that leaks is washed from the sample. But for measurements made in suspension in a cuvette, dye that leaks will be detected as if it were intracellular. For dyes such as fluo-3 and Calcium Green-2, in which the fluorescence is enhanced more than 80-fold upon binding calcium the presence of relatively high levels of calcium in the extracellular medium can result in virtually all of the fluorescence coming from dye that has leaked from the cells, unless extraordinary care is taken to wash the cells. In this case, indicator leakage can usually be detected by a time-dependent increase in fluorescence in cells that are not otherwise active.

3.3   EXAMPLES OF INTRACELLULAR ION INDICATORS

3.3.1   Calcium indicators

By far the most significant group of ion indicators has been those for intracellular calcium. Almost all of these have been fluorescent derivatives of the chelator BAPTA, which is an aromatic analogue of the calcium-selective chelator ethyleneglycol-bis(β-aminoethylether)-N,N,N',N'-tetraacetic acid (EGTA). BAPTA is described along with the prototype calcium indicator, quin-2, in the classic paper by

Table 3.1 Calcium and magnesium indicators.

Name	Ex/Em low ion (nm)	Ex/Em high ion (nm)	Measurement [a]	K_d-Ca^{2+} [b] (nM)	K_d-Mg^{2+} [b] (mM)
Fura-2	362/512	335/505	1	224	
Indo-1	346/495	330/408	2	250	
Fluo-3	506/526	506/526	3	316	
Rhod-2	556/576	553/576	3	565	
Quin-2	352/492	332/492	1	126	
Calcium Green-1	506/534	506/533	3	189	
Calcium Green-2	506/531	506/531	3	574	
Calcium Green-5N	506/531	506/531	3	3 300	
Calcium Orange	554/575	555/576	3	328	
Calcium Crimson	588/611	588/611	3	205	
Fura Red	472/645	436/640	1	133	
Mag-fura-2	370/511	330/491	1	50 000	1.5
Mag-fura-5	369/505	332/482	1	6 500	2.6
Mag-indo-1	349/476	330/417	2		2.7
Magnesium Green	506/532	506/532	3	4 800	0.9

[a] 1 – excitation ratio, 2 – emission ratio, 3 – single wavelength intensity.
[b] Data on dissociation constants comes from several different sources. These values vary considerably depending on temperature, pH ionic strength, viscosity, protein binding, method and equipment for measurement, presence of other ions and other factors.

Dr Roger Tsien (1980). Along with collaborators, he described fura-2 and indo-1 in 1985 (Grynkiewicz *et al.*, 1985) and fluo-3 and rhod-2 in 1989 (Minta & Tsien, 1989). Kuhn and colleagues described Calcium Green (now called Calcium Green-1), Calcium Orange and Calcium Crimson in 1990 and the indicator that Molecular Probes Inc. named Fura Red was described in a patent by DeMarinis *et al.* in 1989 (US Patent No. 4,849,362). Calcium Green-2 and Calcium Green-5N are new and previously unpublished (Haugland, 1992). Their properties and those of other calcium indicators are described in Table 3.1.

More than a thousand papers have been published using fura-2 and, to a lesser extent, indo-1. Several of these are collected in annual issues of the journal *Cell Calcium*, which is an excellent collected source of information on applications and precautions in use of these indicators.* Spectral response curves for most of the indicators are given in the *Handbook of Fluorescent Probes and Research Chemicals* from Molecular Probes Inc. (Haugland, 1992). So far the preferred probes have been fura-2 for ratio imaging applications, indo-1 for flow cytometry and fluo-3 for uses in confocal microscopy and imaging using 'caged probes' such as caged inositoltriphosphate (Kao *et al.*, 1989). Both fluo-3 and Calcium Green-2 are virtually non-fluorescent except in their calcium chelates.

* A collection of papers discussing practical issues associated with using fura-2 for measuring intracellular Ca^{2+} is included in the February/March 1990 issue of *Cell Calcium*. The February/March 1991 issue is devoted to papers on Ca^{2+} oscillations and waves.

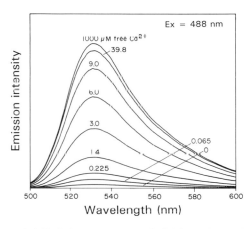

Figure 3.4 Emission spectrum of Calcium Green 5N in increasing concentrations of free calcium.

Calcium Green-1 has the advantage of being more fluorescent in both the Ca^{2+}-bound and Ca^{2+}-unbound states than fluo-3, which makes visualization in the resting cell easier. Calcium Green-5N (Fig. 3.4) and some magnesium indicators, including mag-fura-2 (Delbono & Stefani, 1992), mag-indo-1 and the new Magnesium Green (Haugland, 1992) have calcium ion dissociation constants above $\sim$1 μM. This property makes them useful for detecting spikes in intracellular calcium above 1 μM – a concentration that nearly saturates the response of indicators such as fura-2 – such as occur in smooth muscle (Baylor *et al.*, 1989; Konishi *et al.*, 1991; Kurebayashi, 1992) and central nervous system cells (Connor, 1986; Lipscombe *et al.*, 1988).

3.3.2 Magnesium indicators

All of the current magnesium indicators in Table 3.1 are variants of the calcium indicators in which the BAPTA structure is replaced by a triacetic acid analogue. Mag-fura-2 (which was originally named furaptra by Raju and collaborators in 1989) and the other magnesium indicators respond to Mg^{2+} concentrations in the physiological range of about 0.1–10 mM. As indicated in Table 3.1, they also respond to high levels of intracellular Ca^{2+}. Since their optical responses to Mg^{2+} and to Ca^{2+} are virtually identical, one must carefully discriminate the effects of these two ions. Moreover, the changes that have been observed in intracellular magnesium concentration are far smaller than those of intracellular calcium, resulting in only small fluorescence changes (Raju et al., 1989; Quamme & Rabkin, 1990; Jung et al., 1990).These are best measured using the indicators mag-fura-2 and mag-indo-1, both of which permit ratio measurements. Magnesium Green is a new, long wavelength indicator whose use in cells has not yet been described (Haugland, 1992). Its response to magnesium (Fig. 3.5) is similar to that of Calcium Green-1 to calcium. The indicator has dissociation constants for magnesium and calcium, respectively, of ~0.9 mM and ~4.8 μM.

3.3.3 Sodium and potassium indicators

The only intracellular sodium indicator in common use is SBFI, which was originally described by Tsien and his collaborators (Minta et al., 1989). PBFI is an analogue of SBFI that has a cryptand cavity, allowing it to bind the larger potassium ions. Although their base fluorophore is different, both of these indicators have optical responses to their target ions that mimic that of fura-2 to Ca^{2+} and thus can be used to make ratio imaging measurements. Unfortunately the dissociation constants for both indicators are strongly dependent upon the concentrations of both Na^+ and K^+, resulting in potential interference between the ions (Minta et al., 1989). Furthermore, loading of cells with the acetoxymethyl esters of these indicators has been difficult. The excellent review by Negulescu and Machen (1990) should be consulted when using SBFI-AM. Despite problems in using these indicators, several papers have described using SBFI in cells (Borin & Siffert, 1990; Satoh et al., 1991; Jaffe et al., 1992) and a limited number of papers have described using PBFI (Jezek et al., 1990; Laskay et al., 1992). Like quin-2, these indicators should be considered prototypes for yet undeveloped superior indicators.

3.3.4 Chloride indicators

All of the intracellular chloride indicators in current use are derivatives of 6-methoxyquinolinium dyes. Primary among these is the indicator N-(3-sulphopropyl)-6-methoxyquinolinium (SPQ) (Wolfbeis & Urbano, 1982). SPQ and several of its analogues including MQAE and MQAA, have been exploited by Verkman and colleagues (1989) as intracellular ion indicators. All of the chloride indicators detect chloride through a collisional quenching mechanism (Wolfbeis & Urbano, 1982, 1983; Illsley & Verkman, 1987). This results in fluorescence quenching with no spectral shift. Thus ratio measurements cannot be made using these indicators. In addition, these indicators have at least three other problems:

(1) Low fluorescence output that results primarily from a low absorbance. This necessitates use of a larger amount of indicator and results in low sensitivity. The excitation wavelength is also in the ultraviolet, which may result in high autofluorescence and photodamage to cells.
(2) Rapid leakage of the dye from cells. Unlike the permeant AM and acetate esters of most other indicators, the chloride probes have been loaded into cells by incubating them with a huge excess of the probe and presuming the membranes have a slight permeability to the indicator. Unfortunately the indicator may rapidly leak from cells, often in minutes, making quantitation and calibration difficult.
(3) Variation of the quenching constant with viscosity. Since the quenching is a diffusional process, it shows a high sensitivity to the viscosity of the medium. For instance, it has been reported that SPQ has a quenching constant of 118 M^{-1} in aqueous solution and an intracellular quenching constant of 12 M^{-1} (Krapf et al., 1988).

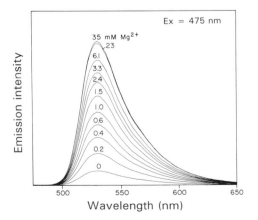

Figure 3.5 Emission spectra for Magnesium Green at various concentrations of Mg^{2+}.

Verkman and his colleagues have attempted to overcome some of the limitations of SPQ by attaching the probe to dextrans (Biwersi *et al.*, 1992). Although these must be microinjected or loaded by some other membrane-disruptive technique, they are less likely to leak from cells or to bind to proteins. Biwersi and Verkman have also described a freely membrane-permeant, chemically reduced form of a 6-methoxyquinolinium dye (Biwersi & Verkman, 1991). Called diH-MEQ, this colourless, non-fluorescent probe is spontaneously oxidized to the indicator inside cells. The susceptibility of diH-MEQ to air oxidation makes it necessary to prepare the probe shortly before use by an easy chemical reduction. This product, however, is still susceptible to leakage from cells and to viscosity artifacts.

3.3.5 pH indicators

Sensitivity of the absorption and emission of some fluorescent dyes to pH has been known for many years. Fluorescein and 7-hydroxy-4-methylcoumarin (β-methylumbelliferone) are well-known examples of this type of dye. In most cases, the pH sensitivity results from ionization of a phenolic moiety on the dye. The pK_a of the dye depends on the chemical structure and substituents of the dye; electron withdrawing groups lower the pK_a. As with the other ion indicators, a pK_a near the pH of the compartment whose pH is to be measured is necessary. In most cases this is a near-neutral pH; however, measurements of pH in endosomes may require the use of a pH indicator whose pK_a is in the range of 4–6.

Because of problems of dye leakage and an inappropriate pK_a for the probe, most of the early attempts to use dyes such as fluorescein diacetate as intracellular pH indicators in single cells were not very successful. Thomas and colleagues introduced carboxyfluorescein diacetate in 1979 and Tsien and colleagues prepared the widely used pH indicator 2′,7′-bis-carboxyethyl-5(6)-carboxyfluorescein (BCECF) in 1982 (Rink *et al.*, 1982). Both of these intrinsically colourless probes use permeabilization groups – acetates and AM esters – to facilitate entry of the dye into the cells, where intracellular esterases cleave the esters to the coloured and highly fluorescent pH indicator. In addition to having a more appropriate pK_a (6.98 versus 6.4), the higher net charge of BCECF compared to fluorescein (−4 to −5 versus −1 to −2) results in considerably improved retention of the dye in cells that have intact membranes.

Almost all pH-sensing dyes respond to protonation by a shift in their absorption spectrum to shorter wavelengths; usually a decrease in absorption intensity also results. This absorption shift – as reflected by a change in the fluorescence excitation spectrum – usually results in little, if any shift in the emission spectrum. Because the acidity of the excited state of phenolic dyes is almost always higher than the ground state, emission usually occurs at the wavelength maximum of the basic form of the dye, even in acidic medium. The lower absorption in acidic medium, however, results in the probe having decreased fluorescence when it is excited in its longest wavelength absorption band. There is often a corresponding rise in a shorter wavelength absorption peak that results in clear isosbestic points. This optical response for pH indicators such as BCECF (Fig. 3.6) enables these dyes to be used in dual-excitation ratiometric measurements. On the other hand, the SNARF and SNAFL pH indicators are a unique exception in that both groups of dyes undergo pH-dependent shifts in both their excitation and emission spectra. The pK_a values of these dyes are in the range 7.4–7.8 (Whitaker *et al.*, 1991a). The quantum yield of the basic form of carboxy-SNARF-1 is greater than that of its acidic form. This compensates for the lower absorption of its basic form at 480–520 nm and results in particularly useful dual-emission optical response (Fig. 3.2). Because carboxy-SNARF-1 can be excited by the argon laser at 488 or 514 nm, it is usually the preferred pH indicator for flow cytometry and laser-excited microscopy. The SNAFL dyes have stronger emission from their acidic forms, which results in this class of dyes being more useful as dual-excitation indicators (Fig. 3.7). SNARF- and SNAFL-dextrans (Thiebaut *et al.*, 1990) as well as BCECF-dextrans (Bright *et al.*, 1987, 1989; Shen & Buck, 1990) have been used to avoid problems of compartmentalization and leakage. As described above, these probes must be microinjected or loaded by relatively disruptive means.

Unlike the cytoplasmic pH indicators, which are

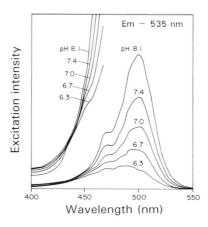

Figure 3.6 The pH-dependent excitation spectra of BCECF. The excitation spectra have been enlarged on the left to reveal BCECF's 439 nm isosbestic point.

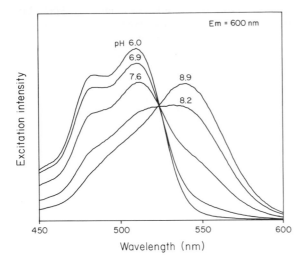

Figure 3.7 The pH-dependent excitation spectra of carboxy-SNAFL-1 with emission monitored at 600 nm.

loaded into cells as permeant esters, indicators for endosomal pH are usually used to follow the uptake and trafficking of protein–indicator conjugates, such as fluorescein transferrin (Murphy *et al.*, 1984; Yamashiro *et al.*, 1984), or other ligands, such as chemotactic peptides (Sklar *et al.*, 1982, 1984), through acidic organelles. Upon internalization into acidic organelles, indicators such as fluorescein undergo strong fluorescence quenching with no emission shift. On the other hand, conjugates of carboxy-SNARF-1 undergo a shift in both their excitation and emission spectra, making carboxy-SNARF-1 the preferred dye for determining whether the conjugate is outside the cell – where its fluorescence properties are determined by the pH of the medium – or internalized. The large pH changes that occur upon internalization of pH-indicating conjugates into endosomes makes this an attractive approach for following ligand uptake and processing. However, to estimate the actual pH of the endosomal compartments or potentially to distinguish between compartments that have different pH values requires fluorescent pH indicators that have a relatively low pK_a of about 4–6. Because the fluorescence of most phenolic pH indicators is quenched by acid and also because most phenolic dyes have pK_a values between 7 and 10, there are few fluorescent probes that can be used to prepare protein conjugates that are useful in this pH range. The pH indicators Cl-NERF and DM-NERF have been briefly described in an abstract (Whitaker *et al.*, 1991b). These have pK_a values of 3.0–3.5 and 5.0–5.5 respectively and long wavelength spectral properties that allow dual-excitation ratio measurements.

3.4 CONCLUSIONS

The numerous applications of the ion indicators described above are beyond the scope of this chapter. The design and synthesis of new indicators is meeting the requirement of an optical response which is specific to the ion, occurs in the expected range of the ion concentration, and is free of artifacts – or where the artifacts are recognized and subject to control or measurement. Development of new fluorescent indicators for ions, such as an intracellular indicator for calcium that would have an optical response mimicking that of carboxy-SNARF-1 (Fig. 3.2), would permit new single cell measurements, particularly in flow cytometry and laser scanning microscopy. Development of optimal fluorescent indicators for intracellular pH is probably close to completion. However, the current sodium, potassium and chloride indicators have recognized limitations that make their applications problematic. Like quin-2, these are prototypical indicators and new generations will hopefully follow from the few laboratories working on synthesis of this type of probe.

REFERENCES

Baylor S.M., Hollingworth S. & Konishi M. (1989) *J. Physiol.* **418**, 69P.

Biwersi J. & Verkman A.S. (1991) *Biochemistry* **30**, 7879.

Biwersi J., Ke Z. & Verkman A.S. (1992) *Biophys. J.* **61**, A35, Abstr. 3104.

Blatter L.A. & Weir W.G. (1990) *Biophys. J.* **58**, 1491.

Blinks J.R., Mattingly P.H., Jewell B.R., Van Leeuwen M., Harrer G.L. & Allen D.G. (1978) *Methods Enzymol.* **57**, 292.

Borin M. & Siffert W. (1990) *J. Biol. Chem.* **265**, 19543.

Bright G.R., Simon J.R. & Taylor D.L. (1987) *J. Cell Biol.* **105**, 1068a.

Bright G.R., Whitaker J.E., Haugland R.P. & Taylor D.L. (1989) *J. Cell Physiol.* **141**, 410.

Bush D.S. & Jones R.L. (1988a) *Eur. J. Cell Biol.* **46**, 466.

Bush D.S. & Jones R.L. (1988b) *Cell Calcium* **8**, 455.

Bush D.S. & Jones R.L. (1990) *Plant Physiol.* **93**, 841.

Connor J.A. (1986) *Proc. Natl. Acad. Sci. USA* **83**, 6179.

Delbono O. & Stefani E. (1992) *Biophys. J.* **61**, A160, Abstr. 916.

DeMarinis R.M., Katerinopoulos H.E. & Muirhead K.A. (1989) US Patent No. 4,849,362.

Di Virgilio F., Steinberg T.H., Swanson J.A. & Silverstein S.C. (1988) *J. Immunol.* **140**, 915.

Di Virgilio F., Steinberg T.H. & Silverstein S.C. (1990) *Cell Calcium* **11**, 57.

Elliott D.C. & Petkoff H.S. (1990) *Plant Sci.* **67**, 125.

Etter E.F., Kuhn M.A., Tuft R.A., Bowman D.S. and Fay F.S. (1992) *Biophys. J.* **61**, A159.

Grynkiewicz G., Poenie M. & Tsien R.Y. (1985) *J. Biol. Chem.* **260**, 3440.

Haugland R.P. (1992) *Handbook of Fluorescent Probes and Research Chemicals*. Molecular Probes, Inc., Eugene, OR.

Illsley N.P. & Verkman A.S. (1987) *Biochemistry* **26**, 1219.

Jaffe D.B., Johnston D., Lasser-Ross N., Lisman J.E., Miyakawa H. & Ross W.N. (1992) *Nature* **357**, 244.

Jezek P., Mahdi F. & Garlid K.D. (1990) *J. Biol. Chem.* **265**, 10522.

Jung D.W., Apel L. & Brierley G.P. (1990) *Biochemistry* **29**, 4121.

Kao J.P.Y., Harootunian A.T. & Tsien R.Y. (1989) *J. Biol. Chem.* **264**, 8179.

Konishi M., Hollingworth S., Harkins A.B. & Baylor S.M. (1991) *J. Gen. Physiol.* **97**, 271.

Krapf R., Berry C.A. & Verkman A.S. (1988) *Biophys. J.* **53**, 955.

Kurebayashi N., Harkins A.B. & Baylor S.M. (1992) *Biophys. J.* **61**, A160, Abstr. 917.

Lakowicz J.R., Szmacinski H. & Johnson M.L. (1992) *J. Fluorescence* **2**, 47.

Laskay G., Varhelyi T., Dale R.E. & Dexter T.M. (1992) *Biophys. J.* **61**, A385, Abstr. 2217.

Lipscombe D., Madison D.V., Poenie M. & Reuter H. (1988) *Neuron* **1**, 355.

Minta A. & Tsien R.Y. (1989) *J. Biol. Chem.* **264**, 8171.

Minta A., Kao J. & Tsien R.Y. (1989) *J. Biol. Chem.* **264**, 19449.

Murphy R.F., Powers S. & Cantor C.R. (1984) *J. Cell Biol.* **98**, 1757.

Negulescu P.A. & Machen T.E. (1990) *Methods Enzymol.* **192**, 38.

Paradiso A.M., Tsien R.Y., Demarest J.R. & Machen T.E. (1987) *Am. J. Physiol.* **253**, C30.

Quamme G.A. & Rabkin S.W. (1990) *Biochem. Biophys. Res. Commun.* **167**, 1406.

Raju B., Murphy E., Levy L.A., Hall R.D. & London R.E. (1989) *Am. J. Physiol.* **256**, C540.

Rink T.J., Tsien R.Y. & Pozzan T. (1982) *J. Cell Biol.* **95**, 189.

Satoh H., Hayashi H., Noda N., Terada H., Kobayashi A., Yamashita Y., Kawai T., Hirano M. & Yamazaki M. (1991) *Biochem. Biophys. Res. Commun.* **175**, 611.

Scarpa A., Brinkley F.J., Tiffert T. & Dubyak G.R. (1978). *Ann. NY Acad. Sci.* **307**, 86.

Shen S.S. & Buck W.R. (1990). *Devel. Biol.* **140**, 272.

Sklar L.A., Jesaitis A.J., Painter R.G. & Cochrane C.G. (1982) *J. Cell. Biochem.* **20**, 193.

Sklar L.A., Finney D.A., Oades Z.G., Jesaitis A.J., Painter R.G. & Cochrane C.G. (1984) *J. Biol. Chem.* **259**, 5661.

Steinberg J.H., Newman A.S., Swanson J.A. & Silverstein S.C. (1987) *J. Cell Biol.* **105**, 2695.

Thiebaut F., Currier S.J., Whitaker J.E., Haugland R.P., Gottesman M.M., Pastan I. & Willingham M.C. (1990) *J. Histochem. Cytochem.* **38**, 685.

Thomas J.A., Buchsbaum R.N., Zimniak A. & Racker E. (1979) *Biochemistry* **18**, 2210.

Tsien R.Y. (1980) *Biochemistry* **19**, 2396.

Tsien R.Y. (1982) *J. Cell Biol.* **95**, 189.

Tsien R.Y. & Pozzan T. (1989) *Methods Enzymol.* **172**, 230.

Verkman A.S., Sellers M.C., Chao A.C., Leung T. & Ketcham R. (1989) *Anal. Biochem.* **178**, 355.

Whitaker J.E., Haugland R.P. & Prendergast F.G. (1991a) *Anal. Biochem.* **194**, 330.

Whitaker J.E., Haugland R.P., Ryan D., Dunn K., Maxfield F.R. & Haugland R.P. (1991b) *Biophys. J.* **59**, 358a, Abstr. 478.

Wolfbeis O.S. & Urbano E. (1982) *J. Heterocyclic Chem.* **19**, 841.

Wolfbeis O.S. & Urbano E. (1983) *Fresenius Z. Anal. Chem.* **314**, 577.

Yamashiro D.J., Tycko B., Fluss S.R. & Maxfield F.R. (1984) *Cell* **37**, 789.

CHAPTER FOUR

Redox Confocal Imaging: Intrinsic Fluorescent Probes of Cellular Metabolism

BARRY R. MASTERS[1] & BRITTON CHANCE[2]

[1] Department of Anatomy and Cell Biology, Uniformed Services University of the Health Sciences, Bethesda, MD, USA

[2] Department of Biochemistry and Biophysics, University of Pennsylvania, Philadelphia, PA, USA

4.1 INTRODUCTION

4.1.1 What is redox fluorometry?

This chapter describes redox imaging of cells, tissues and organs based on intrinsic fluorescent probes of cellular metabolism. Cellular metabolism may be non-invasively interrogated through the 'optical method' based on the fluorescence intensity of intrinsic probe molecules (Chance, 1991). The intrinsic fluorescent probes which report on cellular metabolism are the reduced pyridine nucleotides, NAD(P)H, and the oxidized flavoproteins.

The basis of redox fluorometry is that the quantum yield of the fluorescence, and hence the intensity, is higher for the reduced form of NAD(P) and lower for the oxidized form (Fig. 4.1). The reverse is true for the flavoproteins; the quantum yield, and hence the intensity, is higher for the oxidized form and lower for the reduced form. Measurement of the fluorescence intensity of either NAD(P)H or flavoproteins, or their ratio, is an optical indicator of cellular metabolism.

The main advantage of redox fluorometry is that it is a non-invasive optical method with high spatial and temporal resolution. The temporal resolution is of the order of milliseconds. The spatial resolution is given

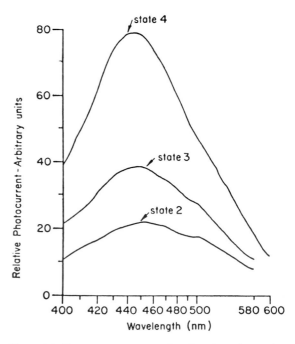

Figure 4.1 Fluorescence spectra of rat liver in various redox states. The excitation is at 366 nm. The upper curve is characteristic of the hypoxic state. The lower curve is characteristic of normoxic tissue.

by the spot size of the excitation light beam, and may be a diffraction-limited spot of a few micrometres in diameter.

Redox fluorometry has numerous applications in cell biology, tissue physiology and organ physiology. Applications include such diverse samples as single cells, muscle tissue and organs such as the eye and the brain. The method is not limited to surface fluorometry of cells and organs; the widespread use of fibre optics in minimally invasive surgery and diagnostics, permits the use of redox fluorometry inside the body. An example of this approach is the use of redox fluorometry to monitor the metabolism of myocardium during open-heart surgery. In this experimental approach, the laser excitation light is transmitted to the *in vivo* myocardium through a fibre optic, and the NAD(P)H fluorescence is transmitted to the redox fluorometer through another fibre optic.

In addition to the use of redox fluorometry to monitor cellular metabolism, it can also be utilized as an optical probe of oxygen concentration in cells, tissues and organs. The intensity of NAD(P)H fluorescence varies with oxygen concentration; this is the basis of non-invasive monitoring of corneal mitochondrial oxygen utilization (Masters, 1988).

Redox fluorometry must be used with proper calibration and an understanding of its technical limitations. As a technique, it has the following limitations: (1) it lacks an absolute scale which relates the fluorescence intensity of the intrinsic probes to a unique index of cellular metabolism; (2) there are contributions of fluorescence from both the mitochondrial space and the cytoplasmic space (in the case of NAD(P)H fluorescence) to the total fluorescence intensity (Sies, 1982); (3) there are several factors which can affect the fluorescence intensity of the NAD(P)H signal; and (4) redox fluorometry based on intensity measurements is dependent on the geometry of the sample and the optical instrument (Rost, 1991).

Recent developments in fluorescent redox imaging based on fluorescent lifetimes can eliminate this geometrical dependence (Lakowicz, 1992). For specific tissues, e.g. the cornea, the NAD(P)H fluorescence intensity from the mitochondrial space dominates the total NAD(P)H fluorescence intensity. When used with proper controls, redox fluorometry provides a simple, non-invasive, optical probe of cellular metabolism.

4.1.2 Chapter scope

This chapter presents a critical, self-contained evaluation of redox confocal imaging. The coverage includes historical developments, biochemical and photophysical basis of the methodology, instrumentation, applications to various tissues and organs, and a comparison with other non-invasive techniques. The subjects of chemiluminescence and bioluminescence, whilst within the subject domain of this chapter, are discussed elsewhere in this book.

The authors have decided to present a critique of the methodology which puts the advantages and disadvantages of the technique in perspective. It is hoped that this approach will enable the investigator to use redox fluorometry in a manner that maximizes its utility and minimizes artifacts. The subjects not covered in depth are cited in the references. The coverage of applications will concentrate on ocular tissue and the emphasis is on two-dimensional confocal redox imaging. Previous reviews have provided extensive coverage of redox applications with other cells, tissues and organs (Masters, 1984a; Chance, 1991).

In summary, this chapter discusses new developments in the methodology. Confocal microscopy provides high resolution optical sectioning yielding two-dimensional images and the concomitant increase in spatial resolution. The application of two-photon confocal fluorescence microscopy to the redox imaging of the eye is a dramatic new development in two-dimensional redox imaging. *In vivo*, real-time confocal microscopy combined with redox fluorometry provides the basis of functional imaging; *in vivo* confocal microscopy provides two-dimensional images of cell and tissue morphology, and redox fluorometry provides the metabolic imaging (Masters, 1993b). The reader of this chapter will find a self-contained review of the theory, methodology and instrumentation.

4.1.3 Previous reviews

An excellent, recent book on spectrofluorometry of cells and tissues is edited by Kohen and Hirschberg (1989). This volume contains comprehensive coverage of the following topics: history of fluorescence microscopy, image spectroscopy of living cells, fluorescence microscopy in three dimensions, frequency-domain fluorescence spectroscopy and fluorescence probes. In addition, cytometry and cell sorting as well as bioluminescence are covered.

Another recent review is the chapter by Balaban and Mandel (1990) on optical methods for metabolic studies of living cells. This review covers the authors' work and details the problem of redox measurements of the surface of a beating heart. They provide a technical solution which consists of an internal fluorescent standard in addition to the NAD(P)H measurements (Balaban & Mandell, 1990).

Reviews which are more specific to studies of ocular tissue include the following: Masters *et al.* (1981, 1982a) and Masters (1984a). The latter review is a comprehensive study of instrumentation and biochemical studies on which redox fluorometry is based.

Much of the instrumentation and techniques that are of use for studies of cells in tissue culture are discussed. The method of freeze-trapping tissues and organs, and mechanical sectioning prior to two-dimensional redox imaging is described. Applications to *in vitro* and *in vivo* organs and tissues are covered in a previous major review of redox fluorometry (Masters, 1984a).

4.2 HISTORY OF THE USE OF INTRINSIC PROBES TO MONITOR CELLULAR METABOLISM

While it is widely acknowledged that the development of the the optical microscope is intimately linked with the development of the science of pathology, the relationship of the spectroscope to medical science is uncommon knowledge. In fact, Leeuwenhoek was able to demonstrate the existence of capillary circulation, first described red blood cells, observed bacteria in dental plaque, and described lenticular fibres in the ocular lens as well as the striations of skeletal muscle fibres. All of these observations were made 350 years ago with his single lens microscope.

The history of intracellular respiration is described in its historical context in a fascinating work by Keilin (1966). The key people in this theme include Harvey, Malpighi, Leeuwenhoek, Hooke, Lower, Mayov, Priestly, Lavoisier, Spallanzani, Pflüger, Ludwig, Berthelot, Liebig, Pasteur, Bernard, Hoppe-Seyler, Berzelius, Traube, Cagniard-Latour and Appert (Keilin, 1966).

In 1880 an important book with the title *The Spectroscope in Medicine* was published in London (MacMunn, 1880) and in 1914, MacMunn published another book, *Spectrum Analysis Applied to Biology and Medicine*. Work on the colour of cells and tissues, and their absorption and fluorescence properties begins with the studies of haemoproteins by MacMunn, continues with the classic work of Keilin on cytochromes, and continues in this century with the modern advances in the development of redox fluorometry.

By the mid-1800s the following three important empirical observations had been made: (1) the fluorescence emission is always at a longer wavelength than the exciting light (known today as Stokes's law); (2) a body must first absorb the light before it can emit fluorescence; and (3) the fluorescent spectrum, which is the intensity of light emitted as a function of wavelength, can be used to characterize specific substances. It is the last point which is the basis of spectral characterization of specific molecules. In 1934 Haitinger coined the term 'fluorochrome' to describe the use of fluorescence dyes which results in fluorescent staining of tissues.

David Keilin had great success with a low-dispersion prism spectroscope fitted to a microscope which was manufactured by Carl Zeiss. This instrument was used by Keilin to study slices of plant and animal tissues as well as suspensions of bacteria and yeast. Keilin was able to observe the oxidation and reduction of cytochrome within living tissues and cells based on changes in the absorption spectrum of visible light. This work then led to the next generation of tissue spectrometers developed in Sweden (Caspersson, 1950, 1954) and in the United States (Chance & Thorell, 1959; Chance, 1951).

The advantage of microspectrographic methods is that cellular and intracellular organelles can be studied in the living state, in real-time with a non-invasive optical technique. This means that cellular organelles do not have to be removed from their natural surroundings or destroyed in the process of studying their metabolism. Prior to 1945, the early work tended to investigate the nucleus and chromosome structure in normal and tumour cells. After the Second World War research shifted towards optical studies of oxidative metabolism in cells and organelles. The studies of intact cells (yeast and bacteria) accelerated, and optical studies of mitochondria became a more active research area.

A historical survey of the origins of modern fluorescence microscopy and fluorescence is useful reading for every fluorescence microscopist (Kasten, 1989). The following listing of historical events is taken from the work of Kasten. Although the term fluorescence was coined by George Stokes in 1852, Kasten credits David Brewster with the first description of the phenomenon of fluorescence in 1838. August Köhler invented the first ultraviolet absorption microscope in Jena in 1904. In 1910, H. Lehmann of Carl Zeiss, Jena used a modification of the liquid filters developed by Robert Wood, a professor of experimental physics, at Johns Hopkins University, and in the next two years developed the first fluorescence microscope. The reader may find it of interest that the light sources were typically Siemens arc lamps with a power of 2000–3000 W!

Han Stübel made the first microscopic observations of autofluorescence in 1911. He investigated the autofluorescence of animal organs, bacteria and protozoa. Further development rapidly followed. In 1929, Ellinger, a pharmacologist at Heidelberg University, developed the intravital fluorescence microscope, which Carl Zeiss, Jena then manufactured. This new instrument utilized vertical or epi-illumination, a water-immersion objective, and a series of filters for excitation wavelength selection which were mounted in a rotating wheel. A steady gravity flow of physiological solutions superfused the living organs for the duration of their microscopic observation.

A microscope which could photographically record the fluorescence spectrum induced on tissues, cells or fluid samples was developed by Borst and Köningsdörffer in 1929. This important technical development was used in numerous studies of porphyrin autofluorescence in patients with porphyria. This important milestone antedated the work of van Euler on riboflavin fluorescence, as well as Caspersson's work on the ultraviolet absorption of nucleic acids and proteins. In 1959, rapid kinetic studies in mitochondria of living cells were performed with a differential microspectrofluorometer (Chance & Thorell, 1959).

Historical and personal perspectives on the development of microspectroscopy and the optical method contain important landmark developments in the field of non-invasive studies of cell and tissue metabolism (Chance, 1989, 1991). The development of the dual-wavelength spectrophotometer and its influence on cell and tissue optical studies is presented here by Dr Chance who was present during this development.

The origins of intracellular microspectrofluorometry started with the work of Warburg who discovered NAD(P)H fluorescence. Dr Bo Thorell and Britton Chance, and Baltscheffsky then discovered that the cell autofluorescence was due to NAD(P)H fluorescence from the mitochondrial matrix space. Together they were able to measure the contribution of the NADH/NADPH in the cytosol to the mitochondrial NAD(P)H signal. These measurements were of fundamental importance in the development of the use of mitochondrial fluorescence as a sensitive indicator of mitochondrial hypoxia in tissues. This study succeeded in measuring the fluorescence from a single mitochondrion within a single cell (Chance & Thorell, 1959).

4.3 THE BIOCHEMICAL BASIS OF INTRINSIC FLUORESCENT PROBES IN LIVING CELLS

There are two intrinsic fluorescent probes which can serve as non-invasive intrinsic probes of cell metabolism: the oxidized flavoproteins and the reduced pyridine nucleotides.

The fluorescence intensity from intrinsic oxidized flavoproteins which are present in mitochondria is a non-invasive measure of cellular metabolic function. The main advantage of measuring the fluorescence intensity from the oxidized flavoproteins is that the fluorescence is localized in the mitochondrial space. Redox imaging based on NAD(P)H fluorescence originates in both the mitochondrial and the cytoplasmic spaces (Chance & Baltscheffsky, 1958; Chance et al. 1978, 1979). Every tissue is different with respect to the ratio between NAD(P)H fluorescence contributions

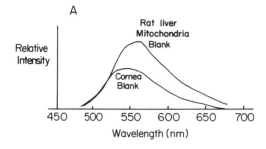

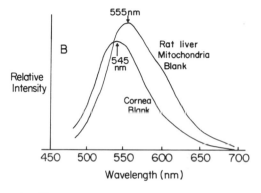

Figure 4.2 Fluorescence spectra of flavoproteins from cornea and rat liver mitochondria. The excitation is at 460 nm.

from the mitochondrial space and from the cytoplasmic space (Chance & Williams, 1955).

The fluorescence intensity from oxidized flavoproteins in cells, e.g. the cornea epithelium, occurs in the region from 520 nm to 590 nm with a broad maximum at 540 nm (Chance et al., 1968). The light absorption of oxidized flavoproteins has a broad maximum at 460 nm and extends from 430 to 500 nm (Fig. 4.2). The fluorescence intensity of the oxidized flavoproteins in cells is usually much lower than the fluorescence intensity from the reduced pyridine nucleotides.

Fluorescence from the intrinsic oxidized flavoproteins in the cornea has been spectroscopically characterized in rabbit corneas frozen to 77K (Chance & Lieberman, 1978). Other evidence that the corneal epithelial cell fluorescence in both in vitro and in vivo corneas is due to oxidized flavoproteins comes from studies by Masters (1984a,b). The distribution of mitochondria, stained with the cationic dye rhodamine 123, in the basal epithelial cells of the rabbit cornea has been studied with a confocal laser scanning fluorescence microscope (Masters 1993b).

The alteration of the fluorescence intensity of the oxidized flavoproteins in the in vivo rabbit corneal epithelial cells as a function of cellular hypoxia has been demonstrated (Masters et al., 1982a). A HeCd laser at 442 nm was used to excite the corneal epithelial cells in a living rabbit and the flavoprotein fluorescence intensity was measured in the wavelength region of 550 nm (Masters *et al.*, 1982b). The fluorescence intensity was reduced in the presence of a flow of hydrated nitrogen (tissue hypoxia) and the effect was reversed in the presence of hydrated air. These studies are in agreement with *in vitro* studies of flavoprotein fluorescence from corneal epithelial cells conducted in both *in vitro* rabbit perfused corneas at 37C, and in freeze-trapped rabbit cornea at 77K.

The fluorescence from the naturally occurring reduced pyridine nucleotides in cells is an indicator of cellular respiration. The fluorescence of reduced pyridine nucleotides is excited with light of 364 nm and has a fluorescence emission in the range of 400–500 nm. Cellular hypoxia is associated with an increased ratio between reduced and oxidized pyridine nucleotides, and therefore increased fluorescence intensity in the region 400–500 nm. Since the quantum efficiency, and thus the fluorescence intensity, of the reduced pyridine nucleotides is significantly greater than that of the oxidized pyridine nucleotides, the fluorescence intensity monitors the degree of cellular hypoxia. This non-invasive technique is called redox fluorometry. We have demonstrated that the 400–500 nm fluorescence excited at 364 nm is due to the reduced pyridine nucleotides (Masters, 1984a). While the fluorescence intensity of the cornea has been investigated using optically sectioning microscopes to monitor the degree of cellular hypoxia, it was not previously possible to obtain single cell images of the reduced pyridine nucleotide fluorescence. However, two-dimensional redox imaging has been demonstrated by other investigators using isolated cardiac myocytes in tissue culture (Eng *et al.*, 1989).

The fluorescence from NAD(P)H is an intrinsic probe which can be used to study cellular metabolism (Chance & Thorell, 1959). The fluorescence intensity from these intrinsic probes provides a non-invasive optical method to monitor cellular respiration. The fluorescence from NAD(P)H has been used to study cellular metabolism in many tissues and organs due to the strong fluorescence intensity. The NAD(P)H fluorescence intensity occurs in two compartments, the mitochondrial and the cytosolic; this complicates the interpretation of the fluorescence studies. However, in some tissues, e.g. rat cardiac myocytes, the NAD(P)H fluorescence is predominantly from the mitochondrial space.

Two-dimensional images of the fluorescence intensity from NAD(P)H have been studied in brain slices and in isolated perfused hearts (Chance, 1991). Kohen *et al.*, (1989) have developed instruments for rapid redox mapping of cells in culture. At the cellular level NAD(P)H imaging of isolated rat cardiac myocytes have been studied with a standard fluorescence microscope (Balaban & Mandel, 1990). These authors

demonstrated that the fluorescence images are mainly due to mitochondrial NAD(P)H fluorescence (Eng *et al.*, 1989). Two-dimensional imaging of the NAD(P)H fluorescence intensity of *in vitro* corneal endothelial cells has been studied with an ultraviolet confocal laser scanning fluorescence microscope (Masters, 1993b).

4.4 INSTRUMENTATION FOR THE USE OF LOW-LIGHT-LEVEL FLUORESCENT IMAGING OF LIVING CELLS AND TISSUES

Masters (1984a) reviewed various instrumental set-ups for redox imaging of organs and tissues. Major advances in instrumentation include confocal microscopy and two-photon confocal microscopy for redox imaging.

The 'gold standard' for two-dimensional redox imaging of freeze trapped specimens is the automated, milling, redox ratio-scanning instrument (Quistorff *et al.*, 1985). This instrument is only used with frozen specimens; however, its spatial resolution and quantitative fluorescence as well as a histogram output of redox state homogeneity is of great value (Figs. 4.3 and 4.4).

The confocal microscope is an optical device which can be used to observe a single focal plane of thick objects with high resolution and contrast as compared to standard microscopes (Wilson, 1990; Masters & Kino, 1990). The confocal microscope used in the fluorescent mode has the excitation at one wavelength, and the fluorescent image is formed at a longer wavelength. This differs from the reflected light mode in which the confocal image is formed at the same wavelength as that of the laser illumination. The depth resolution of fluorescence mode confocal scanning optical microscopes is reduced as compared to that in the reflected imaging mode. The advantages of UV confocal microscopy include increased resolution and a reduced depth of focus as compared to visible light confocal microscopy. These advantages depend on a microscope objective which is corrected for the UV.

The use of the confocal microscope to section the cornea optically has been demonstrated by Lemp *et al.*, (1986). The fine structure of the *in vitro* cornea has been shown with both the one-sided Nipkow disk confocal microscope and with the laser scanning confocal microscope (Masters & Paddock, 1990a). Three-dimensional volume reconstruction from serial confocal optical sections of the *in vitro* cornea has been demonstrated (Masters & Paddock, 1990b).

The laser scanning confocal microscope (Zeiss, UV confocal LSM) permits two-dimensional confocal imaging of the redox fluorescence intensity of corneal

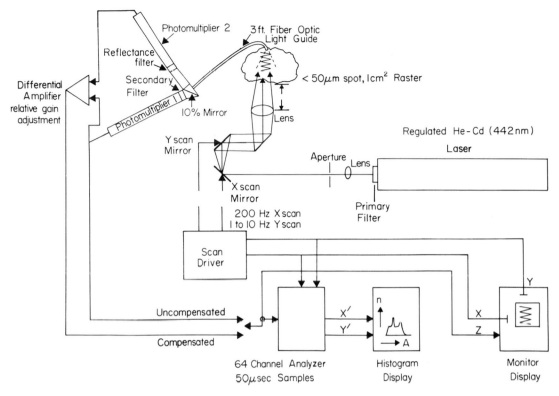

Figure 4.3 A fibre optic laser redox fluorometer for *in vivo* work.

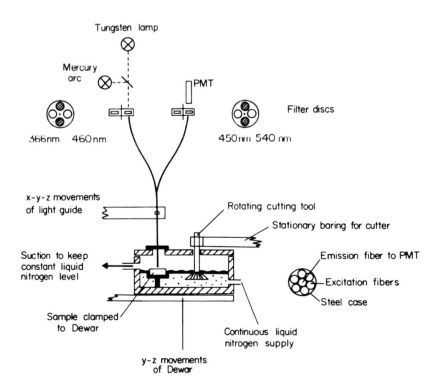

Figure 4.4 A low-temperature scanning, milling redox fluorometer. The freeze-trapped specimen is scanned at liquid nitrogen temperature, then the next layer is milled off the tissue and a new scan is made. The two-dimensional redox scans can be combined to form a three-dimensional image of tissue redox states.

endothelial cells (Kapitza & Wilke, 1988). Thus, a two-dimensional image or map of cellular hypoxia can be obtained. A combination of the reflected light images of cell morphology and the redox fluorescence images of cellular metabolism may be used to construct a multi-modality three dimensional image of cell structure and function.

4.4.1   One-dimensional confocal redox fluorometer

Figures 4.5 and 4.6 illustrate the mechanical and optical components of a confocal redox fluorometer. The device is a confocal microscope because it contains two slits located in conjugate planes, one for the illumination and one for the image plane. The confocal microscope is used in the vertical mode for work on tissue cultures and for studies of *in vitro* eyes, and in the horizontal mode for *in vivo* studies on animals or human subjects. This confocal redox fluorometer can be used to measure fluorescence from NAD(P)H, from oxidized flavoproteins, or from extrinsic probes, i.e. mitochondrial or nuclear stains, and back-scattered light. The unique feature of the confocal redox fluorometer is that the microscope objective is a scanning objective (Masters, 1988). A piezoelectric driver scans the microscope objective which gives rise to a depth profile across the cell or tissue. The spatial

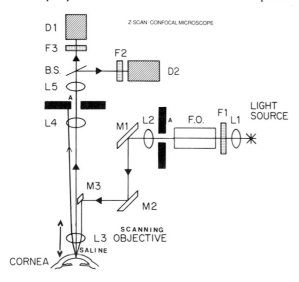

Figure 4.5 Schematic diagram of the scanning one-dimensional confocal microscope showing a light ray path. The light source is either a laser or a mercury arc lamp connected to the microscope by a fibre optic. F1 and F2 are narrow band interference filters to isolate the excitation wavelengths. F3 is a narrow band interference filter to isolate the emission light. M1, M2 and M3 are front surface mirrors, and B.S. is a quartz beam splitter. L3 is the scanning objective 50×, NA 1.00. The piezoelectric driver scans the microscope objective along the optic axis of the eye. This confocal microscope is suitable for use with tissue culture or in the horizontal mode for use with living animal or human subjects.

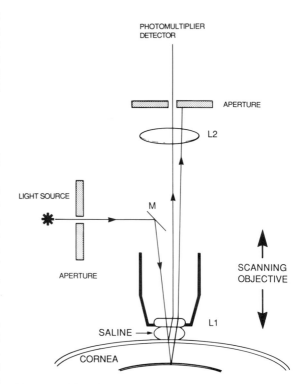

Figure 4.6 The principle of confocal microscopy. Schematic diagram of the *z*-scan confocal microscope developed for optically sectioning the living eye. The light source is connected to the instrument via a quartz fibre optic light guide. There are two conjugate slits which make the microscope a confocal microscope. One slit is imaged onto the object, and the second slit in front of the detector forms an image of itself on the object. The *z*-scan confocal microscope has a scanning microscope objective which moves on an axis under computer control via a piezoelectric driver.

resolution of this system is demonstrated in Fig. 4.7. Figure 4.8 illustrates the *in vivo* use of the confocal redox fluorometer developed by Masters. This example shows the changes in the NAD(P)H fluorescence intensity of the *in vivo* rabbit eye due to contact lens-induced corneal hypoxia.

4.2.2   Confocal redox NAD(P)H imaging

A Zeiss confocal scanning laser microscope (LSM 10 UV, Carl Zeiss, Oberkochen, Germany) has been adapted for UV fluorescence confocal microscopy (Kapitza & Wilke, 1988; Masters, 1993b; Masters *et al.*, 1991). In addition to the argon ion laser (488 and 514 nm) and the HeNe laser (543 nm), another argon laser (364 nm) was added to the microscope. The exact wavelength was 333.6 nm. The UV argon ion laser was a Spectra Physics, Type 2016, with a variable output power of 20–100 mW. In addition to the third laser there were other changes to the microscope; the

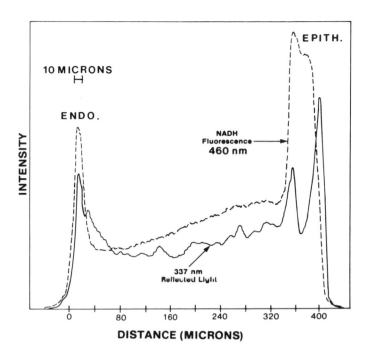

Figure 4.7 One-dimensional *in vivo* redox confocal imaging of the living eye. An optical section through a rabbit cornea illustrating the range resolution for the back-scattered light (solid line) and the NAD(P)H fluorescence emission (broken line). The intensity of the back-scattered light is 10 times that of the fluorescence. The tear film is on the right side of the scan and the aqueous humour is on the left side of the figure.

antireflection coatings on the *x–y* scanner were optimized for UV light.

A Zeiss water-immersion objective of 25X, NA 0.8, corrected for UV, was used to measure the fluorescence from optical sections of a freshly enucleated rabbit eye. The microscope objective was able to focus across the full 400 μm thickness of the *in vitro* rabbit cornea and digital images of 256 × 265 pixels and 256 grey levels were produced. The back-scattered light mode of the microscope at a wavelength of 364 nm was used to locate the endothelial cells which are situated about 400 μm below the anterior surface of the cornea. The microscope was focused about 2μm into the corneal endothelial cells, and a light confocal image formed (Fig. 4.9). Then the microscope was switched to the fluorescence mode (Fig. 4.10). The excitation wavelength was 364 nm and the emission filters collected light in the region of 400–500 nm. Eight images were averaged to improve the signal-to-noise ratio of the final image. The wavelengths used for the laser excitation (364 nm) and the fluorescent emission (400–500 nm) correspond to the excitation and emission wavelengths of the reduced pyridine nucleotides. Several confocal images of the *in vitro* cornea were made using the reflected light mode with the laser source at 488 nm to demonstrate the optical resolution and contrast of the modified microscope.

4.4.3   Two-photon confocal NAD(P)H imaging

Two-photon laser scanning fluorescence microscopy has been used to image the NAD(P)H fluorescence of the basal epithelium of an *in vitro* rabbit cornea (Denk *et al.*, 1990). The technique has two important advantages over single-photon confocal fluorescence imaging. The point spread function varies as *z* to the fourth power, where *z* is the distance from the focal plane. In a standard confocal microscope *z* varies as the second power. Since the illumination is at 700 nm, photobleaching only occurs in the focal plane; with the standard confocal microscope photobleaching occurs in all the planes scanned by the laser beam. The application of two-photon fluorescence confocal microscopy to redox imaging is a major advance. Its disadvantages are complexity and cost; these may change in the future.

4.4.4   Confocal redox flavoprotein imaging

The study by Masters and Kino (1990) used a laser scanning confocal microscope (BioRad MRC-600) to image freshly enucleated rabbit eyes. The microscope uses two scanning mirrors moved by scanning galvanometers to scan the laser beam across the microscope objective. The reflected light retraces the incident path and is collected by a set of curved mirrors and sets of flat mirrors. The scan mirrors are located at conjugate aperture planes. An adjustable pinhole is placed in front of the photodetector.

The freshly enucleated eye was transferred to a black plastic chamber containing Ringer's solution and placed on the stage of the confocal microscope. The bicarbonate Ringer's solution completely immersed

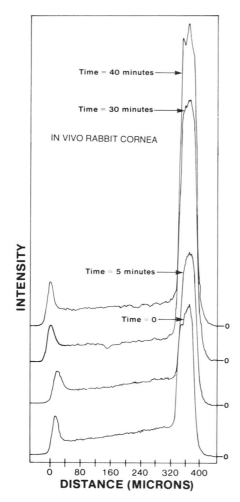

Figure 4.8 *In vivo* redox confocal imaging of the rabbit cornea. Time dependence of the effect of a PMMA contact lens on the NAD(P)H fluorescence intensity of the rabbit cornea. The peaks on the left side are from the corneal endothelium. The larger peaks on the right side are from the corneal epithelium. Time represents the duration that the contact lens was on the eye.

the eye and the tip of the microscope objective. Every 10 min the Ringer's solution was exchanged for fresh aerated solution.

The BioRad laser scanning microscope used an air-cooled 25 mW argon ion laser to provide the 488 nm wavelength. This wavelength produced excitation in the reflected light mode. The low reflectivity of the cornea necessitated maximum amplification of the signal from the detector. At this high amplification, a bright spot of stray light appeared in the centre of each image. A black disk or square was used partially to mask this reflection and to designate its position. Kalman averaging was used to average ten frames to reduce the noise in the final image.

The fluorescence intensity in the wavelength region

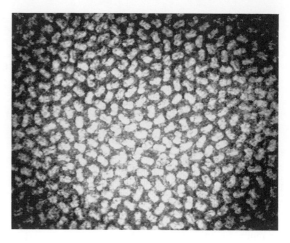

Figure 4.9 A confocal image formed in back-scattered light of the *in vitro* rabbit cornea. The image shows corneal endothelial cells. The bright regions are the cell nuclei. The image was formed with a Zeiss laser scanning confocal microscope.

longer than 515 nm was used to image the flavoprotein fluorescence. The excitation filter centred at 488 nm with a 10 nm band-pass was used to isolate the 488 nm line from the argon ion laser. A dichroic reflector (BioRad DR 510 LP) reflected the 488 nm laser line onto the microscope objective. A BioRad barrier filter CG 515 (yellow) blocked all light with wavelengths shorter than 515 nm from reaching the detector. The emission spectrum of the corneal epithelial cells extends from 500 nm to 600 nm with a maximum at 545 nm. The absorption spectrum of the flavoproteins is in the range 400–500 nm, with a maximum at 550 nm. The argon ion excitation line at 488 nm, while not optimal, is a suitable laser line for fluorescence excitation of the oxidized flavoproteins.

4.5   APPLICATIONS OF INTRINSIC FLUORESCENT REDOX PROBES TO CELLULAR METABOLISM

To illustrate the numerous applications of the technique to *in vitro* and *in vivo* tissues the authors have selected references which illustrate significant studies of cell function. As stated in the Introduction, the applications are focused on redox fluorometry of the cornea.

4.5.1   Redox imaging based on NAD(P)H fluorescence

4.5.1.1   Confocal NAD(P)H redox imaging

In order to demonstrate the advantages of confocal microscopy in monitoring of corneal morphology we

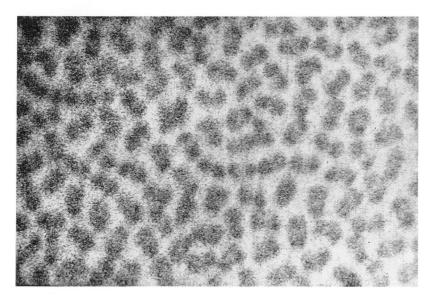

Figure 4.10 A confocal redox image of endothelial cells of the *in vitro* rabbit eye formed from NAD(P)H fluorescence. The bright regions correspond to the NAD(P)H fluorescence. The large dark regions are the endothelial cell nuclei.

have produced images of the *in vitro* cornea both perpendicular to the corneal surface and in the plane of the cornea using 488 nm laser light in the reflected light confocal mode. Figure 4.11 illustrates the quality of *z*-scan of the cornea produced by back-scattered confocal microscopy. Figure 4.12 illustrates a three-dimensional reconstruction of a stack of serial optical sections of the rabbit cornea.

We have found that the image quality of two-dimensional redox images in dependent on the depth within the tissue. Images at the anterior of the cornea have superior resolution compared with those images made 400 μm into the cornea. The image quality at the posterior surface of the cornea, which is made through 400 μm, is shown in Fig. 4.9. The fluorescence of the reduced pyridine nucleotides is shown in Fig. 4.10. These images represent two dimensional maps of cellular metabolic function.

4.5.1.2 Two-photon NAD(P)H redox imaging

To illustrate the superb signal-to-noise ratio obtained with two-photon confocal redox imaging the technique was applied to the *in vitro* cornea. Figure 4.13 is an image of the basal epithelium in the back-scattered light mode. Figure 4.14 shows the two-dimensional

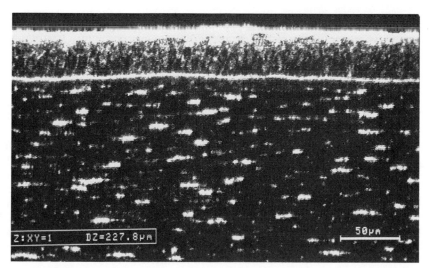

Figure 4.11 A *z*-scan of the cornea of a freshly enucleated rabbit eye obtained using a laser scanning confocal microscope in the back-scattered mode.

Figure 4.12 Three-dimensional volume reconstruction of a serial stack of two-dimensional back-scattered light confocal microscopic images of the full thickness of the cornea from an *in vitro* rabbit eye. The rectangular section is located in the central region of the cornea. The thickness of the cornea is 400 μm. The image was formed in a computer from the stack of two dimensional images using the volume rendering reconstruction technique. The bright line at the top of the figure is the superficial epithelium. The bright line 40 μm below is the reflection from the basal lamina. The horizontal lines are nuclei of stromal keratocytes.

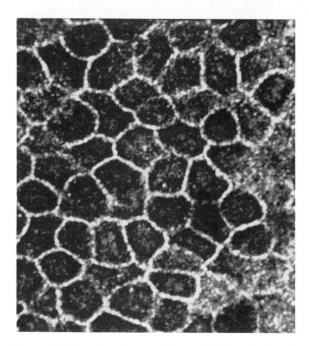

Figure 4.13 A confocal image of basal epithelial cells formed in back-scattered light of the *in vitro* rabbit cornea. The cell borders and the round cell nuclei can be seen.

NAD(P)H imaging of the basal epithelium of the cornea.

4.5.2   Redox imaging based on flavoprotein fluorescence

The confocal microscopic optical section of the basal epithelial cells and their nuclei is shown in Fig. 4.13. The confocal microscope formed the image with light of 488 nm in the back-scattered mode. The contrast of the images is due to local differences in refractive index, e.g. between the cell cytoplasm and the cell nuclei, which generates image contrast. The focal plane was centred at the approximate centre of the height of the basal epithelial cells and in the centre of the cell nuclei. We have used the 488 nm line of the laser scanning confocal microscope to image oxidized flavoprotein fluorescence.

4.5.3   Discussion

In most cells and tissues, the signal from the NAD(P)H is more intense than that from the oxidized flavoproteins. Both signals are subject to photobleaching.

4.5.3.1   NAD(P)H redox imaging

Fluorescent images of pyridine nucleotide fluorescence in basal epithelial cells of the cornea, and cells 400 μm below the surface of the cornea (corneal endothelium) have been shown here. What do the images represent and why are the images of the surface cells different from those of the deeper cells? A comparison of the images obtained using 364 nm reflected light and 364 nm excitation/400–500 nm emission with 364 nm reflected light and 364 nm excitation/400–500 nm emission reveals the following differences. First, the images of the cells on the corneal surface have higher contrast and resolution that those 400 μm below the surface. This was observed for both the reflected light modes and the fluorescent modes. Secondly, the resolution of the confocal images is lower than that of the images made in reflected light. This is consistent with calculations of the microscope point spread function for reflected light imaging and fluorescence light imaging; the latter shows a wider point spread function (Kimura & Munakata, 1989, 1990).

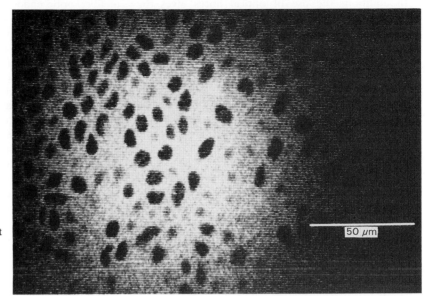

50 μm

Figure 4.14 A two-photon confocal redox image of basal epithelial cells of the *in vitro* rabbit eye formed from NAD(P)H fluorescence. The bright regions indicate NAD(P)H fluorescence. The dark oval regions are cell nuclei.

4.5.3.2 Flavoprotein redox imaging

Metabolic imaging of the flavoproteins has certain advantages over metabolic imaging of the reduced pyridine nucleotides in corneal cells. The oxidized flavoproteins are specifically located within the mitochondria and are not found in the cytoplasmic space of the cells. In contrast, the NAD(P)H is located in both the mitochondria and cytoplasmic spaces, which introduces an extra complexity for quantitative analysis and interpretation.

The previous work on freeze-trapped rabbit cornea showed an emission band with a peak at 545 nm (excitation at 442 nm) which is similar to that observed in mitochondrial preparations. The similarity of the emission spectra for the freeze-trapped rabbit cornea and the isolated mitochondrial provides evidence for the identification of oxidized flavoproteins in the cornea (Chance & Lieberman, 1978).

The fluorescence intensity due to the flavoproteins in the living rabbit corneal epithelium has been investigated (Masters *et al.*, 1982a). The fluorescence emission in the region 550 nm (50 nm band-pass) due to excitation at 442 nm was investigated. The excitation light of the laser grazed the corneal epithelium of a living rabbit and the effect of a gentle flow of either hydrated air or hydrated nitrogen was studied. The nitrogen flow resulted in a decrease of fluorescence intensity which was reversible with the passage of air. These studies did not involve two-dimensional fluorescence imaging. Similar results were observed in mitochondrial suspension from rat liver. This study on the epithelium in a living rabbit and the studies of rabbit corneas conducted in freeze-trapped prepara-

tions supports the conclusion that the fluorescence is due to oxidized flavoproteins. In perfused rabbit corneas the dose–response curves of the effect of mitochondrial respiratory inhibitors at Sites I, II, and III on fluorescence intensity yielded results consistent with the characterization of the fluorescence signal (450 excitation, 550 nm emission) as due to fluorescence from the oxidized flavoproteins (Masters, 1984a).

In order to visualize the functional anatomy of the living cornea we have merged the three-dimensional volume reconstruction with a second volume reconstruction based on redox imaging of flavoproteins. The resulting three-dimensional reconstruction shows the morphology and the mitochondrial function of the cells in the living cornea (Fig. 4.12). The dynamic functional anatomy of the living cornea could be obtained through a time-series of merged three-dimensional functional and morphological volumes.

4.6 COMPARISON WITH OTHER NON-INVASIVE TECHNIQUES

Where are we to place the optical method involving intrinsic fluorescence probes in comparison to other non-invasive techniques (Chance, 1991)? This technique has been developed from the microscopic studies of organelles and cells, to studies on whole organs both *in vitro* and *in vivo*. Applications to the *in vivo* human eye with spatial resolution at the single cell level purport a bright future for the optical method in ophthalmology.

In addition, the use of optical fibres gives the observer a fibre link to the interior of the body, inside vessels of the heart, inside interior cavities, and with the rapid advances of endoscopic diagnostic and surgical techniques this tool promises numerous applications in both research and clinical diagnostics.

Other diagnostic techniques such as CT, PET, SPECT, MRI are either invasive or significantly more complex and costly. These methods involve ionizing radiation or high cost or both.

In times of major financial crisis in the provision of health care cost is a significant factor. The optical method is relatively simple, inexpensive, and can easily be coupled to a variety of endoscopic fibre optic probes. We predict that the optical method using the intrinsic fluorescence probe will continue to contribute to our understanding of cellular function and aid in clinical monitoring (during surgery and during intensive care) and diagnosis of cellular hypoxia.

4.7 SUMMARY AND CONCLUSIONS

Redox fluorometry based on intrinsic fluorescent probes of cellular metabolism is an evolving microscopic technique to investigate tissue hypoxia. Its advantages over other non-invasive methods are that it is a real-time, non-invasive optical technique, and that relatively inexpensive and simple instrumentation is required to perform the measurements.

Several disadvantages associated with the technique include problems associated with calibration and quantitation, and the multivariate nature of tissue fluorescence; metabolic substrate utilization versus oxygen concentration and utilization versus auto-fluorescence of tissue and cells not related to NAD(P)H or flavoproteins. The penetration of ultraviolet light is usually limited to a depth of 1 mm or less, except for transparent tissue such as the cornea or ocular lens of the eye. Therefore, the depth of tissue that can be physically located between the tip of the objective or optical fibre is about 1 mm.

There are several areas which warrant further development. The recent advances in two-dimensional redox imaging of tissues and cells based on laser confocal microscopy should be further developed. Two-photon fluorescence confocal microscopy of intrinsic redox probes, NAD(P)H or flavoproteins is a promising technical development with the advantages of minimal photobleaching of the intrinsic fluorescent probes, and the high spatial resolution and optical sectioning capability of confocal microscopy. More precise tissue localization is now possible with the continued development of confocal microscopes.

Another technological development is the coupling of fibre optic endoscopes to external redox fluorometers. This technology permits redox imaging of internal organs and tissues.

A very important development is the use of solid-state imaging devices (Masters, 1989). Large size (4000 × 4000 pixels) charge-coupled devices (CCDs) are being developed with very high detector quantum efficiencies These back-illuminated, thinned, slow-scanned CCD detectors are useful as area integrating detectors with a large dynamic range. These cooled, slow-scanned CCDs have two major advantages over other imaging detectors such as video cameras: (1) they show no geometric distortions, and (2) once calibrated they are linear photometric detectors, which is a critical requirement for quantitative measurements. For those experimental cases where motion is a problem and integration would result in blurring of the redox image, a rapid line scan which can measure a profile of the fluorescence intensity may be preferable. Newly developed linear diode array detectors permit the rapid acquisition of emission spectral information. All of these instruments discussed so far involve measurements of fluorescence intensity. Further developments of imaging devices based on other fluorescence parameters, such as fluorescence lifetimes, will continue to improve the specificity of redox microfluorescence techniques. The 'optical method' is under vigorous development and continues to provide a non-invasive method to monitor cellular metabolism.

ACKNOWLEDGEMENTS

This work was supported by a grant (BRM) from NIH EY-06958. The authors acknowledge the assistance of Dr W. Webb, and Dr D. Piston, Department of Applied Physics, Cornell University, in obtaining two-photon confocal images of the cornea.

REFERENCES

Balaban R.S. & Mandel L.J. (1990) In *Noninvasive Techniques in Cell Biology*, J.K. Foskett & S. Grinstein, (eds). Wiley-Liss, New York, pp. 213–236.

Caspersson T. (1950) *Exp. Cell Res.* **1**, 595–598.

Caspersson T. (1954) *Exp. Cell Res.* **7**, 598–600.

Chance B. (1951) *Fedn. Proc.* **10**, 171.

Chance B. (1989) In *Cell Structure and Function by Microspectrofluorometry*, E. Kohen & J.G. Hirschberg (eds). Academic Press, New York, pp. 53–69.

Chance B. (1991) In *Annual Review of Biophysics and Biophysical Chemistry*, Vol. 20, D.M. Engelman (ed.). Annual Reviews, Palo Alto, CA, pp. 1–28.

Chance B. & Baltscheffsky H. (1958) *J. Biol. Chem.* **233**, 736–739.

Chance B. & Lieberman M. (1978) *Exp. Eye Res.* **26**, 111–117.

Chance B. & Thorell B. (1959) *J. Biol. Chem.* **234**, 3044–3050.

Chance B. & Williams G.R. (1955) *J. Biol. Chem.* **217**(1), 409–427.

Chance B., Mela L. & Wong D. (1968) In *Flavins and Flavoproteins, 2nd International Congress on Flavins and Flavoproteins*, K. Yagi (ed.). University Park Press, Baltimore, pp. 102–121.

Chance B., Barlow C., Haselgrove J., Nakase Y., Quistoroff B., Matschinsky F. & Mayevsky A. (1978). In *Microenvironments and Metabolic Compartmentation*, (P.A. Sere & R.W. Estabrook (eds). Academic Press, London, pp. 131–148.

Chance B., Schoener B., Oshino R., Itshak F. & Nakase Y. (1979) *J. Biol. Chem.* **254**, 4764–4771.

Denk W., Strickler J.H. & Webb W.W. (1990) *Science*, **248**, 73–76.

Eng J., Lynch R.M. & Balaban R.S. (1989) *Biophys. J.* **55**, 621–630.

Kapitza H.G. & Wilke V. (1988) *Proc. Soc. Photo-Optical Instrument Engineers (SPIE)* **1028**, 173–179.

Kasten F.H. (1989) In *Cell Structure and Function by Microspectrofluorometry*, E. Kohen, J.G. Hirschberg & J.S. Ploem (eds). Academic Press, New York, pp. 3–50.

Keilin D. (1966) *The History of Cell Respiration and Cytochrome.* Cambridge University Press, London.

Kimura S. & Munakata C. (1989) *Appl. Optics* **6**, 1015–1019.

Kimura S. & Munakata C. (1990) *Appl. Optics,* **29**, 489–494.

Kohen E. & Hirschberg J.G. (1989) *Cell Structure and Function by Microspectrofluorometry.* Academic Press, New York.

Kohen E., Kohen C., Hirschberg J.G., Fried M. & Prince J., (1989) *Optical Engng* **28**(3), 222–231.

Lakowicz J.R. (1992) *Principles of Fluorescence Spectroscopy.* Plenum Press, New York.

Lemp M.A., Dilly P.N. & Boyde A. (1986) *Cornea* **4**, 205–209.

MacMunn C.A. (1880) *The Spectroscope in Medicine.* Churchill, London.

MacMunn C.A. (1914) *Spectrum Analysis Applied to Biology and Medicine.* Longmans, Green & Co., London.

Masters B.R. (1984a) In *Current Topics in Eye Research*, Vol. 4, J. Zadunaisky & H. Davson (eds). Academic Press, London, pp. 139–200.

Masters B.R. (1984b) *Curr. Eye Res.* **3**, 23–26.

Masters B.R. (1986) In *The Precorneal Tear Film In Health, Disease and Contact Lens Wear*, F. Holly (ed.). Dry Eye Institute, Lubbock, TX, pp. 966–970.

Masters B.R. (1988) In *The Cornea: Transactions of the World Congress on the Cornea III*, H.D. Cavanagh (ed.). Raven Press, New York, pp. 2810–3860.

Masters B.R. (1989) In *New Methods in Microscopy and Low Light Imaging*, J.E. Wampler (ed.). *Proc. Soc. Photo-Optical Instrument Engineers (SPIE)* **1161**, 350–365.

Masters B.R. (1990a) In *Confocal Microscopy*, T. Wilson (ed.). Academic Press, London, pp. 305–324.

Masters B.R. (1990b) In *Noninvasive Diagnostic Techniques in Ophthalmology*, B.R. Masters (ed.). Springer-Verlag, New York, pp. 223–247.

Masters B.R. (1991) *Machine Vision and Applications* **4**, 227–232.

Masters B.R. (1992) *J. Microsc.* **165**, 159–167.

Masters B.R. (1993a) In *3-D Visualization in Microscopy*, A. Kriete (ed.). VCH, Germany, pp. 183–203.

Masters B.R. (1993b) *Appl. Optics* (in press).

Masters B.R. & Chance B. (1980) *Proc. Int. Soc. for Eye Res.* **1**, 30.

Masters B.R. & Kino G.S. (1989) *Proc. Institute of Physics, Electron Microscopy and Analysis Group*, **98**, 625–628.

Masters B.R. & Kino G.S. (1990) In *Noninvasive Diagnostic Techniques in Ophthalmology*, B.R. Masters (eds). Springer-Verlag, New York, pp. 152–171.

Masters B.R. & Paddock S.W. (1990a) *J. Microsc.* **158**, 267–275.

Masters B.R. & Paddock S.W. (1990b) *Appl. Optics* **29**, 3816–3822.

Masters B.R., Fischbarg J., Chance B. & Lieberman M. (1980) *Invest. Ophthalmol. Vis. Sci.* (Suppl.) **19**, 63.

Masters B.R., Chance B. & Fischbarg J. (1981) *Trends Biochem. Sci.* **6**, 282–284.

Masters B.R., Chance B. & Fischbarg J. (1982a) In *Noninvasive Probes of Tissue Metabolism*, J.S. Cohen (ed.). John Wiley & Sons, New York, pp. 79–118.

Masters B.R., Falk S. & Chance B. (1982b) *Curr. Eye Res.* **1**, 623–627.

Masters B.R., Riley M.V., Fischbarg J. & Chance B. (1983) *Exp. Eye Res.* **36**, 1–9.

Montag M., Kükulies J., Jörgens R., Gundlach H., Trendelenburg M.F. & Spring H. (1991). *J. Microsc.* **163**, 201–210.

Quistorff B., Haselgrove J.C. & Chance B. (1985) *Analyt. Biochem.* **148**, 389–400.

Rost F.W.D. (1991) *Quantitative Fluorescence Microscopy.* Cambridge University Press, Cambridge.

Sies, H. (1982) *Metabolic Compartmentation.* Academic Press, New York.

Wilson T. (1989) *J. Microsc.* **154**, 143–156.

Wilson T. (1990) In *Confocal Microscopy*, T. Wilson (ed.). Academic Press, London, pp. 93–141.

Xiao G.Q., Kino G.S. & Masters, B.R. (1990) *Scanning* **12**, 161–166.

CHAPTER FIVE

Bioluminescent and Chemiluminescent Indicators for Molecular Signalling and Function in Living Cells

ANTHONY K. CAMPBELL & GRACIELA SALA-NEWBY
Department of Medical Biochemistry, University of Wales College of Medicine, Cardiff, UK

5.1 THE NATURAL HISTORY OF BIO- AND CHEMILUMINESCENCE

5.1.1 What are bioluminescence and chemiluminescence?

Chemiluminescence is the emission of light as a result of a chemical reaction (see Campbell, 1988, chapter 1). The enthalpy of the reaction gives rise to an atom or molecule in a vibronically excited state; when the electron decays back to ground state a photon is emitted. Bioluminescence is visible light emission from luminous organisms. All known examples of bioluminescence are biological chemiluminescence. The term bioluminescence is also used to describe reactions extracted, or DNA cloned and engineered, from luminous organisms. All the other chapters in this book are concerned with fluorescence. It is important, therefore, to understand the difference between fluorescence, together with true phosphorescence, and chemiluminescence. The confusion that exists is compounded by the lay term 'phosphorescence' which in the Oxford English Dictionary is still defined as 'the property of shining in the dark'.

The term 'luminescenz', associated with the prefixes photo-, electro-, thermo-, crystallo- and chemi-, was first coined by a German physicist in 1888, Eilhardt Weidemann, to distinguish phenomena which resulted in light emission without requiring the high temperatures necessary for incandescence. Luminescent phenomena do not obey the laws of black body radiation. The 'discovery' of the electron in 1897, together with the development of quantum theory and electron spin, led to a clear understanding of the difference between chemiluminescence and the related phenomenon of photoluminescence. In photoluminescence, usually known either as fluorescence or phosphorescence, the energy for exciting the electron arises from absorption of electromagnetic radiation in the near IR, visible or UV region. In chemiluminescence, however, the energy for exciting the electron comes from the enthalpy of the chemical reaction. Both result in electronically excited states, and can involve intersystem crossing, i.e. to the triplet spin state, when the decay becomes phosphorescence. To produce orange-red light at 600 nm at least 47.6 kcal mol^{-1} (1 orange-red photon = 4.41 × 10^{-19} J) are required, or 63.5 kcal mol^{-1} for blue light at 450 nm. Thus chemiluminescence requires a reaction with an enthalpy of 50–100 kcal mol^{-1}.

There are two important differences between a chemiluminescent and fluorescent compound. First, the actual emitter in chemiluminescence is different chemically from the initial substance. Secondly, and as a consequence of this, a chemiluminescent compound (C) can only produce a photon once. In contrast a fluor (F), in the absence of a photobleaching reaction, is identical chemically to the initial compound and can be excited again and again. Thus:

Fluorescence

$$F \xrightarrow{\text{light}} F^* \longrightarrow F + h\nu \qquad [5.1]$$

Chemiluminescence

$$C + \text{reactants} \xrightarrow{\text{catalyst}} Pr^* \longrightarrow Pr + h\nu \qquad [5.2]$$

Yet fluorescence is associated with chemiluminescence. The initial product (Pr) of the reaction is, by definition, capable of fluorescence, but may only be transiently stable. The initial compound (C) is often also capable of fluorescence, but its fluorescence spectrum will be quite different from Pr, since it is structurally different. Fluors can also act as energy transfer acceptors from Pr*, as they do in several luminous organisms, and in the light sticks commercially available for decoration or use in a golf ball at night. The two most common chemical mechanisms with sufficient energy to generate an excited state are electron transfer and the cleavage of linear or cyclic peroxides.

Five parameters characterize a chemiluminescent reaction: intensity, speed of onset, decay of light intensity, colour, and polarization, if any. Since chemical reactions are much slower than electronic excitation (10^{-15} s) or electronic decay (10^{-9}–10^{-8} s for singlet), the first two of these are dependent on three factors: the rate of the chemical reaction, the efficiency of the chemical reaction in generating electronically excited molecules, and the efficiency of the excited molecules in producing photons:

Thus the overall quantum yield of a chemiluminescent reaction, ϕ_{CL}, is given by:

$$\phi_{CL} = \text{(total number of photons emitted)}/$$
$$\text{(number of molecules reacting)}$$
$$= \phi_C \phi_{EX} \phi_F \qquad [5.3]$$

where ϕ_C = chemical yield, i.e. the fraction of molecules going through the chemiluminescence pathway; ϕ_{EX} = yield of excited state molecules; and ϕ_F = excited state quantum yield. Bioluminescent molecules usually have ϕ_{CL} in the range 0.01–1. For methods of standardization see Hastings and Weber (1963), Hastings and Reynolds (1966), Nakamura (1972), and Campbell (1988, chapter 2).

The analytical application of chemi- and bioluminescence arises from the ability to couple analytes of interest to one of the components of the reaction. The exquisite sensitivity of chemiluminescence analysis, at least down to 10^{-21} mol in a solution of 0.1–1 ml, derives from the low noise from normal solutions. Clean solvents such as H_2O, and cuvette housings painted black, produce no chemiluminescence, but they do fluoresce. Small though this fluorescence may be, it usually means that in the test tube chemiluminescence has superior sensitivity to fluorescence.

5.1.2 Natural occurrence of chemi- and bioluminescence

Chemiluminescence can occur in a liquid, gas or at solid surface of a wide range of inorganic and organic compounds (Harvey, 1952; Campbell, 1988). It occurs in the atmosphere as the 'air glow'. It can follow an electrical discharge as the yellow afterglow, and can be seen over marshy ground as 'Will-o-the-wisp'. A considerable number of chemiluminescence compounds have also been synthesized (Fig. 5.1). Many have high quantum yields (0.01–1), though some are still visible even though ϕ_{CL} may be as low as 10^{-6}–10^{-8}, e.g. Grignard reactions. The oxalate esters, used in light sticks, require a fluor as an energy transfer acceptor to generate visible light.

Luminous organisms have invaded all the major habitats on the earth: the soil (bacteria), caves (New Zealand glowworm), the air (fireflies), vegetation (fungi), mud (scale worms), rock (piddock), the sea (jellyfish). Some 700 genera from 16 phyla contain luminous species (Herring, 1978; Campbell, 1988). These include familiar species such as mushrooms, jellyfish, squid, starfish, worms, fish, and beetles. Yet other familiar groups such as spiders, crabs, amphibians and mammals have no luminous species. Bioluminescence is particularly common in the sea. In the twilight zone, some 900 m below the surface, nearly all the animals are luminous. The colour varies from the deep blue of a searsiid fish, or the blue of jellyfish, to the green of some fungi, or the yellow and orange of some fireflies. Red bioluminescence is very rare but does occur in the deep-sea fish *Malacosteus*, which has two pairs of red-emitting light organs, and two which emit blue light. Similarly, the South American railroad worm, *Phrixothrix* has a pair of red organs on its head, and eleven yellow-green organs on its body. The ability to change the intensity or colour of light emission is an important characteristic for its analytical potential.

There are no luminous mammals, but there are cells in the body which produce an ultraweak chemiluminescence. This is invisible to the naked eye, but detectable by a photomultiplier tube. Ultraweak chemiluminescence is found in organs such as the lung

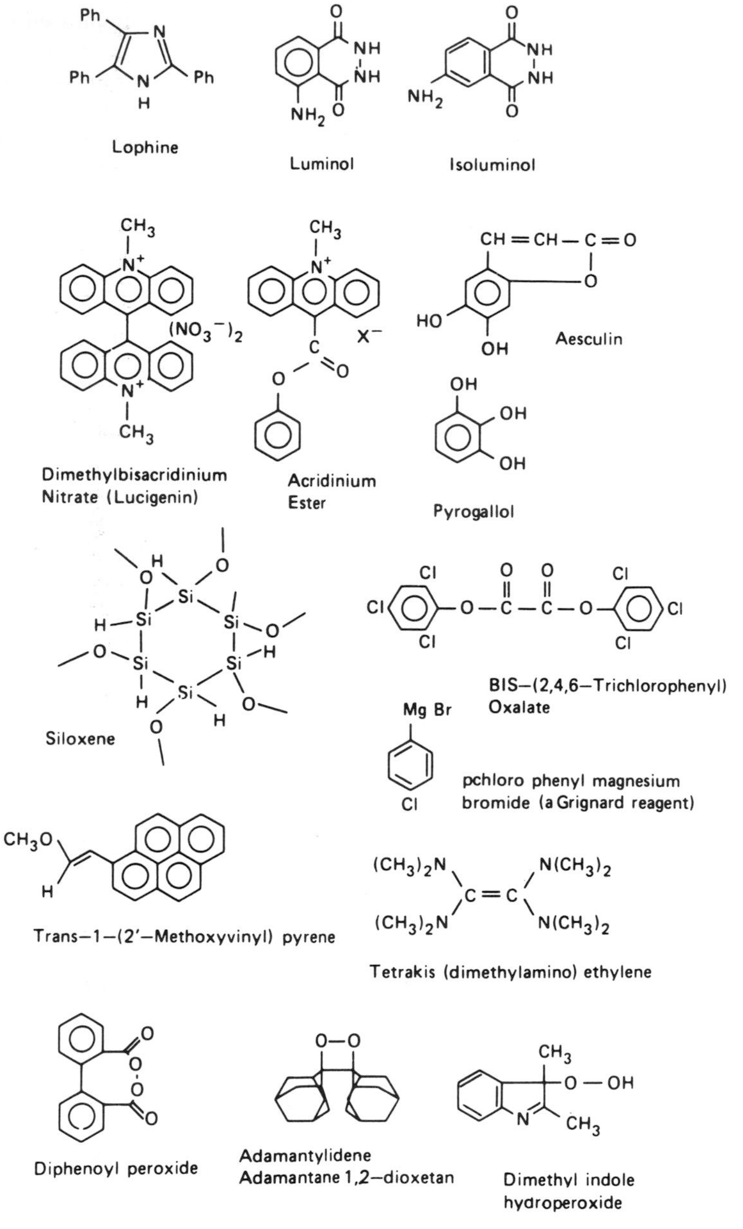

Figure 5.1 Chemiluminescent compounds.

and liver and in isolated cells such as phagocytes, platelets, fertilized eggs, fungi, yeast and plants (Quickenden & Que Hee, 1976; Quickenden *et al.*, 1985; Campbell, 1988, chapter 6) and even in human breath (Williams & Chance, 1983). It is caused by the generation of oxygen metabolites and the oxidation of endogenous substances such as unsaturated fatty acids, and possibly the generation of singlet oxygen, which is itself chemiluminescent (Gorman & Rodgers, 1981).

Thus there are four types of chemiluminescence,

from which assays for biological reactions and substances have been established: synthetic, bioluminescence, ultraweak, and energy transfer.

5.1.3 The chemical reactions

Bioluminescent reactions require a minimum of three components: the chemiluminescent substrate = the luciferin; the catalyst = the luciferase; and oxygen or one of its metabolites. Up to three other components may also be required: a metabolite such as NADH or

(a)

1. ALDEHYDES

$CH_3(CH_2)_n CHO + FMN - OOH_{n>7}$ bacteria

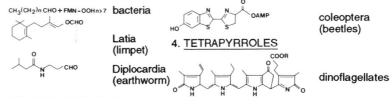

Latia
(limpet)

Diplocardia
(earthworm)

2. IMIDAZOLOPYRAZINES

coelenterates
decapods,
mysids, squid
copepods,
radiolarians
some fish

ostracods

3. BENZOTHIAZOLES

coleoptera
(beetles)

4. TETRAPYRROLES

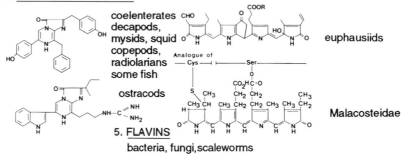

dinoflagellates

euphausiids

Malacosteidae

5. FLAVINS

bacteria, fungi,scaleworms

(b) 1. Bacteria (no role for Ca^{2+}) – blue light

$$NADH + FMN \xrightarrow{\text{reductase}} NAD^+ + FMNH_2$$

$$FMNH_2 + RCHO + O_2 \xrightarrow[\text{luciferase}]{} FMN + RCO_2H + H_2O + h\nu$$

2. Photinus (the firefly) – yellow light

3. Aequorea (a hydrozoan jellyfish): blue light (animal: blue-green light)

4. Latia (a freshwater limpet): pale green light

Latia luciferin

Purple protein – luciferin + O_2 $\xrightarrow{\text{luciferase}}$ light + $\substack{\text{purple} \\ \text{protein}}$ + products

5. Diplocardia (an earthworm) – blue – green light

Diplocardia luciferin

Figure 5.2 (a) Chemical families in bioluminescence. (b) Some chemi- and bioluminescent reactions.

ATP, a cation such as Cu^{2+}, Mg^{2+} or Ca^{2+}, and an energy transfer acceptor. However, even when ATP is required, as in the beetle system, the energy for light comes from oxidation of the luciferin, usually via the cyclic oxygen intermediate dioxetan

$$\left(\begin{array}{c} O-O \\ | \quad | \\ -C-C- \\ | \quad | \end{array} \right) \qquad \left(\begin{array}{c} O-O \\ | \quad | \\ -C-C=O \\ | \end{array} \right)$$

There is insufficient energy in ATP hydrolysis to generate a visible photon. Thus ATP is not required directly for most bioluminescent reactions. It is important to remember that the terms luciferin and luciferase, first introduced by Dubois in 1887, are generic terms. Each luciferase has a unique amino acid sequence, though close homology obviously exists between related species. Five distinct chemical 'families' of luciferin have been identified so far (Fig. 5.2(a)), each resulting in a quite different chemical reaction (Fig. 5.2(b)). Benzothiazole bioluminescence has been found only in luminous beetles, yet imidazolopyrazine bioluminescence is the most common one in the sea, being found in six distinct phyla (Campbell & Herring, 1990). In some of these, namely the radiolarians, cnidarians and ctenophores, the luciferin and oxygen are tightly or covalently bound to the luciferase so that the whole complex can be isolated as a whole (Shimomura et al., 1962; Shimomura & Shimomura 1985). Shimomura, who has carried out such unique and distinguished work on the chemistry of bioluminescence for over 30 years, named this complex photoprotein. Addition of Ca^{2+} to the photoprotein triggers the chemiluminescent reaction (Fig. 5.2(b)), hence it can be used as an indicator for intracellular Ca^{2+}. Once the reaction has taken place the product, the oxyluciferin, is no longer covalently linked to the protein. Dissociation of the Ca^{2+}, aided by addition of a Ca^{2+} chelator, results in its separation from the protein. The apoprotein can then be reactivated to form photoprotein by addition of the 'luciferin', coelenterazine (Campbell et al., 1988; Campbell & Herring, 1990), in the presence of oxygen. Some 30 coelenterazine derivatives have been synthesized by Kishi and colleagues (Shimomura et al., 1988, 1989, 1990). These also reactivate apophotoproteins formed from aequorin or obelin, and result in new photoproteins with interesting and important properties (Fig. 5.3). One has a bimodal spectrum where the ratio of 409/465 is linearly related to free Ca^{2+}, whereas others have changed kinetics, ϕ_{CL} and increased affinity for Ca^{2+}. The affinity of the apoprotein for coelenterazine is very high, just a few nanomolar.

Fish, squid, shrimp, ostracods, copepods and anthozoans which also use imidazolopyrazine bioluminescence, use a simple O_2 luciferin–luciferase

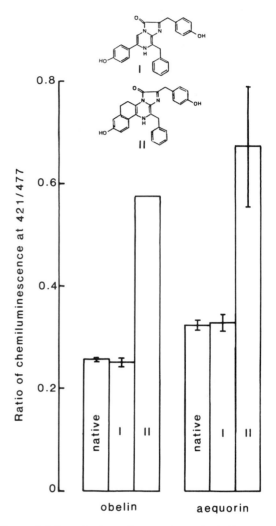

Figure 5.3 Reactivation of aequorin and obelin. Apoaequorin and apo-obelin, prepared as previously described (Campbell et al., 1988; Campbell & Herring, 1990), were incubated for up to 24 h with native coelenterazine(I) and e-coelenterazine(II), kind gift from Professor Kishi, University of Havard. The active photoprotein formed was assayed by recording the chemiluminescence counts in the first 10 s following addition of 25 mM Ca^{2+}. Results are expressed as a percentage of the original photoprotein. (From Campbell et al., 1991, with permission of Wiley, Chichester.)

system, sometimes squirted out into the seawater. They all emit blue light, though the emission maximum can vary from 440 to 490 nm without energy transfer. As with similar changes in the colour of beetle emission from green, green-yellow, orange to red, just a few amino acid differences (Wood et al., 1989a,b; Wood, 1991), usually charged ones, are responsible for these colour shifts. Such spectral differences are not found so dramatically in other luciferin–luciferase systems, e.g. bacteria, unless intermolecular energy transfer occurs. It was this lesson from Nature which led us to select imidazolopyrazine apophotoproteins and benzothiazole luciferases to engineer as indicators for molecular signalling in live cells (see Sections 5.4 and 5.5).

Most synthetic organic chemiluminescent reactions are also oxidations, using O_2, O_2^- or H_2O_2 as oxidant to produce a dioxetan intermediate which then spontaneously generates the excited carbonyl emitter. It is, however, possible to synthesize stable dioxetans (see Adam & Cilento, 1982), which decompose to excited carbonyls on mild heating. One such compound of important application as an immuno-assay and DNA probe is adamantylidene adamantane 1,2-dioxetan (Fig. 5.1). A phosphate derivative of this is not chemiluminescent. However, catalytic removal of the phosphate by alkaline phosphatase results in generation of the chemiluminescent compound, and light emission (Schaap et al., 1989).

5.2 THE ANALYTICAL POTENTIAL OF CHEMILUMINESCENT COMPOUNDS

Chemical light has been exploited by humans for centuries, and in some remarkable ways (Table 5.1). In 1652 Bishop Olaus Magnus reported 'when there is an urgent need people of the far north use an ingenious way of lighting their way through forests in the dark, they place pieces of rotten oak bark that by their glow they may complete their journey'. The luminous fungus in this case was almost certainly honey tuft, *Armillaria mellea*. They had found a way of illuminating the forest without destroying it by fire. Likewise one of the most exciting applications of chemi- and bioluminescence is the ability to light up molecular signals and gene expression in living cells without destroying them (see Sections 5.4 and 5.5). But why haven't chemiluminescent indicators been made obsolete by the inventiveness of chemists producing fluorescent indicators, as reviewed in this book?

The study of cell activation and cell injury, together with the study and detection of microbes, in the research and clinical laboratory requires the identification and quantification of chemical changes, both within the cell and in the surrounding fluid. Substances to measure include substrates and nutrients, metabolites, enzymes and other proteins, hormones and vitamins, ions, drugs and pathogens. The concentrations and amounts of these substances available for analysis can vary over some 11 orders of magnitude, from 140 mM for extracellular Na^+ to a few picomolar for some free hormones, from millimoles of K^+ extracted from a whole liver to a few attomoles of ATP extracted from a single bacterium.

Table 5.1 Exploitation of chemical light by man.

Application	Example
1. Torch	'Lamps' containing luminous bacteria or fireflies Light sticks War-time use for map reading
2. Fishing	Luminous squid as a lure off Portugal Live flashlight fish in bamboo as a lure in Indian ocean Synthetic light lures by anglers
3. Military	Detection of nuclear submarines
4. Ritual	Fireflies on head-dresses of Mexican women Luminous fungi of native ritual, in Ecuador Light sticks adorning night club dancers
5. Fun	Luminous golf balls
6. Scientific	Model for chemistry of the excited state Model for biochemical evolution Analytical probe in biomedicine Probe in genetic engineering Microbial detection in industry and clinical microbiology

See Campbell (1988) chapter 4 for more details.

It is the combination of three features of chemiluminescent compounds which enables us to identify where they should be used; where they have unique potential over absorbing, fluorescent or radioactive probes:

(1) their sensitivity;
(2) the fact that they are non-radioactive; and
(3) their ability to produce a signal from within living cells.

Sensitivity of detection is determined by signal-to-noise ratio. Synthetic chemiluminescent compounds such as isoluminol, acridinium esters and dioxetans can be detected easily in the range $10^{-15}–10^{-19}$ mol, and with enzyme amplification down to 10^{-21} mol. Similarly, firefly luciferase can be detected down to approximately 10^{-18} mol, and obelin or aequorin down to 10^{-21} mol. Isotopes such as ^{14}C and ^3H can be detected only down to $10^{-12}–10^{-15}$ mol, and ^{125}I or ^{32}P to around 10^{-18} mol. Furthermore, radioactive isotopes are hazardous, have a short shelf life if sensitively detectable, often require long counting times, and require a separation step in the procedure to quantify the analyte, i.e. they cannot be used in 'homogeneous' assays. In contrast, chemi- and bioluminescent labels are stable until triggered, apparently safe, can be analysed in seconds, and can

Table 5.2 The major biomedical applications of chemi- and bioluminescence.

Analyte	Chemiluminescent system
1. Metabolites and related enzymes	
ATP	Firefly luciferase
NAD(P)H, FMN	Bacterial luciferase
PAP	*Renilla* luciferase
2. Ions	
Intracellular Ca^{2+}	Aequorin, obelin
Transition metals	Luminol, chaetopterin
Phosphate	Bacterial or firefly
NO, NO_2, NO_3^-	O_2^-/luminol
3. Oxygen	
O_2	Luminous bacteria
O_2^-	Pholasin, polynoidin, lucigenin
H_2O_2	Luminol, earthworm luciferin–luciferase
H_2O_2/peroxidase	Luminol
Singlet oxygen	$^1O_2$ itself, methoxy vinyl pyrenes
OCl^-	Luminol, pholasin
4. Labels in immunoassay or DNA/RNA probes	
Direct label	Isoluminol (ABEI), acridinium ester
Enzyme label	Peroxidase (luminol-enhanced) alkaline phosphatase (adamantylidene adamantane dioxetan)
Bioluminescent label	Bacterial luciferase, aequorin
5. Live microbes	
ATP	Firefly luciferase
Engineered phage	Phage-containing luciferase cDNA
6. Molecular signalling in living cells	
Reporter gene	Bacterial and firefly luciferase DNA
Ca^{2+}	Aequorin cDNA
Cyclic AMP-dependent protein kinase	engineered firefly luciferase cDNA
Organelle Ca^{2+}, ATP, or kinase	Signal targeted aequorin or luciferase

For references see: DeLuca & McElroy (1978, 1987), Campbell *et al.* (1985), Campbell (1988), Sala-Newby & Campbell (1991), Knight *et al.* (1991a,b).

be coupled to homogeneous immunoassays or DNA analysis (Weeks *et al.*, 1983a,b; Campbell *et al.*, 1985; Balaguer *et al.*, 1989, 1991; Geiger *et al.*, 1989; Nelson *et al.*, 1990). This latter feature is essential if a probe is to be used inside a living cell.

Thus synthetic chemi- and bioluminescent assays are widely used for analysis of a range of substances of biological interest (Table 5.2). Firefly luciferase is the method of choice for assaying biomass via ATP. As little as 100 amol can be detected in a solution of 0.1–1 ml. Four chemiluminescent labels are commercially available for a wide range of immunoassays, microbial tests and DNA analysis, including single point mutations and quantification of PCR products. They can also be visualized from Southern, Northern and Western blots. They are:

(1) Peroxidase detected by luminol-enhanced chemiluminescence (Matthews *et al.*, 1985).
(2) Acridinium esters (Weeks *et al.*, 1983a, 1987; Nelson *et al.*, 1990).
(3) Adamantylidene adamantane dioxetan (Hummelen *et al.*, 1987; Geiger *et al.*, 1989; Schaap *et al.*, 1989).
(4) Bacterial luciferase (Balaguer *et al.*, 1990).

But this chapter is focused principally on the application of chemi- and bioluminescence on measurements in live cells, as opposed to fluids and tissue extracts.

5.3 APPLICATION OF CHEMI- AND BIOLUMINESCENCE TO LIVING CELLS

5.3.1 Principles

What do we need to measure and are there appropriate chemi- and bioluminescent probes available?

A full understanding of the mechanisms responsible for activating cells or for cell injury, induced, for example, by components of the immune system in rheumatoid arthritis or by viruses, requires elucidation of the complete molecular sequence, initiated at the plasma membrane or within the nucleus and ending in a cell response (Fig. 5.4). This response begins with the generation of a signal such as Ca^{2+}, IP_3 or cyclic AMP, which then induces a structural or covalent modification in target proteins. If these modifications happen at the right time, in the necessary part of the cell and to the right extent, and if energy phosphate in the form of ATP and GTP is available, the cell undergoes an end response. These end responses include cell movement, secretion, division, transformation, defence by removal of the attacker, apoptosis and lysis. Some years ago (Campbell, 1983, 1988) I proposed that not only these end responses, but also the molecular sequences responsible for determining if and when they occur in a particular cell, are quantal in nature. Each step in the sequence

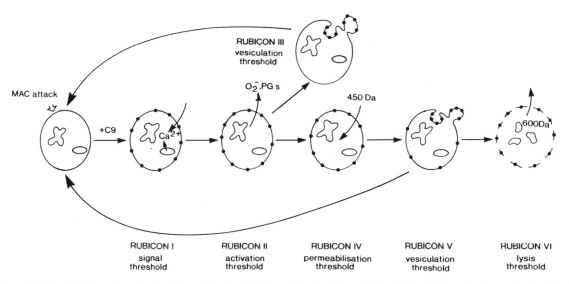

RUBICON IN DEFENCE AGAINST COMPLEMENT ATTACK

Figure 5.4 The Rubicon hypothesis. (From Campbell, 1991, with permission of Cambridge University Press, Cambridge.)

involves a chemico-physical reaction which results in the cell crossing a threshold, or rubicon (Campbell, 1991), which then allows the next chemico-physical process to begin. Single cell analysis using bioluminescent and fluorescent probes has established that this hypothesis is correct. The living cell does not behave like the soup of the biochemist's homogenate.

The timing and magnitude of each step in the sequence varies considerably from cell to cell. Thus ultimately we can only follow the complete sequence using single cell analysis. For example, a calcium cloud, detected using aequorin, has to reach its target before exocytosis in a fertilized fish egg (Gilkey *et al.*, 1978) or gap junction closing in a *Chironomus* salivary gland (Rose & Loewestein, 1976) occur. Using fura-2 imaging for Ca^{2+} and pholasin or 2,7-dichloro-fluorescein for O_2^- and H_2O_2 we have shown that in a population of neutrophils activated by f-Met-Leu-Phe to produce O_2^-, the time-course of O_2^- release is a reflection of individual neutrophils starting to release O_2^-, the threshold, at different times up to 30 min after addition of the stimulus. Furthermore, four types of Ca^{2+} signal have been visualized in different cells (Hallett *et al.*, 1990), a rapid Ca^{2+} signal within 6 s, a delayed Ca^{2+} signal up to 60 s after f-Met-Leu-Phe, an oscillatory Ca^{2+} signal, and no signal at all in 30% of the cells. In contrast the Ca^{2+} signal induced by leukotriene B4 is synchronous, though, once again, not all of the cells respond (Davies *et al.*, 1991a,b, 1992). Under normal conditions the source of the Ca^{2+} is from outside the neutrophil. But in the absence of external Ca^{2+} a highly localized releasable Ca^{2+} store is seen, though in only 30% of the cells. The fraction

of cells having this store doubles in rheumatoid synovial fluid cells, which show a much larger Ca^{2+} cloud, often covering the whole of the cell cytosol. These cells have thus been primed (Davies *et al.*, 1991a,b, 1992).

A clear example of the Rubicon hypothesis in action is seen when nucleated cells respond to membrane pore-forming proteins such as complement (Fig. 5.4). Unlike aged erythrocytes, nucleated cells can protect themselves and do not necessarily die when attacked by the membrane attack complex (MAC) (Campbell & Luzio, 1981; Campbell & Morgan, 1985; Patel & Campbell, 1987; Campbell, 1991). Binding and insertion of the last component, C9, to C5b-8 in the plasma membrane causes the cell to cross the first rubicon, the movement of Ca^{2+} into the cell and the release of Ca^{2+} from internal stores. This Ca^{2+} activates Ca^{2+} dependent responses within the cell and at the rubicon marker, e.g. O_2^- production (rubicon II) and a mechanism where the MACs can be channelled into vesicles and removed by endocytosis and budding (rubicon III). Those cells which do not protect in time go through a second major permeabilization of the plasma membrane to molecules about 450 Da (rubicon IV), but can still protect after this (rubicon V). The remaining cells go through the ultimate rubicon, and lyse. We have shown that even the first rubicon, namely Ca^{2+} release, can vary by at least 30 min from cell to cell. After 20 min we can find lysed cells, cells positive to propidium iodide (450 Da) with raised Ca^{2+} (i.e. fura positive) or cells with resting Ca^{2+}. This mechanism may play an important role in the inflammatory process in rheumatoid arthritis (Patel

& Campbell, 1987; Campbell, 1991), and in demyelination in multiple sclerosis (Scolding *et al.*, 1989).

It is clear from this evidence that cells have not evolved to behave in a graded, synchronous manner like the biochemist's homogenate. Rather, the behaviour, development and survival of the whole tissue and organism in health and disease is dependent on the timing and number of thresholds crossed by particular cells. In order to define these thresholds at the molecular level we need, therefore, methods for quantifying and locating these chemical sequences, not only in live cells but in intact organs. We need to locate not only which cells have undergone a chemical change, and when, but also where in the cell it occurs, i.e. in the centre of the cytosol or close to the mitochondria, plasma membrane, in the endoplasmic reticulum or nucleus, or in the chloroplast or tonoplast in a plant cell. Chemiluminescent probes, in particular extracted and engineered bioluminescent proteins, have the unique potential to do this. The aim is to measure in live cells within whole tissues nine stages in the sequence:

(1) The concentration and movement of intracellular signals, e.g. Ca^{2+}, IP_3, diacylglycerol and cyclic nucleotides.
(2) The concentration and locality of changes in intracellular energy supply and regulatory metabolites, e.g. ATP, GTP and AMP.
(3) The activity of synthetic and catabolic enzymes, e.g. RNA and DNA polymerase, glycogen synthetase, the enzymes of cholesterol biosynthesis.
(4) The covalent modification and activity of regulatory enzymes, e.g. serine, threonine, tyrosine kinases and the enzymes they phosphorylate.
(5) The movement of regulatory proteins and subcellular structures to a target, e.g. calmodulin and secretory vesicles.
(6) The expression of new proteins at their correct site (normally only possible by immunofluorescence on fixed tissue).
(7) The activation or inhibition within the nucleus of proteins, enhancers and promoters, controlling gene expression, cell priming and cell transformation.
(8) Cell and organelle viability and permeability changes in the plasma or organelle membranes.
(9) The end response, e.g. secretion, transformation or death.

5.3.2   The criteria

In order to achieve these objectives we need to find or design chemiluminescent indicators which, ideally, satisfy eight criteria:

(1) It must be possible to couple the substance or process of interest to a chemiluminescent reaction so that it is 'homogeneous', i.e. usable inside a cell without the need for a separation step.
(2) It must be sufficiently sensitive, ultimately for single cell analysis, and specific.
(3) It must distribute uniformly in the required part of the cell, and nowhere else.
(4) It must be possible to quantify and locate the indicator in the cell by a change in light intensity or colour, or possibly polarization, and within subcellular structures and cells from intact organs. It has been argued that for ease of quantification it is better to have a linear relationship between the analyte concentration and the signal. However, a power law relationship, as with aequorin and obelin light emission to free Ca^{2+}, has the advantage of amplification, so that small changes in free Ca^{2+} are more easily detectable. A ratio of intensity at two colours is also advantageous since it is independent of the amount of indicator injected or expressed inside the cell.
(5) It must be possible to incorporate it into the cell.
(6) It must be non-lethal, and not disturb significantly the phenomenon under investigation.
(7) It must be stable enough, and respond with the appropriate kinetics, to follow the phenomenon over the desired time-course, sometimes milliseconds to seconds, sometimes hours to days.
(8) It should be readily available and not too expensive.

Considerable success has already been achieved using proteins extracted from luminous organisms and synthetic chemiluminescent reactions, but, as we shall see in Sections 5.4 and 5.5, the full potential of chemiluminescent indicators can now be realized using controlled expression and engineering of bioluminescent genes.

5.3.3   Detection and visualization of chemi- and bioluminescence

In practice chemi- and bioluminescent indicators produce a lower intensity of photon emission than do fluors, particularly when the latter are excited with a high-intensity lamp or laser. Thus to realize the exquisite sensitivity of chemiluminescence the most sensitive photomultiplier tubes (PMT) available are used (Campbell, 1988, chapter 2). Battery-operated photodiodes are available in chemiluminometers for use in the field. However, their photon sensitivity is usually at least two orders of magnitude lower than a photomultiplier tube. A typical PMT is bialkali with 11 dynodes with a dark current of 0.05 nA at 1000 V, operating in true digital rather than analogue mode, with a recorded background of 10–20 cps. Photomultipliers with 13 dynodes produce a higher absolute signal but more noise. It is important to remember that the peak in spectral sensitivity for all photocathodes available is in the blue. For a bialkali tube some 5–10 more photons are required from firefly luciferase (λ_{max} 565 nm) than from aequorin (λ_{max} 465 nm) to generate the same recorded counts, though it may be possible to improve this slightly by careful selection of the photomultiplier. Taking into account sample geometry, PMT sensitivity (never >25%), discriminator setting and noise, a typical good chemi-luminometer has an efficiency of 0.2–1% and a detection limit of approximately 100 emitted photons s^{-1}, giving a detection limit for a chemiluminescent indicator with $\phi_{CL} = 0.01$ of about 10^{-22} mol.

Several good, true digital chemiluminometers are now available commercially e.g. Berthold, but we have constructed our own for greater flexibility, particularly in the sample housing. This has enabled different size samples, an X–Y table, an O_2 electrode, a dual-wavelength housing and spectrometer to be inter-changed. Photodiode array spectrometers are also available commercially with excellent software and have produced important spectral data from intact luminous organisms (Widder et al., 1983, 1984). However, you usually need to be able to see the light with the naked eye to be able to detect it, unlike a photomultiplier where, if you can see it, the signal is likely to saturate or even damage your detector. The design and characteristics of home-built and commercial luminometers have been reviewed (see Campbell, 1988, chapter 2 for references).

Imaging of bioluminescence was pioneered by Reynolds (1972, 1978), using an EMI four-stage electromagnetically focused imaging intensifier, visualizing aequorin Ca^{2+} waves and clouds in single cells (Rose & Lowenstein, 1976; Eisen & Reynolds, 1985). These are no longer commercially available. Chemiluminescent and antibody probes in Southern or Western blots or immunoassay can be visualized with 20 000 ASA photographic film, with the appro-priate colour sensitivity (Matthews et al., 1985; Kricka & Thorpe, 1987). However, to image individual luminous cells, particularly when observed in a microscope, a good imaging camera is required (Hooper et al., 1990). The three-stage Photonics Science IS1S3M, with photocathode, microchannel plate and CCD produces a video signal which can be processed by a computer, just as for fluorescent dyes. However, it is not sufficiently sensitive for low photon emissions, i.e. down to a few cps for cells. However, addition of a front intensifier stage, results in a photon-counting image intensifer capable of imaging transgenic plants, firefly luciferase in COS

cells (Craig et al., 1991), luminol chemiluminescence from neutrophils, and even ultraweak chemilumine-scence from growing root tips. Image Research have developed a contact imaging camera (Hooper et al., 1990) where a fine fibre optic bundle is in direct contact with a cell culture, enabling an image, albeit a fuzzy one, to be quantified from individual cells without the need of a microscope. This is important for observing cells with very low photon emission, e.g. the resting flow from aequorin at Ca^{2+} <0.1 μM, since at every optical surface in a microscope as many as 50% of the photons may be lost. Contact imaging obviously improves the solid angle considerably.

Liquid N_2-cooled CCD cameras appear to have the necessary sensitivity for imaging chemiluminescence from cells. They have very low noise in the detector, but still introduce noise in the signal processor and often require long accumulation times. This is alright for astronomy but too slow for some biological phenomena. These intensifiers require proper evalua-tion. The four-stage system used by Reynolds, though cumbersome, is still probably the most sensitive detector.

5.3.4 Incorporation into cells

Synthetic chemiluminescent indicators such as luminol and luciferin appear to diffuse into cells, though their location in phagocyte O_2^- production has never been definitively proven. Bioluminescence requires a pro-tein, which cannot diffuse across the plasma mem-brane. The photoprotein aequorin has been injected successfully into giant cells such as barnacle muscle (Ridgway & Ashley, 1967), squid axon (Baker et al., 1971) and many other cells (Blinks et al., 1978, 1982) including photoreceptors, several invertebrate and vertebrate muscle and nerve cells, a variety of eggs, Chironomus salivary gland, slime mould, protozoa and a few plant cells. The elegant studies of Cobbold and his collaborators (Cobbold & Bourne, 1984; Woods et al., 1986; Cobbold & Rink, 1987) have shown that it is also possible to microinject cells just a few micrometres in diameter, including single myocytes, hepatocytes, and adrenal chromaffin cells. The avail-ability of computerized microinjectors makes it pos-sible to inject protein mRNA or cDNA into cells as small as human neutrophils (ca. 7 μm diameter). However, in order to load large numbers of cells so that changes in intracellular Ca^{2+} and ATP can be correlated with the population end response a number of alternative methods have been developed to incor-porate photoproteins and other bioluminescent systems into small cells (see Campbell, 1983, 1988, chapter 7 for details and references). Thus there are currently five methods available, all of which have given new information about cell physiology or pathology:

(1) Microinjection (Blinks *et al.*, 1978; Cobbold & Bourne, 1984);

(2) Transient permeabilization of the plasma membrane:
   (a) ghosting and swelling (Campbell & Dormer, 1978);
   (b) scrape loading;
   (c) ATP;
   (d) electropermeabilization – not well-investigated (Maxwell & Maxwell, 1988);

(3) release from phagocytosed or micropinocytotic vesicles (Campbell *et al.*, 1988; Hallett & Campbell, 1983);

(4) vesicle–cell fusion (Hallett & Campbell, 1982; Campbell & Hallett, 1983);

(5) transformation, transfection or conjugation of genetic material, either mRNA or cDNA, followed by stable or transient expression (Inouye *et al.*, 1985; Campbell *et al.*, 1988; Knight *et al.*, 1991a,b).

It is the last of these which has the most exciting potential as a universal method for controlled expression of bioluminescent indicators in defined cell types. Of course, cells from luminous organisms already have their own indicators and cells from *Obelia*, dinoflagellates and marine bacteria, e.g. in phagocytosis, have been used in this way.

Oxygen is always present, though its K_m for luciferases has been poorly characterized. The luciferin, long-chain aldehyde in the case of bacterial luciferase, has to be added. Aldehydes are lipid-soluble but toxic. Evaporation from a hanging droplet has been used. Coelenterazine is not needed for aequorin or obelin, but is required when starting from mRNA or cDNA. It also appears to penetrate biological membranes readily, but photoprotein reactivation is slow, up to 3–24 h being required for maximal activity (Shimomura *et al.*, 1988; Campbell & Herring, 1990; Fig. 5.3). It is this which determines the kinetics of reactivation of apophotoproteins inside cells, including animals, plants and bacteria (Campbell *et al.*, 1988; Knight *et al.*, 1991a,b). Coelenterazine is very unstable, being susceptible to photooxidation, except when it is bound to photoprotein.

Somewhat surprisingly, firefly luciferin, though charged with a $-CO_2-$ group at physiological pH, penetrates into cells very rapidly. External pH may affect this, but light emission from firefly luciferase in COS cells is immediate after luciferin addition (Craig *et al.*, 1991; Sala-Newby & Campbell, 1992, Fig. 5.5(b)). Light intensity from recombinant luciferase in *E. coli* appears to reach a maximum within 30–60 min (Fig. 5.5(a)). A number of luciferin esters have been synthesized which enable maximum light emission from COS cells to be achieved at concentrations some 10–100 times lower than native luciferin (Craig *et al.*, 1991), which has to be used at 0.5–1 mM for rapid penetration.

5.3.5   Application of extracted bioluminescent and synthetic chemiluminescent indicators in living systems

There have been three major applications in living systems (Table 5.2): measurement of intracellular free Ca^{2+}, oxygen metabolites and ATP. Luminous bacteria were first used to monitor O_2 production during photosynthesis nearly a century ago (Beijerinck, 1902), but the most successful of the bioluminescent indicators so far have been the Ca^{2+}-activated photoproteins, aequorin and obelin (Ashley & Campbell, 1979; Blinks *et al.*, 1982; Campbell, 1983, 1988). Until the successful use of the fluorescent indicators invented by Tsien (Tsien *et al.*, 1982; Grynkiewicz *et al.*, 1985) all the really new information about intracellular free Ca^{2+} in living cells came from studies using the photoproteins. During the 1970s four ranges of free Ca^{2+} were defined, resting cells = 30–300 nM, stimulated cells = <1–5 μM, reversibly injured cells = 5–50 μM and dying cells = >100 μM. Primary stimuli, such as action potentials, transmitters and hormones were identified which worked via a rise in intracellular Ca^{2+}. Ca^{2+}-independent stimuli, e.g. particles on O_2^- in neutrophils (Hallett & Campbell, 1982), were also found even though manipulation of external Ca^{2+} sometimes affected the cell response. Secondary regulators, such as adrenalin in the heart and CO_2 in barnacle muscle, which work by modifying the primary Ca^{2+} transient were defined. Here a major discrepancy between fura-2 and the photoprotein signal has been identified. Obelin in neutrophils has shown that adenosine, which inhibits only the stimuli dependent on a rise in intracellular Ca^{2+}, can abolish the f-Met-Leu-Phe obelin Ca^{2+} signal (Campbell *et al.*, 1988). Whereas when monitored by fura-2 little or no effect was observed. A key role for irreversible cell injury by complement was first identified using obelin (Campbell & Luzio, 1981; Campbell, 1987). The source of intracellular Ca^{2+} for cell activation, external or from an internal store, was first clarified using photoproteins. Ca^{2+} oscillations, Ca^{2+} waves and localized Ca^{2+} transients were first discovered using photoproteins in slime mould, eggs, nerve cells and hepatocytes (see Campbell, 1983; Rose & Lowenstein, 1976; Gilkey *et al.*, 1978; Woods *et al.*, 1987). The availability, ease of use and imaging potential of the fluors has apparently put them now as first choice. Or has it?

Certainly, important new principles as well as much needed details in large numbers of cell types have come from using the fluors (Poinie & Tsien, 1986;

(a)

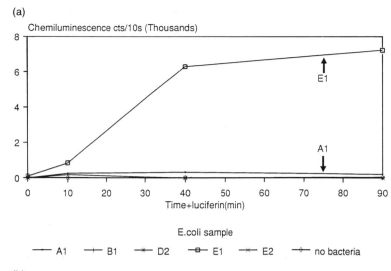

Chemiluminescence cts/10s (Thousands)

E.coli sample

——— A1   —+— B1   —*— D2   —□— E1   —×— E2   —◇— no bacteria

(b)

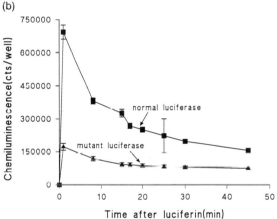

Figure 5.5 Expression of recombinant firefly luciferase in live cells. (a) Expression of firefly luciferase cDNA in *E. coli*. A1 and E1 represent two different plasmids, both containing full length cDNA. (b) Expression of firefly luciferase cDNA in COS cells. Cells incubated at room temperature with 0.75 mM luciferin.

Davies *et al.*, 1991a,b, 1992; see also chapters in this volume). But artifacts can occur. Not all cells load easily with the fluors. It is difficult to use them in bacteria and impossible at present to target them specifically to organelles or sites within the cell. The controlled expression and engineering of photoproteins heralds a new era for their application as indicators of free Ca^{2+} in living systems (see Section 5.4 and 5.5).

Another area, often underrated physiologically and pathologically, where chemi- and bioluminescent indicators have played an important role, has been in identifying cells and organs which produce oxygen metabolites (O_2^-, H_2O_2, OCl^- and $^1O_2$) after exposure to various stimuli or toxins, and in defining the molecular mechanisms responsible (see Adam & Cilento, 1982; Campbell, 1988, chapter 6). O_2 can be measured using live marine bacteria. True endogenous ultraweak chemiluminescence is an indicator of oxidative reactions which may damage DNA, protein or membranes. It has been observed in plant and animal cells, including root tips, phagocytes, platelets, liver, lung, brain and fertilized eggs. It has even been imaged. The infrared chemiluminescence of $^1O_2$ at 1268 nm is a useful criterion of singlet oxygen in biological systems. Addition of a chemiluminescent compound which reacts with one or more of the oxygen metabolites responsible for endogenous ultraweak chemiluminescence results in up to 104 times the light intensity. This indicator-dependent chemiluminescence can also be imaged. Lucigenin, and pholasin from the boring mollusc *Pholas dactylus*, are more selective for O_2^- than luminol, which requires access to peroxidase for maximum light emission. Pholasin is sensitive enough to detect O_2^- release from a single, human neutrophil (Roberts *et al.*, 1987). It has also been injected into myocytes (Cobbold & Bourne, 1984). Indicator-dependent chemiluminescence is a very convenient method for detection and analysis of abnormalities in oxygen metabolite production in clinical samples.

Firefly luciferase has been extensively used to assess

biomass in several industries, including general microbial detection in water, brewing, textiles, milk, fruit juice and food, and in urine. Dead cells have no ATP and also coenzymes are lost. It has even been used to search for life on Mars. These assays are non-specific. *E. coli* cannot be distinguished from *Salmonella* for example. Although apparently very sensitive, ATP from 1 neutrophil (*ca.* 10^{-15} mol) and from 50 bacteria (5×10^{-16} mol) being detectable, clinical analysis requires detection of 10^2 bacteria per ml, which still requires filtration. In food, one *Salmonella* in 25 g must be detected if we are to be certain of avoiding food poisoning. Somewhat surprisingly, there have been few studies reported where firefly luciferase has been microinjected into cells to monitor intracellular ATP. The two main problems are that firefly luciferase is very strongly inhibited by anions, particularly Cl^-, and its K_m for ATP Mg is about 0.25 mM so it should be fully saturated down to 10% of the normal ATP in the cell (i.e. from 5–10 mM in a healthy cell to 0.5–1 mM). Nevertheless, extreme attack to mitochondria and glycolysis will cause a dramatic reduction in luciferase signal. A further problem is the effect of lowering pH to produce a red instead of yellow emission, due to the monoanion form of the oxyluciferin. Since photomultipliers are very insensitive in the red, a decrease in apparent signal will be recorded.

Surprisingly, there have been no reports of using bacterial luciferase to monitor intracellular NAD(P)H, FMN or aldehydes. However, an ingenious application of this bioluminescent system has been reported to detect live bacteria as described in Section 5.4.3.2.

Some years ago we developed a homogeneous chemiluminescence immunoassay for cyclic AMP and cyclic GMP based on resonance energy transfer (Campbell & Patel, 1983; Campbell *et al.*, 1985; Campbell, 1988, chapter 9). ABEI cyclic AMP and cyclic GMP derivatives were synthesized. When bound to fluorescein-labelled antibody, a shift in the colour of light emission from blue to green occurred. This was reversed by displacement with unlabelled cyclic nucleotide. The 'homogeneous' assay quantified cyclic nucleotide by the ratio of light at 460/525 nm over the range 0.1–100 pmol, or 10 nM–10 μM, just right for the cytosol. Experiments with the late Peter Baker injecting squid giant axons at Plymouth resulted in a light signal from within the axon. Unfortunately, ABEI works best at alkaline pH and requires 1 mM H_2O_2 outside the cell! Furthermore, excited states are very susceptible to interference from solvents and other solutes, particularly protein. What was needed, therefore, was a bioluminescent protein label, creating its own 'solvent' within the active centre, and a fluor mimicking the natural energy transfer proteins which

occur in some coelenterates and deep-sea fish. The only way to produce sufficient bioluminescent protein for this purpose was cloning, followed by genetic manipulation.

5.4 BIOLUMINESCENT REPORTER GENES

5.4.1 Isolation of the genes

Several luciferases from organisms belonging to three phyla (Bacteria, Cnidaria and Arthropoda) have been cloned and sequenced. Luminous bacteria are very abundant and widely distributed. They can be found free in seawater or growing on dead fish and meat, in the guts and in the light organs of fish and squid, and as parasites. They belong to three genera: *Vibrio* (sp. *harveyi* and *fischeri*), *Photobacterium* (sp. *phosphoreum*, *leiognathi* and *logei*), and the freshwater or soil species *Xenorhabdus* (*luminescens*) (see Campbell, 1988).

Bacterial luciferases catalyse the following reaction:

$$FMNH_2 + RCHO + O_2 \rightarrow FMN + H_2O + RCOOH + h\nu$$

where R = long aliphatic chain, usually C12–14.

The $FMNH_2$ is generated by an NAD(P)H oxido-reductase present in most pro- and eukaryotic cells.

The luciferase is a 77 kDa chimaeric protein of two non-identical subunits (α and β) coded by adjacent genes *luxA* and *luxB* that form part of the regulated lux operon that also comprises the genes coding for fatty acid reductase (*luxCDE*) and others. The *luxA* and *B* genes and other members of the operons of *V. fischeri* were cloned from a genomic DNA library (Cohn *et al.*, 1983; Baldwin *et al.*, 1989; Foran & Brown, 1988). *V. harveyi* DNA was cloned into Charon 13 that was screened using mixed sequence oligonucleotide probes designed from partial protein sequence data (Cohn *et al.*, 1985; Johnson *et al.*, 1986), *X. luminescens luxAB* genes were cloned and sequenced from plasmid DNA libraries in *E. coli* based on their light emission (Meighen 1991; Frackman *et al.*, 1990; Xi *et al.*, 1991), and *P. leiognathi* genes were cloned in *E. coli* from a genomic library constructed in the transcriptional reporter plasmid pjSDH (Baldwin *et al.*, 1989). *P. phosphoreum* genes were isolated from an *E. coli* genomic library in pBR322 screened with *V. fischeri* lux DNA (Mancini *et al.*, 1988). There is a high degree of homology between the amino acid sequence of the α (54–88%) and β (45–77%) subunits of luciferase among the different species (Meighen, 1991). A fusion protein for *luxAB* has also been constructed (Boylan *et al.*, 1991). The lux operon is regulated by an auto-inducer (Swartzman *et al.*, 1990).

Three proteins which are responsible for coelenterazine bioluminescence have been isolated from *Aequorea victoria*, *Obelia geniculata* and *Renilla reniformis*. Luminescence requires the presence of Ca^{2+} for the first two and not the latter. *Aequorea victoria* luciferase (189/196 amino acids) has been cloned from plasmid cDNA libraries that were screened using oligonucleotide probes designed from partial protein sequence data (Inouye *et al.*, 1985; Prasher *et al.*, 1985, 1987). Clones selected contained open reading frames that started at methionine and coded for seven amino acids more than the mature protein. The sequence of several clones and protein sequence data indicate that 24 residues are heterogeneous with two and sometimes more amino acid variants at each portion. Three EF hands (Ca^{2+} binding domains) were identified, and a hydrophobic domain which may be the coelenterazine binding site (Charbonneau *et al.*, 1985; Cormier *et al.*, 1989). *Renilla reniformis* luciferase (311 amino acids) was cloned and sequenced from a λgt 11 cDNA library screened with degenerate oligonucleotide probes based on peptide sequences. The recombinant protein expressed in *E. coli* is indistinguishable from the native luciferase (Lorenz *et al.*, 1991).

In the marine ostracod *Vargula hilgendorfii*, luminescence is caused by the oxidation of the another imidozolopyrazine luciferin (Fig. 5.2(a)) by molecular oxygen, and is catalysed by a luciferase composed of 555 amino acids, a secreted protein. It has been cloned from a cDNA library in the plasmid pRSVL, screened using degenerate oligonucleotides and expressed in COS cells (Thompson *et al.*, 1989).

Beetle luciferases catalyse the oxidative decarboxylation of a benzothiazole luciferin in the presence of Mg-ATP (Fig. 5.2). The best characterized of the beetle luciferases is that from the firefly *Photinus pyralis* (550 amino acids). It has been cloned from a lantern cDNA library constructed in the expression vector λgt 11, screened with anti-luciferase antibodies. Its gene structure and sequence has been determined (De Wet *et al.*, 1985, and has six small introns. The luciferase from *Luciola cruciata* (548 amino acids) was cloned from a cDNA library in pUC19 screened with the ClaI/EcoRI fragment from *Photinus* cDNA (Tatsumi *et al.*, 1989). From the abdominal light organ of the beetle *Pyrophorus phagiophthalamus* a cDNA library was constructed in the vector ZAP that also converts into a bacterial expression vector. The library was screened with anti-*Photinus* luciferase antibodies also and for light emission from the bacterial colonies. Four luciferases (553 amino acids) that emitted light with colours ranging from green to orange were cloned and sequenced. They are 94–99% identical with each other but only 48% homologous to the *Photinus* luciferase (Wood *et al.*, 1989a).

5.4.2   Uses of bioluminescent genes as reporter genes

The study of gene regulation in higher organisms by *cis*-acting elements (promoters, enhancers, inducers) is carried out in cultured cells and animals or plants transfected with these putative regulatory elements. Their effect on transcription can be determined after analysis of the structure and concentration of the RNA under study, either *in situ* or after its isolation. These direct methods of analysis are time-consuming, destroy the tissue and are not very successful for low abundance mRNAs. They have been steadily replaced by the use of reporter genes. The regulatory DNA element under study is ligated 5' or 3' to the reporter gene, genomic or cDNA. The construct, usually a plasmid that grows in *E. coli*, includes antibiotic resistance genes, bacterial origin of replication, and a polyadenylation signal. Whether one includes a strong promoter with either wide or restricted tissue specificity, sequences for homologous recombination, a selectable marker, a strong transcription termination signal upstream of the promoter to avoid the generation of transcripts initiated at other sites in the vector, will depend on the purpose of the investigation. There are many commercial sources of cassettes tailored to suit different strategies.

The ideal reporter gene should satisfy the following criteria:

(a) Specificity: The protein coded by the reporter gene should be absent from the tissue of interest. Several reporter genes have been used (Table 5.3). These include genes coding for bacterial proteins such as chloramphenicol acetyl transferase (CAT), β-galactosidase (β-gal), β-glucuronidase, bacterial luciferase; as well as eukaryotic proteins such as *Drosophila* alcohol dehydrogenase and several luciferases. Proteins easily distinguishable from tissue proteins can be used if they are present as an isoform or have a very restricted tissue distribution, e.g. globin (Treisman *et al.*, 1983) and placental alkaline phosphatase, or if they are normally secreted into the medium, e.g. human growth hormone and secretory alkaline phosphatase.

Although luminescent mammals do not exist, some cells (e.g. phagocytic) produce weak chemiluminescence that could interfere with the assay of luminescent proteins. As far as specificity is concerned, luminescent proteins are winners. Heat treatment is given to CAT extracts to remove interfering activities. β-Gal activity is expressed by gut epithelial cells (Sambrook *et al.*, 1989) and placental type alkaline phosphatase is expressed in tissues such as lung, testes, cervix and some malignant cells (Alam & Cook 1990). The reporters that code for secretory proteins even if not completely specific are useful to correct for transfection efficiency because they do not

Table 5.3 Reporter genes.

Reporter gene	Type of assay	Tissue preparation	Relative sensitivity[a]	References
Chloramphenicol acetyl transferase	TLC or differential extraction and scintillation counting	Cell extracts	1	Gorman et al. 1982; Sambrook et al., 1989; Alam & Cook, 1990
β-Galactosidase (E. coli lac Z gene product)	Spectrophotometric Histochemical FACS	Cell extracts Tissue In vivo	<1	Sambrook et al., 1989; Nolan et al., 1988
β-Glucuronidase	Spectrophotometric Histochemical Fluorometric FACS	Cell extracts Tissue Cell extracts In vivo	<1 >1	Jefferson et al., 1987; Liu et al., 1990
Human growth hormone	Radio immunoassay	Supernatants	10	Selden et al., 1986
Alkaline phosphatase (or its secretory variant)	Spectrophotometric Luminometric/ scintillation counter Histochemical	Cell extracts Cell extracts/ supernatants Tissue	1 >1	Henthorn et al., 1988; Schaap et al., 1989
Firefly luciferase	Luminometric/ scintillation counting Luminometric/CCD camera/autoradiography	Cell extracts In vivo	30–100	De Wet et al., 1985; Brasier et al., 1989; Nguyen et al., 1988; Gould & Subramani, 1988; Hooper et al., 1990
Aequorin	Luminometric	Cell extracts	1	Tanahashi et al., 1990
Vargula luciferase	Luminometric/ scintillation counter	Supernatants	>1	Thompson et al., 1990
Bacterial luciferase	Luminometric/ scintillation counter/ autoradiography	Cell extracts/ in vivo	>1	Legocki et al.,1986; Olsson et al., 1988; Rogowsky et al., 1987

[a] Relative sensitivity: (% sample needed to measure CAT)/(% sample needed to measure reporter); only measurements for cell extracts or supernatants can be compared.
Relative sensitivity is >1: means that the assay is more sensitive than that for CAT but the exact figure is not known or it is based on the number of molecules of pure reporter protein that can be detected being less than the 10^7–10^8 molecules that can be detected by the CAT assay.
Relative sensitivity is <1: vice versa.

compete for precious transfected cells. To this group belong human growth hormone, secreted alkaline phosphatase and *Vargula* luciferase (Tables 5.3 and 5.4).

(b) Sensitivity: Just a few molecules of the reporter should be necessary to detect its presence. To define precisely the sensitivity of the different reporter genes is difficult. However, their activity can be compared with that of CAT driven by the same promoter in the same cells, and the activity expressed either as the minimum number of cells required in the assay or as the length of time after transfection needed for the reporter protein to be significantly different from

background. Estimation of the minimum number of molecules detected by any assay is only valid when the stability of the reporter gene mRNA or its protein have been evaluated, as well as the effect of the extraction procedure, the presence of inhibitors, and the effect of the incubation temperature. The photo-protein aequorin exemplifies this point. A chemi-luminometer can detect as little as 10^3 molecules of the protein, but when used as a reporter gene it was not better than CAT (Tanahashi et al., 1990), having a limit of detection of at least 10^7 molecules (Alam & Cook 1990). Optimal conditions for extraction and assay have reduced the limit of detection for firefly

Table 5.4 Examples of uses of luminescent proteins as reporter genes.

Luminescent protein	Promoter	Cell/organism	References
Firefly luciferase	1. CaMV35	plants	Ow et al., 1987
	2. SV40, RSV-LTR	CV-1	De Wet et al., 1985
	3. Adenovirus major late	CHO (DHFR–)	Gould & Subramani, 1988
	4. Oestrogen regulatory elements, RSV-LTR	MB231N	Waterman et al., 1988
	5. Vaccinia early and late	BSC-40	Rodriguez et al., 1988, 1989
	6. Interleukin-2	jurkat	Williams et al., 1989
	7. Ferritin H gene	FRTL5	Chazenbalk et al., 1990
	8. HIV-LTR	various lymphoid, monocytic, adherent	Schwartz et al., 1990
	9. Polyhedrin	Sf9	Hasnain & Nakhai, 1990
	10. Vitelogenin	MCF-7	Pons et al., 1990
	11. Polio virus regulatory regions	HeLa	Simoes & Sarnow, 1991
	12. Oxytocin	P19, MCF-7	Adan et al., 1991
Bacterial luciferases	1. nif D, nif H	*Rhizobium*	Legocki et al., 1986
	2. Tetracycline resistance gene	Phytogenic bacteria	Shaw & Kado, 1986
	3. Vi-genes	*Agrobacterium tumefaciens*	Rogowsky et al., 1987
	4. Gene 2′ early and T_R-DNA 1′	Carrot, nicotiana	Koncz et al., 1987
	5. Transposon Tn 172	Bacteria, whole plants	Shaw et al., 1986, 1987
	6. PGK	Yeast	Kirchner et al., 1989
	7. TL-DNA gene 7	Tobacco calli	Olsson et al., 1989
	8. tet	*E. coli*	Rattray et al., 1990
Aequorin	1. SV40, RSV-LTR, HSV-1tk	CV-1, COS, HeLa	Tanahashi et al., 1990
Vargula luciferase	1. SV40, RSV-LTR, hEF, mG-CSF	Mouse, monkey and live cells	Thompson et al., 1990

luciferase to 2000 molecules (Wood, 1991), but in intact cells at least ten times more molecules would be required because anions and pH affect its activity and colour (DeLuca & McElroy, 1978). A yellow emitter has to produce some 5–10 times as many photons as a blue emitter to produce the same number of recorded counts from a photomultiplier.

(c) Detectability: It should be detectable without the destruction of the transfected tissue. One of the most exciting aspects that reporter genes offer is the possibility of detecting low levels of gene expression and how regulatory regions work in individual cells live cells. This would allow examination of the effect of the integration site on expression, the analysis of the variables that control different levels of expression, as well as quantification of cell heterogeneity and isolation of clonal populations. It is here that the fluorescent and bioluminescent reporters have particular application. Fluorogenic assays for β-gal (Nolan et al., 1988) and β-glucuronidase (Jefferson et al., 1987) utilize fluoresceinated substrates which, when cleaved by the respective enzymes, generate

fluorescein inside the cells; this enables FACS analysis to detect and isolate individually expressing cells. Single cells expressing firefly luciferase under the control of a *Vaccinia* promoter have been detected using CCD imaging (Hooper et al., 1990). Firefly luciferase activity has also been detected *in vivo* in bacterial and yeast populations but not in single cells; in plants (Ow et al., 1986) and in transfected cells (Rodriguez et al., 1988; Pons et al., 1990). Some cells required permeabilization to allow in the luciferin (Gould & Subramani, 1988). However, when permeability to luciferin is a problem, it may be possible to solve this by using 0.5–1 mM luciferin or luciferin esters (Craig et al., 1991). Bacterial luciferase has also been detected *in vivo* (Table 5.4).

(d) Amenable to tissue localization: The use of histochemical stains has allowed researchers to look at gene expression in individual cells but only after fixation. β-Gal has been used extensively to study mammalian cells (see Nielsen & Pedersen, 1991 for references) and β-glucuronidase has proved useful in plants (Jefferson et al., 1987). Alkaline phosphatase has also been used but its application is limited to tissues with low levels of endogenous expression or very strong promoters (Henthorn *et al.*, 1988), e.g. *Drosophila* alcohol dehydrogenase (Nielsen & Pedersen, 1991). Luminescent proteins can be detected *in vivo*. Transgenic tobacco plants carrying the firefly luciferase gene showed luciferase activity in various organs and whole plants after soaking with luciferin and exposure to X-ray film (Ow *et al.*, 1986). Bacterial luciferase expresses in plants (Table 5.4) but *in vivo* measurements are hampered by the low levels of $FMNH_2$ (Koncz *et al.*, 1987). The aldehyde is usually added via a hanging droplet.

(e) The assay should be simple, inexpensive, use commercially available reagents, fast to perform and the concentration of the reporter gene product should closely reflect transcriptional activity. CAT is probably the most widely used reporter gene, perhaps because of the widespread availability of constructs. But the assay is cumbersome, requires radioactive substrate and it is expensive. Alam and Cook (1990) have estimated the cost of the most common reporter genes used. Optimal conditions to assay firefly luciferase in extracts (Brasier *et al.*, 1989; Wood, 1991) have been developed. The possession of a luminometer is not essential for the use of bioluminescent reporter proteins except when the photoprotein aequorin is used. All the luciferases and the most sensitive assay for alkaline phosphatase (detects 10^3 molecules) by dioxetan-enhanced chemiluminescence (Schaap *et al.*, 1989) can be measured using a scintillation counter (Nguyen *et al.*, 1988), and sometimes photographic film. However, the real potential of bioluminescent genes as reporters can only be realized by detection using a photon-counting image intensifier preferably with a contact imaging system to by-pass the microscope (Hooper *et al.*, 1990). The use of *Vargula* luciferase is limited because the luciferin is not commercially available.

5.4.3 Other uses of bioluminescent genes

5.4.3.1 As markers of recombinant events

A new cloning vector has been developed based on the loss of bioluminescent phenotype. A multiple cloning site was inserted into the *lucB* gene in a vector containing the *luxA* and *B* genes from *V. harveyi* (pLUM), the *E. coli* ori and phage fori that specifies single stranded DNA synthesis. Transfection into *E. coli* generated luminous bacteria and the insertion of a foreign gene destroyed the luminous phenotype. Light generated by live bacteria was detected on X-ray film. This system seems superior to the IPTG/X–gal/β-gal system (Sevigny & Gossard, 1990).

(a)

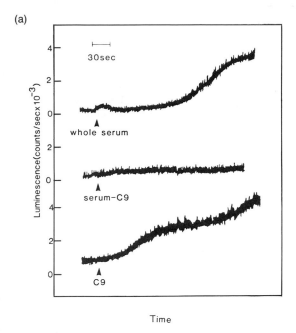

(b)

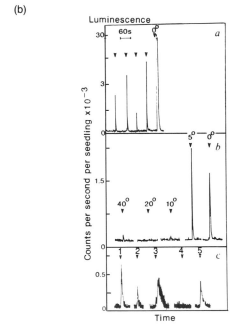

Figure 5.6 Ca^{2+} signals from recombinant aequorin in live cells. (a) Effect of the membrane attack complex of complement on recombinant aequorin expressed in *E. coli*. (From Knight *et al.*, 1991b, with permission of Elsevier Science Publishers.) (b) Effect of *a* touch, *b* cold shock or *c* wind on recombinant aequorin luminescence expressed in a transgenic tobacco plant. (From Knight *et al.*, 1991a, reprinted with permission from *Nature* **352**, 524–526. Copyright (1991) Macmillan Magazines Limited.)

5.4.3.2 In vivo *indicators of cellular functions*

Ca^{2+} spikes have been detected for the first time from an intact tobacco plant, transgenic with aequorin (Knight *et al.*, 1991b, Fig. 5.6(a)), and the effect of antibiotics, complement and phagocytosis together with the cell cycle changes, now becomes possible in *E. coli* (Knight *et al.*, 1991a, Fig. 5.6(b)). The photoprotein aequorin has been expressed in *E. coli* under the control of the P_L promoter (Knight *et al.*, 1991a), in yeast under the control of the promoter of *S. cerevisiae* glyceraldehyde-3-phosphate dehydrogenase (Nakajima-Shimada *et al.*, 1991) and in plants under the control of the CMV 35S promoter (Knight *et al.*, 1991b). Aequorin was reconstituted *in vivo* and used to measure changes in intracellular Ca^{2+} triggered by several agonists (Fig. 5.6). Aequorin has also been expressed in yeast, enabling Ca^{2+} signals associated with the cell cycle to be monitored continuously over periods of hours or even days; this is impossible with the fluors (Nakajima-Shimada *et al.*, 1991a,b). Even if single cells cannot yet be analysed, aequorin has the advantage that it is not sequestered in organelles as is fura-2, and measurements of Ca^{2+} can be carried out constantly for very long periods.

The application of bioluminescence in microbiology for research and industry is a rapidly developing field. The introduction of *lux* genes into bacteriophages followed by infection of a bacterial host leads to the expression of the genes and the generation of the luminescent phenotype. *Salmonella*, *Listeria* and enterics as a group have been targeted (references in Jassim *et al.*, 1990).

Engineered bioluminescent bacteria have been used to monitor for biocides and sublethal injury and recovery because only live bacteria carry the metabolites required for bacterial luciferase luminescence and bacteriophages carrying the *luxAB* genes have been used to screen for virucidal activity (Jassim *et al.*, 1990). The same group engineered promoter elements from *Lactobacillus casei* upstream from the *luxA* and *B* genes from *V. fischeri*. Transfected *Lactobacillus* species became luminescent and have been used to monitor antimicrobial activity in milk and gene expression in lactic acid bacteria in fermentation systems *in situ* (Ahmed & Steward, 1991).

Pathogenic and symbiotic plant bacteria have been given a luminous phenotype by transfection with a plasmid carrying the *lux* operon from *V. fischeri* without decreasing their virulence. Autoradiography was then used to monitor the infection of the plant hours or days before the symptoms became visible. It has also been proposed that genetically engineered microorganisms could be released into the environment tagged with *lux* genes to allow aerial detection (Shaw & Kado, 1986). Bacterial luciferase and luminometry has also been used to detect genetically engineered

microorganisms in soil, 10^2–10^3 cells ml^{-1} being detectable (Rattray *et al.*, 1990).

A recombinant baculovirus containing the firefly luciferase cDNA under the control of the polyhedrin promoter has been used to follow the dissemination of this virus in live caterpillars (Jha *et al.*, 1990). A non-invasive, non-destructive, rapid, *in situ* and population-specific method for naphthalene exposure and biodegradation has also been developed using bacterial *lux* genes. The *lux* genes have been inserted in the *nahG* gene encoding an enzyme induced by naphthalene. Bacteria carrying the plasmid developed luminescence when exposed to naphthalene and soil slurries containing polycyclic aromatic hydrocarbons (King *et al.*, 1991).

Oestrogen receptor positive cell lines, stably transfected with a plasmid that allows expression of firefly luciferase under the control of the oestrogen regulatory element of the vitellogenin A2 gene, have been used to develop an *in vivo* test for oestrogen and anti-oestrogen molecules. Luciferase was detected in extracts and *in vivo* using a CCD camera (Pons *et al.*, 1990).

An assay for the human immunodeficiency virus (HIV) and antiviral drugs has also been developed. A plasmid carrying firefly luciferase cDNA under the control of HIV was transfected into several cell lines. Challenging with HIV resulted in luciferase expression (Schwartz *et al.*, 1990).

5.4.3.3 As labels for immunoassays and DNA

Several groups have coupled aequorin to antibodies or protein A either chemically (Erikaku *et al.*, 1991) or by genetic manipulation (Casadei *et al.*, 1990; Zenno & Inouye, 1990; Lindbladh *et al.*, 1991), or have biotinylated aequorin (Zatta *et al.*, 1991), for use in immunoassays. A simple ELISA method that can detect *Salmonella* contamination in food using biotinylated recombinant aequorin has been reported and compared favourably with other alternatives (Smith *et al.*, 1991).

Biotinylated recombinant firefly luciferase and aequorin have been used as probes to detect proteins and nucleic acids in blots. Firefly luciferase gave the most sensitive results, comparable to using alkaline phosphatase (Stults *et al.*, 1991). A bioluminescent solid phase system using bacterial luciferase has been used to detect DNA and single base mutations obtained by polymerase chain reaction (Balaguer *et al.*, 1991).

Bioluminescent labels for immunoassay or DNA analysis can be at least as sensitive as the best synthetic labels, acridinium esters (Weeks *et al.*, 1983a,b; Woodhead & Weeks, 1991) or adamantane dioxetans + alkaline phosphatase (Schaap *et al.*, 1989). These are superior to luminol-enhanced perioxidase, producing a glow which can be recorded on sensitive (20 000 ASA) photographic film.

The ability to produce 100 mg of recombinant bioluminescent protein from just a few litres of bacterial culture opens the way for exploitation of the exquisite sensitivity of these immunoassay and DNA probes in clinical assays (Zatta *et al.*, 1991).

5.5   ENGINEERING INDICATORS FOR MOLECULAR SIGNALLING IN LIVE CELLS

5.5.1   The objectives

Our aim has been to engineer, initially into firefly luciferase and aequorin cDNA: (a) binding sites for cyclic AMP, cyclic GMP, IP_3, GTP and diacylglycerol, and protein phosphorylation recognition sites, e.g. RRXS for protein kinase A (Kemp *et al.*, 1976); (b) the ATP-Mg binding site in firefly luciferase to reduce its affinity, i.e. to a K_m *ca.* 2–5 mM; (c) targeting peptides to take these indicators into the mitochondria, endoplasmic reticulum (ER) and nucleus. Since the free Ca^{2+} in the ER may be much higher than in the cytosol it will also be important to engineer an aequorin less sensitive to Ca^{2+}, so that the effects of priming agents, secondary regulators and oscillation on ER Ca^{2+} can be determined (Campbell, 1989).

Firefly luciferase and aequorin were selected because nature has already taught us that a few amino acid changes can shift the colour of the emission, and the spectral peak by as much as 70 nm. Luminous beetles can emit green (*ca.* 550 nm), green yellow (555 nm, yellow (565 nm), orange (600 nm) or red (620 nm) light, all with the benzothiazole chemistry (Fig. 5.2(a)1). Luminous radiolarians emit at 440 nm, aequorin at 465 nm, obelin at 475 nm, and ctenophores at 490 nm, all with coelenterazine Ca^{2+}-activated photoproteins.

5.5.2   Engineering strategy

Firefly luciferase cDNA was isolated by cloning in pcDV1 vector and Honjo linker containing the SP6 RNA polymerase promoter (Sala-Newby *et al.*, 1990a,b). Aequorin cDNA was isolated using the reverse transcriptase polymerase chain reaction (PCR), and both genomic DNAs isolated by conventional PCR. A one- or two-stage PCR strategy was established to incorporate a T7 RNA polymerase promoter at the 5′ end, Sac I and Sal I restriction sites at the 3′ end for subcloning, and mutations, protein phosphorylation sites and targeting signals within the proteins or at the N- or C-terminus engineered using specially designed primers. A key part of the protocol was to synthesize mRNA *in vitro* from the normal and engineered cDNAs and then translate this in rabbit reticulocyte lysate to form light-emitting protein (Sala-Newby *et al.*, 1990a,b).

Three criteria were used to confirm a successful PCR:

(1) main (ideally single) ethidium bromide band on agarose gel electrophoresis of the correct size;
(2) synthesis of [^{32}P]mRNA, from [^{32}P]UTP capped using m7 GpppG, of the correct size on glyoxal gel electrophoresis. No mRNA was formed when T7 promoter was absent, or when the SP6 promoter was used instead; and
(3) translation of mRNA to form light-emitting protein, comparing chemiluminescent counts per microgram RNA for various cDNAs.

Six properties of the engineered protein could be characterized from a few tens of microlitres of rabbit reticulocyte translating a few nanograms of RNA, because of the sensitivity of detecting bioluminescent proteins:

(1) [^{35}S]Met produced protein of the correct size on polyacrylamide gel electrophoresis. A number of uncompleted chains were also often seen and could be removed by gel filtration. No protein was formed when the T7 promoter was absent in the mRNA synthesis step.
(2) Specific activity = chemiluminescence counts per mol protein estimated from [^{35}S] met incorporation.
(3) Colour, determined by ratio of light emission at 545/609 for firefly luciferase and 407/470 for aequorin.
(4) pH profile.
(5) K_m for key components; luciferin, O_2, ATP, Ca^{2+}.
(6) Special properties when appropriate:
　　(a) binding to isolated organelles, e.g. mitochondria;
　　(b) effect of phosphorylation and dephosphorylation on intensity and colour.

Each PCR reaction generated several micrograms of cDNA. The efficiency of transcription was up to 20 mol RNA/DNA, but because of the necessity of having the right ratio of GTP to m7 GpppG for capping and thus efficient translation, in most experiments only some 5–10 mol RNA were formed per DNA. The translation efficiency in rabbit reticulocyte lysate was disappointing, being only some 0.2–1 protein molecules per RNA. Up to 20 protein/RNA have been reported (Rhoads, 1985). Increasing the number of bases between the transcription start site, using a Shine–Dalgarno sequence of 12 bp, and the translation start AUG, had little or no effect. A critical factor was the addition of K^+ and Mg^{2+}, to some 80 mM and 2 mM respectively, which had to be titrated for each batch of lysate.

Using this strategy we have established that aequorin gDNA apparently has no introns, as it is the same length as cDNA and translates to form fully active apophotoprotein. Addition of peptides to the N-terminus of either protein or to the C-terminus of luciferase had only a small or no effect on specific activity. However, addition of KDEL to the C-terminus of aequorin appeared to reduce activity.

In an attempt to make large amounts of protein for microinjection and express the proteins in COS cells the engineered cDNAs were inserted into a Sma I/Sal I cut pSV7d plasmid which then transformed *E. coli* strain BL21(DE3) with an IPTG-inducible T7 RNA polymerase.

5.5.3 An indicator for cyclic AMP-dependent protein kinase (protein kinase A)

The consensus recognition sequence for protein kinase A is RRXS. Good artificial substrates include heptapeptides, kemptide LGRRASG (Kemp *et al.*, 1976) found in pyruvate kinase, and malantide (Murray *et al.*, 1990). We had previously reported that chemical coupling of kemptide using disuccinyl suberate (DSS) to luciferase produced a chimaeric protein whose colour shifted to the red when phosphorylated (Jenkins *et al.*, 1990).

Addition of kemptide to the N- or C-terminus had little or no effect on activity, colour or pH profile of firefly luciferase (Sala-Newby & Campbell, 1991) nor was there any detectable effect of protein kinase A catalytic subunit. Firefly luciferase contains two sequences, VRFS (217–220) and RLKS (437–410), where a two-base change alters them to RRFS or RRKS respectively. The latter had no change in specific activity or colour, but the RRFS luciferase had a specific activity some 80% less than the normal protein, and emitted greener light. Phosphorylation by protein kinase A reduced the light intensity by >80%, which was reversed by addition of a phosphatase (Fig. 5.7). This indicator has been expressed in cells (Sala-Newby & Campbell, 1992). Gene expression of this mutant luciferase in cells will provide for the first time an indicator for protein kinase activity in living cells. The specificity for particular kinases and phosphatases remains to be determined.

5.5.4 Targeting

Firefly luciferase contains a peroxisomal targeting signal at the C-terminus, originally thought to reside in the last 12 amino acids (Keller *et al.*, 1987; Gould & Subramani, 1988). Removal of these amino acids using our PCR procedure resulted in >95% loss in activity (Sala-Newby *et al.*, 1990b). However, removal

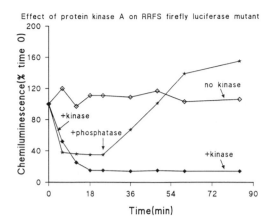

Effect of protein kinase A on RRFS firefly luciferase mutant

Figure 5.7 Effect of phosphorylation and dephosphorylation on the activity of the RRFS mutant firefly luciferase. Partially purified RRFS mutant luciferase was incubated at 30°C with buffer alone, protein kinase A catalytic subunit, kind gift from Dr K. Murray, Smith Kline & Beecham, Welwyn, UK, with ($\Diamond$, $\blacklozenge$) or without ($\star$) protein phosphatase inhibitors. At 24 min to trace ($\downarrow$) alkaline phosphatase and protein kinase inhibitor were added($\star$). Samples were taken at various times up to 90 min, and assayed for luciferase at pH 7.2. Results were expressed as a percentage of chemiluminescence at time 0, and are the mean of two determinations. (From Sala-Newby & Campbell, 1991, with the permission of The Biochemical Society and Portland Press.)

of just the last three amino acids, now thought to be the key signal (Gould & Subramani, 1988), retained full activity. In COS cells expression of the cDNA for luciferase-3 amino acids was uniformly distributed throughout the cell. Firefly luciferase containing the mitochondrial signal synA1 mimicking that for yeast cytochrome oxidase (Allison & Schatz, 1986) bound ten-fold better to isolated mitochondria than normal luciferase (Fig. 5.8). It has been reported using a similar approach that firefly luciferase can be targeted to the mitochondria in live yeast (Aflalo, 1990).

An endoplasmic reticulum (Austin *et al.*, 1984) and a nuclear targeting signal have been added to the N-terminus of aequorin, and the KDEL sequence for retaining proteins in the ER (Munro & Pelham, 1987) to the C-terminus. The latter reduced the activity of aequorin, but manipulation at the N-terminus retained full activity. It has been possible to test the uptake of the ER aequorin into isolated ER, although there is a very large inhibition of protein synthesis in rabbit reticulocyte lysate by these microsome preparations. But recombinant aequorin can be expressed in COS cells. The Ca(II) site has also been mutated, generating an aequorin with a lower affinity for Ca^{2+} and targeted to the ER in live COS cells (Kendall *et al.*, 1992a,b). It has been reported previously that inserting a positively charged amino acid into Ca site I abolishes activity (Tsuji *et al.*, 1986).

SPECIFIC ACTIVITY

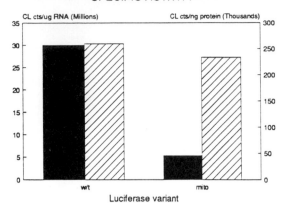

BINDING TO MITOCHONDRIA

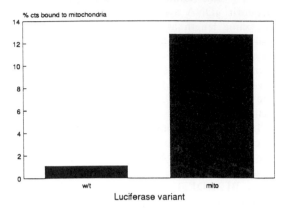

Figure 5.8 Specific activity and binding to rat liver mitochondria of firefly luciferase containing a mitochondrial signal peptide. Isolated mitochondria were incubated with recombinant luciferase, normal (w/t) or one containing the mitochondrial signal syn A1 (Allison & Schatz, 1986) (mito), in the presence of rabbit reticulocyte lysate for 30 min at 37°C. The mitochondria were centrifuged in an Eppendorf for 5 min and washed once in buffer. The luciferase activity associated with the pellet was assessed in the usual way (see Sala-Newby & Campbell, 1991).

5.6 CONCLUSIONS AND FUTURE PROSPECTS

5.6.1 Chemiluminescence versus fluorescence

Chemiluminescent indicators for living cell function, injected into cells, are less susceptible to leakage or intracellular redistribution than the fluors, can detect enzymatic processes as well as concentration changes at indicator concentrations some ten-fold lower than the fluors, and appear to be non-toxic. Precise calibration of both types of indicator has problems. Non-linearity of a chemi- or bioluminescent indicator can be an advantage when amplification of a small change is required to detect it. One must be aware of endogenous ultraweak chemiluminescence, but this is less of a problem than endogenous fluorescence of substances such as NAD(P)H or FMN. It is necessary to ensure that for luciferases O_2 is not limiting, and that the effects of endogenous ions or pH do not affect intensity of colour, and thus apparent intensity.

The two indicators work in symbiosis. The fluors clearly are more easy to use, particularly for single cell imaging, but the engineered indicators have unique potential for measuring covalent modifications, and signalling inside organelles. Furthermore they circumvent the problem of internal accumulation of fluors inside unspecified organelles.

5.6.2 The future

The engineering of bioluminescent proteins heralds a new era for these indicators inside cells. Large amounts will be available for microinjection. For the first time it will be possible to measure components of the molecular signalling system in defined compartments of live cells. By coupling these engineered DNAs to the appropriate promoters they can be expressed in defined cells within a live organ, e.g. a whole beating heart, a liver or a β cell in the pancreas. They also provide a unique tool for following the movement of proteins and organelles in live cells, determining how efficient their targeting systems are, and as an indicator for the activity of promoter and enhancer proteins in cells and in nuclear extracts. The use of phage promoters such as T7 to generate large amounts of mRNA *in vitro* for microinjection will enable these new indicators to be expressed in non-dividing cells, for example human neutrophils (Campbell *et al.*, 1988). The ability to target the engineered bioluminescent proteins to defined sites within the cell means that it will be possible to light up chemical events in particular regions within the cell without the need for expensive imaging devices. There is already sufficient photon emission from transgenic cells to monitor the bioluminescent indicators in individual cells, over a time scale of a few seconds.

Bioluminescence genes thus have considerable potential for studying the control of gene expression in live cells (Table 5.4). Transgenic plants expressing firefly luciferase, bacterial luciferase, or aequorin have been successfully engineered. Attempts to produce transgenic mice and other animals are in progress. This will allow specific cell types to be studied in intact organs for the first time; this is impossible with fluorescence.

The scattered occurrence of bioluminescence, even

within a single family, was even a puzzle to Darwin (see 6th edition of *The Origin of Species*). The occurrence of imidazolopyrazine bioluminescence in six phyla: Sarcomastigophora (Radiolaria), Cnidaria, Ctenophora, Mollusca, Arthropoda and Chordata (Campbell & Herring, 1990), five of which are unrelated, and in which Ca^{2+}-activated photoproteins are found in three, provides a unique model for biochemical and cellular evolution. The central question is how step by step changes in DNA lead to a threshold, an evolutionary Rubicon, at which point, and only then, the phenomenon became susceptible to the forces of Darwinian–Mendelian selection.

Chemi- and bioluminescence are fascinating natural phenomena taking one from the ecology and biology of the deep oceans and a rock pool, through biochemistry and chemistry to the nature of light itself, and perhaps they may be of good use to human life.

ACKNOWLEDGEMENTS

We thank the MRC, AFRC and Arthritis and Rheumatism Council for financial support. We also thank our colleagues Dr Bob L. Dormer, Dr Noor Kalsheker, Dr Rosan Padua, Dr J. Paul Luzio, Dr Marc Knight and Professor Tony Trewavas for helpful and enjoyable discussion, and happy experimental collaboration. Finally, we thank the other members of the Cardiff ARC Molecular Signalling Group for their collaboration.

REFERENCES

Adam W. & Cilento G. (1982) *Chemical and Biological Generation of Excited States*. Academic Press, New York.

Adan R.A.H., Walther N., Cox J.J., Ivell R. & Burbach J.P.H. (1991) *Biochem. Biophys. Res. Commun.* **175**, 117–122.

Aflalo C. (1990) *Biochemistry* **29**, 4758–4766.

Ahmed K.A. & Steward G.S.A.B. (1991) *J. Appl. Bacteriol.* **70**, 113–120.

Alam J. & Cook J.L. (1990) *Anal. Biochem.* **188**, 245–254.

Allison D.S. & Schatz G. (1986) *Proc. Natl. Acad. Sci. USA* **83**, 9011–9015.

Ashley C.C. & Campbell A.K. (eds) (1979) *The Detection and Measurement of Free Ca^{2+} in Cells*. Elsevier/North Holland, Amsterdam.

Atta P.F., Nyame K., Cormier M.J., Mattox S.A., Prieto P.A., Smith D.F. & Cummings R.D. (1991) *Anal. Biochem.* **194**, 185–191.

Austen B.M., Hermon-Taylor J., Kaderbhai A. & Ridd A. (1984) *Biochem. J.* **224**, 317–325.

Baker P.F., Hodgkin A.L. & Ridgway E.B. (1971) *J. Physiol.* **218**, 708–755.

Balaguer P., Terouanne B., Boussioux A.M. & Nicolas J.C. (1989) *J. Biolumin. Chemilum.* **4**, 302–309.

Balaguer P., Terouanne, B., Boussioux A.M. & Nicolas J.C. (1991) *Anal. Biochem.* **195**, 105–110.

Baldwin T.O., Devine J.H., Heckel R.C., Lin J.W. & Shadel G.S. (1989) *J. Biolum. Chemilum.* **4**, 326–341.

Beijerinck M.W. (1902) *Proc. Acad. Sci. Amst.* **4**, 45–49.

Blinks J.R., Mattingly P.H., Jewel B.R., Van Leeuwan M., Harrer G.C. & Allen D.G. (1978) *Methods Enzymol.* **57**, 292–328.

Blinks J.R., Wier, W.G., Hess P. & Prendergast F.G. (1982) *Prog. Biophys. Molec. Biol.* **40**, 1–114.

Boveris A., Cadenas E., Reiter R., Filipkowski M., Nakase Y. & Chance B. (1980) *Proc. Natl. Acad. Sci. USA* **77**, 347–351.

Boylan M.O., Pelletier S., Dhepagnon S., Trudel N., Sonenberg N. & Meighen E.A. (1991) *J. Biolum. Chemilum.* **4**, 310–316.

Brasier A.R., Tate J.E. & Habener J.F. (1989) *BioTechniques* **7**, 1116–1122.

Campbell A.K. (1983) *Intracellular Calcium: Its Universal Role as Regulator*. John Wiley & Sons, Chichester.

Campbell A.K. (1987) *Clin. Sci.* **72**, 1–10.

Campbell A.K. (1988) *Chemiluminescence: Principles and Applications in Biology and Medicine*. Horwood/VCH, Chichester & Weinheim.

Campbell A.K. (1989) British Patent Application 8916806.6. International Patent Application PCT/GB90/01131.

Campbell A.K. (1990) In *Oxygen Radicals and Cellular Damage*, C.J. Duncan (ed.). Cambridge University Press, Cambridge, pp. 189–217.

Campbell A.K. (1991) In *Calcium, oxygen radicals and cellular damage*. SEB Seminar Series 46. C.J. Duncan (ed.) Cambridge University Press, Cambridge, pp. 189–217.

Campbell AK & Dormer R.L. (1978) *Biochem. J.* **176**, 53–66.

Campbell A.K. & Hallett M.B. (1983) *J. Physiol.* **338**, 537–562.

Campbell A.K. & Herring P.J. (1990) *Mar. Biol.* **104**, 219–225.

Campbell A.K. & Luzio J.P. (1981) *Experientia* **37**, 1110–1112.

Campbell A.K. & Patel A. (1983) *Biochem. J.* **216**, 185–194.

Campbell A.K., Roberts P.A. & Patel A.K. (1985) In *Alternative Immunoassays* W.P. Collins (ed.), John Wiley & Sons, Chichester, pp. 153–183.

Campbell A.K., Patel A.K., Razavi, Z.S. & McCapra F. (1988) *Biochem. J.* **252**, 143–149.

Casadei J., Powell M.J. & Kenton J.H. (1990) *Proc. Natl. Acad. Sci. USA* **87**, 2047–2051.

Charbonneau H., Walsh K.A., McCann R.O., Prendergast F.G., Cormier M.J. & Vanaman T.C. (1985) *Biochemistry* **24**, 6762–6771.

Chazenbalk G.D., Wadsworth H.L., Foti D. & Rapoport B. (1990) *Molec. Endocrin.* **90**, 1117–1124.

Cobbold P.H. & Bourne P.K. (1984) *Nature (Lond.)* **312**, 446–448.

Cobbold P.H. & Rink T.J. (1987) *Biochemistry* **248** 313–328.

Cohn D.H., Ogden R.C., Abelson J.N., Baldwin T.O., Nelson K.H., Simon M.I. & Mileham A.J. (1983) *Proc. Natl. Acad. Sci. USA* **80**, 120–123.

Cohn D.H., Mileham A.J., Simon M.I., Nealson K.H.,

Rausch S.K., Bonam D. & Baldwin T.O. (1985) *J. Biol. Chem.* **260**, 6139–6146.

Cormier M.J., Prasher D.C., Longiaru M. & McCann R.O. (1989) *Photochem. Photobiol.* **49**, 509–512.

Craig F.F., Simmonds A.C., Watmore D. & McCapra F. (1991) *Biochem. J.* **276**, 637–641.

Davies E.V., Campbell A.K. & Hallett M.B. (1991a) *FEBS Lett.* **291**, 135–138.

Davies E.V., Campbell A.K., Williams B.D. & Hallett M.B. (1991b) *Brit. J. Rheum.* **30**, 443–448.

Davies E.V., Campbell A.K. & Hallett M.B. (1992) *FEBS Lett.* (in press).

DeLuca M. & McElroy W.D. (1978) *Methods Enzymol.* **57**, 3–15.

DeLuca M.A. & McElroy W.D. (1987) *Methods Enzymol.* **133**.

De Wet J.R., Wood K.V., Helsinki D.P. & De Luca M. (1985) *Proc. Natl. Acad. Sci. USA* **82**, 7870–7873.

Dubois R. (1887) *Comptes Rend. Seanc. Soc. Biol.* **4**, 564–565.

Eisen A. & Reynolds G.T. (1985) *J. Cell. Biol.* **101**, 1522–1527.

Erikaku T., Zenno S. & Inouye S. (1991) *Biochem. Biophys. Res. Commun.* **174**, 1331–1336.

Foran D.R. & Brown W.M. (1988) *Nucleic Acids Res.* **16**, 177.

Frackman S., Anhalt M. & Nealson K.H. (1990) *Bacteriology* **172**, 5767–5773.

French R., Janda M. & Ahlquist P. (1986) *Science* **23**, 1294–1297.

Geiger R., Hauber R. & Miske W. (1989) *Molec. Cell. Probes* **3**, 309–328.

Gilkey J.C., Jaffe L.F., Ridgway E.B. & Reynolds G.T. (1978) *J. Cell Biol.* **76**, 448–466.

Gorman A.A. & Rodgers M.A.J. (1981) *Q. Rev. Chem. Soc.* **10**, 205–231.

Gorman C.M., Moffat L.F. & Howard B.H. (1982) *Molec. Cell. Biol.* **2**, 1044–1051.

Gould S.J. & Subramani S. (1988) *Anal. Biochem.* **175**, 5–13.

Gould S.J., Keller G.A. & Subramani S. (1987) *J. Cell Biol.* **105**, 2923–2931.

Grynkiewicz G., Poenie M. & Tsien R.Y. (1985) *J. Biol. Chem.* **260**, 3440–3450.

Hallett M.B. & Campbell A.K. (1982) *Nature* **295**, 155–158.

Hallett M.B. & Campbell A.K. (1983) *Immunology* **50**, 487–495.

Hallett M.B., Davies E.V. & Campbell A.K. (1990) *Cell Calcium* **11**, 655–663.

Harvey E.N. (1952) *Bioluminescence*. Academic Press, New York.

Hasnain S.E. & Nakhai B. (1990) *Gene* **91**, 135–138.

Hastings J.W. & Reynolds G.T. (1966) In *Bioluminescence in Progress*, F.H. Johnson & Y. Haneda (eds). Princeton University Press, Princeton, pp. 45–50.

Hastings J.W. & Weber G. (1963) *J. Opt. Soc. Am.* **53** 1410–1415.

Henthorn P., Zervoz P. Raducha M., Harris H. & Kadesch T. (1988) *Proc. Natl. Acad. Sci. USA* **85**, 6342–6346.

Herring P.J. (1978) *Bioluminescence in Action*. Academic Press, London & New York.

Hooper C.E., Ansorge R.E., Browne H.M. & Tomkins P. (1990) *J. Biolum. Chemilum.* **5**, 123–130.

Hummelen J.C., Luider T.M. & Wynberg H. (1987) *Methods Enzymol.* **133**, 531–557.

Inouye S., Noguchi M., Sakaki Y., Takagi I., Miyata T., Iwanaga S. & Tsuji F.I. (1985) *Proc. Natl. Acad. Sci. USA* **82**, 3154–3158.

Inouye S., Sakaki Y., Goto T. & Tsuji F.I. (1986) *Biochemistry* **25**, 8425–8429.

Jassim S.A.A., Ellison A., Denyer S.P. & Stewart G.S.A.B. (1990) *J. Biolum. Chemilum.* **5**, 115–122.

Jefferson R.A., Kavanagh T.A. & Bevan M.W. (1987) *EMBO J.* **6**, 3907–3907.

Jenkins T., Sala-Newby G. & Campbell A.K. (1990) *Biochem. Soc. Trans.* **18**, 563–465.

Jha P.K., Nakhai B., Sridhar P., Talwar G.P. & Hasnain S.E.(1990) *FEBS Lett.* **274**, 23–26.

Johnson T.C., Thompson R.B. & Baldwin T.O. (1986) *J. Biol. Chem.* **261**, 4805–4811.

Keller G.A., Gould S., DeLuca M. & Subramani S. (1987) *Proc. Natl. Acad. Sci. USA* **84**, 3264–3268.

Kemp B.E., Benjamani E. & Krebs E.G. (1976) *Proc. Natl. Acad. Sci. USA* **73**, 1038–1042.

Kendall J.M., Sala-Newby G., Ghalaut V., Dormer R.L. & Campbell A.K. (1992a) *Biochem. Biophys. Res. Commun.* **187**, 1091–1097.

Kendall J.M., Sala-Newby G., Ghalaut V., Dormer R.L. & Campbell A.K. (1992b) *Biochem. Biophys. Res. Commun.* **189**, 1008–1016.

King J.M.H., DiGrazia P.M., Applegate B., Buriage R., Sanseverino J., Dunbar P., Larimer F. & Sayler G.S. (1991) *Science* **249**, 778–781.

Kirchner G., Roberts J.L., Gustafson G.D. & Ingolia T.D. (1989) *Gene* **81**, 349–354.

Knight M.R., Campbell A.K., Smith S.M. & Trewavas A.J. (1991a) *Nature* **352**, 524–526.

Knight M.R., Campbell A.K., Smith, S.M. & Trewevas A.J. (1991b) *FEBS Lett.* **282**, 405–408.

Koncz C., Olsson O., Langridge W.H.R., Schell J. & Szalay A.A. (1987) *Proc. Natl. Acad. Sci. USA* **84**, 131–135.

Krika L.J. & Thorpe G.H.G. (1987) *Methods Enzymol.* **133**, 404–420.

Legocki R.P., Legocki M., Baldwin T.O. & Szalay A.A. (1986) *Proc. Natl. Acad. Sci. USA* **83**, 9080–9084.

Lindbladh C., Mosbach K. & Bulow L. (1991) *Immunol. Meth.* **137**, 199–207.

Liu J., Prat S., Willmitzer L. & Frommer W.B. (1990) *Gen. Genet.* **1990**, 401–406.

Lorenz W.W., McCann R.O., Longiaru M. & Cormier M.J. (1991) *Prox. Natl. Acad. Sci USA* **88**, 4438–4442.

Mancini J.A., Boylan M., Soly R.R., Graham A.F. & Meighen E.A. (1988) *J. Biol. Chem.* **263**, 14308–14312.

Matthews J.A., Batki A., Hynds C.C. & Kricka L.J. (1985) *Anal. Biochem.* **151**, 205–209.

Maxwell O.J. & Maxwell F. (1988) *DNA* **7**, 557–562.

Meighen E.A. (1991) *Microbiol. Rev.* **55**, 123–142.

Munro S. & Pelham H.R.B. (1987) *Cell* **48**, 899–907.

Murray K., England P.J., Lynham J.A., Mills D., Schmitz-Peiffer C. & Reeves M.L. (1990) *Biochemistry* **267**, 703–708.

Nakajima-Shimada J.N., Iida H., Tsuji F.I. & Anraku Y. (1991a) *Proc. Natl. Acad. Sci. USA* **88**, 6878–6882.

Nakajima-Shimada J.N., Iida H., Tsuji F.I. & Anraku Y. (1991b) *Biochem. Biophys. Res. Commun.* **174**, 115–122.

Nakamura T. (1972) *Biochemistry* **72**, 173–177.

Nelson N.C., Hammond P.W., Wiese W.A. & Arnold L.J.

(1990) In *Luminescence Immunoassay and Molecular Applications*, K. van Dyke & R. van Dyke (eds). CRC Press, Boston, pp. 293–312.

Nguyen V.T., Morange M. & Bensaude O. (1988) *Anal. Biochem.* **171**, 404–408.

Nielsen L.L. & Pedersen R.A. (1991) *Exp. Zool.* **257**, 128–133.

Nolan G.P., Fiering S., Nicholas J.F. & Herzenberg L.A. (1988) *Proc. Natl. Acad. Sci. USA* **85**, 2603–2607.

Olsson O., Koncz C. & Szalay A.A. (1988) *Molec. Gen. Genet.* **215**, 1–9.

Olsson O., Escher A., Sandberg G., Schell J., Koncz C. & Szalay A.A. (1989) *Gene* **81**, 335–347.

Ow D.W., Wood K.V., De Luca M., De Wet J.R., Helsinki D.R. & Howell S.H. (1986) *Science* **234**, 856–859.

Ow D.W., Jacobs J.D. & Howell S.H. (1987) *Proc. Natl. Acad. Sci USA* **84**, 4870–4874.

Patel A. & Campbell A.K. (1987) *Immunology* **60**, 135–146.

Poenie M. & Tsien R.Y. (1986) *Trends Biochem. Sci.* **11**, 340–355.

Pons M., Gagne D., Nicolas J.C. & Methali M. (1990) *Biotechniques* **9**, 450–459.

Prasher D., McCann R.O. & Cormier M.J. (1985) *Biochem. Biophys. Res. Commun.* **126**, 1259–1268.

Prasher D.C., McCann R.O., Longiaru M. & Cormier M.J. (1987) *Biochemistry* **26**, 1826–1332.

Quickenden T.I. & Que Hee S.S. (1976) *Photochem. Photobiol.* **23**, 201–204.

Quickenden T.I., Comarmond M.J. & Tilbury R.N. (1985) *Photochem. Photobiol.* **41**, 611–615.

Rattray E.A.S., Prosser J.I., Killham K. & Glover L.A. (1990) *Appl. Environ. Microbiol.* **56**, 3368–3374.

Reynolds G.T. (1972) *Q. Rev. Biophys.* **5**, 295–347.

Reynolds G.T. (1978) *Photochem. Photobiol.* **27**, 405–421.

Rhoads R.E. (1985) *Prog. Molec. Subcell. Biol.* **9**, 104–155.

Ridgway E.B. & Ashley C.C. (1967) *Biochem. Biophys. Res. Commun.* **29**, 229–234.

Roberts P.A., Knight J. & Campbell A.K. (1987) *Anal. Biochem.* **160**, 139–148.

Rodriguez J.F., Rodriguez D., Rodriguez J.R., Mcgowan E.B. & Esteban M. (1988) *Proc. Natl. Acad. Sci. USA* **85**, 1667–1671.

Rodriguez D., Rodriguez J.R., Rodrigues J.F., Trauber D. & Esteban M. (1989) *Proc. Natl. Acad. Sci. USA* **86**, 1287–1291.

Rogowsky P.M., Close T.J., Chimera J.A., Shaw J.J. & Kado C.I. (1987) *J. Bacteriol.* **169**, 5101–5112.

Rose B. & Lowenstein W.R. (1976) *J. Memb. Biol.* **28**, 87–119.

Sala-Newby G. & Campbell A.K. (1991) *Biochem. J.* **279**, 727–732.

Sala-Newby B. & Campbell A.K. (1992) *FEBS Lett.* **307**, 241–244.

Sala-Newby G., Kalsheker N. & Campbell A.K. (1990a) *Biochem. Soc. Trans.* **18**, 459–460.

Sala-Newby G., Kalsheker N. & Campbell A.K. (1990b) *Biochem. Biophys. Res. Commun.* **172**, 477–482.

Sambrook J., Frisch, E.F. & Maniatis T. (1989) in *Molecular Cloning*, Vols 1–3. Cold Spring Harbor Laboratory Press, New York, p. 1.

Schaap, A.P. Akhavan H. & Romano L.J. (1989) *Clin. Chem.* **35**, 1863–1864.

Schwartz O., Virelizier J.L., Montagnier L. & Hazan U. (1990) *Gene* **88**, 197–205.

Scolding N.J., Morgan B.P., Houston W.A.J., Linington C., Campbell A.K. & Compston D.A.S. (1989) *Nature* **339**, 690–692.

Selden R.F., Burke Howie K., Rowe M.E., Goodman H.M. & Moore D.D. (1986) *Molec. Cell. Biol.* **6**, 3173–3179.

Sevigny P. & Gossard F. (1990) *Gene* **93**, 143–146.

Shaw J.J. & Kado C.I. (1986) *Biotechnology* **4**, 560–564.

Shaw J.J., Settles L.G. & Kado C.I. (1987) *Molec. Plant–Microbe Interact.* **1**, 39–45.

Shimomura O. (1985) *Soc. Exp. Biol. Symp.* **38**, 351–372.

Shimomura O (1986) *Biochem. J.* **234**, 271–277.

Shimomura O. & Shimomura A. (1985) *Biochem. J.* **228**, 745–749.

Shimomura O., Johnson F.H. & Saiga Y. (1962) *J. Cell Comp. Physiol.* **59**, 223–239.

Shimomura O., Musick B. & Kishi Y. (1988) *Biochem. J.* **251**, 405–410.

Shimomura O., Musicki B. & Kishi Y. (1989) *Biochem. J.* **261**, 913–920.

Shimomura O., Inouye S., Musick B. & Kishi Y. (1990) *Biochem. J.* **270**, 309–312.

Simoes E.A.F. & Sarnow P. (1991) *Virology* **65**, 913–921.

Smith D.F. Stults N.L., Rivera H., Gehle W.D., Cummings R.D. & Cormier M.J. (1991) In *Bioluminescence and Chemiluminescence Current Status*, P.E. Stanley & L.S. Kricka (eds). John Wiley & Sons, Chichester, pp. 529–532.

Stults N.L., Stocks N.A., Cummings R.D., Cormier M.J. & Smith D.F. (1991) In *Bioluminescence and Chemiluminescence Current Status*, P.E. Stanley & L.S. Kicka (eds). John Wiley & Sons, Chichester, pp. 533–536.

Swartzman E., Kapoor S., Graham A.F. & Meighan E.A. (1990) *Bacteriology* **172**, 6797–6802.

Tanahashi H., Ito T., Inouye S., Tsuji F.I. & Sakaki Y. (1990) *Gene* **96**, 249–255.

Tatsumi H., Masuda T., Kajiyama N. & Nakano E. (1989) *J. Biolumin. Chemilum.* **3**, 75–78.

Thompson E.M., Nagata S. & Tsuji F.I. (1989) *Proc. Natl. Acad. Sci. USA* **86**, 6567–6571.

Thompson E.M., Nagata S. & Tsuji F.I. (1990) *Gene* **96**, 257–262.

Treisman R., Green M.R. & Maniatis T. (1983) *Proc. Natl. Acad. Sci. USA* **80**, 7428–7432.

Tsien R.Y., Pozzan T. & Rink T.J. (1982) *J. Cell Biol.* **94**, 325–334.

Tsuji F.O., Inouye S., Goto J. & Sakaki Y. (1986) *Proc. Natl. Acad. Sci. USA* **83** 8107–8111.

Waterman M.L., Adler S., Nelson C., Greene G.L., Evans R.M. & Rosenfeld M.G. (1988) *Molec. Endocrinol.* **2** 14–21.

Weeks I., Beheshti I., McCapra F., Campbell A.K. & Woodhead J.S. (1983a) *Clin. Chem.* **29**, 1474–1479.

Weeks I., Campbell A.K. & Woodhead J.S. (1983b) *Clin. Chem.* **29** 1480–1483.

Weeks I., Sturgess M., Brown R.C. & Woodhead J.S. (1987) *Methods Enzymol.* **133**, 366–387.

Widder E., Latz M.F. & Case J.F. (1983) *Biol. Bull.* **165**, 791–810.

Widder E., Latz M.F. & Herring P.J. (1984) *Science* **225**, 512–514.

Wiedemann E. (1888) *Ann. d. Phsik u. Chemie* **34**, 446–463.

Williams D. & Chance B. (1983) *J. Biol. Chem.* **258**, 3628–3631.

Williams T.M., Bureom J.E., Ogden S., Kricka L.J. & Kant J.A. (1989) *Anal. Biochem.* **176**, 28–32.

Wood K.V. (1991) *J. Biolumin. Chemilum.* **5**, 107–114.

Wood K.V., Lam Y.A. & McElroy W.D. (1989a) *J. Biolumin. Chemilum.* **5**, 31–39.

Wood K.V., Lam Y.A. & McElroy W.D. (1989b) *J. Biolumin. Chemilum.* **5**, 289–301.

Wood K.V., Lam Y.A., Seliger H.H. & McElroy W.D. (1989c) *Science* **244**, 700–702.

Woodhead J.S. & Weeks I. (1991) *Clin. Chem.* **37**, 472.

Woods N.M., Cuthbertson K.S.R. & Cobbold P.H. (1986) *Nature* **319**, 600–602.

Xi L., Cho K.W. & Tu, S.C. (1991) *Bacteriology* **173**, 1399–1405.

Zatta P.F., Nyame K., Cormier M.J., Mattox S.A., Prieto P.A., Smith D.F. & Cummings R.D. (1991) *Anal. Biochem.* **194**, 185–191.

Zenno S. & Inouye S. (1990) *Biochem. Biophys. Res. Commun.* **171**, 169–174.

CHAPTER SIX

Acridine Orange as a Probe for Molecular and Cell Biology*

ALEXANDER V. ZELENIN

Engelhardt Institute of Molecular Biology, Russian Academy of Sciences, Moscow, Russia

6.1 INTRODUCTION

It would be almost impossible not only to name all the stains and dyes used in biology and medicine but even to estimate roughly their number. Many volumes have been written on the subject and almost every day publications appear in which new stains are described or new applications of the known ones are given. Special books and long review papers deal with the history of biological staining (Clark & Kasten, 1983; Kasten, 1989).

Among the thousands of stains and dyes of various different chemical natures, types and complexity one can easily name those that can be regarded as champions from the point of view of their popularity. Among conventional histological stains the nuclear dye haematoxylin, especially in combination with eosin, should be, without any doubt, named as such a champion.

When a closer and more specific field of biological investigation is looked at, such as the use of fluorescent probes, the basic fluorescent dye acridine orange (AO) emerges among the most popular ones.

* To the memory of my teacher Professor Maxim N. Meissel, the founder of fluorescence microscopy in Russia.

The use of AO in different areas of cytochemistry and cell and molecular biology has been reviewed many times (Bertalanffy, 1963; Kasten, 1967; Zelenin, 1967, 1971; Meissel & Zelenin, 1973). The latest reviews (Darzynkiewicz, 1990a; Darzynkiewicz & Kapuscinski, 1990) have, however, mostly dealt with the use of the stain in flow cytometry. To avoid large overlaps and to make this paper complementary to the papers mentioned above, I have tried to cover different fields of AO application and to pay more attention to some other uses besides flow cytometry. Particular attention is paid to the most important works published in Russian and, therefore, scarcely available to Western readers.

6.2 HISTORICAL REMARKS

Acridine orange (AO) was synthesized in 1889 (for review see Albert, 1966; Kasten 1967; Zelenin, 1967, 1971; Darzynkiewicz & Kapuscinski, 1990) but its first biological applications were reported only in 1940 by Bukatsch and Haitinger (1940) and Strugger (1940a,b). Strugger carried out a fundamental investigation on the interaction of AO with plant cells and found that living and dead cells differed in their stainability by

AO. Among those who made substantial contributions to the introduction of AO into biology, M. Meissel, J. Armstrong, L. von Bertalanffy and N. Schummelfeder, in the 1950s, and R. Rigler and Z. Darzynkiewicz, at a later period, should be especially mentioned.

The rapid and successful development of fluorescence microscopy in Russia in general was closely related to the creation of appropriate fluorescence instruments invented and introduced by Dr Eugeny M. Brumberg in Leningrad. In 1948 he suggested the use of a semireflective dichroic mirror in fluorescence micro-scopy (Brumberg & Girshgorin, 1947) and later developed a special opaque-illuminator which allowed the easy conversion of a conventional light microscope into a fluorescence one (Brumberg & Krilova, 1953). In the mid-1950s Brumberg's vertical fluorescence opaque-illuminators with dichroic mirrors began to be produced and sold by the Leningrad Optical Factory (LOMO). Quite soon after that a special fluorescence microscope with easily changeable transmitted and opaque illumination was constructed and, in 1959, LOMO began to offer such microscopes (ML-1) for sale.

The work of Maxim Meissel and his co-workers, carried out in Moscow and published mostly in Russian, requires particular attention. Together with his colleague Dr V. Korchagin Meissel undertook the first investigation of AO binding to nucleic acids *in vitro*. It was shown (Meissel & Korchagin, 1952) that addition of AO to purified DNA and RNA caused their green and red fluorescence, respectively. These works stimulated intensive work on fluorescence microscopy in Russia. AO fluorescence microscopy began to be used on a large scale for the investigation of cancer cells (Meissel & Gutkina, 1953) and X-ray damaged cells (Meissel & Sondak, 1956) (for review see Meissel & Zelenin, 1973). Later, AO was success-fully applied to the investigation of nucleic acid secondary structure *in vitro* (Borisova & Tumerman, 1964, 1965).

6.3   AO AS A FLUORESCENT DYE

Acridine orange (AO), 3,6-dimethylaminoacridine (Fig. 6.1), is a weak base which is readily soluble in water. It is usually prepared and sold as a zinc complex of a hydrochloride salt.

The dye has been shown to exist in aqueous solution in two main forms (Zanker, 1952a,b; Morosov, 1963a,b; Borisova & Tumerman, 1964). The first form is characterized by (1) an absorption maximum at 494 nm, (2) a green fluorescence with a maximum at 530 nm, and (3) a fluorescence lifetime of 2×10^{-9} s.

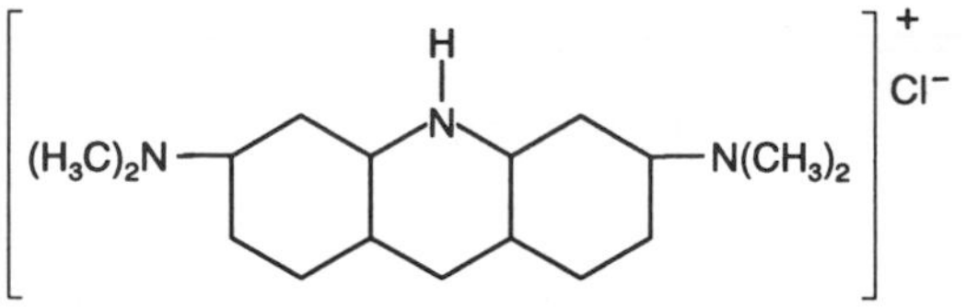

Figure 6.1 The structure of acridine orange.

This form is characteristic of AO in highly diluted solutions (10^{-7} M) in which the stain molecules are presented as single monomers.

The second form is characterized by (1) an absorption maximum at 465 nm, (2) a red fluorescence with a maximum at 640 nm, and (3) by a fluorescence lifetime of 20×10^{-9} s. This form is characteristic of AO in concentrated solutions ($>10^{-2}$ M) in which the stain molecules are presented as dimers.

AO solution of intermediate molarity has mixed spectral properties.

In parallel to the fluorescence, AO in both monomeric and dimeric form has phosphorescence which is rather strong at low temperatures (Zanker et al., 1959; Bradley, 1961; Morosov, 1963a,b; Morosov & Savenko, 1977). This is true for red AO emission both in solutions and in complexes with single-stranded nucleic acids (for details see Darzynkiewicz & Kapuscinski, 1990). Thus, strictly speaking, general AO emission should be called luminescence but not fluorescence. Nevertheless, in most current publications on AO, obviously for the sake of convenience, the term fluorescence is used. We also prefer to use the latter term for all types of AO emission. It must also be taken into consideration that the equipment usually used for the investigation of AO properties, such as fluorometers (Borisov & Tumerman, 1959) or flow cytometers, register only very short signals (shorter than 10^{-6} s).

6.4   SPECTRAL PROPERTIES OF AO IN COMPLEXES WITH NUCLEIC ACIDS AND OTHER BIOPOLYMERS

The first work on the interaction of AO with nucleic acids *in vitro* was performed by Meissel and Korchagin (1952). In this investigation the data on the fluorescence properties of AO complexes with DNA and RNA were obtained, however, only by means of visual fluore-scence microscopy. Later, spectral investigations gave detailed characteristics of these complexes.

Two types of AO complexes with nucleic acids have been described. The A-type complex (Steiner & Beers, 1958, 1959, 1961) arises on interaction of AO with double-stranded DNA or double-stranded regions of RNA with a ratio of the number of nucleotide molecules to number of AO molecules of 4:1 or less. This type of complex is characterized (1) by an absorption maximum at 502–504 nm, (2) by green fluorescence with a maximum at 530 nm, and (3) by a fluorescence yield and lifetime 2–2.5 times higher than that of the AO in a diluted solution when the dye is in the monomeric form.

The AO properties in the A-type complex are similar to those of AO in highly diluted solutions, with the only differences being in the shift of absorption maximum from 594 to 504 nm and in a minor increase in fluorescence lifetime and yield. This means that in the A-type complex AO molecules are located on nucleic acid molecules at such a distance that no dimeric complexes can be formed.

It was suggested about 30 years ago (Lerman, 1961, 1963, 1964) that a complex of this type is formed as a result of the intercalation of an amino acridine molecule between the nucleic acid basepairs (Fig. 6.2). This model was confirmed later by X-ray analysis (Wang, 1974; Berman & Young, 1981; Waring, 1981). For a more recent review see Darzynkiewicz and Kapuscinski (1990).

The B-type complex (Steiner & Beers, 1958, 1959, 1961) differs considerably from the A-type. It is characterized by the following properties: (1) an

absorption maximum at 475 nm, (2) a red fluorescence with maximum at 640 nm, and (3) a certain decrease in the fluorescence yield and sharp (ten-fold) increase in the fluorescence lifetime of the AO molecule (Borisova et al., 1963; Borisova & Tumerman, 1964). A detailed investigation of crystalline acridine complexes with the single-stranded regions of nucleic acids was performed by Neidle et al. (1978).

The properties of this complex are similar to those of AO in dimer form. The B-type complex is formed when AO is combined with single-stranded regions of RNA and denatured DNA with the creation of AO dimers.

AO was shown to form salt-like complexes with acid mucopolysaccharides with red fluorescence (Kuyper, 1957; Appel & Zanker, 1958; Saunders, 1964). At high pH (>7) when the protein carbohydrates are protonized, AO binds to proteins with the formation of the red fluorescing complexes.

6.5 AO IN THE STUDY OF NUCLEIC ACIDS *IN VITRO*

The key work in this area was performed by Borisova and Tumerman (Borisova et al., 1963; Borisova & Tumerman, 1964, 1965). It has been established that investigation of AO properties permits the quantification of the strandedness of nucleic acid in solution. It has been shown (Borisova et al., 1963) that tRNA in solution contains no more than 28–30% single-stranded regions. These data coincide well with the results of direct X-ray analysis of tRNA crystals (Kim et al., 1974).

Later, the properties of pre-mRNA in nuclear particles and in free state in solution were studied. It was found that pre-mRNA in nuclear particles was almost completely single-stranded, whereas in solution about 70% of the same mRNA was double-stranded (Borisova et al., 1981). Similar characteristics (70% double-strandedness) were found for the MS2-phage mRNA in solution (Borisova & Grechko, 1984).

The AO method was also successfully applied to the investigation of parallel double-stranded helices in nucleic acids (Borisova et al., 1991).

6.6 AO IN NUCLEIC ACID CYTOCHEMISTRY

6.6.1 AO as a direct probe for nucleic acids

When used under appropriate conditions (fixation, pH 4.2–4.6, AO concentration 5.10 M, sufficient washing-off of the unbound stain excess), AO acts as an

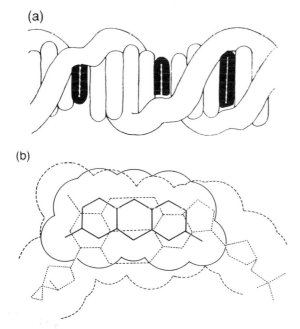

(a)

(b)

Figure 6.2 The structure of DNA-acridine complexes. (a) Acridine molecules intercalated into DNA. (b) A molecule of acridine superposed over a nucleotide pair with the deoxyribose phosphate chain in an extended configuration. (From Lerman, 1964.)

excellent cytochemical probe for nucleic acids. It gives bright-green fluorescence to DNA (nucleus) and deep red fluorescence to RNA (cytoplasm and nucleoli). An important condition for obtaining such a picture is proper fixation. The fixative should preserve nucleic acids and should not contain any denaturing DNA components (formaldehyde, acids, etc.). The AO concentration must be high enough for dimer formation on single-stranded regions of nucleic acids.

Over 30 years' experience allows me to say with complete confidence that the AO staining technique is the best cytochemical approach to nucleic acid investigation in routine work. In our laboratory it is broadly used in all experiments in which quick and reliable screening of the cells for the presence of nucleic acids is required.

The cytochemical interpretation of the fluorescence microscopical picture obtained with AO is based on experiments with cell pretreatments with nucleases (Armstrong, 1956; Bertalanffy & Bickis, 1956; Schummelfeder et al., 1957) as well as on the above-described results on AO binding to nuclei acids in vitro.

Nevertheless, some particular problems of the cytochemical specificity of the obtained pictures should be discussed.

No doubt usually arises about the nature of green fluorescence induced in the cell structures by the AO treatment. A strong A-type complex is specific for AO binding to double-stranded nucleic acids. It should be noted, however, that the sensitivity of the method, at least under visual observation, is not sufficient to detect the green fluorescence of double-stranded mitochondrial DNA in the cytoplasm. At the same time, double-stranded RNAs of some viruses (reovirus) after AO staining acquire green fluorescence which is easily detected microscopically (Gomatos et al., 1962; Gomatos & Tamm, 1963).

In most cases the red fluorescence in AO-treated preparations indicates the presence of RNA. However, red fluorescence is also characteristic of single-stranded DNA of certain phages (Mayor & Hill, 1961) and viruses (Jamison & Mayor, 1965) and is also easily detected cytochemically.

It must also be taken into consideration that AO dimers responsible for the red fluorescence can be formed when AO is bound to any polyanions. Among those, acid mucopolysaccharides should be mentioned first (for details see Section 6.12). At high pH values (>7) AO may form red fluorescing complexes with proteins.

These facts should be always kept in mind when any new type of cell is taken for investigation.

Although RNA has been shown to contain large double-stranded regions in vitro, the whole cytoplasm expresses only red fluorescence, thus indicating that its RNA is completely in the single-stranded form, most likely due to its denaturation under the action of the fixative and AO in the course of staining.

When nucleic acids are investigated visually after AO treatment, this RNA denaturation is usually quite sufficient. For quantitative investigation, additional RNA denaturation is desirable.

The possibility of nucleic acid quantitation with AO requires special discussion. Proteins in nucleoprotein complexes restrict ligand binding to nucleic acids; this is particularly true for the nuclear chromatin. DNA in chromatin is only partially accessible to AO; this accessibility is highly dependent on the DNA–histone interaction and reflects the chromatin functional state. These facts open up the possibility of AO use for the investigation of chromatin functional state (for details see Section 6.9). At the same time they show that for nucleic acid quantitation the proteins restricting AO binding to nucleic acids should be removed or at least their action should be minimized. For that purpose a special technique has been suggested (Darzynkiewicz et al., 1975; Darzynkiewicz, 1990a). The method consists of permeabilization of cells with a detergent in the presence of acid and subsequent staining with AO. Acid treatment results in dissociation and extraction of histones which makes subsequent DNA assay rather quantitative. At the same time AO staining under the suggested conditions causes a complete denaturation of double-stranded RNA regions and the whole RNA thus acquires the red fluorescence. The method has been used successfully for simultaneous DNA/RNA estimation by flow cytometry in different cell systems (Darzynkiewicz et al., 1979a; Darzynkiewicz 1990a). It has been shown that rapid increase in red fluorescence intensity can be detected in quiescent (non-dividing) cells stimulated to proliferate.

6.6.2   Quenching of the fluorescence of AO bound to DNA

6.6.2.1   Actinomycin D as an energy acceptor

When AO is bound to DNA in vitro simultaneously with an energy acceptor, its fluorescence is quenched, partially or completely, owing to the energy transfer. Different substances may be, and are, used for that purpose.

Borisova et al. (1968) showed in experiments in vitro that the antibiotic actinomycin D quenched the fluorescence of AO bound to pure DNA and DNA in chromatin. A method for the estimation of the length of chromatin DNA regions not accessible to ligands was proposed. Approximately half of the DNA in native chromatin was shown to be covered with proteins, and thus not accessible to AO, whereas the other half was easily ligand-accessible and therefore protein-free. The lengths of DNA regions of both types were roughly estimated to be more than 100 basepairs. These early findings coincide closely with the modern concepts of chromatin subdivision into nucleosome and linker parts.

Actinomycin D quenching of the AO fluorescence was later applied to cytochemical investigation of nucleic acids. It was found (Zelenin *et al.*, 1976) that simultaneous treatment of fixed cytochemical preparations with AO and actinomycin D resulted in suppression of fluorescence of AO bound to the DNA-containing cell structures (nucleoplasm). At the same time fluorescence of RNA-containing structures (cytoplasm and nucleoli) remained unchanged. These observations are consistent with selective actinomycin D binding to the DNA guanines.

One possible application of this approach is the distinguishing between double-stranded DNA and RNA. As has been mentioned above, the double-stranded RNA of certain viruses forms green fluorescing complexes with AO. Cytochemical identification of such RNA by conventional methods is rather difficult because it is usually RNAse-resistant (Gomatos & Tamm, 1963). But the fluorescence of such complexes cannot be quenched by actinomycin D. The actinomycin B specificity for guanines also opens up a possibility for application of this approach to the cytochemical detection of long AT-rich regions.

6.6.2.2   *5-Bromodeoxyuridine as an energy acceptor*

5-Bromodeoxyuridine (BrdUrd), incorporated into cellular DNA as a thymidine analogue, quenches fluorescence of various fluorochromes, including AO, applied to these cells. BrdUrd substitution into DNA reduces both fluorescence intensity (Latt, 1976) and phosphorescence of AO (Galley & Purkey, 1972). Labelling of cells with BrdUrd followed by AO staining may thus be used for the investigation of cell cycle. According to Darzynkiewicz *et al.* (1983) all BrdUrd-treated cells, regardless of the cell cycle phase, have shown a decreased fluorescence. The BrdUrd-attributed suppression of the cell fluorescence has been found to be large enough to separate the totally BrdUrd-labelled from unlabelled mitotic cell populations. It has been concluded that the BrdUrd–AO flow cytometry method provides an alternative to the autoradiography technique for obtaining a fraction of labelled mitosis. Elsewhere, BrdUrd–AO flow cytometry was successfully used for discrimination between cycling and non-cycling cells (Darzynkiewicz *et al.*, 1978). In more recent investigations, AO has been mostly replaced, however, by different, more DNA-specific stains (ethidium bromide, bis-benzimides, mithromycin, etc.); for reviews see Bohmer (1990), Crissman and Steinkamp (1990), Poot *et al.* (1990).

6.7   AO STAINING AFTER ACID PRETREATMENTS

Acid treatment, even if mild, brings about at least partial apurinization of DNA and thus changes its stainability by AO. As a result, AO forms dimers at depurinized regions, thus causing the DNA red fluorescence. This fact makes it necessary to avoid the presence of acids in fixatives used for conventional AO fluorescence microscopy of nucleic acids. At the same time it opens up important possibilities for the investigation of the DNA state *in situ*.

It has been shown in several investigations (Schummelfeder *et al.*, 1957; Nash & Plaut, 1964; Roschlau, 1965) that, when preparations containing double-stranded DNA are exposed to a short acid hydrolysis, subsequent staining with AO results in the red or yellow-red fluorescence of DNA-containing structures instead of the standard green one.

These observations have been expanded in two directions:

Darzynkiewicz *et al.*, (1975, 1979b) suggested a flow cytometry technique for measurement of the sensitivity of DNA *in situ* to acid denaturation. The ratio of the red to green fluorescence was taken as a measure of DNA denaturation. Some critical aspects of the procedure (proper fixation, staining at the equilibrium conditions, etc.) should be fulfilled (for details see Darzynkiewicz, 1990b). The method was successfully applied to the investigation of exponentially growing cells, to the investigation of quiescent versus cycling ones, and non-differentiated versus differentiated cells (Darzynkiewicz, 1990b). One of the interesting applications of the method is based on the higher accessibility of mitotic chromatin to the acid treatment compared to that in the interphase nuclei (Darzynkiewicz *et al.*, 1975).

Zelenin *et al.* (1977) used AO staining after acid pretreatment for the investigation of chromatin in *Chironomus* giant chromosomes in cytological preparations. After very mild hydrolysis, as well as in control preparations, all chromosome regions showed green fluorescence. After longer hydrolysis all chromosome bands fluoresced red. Intermediate hydrolysis gave red fluorescence in all transcriptionally inactive bands, including the centomeric ones, whereas transcriptionally active puffing regions were green. The most likely explanation of these observations is that transcriptionally active chromosome regions are less susceptible to acid due to the protective action of the non-histone proteins present there. The method has some advantages over that currently used for the study of chromatin properties (see Section 6.9), as it does not require cytofluorometry, and can be applied to the investigation of cells and their parts in conventional cytological preparations.

Some additional aspects of the application of acid–AO techniques for the investigation of the chromatin functional state are discussed in Section 6.9.

6.8   AO IN THE STUDY OF DNA THERMAL DENATURATION

Experiments *in vitro* have demonstrated applicability of AO spectrofluorometry for distinguishing between single- and double-stranded nucleic acids and, in particular, for the study of the process of thermal DNA denaturation–renaturation (Borisova & Tumerman, 1964). Later, a detailed technique for the investigation of DNA melting *in situ*, i.e. inside undestroyed cells, was developed (Rigler & Killander, 1969; Rigler *et al.*, 1969). The technique consists of the preparations heating and cooling in the presence of formaldehyde, to prevent DNA renaturation, and subsequent estimation of the degree of DNA denaturation based on measuring the ratio of red/green fluorescence by the AO staining (coefficient α). The cell treatment in the presence of formaldehyde is the crucial point of this method. DNA *in situ* is in such a compact form that in the absence of formaldehyde, denatured DNA undergoes immediate and complete renaturation.

Obtained curves of DNA melting *in situ* proved to be similar to those of DNA melting *in vitro*. It was found that DNA accessibility to the action of heat varied in cells with different proliferative activity. This fact was suggested to be due to the protective action of histones in chromatin. The biological results received by this technique are discussed in Section 6.9.

Later, some improvements of the method were suggested, including changing the AO staining conditions (Liedeman & Bolund, 1976b), and the development of a technique for investigation of DNA thermal stability by flow cytometry (Darzynkiewicz *et al.*, 1974, 1975). The latter method, which includes use of various apparatuses as well as different techniques for cell specimen preparation, quickly became broadly used for the study of large cell populations, first of all the peripheral blood cells and spermatozoa. Application of these approaches for the investigation of chromatin functional state is discussed in Section 6.9.

Use of the DNA denaturation–renaturation approach for chromosome banding is described in Section 6.11.

6.9   AO IN THE STUDY OF THE CHROMATIN FUNCTIONAL STATE

6.9.1   Introduction

The chromatin functional state, i.e. the ability of its DNA to serve as a primer in RNA synthesis, is usually

Table 6.1 Main properties of chromatin *in situ*.

- Chromatin in preparations *in situ* is still able to act as a primer in RNA and DNA polymerase reactions catalysed by exogenous polymerases.
- Chromatin *in situ* contains endogenous RNA and DNA polymerases in amounts sufficient to catalyse RNA and DNA synthesis.
- The properties of chromatin *in situ* can be modified in a similar way to those of chromatin *in vitro*.
- Changes in physico-chemical properties of chromatin *in situ* reflect the chromatin changes *in vitro* and *in vivo*.

studied by biochemical (*in vivo*) and molecular biological (*in vitro*) techniques. However, the chromatin *in situ*, i.e. inside the cells in cytological smears, suspension and, sometimes, in histological preparations, retains most characteristics of the chromatin *in vivo*, at least no less than the chromatin *in vitro* (Table 6.1).

A large number of different techniques have been suggested for the investigation of chromatin properties *in situ*. In my review on the subject published 10 years ago more than 20 such techniques were listed (Zelenin, 1982). During the last decade this number has increased at least twice. These techniques are based on different approaches, most of which fall outside the scope of this paper and are discussed in detail elsewhere (Ringertz, 1969; Zelenin, 1977, 1982; Darzynkiewicz, 1990c). Here I will confine myself to the investigation of the chromatin functional state using AO.

Such investigation is based on two different approaches: (1) estimation of direct AO binding to cellular chromatin, and (2) study of the spectral properties of cells stained with AO after different pretreatments.

6.9.2   Direct AO binding to the cellular chromatin

The estimation of AO binding to DNA inside chromatin (Rigler, 1966) was the first and, for a definite period, the most popular approach to chromatin study *in situ* (for reviews see Rigler, 1969; Ringertz, 1969; Auer, 1972; Ringertz & Bolund, 1974; Zelenin, 1977).

The method is based on the assumption that in chromatin AO binds only to DNA regions not covered with proteins. Increase in the cell genome activity accompanied by weakening of the DNA–histone interaction causes an increase in the number of AO molecules attached to DNA and can be detected cytofluorometrically. This fact was first revealed in experiments with peripheral blood lymphocytes exposed to phytohaemagglutinin (Killander & Rigler, 1965, 1969; Rigler, 1966; Rigler & Killander, 1969)

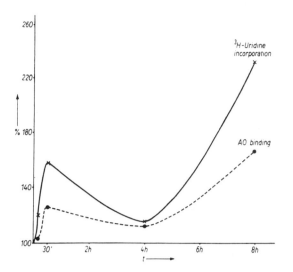

Figure 6.3 AO binding by chromatin and labelled uridine incorporation in RNA of quiescent Chinese hamster cells after growth stimulation. Growth stimulation was achieved in the stationary culture by replacement of nutritive medium. Abscissa: time after stimulation (min). Ordinate: AO fluorescence intensity and radioactive label incorporation (% to the control). A good correlation between both curves is seen. (From Terskikh *et al.*, 1976.)

and then extended to a number of different quiescent cells stimulated to proliferate by different agents: the peripheral blood and lymph node lymphocytes in 'crowded' cultures (Auer *et al.*, 1970; Zelenin & Vinogradova, 1973), hepatocytes of regenerating liver (Kolesnikov *et al.*, 1973; Alvares, 1974; Kushch *et al.*, 1974, 1978), kidney cells induced to grow in culture (Auer *et al.*, 1973), quiescent cultured cells after

addition of serum to the culture medium (Terskikh *et al.*, 1976), hen erythrocytes fused with dividing culture cells (Bolund *et al.*, 1969), etc.

It was shown that removal of histone H1 from the chromatin *in situ* resulted in a substantial increase in AO binding to chromatin (Zelenin & Vinogradova, 1973), whereas cell treatment with 0.3–0.35 M sodium chloride, which presumably removed non-histone proteins, caused a decrease in the AO binding to chromatin of activated cells (Kushch *et al.*, 1974, 1980) and had no influence on AO binding to non-activated cells. In both cases the difference in AO binding to activated and quiescent cells disappeared.

In most of these experiments the results obtained by AO binding were directly or indirectly confirmed by some alternative techniques such as binding of other ligands (actinomycin D, ethidium, etc.), investigation of DNA accessibility to the action of damaging factors (heat, acids) and histone stainability.

A good correlation in many cases between AO cytofluorometry data and the results of *in vivo* DNA template activity studies should be mentioned particularly (Fig. 6.3); for reviews see Zelenin (1977, 1982), Zelenin and Kushch (1985). In our experiments with the liver cells activated by partial hepatectomy (Kolesnikov *et al.*, 1973; Kushch *et al.*, 1974; Sondore *et al.*, 1978), almost complete coincidence between the results *in situ* and *in vitro* (on isolated chromatin) was found (Fig. 6.4).

A concept of chromatin activation was drawn on the basis of these observations (Killander & Rigler, 1965; Rigler, 1966). It was shown later that at least some chromatin properties varied at different stages of this process (Sondore *et al.*, 1978) and three steps of chromatin activation in dormant cells stimulated

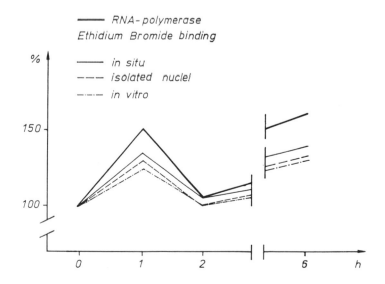

Figure 6.4 Changes in rat liver chromatin properties after partial hepatectomy. Abscissa: time after operation (h). Ordinate: fluorescence intensity and radioactive label incorporation (% to the control). The figure is based on the data of the work of Sondore *et al.* (1978).

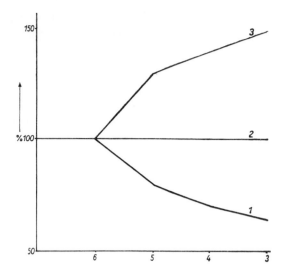

Figure 6.5 Changes in the cytochemical properties of the lymph node lymphocytes during their maturation. Abscissa: cell diameter in arbitrary units. Ordinate: (1) mean values of AO binding; (2) Feulgen DNA staining; (3) histone stainability with an acid dye (primulin). Decrease in the cell diameter is taken as a measure of lymphocyte maturation. Cytochemical parameters for the lymphocytes with the maximal diameter were taken as 100%. (From Manteifel *et al.*, 1973.)

to proliferate were described (Zelenin & Kushch, 1985).

In parallel experiments a marked decrease of AO binding to chromatin in the course of genome inactivation has been described on four cell systems: spermatogenesis (Gledhill *et al.*, 1966), red blood cell (Kernell *et al.*, 1971) and lymphocyte maturation (Fig. 6.5, Manteifel *et al.*, 1973; Zelenin, 1977; Kushch *et al.*, 1980) and contact cell inhibition in culture cells (Zetterberg & Auer, 1970).

At the same time, soon after the creation of Rigler's method there began to appear in the literature evidence indicating that the results obtained by the original Rigler's technique may depend not only on the chromatin functional state but also on the local cell density in a preparation (Bolund *et al.*, 1970; Ringertz & Bolund, 1970; Liedeman *et al.*, 1975).

Attempts to improve Rigler's technique were reported (Kolesnikov *et al.*, 1973; Smets, 1973; Liedeman & Bolund, 1976a,b). Thus in all investigations carried out in our laboratory since the beginning of the 1970s (Zelenin *et al.*, 1974; for review see Zelenin, 1977, 1982; Zelenin & Kushch, 1985) a modification of the original Rigler's technique has been used. In this modification (Kolesnikov *et al.*, 1973) the acetylation step is omitted and after staining in a high concentration of AO (10^{-4} M) the preparations are washed in a low concentration (10^{-6} M). For the subsequent cyto-

fluorometry the cell preparations are also placed in the latter solution. This modification, which may be called semi-equilibrium staining, is similar to the cell staining in equilibrium conditions proposed by Liedeman (Liedeman *et al.*, 1975; Liedeman & Bolund, 1976a).

At the same time the flow cytometric techniques for estimation of the chromatin functional state began to be developed. These techniques were mainly based on non-direct approaches such as the study of the chromatin susceptibility to the action of heat or acids with subsequent AO staining (for details see Section 6.9.3) and enabled quick investigation of vast cell populations. Simultaneously new flow cytofluoro-meters were developed and became increasingly available. As a result, from the early 1980s in many laboratories the estimation of the amount of AO directly bound to DNA for the study of chromatin properties was gradually replaced by the flow cyto-metrical indirect methods (Darzynkiewicz, 1990c; Darzynkiewicz & Kapuscinski, 1990).

Our long experience proved the direct estimation of AO binding to the nuclear chromatin to be a reliable approach to the estimation of its functional state (i.e. the total chromatin template activity).

This conclusion is based on the results obtained on different (from lymphocytes) cell systems for which the local cell density in a preparation is irrelevant, and on a correlation of the AO binding with the chromatin template activity. In addition to the data given above, the results of our more recent investigations (Smol'yaninova *et al.*, 1991) may be given (Fig. 6.6). The data cited above from many works, including ours, show a good coincidence of the results obtained by AO cytofluorometry with those from alternative techniques such as chromatin accessibility to the damaging factors, binding of different ligands includ-ing non-intercalating ones, histone stainability, etc.

It is necessary, however, to remember that only qualitative or semiquantitative conclusions may be drawn from the results obtained by these techniques. Thus, from the data presented in Fig. 6.6 one may certainly conclude that the chromatin is more active 45 min after laser irradiation than before it but it is impossible to say that its activity increased a certain number of times.

6.9.3 Investigation of the spectral properties of the chromatin stained with AO after different pretreatments

These methods are based on the assumption that the DNA–protein interactions or the chromatin super-structure change the DNA susceptibility to heat or acid, and that the DNA denaturation in chromatin is different in cells with different genome activities.

Figure 6.6 AO binding by chromatin and labelled uridine incorporation in the RNA of human peripheral blood lymphocytes after phytohaemagglutinin (PHA) treatment and HeNe laser irradiation. Abscissa: time after stimulation (h). Ordinate: AO fluorescence intensity and radioactive label incorporation (% to the control). (1) AO binding after PHA treatment; (2) AO binding after laser irradiation; (3) [14C]-uridine incorporation after PHA treatment; (4) [14C]-uridine incorporation after laser irradiation. A good correlation between AO binding and RNA synthesis after both treatments is seen. The figure is based on data from the work of Smol'yaninova et al., (1991).

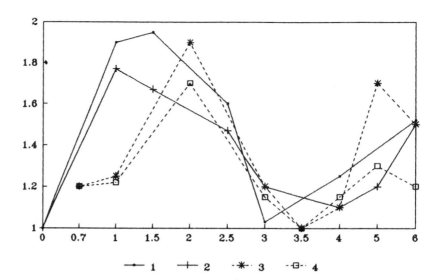

Subsequent AO staining allows the degree of DNA denaturation to be registered (for details see Sections 6.7 and 6.8). The advantage of this approach, compared with the method based on direct AO binding, is the possibility of receiving the characteristics for a single cell expressed in the form of a ratio between the red and green fluorescence (coefficient α) without the need to subtract the background. The detailed techniques based on static (Rigler et al., 1969; Liedeman & Bolund, 1976b) and flow (Darzynkiewicz et al., 1974, 1975, 1979b; Darzynkiewicz, 1990c) fluorometry were developed and became widely used for the investigation of chromatin properties in single cells and vast cell populations.

There are still many contradictions connected with these methods. The flow cytometry techniques in most cases reveal a decreased susceptibility of active chromatin to damaging factors, compared to the inactive one. This phenomenon was assumed to be unrelated to the histone H1, to the modification (phosphorylation, acetylation, etc.) of core histones, as well as to the presence of HMG proteins, but dependent on the higher order of the chromatin structure (Darzynkiewicz, 1990c).

At the same time, in work performed on cytological preparations, the opposite regularity – an increased accessibility of the active chromatin to the action of damaging factors – was described (Rigler et al., 1969, Ringertz, 1969; Zelenin, 1977, 1982). Experiments in which the preparations were treated with salt solution that removed H1 and non-histone proteins, demonstrated the involvement of these proteins in the DNA thermal stability in chromatin (Zelenin & Vinogradova, 1973; Kushch et al., 1974). For an attempt to explain these divergences see Darzynkiewicz (1990c) and Darzyinkiewicz and Kapuscinski (1990).

It may be just concluded here that when used under strict conditions these techniques (both flow and static) give us reliable qualitative and semiquantitative information about the functional state of chromatin in situ. Quantitative conclusions on the subject should, however, be made with great caution (Liedeman & Bolund, 1976b).

6.9.4 Concluding remarks: practical uses of chromatin cytochemistry

There are two sets of AO methods which allow us to judge the functional state of chromatin in situ based on different approaches to the preparation of the cell specimens (on slides and in suspension), registration of fluorescence properties (static and flow cytometry) and probably reflecting different features of chromatin structure (the DNA–histone interactions and the higher order of chromatin structure). These circumstances should be taken into account when results obtained by the alternative techniques are compared or the choice for a particular technique is being made.

Flow cytometry is more convenient and more applicable for chromatin investigation when large cell populations are being studied and suitable material is available. At the same time chromatin smear cytochemistry is useful in cases when investigation of different cells in cytological preparations is required; for example, in cell hybrids. These methods should, therefore, be regarded as complementary, with their particular fields of application.

AO chromatin cytochemistry has been successfully applied to many problems of medicine and agriculture including cancer diagnosis (Kunicka et al., 1987), diagnosis and prognosis of septicaemia in children

(Karachunski *et al.*, 1987) as well as to the investigation of the pea root meristematic cell germination (Troyan *et al.*, 1984); for reviews see Zelenin & Kushch (1985) and Darzynkiewicz (1990c).

6.10   FLUORESCENCE POLARIZATION OF AO BOUND TO DNA

This line of investigation was started by MacInnes and Uretz (1966) who studied the degree of polarization of AO bound to the polytene chromosomes of *Drosophila virilis*. As a result of their investigation the authors suggested that the interband DNA was not supercoiled but lay parallel to the chromosome axis. Such an organization of the interband DNA was later demonstrated by direct electron microscope observations (Ananiev & Barsky, 1985).

In other works polarized fluorescence microscopy was applied to the study of DNA organization in the sperm of different organisms (MacInnes & Uretz, 1968; Vinogradov *et al.*, 1980). Using a two-channel polarization microfluorometer Zotikov (1982) observed changes in the polarization of AO bound to *Tetrachimena periformis* DNA during the cell division cycle. Developing this approach Zotikov and Zelenin (1987) found a marked decrease in AO polarization degree after histone H1 removal from the chromatin by the fixed cell treatment with 0.6 M sodium chloride.

The above cited results suggest that measurement of the degree of polarization of the AO fluorescence provides useful information about the structural organization of DNA *in situ*.

6.11   AO IN CHROMOSOME BANDING

At least three methods of chromosome banding based on the use of AO have been suggested:

(1) The R-banding technique (Bobrov & Modan, 1973; Comings *et al.*, 1973; Verma & Lubs, 1975) includes chromosome preparation pretreatment in DNA denaturing conditions (heat denaturation, formaldehyde treatment, trypsin treatment, etc.) with subsequent AO staining. The regions between Q(G)-bands acquire orange-red fluorescence, while the other chromosome parts are weakly green. The picture is the reverse of the conventional Q/G-banding pattern, thus explaining the name of the technique ('R' – reverse banding). Some minor differences in the AO picture may appear, depending on the type of pretreatment (Bobrov & Modan, 1973). The cytochemical nature of the

method is rather obscure but the technique certainly enlarges the possibilities for chromosome analysis.

(2) A technique which reveals the fast and slowly reassociating chromosome regions was suggested by Stockert and Lisanti (1972). The chromosomes are denatured by heat, then treated in conditions which allow their partial renaturation (a short time of reassociation, cooling in the presence of formaldehyde) and after that stained with AO. As a result the centromeric regions containing highly repetitive DNA acquire green and other parts of chromosomes red fluorescence. The banding pattern resembles that of the conventional C-banding technique.

(3) The T-banding technique (Dutrillaux, 1973) is also based on chromosome thermal denaturation and AO staining. The telomeric regions become green whereas the other parts of chromosomes acquire red fluorescence. The bright green fluorescence of the telomeres – hence the name of the technique – seems to be due to DNA renaturation based on the presence of repetitive sequences.

Although all three techniques are similar, the banding patterns obtained by them are quite different, probably due to some minor variations in experimental conditions.

6.12   AO IN ACID MUCOPOLYSACCHARIDE HISTOCHEMISTRY

Acridine diamino derivatives, including AO as a bis-diamine, form salt-like complexes with acid mucopolysaccharides characterized by bright red fluorescence (Kuyper, 1957; Appel & Zanker, 1958) and a shift in absorption typical for the AO dimer form (Saunders, 1962).

The ability of AO to induce red fluorescence in cell structures containing acid mucopolysaccharides was described in all the early publications on the use of AO in nucleic acid cytochemistry (Armstrong, 1956; Bertalanffy & Bickis, 1956; Schummelfeder *et al.*, 1957) but mostly from the point of distinguishing between these two types of biopolymers. Details of the fluorescence cytochemistry of acid mucopolysaccharides were developed by Saunders (1962, 1964) and Zelenin and Stepanova (1968).

The method suggested by Saunders (1962, 1964) is based on two approaches:

(1) prestaining preparations with AO (0.1%), followed by washing in solutions of progressively increasing concentrations of sodium chloride; and

(2) pretreating preparations with cetyltrimethylammonium chloride, followed by AO staining.

A combination of these methods allowed the author to distinguish between different types of acid mucopolysaccharides: washing with 0.3 M NaCl removed the hyaluronic acid staining, washing with 0.6 M NaCl eliminated staining of hyaluronic acid and chondroitinsulphuric acid, and washing with 0.8 M NaCl removed the staining of both these mucopolysaccharides as well as heparin. After the cetyltrimethylammonium treatment hyaluronic acid acquired red fluorescence.

Zelenin and Stepanova (1968) suggested staining with AO at different pHs to distinguish between different types of acid mucopolysaccharides. At pH 4.5 all acid mucopolysaccharides, including hyaluronic acid, obtained the red fluorescence; however, at pH 2.0 this fluorescence was obtained only by the sulphonated ones (heparin and chondroitin sulphates). If necessary, RNAse or different hyaluronidases could be applied.

Our experience (Zelenin, 1967) allows us to recommend combinations of both Saunders's and our approaches. To eliminate a false positive staining which may be present in epithelial cells, highly diluted AO solutions can be used since chondroitin sulphuric acid and heparin still fluoresce red under these conditions (Saunders, 1962).

The approaches discussed above are all suggested for fixed preparations and there are few reports about the applicability of AO for the investigation of acid mucopolysaccharides *in vivo*. However, when a preparation of the subcutaneous connective tissue is treated with AO in vital or supravital staining (see Section 6.13.1) a number of big, brightly red fluorescent heparin granules can be observed in the mast cells. Other cells in the preparation (fibroblasts, leukocytes, etc.) give the typical picture of living cells. It thus seems that at least in some cases (mast cells) AO reveals acid mucopolysaccharides *in vivo*.

6.13   AO BINDING TO A LIVING CELL

6.13.1   General regularities. Comparison of pictures of fixed and living cells

AO was applied to the investigation of living cells in the very first work on the biological use of this fluorescent stain (Strugger, 1940a,b). Among the long list of authors who have since used AO for this purpose Meissel in Moscow should be named particularly. Reviews on *in vivo* use of AO can be found in Kasten (1967), Zelenin (1971), Meissel and Zelenin (1973), Darzynkiewicz and Kapuchinski (1990).

The picture of a living cell stained with AO differs drastically from that of a fixed one. A fixed cell stained with AO has a bright-green nucleus with orange-red nucleoli inside it. The cytoplasm has intense deep red fluorescence. In a living cell the weakly fluorescing green nucleus can usually be distinguished; some authors claim, however, that when AO is used in very low concentrations the nucleoplasm is undistinguishable from the background (Delic *et al.*, 1991). More bright-green fluorescence of nucleoli is seen in all cases. The fluorescence of the cytoplasm is so weak that it is almost undistinguishable from the background. At the same time there are a number of small granules in the cytoplasm with bright-red fluorescence. These granules present the most striking feature of a living cell stained with AO.

There are a number of different approaches to staining living cells with AO: a preparation of cultured cells on a coverslip or a smear of different animal and plant cells can be placed into the AO solution; AO can be added to the cell suspension; or AO can be injected intravenously or intraperitoneally into a living animal.

The fluorescence microscopical picture is highly dependent on the staining conditions, especially on the AO concentration. The typical picture described above is usually obtained when AO is used at a concentration of about 5×10^{-6} M. This concentration varies, however, depending on the ratio of cells to the volume of the stain solution. The pH of the staining solution is also very important; it should be about neutral or a bit higher (7.2–7.4); the staining medium should not contain any other stains or chemicals which may fluoresce themselves or quench the AO fluorescence. At the same time addition of glucose to the staining solution is desirable (Zelenin, 1971).

Staining under these conditions is usually called vital. Strictly speaking, the term should mean that the studied cell retains all the properties of an unstained cell. This is, however, hardly possible in any 'vital' stain, including AO. AO in a living cell binds with nucleic acids, thus interfering with their synthesis and inhibiting protein synthesis and mitotic activity. If applied for a long time, it reduces the size of nucleoli and alters the morphology of red fluorescent granules (for details see Section 6.13.3). Still, the main features of a living cell are preserved under these conditions. Therefore, such staining is often called 'supravital'.

If AO is used in very high concentrations (10^{-2} M–10^{-3} M) or for too long, the cell can be damaged or even killed by the action of the dye. The first feature of such damage is the appearance of orange-red nucleoli. Later the nucleoplasm acquires bright-green fluorescence and, as the last step of such 'mortal' cell treatment, the cytoplasm appears diffuse red. The fluorescence picture of such a cell is quite similar to the picture of a fixed one. As a typical example of mortal staining, the work of Bertalanffy and Bickis (1956) may be taken. Sometimes such staining is not quite mortal and a cell can be reversed at least into a 'supravital' state if placed into a fresh culture medium, in particular one containing an excess of glucose.

6.13.2 Cytochemical interpretation of microscopical pictures

Such an interpretation is rather difficult for a living cell because standard cytochemical approaches such as enzyme pretreatments are hardly applicable to it. Circumstantial evidence is therefore mostly used for that purpose.

Important information has been obtained through spectroscopical investigations. It has been shown that the absorption spectrum of AO shifts after staining of a living cell in the same way as when AO is bound to the nucleic acids *in vitro* (Loeser *et al.*, 1960). Instead of one peak of absorption characteristic of AO in highly diluted solutions (λ_{abs} 494 nm), two new peaks appear, with λ_{max} 502 nm and 470 nm respectively. As was mentioned previously, the 502 nm peak is characteristic of AO bound to the double-stranded nucleic acids. Thus it can be assumed that the green fluorescence of a living cell stained with AO reflects its binding to the nucleic acids.

This assumption in connection to the nuclear fluorescence is confirmed by the following observations:

(1) Like other acridine amino derivatives, AO inhibits RNA synthesis *in vivo* (Goldberg *et al.*, 1963; Zelenin & Liapunova, 1966; Zelenin, 1971). These data may be assumed to indicate AO binding to DNA.

(2) It is well-known that cell treatment with low doses of actinomycin D results in a gradual reduction of the RNA component of nucleoli. The AO vital staining reveals in such cells a marked reduction of green fluorescent nucleoli (Zelenin, 1971).

In contrast to the data described above, Delic *et al.* (1991) recently concluded that AO does not intercalate into the nuclear DNA of a living cell. Intensified fluorescence microscopy coupled with a digital imaging system was used in this work. It allowed the authors to receive good fluorescence pictures at low AO concentrations (10^{-6} M) and very low levels of excitation. Under the conditions of these experiments the only fluorescence registered in the cell was green fluorescence of nucleoli and red fluorescence of cytoplasmic granules. It is quite possible that under these conditions AO does not bind to DNA and thus less injures the cell. Such staining is possibly closer to the really 'vital'.

Interpretation of the cytoplasmic pictures is more difficult. It has been shown that AO specifically inhibits protein synthesis in cells, that effect being due to its binding to cytoplasmic RNAs, tRNA in particular (Zelenin & Liapunova, 1964a; Zelenin, 1971). It can be assumed that AO binds to the cytoplasmic RNAs and is responsible for the weak green fluorescence of the cytoplasm usually seen in AO-treated cells.

But the most interesting story is connected with the red cytoplasmic granules which were shown to be lysosomes which accumulated AO.

6.13.3 *In vivo* lysosome investigation

The presence of red cytoplasmic granules is the most striking feature of a living cell exposed to the action of AO. These granules were first found by Vonkennel and Wiedemann in 1944 and later described in many animal cells (see Bertalanffy & Bickis, 1956; Meissel & Zelenin, 1973). Their cytochemical and morphological nature was for many years a subject of some controversy. At first they were regarded as a result of a complex formation between AO and the cytoplasmic RNA. This assumption was based on the *in vitro* data on the red fluorescence of AO in complex with RNA. Some authors regarded the granules as complexes between AO and acid mucopolysaccharides or proteins, or as AO-stained mitochondria (for review, see Zelenin, 1967, 1971; Meissel & Zelenin, 1973).

Later, however, these granules were proved to be lysosomes and lysosome-related structures that had accumulated AO (Koenig, 1963; Robbins *et al.*, 1964; Zelenin & Liapunova, 1964b; Zelenin *et al.*, 1965; Zelenin, 1966; Dingle & Barrett, 1967). It was shown that even isolated lysosomes retained the capacity to concentrate AO inside them (Dingle & Barrett, 1967; Cononico & Bird, 1969). AO was therefore named as an excellent non-enzymatic lysosome marker (Cononico & Bird, 1969).

Almost immediately AO* became widely used for *in vivo* lysosome investigations of primary and secondary lysosomes (Allison & Young, 1964, 1969; Blume *et al.*, 1969) and in special lysosome analogues such as acrosomes (Allison & Hartee, 1970).

The bright red fluorescence of AO lysosomes shows that AO is there in the dimer form, whereas in other cellular structures which fluoresce green it is in a monomer form. This means that the AO concentration

---

* In some early papers (Dingle & Barrett, 1967; Blume *et al.*, 1969; Unanue *et al.*, 1969; Allison & Hartee, 1970) a fluorochrome called euchrysine 3R was reported to be used for that purpose. As we noted some years ago (Zelenin, 1971), euchrysine 3R is just one of numerous trade marks of AO (*Colour Index*, 1956). Allison and Young (1969) themselves showed the chromatographic identity of euchrysine 3R with AO.

inside the lysosomes is at least a thousand times higher than in the cytosol. Preferential accumulation of AO in lysosomes was demonstrated directly by Dingle and Barrett (1968) in experiments with tritium-labelled AO.

The cytophysiological mechanism of AO accumulation in lysosomes merits special discussion.

It was shown that the capacity of lysosomes to concentrate AO is connected with the presence on their membrane of a proton pump responsible for the maintenance of low pH inside lysosomes (De Duve *et al.*, 1974; Yamashiro *et al.*, 1983; Moriyama *et al.*, 1982, 1984). Low pH causes accumulation inside lysosomes of different cationic compounds including AO. It is, however, possible that red lysosome fluorescence is due not only to the high concentration of AO but also to its binding to some acid substrate (polymer) which is capable of binding AO and thus facilitates formation of dimers. One candidate for such a substrate is a strongly acidic component capable of binding cations, which was described in lysosomes by Barrett and Dingle (1967).

AO vital and supravital staining may be used for detection of lysosomes, investigation of their distribution, localization and morphology. It is very useful for the investigation of secondary lysosomes and for the study of the process of lysosome fusion, for example with phagosomes. It should be taken into account that only acidified lysosome-like vesicles acquire red fluorescence after AO staining. If applied in low concentration for a long time (days) AO staining reveals the general accumulative capacity of the cellular lysosomes.

In parallel, lysosome staining with AO opens up important possibilities in the study of cell physiology. The capacity of a cell to concentrate AO in its lysosomes indicates that its proton pump functions normally. This requires an adequate energy supply. It was shown many years ago that the AO accumulation inside lysosomes is an energy-dependent process (Weissmann & Gilgen, 1956; Kirianova & Zelenin, 1970; Zelenin, 1971). Thus the presence of the red cytoplasmic granules in the AO-treated cells indicates that the energy-supplying mechanisms function normally in these cells.

Although the AO method of lysosome investigation was suggested and developed a long time ago it is still widely used. This approach was recently successfully applied by Del Bing *et al.* (1991) to the investigation of the cytotoxic action of the DNA topoisomerase I inhibitor camptophecin. The inhibitor had no effect on AO uptake into lysosomes. This allowed the authors to conclude that the cytotoxic effect of this substance was not connected with its action on the maintenance of the proton pump.

In the works analysed above the AO-stained lysosomes were studied by conventional (static) fluorescence microscopy. In parallel to these, AO has also been widely used in flow cytometry. The most common application of this approach is the study of cell viability (see Section 14). Lysosome staining with AO has been successfully applied in the flow cytometric investigation of living blood cells (Melamed *et al.*, 1972, 1974) as well as for the distinguishing between different types of lung cells (Wilson *et al.*, 1986).

6.14 AO IN THE STUDY OF CELL VIABILITY

In the very first publication of the biological application of AO, Strugger (1940b, 1942) demonstrated the possibility of its use for distinguishing between dead and living cells. In his experiments dead cells stained with AO fluoresced red, whereas living cells fluoresced green. These findings were carefully studied in many works. Some authors confirmed these observations but others failed to do so (see reviews by Bertalanffy & Bickis, 1956; Zelenin, 1967). It is difficult to give a complete explanation of Strugger's observations. It is, however, possible that in successful experiments when the 'Strugger effect' was observed, the staining had been carried out with such AO concentrations and under such conditions as to cause the red fluorescence of the ribosomal RNA of the dead or damaged cells. At the same time, the cells had very few or no lysosomes (this is typical for the plant, yeast and bacterial cells mostly used in those early experiments), and thus only acquired green fluorescence after staining *in vivo*.

After the discovery that AO could be actively concentrated in the lysosomes of living cells it began to be used for the study of cell viability. As pointed out above (Section 6.13.3), the presence of the red fluorescing granules in vitally (supravitally) stained cells is a reliable indication of the normal state of their lysosomes and the active function of the energy-supply mechanisms. In our laboratory we use AO staining as a primary conventional test of the viability of culture cells.

More sophisticated applications of AO for that purpose have been described.

Singh and Stephens (1986) developed a two stain method for the determination of cell viability. They applied two fluorescent dyes – AO and Hoechst 33258 for that purpose. Under the conditions used AO penetrated only into the living cells whereas H 33258 penetrated only into the dead ones. When these dyes were used in combination, the dead cells fluoresced brilliant blue and the living ones green.

Singh and Stephens (1986) used in their work a

conventional fluorescence microscope. Other authors applied flow cytometry. This approach is now widely used in such investigations (Darzynkiewicz & Kapuchinscki, 1990; Pollack & Ciancio, 1990). Bohmer (1984, 1985) combined AO with ethidium bromide to detect living and dead cells. Three groups of cells treated with agents which caused complete cell cycle arrest were found: vital cells stained only with AO, dead cells in which ethidium stained only the cytoplasm, and dead cells in which ethidium stained both cytoplasm and nucleus. The method proved itself to be more sensitive than the conventional trypan blue exclusion test. Two steps in the cell death process were revealed.

6.15   AO IN FLOW CYTOMETRY

The use of AO in flow cytometry has been reviewed many times and at least twice during 1990 (Darzynkiewicz, 1990a; Darzynkiewicz & Kapuchinski, 1990). Therefore, in this section the most important directions of these investigations will just be listed.

(1) Determination of the RNA/DNA ratio in the cell cycle study, in study of the activation of proliferation of the quiescent cells, in the investigation of the changes in cell transcriptional activity, etc.
(2) Investigation of chromatin properties after acid or thermal denaturation in cells with different genome activity (stimulation of proliferation in quiescent cells, determination of mitotic activity, etc.).
(3) Cell cycle analysis by the BrdUrd-suppressed AO fluorescence, including discrimination between cycling and non-cycling cells.
(4) Investigation of cell viability.

6.16   OTHER APPLICATIONS OF AO

There are a number of AO applications in which it is used just for obtaining a fluorescence microscopical picture without any cytochemical analysis. It is difficult not only to cite these works but even to classify them. Some old publications in this field are given in Gurr's *Encyclopedia* (1960). Recent ones can be found in the current literature.

AO has been used for the electron microscopical investigation of chromatin and mitotic chromosomes. It was found that treatment of the living cell with AO results in a change of chromatin structure (Frenster, 1971) and suppression of chromosome condensation (Matsubara & Nakagone, 1983). The latter fact allows the use of AO to increase the resolution of chromosome banding, similar to ethidium (Ikeuchi, 1984; Chiryaeva *et al.*, 1989).

Among new and very promising intercalators, related to AO, the bis- and tris-acridines should be mentioned (Chen *et al.*, 1978; Gaugain *et al.*, 1984; Denny *et al.*, 1985).

REFERENCES

Albert A. (1966) *The Acridines*. E. Arnold, London.
Allison A.G. & Hartee E.F. (1970) *J. Reprod. Fertility* **21**, 501–515.
Allison A.C. & Young M.R. (1964) *Life Sci.* **3**, 1407–1414.
Allison A.C. & Young M.R. (1969) In *Lysosomes in Biology and Pathology*, Vol. 2, J.T. Dingle & H.B. Fell (eds). North-Holland, Amsterdam & London, pp. 603–628.
Alvarez M.R. (1974) *Exp. Cell Res.* **83**, 225–230.
Appel W. & Zanker V. (1958) *A. Naturforsch.* **13b**, 126–134.
Ananiev E.V. & Barsky V.E. (1985) *Chromosoma* **19**, 104–112.
Armstrong J.A. (1956) *Exp. Cell Res.* **11**, 640–643.
Auer G. (1972) *Cytochemical Properties of Chromatin Related to Changes in Cellular Growth Activity*. Almqvist and Wikseli Informationindustri AB, Uppsala, 1–78.
Auer G., Zetterberg A. & Killander D. (1970) *Exp. Cell Res.* **62**, 32–38.
Auer G., Moore G.P.M., Ringertz N.R. & Zetterberg A. (1973) *Exp. Cell Res.* **76**, 229–233.
Barret R.E. & Dingle J.T. (1967) *Biochem. J.* **105**, 20p.
Berman H.M. & Young P.R. (1981) *Ann. Rev. Biophys. Bioeng.* **10**, 87–114.
Bertalanffy L. von (1963) *Protoplasma* **57**, 51–83.
Bertalanffy L. von & Bickis I. (1956) *J. Histochem. Cytochem.* **4**, 481–493.
Blume R.S., Clade P.R. & Chessin L.N. (1969) *Blood* **33**, 87–99.
Bobrov M. & Modan K. (1973) *Cytogenet. Cell Genet.* **12**, 145–156.
Bohmer R.M. (1984) *Cell Tissue Kinet.* **17**, 593–600.
Bohmer R.M. (1985) *Cytometry* **6**, 215–218.
Bohmer R.M. (1990) In *Methods in Cell Biology*, Vol. 33, *Flow Cytometry*, Z. Darzynkiewicz & H.A. Crissman (eds). Academic Press, London, pp. 173–184.
Bolund L., Ringertz N.R. & Harris H. (1969) *J. Cell Sci.* **4** 71–87.
Bolund L., Darzynkiewicz Z. & Ringertz N.R. (1970) *Exp. Cell Res.* **62**, 76–81.
Borisov A.Yu. & Tumerman L.A. (1959) *Izvestiya Akad. Nauk SSSR, ser. phys.* **23**, 97–101 (Moscow).
Borisova O.F. & Grechko V.V. (1984) *Studia Biophys.* **101**, 133–134.
Borisova O.F. & Tumerman L.A. (1964) *Biofisika* **9**, 537–544 (Moscow).
Borisova O.F. & Tumerman L.A. (1965) *Biofiskia* **10**, 32–36 (Moscow).
Borisova O.F., Kisselev L.L. & Tumerman L.A. (1963) *Dokl. Akad. Nauk SSSR* **152**, 1001–1004 (Moscow).

Borisov O.F., Horachek P., Gursky G.V., Minyat E.E. & Tumanyan V.G. (1968) *Izvestiya Akad. Nauk SSSR, ser. phys.* **32**, 1317–1324 (Moscow).

Borisova O.F., Krichevskaya A.A. & Samarina O.P. (1981) *Nucleic Acids Res.* **9**, 663–681.

Borisova O.F., Golova Yu.B., Gotthich B.P., Zibrov A.S., Il'icheva Yu.P., Lysov Yu.P., Mamaeva O.K., Chernov B.K., Chernyi A.K., Shchelkina A.K. & Florentiev V.L. (1991) *J. Biomol. Struct. Dyn.* **8**, 1187–1201.

Bradley D.F. (1961) *Trans. NY Acad. Sci. Ser II* **24**, 64–74.

Brumberg E.M. & Girshgorin A.S. (1947) An opack-illuminator, primarily for luminescence microscopy. USSR Patent, Klass 42, 14/05, No. 78635.

Brumberg E.M. & Krilova G.H. (1953) *Jurnal obshchey biologii* **14**, 461–464 (Moscow).

Bukatsch F. & Haitinger M. (1940) *Protoplasma* **34**, 515–523.

Chen T.K., Fico R. & Canellakis E.S. (1978) *J. Med. Chem.* **21**, 868–874.

Chiryaeva O.G., Amosova A.V., Efimov A.M., Smirnov A.F., Kaminir L.B., Yakovlev A.F. & Zelenin A.V. (1989) In *Cytogenetics of Animals*, C.R.E. Halnan (ed.). CAB International, Wallingford, UK, pp. 211–219.

Clark G. & Kasten F.H. (1983) *History of Staining* (3rd edition). Williams & Wilkins, Baltimore & London.

Colour Index (1956) in 4 volumes. Bradford Society of Dyers and Colourists, Lowell.

Comings D.E., Aveling E., Okada T.A. & Wyandt H.E. (1973) *Exp. Cell Res.* **77**, 469–493.

Cononico P.G. & Bird J.W.C. (1969) *J. Cell Biol.* **43**, 367–371.

Crissman H.A. & Steinkamp J.A. (1990) *Methods Cell Biol.* **33**, 199–206.

Darzynkiewicz Z. (1990a) *Methods Cell Biol.* **33**, 285–298.

Darzynkiewicz Z. (1990b) *Methods Cell Biol.* **33**, 337–352.

Darzynkiewicz Z. (1990c) In *Flow Cytometry and Sorting*, M.R. Melamed, T. Lindmo & M.L. Mendelson (eds). Wiley-Liss, New York, pp. 315–340.

Darzynkiewicz Z. & Kapuscinski J. (1990) In *Flow Cytometry and Sorting*, M.R. Melamed, T. Lindmo & M.L. Mendelson (eds). Wiley-Liss, New York, pp. 291–314.

Darzynkiewicz Z., Troganos F., Sharpless T. & Melamed M.R. (1974) *Biochem. Biophys. Res. Commun.* **59**, 392–399.

Darzynkiewicz Z., Troganos F., Sharpless T. & Melamed M.R. (1975) *Exp. Cell Res.* **90**, 411–428.

Darzynkiewicz Z., Andreeff M., Troganos F., Sharpless T. & Melamed M.R. (1978) *Exp. Cell Res.* **115**, 31–35.

Darzynkiewicz Z., Evenson D., Stainano-Coico L., Sharpless, T. & Melamed M.R. (1979a) *J. Cell Physiol.* **100**, 425–438.

Darzynkiewicz Z., Troganos F., Andreef M., Sharpless T. & Melamed M.R. (1979b) *J. Histochem. Cytochem.* **27**, 478–485.

Darzynkiewicz Z., Troganos F. & Melamed M.R. (1983) *Cytometry* **3**, 345–348.

DeDuve C., de Barsy T., Poole B., Trouet A., Tulkens P. & Van Hoof F. (1974) *Biochem. Pharmacol.* **23**, 2495–2531.

Del Bing G., Lassota P. & Darzynkiewicz Z. (1991) *Exp. Cell Res.* **193**, 27–35.

Delic J., Coppey J., Magdelenat H. & Coppey-Moisan M. (1991) *Exp. Cell Res.* **194**, 147–153.

Denny W.A., Atwell G.J., Baquley B.C. & Walelin L.B.G. (1985) *J. Med. Chem.* **28**, 1568–1574.

Dingle J.T. & Barrett A.J. (1967) *Biochem. J.* **105**, 19p–20p.

Dingle J.T. & Barrett A.J. (1968) *Biochem. J.* **109**, 19p.

Dutrillaux B. (1973) *Chromosoma* **41**, 395–402.

Frenster J.H. (1971) *Cancer Res.* **31**, 1128–1133.

Galley W.C. & Purkey R.M. (1972) *Proc. Natl. Acad. Sci. USA* **69**, 2198–2202.

Gaugain B., Markovits J., LePecq J.-B. & Roques B.R. (1984) *FEBS Lett.* **169**, 123–126.

Gledhill B.L., Gledhill M.P., Rigler R. & Ringertz N.R. (1966) *Exp. Cell Res.* **41**, 652–665.

Goldberg I.H., Reich E. & Rabinowitz M.M. (1963) *Nature* **199**, 44–46.

Gomatos P.J. & Tamm J. (1963) *Proc. Natl. Acad. Sci. USA* **49**, 707–714.

Gomatos P.J., Tamm J., Dales S. & Franklin R.M. (1962) *Virology* **17**, 441–454.

Gurr E. (1960) *Encyclopedia of Microscopic Stains*. Leonard Hill, London, pp. 1–498.

Ikeuchi T. (1984) *Cytogen. Cell Genet.* **38**, 56–63.

Jamison R.M. & Mayor H.D. (1965) *J. Bacteriol.* **90**, 1486–1488.

Karachunski A.I., Volodin N.N. & Zelenin A.V. (1987) *Pediartya* No. 2, 67–72 (Moscow).

Kasten F.H. (1967) *Int. Rev. Cytol.* **21**, 141–202.

Kasten F.H. (1989) In *Cell Structure and Function by Microspectrofluorometry*, E. Kohen (ed.). Academic Press, London, pp. 3–51.

Kernell A.M., Bolund L. & Ringertz N.R. (1971) *Exp. Cell Res.* **65**, 1–6.

Killander D. & Rigler R. (1965) *Exp. Cell Res.* **39**, 701–712.

Killander D. & Rigler R. (1969) *Exp. Cell Res.* **54**, 163–170.

Kim S.H., Suddath F.L., Quegley G.J., McPherson A., Susman J.L., Wang A.H.J., Seeman N.C. & Rich A. (1974) *Science* **185**, 435–440.

Kirianova E.A. & Zelenin A.V. (1970) *Dokl. Akad. Nauk SSSR* **190**, 451–454 (Moscow).

Koenig H. (1963) *J. Cell Biol.* **19**, 87A.

Kolesnikov V.A., Kushch A.A. & Zelenin A.V. (1973) *Dokl. Akad. Nauk SSSR* **212**, 489–501 (Moscow).

Kunicka J.E., Darzynkiewicz Z. & Melamed M.R. (1987) *Cancer Res.* **47**, 3942–3947.

Kushch A.A., Kolesnikov V.A. & Zelenin A.V. (1974) *Exp. Cell Res.* **86**, 419–422.

Kushch A.A., Terskikh V.V., Kolesnikov V.A., Niyazmatov A.A., Koslov Yu.V. & Zelenin A.V. (1978) *Cytobiologie* **16**, 161–170.

Kushch A.A., Niyazmatov A.A. & Zelenin A.V. (1980) *Cell Differ.* **9**, 291–304.

Kuyper Ch.M.A. (1957) *Exp. Cell Res.* **13**, 198–200.

Latt S.A. (1976) In *Chromosomes To-day*, Vol. 5, P. Pearson & K. Lewis (eds). Wiley, New York, pp. 367–372.

Lerman L.S. (1961) *J. Molec. Biol.* **3**, 18–31.

Lerman L.S. (1963) *Proc. Natl. Acad. Sci. USA* **49**, 94–102.

Lerman L.S. (1964) *J. Cell. Compar. Physiol.* **64**, 1–18.

Liedeman R.R. & Bolund L. (1976a) *Exp. Cell Res.* **101**, 164–174.

Liedeman R.R. & Bolund L. (1976b) *Exp. Cell Res.* **101**, 175–183.

Liedeman R.R., Matveyeva N.P., Vosrticova S.A. & Prilipko L.L. (1975) *Exp. Cell Res.* **90**, 105–110.

Loeser C.N., West S.S. & Schoenberg M.D. (1960) *Anatom. Rec.* **138**, 163–187.

MacInnes J.W. & Uretz R.B. (1966) *Science* **151**, 689–691.

MacInnes J.W. & Uretz R.B. (1968) *J. Cell Biol.* **38**, 426–436.

Manteifel V.M., Vinogradova N.G. & Zelenin A.V. (1973) *Izvestija Akad. Nauk SSSR, Ser. Biol.* **6**, 740–743 (Moscow).

Mayor H.D. & Hill N.O. (1961) *Virology* **14**, 264–266.

Matsubara T. & Nakagome S. (1983) *Cytogen. Cell Genet.* **35**, 148–151.

Meissel M.N. & Gutkina A.V. (1953) *Dokl. Akad. Nauk SSSR* **91**, 647–650 (Moscow).

Meissel M.N. & Korchagin V.A. (1952) *Bull. Exp. Biol. Mediz.* No. 5, 49–51 (Moscow).

Meissel M.N. & Sondak V.A. (1956) *Biofisika* **1**, 262–273 (Moscow).

Meissel M.N. & Zelenin A.V. (1973) In *Unity Through Diversity*, Vol. 9, Part II, *A Festschrift for Ludwig von Bertalanffy* W. Gray & N.D. Rizzo (eds). Gordon & Breach, New York, London, Paris, pp. 701–738.

Melamed M.R., Adams L.R., Traganos F., Zimring A. & Kamentsky L.A. (1972) *Am. J. Clin. Pathol.* **57**, 95.

Melamed M.R., Adams L.R., Traganos F. & Kamentsky L.A. (1974) *J Histochem. Cytochem.* **22**, 526–530.

Moriyama Y., Takano T. & Ohkuma S. (1982) *J. Biochem. (Tokyo)* **92**, 1333–1336.

Moriyama Y., Takano T. & Ohkuma S. (1984) *J. Biochem. (Tokyo)* **93**, 927–930.

Morosov Yu.V. (1963a) *Biofisika* **8**, 167–171 (Moscow).

Morosov Yu.V. (1963b) *Biofisika* **8**, 331–334 (Moscow).

Morosov Yu.V. & Savenko A.K. (1977) *Molec. Photochem.* **8**, 1–43.

Nash D. & Plaut W. (1964) *Proc. Natl. Acad. Sci. USA* **51**, 731–735.

Neidle S., Taylor G. & Saunderson M. (1978) *Nucleic Acids Res.* **5**, 4417–4422.

Pollack A. & Ciancio G. (1990) *Methods Cell Biol.* **33**, 19–24.

Poot M., Kubbies M., Hoehn H., Grossmann A., Chen Y. & Rabinovich P. (1990) *Methods Cell Biol.* **33**, 185–198.

Rigler R. (1966) *Acta physiol. Scand.* **67**, suppl. 267, 1–122.

Rigler R. (1969) *Ann. NY Acad. Sci.* **157**, 211–224.

Rigler R. & Killander D. (1969) *Exp. Cell Res.* **54**, 171–179.

Rigler R., Killander D., Bolund L. & Ringertz N.R. (1969) *Exp. Cell Res.* **55**, 215–224.

Ringertz N.R. (1969) In *Handbook of Molecular Cytology*, A. Lima-de-Faria (ed.). North-Holland, Amsterdam, pp. 656–687.

Ringertz N.R. & Bolund L. (1970) *Exp. Cell Res.* **55**, 205–214.

Ringertz N.R. & Bolund L. (1974) *Int. Rev. Exp. Pathol.* **8**, 83–116.

Robbins E., Marcos P.J. & Ganatos N.K. (1964) *J. Cell Biol.* **21**, 49–62.

Roschlau G. (1965) *Histochemie* **5**, 396–406.

Saunders A.M. (1962) *J Histochem. Cytochem.* **10**, 683–684.

Saunders A.M. (1964) *J Histochem. Cytochem.* **12**, 164–170.

Schummelfeder N., Ebschner K.-J. & Krogh E. (1957) *Naturwissenschaften* **44**, 467–468.

Singh N.P. & Stephans R.I. (1986) *Stain Technol.* **61**, 315–318.

Smets L.A. (1973) *Exp. Cell Res.* **79**, 239–243.

Smol'yaninova N.K., Karu T.I., Fedoseeva G.E. & Zelenin A.V. (1991) *Biomed. Sci.* **2**, 121–126.

Sondore O.Yu., Fedoseeva G.E., Kadykov V.A. & Zelenin A.V. (1978) *Molec. Biol. Rep.* **4**, 137–141.

Steiner R.F. & Beers R.F. (1958) *Science* **127**, 335–339.

Steiner R.F. & Beers R.F. (1959) *Arch. Biochem. Biophys.* **18**, 75–92.

Steiner R.F. & Beers R.F. (1961) In *Polynucleotides, Natural and Synthetic Nucleic Acids*. Elsevier, Amsterdam & London, pp. 301–323.

Stockert J.C. & Lisanti J.A. (1972) *Chromosoma* **37**, 117–130.

Strugger S. (1940a) *Dtsch. tieararztzl. Wochenschr.* **48**, 645–692.

Strugger S. (1940b) *Jenaische Z. Naturforsch.* **73**, 97–134.

Strugger S. (1942) *Dtsch. tieararztzl. Wochenschr.* **50**, (5/6), 51–61.

Terskikh V.V., Kirianova E.A., Zosimovskaya A.I. & Zelenin A.V. (1976) *Tsitologia* **18**, 1085–1089 (Leningrad).

Troyan V.M., Kolesnikov V.A., Kalinin F.L. & Zelenin A.V. (1984) *Plant Sci. Lett.* **33**, 213–219.

Unanue E.R., Askonas B.A. & Allison A.C. (1969) *J. Immunol.* **103**, 71–78.

Verma R.S. & Lubs H.A. (1975) *Am. J. Hum. Genet.* **27**, 110–117.

Vinogradov A.E., Rosanov Ju.M. & Barsky I.Ya. (1980) *Tsitologia* **22**, 1314–1322 (Leningrad).

Vonkennel & Wiedemann. (1944) *Dtsch. med. Wochenschr.* **70**, 27–28, 529–532.

Wang J.C. (1974) *J. Molec. Biol.* **89**, 783–801.

Waring J.M. (1981) *Ann. Rev. Biochem.* **50**, 159–192.

Weissmann G. & Gilgen A. (1956) *Z. Zellforsch.* **44**, 292–299.

Wilson J.S., Steinkamp J.A. & Lehnert B.E. (1986) *Cytometry* **7**, 157–162.

Yamashiro D.J., Fluss S.R. & Maxfield F.R. (1983) *J. Cell Biol.* **97**, 929–934.

Zanker V. (1952a) *Z. Phys. Chem.* **199**, 255–258.

Zanker V. (1952b) *Z. Phys. Chem.* **200**, 250–292.

Zanker V., Held M. & Rammensee H. (1959) *Z. Naturforsch.* **146**, 789–801.

Zelenin A.V. (1966) *Nature* **212**, 425–426.

Zelenin A.V. (1967) *Luminescence Cytochemistry of nucleic acids*. Nauka, Moscow.

Zelenin A.V. (1971) *Interaction of Aminoacridines with a Cell*. Nauka, Moscow.

Zelenin A.V. (1977) *Biol. Zentralblad* **86**, 407–422.

Zelenin A.V. (1982) *Acta Histochem.* Suppl.-Band **XXVI**, 179–187.

Zelenin A.V. & Kushch A.A. (1985) *Molec. Biol.* **19**(1), part 2, 242–250.

Zelenin A.V. & Liapunova E.A. (1964a) *Nature* **204**, 45–46.

Zelenin A.V. & Liapunova E.A. (1964b) In *Second International Congress Histo- and Cytochemistry*, Abstracts, Frankfurt/Main, p. 217.

Zelenin A.V. & Liapunova E.A. (1966) *Farmakol. toksikol.* **4**, 481–483 (Moscow).

Zelenin A.V. & Stepanova N.G. (1968) *Arch. anat. gistol.* **54**, 82–88 (Leningrad).

Zelenin A.V. & Vinogradova N.G. (1973) *Exp. Cell Res.* **82**, 411–414.

Zelenin A.V., Biriuzova V.I., Vorotnitskaya N.E. & Liapunova E.A. (1965) *Dokl. Acad. Nauk. SSSR* **162**, 925–927 (Moscow).

Zelenin A.V., Shapiro, I.M., Kolesnikov V.A. & Senin V.M. (1974) *Cell Differ.* **3**, 95–101.

Zelenin A.V., Kirianova E.A., Kolesnikov V.A. & Stepanova N.G. (1976) *J. Histochem. Cytochem.* **24**, 1169–1172.

Zelenin A.V., Stepanova N.G. & Kiknadze I.I. (1977) *Chromosoma* **64**, 327–335.

Zetterberg A. & Auer G. (1970) *Exp. Cell Res.* **62**, 262–270.

Zotikov A.A. (1982) *Acta histochem.* Suppl. Band **26**, 219–221.

Zotikov A.A. & Zelenin A.V. (1987) *Tsitologia* **29**, 1398–1401 (Leningrad).

CHAPTER SEVEN

Fluorescent Lipid Analogues: Applications in Cell and Membrane Biology

JAN WILLEM KOK & DICK HOEKSTRA
Laboratory of Physiological Chemistry, University of Groningen, The Netherlands

7.1 INTRODUCTION

Recent advances in studies involving the structure and dynamics of membranes have shown that fluorescent lipid probes have become important and, frequently, indispensable tools in this area of research. These probes are applied in investigations as diverse as those dealing with biophysical aspects of membranes, including lateral mobility, phase transitions and phase separations, but also in studies of the cell biology of membranes, including membrane flow and lipid trafficking.

The type of probe to be used for a particular study depends very much on the sort of research involved. For example, to characterize overall biophysical properties of membranes, the probe to be used need not necessarily resemble that of a lipid. Essential is usually its lipid-like properties and its incorporation in the hydrophobic core or lipid phase of the membrane. A typical and classical probe in this respect is 1,6-diphenyl-1,3,5-hexatriene (DPH) and its derivatives (Shinitzky & Barenholz, 1978; Loew, 1988). By contrast, in studies aimed at revealing the cell biology of lipids, including intracellular processing and metabolism, fluorescent lipid probes are desirable that much more closely resemble the structure of the natural lipids. Therefore, in this case a particular lipid analogue can be studied, which is derived from the natural lipid by derivatization with a fluorescent tag (Pagano, 1989; Pownall & Smith, 1989). Especially these latter studies have added another element to the application of a fluorescent probe, other than just as an alternative for radiochemical or spin probes. This element involves fluorescence microscopy, which in studies of intracellular lipid trafficking, allows direct visualization of the probe without perturbation of cellular integrity.

In the following sections we will describe these various aspects of the application of fluorescent lipid probes. The properties of some frequently employed lipid probes, in particular nitrobenzoxadiazole-derivatized-lipids (NBD-lipids), will be summarized. Some attention will also be paid to 'lipid-like' probes and in this case we will limit the discussion to the application of fluorescent compounds that are attached to an acyl or alkyl chain ('fatty acid' probes). Practical aspects and various examples will be given as to the fruitful application of these fluorescent lipid and lipid-like probes in areas such as lipid trafficking, lipid metabolism, membrane fusion, lipid transfer and in issues dealing with the polarized and asymmetric distribution of lipids.

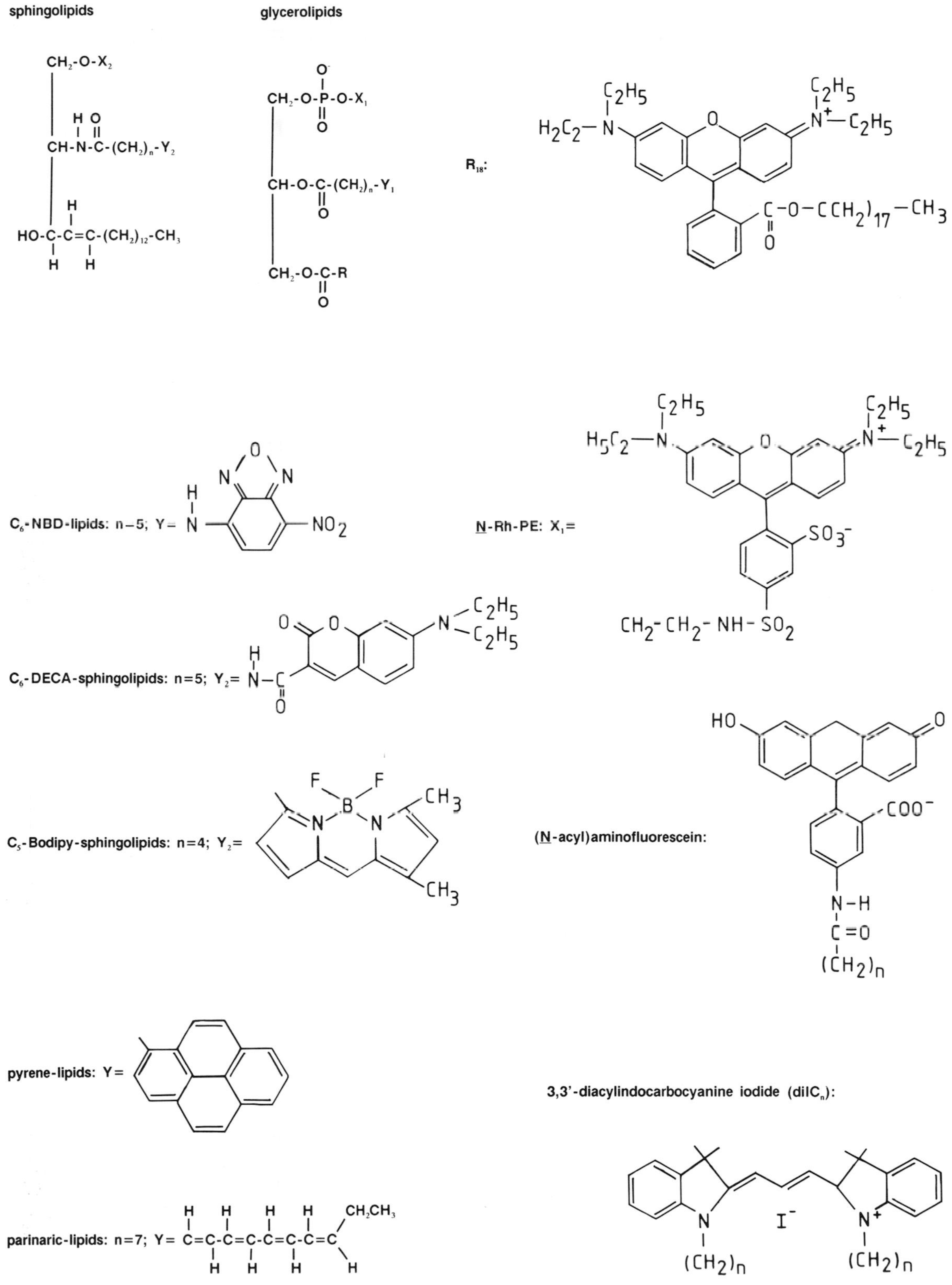

sphingolipids   glycerolipids

CH_2-O-X_2

$CH-N-C-(CH_2)_n-Y_2$

$HO-C-C=C-(CH_2)_{12}-CH_3$

$CH_2-O-P-O-X_1$

$CH-O-C-(CH_2)_n-Y_1$

$CH_2-O-C-R$

R_{18}:

C_6-NBD-lipids: n−5; Y= N

N-Rh-PE: X_1=

C_6-DECA-sphingolipids: n=5; Y_2=

C_5-Bodipy-sphingolipids: n=4; Y_2=

(N-acyl)aminofluorescein:

pyrene-lipids: Y=

3,3'-diacylindocarbocyanine iodide (diIC$_n$):

parinaric-lipids: n=7; Y=

Figure 7.1 Structures of fluorescent lipid analogues.

7.2 FLUORESCENT LIPID ANALOGUES

7.2.1 Structure, availability and synthesis

Typical fluorescent lipid analogues that have found their application in studies as diverse as membrane flow, intracellular lipid trafficking, membrane fusion, lipid phase separations and transitions are the so-called NBD-lipids. These lipid analogues contain a 7-nitrobenz-2-oxa-1,3-diazol-4-yl (NBD) group. The fluorophore can be attached via a short chain (C_6) or longer spacer (C_{12}) to the lipid backbone, which can be mono-acylglycerol (in NBD-phospholipids) or sphingosine (in NBD-sphingolipids; Fig. 7.1). Alternatively, NBD can be attached to the headgroup of phosphatidylethanolamine (PE) via an amide bond. Thus, in this case a typical headgroup-labelled lipid analogue is obtained.

Several NBD-phospholipids are commercially available (Avanti; Molecular Probes). These include the fluorescent derivatives of phosphatidylcholine (PC), PE, phosphatidylglycerol (PG) and phosphatidic acid (PA), all of which are available with C_6 and C_{12} spacers. C_6-NBD-PS is not available but can be synthesized from C_6-NBD-PC by a base-exchange reaction (Comfurius *et al.*, 1990). Of the C_6-NBD-sphingolipids C_6-NBD-ceramide and C_6-NBD-sphingomyelin are now available (Molecular Probes), but all the glycosphingolipids must be synthesized in one's laboratory.

Glycosphingolipids consist of a long-chain sphingosine backbone, to which a fatty acid is attached via an amide linkage. This makes up the hydrophobic moiety, called ceramide. The carbohydrate moiety, which can vary from a single hexose (usually glucose or galactose) to a complex structure of many linked hexose units, is attached at the primary hydroxy group of the sphingoid base (Wiegandt, 1985).

To synthesize a C_6-NBD-(glyco)sphingolipid, the fatty acid is to be replaced by C_6-NBD-hexanoic acid. Thus one needs the deacylated (lyso-) form of the parent lipid. Several deacylated (glyco)sphingolipids are commercially available (Sigma). These include sphingosine (deacylated ceramide), sphingosylphosphorylcholine (deacylated sphingomyelin), psychosine (deacylated galactosylceramide), glucopsychosine (deacylated glucosylceramide) and lysosulphatide. When a deacylated sphingolipid is not available (or too expensive), as in the case of lactosylceramide, the parent lipid can be deacylated according to a modified procedure of Goda *et al.* (1987). In our laboratory we routinely apply the following procedure: 10 mg of lactosylceramide (Sigma) is added to 9 ml *n*-butanol/ 1 ml 10 N KOH in a round-bottom flask. The mixture is stirred and heated in a heating mantle for 6 h at 117°C (boiling point of *n*-butanol), while refluxing. Thereafter the solvent is evaporated under a nitrogen flow, while keeping the temperature at about 40°C to speed up the evaporation. The dried material is then dissolved in 16 ml of $CHCl_3/CH_3OH/H_2O$ (2:2:1), shaken vigorously and centrifuged to separate phases. The lower chloroform phase is washed with 4 ml of water and then evaporated. The dried material is separated on preparative TLC using $CHCl_3/CH_3OH/ CaCl_2$ (0.2 %) in H_2O (50:42:11) as the solvent system. Lactosylsphingosine (R_f 0.39) is well-separated from lactosylceramide (R_f 0.68) and also from the degradation products glucosylsphingosine (R_f 0.44) and glucosylceramide (R_f 0.81). The lactosylsphingosine band is scraped from the TLC plate and washed with 20 ml $CHCl_3/CH_3OH$ (1:1) followed by 20 ml methanol to remove the lipid from the silica. After evaporation the product is tested for purity on TLC, using orcinol staining (Kundu, 1981) and, if necessary, subjected to another preparative TLC procedure.

The reacylation with C_6-NBD-hexanoic acid (Sigma) is performed as described by Kishimoto (1975) with slight modifications. The C_6-NBD-sphingolipid is synthesized from free fatty acid and deacylated sphingolipid by an oxidation–reduction condensation with triphenylphosphine and 2,2′-dipyridyldisulphide, a method originally devised for peptide synthesis (Mukaiyama *et al.* 1970). Typically, 10 μmol (6.2 mg in case of lactosylsphingosine) of the deacylated sphingolipid is mixed with 10 μmol (2.9 mg) of C_6-NBD, 20 μmol (5.2 mg) of triphenylphosphine and 20 μmol (4.4 mg) of 2,2′-dipyridyldisulphide and dissolved in 200 μl of chloroform. The mixture is stirred overnight at room temperature under an atmosphere of nitrogen. The material is then applied on preparative TLC plates, which are run in a solvent system consisting of $CHCl_3/CH_3OH/20\%$ (wt/vol.) NH_4OH (70:30:5) for C_6-NBD-(glyco)sphingolipids, or $CHCl_3/CH_3OH/HAc$ (90:2:8) for C_6-NBD-ceramide. C_6-NBD-lactosylceramide (R_f 0.16) is well-separated from C_6-NBD (R_f 0.30) and from high-R_f-value side-products, as well as from the other reactants (Fig. 7.2) The NBD-band of interest is scraped from the plate and washed with 20 ml $CHCl_3/CH_3OH$ (1:1) followed by 20 ml of methanol to remove the lipid from the silica. Purity of the product is tested in several TLC solvent systems (basic, neutral, acidic) and repurified on TLC if necessary to finally yield a single spot. The final yield of the deacylation/reacylation procedure varies from 10 to 25%, depending on the type of lipid. The above described procedure has the advantage that an extra (preceding) step of derivatization of the fatty acid to, for instance, an *N*-hydroxysuccinimide ester is not necessary. In the case of C_6-NBD-hexanoic acid, however, the succinimidyl ester is available (Molecular Probes), which renders

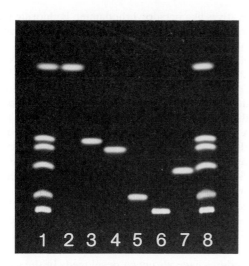

Figure 7.2 Separation of C_6-NBD-sphingolipids by thin-layer chromatography. Five different C_6-NBD-sphingolipids were synthesized from the corresponding deacylated parent lipids and C_6-NBD-hexanoic acid. The NBD-lipids were separated on a TLC plate employing $CHCl_3$/CH_3OH/20 % (wt/vol.) NH_4OH (70:30:5) as the running solvent system. (Individual C_6-NBD-sphingolipids from left to right are: C_6-NBD-ceramide (lane 2); C_6-NBD-glucosylceramide (lane 3); C_6-NBD-galactosylceramide (lane 4); C_6-NBD-lactosylceramide (lane 5) and C_6-NBD-sphingomyelin (lane 6); in lane 7 C_6-NBD-hexanoic acid was spotted.

this reaction a good alternative to the above described condensation reaction, with similar yields. Typically, 10 µmol of the deacylated sphingolipid and 15 µmol (5.9 mg) of succinimidyl 6 (7 nitrobenz 2 oxa-1,3-diazol-4-yl)aminohexanoate (Molecular Probes) are dissolved in 200 µl of $CHCl_3$/CH_3OH/$(C_2H_5)_3N$ (8:2:0.1) and allowed to react overnight at room temperature and under nitrogen (cf. Gardam *et al.*, 1989).

In the case of gangliosides, which are acidic glycosphingolipids containing one or more N-acetyl-neuraminic acid (NANA) residues in the carbohydrate moiety, the NBD labelling procedure becomes more complex. This is due to a concomitant deacetylation of the NANA during the deacylation of the lipid. After the labelling procedure the ganglioside has to be subjected to re-N-acetylation with acetic anhydride. Furthermore, the deacylation is performed with tetramethylammonium hydroxide instead of KOH and the reacylation with either the succinimide ester (Schwarzmann & Sandhoff, 1987) or the mixed anhydride of the fatty acid and chlorocarbonate, thus involving a fatty acid activation step (for details see Sonnino *et al.*, 1986; Acquotti *et al.*, 1986).

Fluorescently labelled lipids offer the great advantage that their intracellular trafficking can be monitored directly in living cells with fluorescence microscopy. When combined with the use of fluorescently tagged protein (Dunn & Maxfield, 1990), this offers the possibility to study lipid trafficking in relation to the routes by which protein ligands like transferrin are processed, which are well-established (Van Renswoude *et al.*, 1982; Chiechanover *et al.*, 1983). Thus one can perform co-internalization studies of C_6-NBD-lipids and proteins tagged with fluorophores such as those of, for example, the rhodamine group (lissamine rhodamine B (LR), (tetramethyl)rhodamine (T)RITC, Texas Red (TR)) (Kok *et al.*, 1989, 1990; Koval & Pagano, 1989).

The labelling of lipids with fluorophores other than NBD, displaying different spectral properties, makes it also possible to perform double-labelling studies with different types of lipids (Uster & Pagano, 1986; Kok *et al.*, 1991). Using different combinations of fluorophores, even triple-labelling studies can be carried out. One has to take into account, however, that the overall physico-chemical properties of the lipid analogues can be influenced by the (type of) fluorescent tag, as discussed below. An example of the double-labelling strategy involving lipids is the use of 7-diethylaminocoumarin-3-carboxylic acid (DECA) as an alternative for (C_6-)NBD (Gardam *et al.*, 1989), which emits blue fluorescence instead of the green fluorescence of NBD. Microscopically, DECA can be separated from NBD by using appropriate filter sets. C_6-DECA-sphingolipids can be synthesized by first coupling an activated DECA fluorophore to hexanoic acid followed by linking this C_6-DECA to the deacylated sphingolipid with a condensation reaction (Kishimoto, 1975) as described above for the C_6-NBD-sphingolipids. The C_6-DECA can be synthesized as follows: 6-amino-n-hexanoic acid (Sigma) is dissolved in methanol at a concentration of 20 mg ml^{-1}. Then 100 µl (2 mg or 15 µmol) of this solution is added to 200 µl $CHCl_3$/CH_3OH/$(C_2H_5)_3N$ (8:2:0.1) containing 8.8 mg (20 µmol) of DECA-succinimidyl ester and incubated overnight, under nitrogen and at room temperature. Thereafter the C_6-DECA is purified by preparative TLC employing a solvent system consisting of $CHCl_3$/CH_3OH/20% (wt/vol.) NH_4OH (70:30:5), followed by scraping of the C_6-DECA band (R_f is 0.46, i.e. well-separated from DECA-succinimidyl ester (R_f 0.84) and washing as described above.

A disadvantage of the NBD-fluorophore is that it rapidly bleaches during observations with the fluorescence microscope. Therefore, photography is sometimes difficult and often tedious, while long-time or repeated observation of single cells at high intensity is virtually impossible. Recently, a new fluorophore, boron dipyrromethane difluoride (Bodipy, Molecular Probes), has become available, which has a higher quantum yield, an improved photostability and which can be linked to (sphingo)lipids, analogous to NBD. A highly interesting and exciting property of the probe is that it shows a density-dependent fluorescence emission shift from 515 to 620 nm. This property makes it possible to monitor the generation or elimination of high local concentrations of probe-carrying lipids in membranes of different cellular compartments (Pagano et al., 1991). A disadvantage of the use of Bodipy-labelled lipids is that due to the less polar character of the fluorophore (in comparison to NBD), the lipid cannot be fully recovered from the plasma membrane by a so-called 'back-exchange', as discussed below.

Other fluorescent lipid analogues that have frequently been used in studies of the dynamics of membranes are analogues, labelled synthetically or biosynthetically with pyrene or parinaric acids. These probes are primarily used in experiments involving fluorometric measurements in the fluorometer (for a review see Pownall & Smith, 1989), while their spectral properties preclude their fruitful application in fluorescence microscopy. Although parinaric acids are most similar to natural fatty acids, these probes are particularly difficult to work with because of their rapid decomposition, due to their air-sensitivity. A spectral inert gas atmosphere is therefore necessary. The reader is referred to several excellent reviews on these lipid analogues, that were recently published elsewhere (Loew, 1988; Pownall & Smith, 1989).

7.2.2 Some properties and practical implications

C_6-NBD-lipids can easily be inserted into the plasma membrane of cells and, inversely, also recovered from the plasma membrane by a 'back-exchange'. This is due to the fact that these lipid analogues are less hydrophobic than their natural counterparts and are therefore able to transfer spontaneously from a suitable donor system to a cellular membrane. C_6-NBD-PC has been shown to transfer between vesicle populations (Pagano et al., 1981), between vesicles and cells (Struck & Pagano, 1980) and between biological membranes (Kok et al., 1990). This is not a typical feature of C_6-NBD-labelled lipids. In general, the transfer rate of fluorescent lipid analogues is dependent on the length of the acyl chain to which the probe is attached. For pyrene-labelled lipids, it has been shown that when the number of methylene units decreases, the transfer rate increases. This appears to be directly related to the increased solubility of the analogue in water (Pownall & Smith, 1989; Chattopadhyay, 1990). In contrast, the tendency of exogenously incorporated acyl or alkyl derivatives to leave the bilayer decreases with increasing chain length (Frank et al., 1983; Nichols & Pagano, 1983). Finally, headgroup-labelled lipid derivatives, such as N-NBD-PE and N-(lissamine rhodamine B sulphonyl) phosphatidylethanolamine (N-Rh-PE) or lipid-like

derivatives like octadecyl rhodamine B chloride (R_{18}) behave *in this respect* like natural lipids, showing monomeric transfer kinetics in the order of days (Keller et al., 1977; Struck et al., 1981; Hoekstra, 1982a; Nichols & Pagano, 1983; Hoekstra et al., 1984).

The proper intercalation of a fluorescent lipid analogue in the lipid phase of the plasma membrane of the cells can be shown (Struck and Pagano, 1980; Kok et al., 1990) by measuring lateral diffusion rates of the lipid with a technique based on fluorescence recovery after photobleaching (FPR; Axelrod et al., 1976; Jacobsen et al., 1976). When properly inserted, fluorescent lipid analogues display diffusion rate constants of approximately $10^{-8}\,cm^2\,s^{-1}$ (Edidin, 1992).

Apart from using vesicles as donor membranes for lipid probe insertion into acceptor membranes via monomeric transfer through the aqueous phase, other procedures have also been developed. In the case of C_6-NBD-ceramide a method for introduction into cellular membranes has been described (Pagano & Martin, 1988), that makes use of NBD-lipid/BSA complexes. The lipid is dissolved in ethanol and injected under vortexing into a defatted BSA-containing buffer, followed by dialysis. The complex is then added to the cells resulting in transfer of the lipid from the BSA complex to the cell membrane.

A generally applicable method for the insertion of C_6-NBD-lipids into the plasma membrane of cells involves the use of ethanol micelles (Kok et al., 1989, 1990). Appropriate amounts of C_6-NBD-lipid are dried under nitrogen and subsequently solubilized in absolute ethanol. An aliquot of the ethanolic solution (0.5% (v/v), final concentration) is injected into the appropriate buffer under vigorous vortexing. This solution (usually containing 4–5 μM NBD-lipid) is then added to the cells. It turns out that with this method it is also possible to insert N-Rh-PE into the plasma membrane of cells (Kok et al., 1990). This analogue is a fluorescent PE derivative (but no longer has specific PE properties, see below) which has the fluorophore (rhodamine) attached to the amino group of the headgroup of the lipid. N-Rh-PE is defined as a non-exchangeable lipid analogue, based on observations that the probe, when incorporated into liposomal bilayers, does not spontaneously transfer between the labelled liposomes and non-labelled artificial or biological membranes (Pagano et al., 1981; Struck et al., 1981; Hoekstra et al., 1988). However, when baby hamster kidney (BHK) cells are incubated at 2°C with N-Rh-PE (dispersed in ethanol), the lipid is inserted into the plasma membrane, as indicated by several criteria including FPR measurements (Kok et al., 1990).

The exchangeable NBD-lipids can be removed from the outer leaflet of the plasma membrane ('back-exchange') by incubating the cells with

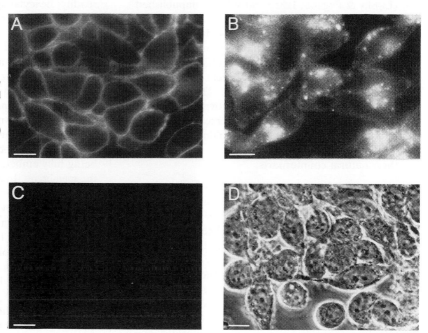

Figure 7.3 Plasma membrane insertion, endocytosis, and back-exchange of a C$_6$-NBD-lipid. Baby hamster kidney cells were labelled at 2°C with C$_6$-NBD-glucosylceramide (5 mol %)-containing DOPC liposomes for 30 min (A). Subsequently either a back-exchange was performed with 5% BSA/Hank's for 30 min at 2°C (C and D), or the labelled cells were first incubated at 37°C for 30 min, followed by a back-exchange (B). Note that the back-exchange removes all plasma membrane-inserted C$_6$-NBD-lipid. Similar results were obtained when the cells were labelled with the ethanol-injection method or when the back-exchange was performed with DOPC SUV (bars = 10 μm). (For details, see Kok *et al.*, 1989.)

either dioleoylphosphatidylcholine (DOPC) small unilamellar vesicles (SUV) (Struck & Pagano, 1980) or (bovine) serum albumin (Van Meer *et al.*, 1987; Kok *et al.*, 1989). The vesicles or the albumin function as acceptors for the NBD-lipid molecules. If a sufficient amount of these acceptors is added to the cells (usually two consecutive washes), the NBD-lipid can be quantitatively removed from the outer leaflet of the plasma membrane (Fig. 7.3). As anticipated, this back-exchange procedure does not work for the non-exchangeable N-Rh-PE (Kok *et al.*, 1990).

When the C$_6$-NBD-lipids are compared to their natural counterparts in cells, several arguments exist in favour of a good analogy between the probe and the natural lipid. First, C$_6$-NBD-PA is metabolized in cells initially to C$_6$-NBD-DG, followed by synthesis of C$_6$-NBD-TG and C$_6$-NBD-PC. Furthermore, the cell recognizes these products and sorts them to different intracellular compartments, since only C$_6$-NBD-TG becomes associated with lipid droplets (Pagano & Sleight, 1985; Pagano & Longmuir, 1985). Secondly, analogous to radioactively labelled PE, C$_6$-NBD-PE undergoes transbilayer movement at the plasma membrane (Pagano & Sleight, 1985; Sleight & Pagano, 1985). Similarly, when C$_6$-NBD-PS is supplied to erythrocytes, it inserts into the *outer* leaflet of the membrane followed by translocation to the *inner* leaflet, resulting in an asymmetric distribution among the leaflets. By contrast, C$_6$-NBD-PC does not translocate (Connor & Schroit, 1987). These results are quite compatible with the naturally occur-

ring situation, where aminosphopholipids preferentially reside in the *inner* leaflet, while the choline phospholipids are predominantly localized in the *outer* leaflet of the red blood cell membrane (Verkleij *et al.*, 1973; Gordesky *et al.*, 1975). Thirdly, the metabolism and intracellular transport of C$_6$-NBD-ceramide seem to reflect the behaviour of endogenous ceramide in the following ways.

(1) C$_6$-NBD-ceramide is metabolized to C$_6$-NBD-sphingomyelin and C$_6$-NBD-glucosylceramide (Lipsky & Pagano, 1983, 1985a; Van Meer *et al.*, 1987) and to other glycosphingolipids (Kok *et al.*, 1991), indicating that the respective biosynthetic enzymes are capable of recognizing the NBD-labelled substrate. Furthermore, when radiolabelled ceramide or C$_6$-NBD-ceramide are incubated with isolated subcellular fractions prepared from cultured fibroblasts, no preferential metabolism of the different ceramides is seen (Lipsky & Pagano, 1985a).

(2) The half-life for transport of C$_6$-NBD-sphingomyelin and C$_6$-NBD-glucosylceramide from the Golgi apparatus to the plasma membrane (Lipsky & Pagano, 1983, 1985a) is consistent with the half-life for the appearance of newly synthesized, radiolabelled neuronal gangliosides at the plasma membrane (Miller-Podraza & Fishman, 1982). Furthermore, monensin inhibits the appearance of both isotopically labelled glycosphingolipids (Saito *et al.*, 1984) and C$_6$-NBD-glucosylceramide

(Lipsky & Pagano, 1985a and our own unpublished results) at the cell surface.

(3) In polarized Madin-Darby canine kidney (MDCK) cells C_6-NBD-glucosylceramide is sorted from C_6-NBD-sphingomyelin and preferentially delivered to the apical surface (Van Meer *et al.*, 1987; Van't Hof & Van Meer, 1990). This polarized delivery is consistent with the known enrichment of glycosphingolipids in the apical membrane, thus indicating that the C_6-NBD-sphingolipids are sorted and transported similarly as compared to their natural counterparts.

Finally, in the endocytic uptake pathway C_6-NBD-sphingolipids can be recycled to the plasma membrane (Kok *et al.*, 1989; Koval & Pagano 1989) and sorted from each other (Kok *et al.*, 1991). The recycling of intact C_6-NBD-glucosylceramide molecules to the cell surface after their initial uptake by endocytosis (Kok *et al.*, 1989) is consistent with the notion that glycolipids are thought to reside mainly in the outer leaflet of the plasma membrane, indicating that this class of lipids in particular is subject to sorting (Hakomori, 1981). In liver, gangliosides can be resynthesized from recycled glucosylceramide (Trinchera *et al.*, 1990), indicating that the lipid is sorted during inbound cellular traffic, thereby avoiding degradation in the lysosomal compartment. The sorting of C_6-NBD-glucosylceramide from other (glyco)sphingolipids, like C_6-NBD-sphingomyelin during endocytosis in HT29 cells, indicates that also in the endocytic pathway C_6-NBD-sphingolipids are recognized by the cellular sorting and transport machinery (cf. (3) above).

Taken together, valuable insight into the cell biology of lipids has been obtained from the use of C_6-NBD analogues of lipids. However, there are some indications that the use of these analogues is restricted and should always be critically evaluated.

(1) Studies aimed at investigating the role of fatty acid heterogeneity are, by definition, impossible, especially in the case of the C_6-NBD-sphingolipids. Reacylation of lipids cannot be monitored, since the marker (C_6-NBD) is lost after deacylation and is not reused for synthesis. On the one hand, this property offers the advantage of a relatively simple interpretation since the fate of only one labelled lipid derivative is usually monitored. On the other hand, however, it clearly distinguishes NBD-lipid analogues from pyrene and parinaric fatty acid-labelled lipids. The latter derivatives even more closely resemble natural lipids as both fatty acid probes can be biosynthetically incorporated into cellular lipids (Loew, 1988; Pownall & Smith, 1989). In contrast to C_6-NBD-PC and other NBD-phospholipids, which are

readily deacylated (Sleight & Pagano, 1984; Moreau, 1989; Kok *et al.*, 1990), the C_6-NBD-sphingolipids do not seem to be subject to this type of catabolism (Kok *et al.*, 1989; Koval & Pagano, 1989), although it is known that their natural counterparts are (Trinchera *et al.*, 1990).

Another reason for caution in studies involving acyl chain structure stems from studies by Chattopadhyay and co-workers (1988, 1990). They show that the NBD group of C_6-NBD-lipids loops back to the polar region of the membrane, instead of being fully embedded in the hydrophobic core of the membrane. In a recent review (Chattopadhyay, 1990) the use of NBD-labelled lipids as analogues of natural lipids in membrane and cellular processes is critically evaluated. The conclusion is reached that in spite of the perturbed acyl chain conformation the NBD-labelled lipids have served as reasonably good analogues for these studies.

(2) With regard to lipid anabolism, the more complex glycolipids and gangliosides are not synthesized from the C_6-NBD-ceramide precursor. Apparently the biosynthetic enzymes for glucosylceramide and galactosylceramide are able to recognize the NBD-labelled ceramide substrate. Furthermore, C_6-NBD-glucosylceramide can also serve as a substrate for lactosylceramide synthesis. However, C_6-NBD-GM3 synthesis is not observed in BHK cells, although these cells do contain the natural counterpart of this lipid (Kok *et al.*, unpublished observations).

(3) The ratio of (glyco)sphingolipids synthesized from fluorophore-labelled ceramide may depend on the type of fluorescent probe (C_6-NBD vs. C_6-Bodipy and C_6-DECA) (Pagano *et al.*, 1991; Kok *et al.*, unpublished observations).

In conclusion, although NBD-labelled lipid analogues have shown their value in numerous cell biological applications, one should be careful in interpreting results from metabolic studies employing these fluorescent lipids, especially regarding absolute amounts of synthesis. Also, the suitability of the C_6-NBD label for the study of complex glycolipids (gangliosides) is questionable.

To study dynamic membrane properties such as membrane fusion, lipid translocation, lipid phase transitions and separations, advantage can also be taken of the specific fluorescent properties of certain lipid probes. These properties include concentration-dependent self-quenching of the probes and their ability to engage in resonance energy transfer (RET). In Section 6.2 several examples of experiments that rely on these principles will be discussed. Given the versatile use of approaches that rely on resonance energy transfer, it is useful to outline briefly here the basic features of such an approach. In essence, RET involves the interactions that may occur between two different fluorophores, provided that the emission band of one fluorophore, the energy donor, overlaps with the excitation band of the second fluorophore, the energy acceptor. Energy transfer takes place when the probes are in close physical proximity. It involves the transfer of excited state energy from the donor to the acceptor. This energy is derived from a photon absorbed by the energy donor upon excitation. The acceptor will then fluoresce as though it had been excited directly. The energy transfer efficiency depends on the extent of overlap of the donor's emission spectrum and the acceptor's absorption spectrum and the distance between the probes. The energy transfer efficiency, E, is given by the equation $E = (1-F/F_0) \times 100\%$, where F is the fluorescence measured in the presence of the acceptor and F_0 represents the fluorescence read in the absence of acceptor (usually obtained by reading fluorescence after addition of detergent, i.e. at infinite dilution, see below). With regard to distance, the efficiency of energy transfer is proportional to the inverse of the sixth power of the distance between the donor and acceptor (Fung & Stryer, 1978). The distance between two fluorophores can then be calculated, using the equation:

$$R = R_0[(0.5/E)^{1/6}]$$

in which R_0 is the so-called Förster distance, i.e. the distance between donor and acceptor at which E equals 50%.

A convenient energy transfer couple that has found wide application is that involving NBD-labelled lipid as energy donor, and N-Rh-PE as energy acceptor. Studies of membrane fusion (Struck et al., 1981; Hoekstra, 1982a), protein-mediated and spontaneous exchange and transfer of lipids (Nichols & Pagano, 1983) and (transbilayer) flip-flop studies (Connor & Schroit, 1987; Pagano, 1989) have all benefited from employing this particular couple. In this system, transfer efficiency can be translated in the ability of rhodamine to quench NBD fluorescence. (Note that the transfer efficiency also depends on the concentration of the energy acceptor.) In practice, this means that the reaction (lipid transfer, flip-flop or fusion) is monitored at the NBD emission maximum of approximately 530 nm, while exciting the donor at 475 nm. It should be noted that energy transfer can also be measured between suitably labelled protein molecules and a lipid probe. In the following section, various examples will be discussed of the use of fluorescent lipid probes in cell and membrane biology.

7.3 APPLICATIONS

7.3.1 Applications in cell biology

C_6-NBD-lipids are very useful tools in studying the trafficking of individual types of lipids (for reviews see Pagano & Sleight, 1987; Simons & Van Meer, 1988; Van Meer, 1989; Hoekstra et al., 1989; Schwarzmann & Sandhoff, 1990; Koval & Pagano, 1991). The fate of the lipid can be followed directly with fluorescence microscopy. Furthermore, these observations can be combined with TLC analysis and sensitive fluorometric quantitation, providing information about the localization of the lipid in relation to its metabolism.

For microscopical observation of endocytic trafficking pathways, cells are grown on glass coverslips (coated with, for instance, collagen if necessary, depending on the cell type) in Petri dishes, till confluency. Before labelling with the desired lipid probe, the cells are cooled to 2°C. At this temperature all trafficking processes cease. This allows the fluorescent lipid to be inserted into the plasma membrane of the cell (see Section 7.2.2) and 'synchronized' endocytosis to be triggered, which is done by elevating the temperature to 37°C (Fig. 7.3). When a comparison is to be made to the trafficking of a protein ligand, for example transferrin (Tf), the cells are also incubated with LR-Tf at 2°C, allowing the protein to bind to its receptor. After washing away free ligand, the cells are warmed instantly to 37°C by adding prewarmed (Hank's) buffer, allowing the cells to internalize the lipid and the protein for a defined period of time. Thereafter the cells are cooled again to 2°C to 'fix' the trafficking processes. Subsequently, a back-exchange is carried out to remove the residual plasma membrane NBD-lipid pool. Since only a small (10–20%) fraction of the original plasma membrane (NBD-)lipid pool is internalized, visualization of intracellular fluorescence is difficult without such a back-exchange. Employing the foregoing protocol we have shown that only 2 min of incubation at 37°C is necessary for C_6-NBD-glucosylceramide to reach early endosomes (Fig. 7.4), as indicated by colocalization with LR-Tf, which by definition labels early endosomes, seen as peripheral vesicles, after 2 min of receptor-mediated endocytosis (Kok et al., 1989). When studying such short time (2 min) trafficking with microscopy, it is necessary to somehow fix the lipid inside the intracellular compartment. This can be achieved by cooling the microscope stage to 2°C or by using chemical fixatives. A fixation method has been developed for glycolipids (Kok et al., 1989). The fixative, containing periodate, lysine and paraformaldehyde, stabilizes carbohydrate moieties. Stabilization involves the carbohydrates being oxidized

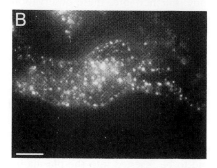

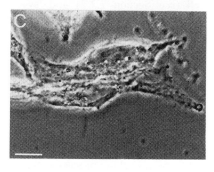

Figure 7.4 Co-internalization of C_6-NBD-glucosylceramide and LR-Tf. Baby hamster kidney cells were labelled at 2°C with C_6-NBD-glucosylceramide (5 mol %)-containing liposomes for 30 min, washed, and labelled with LR-Tf (30 min at 2°C; 0.2 mg ml^{-1}). Subsequently, the cells were incubated at 37°C for 2 min, immediately cooled to 2°C, and subjected to a back-exchange with a 5% BSA/Hank's solution for 30 min at 2°C. The cells were fixed at room temperature (see Section 6.2.1) and examined by fluorescence microscopy, using appropriate filter sets for visualizing either NBD or rhodamine fluorescence. (A) NBD(-lipid); (B) LR(-Tf); (C) corresponding phase-contrast image. Note that the lipid and the protein colocalize in early endosomes (bars = 10 μm). (For further details, see Kok *et al.*, 1989.)

by periodate and cross-linked by lysine (McLean & Nakane, 1974). In practice, cells are fixed for 30 min at room temperature with 2% (wt/vol.) formaldehyde in a buffer, containing 20 mM sodium phosphate (pH 7.4), 100 mM lysine, 60 mM sucrose, and 100 mM sodium periodate, and are post-fixed sequentially with 4% formaldehyde and 6% formaldehyde, both in the same buffer, for 5 and 10 min respectively.

The intracellular endocytic pathways of sphingolipids have been carefully studied in fibroblasts (Kok *et al.*, 1989; Koval & Pagano, 1989, 1990). One aspect of lipid trafficking that has become evident is that, like proteins, lipids can also be subject to recycling, i.e. after their initial uptake into the cell by endocytosis, original molecules return to the cell surface. In this respect, it is important to note the intramembrane topology of the NBD-lipids, that are not subject to transbilayer movement (Fig. 7.5). The NBD-lipid molecules are inserted into the outer leaflet of the plasma membrane, which means that they will be trapped inside an endocytic vesicle that buds from the membrane. During its consecutive trafficking inside the cell the lipid will remain locked up inside organelles and when recycling occurs it will reappear on the outer leaflet of the plasma membrane. Thus, after recycling, the lipid is susceptible to back-exchange, allowing quantification of the extent of recycling (as described below). Recycling of C_6-NBD-glucosylceramide in BHK cells occurs from both early and late endosomes, while another fraction of the lipid cycles via the Golgi area, possibly involving the trans-Golgi network (TGN) (Kok *et al.*, 1989).

The ability to undergo transbilayer movement ('flip-flop') depends on the type of lipid (see also Section 7.3.2). Among the (NBD-)lipids that are not subject to flip-flop are the choline phospholipids (PC, SM) (Sleight & Pagano, 1984; Koval & Pagano, 1989) and the glycolipids (Kok *et al.*, 1989, 1991). However, the amino-phospholipids PE and PS are involved in rapid transmembrane movement (Pagano & Longmuir, 1985; Connor & Schroit, 1987), probably involving a protein translocator (Connor & Schroit, 1988). This translocation occurs at temperatures around 7°C and above, and may lead to subsequent labelling of intracellular organelles by non-vesicular movement. The lipid precursors C_6-NBD-DG and C_6-NBD-ceramide already translocate at 2°C and label all intracellular membranes. At 37°C coordinated transport to specific organelles and metabolism occur. In the case of C_6-NBD-ceramide accumulation takes place in the Golgi apparatus. This is accompanied by metabolism to C_6-NBD-glucosylceramide, C_6-NBD-sphingomyelin and other glycosphingolipids (see Section 7.2.2). Because of this accumulation the ceramide can be used as a marker for the Golgi apparatus, both for fluorescence and electron microscopy (Lipsky & Pagano, 1985b; Pagano *et al.*, 1989, 1991). The metabolic products are subsequently transported to the plasma membrane by vesicular carriers. That transport is indeed vesicular – although the carrier vesicles have never been visualized – is supported by the kinetics of transfer, the inhibition by monensin (Lipsky & Pagano, 1985a), and the absence of transport in mitotic cells. By contrast, the transport

Figure 7.5 Membrane topology of
C$_6$-NBD-lipids that do not flip-flop
during trafficking. When cells are
labelled with C$_6$-NBD-
glucosylceramide (or any other
NBD-lipid that is not subject to
flip-flop), the lipid resides in the
outer leaflet of the plasma
membrane (I). Upon
internalization the NBD-lipid is
'locked up' inside the endocytic
vesicle (III), which is formed after
budding (II) from the plasma
membrane. In the absence of flip-
flop, the lipid will remain in the
inner leaflet of all intracellular
organelles, reached by vesicular
trafficking and fusion of the
vesicles. Eventually the lipid
analogue may reappear in the *outer*
leaflet of the plasma membrane
after recycling.

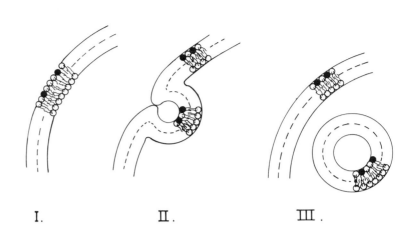

I.   II.   III.

of PE is thought not to be mediated by vesicular carriers (Kobayashi & Pagano, 1989).

In conclusion, vesicular trafficking of sphingolipids can be studied both in the endocytic pathway (uptake of plasma membrane-localized (glyco)sphingolipids) and the biosynthetic pathway (transport of newly synthesized (glyco)sphingolipids). However, the outbound traffic to the plasma membrane is more difficult to compare directly to that of proteins, since methods are not available to introduce fluorophores into proteins during their biosynthesis.

The studies on (sphingo)lipid trafficking in fibroblasts have been further extended to other cell types, including polarized cells (Van Meer *et al.*, 1987; Van't Hoff & Van Meer, 1990; Crawford *et al.*, 1991; Kok *et al.*, 1991). In both MDCK and Caco-2 polarized epithelial cells sorting of the sphingolipids C$_6$-NBD-glucosylceramide and C$_6$-NBD-sphingomyelin after their synthesis in the Golgi apparatus occurs, glucosylceramide being preferentially delivered to the apical membrane of the cell. The cells are grown as tight monolayers on filter supports. The basolateral domain of the cells, which faces the filter, is separated from the apical domain by tight junctions. These junctional complexes block the lateral diffusion of lipids in the outer leaflet of the plasma membrane (see Section 7.3.2). Since BSA can pass the pores in the filter the apical and basolateral membrane domains can thus be analysed separately for NBD-lipid content by a back-exchange. In this way, one can separate not only the plasma membrane NBD-lipid pool from the intracellular pool, but, in addition, pools of different plasma membrane domains (Van Meer *et al.*, 1987). In HT29 cells C$_6$-NBD-glucosylceramide is sorted from other sphingolipids (including sphingomyelin) in the endocytic pathway, being preferentially trans-

ported to the Golgi apparatus. Furthermore, this sorting phenomenon is only observed in the undifferentiated HT29 cell type and not in a differentiated line that was obtained by clonal selection (Kok *et al.*, 1991). Thus sphingolipid trafficking has been studied in relation to cell polarity and differentiation. The relevance of these kind of studies is clear when considering the involvement of (glyco)sphingolipids in processes such as cellular interaction, differentiation and oncogenesis, as reviewed by Hakomori (1981, 1984).

The observed sorting of sphingolipids both in the endocytic and biosynthetic pathways suggests that mechanisms are operative in cells that recognize, select and target specific sphingolipids. It is tempting to assume that proteins are involved in these kinds of processes. A promising approach to identify such putative sorting proteins is the use of 5-[^{125}I]iodonaphthyl-1-azide (INA). This is a photoactivatable probe which, upon activation with UV light, covalently couples to adjacent molecules, which thus become labelled with [125]I. Interestingly, the INA can also be indirectly activated through excitation of adjacent fluorophores (photosensitizers) (Raviv *et al.*, 1987). Conveniently also the NBD-labelled lipids can serve as appropriate photosensitizers for the INA (Rosenwald *et al.*, 1991; and our own unpublished observations). Thus one can, in principle, identify proteins that are in the vicinity of the NBD-lipids and therefore may be functionally related to these lipids. This approach offers the possibility of combining biochemical data on proteins involved in lipid trafficking with the (known) intracellular fate of the lipid, as revealed by fluorescence microscopy. This approach is currently employed in our laboratory and is compared to an approach involving the use of photoaffinity labelled (glyco)sphingolipids, that have a photoactivatable group in the fatty acid moiety (cf. Sonnino *et al.*, 1989).

A step nearer to the *in situ* situation in the organism is the use of primary cultured cells, like hepatocytes. Hepatocytes are polarized cells with bile canalicular membrane domains that are structurally and functionally separate from sinusoidal membrane domains. From liver so-called hepatocyte couplets can be isolated. A couplet is the smallest functioning hepatobiliary unit (Gautam *et al.*, 1987), that consists of two hepatocytes that enclose a bile canaliculus in between them. In this system biliary secretion of lipids can be studied in relation to intracellular trafficking of these lipids. Recently C_6-NBD-ceramide was used to study the intracellular effects of bile salts connected to their capacity to induce biliary lipid excretion (Crawford *et al.*, 1991). Taurocholate induces translocation of the Golgi apparatus to a peri-canalicular location after which excretion occurs of the metabolic products C_6-NBD-glucosylceramide and C_6-NBD-sphingomyelin in bile. One practical disadvantage of the couplet system in microscopical studies is the fact that these cells have a spherical shape, which makes focusing of intracellular structures containing fluorescent lipids difficult. However, this problem can be overcome by using confocal laser scanning microscopy to optically section the cells, which drastically improves the focus in the *z*-dimension. We are currently using this technique to study the exact localization of intracellular endocytic organelles in relation to the canaliculus between the couplet-forming cells.

Cellular processing of NBD-lipids can be easily analysed and quantified. In order to quantify trafficking or analyse metabolism, $1–5 \times 10^7$ cells are incubated with the lipid (precursor) at 37°C for the desired time. A back-exchange can then be performed to separate the plasma membrane NBD-lipid pool from the intracellular pool. Thereafter, both the back-exchange fraction and the cellular fraction are subjected to lipid extraction according to Bligh and Dyer (1959). Briefly, the volume to be extracted is mixed with an equal volume of methanol and two volumes of chloroform. The mixture is vortexed vigorously and centrifuged to separate phases. The lower chloroform phase is collected and the upper methanol/water phase re-extracted with an equal volume of chloroform. The two chloroform phases are combined and dried. The dried lipid is either directly measured (as described below) after dissolving it in 1% (vol./vol.) Triton X-100, or run on TLC for analysis of metabolism. For TLC analysis the dried lipid is taken up in a small (200 μl) volume of $CHCl_3/CH_3OH$ (1:1) and applied on high-performance (HP) TLC plates (Merck). NBD-lipids can be run in one dimension, or if this does not

allow sufficient separation, in two dimensions. A system employing the solvent $CHCl_3/CH_3OH/20\%$ (wt/vol.) NH_4OH (70:30:5) in the first and $CHCl_3/CH_3OH/HAc/H_2O$ (90:40:12:2) in the second dimension separates most of the neutral NBD-(glyco)sphingolipids, which can be identified by comparison to synthesized standards. The individual spots are quantified as follows. The spots are scraped from the plate and the silica is suspended in a 1% (vol./vol.) Triton X-100 solution. This solution is then shaken vigorously at 37°C for 1 h, followed by spinning down the silica. The supernatant is measured in a fluorometer against a standard curve of C_6-NBD-PC in 1% Triton X-100. Corrections should be made for differences in efficiency of removal of the NBD-lipids from the silica. (In the case of sphingomyelin the efficiency is 10% lower than for the other sphingolipids, independent of the absolute amounts.)

In the endocytic pathway (NBD-)lipids are catabolized. Examples are the breakdown of C_6-NBD-PA to C_6-NBD-DG (Pagano & Longmuir, 1985) and the deacylation of C_6-NBD-PC (Sleight & Pagano, 1984), both occurring in the plasma membrane. NBD-(glyco)sphingolipids are not deacylated during their intracellular trafficking. De/reacylation with C_6-NBD can be excluded since C_6-NBD, once liberated, is not reused for synthesis. Although deacylation does occur with radiolabelled sphingolipids (Trinchera *et al.*, 1990; and our own observations), it is convenient that trafficking and processing of NBD-(glyco)sphingolipids can be followed without losing the NBD label. The NBD-sphingolipids are degraded to NBD-ceramide, probably in the lysosomes (Koval & Pagano, 1989, 1990). However, during every cycle of the lipid from the plasma membrane into the cell and back to the plasma membrane (recycling) only a small fraction is degraded (Koval & Pagano, 1990; Kok *et al.*, 1989). The formed NBD-ceramide can move from the lysosomes to the Golgi apparatus, where it can be reused for the synthesis of sphingolipids (see Section 6.2.2). Thus, these lipids can be remodelled during their trafficking inside the cell according to the demands of the cell. Extensive remodelling is a long-term process (hours to days) compared to the timescale of trafficking, where an equilibrium situation is usually reached after *ca.* 30 min of incubation at 37°C. In HT29 cells equilibrium levels of sphingolipids synthesized from C_6-NBD-ceramide are reached after *ca.* 24 h of incubation at 37°C. A similar composition is reached when the cells are incubated with C_6-NBD-glucosylceramide or C_6-NBD-sphingomyelin, indicating that in the long term, complete remodelling occurs. In HeLa cells the more complex glycolipid GM1, when labelled with pyrene in the fatty acid moiety, gives rise to catabolic (GM2 and GM3) and anabolic (GD1a) derivatives after a 24 h incubation at 37°C, whereas after 2 h metabolic processing has not yet occurred (Masserini *et al.*, 1990).

In HT29 cells the composition of NBD-(glyco)-sphingolipids is changing upon differentiation of the cells (Babia *et al.*, manuscript submitted). These long-term metabolic differences related to differentiation are reflected in short-term differences in trafficking (see above). Since it is known that the ratio of the various sphingolipids synthesized from the precursor ceramide depends on the type of fluorescent probe used (as discussed in Section 7.2.2), one should not base firm conclusions on absolute amounts of synthesis. However, in the HT29 system, when studying metabolism related to differentiation, this problem is overcome, since only relative differences in metabolism are considered. Therefore, these studies do provide insight into the changing sphingolipid demands of the cell during its differentiation.

So far we have discussed the specific trafficking of individual lipid molecules, probed with fluorophores in the fatty acid moiety. These lipids do not monitor the overall membrane flow during, for example, endocytosis. However, the PE derivative described above, labelled with rhodamine in the headgroup (*N*-Rh-PE), can serve this purpose. Since the lipid in this case has a modified headgroup, it is no longer comparable to the natural PE, and is not recognized as such by the cell. This implies, for example, that the lipid derivative is not translocated across the membrane, as observed for acyl chain labelled PE (Section 7.3.2). It appears, however, that *N*-Rh-PE follows the general flow of membranes during endocytic membrane uptake, after its insertion into the plasma membrane (see Section 7.2.2). In fact, it turns out that *N*-Rh-PE follows the membrane flow in the fluid-phase endocytic pathway (Kok *et al.*, 1990), since (1) it is delivered to the lysosomes, as shown by microscopy and cellular fractionation; (2) it is not found in early endosomes that are labelled by FITC-Tf, a marker for the receptor-mediated endocytic pathway (unpublished observation); (3) it colocalizes with Lucifer yellow (a fluid-phase marker) throughout its intracellular trafficking pathway. Thus, *N*-Rh-PE can be used to monitor this pathway and as a marker for lysosomes, where it has accumulated after a 1 h incubation at 37°C.

The most commonly used fluorescent markers for fluid-phase endocytosis are Lucifer yellow and FITC-dextran (for a review see Swanson, 1989). However, Lucifer yellow appears to cross intracellular membranes in some cell types (Swanson, 1989), whereas the use of FITC-dextran of commercial sources may be frustrated by the presence of low molecular weight impurities that easily penetrate the cell (Preston *et al.*, 1987) (purification protocols now exist to remove these impurities (Cole *et al.*, 1990)). Thus, the intracellular

labelling observed with these fluid-phase markers may, in part, be a consequence of aspecific exchange instead of vesicular trafficking. Using *N*-Rh-PE, which is a non-exchangeable membrane-associated probe, this problem cannot arise. Furthermore, fluid-phase uptake also occurs as a consequence of receptor-mediated endocytosis, so fluid-phase probes do not necessarily discriminate between the pathways of pinocytosis and receptor-mediated endocytosis. The fluorescent phospholipid analogue, however, seems to do so, since it is not found to colocalize with ligands (Tf) probing the receptor-mediated pathway. Finally, because of its rhodamine fluorophore *N*-Rh-PE is a convenient probe for performing collabelling studies with NBD-lipids, to monitor their possible trafficking along the fluid-phase endocytic pathway.

7.3.2 Applications in membrane biology

Apart from following the intracellular fate of a defined species of a phospho- or glycosphingolipid analogue, there are numerous applications of fluorescent lipid and lipid-like derivatives, aimed at investigating the dynamics of membranes in general, including overall biophysical properties. It is virtually impossible to present even a far-from-complete overview. To be explicit, therefore, only a few typical examples are mentioned here. Other applications and further details can be found in several recent fluorescence textbooks (Laskowicz, 1983; Loew, 1988).

The occurrence of phase separations, phase transitions and the relative 'fluidity' of membranes has been determined with fluorescent probes such as diphenylhexatriene (DPH). The probe can also be coupled to phospholipids (Parente & Lentz, 1985) and the trimethylammonium salt derivative (TMA-DPH) has been employed in measuring membrane fusion during exocytosis (Bonner *et al.*, 1986; see below). DPH displays fluidity-sensitive fluorescence polarization characteristics (Shinitzky & Barenholz, 1978). In aqueous solutions, its fluorescence is negligible. Membranes are labelled by adding the probe, solubilized in tetrahydrofuran, to the system of interest at a relatively low concentration (micromolar). DPH fluorescence polarization, *p*, is calculated after measuring fluorescence (excitation and emission wavelengths for DPH are 360 and 428 nm respectively) with polarizers in crossed and parallel positions, using the equation:

$$P = \frac{I_{II} - I_{\perp}}{I_{II} + I_{\perp}}$$

where I_{II} and $I_{\perp}$ represent fluorescence intensities detected with the polarizers oriented in parallel and perpendicular positions, respectively, to the direction

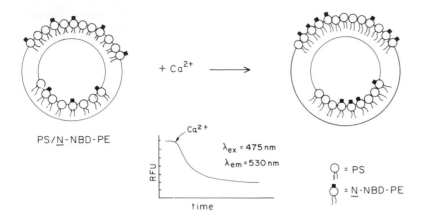

PS/<u>N</u>-NBD-PE

+ Ca²⁺ ⟶

Ca²⁺

λ_{ex} = 475 nm
λ_{em} = 530 nm

RFU

time

= PS

= <u>N</u>-NBD-PE

Figure 7.6 Phase separation in membranes, monitored by fluorescence self-quenching. When Ca^{2+} is added to N-NBD-PE (5 mol %)-containing PS liposomes, phase separation occurs, leading to a local increase of the concentration of the fluorescent lipid analogue, with a concomitant fluorescence self-quenching. This self-quenching can be continuously monitored in a fluorometer as a decrease in the NBD fluorescence signal.

of polarization of the excitation light. $I\perp$ is corrected for the intrinsic polarization of the instrument.

A typical feature of many fluorescent probes is their concentration-dependent self-quenching. NBD-labelled lipid analogues also show this prominent behaviour which has been exploited, among others, to monitor the occurrence and kinetics of lipid phase separations in membranes (Hoekstra, 1982a; Silvius & Gagne, 1984). The principle of the assay is shown in Fig. 7.6. It involves the incorporation of an acyl (C_6-NBD) or headgroup-labelled (N-NBD-PE) phospholipid analogue in lipid bilayers at a concentration of approximately 5 mol % with respect to total lipid. The occurrence of phase separation leads to an increase in the apparent concentration of the fluorescent lipid analogue as it is 'squeezed out' of the separating phases. The phase separation is thus revealed as a quenching of fluorescence, the kinetics of which can be monitored continuously in a fluorometer. DPH- and pyrene-labelled phospholipid analogues can similarly be used for this purpose. Relative clustering of monomers will enhance excimer fluorescence (Kido *et al.*, 1980; Parente & Lentz, 1986). Among others, these types of experiments have been of relevance in analysing and interpreting the role of lipid-phase separations in mechanisms involved in membrane fusion of artificial membrane systems (Hoekstra, 1982b; Silvius & Gagne, 1984).

NBD-labelled lipid analogues have also been employed in studies involving polymorphic transitions of certain phospholipids (Hong *et al.*, 1988; Stubbs *et al.*, 1989). In general, the fluorescent properties of a probe are quite sensitive to the environment of the probe. For example, when N-NBD-PE is incorporated in a bilayer-forming phospholipid system, the headgroup protrudes into the aqueous bilayer/water interface, sensing a polar environment (see, for example, Chattopadhyay & London, 1988; Chattopadhyay, 1990). In an apolar environment – which can be simulated by determining spectral properties of the fluorescent probe in an apolar environment using defined organic solvents of distinct polarity (see, for example, Hoekstra *et al.*, 1984) – the fluorescence quantum yield commonly increases while there is a blue shift in the emission maximum (Edidin, 1981, 1992). The latter phenomenon is seen when the N-NBD-PE analogue is incorporated at relatively low concentrations (<1 mol %) into lipid bilayers that contain lipids which can adopt non-bilayer structures, such as a hexagonal H_{II}-phase. Thus a transition from bilayer to non-bilayer structure can be reported by such labelled lipid analogues as an increase in fluorescence, occurring around the temperature at which such a transition can take place. The underlying mechanism of these fluorescence changes involves, however, a dehydration of the lipid phase (Stubbs *et al.*, 1989). Thus, this feature is not 'typical' for an H_{II} transition and it is evident, therefore, that one cannot solely rely on shifts in quantum yield and emission wavelength maxima when determining non-bilayer lipid transitions. Thus far, the application has only been reported in artificial membrane systems.

Lipid exchange or transfer is a prominent aspect of studies involving the dynamics of lipids and membranes (Fig. 7.7). In studies of membrane fusion, such events might interfere with assays that rely on lipid mixing and, hence, should be excluded in those particular cases (see below). However, biologically relevant lipid transfer processes have been claimed, catalysed by specific and non-specific lipid exchange proteins. Fluorescent lipid probes have been elegantly applied to measure the kinetics of protein-mediated phospholipid exchange between artificial bilayers (Nichols & Pagano, 1983). For this purpose, the approach was based on the principle of resonance energy transfer using N-Rh-PE as the energy acceptor and various NBD-labelled phospholipid species. Donor vesicles were labelled with 1 mol % each of N-Rh-PE and the NBD-labelled lipid of interest. In this case C_{12}-NBD lipid derivatives were used which,

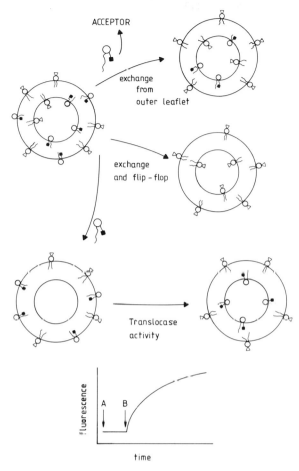

Figure 7.7 Lipid exchange, flip-flop and translocation. Starting from the upper left vesicle, which contains both a fatty-acyl labelled NBD(■)-lipid and the headgroup labelled N-rhodamine($\triangle$)-PE in both leaflets, the NBD-lipid can be exchanged from the outer leaflet, either spontaneously (C_6-NBD-analogues) or by an exchange protein (C_{12}-NBD-analogues). This results in the upper right situation. When the NBD-lipid can also flip-flop (for instance, C_6-NBD-ceramide) from the inner to the outer leaflet, the probe can be completely removed from the starting vesicle, provided that a sufficient amount of acceptor (membrane) is present.

In the lower part of the figure the insertion of a NBD-lipid in the outer leaflet (already containing N-Rh-PE) of a vesicle is depicted, followed by its translocation to the inner leaflet by a translocase activity. Insertion leads to efficient energy transfer (level A in kinetic curve).

In the case of both exchange and translocation a decrease of RET between the NBD-lipid and the N-Rh-PE occurs (B), that can be measured in the fluorometer as an increase of the NBD fluorescence signal.

compared to C_6-NBD derivatives (see previous section), show a relatively low spontaneous intermembrane transfer rate, i.e. negligible on the time-scale of the exchange experiments. Upon addition of unlabelled acceptor vesicles and the specific bovine liver PC-transfer protein or the bovine liver non-

specific lipid transfer protein, the occurrence and lipid specificity of transfer were monitored. This was revealed by the relief of NBD fluorescence quenching, when energy transfer with N-Ph-PE, occurring in the original doubly labelled bilayers, is eliminated. The latter occurred when NBD-lipid, but not N-Rh-PE, transferred to unlabelled acceptor membranes, catalysed by the transfer protein. With the PC-specific exchange protein, the headgroup specificity of the various NBD analogues was maintained, while the acceptor membrane was found to require PC. Furthermore, with the non-specific transfer protein the same order of lipid specificity was seen as that of natural lipids (PA>PC>PE). The authors concluded, therefore, that the data obtained with NBD-lipids are fully consistent with data obtained by other techniques (see Nichols & Pagano, 1983). Evidently, in this case applications of fluorescent lipid analogues offers the advantage over radiolabelled assays that separation of reactants and products is not required. Pyrene-labelled phospholipids have also been used in protein-mediated lipid transfer studies. Density-dependent changes in excimer fluorescence were used as a measure for lipid transfer (Massey *et al.*, 1985; Somerharju *et al.*, 1987). This procedure is less sensitive than the RET approach and requires, therefore, a higher lipid probe concentration.

In solving issues related to protein structure and function, resonance energy transfer measurements are frequently applied to determine intra- and inter-molecular distances. A similar approach was recently applied in studies aimed at understanding the functioning of specific translocases in biological membranes that mediate (transbilayer) flip-flop of certain lipids (Connor & Schroit, 1987). It is assumed that phospholipids are asymmetrically distributed in biological membranes, aminophospholipids such as phosphatidylserine (PS) and phosphatidylethanolamine (PE) residing in the inner leaflets of natural membranes, while choline-containing lipids such as PC and SM are found mainly in the outer leaflet (Op den Kamp, 1979). Various experimental approaches have indeed shown that exogenously added PS and PS analogues, including spin-labelled or fluorescently tagged derivatives, translocate in an ATP-dependent and protein-mediated manner from initial site of insertion in the outer leaflet to the inner leaflet, thereby adopting an asymmetric distribution (Seigneuret & Deveaux, 1984; Daleke & Huestis, 1985; Tilley *et al.*, 1986; Zachowski *et al.*, 1987; Martin & Pagano, 1987; Connor & Schroit, 1987). When adding C_6-NBD derivatives of PE and PS to cultured cells (fibroblasts), it can be seen that in contrast to the PC analogue, the former lipids rapidly internalize at a threshold temperature of about 7°C, i.e. at conditions where commonly operating endocytic mechanisms are not yet active

(Martin & Pagano, 1987). The transbilayer movement was found to be stereospecific. These studies indicated that intracellular membranes became labelled as a result of ATP-dependent flip-flop of the exogenously supplied lipids, followed by diffusion of the lipid analogues through the cytosol between inner leaflet and other intracellular membranes. In erythrocyte membranes – in which the translocase activity has been most extensively characterized thus far (see for example, Connor & Schroit, 1988), a resonance energy transfer technique has been applied showing the occurrence of the selective translocation of exogenously supplied C_6-NBD-PS to the inner leaflet, whereas C_6-NBD-PC remained in the outer leaflet (Connor & Schroit, 1987). As energy acceptor N-Rh-PE was used. The principle of the approach and the interpretation of the results were based upon the rationale that a symmetric distribution of the probes should give rise to a strong energy transfer efficiency whereas relatively weak energy transfer should be seen when the probes are in opposing leaflets. These qualitative distinctions can be readily revealed by this approach, but quantitative calculations as to the exact distances between donor and acceptor lipid are, in all probability, hampered by the fact that acyl chain-labelled NBD, due to its polar character, may have a tendency to loop back to the bilayer/water interface (Chattopadhyay & London, 1988), while exogenous insertion of N-Rh-PE (as done in this study) may lead to a non-random and clustered distribution of the probe (Kok *et al.*, 1990). Since energy transfer depends on the concentration of the energy acceptor this may lead to an erroneous outcome. This may be particularly the case when the calculations are also based upon and compared with model systems, where probe randomization occurs when preparing the bilayer. In spite of this quantitative limitation the RET approach may give qualitative insight as further supported by several independent control experiments, described in the study of Connor and Schroit (1987). It was demonstrated that the exogenously inserted NBD-PS was inaccessible to back-exchange. Also, the lipid analogue is not derivatized by TNBS. TNBS derivatization was seen when the PS transporter activity was inhibited by using inhibitors such as sulphydryl-reactive compounds.

Finally, the occurrence of *spontaneous* flip-flop can also be demonstrated to occur, using RET. As noted above (previous section), NBD-ceramide shows a remarkable tendency to flip-flop, i.e. the analogue rapidly gains intracellular access when endocytic internalization is not operative. In model membranes, ceramide and N-Rh-PE were symmetrically incorporated under conditions of efficient energy transfer (usually 1 mol % each). Subsequently, a large excess of unlabelled membranes was added. A reduction in energy transfer efficiency of about 50% would be anticipated when only the outer leaflet lipid would be accessible to transfer/exchange. On the other hand, complete removal and hence flip-flop ability of the NBD-ceramide should be reflected by a complete elimination of resonance energy transfer. Indeed, the latter was observed (Pagano, 1989), indicating and confirming the high susceptibility of NBD-ceramide to engage in spontaneous flip-flop. In this regard it should be noted that spontaneous flip-flop of natural lipids is usually in the order of days. The kinetics of exchange/transfer are probably codetermined by the chemical and physical nature of the derivative, including polarity properties. In fact, the probe-dependent metabolic fate of the analogues further exemplifies the point (see, for example, Pagano & Martin, 1988).

The artificial nature of certain lipid(-like) probes and the ability to undergo flip-flop as a result of that nature, has been of relevance in defining the role of tight junctions in maintaining distinct lipid domains in polarized cells. In such cells, the lipid compositions of the apical and basolateral membranes are different, the apical membrane being particularly enriched in (glyco)sphingolipids. By adding various fluorescent lipid analogues to the apical or basolateral plasma membrane domains of polarized epithelial cells it has been shown that tight junctions prevent randomization of lipids between the two membrane domains, provided that the lipids are restricted to the outer leaflet of the plasma membrane bilayer (Dragsten *et al.*, 1981; Spiegel *et al.*, 1985; Van Meer & Simons, 1986). In contrast, lipids present in the inner leaflet of the plasma membrane are free to mix by lateral diffusion.

The experiments were carried out with confluent monolayers of epithelial cells, grown in such a way (among others, by growing the cells on filters) that it enabled selective labelling of apical membranes. Dragsten *et al.* (1981) selected four lipid(-like) probes (Fig. 7.1) that were inserted into the apical membrane by adding them to medium bathing the apical surface of a cell monolayer. Fluorescence microscopy was employed to visualize the distribution of the probes. Restriction of the probe to the apical membrane can be seen as incomplete rings or arcs of fluorescence obtained by focusing the microscopy part way up a dome of cells so that apical and basolateral surfaces of individual cells are in focus simultaneously. Thus labelling the cells with C_6-NBD-PC, N-hexadecanoyl amino fluorescein or rhodamine-labelled gangliosides (Spiegel *et al.*, 1985) showed a fluorescence distribution typical of probes exclusively restricted to the apical surface displaying lateral diffusion coefficients of *ca.* 10^{-8} cm^2 s^{-1}, similar to other lipid probes (Edidin, 1992).

However, both $diIC_{16}$ (a hexadecyl carbocyanine)

and dodecanoyl amino fluorescein labelled the entire cell membrane, in spite of their initial insertion in the outer leaflet of the apical membrane only. This suggests their passage over the tight junction. By using agents capable of quenching the fluorescence of the lipid probe specifically it could be shown that $diIC_{16}$ and dodecanoyl amino fluorescein were capable of flipping across the apical membrane, thus gaining access to the inner leaflet of that domain. It was suggested, therefore, that the tight junction apparently presents an impassable barrier only to lipid components that are exclusively present in the outer leaflet of the plasma membrane. Lipids that have gained access to the inner leaflet diffuse freely, leading to intermixing of inner leaflet apical and basolateral lipids.

Evidently, proof for such a hypothesis would be obtained by directly inserting one and the same lipid either exclusively in the outer leaflet of the apical membrane or alternatively in either the inner or both leaflets, which would reveal the distinct differences in subsequent lipid distribution. Only in the latter case should randomization over all membrane domains occur. Such an experiment became possible when it was recognized that certain viruses can bud from specific epithelial membrane domains, either apical or basolateral (Hoekstra, 1990b; Hoekstra & Nir, 1991). The experiment (Van Meer & Simons, 1986) involved the use of cells specifically expressing the influenza fusion glycoprotein HA at the apical surface obtained by infecting the cells with influenza virus. Subsequently, lipid vesicles were prepared with a symmetric or asymmetric distribution of N-Rh-PE. The asymmetric localization of N-Rh-PE (in the outer leaflet of the liposome) was obtained when preparing the vesicles by a reconstitution procedure using octyl-glucoside. In addition, the vesicles contained a ganglioside acting as an anchorage site for the viral HA protein, which will also cause the vesicles to fuse with the cell surface. By this procedure N-Rh-PE could be inserted almost exclusively in the outer leaflet of the apical membrane (with asymmetrically labelled vesicles) or in both leaflets using the symmetrically labelled vesicles. In the former case, the lipid probe did not pass the tight junction whereas in the latter case rapid diffusion to the basolateral plasma membranes also occurred.

These results thus elegantly confirmed the results by Dragsten *et al.* (1981) that the tight junction forms an exclusive diffusion barrier in the *exoplasmic* leaflet of the plasma membrane, while lipids in the cytoplasmic leaflet appear to be capable of freely diffusing between both domains. With regard to the application of fluorescent lipid probes it should be emphasized that in this particular case physico-chemical properties determine localization of the probe rather than specific sorting and/or recognition phenomena (see previous section). When inserted in the apical membrane, any fluorescent lipid incapable of flip-flop will remain in the apical domain. Thus, the rhodamine-labelled ganglioside (Spiegel *et al.*, 1985) being restricted to the apical surface is, by itself, no proof for enrichment of glycosphingolipids in the exoplasmic apical membrane.

In the past decade, considerable progress has been made in understanding the mechanism(s) of membrane fusion, an event crucial to numerous cell biological phenomena such as endocytic and biosynthetic transport processes, the infectious entry of viruses into cells and fertilization (Ohki *et al.*, 1988; Wilschut & Hoekstra, 1990). An important aspect of these studies has been the ability to reveal the occurrence of fusion in a simple and reliable manner and to monitor this process in a continuous fashion so that parameters affecting the fusion event (thus providing insight as to how fusion is accomplished) can be readily revealed. The application of appropriate fluorescent lipid(-like) probes has played a prominent role in this respect (Düzgüneş & Bentz, 1988; Hoekstra, 1990a; Hoekstra & Klappe, 1993). The type of lipid probes that are crucial to fusion assays are those that remain associated with the membranes with which they were first associated. Thus the lipid probes of choice should be essentially non-exchangeable. This implies, therefore, that those probes that contain short-chain fatty acids (such as C_6-NBD-lipids) and which show a relatively enhanced water-solubility, are not suitable for application in assays reporting membrane fusion. Rather, typical probes for a lipid mixing assay are those that have the fluorophore coupled to lipid headgroups, such as the N-NBD-PE/N-Rh-PE couple (Struck *et al.*, 1981) or to relatively long acyl or alkyl chains (as in the lipid-like probe, octadecyl rhodamine B chloride, R_{18} (Hoekstra *et al.*, 1984) or in case of the RET couple (12-CPS)-18PC/(12-DABS)-18-PC, in which CPS[[N-[4-[7-(diethylamino)-4-methylcoumarin-3-yl]phenyl]carbamoyl]methyl]thio and DABS (4-(4-(dimethylamino)-phenyl)sulphonyl)methyl amino are coupled to the glycerol backbone via C_{12} spacers (Silvius *et al.*, 1987). As noted above, the effective length of the acyl chain will ensure the probe's proper intercalation in and association with the membrane, as observed for natural lipids.

Application of fluorescent lipid probes in studies of membrane fusion has generally followed the principle of monitoring the relief of fluorescence 'quenching', either as a result of resonance energy transfer or self-quenching, occurring when labelled membranes fuse with non-labelled membranes (Fig. 7.8). Over other assays, the use of fluorescence assays to monitor fusion offers a number of advantages. These include, among

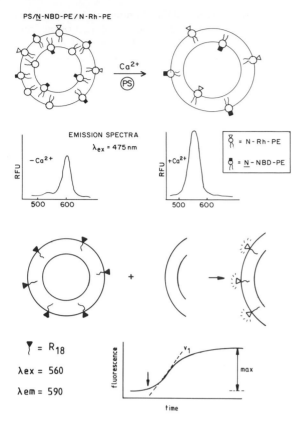

PS/*N*-NBD-PE/*N*-Rh-PE

EMISSION SPECTRA
λ_{ex} = 475 nm

$-Ca^{2+}$

$+Ca^{2+}$

= N - Rh - PE

= N - NBD - PE

Υ = R_{18}

λex = 560

λem = 590

Figure 7.8 Lipid-mixing assays to monitor membrane fusion. Membrane fusion can be measured by employing the *N*-NBD-PE/*N*-Rh-PE couple. Fusion is monitored by measuring the increase in fluorescence emission of the NBD fluorophore (see emission spectra). This increase is a result of the decrease in quenching of the NBD by the rhodamine (due to resonance energy transfer) upon dilution of both probes after mixing of the lipids with those of the unlabelled fusion target membrane.

In the lower part of the figure, the principle of relief of self-quenching is depicted, which can be employed to measure fusion using R_{18} as a probe. When a vesicle containing a high (self-quenching) concentration of R_{18} fuses with an unlabelled vesicle, dilution of the R_{18} probe occurs, resulting in an increase of rhodamine fluorescence (see kinetic curve). With both assays fusion can be continuously monitored, allowing determination of parameters such as the initial fusion rate (V_1) and the extent of fusion, which are frequently used to describe the fusion process.

others, its relatively high sensitivity, its versatility and convenience, its ability to detect and monitor the onset and kinetics of the fusion process, and the relative ease with which quantitative data are required. In addition, in cases in which fusion of biological membranes is involved, it is usually possible to examine the fusion event by fluorescence microscopy.

Pyrene-labelled lipid probes have been widely used to study various aspects of membrane dynamics, including membrane fusion (Schenkman *et al.*, 1981; Pal *et al.*, 1988). The probe is usually attached to the

acyl chain (of at least 16 carbons in length, to avoid spontaneous transfer), thus substituting for one of the lipid's fatty acids. Since insertion as such in a biological membrane is difficult to accomplish, the derivatized lipid is commonly used in a semi-artificial system, involving the interaction of artificial bilayers with biological membranes or in a fusion model system consisting of reconstituted biological membranes, as, for example, in the case of reconstituted viral envelopes (Anselem *et al.*, 1986; Hoekstra, 1990b). A particular advantage of pyrene-fatty acid probes is, however, that they can become metabolically incorporated into 'natural' phospholipids. This approach has been taken to study the fusion properties of an enveloped virus (vesicular stomatitis virus), which infects a cell by fusion with intracellular membranes after receptor-mediated endocytic internalization (Pal *et al.*, 1988). The principle of measuring fusion with pyrene-labelled lipids involves the (discontinuous) determination of changes in the so-called excimer/monomer fluorescence intensity ratio, occurring when pyrene-labelled membranes fuse, and thus dilute, with non-labelled membranes (cf. Hoekstra & Klappe, 1993). The fluorescence emission spectrum of pyrene is sensitive to the concentration of the probe. At relatively high (5–10 mol %) concentrations, the emission fluorescence shows a typical 'excimer' emission at 470 nm (excitation wavelength is 330 nm). The excimeric state of excited pyrene dimers reduces in number when the concentration decreases (Pownall & Smith, 1989), occurring when labelled membranes fuse with non-labelled membranes. This reduction is correlated to the emitted fluorescence derived from monomers, which is detected at an emission wavelength of 385 nm. Since the *E/M* ratio is linearly dependent on the concentration of probe up to *ca.* 7 mol %, the change in ratio can be directly correlated to the extent of fusion.

The fluorescent lipid-like probe octadecyl rhodamine B chloride (R_{18}) is extensively used to study fusion between intact biological membranes, in particular the fusion between viruses and cells (Fig. 7.8). The principle relies on the surface density-dependent self-quenching properties of the dye, implying that the fluorescence intensity increases when the density of the probe decreases and vice versa (Hoekstra *et al.*, 1984). The approach involves the labelling of one membrane preparation at an R_{18} concentration such that fluorescence self-quenching occurs. Fusion is then monitored continuously, at excitation and emission wavelengths of 560 and 590 nm, respectively, in a fluorometer. Essentially, the increase in rhodamine fluorescence is followed, occurring when the labelled membrane merges with the unlabelled membrane.

In practice, the procedure can be carried out as

follows. Membrane labelling is done by rapidly injecting an ethanolic solution (final concentration <1% (v/v)) of R_{18} in the membrane mixture, while vigorously vortexing. The suspension is left in the dark for approximately 20–30 min, although shorter incubation periods suffice. Non-incorporated probe – upon labelling of Sendai virus some 20–30% – is removed by gel filtration (Sephadex G75, 1×15 cm), and adsorbs on top of a column. The final concentration of the probe in the membrane that is labelled should be 6–7 mol %, so as to obtain a self-quenching of approximately 70–80%. This is determined by adding Triton X-100 (1% (v/v)) which dilutes the probe infinitely and does not affect its quantum yield. The detergent is also used to calibrate the fluorescence scale by setting the fluorescence of the labelled membrane preparation at the 'zero' level (no fusion) while 100% is obtained after addition of Triton X-100, and a correction for sample dilution. Fluorescence dequenching is thus used as a reflection of the occurrence of fusion, while the rate of dequenching corresponds to the rate of fusion. In this respect, it is important to establish that probe redistribution is not rate-limiting, i.e. the kinetics of dequenching should accurately reflect the kinetics of the fusion reaction. In spite of the *assumption* that probe mixing is very fast and not rate-limiting in the overall fusion event, the reasonability of this assumption has only been proven in a few cases (Rubin & Chen, 1990). Thus, redistribution of R_{18} is not rate-limiting in virus–cell fusion but is close to rate-limiting in viral spike–protein-mediated cell–cell fusion.

Apart from fluorometric measurements, appropriate fluorescent probes also offer the possibility to study fusion events by advanced fluorescence microscopic techniques. Recently, a technique has been developed that employs digital fluorescence imaging to detect and quantify fusion of viruses with cells by sequential imaging obtained at low virus/cell ratios (Georghiu et al., 1989). In this case, single cells can be analysed in contrast to an average response of an entire cell population, as obtained by fluorometric measurements. Single cell analysis has also been carried out by application of a technique involving fluorescence recovery after photobleaching (Aroeti & Henis, 1986). In the latter case the lateral mobility of a fluorescent derivative such as N-NBD-PE has been monitored after interaction of reconstituted viral envelopes, containing the probe, with cells. The occurrence of fusion is then determined by measuring the mobile fraction of the cell-associated fluorophore. This is determined by following the recovery of fluorescence after bleaching of a relatively small spot on the cell surface with a high-energy laser beam. Note that this approach is similar to that used to measure diffusion rate constants for membrane-inserted lipid probes (see Section 7.2.2).

In principle, the N-NBD-PE, as an energy donor, can also be employed in conjunction with an energy acceptor, such as N-Rh-PE to measure fusion by resonance energy transfer, at probe concentrations low enough (usually less than 1 mol % with respect to total lipid) to avoid self-quenching (Fig. 7.8). The principle of resonance energy transfer has been described above. Numerous combinations of probes fulfilling the typical features of an overlap between donor emission spectrum and excitation spectrum of the acceptor have been employed for the purpose of determining and monitoring fusion (see, for example, Gibson & Loew, 1979; Vanderwerf & Ullman, 1980; Uster & Deamer, 1981; Loew, 1988; Düzgünes & Bentz, 1988). Usually, the relief of energy transfer is followed by measuring the increase in donor fluorescence as a function of time (Fig. 7.8), occurring when labelled membranes fuse with unlabelled target membranes.

Assays based on resonance energy transfer have been applied primarily in fusion involving artificial membranes. This is due to the fact that spontaneous insertion of the RET probes in a manner suitable for efficient energy transfer is difficult to accomplish by exogenous addition of the probes, as would be required for incorporation in biological membranes. However, liposomes or reconstituted biological membranes allow rapid incorporation, since the fluorescent lipid probes are readily incorporated during the procedure of preparing these membrane preparations (see, for example, Hoekstra & Klappe, 1993).

Finally, as already noted above, particularly in studies involving the use of fluorescent lipid probes as reporters of membrane fusion, it is essential to exclude that lipid mixing has been accomplished by spontaneous transfer or flip-flop of the *probes* between membranes. It is therefore crucial that with any probe or combination of lipid probes used for this purpose and in the system in which fusion is studied, appropriate control experiments are carried out. This can be done, for example, by comparing different assays in the same system or by using a particular assay in a system at conditions where fusion is known not to occur as by using specific inhibitors or by proteolytically cleaving proteins as in the case of viruses. Provided that such controls are carried out properly, these assays can be valuable tools for analysing mechanisms of membrane fusion, adding another versatile aspect to employment of fluorescent lipid probes in the (cell) biology of membranes.

ACKNOWLEDGEMENTS

Work cited in this paper and carried out in the authors' laboratory was supported by NIH Grant AI

255534 and by The Netherlands Organization for Scientific Research (NWO/SON). The secretarial assistance of Mrs Rinske Kuperus is gratefully acknowledged.

REFERENCES

Acquotti D., Sonnino S., Masserini M., Casella L., Fronza G. & Tettamanti G. (1986) *Chem. Phys. Lip.* **40**, 71–86.

Amselem S., Barenholz Y., Loyter A., Nir S. & Lichtenberg D. (1986) *Biochim. Biophys. Acta* **860**, 301–313.

Aroeti B. & Henis Y.I. (1986) *Biochemistry* **25**, 4588–4596.

Axelrod D., Koppel D.E., Schlessinger J., Elson E. & Webb W.W. (1976) *Biophys. J.* **16**, 1055–1069.

Bligh E.G. & Dyer W.J. (1959) *Can. J. Biochem. Physiol.* **37**, 911–917.

Bronner C., Landry Y., Fonteneau P. & Kuhry J.G. (1986) *Biochemistry* **25**, 2149–2154.

Chattopadhyay A. (1990) *Chem. Phys. Lip.* **53**, 1–15.

Chattopadhyay A. & London E. (1988) *Biochim. Biophys. Acta* **938**, 24–34.

Chiechanover A., Schwartz A.L., Dautry-Varsat A. & Lodish H.F. (1983) *J. Biol. Chem.* **258**, 9681–9689.

Cole L., Coleman J., Evans D. & Hawes C. (1990) *J. Cell. Sci.* **96**, 721–730.

Comfurius P., Bevers E.M. & Zwaal F.A. (1990) *J. Lipid Res.* **3**, 1719–1721.

Connor J. & Schroit A.J. (1987) *Biochemistry* **26**, 5099–5105.

Connor J. & Schroit A.J. (1988) *Bochemistry* **27**, 848–851.

Crawford J.M., Vinter D.W. & Gollan J.L. (1991) *Am. J. Physiol.* **260**, G119–G132.

Daleke D.L. & Huestis W.H. (1985) *Biochemistry* **24**, 5406–5416.

Dragsten P.R., Blumenthal R. & Handler J.S. (1981) *Nature* **294**, 718–722.

Dunn K.W. & Maxfield F.R. (1990) In *Noninvasive Techniques in Cell Biology*, J.K. Foskett & S. Grinstein (eds). Wiley-Liss, New York, pp. 153–176.

Düzgüneş N. & Bentz J. (1988) In *Spectroscopic Membrane Probes*, L.D. Loew (ed.). CRC Press, Boca Raton, pp. 117–159.

Edidin M. (1981) In *New Comprehensive Biochemistry*, Vol. I, A. Neuberger & L.L.M. van Deenen (eds). Elsevier, Amsterdam, pp. 37–82.

Edidin M. (1992) In *The Structure of Biological Membranes*, P.L. Yeagle (ed.). CRC Press, Boca Raton, pp. 539–572.

Foster T. (1948) *Ann. Phys. (Leipzig)* **2**, 55–75.

Frank A., Bazenholz Y., Lichtenberg D. & Thompson T.E. (1983) *Biochemistry* **22**, 5647–5651.

Fung B.K.-K. & Stryer L. (1978) *Biochemistry* **17**, 5241–5248.

Gardam M.A., Itovitch J.J. & Silvius J.R. (1989) *Biochemistry* **28**, 884–893.

Gautam A., Ng O.-C. & Boyer J.L. (1987) *Hepatology* **7**, 216–223.

Georghiu G., Morrison I.E.G. & Cherry R.J. (1989) *FEBS Lett.* **250**, 487–492.

Gibson G.A. & Loew L.M. (1979) *Biochem. Biophys. Res. Commun.* **88**, 135–140.

Goda S., Kobayashi T. & Goto I. (1987) *Biochim. Biophys. Acta* **920**, 259–264.

Gordesky S.E., Marinetti G.V. & Love R. (1975) *J. Membr. Biol.* **20**, 111–132.

Hakomori S.-I. (1981) *Ann. Rev. Biochem.* **50**, 733–764.

Hakomori S.-I. (1984) *Trends Biochem. Sci.* **9**, 453–458.

Hoekstra D. (1982a) *Biochemistry* **21**, 1055–1061.

Hoekstra D. (1982b) *Biochemistry* **21**, 2833–2840.

Hoekstra D. (1990a) *Hepatology* **12**, 615–665.

Hoekstra D. (1990b) *J. Bioenerg. Biomembr.* **22**, 121–155.

Hoekstra D. & Klappe K. (1993) *Methods Enzymol.* (in press).

Hoekstra D. & Nir S. (1991) In *The Structure of Biological Membranes*, P.L. Yeagle (ed.), CRC Press, Boca Raton, pp. 949–996.

Hoekstra D., De Boer T., Klappe K. & Wilschut J. (1984) *Biochemistry* **23**, 5675–5681.

Hoekstra D., Klappe K., Stegmann T. & Nir S. (1988) In *Molecular Mechanisms of Membrane Fusion*, S. Ohki, D. Doyle, S.W. Hui & E. Mayhew (eds). Plenum Press, New York, pp. 399–412.

Hoekstra D., Eskelinen S. & Kok, J.W. (1989) In *Organelles in Eukaryotic Cells. Molecular Structure and Interactions*, J.M. Tager, A. Azzi, S. Papa & F. Guerreri (eds). Plenum Press, New York, pp. 59–83.

Hong K., Baldwin P.A., Allen T.M. & Papahadjopoulos D. (1988) *Biochemistry* **27**, 3947–3955.

Jacobsen K., Derzko Z., Wu E.S., Hou Y. & Poste G. (1976) *J. Supramol. Struct.* **5**, 565–576.

Keller P.M., Person S. & Snipes W. (1977) *J. Cell Sci.* **28**, 167–177.

Kido N., Tanaka F., Kaneda N. & Yagi K. (1980) *Biochim. Biophys. Acta* **603**, 255–265.

Kishimoto Y. (1975) *Chem. Phys. Lip.* **15**, 33–36.

Kobayashi T. & Pagano R.E. (1989) *J. Biol. Chem.* **264**, 5966–5973.

Kok J.W., Eskelinen S., Hoekstra K. & Hoekstra D. (1989) *Proc. Natl. Acad. Sci. USA* **86**, 9896–9900.

Kok J.W., ter Beest M., Scherphof G. & Hoekstra D. (1990) *Eur. J. Cell Biol.* **53**, 173–184.

Kok J.W., Babia T. & Hoekstra D. (1991) *J. Cell Biol.* **114**, 231–239.

Koval M. & Pagano R.E. (1989) *J. Cell Biol.* **108**, 2169–2181.

Koval M. & Pagano R.E. (1990) *J. Cell Biol.* **111**, 429–442.

Koval M. & Pagano R.E. (1991) *Biochim. Biophys. Acta* **1082**, 113–125.

Kundu S.K. (1981) *Meth. Enzymol.* **72**, 185–204.

Laskowicz J.R. (ed.) (1983) *Principles of Fluorescence Spectroscopy*. Plenum Press, New York.

Lipsky N.G & Pagano R.E. (1982) *Proc. Natl. Acad. Sci. USA* **80**, 2608–2612.

Lipsky N.G. & Pagano R.E. (1985a) *J. Cell Biol.* **100**, 27–34.

Lipsky N.G. & Pagano R.E. (1985b) *Science* **228**, 745–747.

Loew L.M. (ed.) (1988) *Spectroscopic Membrane Probes*, Vol. 1-III, CRC Press, Boca Raton.

McLean I.W. & Nakane P.K. (1974) *J. Histochem. Cytochem.* **22**, 1077–1083.

Martin O.C. & Pagano R.E. (1987) *J. Biol. Chem.* **262**, 5890–5898.

Masserini M., Giuliani A., Palestini P., Acquotti D., Pitto M., Chigorno V. & Tettamanti G. (1990) *Biochemistry* **29**, 697–701.

Massey J.B., Hickson-Bick D., Via D.P., Gotto Jr. A.M. & Pownall H.J. (1985) *Biochim. Biophys. Acta* **835**, 124–131.

Miller-Podraza H. & Fishman P.H. (1982) *Biochemistry* **21**, 3265–3270.

Moreau R.A. (1989) *Lipids* **24**, 691–699.

Mukaiyama T., Matsueda R. & Suzuki M. (1970) *Tetrahedron Lett.* **22**, 1901–1904.

Nichols J.W. & Pagano R.E. (1983) *J. Biol. Chem.* **258**, 5368–5371.

Ohki S., Doyle D., Flanagan T., Hui S. & Mayhew E. (eds) (1988) *Molecular Mechanisms of Membrane Fusion.* Plenum Press, New York.

Op den Kamp J.A.F. (1979) *Ann. Rev. Biochem.* **48**, 47–71.

Pagano R.E. (1989) *Methods Cell Biol.* **29**, 75–85.

Pagano R.E. & Longmuir K.J. (1985) *J. Biol. Chem.* **260**, 1909–1916.

Pagano R.E. & Martin O.C. (1988) *Biochemistry* **29**, 4439–4445.

Pagano R.E. & Sleight R.G. (1985) *Science* **229**, 1051–1057.

Pagano R.E., Martin O.C., Schroit A.J. & Struck D.K. (1981) *Biochemistry* **20**, 4920–4927.

Pagano R.E., Longmuir K.J. & Martin O.C. (1983) *J. Biol. Chem.* **258**, 2034–2040.

Pagano R.E., Sepanski M.A. & Martin O.C. (1989) *J. Cell Biol.* **109**, 2067–2079.

Pagano R.E., Martin O.C., Kang H.C. & Haughland R.P. (1991) *J. Cell Biol.* **113**, 1267–1279.

Pal R., Barenholz Y. & Wagner R.R. (1988) *Biochemistry* **27**, 30–36.

Parente R.A. & Lentz B.R. (1985) *Biochemistry* **24**, 6178–6185.

Parente R.A. & Lentz B.R. (1986) *Biochemistry* **25**, 1021–1026.

Pownall H.J. & Smith L.C. (1989) *Chem. Phys. Lip.* **50**, 191–211.

Prendergast F.G., Haughland R.P. & Callahan P.J. (1981) *Biochemistry* **20**, 7333–7338.

Preston R.A., Murphy R.F. & Jones E.W. (1987) *J. Cell Biol.* **105**, 1981–1987.

Raviv Y., Salomon Y., Gitler C. & Bercovici T. (1987) *Proc. Natl. Acad. Sci. USA* **84**, 6103–6107.

Rosenwald A.G., Pagano R.E. & Raviv Y. (1991) *J. Biol. Chem.* **266**, 9814–9821.

Rubin R.J. & Chen Y. (1990) *Biophys. J.* **58**, 1157–1167.

Saito M., Saito M. & Rosenberg A. (1984) *Biochemistry* **23**, 1043–1046.

Schenkman S., Aranjo P.S., Dijkman R., Quina F.H. & Chaimovich H. (1981) *Biochim. Biophys. Acta* **649**, 633–641.

Schwarzmann G. & Sandhoff K. (1987) *Methods Enzymol.* **138**, 319–341.

Schwarzmann G. & Sandhoff K. (1990) *Biochemistry* **29**, 10865–10871.

Seigneuret M. & Devaux P.F. (1984) *Proc. Natl. Acad. Sci. USA* **81**, 3751–3755.

Shinitzky M. & Barenholz Y. (1978) *Biochim. Biophys. Acta* **515**, 367–394.

Silvius J.R. & Gagne J. (1984) *Biochemistry* **23**, 3241–3247.

Silvius J.R., Leventis R., Brown P.M. & Zuckermann M. (1987) *Biochemistry* **26**, 4279–4287.

Simons K. & Van Meer G. (1988) *Biochemistry* **27**, 6197–6202.

Sleight R.G. (1987) *Ann. Rev. Physiol.* **49**, 193–208.

Sleight R.G. & Pagano R.E. (1984) *J. Cell Biol.* **99**, 742–751.

Sleight R.G. & Pagano R.E. (1985) *J. Biol. Chem.* **260**, 1146–1154.

Somerharju P.J., Van Loon D. & Wirtz K.W.A. (1987) *Biochemistry* **26**, 7193–7199.

Sonnino S., Acquotti D., Riboni L., Giuliani A., Kirschner G. & Tettamanti G. (1986) *Chem. Phys. Lip.* **42**, 3–26.

Sonnino S., Chigorno V., Acquotti D., Pitto M., Kirschner G. & Tettamanti G. (1989) *Biochemistry* **28**, 77–84.

Spiegel S., Blumenthal R., Fishman P.H. & Handler J.S. (1985) *Biochim. Biophys. Acta* **821**, 310–318.

Struck D.K. & Pagano R.E. (1980) *J. Biol. Chem.* **255**, 5405–5410.

Struck D.K., Hoekstra D. & Pagano R.E. (1981) *Biochemistry* **20**, 4093–4099.

Stryer L. (1978) *Ann. Rev. Biochem.* **47**, 819–832.

Stubbs C.D., Williams B.W., Boni L.T., Hoek J.B., Taraschi T.F. & Rubin E. (1989) *Biochim. Biophys. Acta* **986**, 89–96.

Swanson J. (1989) *Methods Cell Biol.* **29**, 137–151.

Tilley L., Cribier S., Roelofsen B., Op den Kamp J.A.F. & Van Deenen L.L.M. (1986) *FEBS Lett.* **194**, 21–27.

Trinchera M., Ghidoni R., Sonnino S. & Tettamanti G. (1990) *Biochem. J.* **270**, 815–820.

Uster P.S. & Deamer D.W. (1981) *Arch. Biochem. Biophys.* **209**, 385–395.

Uster P.S. & Pagano R.E. (1986) *J. Cell Biol.* **103**, 1221–1234.

Vanderwerf P. & Ullman E.F. (1980) *Biochim. Biophys. Acta* **596**, 302–314.

Van Meer G. (1989) *Ann. Rev. Cell Biol.* **5**, 247–275.

Van Meer G. & Simons K. (1986) *EMBO J.* **5**, 1455–1464.

Van Meer G., Stelzer E.H.K., Wijnaendts-Van-Resandt R.W. & Simons K. (1987) *J. Cell Biol.* **105**, 1623–1635.

Van Renswoude J., Bridges K.R., Hartford J.B. & Klausner R.D. (1982) *Proc. Natl. Acad. Sci. USA* **79**, 6186–6190.

Van't Hof W. & Van Meer G. (1990) *J. Cell Biol.* **111**, 977–986.

Verkleij A.J., Zaal R.F.A., Roelofsen B., Comfurius P., Kastelijn D. & Van Deenen L.L.M. (1973) *Biochim. Biophys. Acta* **323**, 178–193.

Wiegandt H. (1985) *New Comprehensive Biochemistry*, Vol. 10. Elsevier, Amsterdam.

Wilschut J. & Hoekstra D. (eds) (1990) *Membrane Fusion.* Marcel Dekker, New York.

Zachowski A., Herrmann A., Paraf A. & Devaux P.F. (1987) *Biochim. Biophys. Acta* **897**, 197–200.

CHAPTER EIGHT

Probes for the Endoplasmic Reticulum

MARK TERASAKI

National Institutes of Health, Bethesda, MD, USA

8.1 INTRODUCTION

The endoplasmic reticulum (ER) is a site of protein synthesis, lipid synthesis and calcium regulation, and has other functions, such as detoxification. There are, therefore, many reasons why it might be interesting to observe the distribution of the ER in a living cell.

An ideal probe for the ER would label all the ER and only the ER. At present, there is no such probe available. Instead, the most often used probe is a dye that probably stains all membranes. This dye, $DiOC_6(3)$, is effective because the ER has a distinctive morphology; among the intracellular organelles, the ER has very extensive continuity. In practice then, $DiOC_6(3)$ is useful when the continuity of membranes can be observed; $DiOC_6(3)$ has not been useful in conditions where organelles cannot be distinguished from each other spatially.

Staining with $DiOC_6(3)$ was reviewed earlier (Terasaki, 1989). Since that time, several papers have reported use of the dye. The techniques for staining with $DiOC_6(3)$ are essentially the same as described earlier, so the techniques will be described more briefly, with more emphasis on recent work. Recent work with another dye, DiI, is briefly described. In addition, procedures for labelling ER in fixed cells and in cell-free preparations are described.

8.2 MECHANISM OF STAINING BY $DiOC_6(3)$

$DiOC_6(3)$ is a dicarbocyanine dye. Its molecular weight is 572. It is a positively charged molecule that passes through the plasma membrane, after which it does not seem to be metabolized or chemically altered by cells. At relatively low doses, $DiOC_6(3)$ accumulates in mitochondria, while at higher doses it accumulates in other membranous organelles, including the endoplasmic reticulum.

$DiOC_6(3)$ was the brightest and most photostable of several dyes that were found to have similar staining properties (Terasaki et al., 1984). These dyes were $DiOC_2(3)$, $DiOC_3(3)$, $DiOC_4(3)$, $DiOC_2(5)$, $DiOC_6(5)$, $DiIC_1(3)$, $DiIC_3(3)$, $DiIC_4(3)$ and $DiIC_6(3)$, $DiSC_6(3)$ and rhodamine 6G and rhodamine 3B. All of these dyes have a positive charge. They consist of a polycyclic fluorescent portion with some hydrocarbon (i.e. $-CH_2-$ or $-CH_3$) groups attached. The hydrocarbon groups seems to increase the association of the dyes with membranes. To test this, the staining of two similar dyes was compared. Rhodamine B is a positively charged molecule with no attached hydrocarbon groups and does not stain internal membranes. Chemists at Molecular Probes attached a 6 carbon chain to it, forming hexyl ester rhodamine B, a dye that does stain internal membranes. When the hydrocarbon chains are longer, the dyes still associate with membranes but they do not permeate through them. For instance, $DiIC_{12}(3)$ stains only the plasma membrane when applied from outside the cell.

The most straightforward explanation of the interaction of $DiOC_6(3)$ and related dyes with the cell is that the dyes pass through the plasma membrane and partition into the lipid phase of all intracellular membranes. At low doses, the positive charge on the dyes cause them to accumulate in mitochondria under influence of the negative mitochondrial membrane potential. In the original publication on $DiOC_6(3)$ (Terasaki et al., 1984), it was claimed that the dye did not stain the Golgi apparatus or endosomes, but this author now thinks otherwise (Terasaki, 1989; Terasaki & Reese, 1991).

8.3 METHODS

$DiOC_6(3)$ can be obtained from several companies, such as Molecular Probes and Eastman Kodak. Its full name is 3,3'-dihexyloxacarbocyanine iodide. A convenient way to store it is as a stock solution of 0.5 mg ml^{-1} in ethanol in 10 ml in a scintillation vial wrapped with aluminium foil and kept at room temperature. To view its fluorescence, conventional fluorescein filters are used. Hexyl ester rhodamine B (obtainable from Molecular Probes) or rhodamine 6G have similar staining properties but are viewed with conventional rhodamine filters. Less work has been done with these dyes, but the doses for staining etc. are probably similar to those for $DiOC_6(3)$. One complication of hexyl ester rhodamine B is that fluorescence illumination apparently can cause breakdown products whose fluorescence is visible in the fluorescein filter.

8.3.1 ER in living cells

In the original publication, cells were stained in media containing 0.5 µg ml^{-1} $DiOC_6(3)$ for 10 min (Terasaki et al., 1984). In practice, there is a fair amount of variability in staining of live cells, so some experimentation was recommended for each different preparation (Terasaki, 1989). The following is a compilation of staining conditions reported by investigators using $DiOC_6(3)$ to stain ER.

Lee and Chen (1988) and Lee et al. (1989) stained CV-1 cells (an African green monkey kidney cell line) for 5 min in 2.5 µg ml^{-1} $DiOC_6(3)$ in culture medium. Dailey and Bridgman (1989) stained explants of embryonic day 21 rat superior cervical ganglia after 12–24 h in culture with 0.3 µg ml^{-1} $DiOC_6(3)$ for

10 min. Sanger et al. (1989) stained PtK–2 cells (a rat kangaroo epithelial line) and myotubes in culture that had fused from myoblasts isolated from chick or quail embryos. These cells were stained for 5–30 min with 0.5 µg ml^{-1} $DiOC_6(3)$ in culture medium. A6 cells (a frog kidney cell line) have been stained by mounting the cells in 0.5 µg ml^{-1} and observing the cells after 5 min (Terasaki, unpublished). Caulonemata of the moss *Funaria hygrometrica* were labelled with 2.5–5 µg ml^{-1} $DiOC_6(3)$ in water or Laetsch medium for 10–20 min followed by 5–10 min wash (McCauley & Hepler, 1990). Onion bulb epidermis cells were stained with 2–5 µg ml^{-1} $DiOC_6(3)$ in water (time not specified) (Quader et al., 1987).

8.3.2 ER in cell-free preparations

Generally, it is faster and easier to stain ER in cell-free preparations than in live cells. Dabora and Sheetz (1988) stained a cell-free preparation of tubular membranes with 2.5 µg ml^{-1} $DiOC_6(3)$ for 10 s. ER membranes in a squash preparation of squid axoplasm and in the sea urchin cortex preparation were labelled with 2.5 µg ml^{-1} $DiOC_6(3)$ for 10 s (Terasaki et al., 1987, 1991).

8.3.3 Labelling in fixed cells

Glutaraldehyde-fixed cells can be stained with $DiOC_6(3)$ very well (Terasaki et al., 1984); fixed cells are stained for 10 s with 2.5 µg ml^{-1} $DiOC_6(3)$. This staining is much less variable than staining of living cells. In contrast to living cells, there is no low dose at which only mitochondria are stained. In one experiment, living cells stained with $DiOC_6(3)$ were photographed, then fixed in glutaraldehyde, re-stained and then rephotographed (Terasaki & Reese, 1992). The ER patterns were identical, indicating that glutaraldehyde fixation preserved the ER distribution well.

In glutaraldehyde-fixed cells, blebs from the plasma membrane form with time; also, large stained vesicles slowly become brighter with time, and autofluorescence due to the glutaraldehyde fixation slowly develops. For these reasons, staining in glutaraldehyde-fixed cells is optimal for the first 10–20 min after staining. Formaldehyde fixation often vesiculates the ER, and methanol fixation extracts the ER membranes so that no pattern can be seen at all.

8.3.4 Labelling of cell-free preparations by DiI

$DiOC_6(3)$ is not useful in some cell-free preparations because it stains all the membranes and the ER cannot be easily observed among them. Another dicarbocyanine dye, DiI, has been useful in some of these preparations.

DiI refers to either $DiIC_{16}(3)$ or $DiIC_{18}(3)$; both

molecules behave the same in the work described here. DiI is similar to $DiOC_6(3)$ except that it has two long carbon chains which give it different staining properties. Like $DiOC_6(3)$, DiI associates with membranes, but the long carbon chains prevent DiI from permeating through membranes. Once DiI becomes incorporated into a membrane, it remains in that membrane. It is free to diffuse in the membrane bilayer, and thus spreads throughout any continuous membrane into which it has incorporated. DiI has been used to label the complex shapes of plasma membranes of neurons (Honig & Hume, 1986) and muscle (Flucher et al., 1991).

DiI was used to label the cortical ER on isolated sea urchin egg cortices (Henson et al., 1989). It was shown that DiI labels the ER by a random hit mechanism involving DiI aggregates (Terasaki et al., 1991). When diluted from an ethanolic stock into aqueous solution, DiI forms microscopic aggregates that are visible by fluorescence microscopy. DiI aggregates collide with membranes on the isolated cortices, labelling occasional cortical granules but spreading in the continuous network of the ER. In this way, the cortical ER is labelled so it can be easily distinguished from the other membranes. $DiOC_6(3)$ is not useful in this preparation because it labels all of the cortical granules as well as the ER. To label the cortical ER, DiI was diluted from a 2.5 mg ml^{-1} ethanolic stock to 8 µg ml^{-1} in an intracellular buffer, and then applied to the cortices for 1 min. DiI has been used to label cortical ER of ascidian eggs (Gualtieri & Sardet, 1989) and of frog eggs (Houliston & Elinson, 1991). In the second case, a 1/200–1/500 dilution of 2.5 mg ml^{-1} ethanolic stock of DiI was used to stain for 1–2 min.

Baumann et al. (1990) applied a DiI suspension to the dissected labial adductor muscle of the honey bee Apis mellifera. In some of the muscle fibres, DiI labelled the sarcoplasmic reticulum. The authors proposed that the dye entered the muscle cell through sites of mechanical damage produced during the dissection. A stock of 10 mg ml^{-1} DiI in ethanol or DMSO was diluted to 25–100 mg ml^{-1} and used to stain for 30–60 min. The mechanism of staining is probably the same as that described in the sea urchin cortex preparation.

8.3.5   Labelling of living cells by DiI

DiI was recently used to label the ER in a living cell (Terasaki & Jaffe, 1991). A saturated solution of DiI in soybean oil (Wesson cooking oil) was microinjected into sea urchin eggs. In a period of 30 min, the dye spread in the ER throughout the egg. The dye did not label the abundant yolk platelets in the interior of the egg or the cortical granules. Using confocal microscopy,

the organization of the ER was observed in the unfertilized egg, and a striking reorganization was observed at fertilization.

This technique has worked in other eggs, such as starfish and ascidian eggs. The dye, however, did not spread when microinjected into the squid giant axon. It is possible that alterations in the technique are required for it to work in different organisms and tissues. It has been seen that DiI does not spread as well from corn oil than from soybean oil in ascidian eggs! So the oil carrier may make a difference. The long-chain dye may make a difference as well; dyes that could be tried are $DiOC_{18}(3)$, octadecylrhodamine, or other new dyes currently being developed by Molecular Probes.

8.4   IDENTIFICATION OF ER

$DiOC_6(3)$ and DiI are not specific probes for the ER. To use them, one observes the cells and identifies continuous interconnected membranes as ER. A potential source of difficulty is that the trans-Golgi apparatus (Pagano, 1989) and tubular lyosomes (Swanson, 1989) take on tubular network distributions in some cells. These organelles, though, have a characteristic type of distribution and can usually be eliminated through using specific probes for these organelles. The most difficult problem in using $DiOC_6(3)$ and DiI is related to uncertainties about the degree of continuity of ER membranes.

It is not known how continuous the ER membranes are. It is possible that the ER is a single interconnected membrane, or it is possible that some or many parts of the ER are normally discontinuous with each other. Since the degree of continuity of ER membranes is not known, there is some uncertainty in how to interpret $DiOC_6(3)$ and DiI staining. $DiOC_6(3)$ probably labels all membranes, so it probably labels all of the ER. However, it is not clear that some parts of the ER are discontinuous with the main network and cannot therefore be identified by the morphological criterion of continuity. DiI appears to be spreading in a continuous membrane, but it is not known if there are discontinuous parts of the ER that are not labelled then by DiI.

In spite of this difficulty, $DiOC_6(3)$ and DiI have been demonstrated in two cases to stain membranes that have been identified as ER. BiP, a protein that is known to be present in the ER, co-distributes identically with the network labelled by $DiOC_6(3)$ in a cultured kidney cell line (Terasaki & Reese, 1992). Also, a calsequestrin-like protein isolated from sea urchin egg microsomes co-distributes identically with the network labelled by DiI on isolated sea urchin egg cortices (Terasaki *et al.*, 1991). In both of these cases, the prominent interconnected membrane staining by $DiOC_6(3)$ or DiI is part of the ER. Thus, in any given cell on which the dyes are tried, the continuous, extensive membrane system that they label is very likely to be a part of the ER; in cases where an antibody to a protein in the ER is available, it would be useful to compare the immunofluorescence pattern to verify that the dye-stained membranes are part of the ER.

REFERENCES

Baumann O., Kitazawa T. & Somlyo A.P. (1990) *J. Struct. Biol.* **105**, 154–61.

Dabora S.L. & Sheetz M.P. (1988) *Cell* **54**, 27–35.

Dailey M.E. & Bridgman P.C. (1989) *J. Neurosci.* **9**, 1897–1909.

Flucher B.E., Terasaki M., Chin H., Beeler T.J. & Daniels M.P. (1991) *Devel. Biol.* **145**, 77–90.

Gualtieri R. & Sardet C. (1989) *Biol. Cell.* **65**, 301–304.

Henson J.H., Begg D.A., Beaulieu S.M., Fishkind D.J., Bonder E.M., Terasaki M., Lebeche D. & Kaminer B. (1989) *J. Cell Biol.* **109**, 149–161.

Honig M.G. & Hume R.I. (1986) *J. Cell Biol.* **103**, 171–187.

Houliston E. & Elinson R.P. (1991) *J. Cell Biol.* **114**, 1017–1028.

Lee C. & Chen L.B. (1988) *Cell* **54,** 37–46.

Lee C., Ferguson M. & Chen L.B. (1989) *J. Cell Biol.* **109**, 2045–2055.

McCauley M.M. & Hepler P.K. (1990) *Development* **109**, 753–764.

Pagano R.E. (1989) *Methods Cell Biol.* **29**, 75–85.

Quader H., Hofmann A. & Schnepf E. (1987) *Eur. J. Cell Biol.* **44**, 17–26.

Sanger J.M., Dome J.S., Mittal B., Somlyo A.V. & Sanger J.W. (1989) *Cell Motil Cytoskel.* **13**, 301–319.

Swanson J. (1989) *Methods Cell Biol.* **29**, 137–151.

Terasaki M. (1989) *Methods Cell Biol.* **29**, 125–135.

Terasaki M. & Jaffe L.A. (1991) *J. Cell. Biol.* **114**, 929–940.

Terasaki M. & Reese T.S. (1992) *J. Cell Sci.* **101**, 315–322.

Terasaki M., Song J.D., Wong J.R., Weiss M.J. & Chen L.B. (1984) *Cell* **38**, 101–108.

Terasaki M., Gallant P. & Reese T.S. (1987) *J. Cell Biol.* **105**, 128a.

Terasaki M., Henson J., Begg D., Kaminer B. & Sardet C. (1991) *Devel. Biol.* **148**, 398–401.

NOTE ADDED IN PROOF

An article that was inadvertently omitted describes the successful use of $DiOC_6(3)$ to label plant cell ER [Knebel W., Quadar H. & Schnepf E. (1990) *Eur J. Cell Biol.* **52**, 328–340]. Two recent articles give details of the use of $DiOC_6(3)$ (referred to as DHCC and DECC respectively) as a general membrane dye in cell-free preparations to observe nuclear envelope formation and breakdown [Boman A.L., Delannoy M.R. & Wilson K.L. (1992) *J. Cell Biol.* **116**, 281–294; and Newport J. & Dunphy W.J. (1992) *Cell Biol.* **116**, 295–306]. Lastly, $DiOC_6(3)$ has been successfully used to observe organelle dynamics in living yeast cells [Koning A.J., Lum P.Y., Williams J.M. & Wright R. (1993) *Cell Mot. and Cytoskel.*, in press].

CHAPTER NINE

Probing Mitochondrial Membrane Potential in Living Cells by a J-Aggregate-Forming Dye

LAN BO CHEN & STEPHEN T. SMILEY

Dana-Farber Cancer Institute, Harvard Medical School, Boston MA, USA

9.1 INTRODUCTION

Mitochondria are known to have proton pumps that generate a proton gradient across their inner membranes (Mitchell, 1979). This gradient has two components: membrane potential and pH gradient. In order to support a high rate of ATP synthesis, mitochondria must maintain a relatively high electrochemical gradient. In principle, it can be as high as 240 mV across the 5-nm-thick inner mitochondrial membrane (equivalent to 480 000 V across 1 cm). Although both membrane potential and pH gradient drive the synthesis of ATP, each is involved in additional biochemical events. For example, the uptake of pyruvate and glutamate by mitochondria is proportional to the pH gradient, whereas the import of mitochondrial enzymes from the cytoplasm, the uptake of calcium, and the maintenance of mitochondrial protein synthesis are dependent upon membrane potential.

Lipophilic cations such as rhodamine 123 and cyanines have been used to monitor mitochondrial membrane potential in living cells (Cohen & Salzberg, 1978; Bashford & Smith, 1979; Waggoner, 1979; Rottenberg, 1979; Freedman, 1981; Chen, 1988). One of the interesting findings from these studies has been the variation in mitochondrial membrane potential between cell types (for example, very high in cardiac muscle cells but very low in bladder epithelial cells), between differentiated states (high in myotubes but low in myoblasts), between cells transformed with different oncogenes (high in *fos*-transformed cells but low in *fes*-transformed cells), between different cancers (high in adenocarcinomas but low in oat cell carcinoma), between descendants of a recently cloned cell, and even between two daughter cells of a mitotic division. These findings stand in contrast to the generally accepted belief that mammalian mitochondria in different cell types have similar membrane potentials. Several long-standing issues, however, have not been resolved by the previously used lipophilic cationic dyes. For example, do all mitochondria within one cell necessarily adopt identical membrane potentials? Can a localized higher membrane potential exist within certain regions of a single mitochondrion?

It has been known for over 50 years that some dyes form J-aggregates in certain environments (Jelley, 1937; Scheibe, 1937). The formation of J-aggregates is often accompanied by dramatic shifts in both absorption and fluorescence maxima. A unique feature of these dyes, which is potentially useful for cell biological studies, is their propensity to form J-aggregates locally and instantaneously. Such dyes may be useful as reporter molecules for localized biochemical events. With the increasing realization that the location is often more important than the magnitude of an event, J-aggregate-forming dyes may find a variety of applications in biology and medicine. Surprisingly, this remarkable phenomenon has yet to be purposefully exploited by biologists.

Factors that may influence the formation of J-aggregates have previously been characterized by photographic scientists (Bird *et al.*, 1968; Gray *et al.*, 1970; Norland *et al.*, 1970; Ballard & Gardner, 1971; Beretta & Jaboli, 1974; Collier, 1974; Hada *et al.*, 1977, 1985). Within the context of living cells, the most important factors are the concentration of the dye, and the pH and ionic strength of the environment (Kay *et al.*, 1964a, b). Certain J-aggregate-forming dyes are lipophilic cations. Since it has been previously established that lipophilic compounds with delocalized positive charges are taken up by mitochondria in accordance with the Nernst equation, we have explored the possibility of using J-aggregate-forming lipophilic cations to study mitochondria in living cells. Can J-aggregates be formed in living cells? Will mitochondria be the major sites for J-aggregate formation? Can J-aggregates act as reporter molecules for localized mitochondrial membrane potentials?

9.2 CELL CULTURE

'Normal' African green monkey kidney cell line CV-1 obtained from American Type Culture Collection (ATCC), normal human fibroblast strain FS-2 obtained from Dr R. Sager (Dana-Farber Cancer Institute), and human breast carcinoma cell line MCF-7 obtained from Michigan Cancer Foundation were grown in Dulbecco's modified Eagles' medium (GIBCO) supplemented with 10% calf serum (M A Bioproducts) Human colon carcinoma cell line CX-1 obtained from Dr S. Bernal (Dana-Farber Cancer Institute), pancreatic carcinoma cell line PaCa-2 obtained from ATCC, bladder transitional cell carcinoma line EJ obtained from Dr I.C. Summerhayes (New England Deaconess Hospital) were grown in 50% Dulbecco's modified Eagles' medium and 50% RPMI 1640 medium (GIBCO) supplemented with 5% calf serum and 5% NuSerum (Collaborative Research). Bovine kidney epithelial cell line CCL22 obtained from ATCC, normal mouse bladder epithelial cells prepared by the procedures of Summerhayes and Franks (1979) were grown in F12 medium (GIBCO) supplemented with 10% fetal bovine serum (GIBCO). All cells were maintained at 37°C, 5% CO_2 and 100% humidity.

9.3 STAINING OF CELLS FOR MICROSCOPY

All cells grown on 12-mm-square glass coverslips (Bradford Scientific, Epping, NH) were stained with 50 µl of 5,5',6,6'-tetrachloro-1,1',3,3'-tetraethylbenzimidazolocarbocyanine iodide (JC-1, 10 µg ml^{-1}, Polaroid Co., Cambridge, MA) in Dulbecco's modified Eagles' medium for 10 min in cell culture incubator. Cells were rinsed in dye-free culture medium and mounted in a living cell chamber made of 0.7-mm-thick silicon rubber (N.A. Reiss, Belle Mead, NJ) as described previously (Johnson *et al.*, 1980).

9.4 FLUORESCENCE MICROSCOPY

A Zeiss Axiophot Microscope or a Zeiss Photomicroscope III equipped with epifluorescence optic was used. Objective lenses used included Planapo 40× (NA 1.3), Planapo or Neofluar 100× (NA 1.2). A 100 W mercury bulb was used for either microscope. Microscopic images were recorded on Kodak Professional Ektamatic P800/1600 positive films at E.I. 800 and developed by E-6 process at Push 1. Colour photographs were made with Ilford Cibachrome A-II papers developed by a Cibachrome automatic processor.

9.5 SPECTROPHOTOMETRIC ANALYSIS

JC-1 (10 µg ml^{-1}) in 40% dimethylsulphoxide (DMSO) in double distilled water at pH 7.2 or 1% DMSO in high K^+ buffer (3.6 mM NaCl, 137 mM KCl, 0.5 mM $MgCl_2$, 1.8 mM $CaCl_2$, 4 mM HEPES, 1 mg ml^{-1} dextrose, and 1% modified Eagles' medium amino acid solution (100×, GIBCO), pH 7.2) was placed in a 1 cm quartz cuvette and examined by a Beckman DU-70 spectophotometer. JC-1 was also dissolved in 50 mM Tris–HCl (pH 8.2) containing 1% DMSO, mixed thoroughly for 10 min in a 1 cm quartz cuvette equipped with a magnetic stirrer, and examined by a Kontron SFM25 fluorescent spectrophotometer. Fluorescence scans were made by synchronously varying both the emission and excitation wavelengths with a constant differential of 15 nm. Slit width was 10 nm.

Uptake of JC-1 by human colon carcinoma cell line CX-1 was also examined by fluorescent spectrophotometer. Cells were washed with (5 ml) and incubated in (1 ml) low K^+ buffer (137 mM NaCl, 3.6 mM KCl, 0.5 mM $MgCl_2$, 1.8 mM $CaCl_2$, 4 mM HEPES, 1 mg ml^{-1} dextrose, and 1% modified Eagles' medium amino acid solution (100×, GIBCO), pH 7.2) for 10 min. Cells were then washed three times with (2 ml each) and left in (1 ml) trypsin (1× M.A. Bioproducts) in low K$^+$ buffer for 5 min. About 0.8 ml of cell suspension was mixed with 1.2 ml of low K$^+$ buffer in a 1 cm quartz cuvette for 5 min. Recordings of spectra from 550 to 620 nm were repeated at a higher detector sensitivity.

9.6   DRUGS AND AGENTS

p-Trifluoromethoxyphenylhydrazone (FCCP), nigericin, dinitrophenol, sodium azide and oliomycin B, were from Sigma (St Louis, MO), valinomycin was from Calbiochem–Behring Corp. (La Jolla, CA), nocodazole was from Janssen Pharmaceutica (Piscataway, NJ).

9.7   J-AGGREGATE-FORMING LIPOPHILIC CATIONS

There are numerous lipophilic cations that form J-aggregates. Many of them have been used as photographic sensitizers for silver halide emulsions. Some of them are not suitable for biological studies because of their solubility, pKa, toxicity, requirement for non-physiological ionic strength, or fluorescence maxima which are inconvenient for the filters commonly used in fluorescence microscopy. It is necessary, therefore, to first identify dyes that can form J-aggregates at an ionic strength and pH which are compatible with the intramitochondrial environment, and at a concentration that can readily be attained by mitochondria in response to the Nernst potentials.

Among the numerous J-aggregate-forming dyes we screened, one was found to be quite suitable for probing mitochondria in living cells, namely 5,5',6,6'-tetrachloro-1,1',3,3'-tetraethylbenzimidazolocarbocyanine iodide (Fig. 9.1). This is the first J-aggregate-forming cationic dye which is of use to us and will be denoted JC-1. The effects of ionic strength,

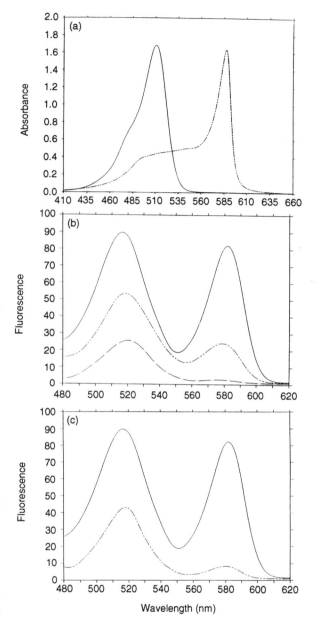

Figure 9.2 Effects of ionic strength, pH and concentration on JC-1. (a) The absorption spectra of JC-1 (10 μg ml^{-1}) in a 1 cm quartz cuvette analysed by a Beckman DU-70 spectrophotometer. Solid line is in 40% dimethylsulphoxide (DMSO) in double distilled water at pH 7.2; dashed line is 1% DMSO in high K$^+$ buffer. As shown, J-aggregate formation is favoured by a buffer with ionic strength comparable to that inside the cells. (b) The fluorescence spectra of JC-1 at various concentrations in 50 mM Tris–HCl (pH 8.2) containing 1% DMSO. Solid line is 200 ng ml^{-1}; dashed line is 100 ng ml^{-1}; dotted line is 50 ng ml^{-1}. As shown, J-aggregate formation is highly concentration-dependent. (c) The fluorescence spectra of JC-1 (200 ng ml^{-1} in 50 mM Tris-HCl containing 1% DMSO at pH 8.2 (solid line) or pH 7.2 (dashed line). As shown, J-aggregate formation is strongly favoured by pH 8.2, the intramitochondrial pH.

Figure   9.1 5,5',6,6'-Tetrachloro-1,1',3,3'-tetraethylbenzimidazolocarbocyanine iodide (JC-1).

pH and dye concentrations on the spectra of JC-1 are shown in Fig. 9.2. Previous reports have identified the first peak (absorption maximum = 510 nm and fluorescence maximum = 520 nm) as the monomeric dye species, and second peak (absorption maximum = 585 nm and fluorescence maximum = 585 nm) as the J-aggregate (Smith & Luss, 1972; Hada *et al.*, 1977). These findings indicate that the intramitochondrial environment should permit the formation of J-aggregates of JC-1. Also, the effect of JC-1 concentration on J-aggregate formation should allow us to monitor sensitive differences in the Nernst potentials.

9.8 J-AGGREGATES IN LIVING CELLS

When cultured cells were stained with JC-1 at 10 µg ml^{-1} in culture medium at 37°C for 10 min and examined by standard epifluorescence microscopy, either green fluorescence (under blue excitation) or red fluorescence (under green excitation) was detected. By use of a long-pass filter system which allows the visualization of red and green fluorescence simultaneously, orange fluorescence was detected in regions where both fluorescences coexisted. Plate 9.1 illustrates how these three colours may be detected in human breast carcinoma MCF-7 cells stained with JC-1. Visualization under green excitation with a narrow band-pass filter produced red fluorescence (A), under blue excitation with a narrow band-pass filter produced green fluorescence (B), under light-blue excitation with a long-pass filter produced orange fluorescence (C). Although most mitochindria shown in Plate 9.1 (C) display orange fluorescence, there are also a few mitochondria with only green fluorescence. For the rest of this paper, only the photographs that we generated by using the long-pass filter will be shown. Orange regions indicate the presence of both green and red fluorescence, whereas green regions have no red fluorescence.

To establish that in living cells the green fluorescence represents the monomer and the red fluorescence the J-aggregate, human colon carcinoma CX-1 cells were incubated with JC-1, trypsinized, transferred to a cuvette and analysed by fluorescence spectrophotometer. Figure 9.3(a) shows the fluorescence spectrum. The two peaks at 520 nm and 585 nm correspond to the monomer fluorescence and the J-aggregate fluorescence, respectively (as demonstrated in Fig. 9.2). Thus, the green fluorescence in living cells represents the monomer and the red fluorescence the J-aggregate.

To confirm that J-aggregate formation is critically dependent on the concentration of JC-1 attained by mitochondria, the effect of extracellular dye concentration on the formation of J-aggregates inside the

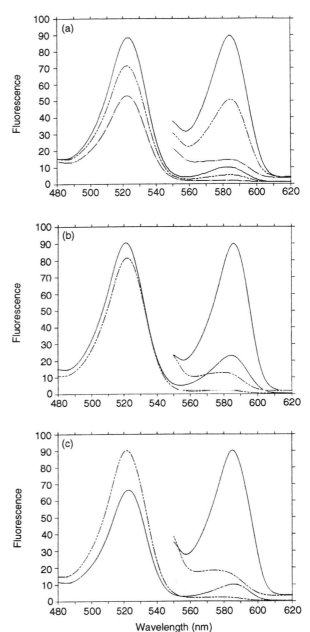

Figure 9.3 Effect of mitochondrial depolarization and JC-1 concentration. (a) Uptake of JC-1 in low K$^+$ buffer by CX-1 cells: 5 µg ml^{-1} (solid line); 2.5 µg ml^{-1} (dashed line); 1.25 µg ml^{-1} (dotted line). (b) CX-1 cells were incubated with JC-1 (10 µg ml^{-1}) in the presence of 5 µM FCCP and 0.5% ethanol (solid line) or 0.5% ethanol (dashed line). (c) CX-1 cells were first incubated with JC-1 (10 µg ml^{-1}), then treated with FCCP (5 µM with 0.5% ethanol, dashed line) or 0.5% ethanol (solid line).

mitochondria was examined. After a 10-min incubation of CX-1 cells with JC-1 at 1.25 µg ml^{-1} in culture medium, a very small amount of J-aggregate was observed; at 2.5 µg ml^{-1} more was generated; at 5 µg ml^{-1} the amount of J-aggregate greatly increased. To ensure an excess of JC-1 in culture medium for cellular uptake, we have used JC-1 at a concentration of 10 μg ml^{-1} for the rest of this work.

9.9 J-AGGREGATE FORMATION IS MEMBRANE POTENTIAL DEPENDENT

JC-1 is a lipophilic permeant with a delocalized positive charge. The uptake of such cations is expected to be driven by membrane potential. To extend such an expectation to J-aggregate-forming dyes, the effects of a variety of drugs and ionophores were tested. Figure 9.3(b) shows that in the presence of FCCP, a proton ionophore that abolishes the electrochemical gradient, very little J-aggregate was detected. This suggests that the formation of J-aggregates is dependent on the presence of an electrochemical gradient. When cells were allowed to form J-aggregates, and then placed in medium containing FCCP, the J-aggregates rapidly disappeared (Fig. 9.3(c)). The maintenance of J-aggregates in mitochondria is therefore also dependent upon an electrochemical gradient. Other agents known to abolish the mitochondrial electrochemical gradient (including CCCP, dinitrophenol, azide plus oligomycin, antimycin A plus oligomycin, and rotenone plus oligomycin) not only prevented the formation of J-aggregates but also disintegrated preformed J-aggregates.

To identify the component of the electrochemical gradient that is responsible for the formation and maintenance of J-aggregates, the effects of two ionophores were investigated: valinomycin, a K$^+$ ionophore that dissipates the membrane potential but not the pH gradient, and nigericin, a K$^+$/H$^+$ ionophore that abolishes the pH gradient but induces a compensatory increase in membrane potential with continued respiration. Plate 9.2(B) shows that nigericin (5 μg ml^{-1} for 30 min) in the presence of ouabain (5 μg ml^{-1}; to inhibit hyperpolarization of the plasma membrane) dramatically increased the formation of J-aggregates in kidney epithelial cells such that every mitochondrion had a detectable amount of J-aggregate. On the other hand, valinomycin completely prevented the uptake of JC-1 and J-aggregate formation. When cells were prestained with JC-1 and placed in valinomycin or nigericin in the absence of dye, the former abolished the orange fluorescence and the latter had no observable effect. Thus, the pH gradient is not required either for the uptake of JC-1 and subsequent formation of J-aggregates or for the maintenance of preformed J-aggregates. The component of the electrochemical gradient responsible for the formation and

maintenance of J-aggregates in mitochondria is thus likely to be the membrane potential.

Oligomycin (an inhibitor of mitochondrial ATPase) is known to partially increase the membrane potentials of mitochondria having low membrane potentials. As expected, it increased the formation of J-aggregates. Chloramphenicol (an inhibitor of mitochondrial protein synthesis), cycloheximide (an inhibitor of cellular protein synthesis), and actinomycin D (an inhibitor of RNA synthesis) did not affect the mitochondrial membrane potential; nor did they affect the formation of J-aggregates.

In living cells, mitochondria are surrounded by the plasma membrane, whose potential has been shown to have a pre-concentrating effect on the accumulation of lipophilic cations by mitochondria (Davis *et al.*, 1985). If J-aggregate formation is largely membrane potential dependent, a reduction in the plasma membrane potential should also lead to a reduction in J-aggregate formation. Indeed, by incubating CX-1 cells in high K$^+$ buffer that dissipates the plasma membrane potential, the amount of J-aggregate formation is reduced.

Effects of temperature and time on the uptake of JC-1 and J-aggregate formation were also investigated. No J-aggregate was detected when MCF-7 cells were incubated at 4°C with JC-1 (10 μg ml^{-1}) in culture medium for 10 min; a small amount was detected at 25°C; a large amount was detected at 37°C. MCF-7 cells were then mounted in a live cell chamber containing 10 μg ml^{-1} of JC-1 in culture medium and maintained at 37°C on the microscope stage with an air curtain. The uptake of JC-1 and formation of J-aggregates were monitored at 1 min intervals by fluorescence microscopy. After 3 min, green fluorescence with a few speckles of red fluorescence was detected in mitochondria. At 5 min, the intensity of green fluorescence significantly increased and rod-like structures with red fluorescence were detected. At 7 min, almost every mitochondrion exhibited red fluorescence. After 10 min, all mitochondria were intensely illuminated with red fluorescence.

Taken together, the above results strongly suggest that in living cells the uptake of JC-1 by mitochondria and the subsequent formation and maintenance of J-aggregates are driven by membrane potentials. J-aggregate formation requires higher concentrations of JC-1, which in turn can only be provided by higher mitochondrial membrane potentials.

9.10 J-AGGREGATES IN MITOCHONDRIA OF VARIOUS CELL TYPES

A variety of cell types and cell lines were stained with JC-1 at 10 μg ml^{-1} in culture medium for 10 min.

The results show that many cell types have two populations of mitochondria: one with only green fluorescence and the other with orange fluorescence. As an example, a single cell in a sparse culture of African green monkey kidney cell line CV-1 cells is shown in Plate 9.3. (In confluent cultures of CV-1 cells, mitochondria with orange fluorescence were rarely detected.) More remarkably, in some cell types/lines, within a single mitochondrion there are often some regions with only green fluorescence and other regions with orange fluorescence. Plate 9.4 shows such mitochondria in normal human FS-2 fibroblasts.

9.11 DISCUSSION

The aggregation of cyanine molecules is accompanied by large shifts in absorption spectra, which lead to violations of Beer's law. The spectrum of a monomer usually consists of a broad peak with a vibrational shoulder at the shorter wavelength side (Fig. 9.2(a)). This peak has been called the M-band (for monomer). Dye aggregation may lead to a shift of the absorption maximum to a shorter wavelength (called H-aggregates or H-bands, for hypsochromic), or to a longer wavelength (called J-aggregates or J-bands for its discoverer, Jelley). H-aggregates do not fluoresce and this feature has been previously exploited for the measurement of membrane potentials (Cohen & Salzburg, 1978; Bashford & Smith, 1979; Waggoner, 1979; Freedman, 1981). In contrast J-aggregates are often intensely fluorescent. The wavelength of such fluorescence is very similar to the absorption wavelength of the J-aggregates, and this lack of a Stokes's shift is termed 'resonance fluorescence'.

Several mathematical models have been developed to explain the spectrophotometric changes that accompany cyanine dye aggregation (Sturmer & Heseltine, 1977). A widely accepted model uses molecular exciton theory and assumes that the aggregation is the result of Coulombic coupling between the electrons of neighbouring molecules. This model characterizes aggregates by the slip angle between the long axis of a monomer and a line through the centres of the aggregated molecules. If the slip angle is small, for example less than approximately 35°, J-aggregates are produced. If the angle is larger, H-aggregates develop. The varied lengths of the polymer chains, as well as the existence of intertwined chains, complicate the mathematics and make it difficult to give exact predictions of the shift magnitudes. Nonetheless, the model satisfactorily explains much of the spectrophotometric and X-ray diffraction data.

JC-1 has been extensively used and studied as a sensitizer for silver halide-based photographic emulsion. Like most cyanine dyes, JC-1 in aqueous solution may exist as three distinct molecular species; monomer, H-aggregates, and J-aggregates. Each species may be characterized by its unique absorption/fluorescence spectra. The extent to which each species is present is governed by two distinct and reversible equilibria. One of these is governed by pH. The apparent pK_a of JC-1 is 7.9 (Beretta & Jaboli, 1974; Herz, 1974a, b). Above pH 7.9, the majority of JC-1 molecules have a single delocalized positive charge. Other resonance forms may be drawn in which the double bonds are shifted such that the positive charge falls on one of the other four nitrogen atoms. The conjugated electron system allows this species of JC-1 to absorb energy corresponding to visible blue-green wavelengths. Subsequent release of the absorbed energy results in the emission of green light. Below pH 7.9, most of the JC-1 molecules are protonated. Nuclear magnetic resonance studies have confirmed that the molecules are protonated at a carbon of the methine chain adjacent to the heterocyclic nuclei (West & Pearce, 1965; Feldman et al., 1968; Smith & Luss, 1972). Thus JC-1 and other cyanines are carbon acids rather than the more common nitrogen acids. Protonation at this position has two major consequences. First, the molecule now has an overall charge of +2. Secondly, the conjugated methine chain has been disrupted, resulting in the loss of absorbance and fluorescence of visible light. Unlike thia-, indo-, oxa-, or classic cyanines, the positive charges on the carbon acids of imidazolocyanines like JC-1 remain delocalized – resonance forms can still be drawn which place the positive charges on either nitrogen of each heterocyclic nucleus. Thus, these molecules remain lipophilic in their acid form.

A second equilibrium is one between monomers and aggregates. The monomeric species are favoured in organic solvents, in dilute aqueous solutions, in the presence of organic deaggregants, and at high temperatures. Aggregates may consist of dimers, trimers, or higher polymers. Factors favouring aggregation in aqueous environments include increased dye concentration and/or ionic strength.

When many cell types/lines were stained with JC-1, mitochondria were observed with simultaneous red fluorescence and green fluorescence. The two equilibria discussed above may both be relevant to the formation of red and green mitochondria. When JC-1 is diluted from a stock solution into physiological buffer of pH 7.2, much of the JC-1 should exist as the uncoloured, doubly positively charged carbon acid (the apparent pK_a of the dye is 7.9). This species of JC-1 should be taken up by the cells in response to their Nernst potentials, since the molecule is a delocalized lipophilic cation. Once inside the mitochondria, however, this species of JC-1 will revert back to its basic, monomeric

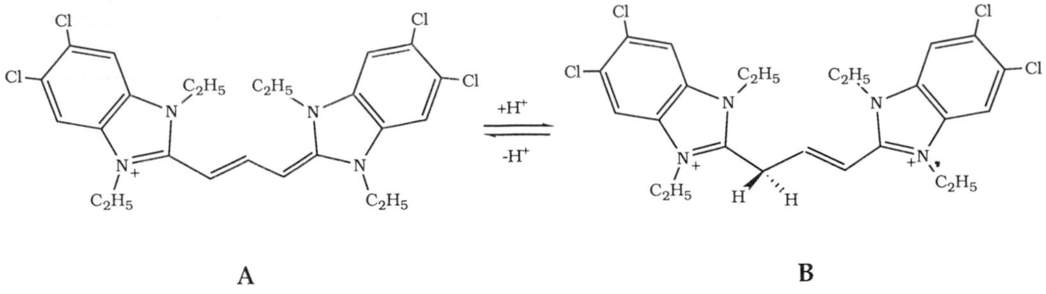

Figure 9.4 Protonation of JC-1.

form and fluoresce green because the intramitochondrial pH is known to be 8.2. With continuous uptake of JC-1, eventually J-aggregates will form when the concentration of monomer reaches nadir.

To confirm that the uptake of JC-1 and subsequent formation of J-aggregates in living cells is caused principally by the mitochondrial membrane potential, a variety of agents known to inhibit normal mitochondrial functions were tested. Only agents that abolish or prevent formation of a potential across the mitochondrial membrane inhibit the formation of J-aggregates. Most noteworthy among our results is that the pH gradient across mitochondria is not directly responsible for J-aggregate formation. Valinomycin, an ionophore that abolishes the membrane potential but not the pH gradient, inhibited the uptake of JC-1 and formation of J-aggregates. On the other hand, nigericin, an ionophore that converts the pH gradient into membrane potential with continued respiration, increased the uptake of JC-1 and the formation of J-aggregates. Moreover, cell types/lines that were found previously to have low mitochondrial membrane potentials, such as CV-1, the mouse bladder-derived cells shown here, and feline sarcoma virus-transformed mink cells (C. Hartshorn, M. Lin, S. Smiley, & L.B. Chen, unpublished results), permit only a small amount of J-aggregate to form. In contrast, cell types/lines known to have high mitochondrial membrane potentials, including carcinoma-derived cells (MCF-7 and CX-1) shown here and cardiac muscle cells (T.L. Lampidis, S. Smiley & L.B. Chen, unpublished results), allow most of their mitochondria to form a large amount of J-aggregate. In the course of our screening of J-aggregate-forming dyes, none of the neutral or anionic dyes (including closely related analogues of JC-1) were taken up by mitochondria. It is difficult to explain this observation without the following hypothesis: only the positively charged JC-1 (and not the neutral or negatively charged analogues) may be electrophoresed into mitochondria because their interior is negative as a result of membrane potential. It appears most probable that intramito-

chondrial J-aggregate formation is a consequence of the high membrane potential across mitochondria in living cells.

What might be the magnitude of difference in mitochondrial membrane potentials between green and red mitochondria? Mitochondria isolated from confluent cultures of CV-1 cells *in vitro* are known to have a membrane potential of approximately 100 mV (Modica-Napolitano & Aprille, 1987). If one assumes that the mitochondrial membrane potential in living cells is preserved in isolated mitochondria *in vitro*, the green mitochondria may represent a membrane potential of approximately 100 mV, since almost all mitochondria in confluent cultures of CV-1 cells are stained green. The results in Figs 9.2(b) and 9.3(a) show that approximately a four-fold increase in JC-1 concentration (at certain ranges) can lead to the conversion from a J-aggregate-scarce state to a J-aggregate-abundant state. Under ideal conditions, the Nernst equation would predict a four-fold increase to be caused by a hyperpolarization of approximately 40 mV. One might then speculate that mitochondria with membrane potentials greater than 140 mV should accumulate enough JC-1 to form J-aggregates. This prediction is supported by the fact that mitochondria of CX-1 cells readily permit J-aggregate formation and are known to have a membrane potential of about 160 mV *in vitro* (Modica-Napolitano & Aprille, 1987).

Based on the results described here, mitochondria within the same cell may not have identical membrane potentials. Why have rhodamine 123 and the cyanines previously used failed to reveal any heterogeneity in fluorescent intensity among mitochondria within a single cell? The human eye, photography, and video imaging all have limited ranges for a linear response to increasing light intensity. Consequently, once the fluorescent intensity reaches a certain nadir, detection systems fail to respond to further increases in fluorescence. When cells are incubated for 10 min in rhodamine 123 (10 µg ml^{-1}), all mitochondria stain intensely, even those with low membrane potentials (Johnson *et al.*, 1980, 1981, 1982). Mitochondria with higher potentials in the same cell are difficult to distinguish from those with lower potentials because they are already brightly stained. Furthermore, most cyanines previously used form H-aggregates rather than J-aggregates. As discussed above, H-aggregates quench the dye fluorescence. Thus, an increase in dye uptake as a result of higher mitochondrial membrane potential may not necessarily lead to brighter fluorescence; it may even reduce the fluorescence to an extent that such mitochondria become undetectable.

For the majority of cell types/lines examined thus far, however, it is possible that all mitochondria in the same cell need not maintain the same membrane potential. With increasing certainty that mitochondria are descendants of bacteria and that all mitochondria within one cell are not interconnected as are the endoplasmic reticula, it is conceivable that some degree of autonomy and individuality may be maintained by mitochondria within a single cell. Mitochondria have the capacity to vary enormously in morphology and density within a cell. Electron micrographs show that even cristae densities can vary among mitochondria within the same cell. In large cells such as skeletal muscle fibres or neurons, two mitochondria can be quite far apart. It is difficult to imagine that such mitochondria will have identical microenvironments. Possibly, each mitochondrion responds to its immediate environment and retains the autonomy to set its own membrane potential. For example, when a mitochondrion migrates to a region with a higher Ca^{2+} concentration, it may be compelled to take up the excess Ca^{2+}. By doing so, its membrane potential may be reduced. The idea that local Ca^{2+} concentrations may be one of the determining factors of mitochondrial membrane potential is supported by the observation that in cells lacking detectable P-glycoprotein and having low mitochondrial membrane potentials (such as CV-1 cells), a calcium channel blocker like verapamil significantly increases J-aggregate formation (C. Hartshorn, M. Lin, S. Smiley & L.B. Chen, unpublished results). Since verapamil does not detectably increase J-aggregate formation in cells with high mitochondrial membrane potentials (such as MCF-7 and CX-1 cells), it is conceivable that these mitochondria are not actively using the electrogenic calcium pump, and are thus insensitive to verapamil. Such a scenario may also explain why all mitochondria in cardiac muscle cells are intensely stained with red fluorescence. Cardiac muscle cells may use primarily the sarcoplasmic reticulum for regulating cytoplasmic free Ca^{2+}. Their mitochondria would not be required to use the electrogenic calcium pump and, consequently, a relatively high mitochondrial membrane potential would be maintained in such cells.

The possibility of a regional heterogeneity in membrane potentials within one mitochondrion is provocative. Why have rhodamine 123 and cyanines previously failed to reveal this possibility? If there is regional heterogeneity in membrane potential within a single mitochondrion, regions that take up more dye may quickly equilibrate with regions that take up less dye. Thus, the entire mitochondrion would fluoresce with equal intensity. Also, fluorescence quenching and difficulty in obtaining a linear response described above may have compounded prior inabilities to detect heterogeneity within a mitochondrion. Since J-aggregate polymers form quickly and are readily detectable, the heterogeneity has now been visualized for the first time.

Debate on the proton circuit in mitochondria continues. One consequence of the localized proton circuit theory may be that the membrane potential may not be identical throughout the inner membrane of a mitochondrion. Neurobiologists have established that a membrane need not maintain the same electrical potential throughout. If the plasma membrane potential of a neuron can have regional variability, there is no a priori reason why the inner membrane of a mitochondrion cannot behave in the same manner. One simple mechanism that can account for such heterogeneity involves the uneven distribution of Ca^{2+} along the surface of a mitochondrion. Using the arguments described above, one may propose that regions of green fluorescence have lower membrane potentials because their electrogenic calcium pumps are being utilized to take up free Ca^{2+}, whereas regions of red fluorescence have higher membrane potentials because no such utilization is required.

REFERENCES

Ballard R.E. & Gardner B.J. (1971) *J. Chem. Soc. (B)* 736–738.

Bashford C.L. & Smith J.C. (1979) *Methods Enzymol.* **55**, 569–586.

Beretta P. & Jaboli A. (1974) *Photogr. Sci. Engng* **18**(2), 197–207.

Bird G.R., Norland K.S., Rosenoff A.E. & Michaud H.B. (1968) *Photogr. Sci. Engng* **12**(4), 196–206.

Brand M.D. & Felber S.M. (1984) *Biochem. J.* **217**, 453–459.

Chen L.B. (1988) *Ann. Rev. Cell Biol.* **4**, 155–181.

Cohen L.B. & Salzberg B.M. (1978) *Rev. Physiol. Biochem. Pharmacol.* **83**, 35–88.

Collier S.S. (1974) *Photogr. Sci. Engng* **18**(4), 430–440.

Davis S., Weiss M.J., Wong J.R., Lampidis T.J. & Chen L.B. (1985) *J. Biol. Chem.* **260**, 13844–13850.

Emaus R.K., Grunwald R. & Lemasters J.J. (1986) *Biochim. Biophys. Acta* **850**, 436–448.

Feldman L.H., Herz A.H. & Regan T.H. (1968) *J. Phys. Chem.* **72**(6), 2008–2013.

Freedman J.C. (1981) *Int. Rev. Cytol.* **12**, 177–246.

Gray W.E., Brewer W.R. & Bird G.R. (1970) *Photogr. Sci. Engng* **14**(5), 316–320.

Hada H., Honda C. & Tanemura H. (1977) *Photogr. Sci. Engng* **21**(2), 83–91.

Hada H., Hanawa R., Haraguchi A. & Yonezawa Y. (1985) *J. Phys. Chem.* **89**, 560–562.

Herz A.H. (1974a) *Photogr. Sci. Engng* **18**(2), 207–215.

Herz A.H. (1974b) *Photogr. Sci. Engng* **18**(3), 323–335.

Jelley E.E. (1937) *Nature* **139**, 631–632.

Johnson L.V., Walsh M.L. & Chen L.B. (1980) *Proc. Natl. Acad. Sci. USA* **77**, 990–994.

Johnson L.V., Walsh M.L., Bockus B.J. & Chen L.B. (1981) *J. Cell Biol.* **88**, 526–535.

Johnson L.V., Summerhayes I.C. & Chen L.B. (1982) *Cell* **28**, 7–14.

Kay R.E., Walwick E.R. & Gifford C.K. (1964a) *J. Phys. Chem.* **68**(7), 1896–1906.

Kay R.E., Walwick E.R. & Gifford C.K. (1964b) *J. Phys. Chem.* **68**(7), 1907–1916.

Mitchell P. (1979) *Science* **206**(7), 1148–1159.

Modica-Napolitano J.S. & Aprille J.R. (1987) *Cancer Res.* **47**, 4361–4365.

Nadakavukaren K.K., Nadakavukaren J.J. & Chen L.B. (1985) *Cancer Res.* **11**, 541–556.

Norland K., Ames A. & Taylor T. (1970) *Photogr. Sci. Engng* **14**(5), 295–307.

Rottenberg H. (1979) *Methods Enzymol.* **55**, 547–569.

Scheibe G. (1937) *Z. Angew. Chem.* **50**, 212.

Smith D.L. & Luss H.R. (1972) *Acta Cryst.* **B28**, 2793–2806.

Sturmer D.M. & Heseltine D.W. (1977) In *The Theory of the Photographic Process*, T.H. James (ed.) MacMillan, New York, pp. 194–234.

Summerhayes I.C. & Franks (1979) *J. Natl Cancer Inst.* **62**, 1017–1023.

Summerhayes I.C., Lampidis T.J., Bernal S.D., Nadakavukaren J.J., Nadakavukaren K.K., Shepherd E.L. & Chen L.B. (1982) *Proc. Natl. Acad. Sci. USA* **79**, 5292–5296.

Waggoner A.S. (1979) *Ann. Rev. Biophys. Bioengng* **8**, 47–68.

West W. & Pearce S. (1965) *J. Phys. Chem.* **69**(6), 1894–1903.

CHAPTER TEN

Optical Probes for Cyclic AMP

STEPHEN R. ADAMS,[1] BRIAN J. BACSKAI,[1] SUSAN S. TAYLOR[2]
AND ROGER Y. TSIEN[1]

[1] Howard Hughes Medical Institute 0647, University of California San Diego, La Jolla, CA, USA
[2] Department of Chemistry 0654, University of California San Diego, La Jolla, CA, USA

10.1 RATIONALE FOR CREATING OPTICAL PROBES FOR CYCLIC AMP

Cyclic 3′,5′-adenosine monophosphate (usually abbreviated to cyclic AMP or cAMP) was the first molecule to be explicitly recognized as a 'second messenger', or mediator between extracellular stimuli ('first messengers') and intracellular biochemical responses (Rall & Sutherland, 1961; Robison *et al.*, 1971). For several years after the initial delineation of its role in transducing hormone signals, the involvement of cAMP in nearly every imaginable example of signal transduction was hypothesized and tested. Eventually it became clear that cAMP is only one member, albeit one of the most important, of a variety of second messengers and transduction pathways.

The longest-established roles of cAMP, mediating the immediate intracellular actions of numerous peptide hormones and transmitter substances, are still of tremendous clinical and pharmacological importance. In addition, new roles for cAMP under intense investigation include vertebrate olfaction (Nakamura & Gold, 1987), invertebrate neuronal plasticity (Schacher *et al.*, 1988), signalling in yeast (*Saccharomyces*; Broach & Deschenes, 1990) and slime moulds (*Dictyostelium*; Van Haastert, 1991), and the control

of gene expression (Karin, 1989). Also, rapid progress continues to be made in the biochemistry and molecular biology of the proteins involved in the cAMP signalling cascade, e.g. β-adrenergic receptors (O'Dowd *et al.*, 1989), G-proteins (Bourne *et al.*, 1991), adenylyl cyclase (Gilman, 1990), cyclic-AMP-dependent protein kinase (A-kinase; Taylor *et al.*, 1990a,b), cAMP phosphodiesterases (Beavo & Reifsnyder, 1990), and kinase substrates such as the cAMP-response-element binding protein (CREB) mediating gene activation (Montminy & Bilezikjian, 1987; Meinkoth *et al.*, 1991).

However, one area where cAMP research has still been lagging is the detailed exploration of the spatial and temporal dynamics of cAMP signalling. This gap is particularly noticeable in comparison with the study of intracellular Ca^{2+}, the only other second messenger known to be of comparable ubiquity and importance to cAMP. Optical probes for Ca^{2+} have revealed that Ca^{2+} signalling is often remarkably complex in fine structures such as heterogeneity of nominally identical neighbouring cells, spatial gradients of cytosolic $[Ca^{2+}]_i$ in individual cells, and temporal oscillations in response to steady stimuli (Tsien & Poenie, 1986; Berridge *et al.*, 1988; Tsien & Tsien, 1990; Meyer & Stryer, 1991). Their probable function is to enable ensembles of cells to generate multivariate or vectorial responses, which are too complex to be coded by a single, slowly varying, spatially uniform scalar concentration of a messenger substance. Somewhat analogous intracellular compartmentation of cAMP has long been suspected on indirect grounds such as discrepancies between the total cAMP, measured after destruction of the tissue, and physiological functions believed to be controlled by cAMP (Terasaki & Brooker, 1977; Hayes & Brunton, 1982; Greenberg *et al.*, 1987; Bode & Brunton, 1988; Aass *et al.*, 1988; Akil & Fisher, 1989; Murray *et al.*, 1989). The development of non-destructive optical imaging of cAMP in living cells would at long last enable direct exploration of the spatio-temporal intricacies of cAMP signalling. A further benefit would be the ability to measure cAMP in single cells, particularly valuable when the cells are scarce or embedded amongst other cell types or when the cell's electrophysiological behaviour is being simultaneously recorded.

10.2   PREVIOUS METHODS FOR MEASURING cAMP OR IMAGING RELATED MOLECULES

10.2.1   Assay in cell lysates

By far the most common existing method for measuring cAMP is to lyse the tissue by acid quenching, freezing, or other destructive methods, then to measure cAMP in the soluble supernatant by various binding or enzymatic assays, the most popular of which is radioimmunoassay. Because these methods are well-established and described (Brooker *et al.*, 1979; Brooker, 1988), they will not be further discussed here, except to note that total cAMP is measured, any compartmentation (Hayes & Brunton, 1982; Murray *et al.*, 1989) will be overriden, a significant number of cells must be destroyed for each time point, and the amount of tissue needs to be measured (typically as milligrams of protein) if the amount of cAMP (typically measured as picomoles) is to be normalized into concentration-type units. Thus such assays are analogous to measuring cellular Ca^{2+} by atomic absorption spectrophotometry or $^{45}Ca^{2+}$ equilibration.

10.2.2   Immunocytochemistry

Attempts have been made to use immunocytochemistry to localize cAMP in fixed or frozen tissues (Cumming, 1981). A basic problem is that cAMP should be reasonably diffusable, so that it is vulnerable to smearing or leaching from the tissue during the histological processing. These steps must involve permeabilization of the membranes and introduction of antibodies, so that relatively few positive results have been obtainable. Recently Barsony and Marx (1990) have reported that microwave fixation under very specific conditions traps cAMP immunoreactivity in interesting spatial patterns. Since it is unknown how microwaves could cross-link cAMP to tissue macromolecules and what sorts of cross-links would still allow recognition by the antibody, it is difficult to judge whether the cAMP distributions reflect the true pattern before fixation or whether they are somehow created or accentuated by fixation.

Considerably more success has been reported for immunocytochemistry of the macromolecules related to cAMP signalling, particularly of A-kinase subunits, since they are conventional proteins with respect to fixation. Detection can be accomplished either with antibodies or with exogenous fluorescein-labelled catalytic subunits (to detect free regulatory subunits; Fletcher *et al.*, 1988) or fluorescein-labe led protein kinase inhibitor (to detect free catalytic subunits; Byus & Fletcher, 1988). Several reviews may be consulted for further information (Cumming, 1981; Lohmann & Walter, 1984; Nigg, 1990).

10.2.3   Fluorescent analogues of cyclic nucleotides

Several fluorescent analogues of cyclic AMP have been synthesized, for example, $1,N^6$-etheno-cAMP and $2'$-O-(N-methylanthraniloyl)-cAMP (Hiratsuka, 1982). The latter contains an environmentally sensitive fluorophore and is about 20% as effective as unmodified cAMP as a substrate for beef heart phosphodiesterase. Despite such utility as substrate analogues for *in vitro* biochemical studies, it is not clear how such derivatives could be used to monitor endogenous unlabelled cAMP.

10.2.4   Fluorescently labelled cAMP receptor protein from *E. coli*

The cAMP receptor protein (CRP) from *E. coli* is a transcription factor that directly binds cAMP. Wu *et al.* (1974) found that CRP could be labelled with either of two environmentally sensitive fluorophores, acetamidoethylnapthalenesulphonate (AENS) or dansyl. The fluorescences of the resulting conjugates were somewhat sensitive to the binding of cAMP, which increased the emission intensity of the AENS label by 30%, whereas the dansyl derivative showed a 10% decrease. Meanwhile, the emission wavelengths were hardly changed by cAMP. The apparent dissociation constant of the AENS derivative for cAMP was 10 μM, which is considerably higher than the concentrations for half-maximal activation of A-kinase, the mammalian cAMP-sensor. The AENS-labelled protein was used for temperature-jump studies of the rate of reaction with cAMP (Wu & Wu, 1974); no attempt was reported to try it in intact cells.

With modern imaging equipment and perhaps better labels, CRP might justify re-examination as the basis for a fluorescent probe for cAMP, especially for the upper range of cAMP concentrations. Unfortunately, there is not yet enough information to predict rationally what fluorescence change would be produced by cAMP, especially how one might generate wavelength shifts or dual-wavelength ratio changes, which are much more reliably quantified than mere intensity changes (Tsien & Poenie, 1986; Bright *et al.*, 1989; also see other chapters in this book).

10.3 ALTERNATIVE cAMP BINDING SITES

Before discussing A-kinase holoenzyme, the basis for our current cAMP probe, it is worthwhile discussing yet other sources of cAMP binding sites that might also be made into optical indicators.

10.3.1 Abiotic, totally synthesized receptors

Recently chemists have made remarkable progress in designing totally artificial binding sites for nucleotides, including cAMP (Deslongchamps *et al.*, 1992). A carbazole ring system provides π-stacking interaction with the adenine, bicyclic imides provide Watson–Crick and Hoogsteen basepairing, and a guanidinium group is poised to ion-pair with the phosphate. At present, selectivity for cAMP over cGMP is only modest, though this would normally not pose a biological problem since usually cAMP considerably exceeds cGMP in concentration. However, binding was assessed only as the ability to extract cAMP from an 0.11 M aqueous solution into dichloromethane. Therefore, it is unclear what affinity might be observed in aqueous media at more physiological concentrations of cAMP and whether any optical changes can be coupled to binding. Nevertheless, this approach offers much promise, since it offers about the only approach to probes that would not need microinjection. Also, even without optical sensitivity, artificial cAMP binding molecules could be very valuable as buffers or antagonists of cAMP signalling.

10.3.2 Regulatory subunit of A-kinase

Because the cAMP binding sites for A-kinase are on the regulatory (R) subunit, the latter would be an obvious target of labelling. However, there are at least three theoretical objections to using labelled R alone as a cAMP probe: (1) In the absence of the catalytic (C) subunit, the affinity of R for cAMP is extremely

high because the dissociation rate becomes kinetically very slow (Døskeland, 1978; Rannels & Corbin, 1980). Therefore, exogenous R subunits in excess of C could well remain saturated with cAMP even at basal intracellular levels at which the endogenous holoenzyme would be largely inactive. (2) If under basal conditions the R subunits were *not* saturated with cAMP, they would tend to compete for any increase in cAMP and thereby inhibit the very transduction cascade being monitored. (3) Just as with bacterial CRP, there is no rational way to design a usable fluorescence change upon cAMP binding; one would simply have to rely on random trials of different labels.

10.3.3 *Dictyostelium* receptors for cAMP

The slime mould *Dictyostelium discoideum* uses extracellular cAMP signals to trigger aggregation and a new developmental programme during nutritional starvation (recently reviewed by Van Haastert, 1991). The cAMP concentrations needed for activation probably overlap those used intracellularly in animal cells. Unfortunately, the *Dictyostelium* cAMP receptors are integral membrane proteins belonging to the family with seven transmembrane helical domains and which act through G-proteins (Klein *et al.*, 1988). Therefore, it would be very difficult to get soluble injectable protein. Genetic incorporation would be useless unless the extracellular-facing binding site could be turned around to face the cytoplasm of the target cell.

10.4 PROPERTIES OF A-KINASE

This enzyme is the major intracellular receptor of cAMP in eukaryotic cells. The inactive holoenzyme is a tetramer consisting of two regulatory (R) subunits (present as a dimer) and two catalytic (C) subunits (Beebe & Corbin, 1986). Binding of two cAMP molecules to each R subunit results in the reversible dissociation to an R_2 dimer and free C subunits, according to the following equation:

$$R_2C_2 + 4\,cAMP \rightleftharpoons R_2(cAMP)_4 + 2\,C$$
$$\text{(inactive)} \qquad\qquad\qquad \text{(active)}$$

We have exploited this unusual activation mechanism for A-kinase to make a fluorescent sensor for cAMP, simply by labelling each of the R and C subunits with two different fluorophores which are capable of fluorescence resonance energy transfer (FRET) in the holoenzyme (Adams *et al.*, 1991). Upon binding cAMP, the subunits dissociate to effectively infinite distance, thus preventing FRET (Fig. 10.1).

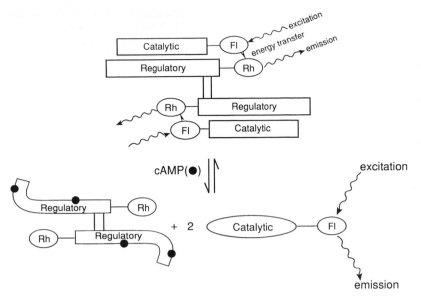

Figure 10.1 Diagram depicting the method for detecting cAMP using fluorescence resonance energy transfer (FRET) between the subunits of cAMP-dependent protein kinase (Adams *et al.*, 1991). When the cAMP concentration is low, most of the A-kinase is in the holoenzyme state, which permits FRET to occur between the fluorescein-labelled catalytic (C) subunits and the rhodamine-labelled regulatory (R) subunits. In high cAMP, the kinase is mostly dissociated and prevents energy transfer, so excitation of fluorescein results in fluorescein emission instead of rhodamine re-emission. (From Adams *et al.*, 1991, with permission.)

10.4.1 Isozymes

In mammalian cells, there are two major types of A-kinase (types I and II), which are differentiated by their R subunits, R^I or R^{II}; the C subunits are identical (Beebe & Corbin, 1986; Taylor *et al.*, 1990a,b). R^I and R^{II} differ in amino acid sequence, the ability of R^{II} to undergo autophosphorylation, which may decrease its rate of recombination with C (Rangel-Aldao & Rosen, 1977; Rymond & Hofmann, 1982), and by the high-affinity binding of MgATP to type I holoenzyme. Several isoforms of each subunit have been described, which differ in their tissue distribution. α-Forms are generally more widespread, whereas the β-form is restricted to neural and reproductive tissues. Activation of the holoenzyme composed of R^I_β is believed to occur at 3–7 times lower concentrations of cAMP than the R^I_α-containing isoform (Cadd *et al.*, 1990). The crystal structure of the C subunit complexed with a peptide inhibitor was recently solved to 2.7 Å resolution (Knighton *et al.*, 1991a,b), revealing a bilobal ellipsoid ($65 \times 45 \times 45$ Å) with a deep dividing cleft containing the catalytic site. Both MgATP and peptide substrate bind to the cleft. No detailed structure of the R subunits or holoenzyme are yet available, although each R subunit contains a substrate-like autoinhibitor site that occupies the consensus peptide binding site in the cleft.

The apparent dissociation constant (of the subunits) for the holoenzyme (type I or II) is 0.2–0.7 nM (Hofmann, 1980); this increases by four orders of magnitude upon binding cAMP (Granot *et al.*, 1980). The intracellular concentration of A-kinase has been reported to be in the range of 0.2–2 μM in a variety of tissues, almost in the same concentration range as its activator, cAMP (Beavo *et al.*, 1974; Lohmann & Walter, 1984).

10.4.2 Sources

The traditional sources of the holoenzyme or the subunits have been bovine heart for the type II and rabbit or porcine skeletal muscle for type I (Beebe & Corbin, 1986). The two forms may be separated (and in fact were named) by their elution from DEAE-cellulose at different salt concentrations. The subunits may be isolated by cAMP-affinity chromotography, which retains the R subunits which can then be eluted with cAMP.

More recently, high-yielding *E.coli* expression systems for the recombinant subunits have been used. We use the recombinant murine C_α (Slice & Taylor, 1989) and bovine R^I_α (Saraswat *et al.*, 1986); purification of the latter involves an ammonium sulphate precipitation and elution from DEAE-cellulose. C subunit requires phosphocellulose chromatography and ammonium sulphate precipitation followed by gel-filtration. Yields range from 5 to 10 mg pure protein per litre of culture broth; each run produces tens of milligrams for approximately 1 weeks' labour. The recombinant subunits appear to be similar to native enzyme in their capacity to form holoenzyme and be activated by cAMP. The recombinant C subunit lacks an *N*-myristoyl group and is slightly more sensitive to heat denaturation than the mammalian enzyme. However, an expression system that includes *N*-myristoylation has been devised (Duronio *et al.*, 1990). The purified R^I subunit contains two tightly bound cAMP molecules, whose removal requires strongly denaturing conditions (8 M urea). A recombinant system for expressing the R^{II}_{α} subunit has also been described (Scott *et al.*, 1990).

10.4.3 Subcellular distribution and function

Specific tissues usually contain predominantly one type of A-kinase or a mixture of both, although this may differ between species. For example, bovine heart contains predominantly type II, rat and mouse heart contain type I, and human and rabbit heart contain a mixture of both types (Beebe & Corbin, 1986). The rationale for such a variety is still obscure, although differences in sensitivity to cAMP (Cadd *et al.*, 1990), selective activation of the two types by hormones (see Beebe *et al.*, 1988 for review; Cho-Chung *et al.*, 1991; Jones *et al.*, 1991; Boshart *et al.*, 1991), and homone inducibility (Gross *et al.*, 1990) have been described. R^{II} has been reported to bind specifically to several intracellular proteins (e.g. Carr *et al.*, 1991; Lohmann *et al.*, 1988; Scott, 1991) and such binding probably localizes the kinase to specific sites. This binding, for example, may account for membrane- or cytoskeletal-associated A-kinase and influence substrate availability. The possibility that the R^{II} subunit could play a separate role from the C subunit in regulating gene transcription following an increase in intracellular cAMP levels, has been dismissed (Büchler *et al.*, 1990; reviewed by Nigg, 1990).

More recently, direct microinjection of fluorescently labelled C subunit (or holoenzyme) has directly confirmed the numerous (but often conflicting) reports (reviewed by Nigg, 1990) of translocation of the C subunit from the cytoplasm to the nucleus upon cAMP elevation. The injected holoenzyme and free R^{I} subunit appear to stay in the cytoplasm (Meinkoth *et al.*, 1990).

10.5 FLUORESCENT LABELLING OF A-KINASE

10.5.1 Fluorescence resonance energy transfer

The transfer of excitation energy of a donor fluorophore through space (non-radiatively) to a nearby acceptor chromophore or fluorophore is called fluorescence resonance energy transfer or FRET (Stryer, 1978; Lakowicz, 1983; Jovin and Arndt-Jovin, 1989; Herman, 1989). This process depends upon spectral overlap between the donor emission spectrum and the acceptor excitation (or absorbance) spectrum, the distance between the donor–acceptor pair (generally 2–7 nm), and the orientation of their respective dipoles. The efficiency of energy transfer is inversely proportional to the sixth power of the distance between donor and acceptor. FRET can be experimentally observed by a decrease in donor fluorescence in the presence of the acceptor, by sensitized acceptor re-emission when excited at the donor wavelength, and by a decrease in the fluorescence lifetime of the donor. There have been many attempts to use this phenomenon as a 'spectroscopic ruler' in biology (Stryer, 1978; Lakowicz, 1983).

10.5.2 Choice of fluorophores

Our method (Adams *et al.*, 1991) for converting A-kinase to a fluorescent cAMP sensor relies upon the reversible dissociation of the tetrameric holoenzyme to subunits upon binding cAMP. In the holoenzyme, the fluorophores are initially in close proximity. However, upon dissociation of the subunits by cAMP, the fluorophores become separated by effectively an infinite distance. Because the extent of energy transfer in the holoenzyme can be empirically determined, and the variable of interest is the percentage of holoenzyme left, it is not necessary to label unique sites on each subunit. This contrasts with the many applications of FRET in which a precise distance between two specified sites is being sought.

In general, the choice of donor acceptor pair must at least partially fulfil certain criteria:

(1) The excitation spectrum of the acceptor should maximally overlap the emission spectrum of the donor.
(2) The fluorescent quantum yields of the donor and acceptor should be as high as possible as this results in the optimal acceptor re-emission.

More specifically for the cAMP sensor:

(3) The fluorophores should be environmentally insensitive so changes in quantum yield and emission wavelength do not occur upon minor conformational changes in the protein, or upon binding to other macromolecules within a cell. Thus any changes in fluorescence seen will result only from changes in the association and dissociation of the C and R subunits.

Our first choice of donor fluorophore was fluorescein because of its high quantum yield (0.90 in water, although this is generally reduced on protein surfaces; Tsien & Waggoner, 1990), excitation and emission at convenient visible wavelengths, availability with a variety of coupling chemistries, and relatively environmentally insensitive fluorescence. Disadvantages include a small Stokes's shift (the difference between absorption and emission wavelengths), relatively poor photostability (compared to rhodamine), and pH-sensitivity. Possible acceptor fluorophores include substituted fluoresceins, rhodamines, cyanines and phycobiliproteins. Rhodamines are the classical acceptor for fluorescein, with good photostability and spectral overlap (particularly tetramethylrhodamine or rhodamine B, but less ideal with rhodamine X or Texas Red chromophores). However, the quantum yields tend to be quenched when coupled to proteins (Tsien & Waggoner, 1990). Cyanines (Southwick *et al.*, 1990), borate-dipyrromethenes or Bodipy (Haugland, 1989) and substituted fluoresceins are more recently described alternatives which may offer some advantages. Phycobiliproteins (Glazer & Stryer, 1990), the fluorescent proteins from cyanobacteria, are highly fluorescent, show large Stokes's shifts and little photobleaching (White & Stryer, 1987) but their high molecular weight dwarfs the subunits of A-kinase and might very likely hinder holoenzyme formation and subunit diffusion in cells.

10.5.3 Coupling chemistry

The most chemically reactive groups in proteins are generally amines (from lysines and the N-terminal) and sulphydryls (cysteines). The latter are generally much less frequent in occurrence but more reactive and offer the ability to label specific sites. Of the two exposed cysteines in the C subunit, the most reactive is Cys^{199} near the active site which is essential for kinase activity. Cysteine modifications have been reviewed by Bramson *et al.* (1984). First and Taylor (1989) devised a procedure for selective labelling of the least reactive thiol, Cys^{343}, which requires temporary protection of the most reactive. The R^I subunit has no readily reactive cysteines (although R^{II} has several; Nelson & Taylor, 1983) so we therefore initially concentrated on labelling the more ubiquitous amines. Amines are far more common in proteins and hence it is harder to selectively modify one residue, but such selectivity is not a necessity for the cAMP sensor because the subunits undergo complete dissociation upon binding cAMP. Random labelling at a number of sites may yield more efficient energy transfer because of an average closer distance between the donor and acceptor fluorophores. The most common 'amine-selective' groups are isothiocyanates, N-hydroxysuccinimidyl esters of carboxylic acids, sulphonyl halides and dichlorotriazinylamino (DTA) derivatives. From a chemical point of view, N-hydroxysuccinimidyl esters are probably the best, since they are the most selective reagents, are reactive at pH values close to 7, and form stable amides as products (Anjaneyulu & Staros, 1987). Isothiocyanates react more rapidly but reversibly with sulphydryl groups than with amines (Drobnica *et al.*, 1977), and sulphonyl halides and DTA derivatives are less reactive at lower pH (Haugland, 1989). Because of these complexities, we empirically tried a wide range of combinations of fluorophore and coupling chemistry.

10.5.4 Labelling strategy

Table 10.1 summarizes the results of our labelling attempts of A-kinase holoenzyme and subunits. All experiments exploit protection by MgATP of the catalytic subunit activity. The cAMP binding domains of the isolated R subunits (but not in the holoenzyme) are probably protected by bound cAMP. Initial attempts to label isolated C and R^I subunits with N-hydroxysuccinimidyl esters (OSu) were unsuccessful as holoenzyme could not be reformed. Unlabelled C could form holoenzyme with labelled R^I, suggesting we were labelling some site on C important in binding R^I. To protect these sites, we tried labelling the holoenzyme, separating the subunits by cAMP-affinity chromatography, and recombining with similarly prepared subunits labelled with a different fluorophore. Although this strategy appeared to work, it involved a moderately difficult separation and more steps than simply labelling the subunits. Interestingly, Leathers *et al.* (1990) have recently reported the successful labelling of C with carboxyfluorescein succinimidyl ester (FOSu) but only by labelling holoenzyme followed by separation. We also tried the isothiocyanates of fluorescein and tetramethylrhodamine (FITC and TRITC) and immediately found moderate energy transfer. Fortunately, these labels seemed to work as well on the isolated subunits, so that the complications of labelling the holoenzyme could be avoided. This combination of C labelled with FITC and R^I with TRITC has proven to be the most successful for energy transfer within type I holoenzyme, despite intensive testing of other fluorophores and coupling chemistries. Previous preparations of C-FITC as a cytochemical probe for R subunits (Fletcher *et al.*, 1988) stressed the need to label the intact holoenzyme and then separate subunits by DEAE-cellulose chromatography, but in our hands MgATP was sufficient to protect the C subunit.

The currently favoured labelling conditions for forming type I holoenzyme are as follows: recombinant C_α and R^I_α subunits (approximately 0.5–1.0 mg, as 0.5–2 mg ml^{-1}) were dialysed (Spectra/Por, MWCO 12 000–14 000; Spectrum Medical Industries, Inc., Los Angeles, CA) for 4 h at 4°C against 25 mM bicine, 0.1 mM EDTA, titrated to pH 8 with KOH, with several buffer changes. Following measurement of the absorbance of each solution at 280 nm (to determine approximate protein concentration using extinction coefficients of 45 000 and 48 000 M^{-1} cm^{-1} for C and R^I respectively), the solution of C was made 8 mM in MgCl$_2$ and 5 mM in ATP (from a stock solution titrated to about pH 8). R^I was labelled with 0.5 mM tetramethylrhodamine 5-isothiocyanate (TRITC) and C was labelled with 0.3 mM fluorescein 5-isothiocyanate (FITC) for 30 min at room temperature. Both dye reagents were from Molecular Probes (Eugene,

Table 10.1 Fluorescent labelling of cAMP-dependent protein kinase.

Fluorescent label		Holoenzyme formation	Energy transfer	Comments
Catalytic	R^I			
F.OSu[a]	TR.OSu[b]	—	—	
F.OSu[a]	TCF.OSu[c]	—	—	R^I cross-linked
F.OSu[a]	Coum.OSu[d]	—	—	R^I unlabelled
F.OSu[a]	Unlabelled	—	—	
TR.OSu[b]	F.OSu[a]	—	—	
Unlabelled	F.OSu[a]	+	—	
Coum.OSu[d]	F.OSu[a]	+	—	C unlabelled
TCF.OSu[c]	F.OSu[a]	—	—	C cross-linked
Cy3.18.OSu[e]		nd	nd	C cross-linked
Resos.OSu[f]		nd	nd	C reacts with -SH
	Rhod6G.OSu[g]	nd	nd	R^I cross-linked
FITC[h]	XRITC[i]	+	+	40% ratio change
FITC[h]	TRITC[j]	+	+	34–60% ratio change
FITC[h]	EosinITC[k]	+	+	35% FITC quench
FITC[h]	ErythrosinITC[l]	—	—	
FITC[h]	RhodBITC[m]	+	+	
FITC[h]	TR.OSu[b]	—	—	
FITC[h]	TCFITC[n]	+	+	TCFITC quenched
FITC[h]	Unlabelled	+	—	
FITC[h]	RhodB.SO$_2$Cl[o]	—		RhodB.SO$_2$Cl cross-links
TRITC[j]	FITC[h]	—	—	
TRITC[j]	F-C$_6$.OSu[p]	+	+	<10% ratio change
XRITC[i]	FITC[h]	—	—	C precipitated
EosinITC[k]	FITC[h]	—	—	
ErythrosinITC[l]	FITC[h]	—	—	
TRITC[j]	Bodipy.OSu[q]	+	—	Bodipy quenched
Bodipy.OSu[q]	TRITC[j]	+	—	Bodipy quenched
DTAF[r]	TRITC[i]	+	+	40% ratio change
	Type I holoenzyme			
	TCF.OSu[c]	nd	nd	Cross-linked
	F.OSu[a]	nd	nd	
	FITC[h]	+	+	With TRITC
	XRITC[i]	—	—	Precipitated
	TRITC[j]	+	+	With FITC
Catalytic	R^{II}			
FITC[h]	TRITC[j]	—	—	
FITC[h]	TR.OSu[b]	+	+	75–100% ratio change
FITC[h]	RhodB.SO$_2$Cl[o]	+	+	35% ratio change

nd, not determined.

Labels (supplier, catalogue number):

[a] 5(6)-Carboxyfluorescein succinimidyl ester (MP, C–1311).
[b] 5(6)-Carboxytetramethylrhodamine succinimidyl ester (MP, C–1171).
[c] 5(6)-Carboxy-2′,4′,5′,7′-tetrachlorofluorescein succinimidyl ester (Adams & Tsien, unpublished results).
[d] Coumarin 343, succinimidyl ester (MP, C–1419).
[e] Cy3.18, succinimidyl ester (gift of A. Waggoner).
[f] N-(Resorufin-4-carbonyl)-piperidine-4-carboxylic acid, N'-hydroxy succinimide ester (BM, 1042 653).
[g] 5(6)-Carboxyrhodamine 6G succinimidyl ester (gift of R. Haugland).
[h] Fluorescein 5-isothiocyanate (MP, F–143).
[i] Rhodamine X isothiocyanate (MP, R–491).
[j] Tetramethylrhodamine-5-isothiocyanate (MP, T-1480).
[k] Eosin 5-isothiocyanate (S, E-2005).
[l] Erythrosin B isothiocyanate isomer II (A, 32,380–2).
[m] Rhodamine B isothiocyanate (S, R–1755).
[n] 2′,4′,5′,7′-Tetrachlorofluorescein 5-isothiocyanate (Adams & Tsien, unpublished results).
[o] Lissamine rhodamine B sulphonyl chloride (MP, L–20).
[p] Fluorescein 5(6)-carboxamido-6-hexanoic acid, succinimidyl ester (MP, F–2181).
[q] Bodipy propionic acid, succinimidyl ester (MP).
[r] 5-(4,6-Dichlorotriazinyl)aminofluorescein (MP, D–16).

MP, Molecular Probes, Eugene OR; BM, Boehringer-Mannheim, Indianapolis, IN; S, Sigma Chemical Company, St Louis, MO; A, Aldrich Chemical Company, Inc., Milwaukee, WI.

OR) and were freshly prepared as 100 mM stocks in dry dimethyl formamide. The labellings were quenched by the addition of 5 mM glycine (pH 8) for 10–15 min. Excess dye was removed by passing each protein solution through a Sephadex G-25 column (3 ml), eluting with 25 mM potassium phosphate pH 6.8, 2 mM EDTA, 5 mM 2–mercaptoethanol and 5% glycerol at 4°C. The first coloured band was collected and dialysed (Spectra/Por 1, MWCO 6000–8000) overnight against the same buffer (2 litres) at 4°C. The dye:protein stoichiometries were determined by absorbance spectrometry (at pH 7.2) assuming the following extinction coefficients (in M^{-1} cm^{-1}) of 65 000 and 11 000 for protein-bound fluorescein (Haugland, 1989) at 495 and 280 nm, and 72 000 and 18 000 for protein-bound tetramethylrhodamine (Johnson *et al.*, 1984) at 550 and 280 nm respectively. Covalent attachment of most of the dye was verified by gel electrophoresis under denaturing conditions (SDS-PAGE). The subunits were then mixed at equal concentrations by weight (assessed by absorbance at 280 nm and Coomassie Blue staining of SDS-PAGE), typically 0.5 mg ml^{-1}, and dialysed against several changes of 25 mM potassium phosphate pH 6.7, 0.5 mM $MgCl_2$, 0.1 mM ATP, 5 mM 2–mercaptoethanol, 5% glycerol for 3–5 days at 4°C. Formation of holoenzyme was monitored by kinase assay (Cook *et al.*, 1982). For storage, the labelled holoenzyme was concentrated (using Centricon 30 microconcentrators; Amicon, Beverly, MA) to about 10 mg ml^{-1} in injection buffer (25 mM potassium phosphate pH 7.2, 1 mM EDTA, 0.5 mM 2–mercaptoethanol, and 2.5% glycerol). It has also been successfully stored frozen in buffer containing 50% glycerol.

More recently we have also labelled the R^{II} subunit (unpublished results). In contrast to R^I, labelling R^{II} with tetramethylrhodamine OSu (R^{II}-TROSu) gave the best energy transfer with C-FITC, whereas R^{II}-TRITC gave the least (Table 10.1). Recombinant R^{II} behaves similarly to R^{II} isolated from pig heart. Independently, Leathers *et al.* (1990) have prepared C-FOSu and R^{II}-Texas Red, reconstituted holoenzyme and detected energy transfer.

Our current protocol for preparing labelled type II holoenzyme is similar to that for type I described above except that 5(6)-carboxytetramethylrhodamine succinimidyl ester (TROSu; obtained from Molecular Probes) replaces TRITC as the label for R^{II} subunit. A lower concentration (0.1 mM) of TROSu is required in the labelling reaction of R^{II} (about 1 mg ml^{-1}) with a similar 30 min reaction time. The C labelled with FITC used for the type I holoenzyme is also used for the type II. The remaining procedures are the same for type II as type I except that MgATP is not required in the buffer for forming holoenzyme (2 mM EDTA replaces it) and only 1–2 days of dialysis are required.

Holoenzyme formation can be monitored by a modified kinase assay that uses only 10 mM potassium MOPS and no KCl, since preliminary results indicate the labelled type II holoenzyme partially dissociates at the low protein concentrations used (1–10 nM) if normal salt levels are present.

10.6   PROPERTIES OF FlCRhR

As described above, the best combination for FRET in A-kinase has been C-FITC with R^I-TRITC or R^{II}-TROSu, typically with about one dye molecule per subunit. These are the only cAMP sensors we have used in living cells to measure intracellular free cAMP levels (Adams *et al.*, 1991; Gurantz *et al.*, 1991; Sammak *et al.*, 1992). Therefore, only the properties of these two preparations will be described in this section. For brevity we propose calling the labelled type I and II holoenzymes FlCRhR^I and FlCRhR^{II}, respectively, as acronyms for *Fl*uorescein-labelled *C*atalytic and *Rh*odamine-labelled *R*egulatory subunit type I or II (pronounced 'flicker').

10.6.1   Energy transfer

The change in emission spectrum of FlCRhR^I from low to high cAMP in a cuvette is shown in Fig. 10.2. In the holoenzyme (zero cAMP), the fluorescein

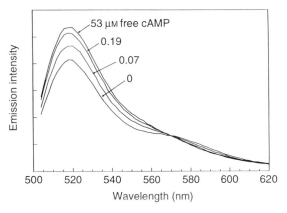

Figure 10.2 The response of the emission spectra of FlCRhR^I upon titration with cAMP when the fluorescein label is excited at 495 nm. With increasing cAMP, the emission at 520 nm (from fluorescein) increases while the emission at 580 nm (mostly from tetramethylrhodamine) decreases. For clarity, only four concentrations of cAMP are shown. The cuvette contained 3 nM FlCRhR^I in 130 mM KCl, 5 mM $MgCl_2$, 3 mM ATP, 10 mM MOPS, pH 7.2 at 22°C. Free cAMP was calculated by subtracting the calculated bound cAMP from the total cAMP assuming that two molecules of cAMP must bind to R^I to release C. Excitation and emission bandwidths were 1.8 and 4.6 nm respectively. (Reproduced from Adams *et al.*, 1991, with permission.)

emission at 520 nm is quenched by about 31% and the emission at 580 nm, mostly but not entirely due to rhodamine, is enhanced by about 13% compared to the respective amplitudes after full dissociation. Control experiments indicate that fluoresceinated catalytic subunit exhibits negligible change in fluorescence upon combination with unlabelled R^I, so that the partial quenching seen with rhodamine-labelled R^I represents energy transfer and not a conformational or environmental effect. The energy transfer efficiency in the holoenzyme is therefore about 0.31, which corresponds to an average distance of 5.8 nm between the donor and acceptor fluorophores (assuming random orientations for their transition moments). Although large, this distance is consistent with the known dimensions of the holoenzyme, whose Stokes's radius is 4–5 nm (Zoller et al., 1979) and whose distance between adjacent C active sites is >5 nm (E. First, D. Johnson & S. Taylor, unpublished results). Possibly the ability of isothiocyanates to label sites distant to the C and R binding domains explains why these labels were the most compatible with reformation of holoenzyme. We have not yet determined if the labelling is truly random or (at least partly) site-specific. One fluorescein molecule appears to be incorporated faster than further ones into the C subunit, suggesting some selectivity (unpublished results).

Upon binding cAMP, the emission of $FICRhR^I$ (when the fluorescein is excited at 495 nm) increases at 520 nm and decreases at 580 nm (Fig. 10.2). The ratio of emissions at 520 to 580 nm increases by a factor of 1.4 to 1.65, depending on the particular batch, upon changing the cAMP concentration from 0 to 50 μM. FICRhR is therefore a ratiometric indicator, which is of particular advantage in intracellular measurements by fluorescence microscopy because ratioing allows correction for variations in probe concentration, optical pathlength, lamp intensity and emission collection efficiency (Tsien & Poenie, 1986; Bright et al., 1989). The reason why the apparent change in emission intensity at 580 nm is smaller than that at 520 nm is due to at least two reasons: the rhodamine does not re-emit a photon every time it quenches a fluorescein excited state, and the emission at 580 nm contains a component due to the tail of the fluorescein emission spectrum, which is affected oppositely from the rhodamine.

$FICRhR^{II}$ shows a larger change than $FICRhR^I$ in the emission ratio of 520–580 nm, with changes of 1.6–2.2-fold being obtained in calibrations and in living cells. One price to be paid for this enhanced magnitude of response to cAMP is the slow dissociation of the holoenzyme when diluted to submicromolar concentrations even in the absence of cAMP. Perhaps in $FICRhR^{II}$ the rhodamine label is closer

to the R^{II}–C interaction site and weakens their interaction.

10.6.2 cAMP sensitivity and kinase activity

The optical sensitivity of FICRhR to cAMP can be measured in vitro simply by recording the changes in fluorescence emission upon titration with cAMP as in Fig. 10.2. The free cAMP concentration can then be calculated for each curve by measuring the percentage activation and subtracting the calculated amount of cAMP bound to the holoenzyme. This requires knowing the holoenzyme concentration in the cuvette and assumes that two cAMP molecules bind per R subunit.

The kinase activity of FICRhR and unlabelled holoenzyme can be measured by ^{32}P incorporation into a peptide substrate, Kemptide, Leu-Arg-Arg-Ala-Ser-Leu-Gly, or more conveniently by coupling the decrease in ATP (again with Kemptide) to a change in NADH using pyruvate kinase and lactate dehydrogenase (Cook et al., 1982).

When the ratio of emissions at 520 nm to that at 580 nm of $FICRhR^I$ is measured as a function of free cAMP concentration (Fig. 10.3, taken from Adams et al., 1991), the resulting calibration curve is similar to the plot of kinase activity vs. cAMP for both the labelled and unlabelled enzyme. The activation constants, i.e. free cAMP concentrations that give half-maximal emission ratio or kinase activity, are all about 0.1 μM with slight positive cooperativity (Hill coefficient 1.2–1.8). These values, measured at vertebrate ionic strength in the presence of MgATP, vary slightly between batches but show no systematic difference between labelled and unlabelled holoenzyme. The kinase activity of C measured with Kemptide appears to be almost unchanged by labelling with FITC in the presence of MgATP. The advantage of using enzymically active FICRhR in intracellular experiments is that cAMP binding to the exogenous enzyme should still elicit normal patterns of phosphorylation, whereas if the sensor were catalytically inactive, its binding of cAMP would detract from the downstream physiological response. If enzymically inactive FICRhR were desired, MgATP could be omitted during the labelling of C, or C could be subsequently reacted with sulphydryl reagents.

The emission ratio of $FICRhR^{II}$ can be calibrated in a similar way to $FICRhR^I$ except that at the enzyme concentrations used (typically <10 nM) preliminary experiments suggest that the labelled holoenzyme slowly dissociates in the absence of cAMP (unpublished results). By contrast, unlabelled type 2 holoenzyme has been reported to be 50% dissociated when diluted to 0.2–0.3 nM (Hofmann, 1980). However, when microinjected into cells, $FICRhR^{II}$ shows a stable emission

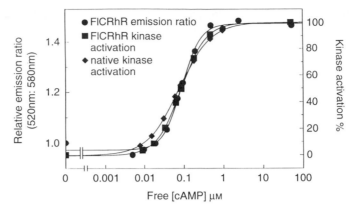

Figure 10.3 Typical calibration curve of FlCRhR[I] relating the ratio of emissions at 520 nm to 580 nm to the free cAMP concentration (circles, left-hand scale). Also depicted is the activation of kinase activity of FlCRhR[I] and unlabelled A-kinase holoenzyme by cAMP (squares and diamonds respectively, right-hand scale). The kinase activity was measured using the assay of Cook *et al.*, (1982) with the synthetic peptide substrate Kemptide. The *y*-axis are normalized to emission ratio or percentage kinase activation at zero cAMP for clarity. The data were fitted by least-squares to the Hill equation (using Graphpad Inplot 3.0, San Diego, CA); the cAMP concentrations giving half-maximal response were 88 nM, 86 nM and 76 nM, and the Hill coefficients were 1.8, 1.4 and 1.1 for emission ratio, labelled kinase, and native kinase respectively. The minor differences within these values are within the range shown by different batches of protein. (From Adams *et al.*, 1991, with permission.)

ratio in unstimulated cells (see later) suggesting dissociation is not a problem at these higher protein concentrations, micromolar or greater. Calibration in a cuvette at these higher holoenzyme concentrations is inaccurate because of errors in converting from total cAMP to the more biologically relevant free cAMP as this requires subtracting two large, imprecise numbers. A better method is to calibrate in a microdialysis capillary on the fluorescence microscope (see Section 10.8.4). An alternative is to calibrate quickly in a cuvette before significant dissociation can occur. Both methods give similar values for activation constant, Hill coefficient, and percentage ratio change for the same batch of protein (0.4 μM, 1.0 and 70–80% respectively, measured at high ionic strength mimicking marine invertebrate cytoplasm).

10.6.3   Stoichiometry of labelling and subunits

The dye-to-protein ratio can be estimated from the absorbance spectrum of the labelled C or R subunit and the extinction coefficients for subunit and dye at 280 nm and dye absorbance maximum (Adams *et al.*, 1991). With FlCRhR[I], there is an average of one fluorophore per subunit, though some subunits may be unlabelled and some multilabelled. Optimal energy transfer would probably occur with as many labels per subunit as possible but empirically 1:1 appears to be the best. For example, overlabelling the R[I] subunit with TRITC increases its susceptibility to precipitate and diminishes its ability to reform holoenzyme. Fluorophores that are in too close proximity (<2 nm)

can quench each other by mechanisms other than FRET (Tsien & Waggoner, 1990).

When recombining the labelled subunits to form FlCRhR, we usually balance the subunits by determining concentrations by absorbance or by protein staining following SDS-PAGE, and by loss of kinase activity upon dialysis. Usually a slight excess of R over C is preferred to minimize free C-FITC, which can interfere in experiments studying nuclear translocation of C. Neither method is foolproof as some R may be unable to form holoenzyme because of labelling. A slight excess of R subunit appears to be less detrimental than free C, perhaps because R already contains bound cAMP and seems not to perturb cAMP signalling (Sammak *et al.*, 1992).

10.6.4   Concentration and storage

The holoenzyme is usually re-formed at protein concentrations of about 1 mg ml[-1] or 6 μM in holoenzyme. For microinjection, 10–fold higher concentrations are usually required in the pipette so that injections of 5–10% of cell volume give final intracellular levels of 3–6 μM FlCRhR. FlCRhR solutions are concentrated by using centrifugal ultrafiltration (e.g. Centricon filters, Amicon Corp.), which causes minimal protein loss through adsorption to the membrane, although some protein precipitation may occur. We also change the buffer to one more suitable for microinjection and storage, e.g. 25 mM potassium phosphate, pH 7.3, 1 mM EDTA, 0.5 mM 2-mercaptoethanol, and 2.5% glycerol. The last two components are to help stabilize the protein; the

EDTA is to discourage bacterial contamination, which nevertheless occurs eventually after dipping many non-sterile pipette tips and removal of many aliquots from one vial. The protein seems to be quite stable just kept at 4°C in a sealed vial protected from light; samples over 6 months old have still been responsive to cAMP in cells. Deterioration can be recognized by increasing difficulty in microinjection, even after extensive centrifugation at 16 000 × **g** at 4°C to remove any precipitated protein and particulates, and decreased ratio change in cells. Proteolysis is visualized by polyacrylamide gel electrophoresis under denaturing conditions.

10.7 INTRODUCTION OF FlCRhR INTO CELLS

10.7.1 Microinjection

A necessary step for biological applications of FlCRhR is the introduction of the protein into the cytosol of living cells. Because FlCRhR is a protein complex with an aggregate molecular weight of 167 or 172 kDa (type I or II respectively), it cannot be loaded via acetoxymethyl esters or iontophoresis as can low-molecular-weight cation indicators. For all applications to date we have relied on pressure microinjection into individual cells (Wang *et al.*, 1982). Microinjection is quick, direct and economical of protein. In recent years it has also become much easier and more widely applicable, thanks to automated micromanipulators, pressure pulse regulators, and piezoelectric stabbers.

Pressure injection of single cells requires micropipettes for penetrating cells and introducing the protein, a closed system for delivering pressure, and a means for monitoring injection volume. Perhaps the most important component for this technique is the micropipette. We use 1.0 mm thin-walled borosilicate glass which contains a microfilament along the inner surface. The microfilament provides a convenient and effective means of capillary back-filling the pipette with the minimum volume necessary. With these electrodes, volumes of <0.5 μl of the concentrated protein solution are more than sufficient. A reproducible and reliable pipette puller is also important. We use a Flaming-Brown type puller (Sutter Instruments, Inc., P80–PC). The actual shape and size of the pipette may depend on the type of cell being injected. For general-purpose injection of cells >20 μm in diameter, pipettes with tips of about 0.5 μm are used. Finer tipped pipettes would be preferable except that the concentrated high-molecular-weight protein tends to clog very small pipettes much more easily. For very delicate cells, the pipette tips may require bevelling as well.

Once pulled and filled, the pipettes are placed in a pipette holder (E.W. Wright, Guilford, CT) which contains a side-port to attach tubing. The holder is attached to a three–axis hydraulic micromanipulator (Narishige Corp. or Newport Corp.) in order to position the pipette stably and precisely on the cells. Most often, some means for isolating vibrations is necessary, e.g. an air suspension table. Occasionally, movement along the oblique axis of the pipette is useful, and can be achieved either with an additional hydraulic positioner or better yet with a piezoelectric manipulator (Inchworm no. IW-700, Burleigh Instruments). The latter allows rapid and reproducible axial movements of the pipette, valuable for fast penetrations of tough cell membranes. To deliver pressure, we have used various systems ranging from a 50 ml disposable syringe to regulated, high-precision automated pressure injection systems (Eppendorf no. 5242, or Medical Systems Corp., PLI-100). The automated systems generally provide three or more available pressures; a constant balance pressure, a timed injection pressure, and a clearing pressure for clogged tips. These systems provide reliable and reproducible injections with a touch of a button.

Monitoring the injection volume can be difficult. With a precision pressure injector, one can calibrate the time of injection pulse and the pressure required for a reproducible pipette geometry to obtain a known volume per pulse. Others have monitored the movement of the meniscus within the shaft of the pipette to estimate injection volume (Castellucci *et al.*, 1980). It is also possible to use the fluorescence of the probe itself to estimate the injection volume under constant imaging conditions. That is, one can compare the brightness of the fluorescence within the cell with *in vitro* samples of the protein at known concentrations. In this way, one can measure the dilution of the fluorescence after injection into the cell, and with an estimate of the volume of the cell, can calculate the volume introduced. This is perhaps the most practical approach, since the minimum volume that must be injected is that which provides a bright enough fluorescent signal for optical measurements. Therefore, we inject the minimum volume of protein for a good signal-to-noise ratio, which is readily monitored on the imaging microscope.

Two general approaches are used for loading cells. For flat, well adherent, easily penetrable cultured cells, a small constant pressure is applied to the pipette so that the pipette solution is continuously flowing while the pipette tip is lowered vertically into contact with the cell. The injection volume then depends on the amount of time the pipette is kept intracellularly. In the other protocol, a pressure pulse is applied to the pipette only after cell penetration. This approach requires more elaborate equipment but is best for cells that can only tolerate very limited injection volumes or are difficult to penetrate because they are not rigidly anchored everywhere and would tend to back away from a continuous outflow.

10.7.2   Alternatives to microinjection

There will be applications in which microinjection of individual cells is not applicable, for example in the smallest or most fragile cell types or when biochemical responses of a bulk population must be studied. Electroporation would seem to be the most promising approach (see, for example, Liang *et al.*, 1988). We have not tried this yet, mainly because electroporation in commercial equipment wastes relatively large amounts of protein merely to fill the chamber. However, protocols may well be developed once the protein or microscale electroporation apparatus become more easily available. Other techniques (see review by McNeil, 1989) such as scrape loading, hypotonic shock/resealing, or vesicle fusion (Straubinger *et al.*, 1985) have not been tried but may prove helpful.

10.8   IMAGING OF FlCRhR AND FREE cAMP

10.8.1   Imaging modes

Observation of FlCRhR in single cells requires a fluorescence microscope equipped for single-excitation (490 nm) and dual-emission (520 and 580 nm) wavelength recording. Dual-wavelength recording is essential because the intensity changes at any one wavelength are modest in amplitude, relatively slow (seconds to minutes), and because bleaching of the fluorescein is a significant concern. In excitation ratioing, one would illuminate with fluorescein and rhodamine excitation wavelengths in alternation, while continually monitoring the emission at rhodamine wavelengths. Unfortunately, the former signal decreases only 10–20% upon cAMP binding (as explained above in Section 10.6.1) and the latter signal changes negligibly. Therefore excitation ratioing would be very insensitive. By contrast, emission ratioing (with constant excitation of the fluorescein) makes use of both the 30–50% enhancement of fluorescein emission and 10–20% decrease in rhodamine emission, giving 35–65% increases in FlCRhR[1]ratio upon binding of cAMP, and is therefore far superior.

10.8.2   Apparatus for emission ratioing

The simplest and perhaps most sensitive approach for dual emission detection uses an image-plane mask to isolate the region of interest. The emitted light from this region is split into two wavelength bands by a dichroic mirror and sent to two detectors whose outputs undergo background subtraction and then ratioing. Such simple, relatively inexpensive, and commercially available apparatus (e.g. Photoscan, Nikon Corp.) provides a time-course of spatially averaged fluorescence. However, to resolve the spatial segregation of the probe or its ratio changes, or to track moving cells, an imaging system must be used. Most ratio imaging systems have been set up for excitation ratioing because that mode best suits many popular probes for cations such as fura-2, BCECF, and SBFI (Tsien, 1989). Dual-emission ratio imaging is optically more troublesome than dual-excitation ratio imaging because emission-selecting optics must transmit as many photons as possible, preserve the optical quality of the image, and maintain accurate registration between the images at two wavelengths. By contrast, excitation wavelength-selecting optics can be relatively wasteful of photons and need handle only diffuse illumination. Dual-emission ratio imaging on a conventional microscope is generally accomplished either by mechanically switching barrier filters in front of a single imaging camera, or by splitting the fluorescence emission with a dichroic mirror and directing the separate emission wavelengths to two identical cameras. Both of these techniques suffer from the difficulty of accurately registering pixel information at each of the two wavelengths. The single-camera approach wastes half the photons, takes twice as long per wavelength pair, and is vulnerable to motion artifacts if the cell moves significantly between the two exposures. Further registration artifacts result from small spatial irregularities in barrier filter construction and inaccuracies of filter positioning from cycle to cycle. The use of two imaging cameras is costly and requires perfect matching of magnifications, freedom from geometrical distortions, and careful translational and rotational alignment to achieve registration. Adaptive warping algorithms have been described to correct digitally for misregistered images (e.g. Jerićevic *et al.*, 1989; Takamatsu & Wier, 1990), but they are still computationally complex and far from routine.

10.8.3   Confocal microscopy

Our preferred strategy to achieve dual-emission image registration is the use of a confocal laser scanning microscope (CLSM) equipped with two emission detectors (Tsien, 1990). Registration is automatically achieved by using an achromatic scanner both to deflect the exciting laser beam and to de-scan the returning emission. After the emission has passed back through the scanner and the confocal pinhole, the spatial properties of the beam are no longer of interest, so the light can simply be split with a dichroic mirror and fed to two non-imaging detectors, whose outputs inherently correspond to the same point of the raster scan (Tsien & Waggoner, 1990). Notice that such registration only holds for scanning with non-imaging detectors. Scanning-disk confocal microscopes that use cameras as detectors will have the same problems of registration as conventional wide-field microscopes. An independent advantage of confocal microscopy is that its optical sectioning capability helps in showing where the R and C subunits have migrated within the cell, as mentioned below (Section 10.9).

10.8.4 Calibrations *in vitro*

Ratio imaging of FlCRhR in single cells allows measurements of relative changes in A-kinase activity independently of the concentration of the probe, optical path length of the imaging system, and illumination intensity. One can readily derive that the calibration of an observed ratio R in terms of an absolute concentration of free cAMP [cAMP] is given by the following equation:

$$[cAMP] = K'_d \left(\frac{R - R_{min}}{R_{max} - R} \right)^{1/n}$$

Where R_{min} and R_{max} are the ratios observed at zero and saturating [cAMP] respectively; n is the Hill coefficient describing the cooperativity of cAMP binding to FlCRhR; and K'_d is an apparent dissociation constant corresponding to the cAMP concentration at which R is midway between R_{min} and R_{max}. K'_d is also given by the following expression:

$$K'_d = K_d \left(\frac{S_{f2}}{S_{b2}} \right)^{1/n}$$

Here K_d is the true dissociation constant for cAMP binding to the probe, and (S_{f2}/S_{b2}) is the ratio of emission intensities of the probe without and with cAMP measured at the denominator wavelength, i.e. at rhodamine emission wavelengths. These equations are closely analogous to the standard ones used for calibrating cation indicators with 1:1 binding stoichiometry (Grynkiewicz *et al.*, 1985), except that they have been modified here to allow for cooperativity, i.e. a Hill coefficient n that can take values other than 1.

In vitro, values for R_{min}, R_{max}, n, and K'_d are readily obtainable by direct titration of the probe with cAMP, providing that one has a way of controlling or at least determining free cAMP. Unfortunately, no reliable buffer systems for cAMP are available at present. The simplest method is to put FlCRhR at a known low concentration in a cuvette and add successive aliquots of cAMP between measurements of the emission spectrum, as in Fig. 10.3. Free cAMP is retrospectively calculated for each known concentration of total

cAMP by measuring the percentage saturation of the probe and assuming that 4 moles of cAMP are bound per mole of holoenzyme that has been dissociated. This approach has a number of limitations, so we now prefer to suck a tiny droplet of FlCRhR into a microdialysis capillary that is permeable to solutes of molecular weight below a few kilodaltons (e.g. Spectra/Por hollow fibre microdialysis tubing, 150 μm diameter, 9 kDa cut-off). This capillary is then immersed in a chamber containing cAMP solutions mounted on the microscope. This approach has many advantages. Free cAMP inside the capillary is directly controlled by the external cAMP, so that the concentration of protein can be relatively high (as required for FlCRhR[II]) yet no assumptions need to be made about the concentration or stoichiometry of binding sites. [cAMP] can be lowered as well as raised, so that a single sample of protein can be checked for hysteresis, stability, self-modification perhaps by auto-phosphorylation, or alterations due to ionic strength, ATP concentration, sulphydryl reductants, etc. Finally, the response is directly measured in the same microscope used for the imaging in cells, so that no corrections are required for differences between spectrofluorometer and microscope in spectral sensitivities and bandwidths.

10.8.5 Calibration in intact cells

In intact cells, methods have not yet been worked out to control the free cAMP concentration at precise intermediate concentrations as required to perform an *in situ* calibration. Perhaps selective permeabilization of the cells to low-molecular-weight solutes, together with inactivation of adenylate cyclase and phosphodiesterases, might be sufficient to clamp intracellular cAMP equal to extracellular. Until such methods are validated, the values for K'_d and Hill coefficient n can only be assumed to equal *in vitro* values. R_{min} and R_{max} are somewhat more accessible. R_{min} should be the ratio observed when no stimuli to raise cAMP are present and a pharmacological antagonist of cAMP is added. The only such antagonist currently available is R_pcAMPS (BioLog Life Science Institute, La Jolla, CA). These phosphorothioate analogues of cAMP (Botelho *et al.*, 1988) are supposed to be somewhat membrane-permeant, but to conserve expensive material and ensure delivery, we have mostly injected them directly into cells. In most cells that we have examined, before stimulation the emission ratio of FlCRhR is unaffected by addition of the antagonist, suggesting that the basal level of free cAMP is essentially zero. Likewise, R_{max} can be assessed in intact cells by saturating them with cAMP or an effective analogue, for example using forskolin to stimulate adenylate cyclase directly, IBMX to inhibit phosphodiesterase activity, permeant non-hydrolysable cAMP analogues like dibutyryl cAMP, direct injection of cAMP, or combinations thereof. Our impression is that R_{min} and R_{max} are up to 30% lower inside cells than in the microdialysis capillary, but that they are both changed by about the same factor.

10.8.6   Potential artifacts

10.8.6.1   Bleaching

Excessive illumination of FlCRhR at fluorescein excitation wavelengths causes a systematic fall in the ratio of fluorescein to rhodamine emissions, simulating a decrease in [cAMP]. This effect seems due to photobleaching of the fluorescein, which suffers more than the rhodamine both because the input light is primarily received by the fluorescein and because fluorescein is inherently more easily bleached. If the emission at rhodamine wavelengths were due only to FRET from fluorescein or the long-wavelength tail of fluorescein itself, fluorescein bleaching would affect the emissions at both wavelengths proportionately and cancel out in the ratio. However, the emission at rhodamine wavelengths includes a significant component due to direct excitation of rhodamine via the short-wavelength tail of its absorbance spectrum. This component does not get bleached, so the ratio decreases somewhat as fluorescein bleaches. The most obvious remedy is to decrease the frequency, duration or intensity of illumination periods. We generally have been able to collect several tens to hundreds of ratio images without too much bleaching. Obviously a less bleachable fluorophore would be desirable in place of fluorescein, but trials with alternatives with supposedly greater photostability, such as Bodipy or cyanines, have thus far been unsuccessful (see Table 10.1). Another approach might be to increase the spectral separation between the fluorophores to reduce direct excitation of the acceptor. Some small improvement might be obtained this way, but too great a spectral separation would decrease FRET efficiency.

10.8.6.2   Scrambling with endogenous kinase

The presence of endogenous unlabelled kinase sub-units within the cell raises the concern that a transient elevation of cAMP would cause scrambling and reassociation of exogenous labelled subunits with unlabelled endogenous subunits. FRET would not be reconstituted in these hybrid holoenzyme complexes, so [cAMP]$_i$ would seem to remain elevated even when it had really fallen. Such an effect can be demonstrated in the microdialysis capillary by deliberately mixing FlCRhR with unlabelled holoenzyme. Nevertheless, in real cells FlCRhR has usually proven able to

reconstitute energy transfer and to indicate a decline in apparent [cAMP]$_i$, often back to levels indistinguishable from the prestimulus state. The probable explanation is that the exogenous kinase is in considerable excess over the endogenous, so statistics favour the desired reassociation of FlC with RhR. However the possibility of scrambling must be kept in mind whenever apparent [cAMP]$_i$ fails to decrease as expected. It would be interesting to try cross-linking FlC and RhR with a flexible tether; this would discourage scrambling, but might also prevent cAMP from dissociating the subunits at all.

10.8.6.3   Perturbation of the cAMP response pathway by exogenous holoenzyme

The realization that microinjected FlCRhR is probably in considerable excess over endogenous kinase leads to concern that FlCRhR might buffer [cAMP]$_i$. Some such effect is probably unavoidable; its seriousness should be testable by deliberately introducing different amounts of FlCRhR, and minimizable by using the lowest concentrations that give adequate fluorescence and permit reconstitution of FRET upon reduction of [cAMP]$_i$. There is some evidence in hepatocytes (Corbin *et al.*, 1985) that kinase activation exerts negative feedback in cAMP levels; such a homeostatic mechanism might render [cAMP]$_i$ more resistant to perturbation by buffering. We have usually observed that strong stimulation of cAMP-linked receptors generates enough endogenous cAMP to saturate at least a few micromolar FlCRhR. In one cell type, angelfish melanophores, deliberate injections of tens of micromolar FlCRhR have been tried; at this level of enzyme, apparent [cAMP]$_i$ levels were depressed (Sammak *et al.*, 1992) compared to those observed with standard injections.

A distinct question is whether exogenous FlCRhR promotes or inhibits downstream events mediated by kinase activation. The strategy of using the native kinase was meant to minimize such perturbations, but may not be perfectly successful. If the cell can only make a limited amount of cAMP, then provision of extra kinase might be expected to maximize the efficacy of each cAMP molecule, resulting in accentuation of distal effects. But if the added FlCRhR is not quite as active as endogenous kinase or is in the wrong location with respect to substrates, it could divert cAMP into less productive binding and reduce distal phosphorylations. The fact that R subunits need two cAMP molecules to release C might also favour inhibition, since excess kinase would statistically favour non-productive single occupancy of R subunits at the expense of productive double occupancy. Obviously, it will be important to monitor downstream cell responses whenever possible. So far, responses such as rounding up in fibroblasts, aggregation and dispersal of pigment granules in melanophores (Sammak *et al.*, 1993), and increased excitability of *Aplysia* sensory neurons appear to be qualitatively unaffected by the usual levels of a few micromolar FlCRhR. However, tens of micromolar FlCRhR caused permanent aggregation of melanophore granules consistent with depression of [cAMP]$_i$ and overall kinase activity.

10.9 APPLICATIONS

To date (September 1991) we have successfully obtained cAMP recordings from REF52 fibroblasts, BC3H-1 smooth muscle-derived cells (Adams *et al.*, 1991), PC-12 pheochromocytoma cells, neonatal cardiac myocytes, primary and transformed osteoblasts, chick ciliary ganglion neurons (Gurantz *et al.*, 1991), rat Schwann cells, MA-10 Leydig tumour cells, fish melanocytes (Sammak *et al.*, 1992), and cultured *Aplysia* sensory neurons. Space does not permit detailed description of the results obtained, which have been or will be reported in appropriate primary research papers. However, one sample application can be described here, the visualization of the spread of cAMP within MA-10 Leydig tumour cells from sites of relatively localized hormone stimulation (Plate 10.1). This experiment, an extension of previous morphological observations (Podesta *et al.*, 1991), was performed by Brian Bacskai and Clotilde Randriamampita in collaboration with José Lemos, Ernesto Podesta and Mario Ascoli. Luteinizing hormone (LH) was locally applied from a puffer pipette placed sequentially at two locations near a cluster of four MA-10 cells injected with FlCRhR. Elevations in intracellular cAMP can be seen to spread from each site of local application. Further details can be found in the plate caption.

Besides the ability of FlCRhR to monitor dynamic changes in free cAMP concentrations in single cells, it is also possible to use the dual labelling of the protein subunits to independently track their positions within a cell. Spatial segregation and dynamic translocation of proteins within cells can crucially influence their biological effectiveness. In general, FlCRhR injected in the cytoplasm remains there and seems not to enter the nucleus as long as cAMP is low. After cAMP-induced dissociation of the holoenzyme, the C subunit translocates into or equilibrates with the nucleus, whereas the R subunits seem to remain in the cytoplasm (Meinkoth *et al.*, 1990). Upon lowering cAMP, the fluorescein-labelled C subunit leaves the nucleus and re-forms holoenzyme with the R subunit in the cytoplasm (Adams *et al.*, 1991). Labelled holoenzyme or subunits should also be a powerful tool for assessing and visualizing binding to cytoskeletal elements or other organelles.

Results such as these suggest that imaging of FlCRhR in single cells can provide unprecedented spatial and temporal resolution to study the distribution of cAMP and A-kinase and thereby to analyse the biochemical control and physiological function of this most important second messenger pathway. Furthermore, the ability of fluorescence resonance energy transfer to monitor protein–protein and protein–DNA interactions non-destructively in living cells might well be extended to many other analogous supramolecular combinations. Just within the field of signal transduction, a few of the obvious examples would include complexes of receptor tyrosine kinases with effector proteins, G-protein α subunits with βγ subunits or with 7–transmembrane-segment receptors or effector proteins (e.g. Erickson & Cerione, 1991), calmodulin with its targets, transcription factor complexes (e.g. Patel *et al.*, 1990), and so on.

Enquiries regarding the commercial availability of FlCRhR should be directed to Atto Instruments, 1500 Research Boulevard, Rockville, MD 20850 USA, which has been licensed by the University of California to produce the material.

REFERENCES

Aass H., Skomedal T. & Osnes J.-B. (1988) *J. Molec. Cell Cardiol.* **20**, 847–860.

Adams S.R., Harootunian A.T., Buechler Y.J., Taylor S.S., & Tsien R.Y. (1991) *Nature* **349**, 694–697.

Akil M. & Fisher S.K. (1989) *J. Neurochem* **53**, 1479–1486.

Anjaneyulu P.S.R. & Staros J.V. (1987) *Int. J. Peptide Protein Res.* **30**, 117–124.

Ascoli M. (1981) *Endocrinology* **108**, 88–95.

Barsony J. & Marx S.J. (1990) *Proc. Natl. Acad. Sci. USA* **87**, 1188–1192.

Beavo J.W. & Reifsnyder D.H. (1990) *Trends Pharmacol. Sci.* **11**, 150–155.

Beavo J.W., Bechtel P.J. & Krebs E.G. (1974) *Proc. Natl. Acad. Sci. USA* **71**, 3580–3583.

Beebe S.J. & Corbin J.D. (1986) In *The Enzymes*, Vol. 17, *Control by Phosphorylation*, Part A, P.D. Boyer & E.G. Krebs (eds). Academic Press, New York, pp. 43–111.

Beebe S.J., Blackmore P.F., Chrisman T.D. & Corbin J.D. (1988) *Methods Enzymol.* **159**, 118–189.

Berridge M.J., Cobbold P.H. & Cuthbertson K.S.R. (1988) *Phil. Trans. R. Soc. Lond. B* **320**, 325–343.

Bode D.C. & Brunton L.L. (1988) *Molec Cell. Biochem.* **82**, 13–18.

Boshart M., Weih F., Nichols M. & Schütz G. (1991) *Cell* **66**, 849–859.

Botelho L.H.P., Rothermel J.D., Coombs R.V. & Jastorff B. (1988) *Methods Enzymol.* **159**, 159–172.

Bourne H.R., Sanders D.A. & McCormick F. (1991) *Nature* **349**, 117–127.

Bramson, H.N., Kaiser E.T. & Mildvan A.S. (1984) *CRC Crit. Rev. Biochem.* **15**, 92–124.

Bright G.R., Fisher G.W., Rogowska J. & Taylor D.L. (1989) *Methods Cell Biol.* **30**, 157–192.

Broach J.R. & Deschenes R.J. (1990) *Adv. Cancer Res.* **54**, 79–139.

Brooker G. (1988) *Methods Enzymol.* **159**, 45–50.

Brooker G., Harper J.F., Terasaki W.L. & Moylan R.D. (1979) *Adv. Cyc. Nucl. Res.* **10**, 1–33.

Büchler W., Meinecke M., Chakraboty T., Jahnsen T., Walter U. & Lohmann S.M. (1990) *Eur. J. Biochem.* **188**, 253–259.

Byus C.V. & Fletcher W.H. (1988) *Methods Enzymol.* **159**, 236–254.

Cadd G.G., Uhler M.D. & McKnight G.S. (1990) *J. Biol. Chem.* **265**, 19502–19506.

Carr D.W., Stofko-Hahn R.E., Fraser I.D.C., Bishop S. M., Acott T.S., Brennan R.G. & Scott J.D. (1991). *J. Biol. Chem.* **266,** 14188–14192.

Castellucci V.F., Kandel E.R., Schwartz J.H., Wilson F.D., Nairn A.C. & Greengard P. (1980) *Proc. Natl. Acad. Sci. USA* **77**, 7492–7496.

Cho-Chung Y.S., Clair T., Tortora G. & Yokozaki H. (1991) *Pharmac. Ther.* **50**, 1–33.

Cook P.F., Neville M.E. jr., Vrana K.E., Hartl F.T. and Roskoshi R. jr. (1982) *Biochemistry* **21**, 5794–5799.

Corbin J.D., Beebe S.J., Blackmore P.F. (1985) *J. Biol. Chem.* **260**, 8731–8735.

Cumming R. (1981) *Trends Neurosci.* **4**, 202–204.

Deslongchamps G., Galàn A., de Mendoza G. & Rebek J. Jr (1992) *Angew. Chem. Intl Ed.* **31**, 61–63.

Døskeland S.O. (1978) *Biochem. Biophys. Res. Commun* **83**, 542–549.

Drobnica Ľ., Kristián P., & Augustín J. (1977) In *The Chemistry of Cyanates and Their Thio Derivatives*, Part 2, S. Patai (ed.). J. Wiley, New York, pp. 1003–1221.

Duronio R.J., Jackson-Machelski E., Heuckeroth R.O., Olins P.O., Devine C.S., Yonemoto W., Slice L.W., Taylor S.S. & Gordon J.I. (1990) *Proc. Natl. Acad. Sci. USA* **87**, 1506–1510.

Erickson J.W. & Cerione R.A. (1991) *Biochemistry* **30**, 7112–7118.

First E.A. & Taylor S.S. (1989) *Biochemistry* **28**, 3598–3605.

First E.A., Johnson D.A. & Taylor S.S. (1989) *Biochemistry* **28**, 3606–3613.

Fletcher W.H., Ishida T.A., Van Patten S.M. & Walsh D.A. (1988) *Methods Enzymol.* **159**, 255–267.

Gilman A.G. (1990) *Adv. Sec. Mes. Phos. Prot. Res.* **24**, 51–57.

Glazer A.N. & Stryer L. (1990) *Methods Enzymol.* **184**, 188–194.

Granot J., Mildvan A.S., Hiyama K., Kondo H. & Kaiser E.T. (1980) *J. Biol. Chem.* **255**, 4569–4573.

Greenberg S.M., Bernier L. & Schwartz J.H. (1987) *J. Neurosci.* **7**, 291–301.

Gross R.E., Lu X. & Rubin C.S. (1990) *J. Biol. Chem.* **265**, 8152–8158.

Grynkiewicz G., Poenie M. & Tsien R.Y. (1985) *J. Biol. Chem.* **260**, 3440–3450.

Gurantz D., Harootunian A.T., Tsien R.Y., Dionne V.E. & Margiotta J.F. (1991) *Soc. Neurosci. Abstr.* **17**, 959 (Abstr. 384.5).

Haugland R.P. (1989) *Handbook of Fluorescent Probes and Research Chemicals.* Molecular Probes, Eugene, Oregon.

Hayes J.S. & Brunton L.L. (1982) *J. Cyc. Nucl. Res.* **8**, 1–16.

Herman B. (1989) *Methods Cell Biol.* **30**, 219–243 .

Hiratsuka T. (1982) *J. Biol. Chem.* **257**, 13354–13358.

Hofmann F. (1980) *J. Biol. Chem.* **255**, 1559–1564

Jerićevic Ž., Wiese B., Bryan J. & Smith L.C. (1989) *Methods Cell Biol.* **30**, 47–83.

Johnson D.A., Voet J.G. & Taylor P. (1984) *J. Biol. Chem.* **259**, 5717–5725.

Jones K.W., Shapero M.H., Chevrette M. & Fournier R.E.K. (1991) *Cell* **66**, 861–872.

Jovin T.M. & Arndt-Jovin D.J. (1989) In *Cell Structure and Function by Microspectrofluorometry*, E. Kohen & J.G. Hirschberg (eds). Academic Press, San Diego, pp. 99–117.

Karin M. (1989) *Trends Genetics* **5**, 65–67.

Klein P.S., Sun T.J., Saxe C.L. III, Kimmel A.R., Johnson R.L. & Devreotes P.N. (1988) *Science* **241**, 1467–1472.

Knighton D.R., Zheng J., Ten Eyck L.F., Ashford V.A., Xuong N-h., Taylor S.S. & Sowadski J.M. (1991a) *Science* **253**, 407–414.

Knighton D.R., Zheng J., Ten Eyck L.F., Xuong N-h., Taylor S.S. & Sowadski J.M. (1991b) *Science* **253**, 414–420.

Lakowicz J.R. (1983) *Principles of Fluorescence Spectroscopy.* Plenum Press, New York, pp. 303–339.

Leathers V.L., Fletcher W.H. & Johnson D.A. (1990) *J. Cell. Biol.* **111**, 90a (Abstr. 393).

Liang H., Purucker W.J., Stenger D.A., Kubinec R.T. & Hui S.W. (1988) *BioTechniques* **6**, 550–558.

Lohmann S.M. & Walter U. (1984) *Adv. Cycl. Nucl. Prot. Phos. Res.* **18**, 63–117.

Lohmann S.M., De Camilli P. & Walter U. (1988) *Methods Enzymol.* **159**, 183–193.

McNeil P.L. (1989) *Methods Cell Biol.* **29**, 153–173.

Meinkoth J.L., Ji Y., Taylor S.S. & Feramisco J.R. (1990) *Proc. Natl. Acad. Sci. USA* **87**, 9595–9599.

Meinkoth J.L., Montminy M.R., Fink J.S. & Feramisco J.R. (1991) *Molec. Cell Biol.* **11**, 1759–1764.

Meyer T. & Stryer L. (1991) *Ann. Rev. Biophys. Biophys. Chem.* **20**, 153–174.

Montminy M.R. & Bilezikjian L.M. (1987) *Nature* **328**, 175–178.

Murray K.J., Reeves M.L. & England, P.J. (1989) *Mol. Cell. Biochem.* **89**, 175–179.

Nakamura T. & Gold G.H. (1987) *Nature* **325**, 442–444.

Nelson N.C. & Taylor S.S. (1983) *J. Biol. Chem.* **258**, 10981–10987.

Nigg E.A. (1990) *Adv. Cancer Res.* **55**, 271–310.

O'Dowd B.F., Lefkowitz R.J. & Caron M.G. (1989) *Ann. Rev. Neurosci.* **12**, 67–83.

Patel L., Abate C. & Curran, T. (1990) *Nature* **347**, 572–575.

Podesta E.J., Solano A.R. & Lemos J.R. (1991) *J. Mol. Endocrinol.* **6**, 269–279

Rall T.N. & Sutherland E.W. (1961) *Cold Spring Harbor Symp. Quant. Biol.* **26**, 347–354.

Rangel-Aldao R. & Rosen O.M. (1977) *J. Biol. Chem.* **252**, 7140–7145.

Rannels S.R. & Corbin, J.D. (1980) *J. Biol. Chem.* **255**, 7085–7088.

Robison G.A., Butcher R.W. & Sutherland E.W. (1971) *Cyclic AMP*. Academic Press, New York.

Rymond M. & Hofmann F. (1982) *Eur. J. Biochem.* **125**, 395–400.

Sammak P.J., Adams S.R., Harootunian A.T., Schliwa M., Tsien R.Y. (1992) *J. Cell Biol.* **117**, 57–72.

Saraswat L.D., Filutowicz M. & Taylor S.S. (1986) *J. Biol. Chem.* **261**, 11091–11096.

Schacher S., Castellucci V.F. & Kandel E.R. (1988) *Science* **240**, 1667–1669.

Scott J.D. (1991) *Pharmac. Ther.* **50**, 123–145.

Scott J.D., Stofko R.E., McDonald J.R., Comer J.D., Vitalis E.A. & Mangili J.A. (1990) *J. Biol. Chem.* **265**, 21561–21566.

Slice L.W. & Taylor, S.S. (1989) *J. Biol. Chem.* **264**, 20940–20946.

Southwick P.L., Ernst L.A., Tauriello E.W., Parker S.R., Mujumdar R.B., Mujumdar S.R., Clever H.A. & Waggoner A.S. (1990) *Cytometry* **11**, 418–430.

Straubinger R.M., Düzgünes N. & Papahadjopolous D. (1985) *FEBS Lett.* **179**, 148–154.

Stryer L. (1978) *Ann. Rev. Biochem.* **47**, 819–846.

Takamatsu T. & Wier W.G. (1990) *Cell Calcium* **11**, 111–120.

Taylor S.S., Buechler J.A. & Knighton D.R. (1990a) In *Peptides & Protein Phosphorylation* B.E. Kemp (ed.). CRC Press, Boca Raton pp. 1–41.

Taylor S.S., Buechler J.A. and Yonemoto W. (1990b) *Ann. Rev. Biochem.* **59**, 971–1005.

Terasaki W.L. & Brooker G. (1977) *J. Biol. Chem.* **252**, 1041–1050.

Tsien R.Y. (1989) *Methods Cell Biol.* **30**, 127–156.

Tsien R.Y. (1990) *Proc. Roy. Microsc. Soc.* **25**, S53 (Micro '90 Supplement).

Tsien R.Y. and Poenie M. (1986) *Trends Biochem. Sci.* **11**, 450–455.

Tsien R.Y. & Tsien R.W. (1990) *Annu. Rev. Cell Biol.* **6**, 715–760.

Tsien R.Y. & Waggoner A. (1990) In *Handbook of Biological Confocal Microscopy*, J. Pawley (ed.). Plenum Press, New York, pp. 169–178.

Van Haastert P.J.M. (1991) *Adv. Sec. Mess. Phos. Prot. Res.* **23**, 185–226.

Wang K., Feramisco J.R. & Ash J.F. (1982) *Methods Enzymol.* **85**, 514–562.

White J.C. & Stryer L. (1987) *Anal. Biochem.* **161**, 442–452.

Wu C.-W. & Wu F.Y.-H. (1974) *Biochemistry* **13**, 2573–2578.

Wu F.Y.-H., Nath K. & Wu C.-W. (1974) *Biochemistry* **13**, 2567–2572.

Zoller M.J., Kerlavage A.R. & Taylor S.S. (1979) *J. Biol. Chem.* **254**, 2408–2412.

CHAPTER ELEVEN

Potentiometric Membrane Dyes

LESLIE M. LOEW

Department of Physiology, University of Connecticut Health Centre, Farmington, CT, USA

11.1 INTRODUCTION

Dye indicators of membrane potential have been employed in numerous studies of cell physiology ever since their introduction in the early 1970s. L.B. Cohen and his co-workers (Davila *et al.*, 1973; Cohen *et al.*, 1974; Ross *et al.*, 1977; Gupta *et al.*, 1981) have pioneered this effort leading to the discovery of a large number of organic dyes whose spectral properties are sensitive to changes in membrane potential. This enterprise was motivated by studies requiring multisite mapping of electrical activity in complex neuronal systems; a variety of applications in both cell biology and neuroscience have emerged, however, and were reviewed in a series of chapters in a recent book (Loew, 1988). The aim of this chapter is to review the methodology in sufficient detail to provide the reader with an appreciation of the factors involved in appropriate choice of potentiometric molecular probes and implementation of the technology in particular applications. The size of the voltage-dependent signal, while certainly important, is by no means the only factor to be considered in choosing a dye. The intent here is to identify the important parameters which have to be understood about the chemistry of a dye if it is to be a useful and practical indicator of potential.

Also introduced below are some brief generalities about the structures and physical properties of the various dye chromophores which have formed the backbones of potentiometric dyes. This information should help to take some of the mystery out of the chemistry of potentiometric dyes and provide guidance toward the appropriate dye selection.

My laboratory has been engaged in the design of membrane staining styryl dyes based on the theoretical expectation of an electrochromic mechanism for the potentiometric response (Platt, 1956; Loew *et al.*, 1978). Fluorescent indicators which operate via a potential-dependent redistribution across the membrane have also emerged from our efforts at dye development (Ehrenberg *et al.*, 1988; Farkas *et al.*, 1989). Particular emphasis, therefore, will be placed on these sets of potentiometric probes. The very different properties of these classes of dyes will help to illustrate the variety of considerations required for appropriate and successful application of this complex methodology. I will also present applications to problems of interest in this laboratory which demonstrate a merger of the probe technology with quantitative fluorescence microscopy. A primary application of potentiometric dyes is, of course, mapping of electrical activity in complex preparations of excitable tissue; this is described in this volume in Chapter 30.

11.2 OPTIMIZATION OF DYE INDICATOR SENSITIVITY

The first attempts to develop potentiometric dyes relied on serendipity and trial and error. By screening large numbers of dyes on the squid giant axon, Cohen and his colleagues (Cohen *et al.*, 1974; Ross *et al.*, 1977; Gupta *et al.*, 1981) were able to discover a series of highly sensitive dyes as well as develop a large database of information. The resultant database revealed several broad rules which could be used to help design new generations of probes. For example, it became clear that the large class of azo dyes were not going to be particularly suitable because of their photo-instability and their propensity for toxic and photodynamic damage to biological preparations. The dyes had to have some hydrophobic appendages to promote interaction with the membrane, but alkyl groups longer than about eight carbons imparted too much insolubility for some applications. It became apparent that several different molecular mechanisms were employed by different dyes to produce potential-dependent spectral changes. It also soon became clear that the same dye could behave very differently in different model and biological preparations (Ross & Reichardt, 1979; Loew *et al.*, 1985b).

Cohen and Salzberg (Cohen & Salzberg, 1978) divided potentiometric dyes into two simple classes – slow and fast. 'Fast' dyes are able to follow changes in the millisecond range – fast enough to monitor individual electrical events in excitable cells and tissue. They are membrane stains which are subtly perturbed by the transmembrane voltage resulting in a small spectral response. 'Slow' dyes can measure voltage changes that may accompany hormonal responses in non-excitable cells or the level of activity in energy-transducing organelles. They are charged and are pulled from one environment to another by the potential difference between them; the environmental dependence of their spectra (e.g. aqueous vs. membrane-bound) underlies the typically large potentiometric response. Therefore, in addition to identifying the range of applications accessible to potentiometric indicators, mechanism and sensitivity are, to some extent, classified by the 'fast' and 'slow' designations.

11.2.1 Fast dyes

Generally, the mechanisms underlying the fast dye responses involve potential-dependent intramolecular rearrangements or small movements of the dye from one chemical environment to another. These reactions have the requisite speed but usually do not produce very large changes in the spectra of the dye. The largest reported fluorescence response to an action potential was greater than 20% per 100 mV for the dye RH421 in a neuroblastoma cell preparation (Grinvald, *et al.*, 1983). This dye is depicted in Fig. 11.1 together with several other examples of good fast dyes. Typically good fast dyes respond with fluorescence changes of only 2–10% per 100 mV. It should also be stressed again that a fast dye may give a large response in one biological preparation with a given set of experimental conditions but may be totally insensitive to potential changes when applied to a different preparation.

An effort to develop fast dyes which employ an

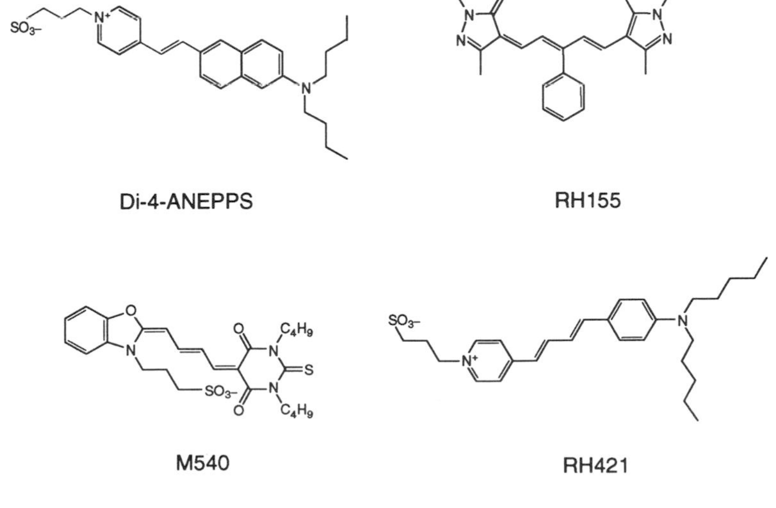

Figure 11.1 A representative sampling of fast dyes. Di-4-ANEPPS is a versatile fluorescence and absorbance indicator (Fluhler *et al.*, 1985; Loew *et al.*, 1992). RH155 is a good absorbance dye (London *et al.*, 1987). M-540 is one of the original and most thoroughly studied potentiometric dyes (L.B. Cohen *et al.*, 1974). RH421 displays a 21% per 100 mV fluorescence change on a neuroblastoma cell preparation (Grinvald *et al.*, 1983).

Di-4-ANEPPS

RH155

M540

RH421

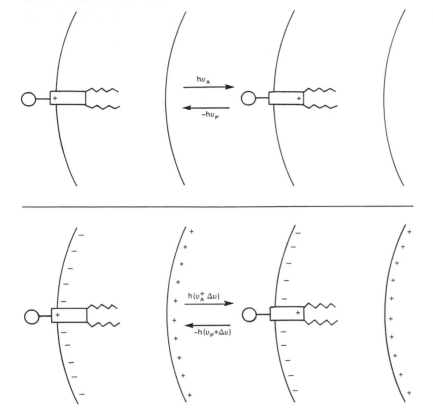

Figure 11.2 An electrochromic dye undergoes an internal charge migration upon excitation (top). In the presence of the electric field produced by a potential drop across the membrane, the energy required for the excitation is altered (bottom), resulting in a spectral shift.

electrochromic (Platt, 1956; Liptay, 1969; Loew *et al.*, 1978) mechanism has been a major focus of this laboratory. Briefly, electrochromism is possible if there is a large shift in electronic charge when a chromophore is excited from the ground to the first excited state; if the direction of charge movement lies parallel to an electric field, the energy of the transition will be sensitive to the field amplitude. Thus, if the chromophore is oriented so that the charge redistribution is perpendicular to the membrane surface, an electrochromic dye should be an indicator of membrane potential. These ideas are summarized in Fig. 11.2. The reason this mechanism was attractive to us was that it lent itself to a theoretical molecular orbital treatment (Loew *et al.*, 1978) which could aid in the design of appropriate targets for organic synthesis. The aminostyrylpyridinium chromophore best fits the criterion of a large charge shift upon excitation and could easily be modified with sidechains that would assure an appropriate orientation in the membrane. Variations on this chromophore have been the targets of synthetic efforts (Loew *et al.*, 1979; Grinvald *et al.*, 1982; Grinvald *et al.*, 1983; Hassner *et al.*, 1984; Fluhler *et al.*, 1985). Figure 11.1 includes two examples: di-4-ANEPPS (Fluhler *et al.*, 1985) and

RH421 (Grinvald *et al.*, 1983), which have been particularly successful in a variety of applications. Among their special attributes are relatively good photostability, a high fluorescence quantum yield for the membrane-bound dyes and almost no fluorescence from aqueous dye. It should be noted that the potentiometric responses of these dyes in different biological preparations may be complex superpositions of several mechanisms with only a minor contribution from electrochromism (Loew *et al.*, 1985b). Nevertheless, the styryl dyes have emerged as the most popular fast fluorescent potentiometric indicators.

Di-4-ANEPPS is a good, general-purpose fast probe which has several beneficial attributes and has been used on a large variety of preparations (Fluhler *et al.*, 1985; Gross *et al.*, 1986; Ehrenberg *et al.*, 1987; Lojewska *et al.*, 1989; Müller *et al.*, 1989; Chien & Pine, 1991; Rosenbaum *et al.*, 1991; Davidenko *et al.*, 1992; Loew *et al.*, 1992). It has been reasonably consistent in giving a relative fluorescence change of $\approx$10% per 100 mV in a number of different cell types and experimental protocols. It can be used in a dual-wavelength ratiometric excitation mode to normalize out artifacts due to uneven staining or photobleaching

(Montana *et al.*, 1989). As with other styryls, only the membrane-bound dye displays appreciable fluorescence. In several experimental protocols, it has been persistent for hours under continuous perfusion with dye-free medium.

Rational modifications can be made to optimize probe characteristics for particular experimental demands. Decreasing the size of the chromophore as in di-4-ASPPS (Loew & Simpson, 1981; Fluhler *et al.*, 1985) results in a blue shift of the spectral characteristics of the response of about 30 nm; unfortunately, a 40% reduction in sensitivity also results. Decreasing the chain length of the hydrocarbon tails as in di-2-ASPSS increases the water solubility of the dye. This is necessary for thick tissue preparations where the dye must penetrate through many cell layers; the more water-soluble dyes, of course, will give less persistent staining of the preparation. The opposite situation will pertain to a dye like di-8-ANEPPS (Ehrenberg *et al.*, 1990; Bedlack *et al.*, 1992), where low solubility requires a high-molecular-weight surfactant like Pluronic F127 (BASF Corp.) to promote staining (Davila *et al.*, 1973; Lojewska & Loew, 1987). In addition to persistence, we have discovered that another attribute of di-8-ANEPPS is a very slow rate of internalization. In several cell types, di-4-ANEPPS was found to be incorporated into intracellular organelles over times as short as 10 min; di-8-ANEPPS is retained exclusively on the plasma membrane over periods of hours. Varying headgroup charge can subtly change the location of bound dye in the membrane with the effect of sometimes significant improvements in sensitivity. Also, it seems generally true that positively charged dyes are especially well-suited for experiments requiring microinjection to localize dye to just one cell in a complex preparation (Grinvald, A. *et al.*, 1987). Many analogues of styryl dyes have been synthesized to meet these various experimental demands in both this laboratory by Joseph Wuskell and in the laboratory of A. Grinvald by Rena Hildesheim.

Merocyanine and oxonol chromophores have formed the basis of potentiometric dyes which respond to voltage via changes in absorbance rather than fluorescence. Absorbance measurements are often desirable, especially for complex preparations with non-excitable satellite cells. This is because the large background fluorescence from the membranes of these uninvolved cells would attenuate the relative fluorescence response with a corresponding degradation of signal-to-noise; a transmitted light signal is used in absorbance and is not significantly affected by background staining. The oxonol class of dyes has now emerged as the most sensitive for detecting fast potential-dependent absorbance changes, although some merocyanine dyes, the earliest class of successful fast indicators, are also still in use. The oxonol chromophore is defined by its delocalized negative charge. The mechanism for the potential-dependent response has been determined for several of these probes; it involves a movement between a binding site on the membrane surface and an aqueous region adjacent to the membrane (Waggoner *et al.*, 1977; George *et al.*, 1988). An example of an oxonol dye, RH155, is given in Fig. 11.1.

The merocyanine chromophore is characterized by neutral and zwitterionic resonance structures; merocyanines are often highly solvatochromic and can theoretically respond to membrane potential via electrochromism. However, the only thorough experimental investigations of merocyanine mechanisms performed to date have been with just one dye, M540 (Fig. 11.1); these studies indicated that dye reorientation possibly coupled with aggregation at the membrane surface underlies the potentiometric response of this early member of the merocyanine family of potentiometric dyes (Ross *et al.*, 1977; Dragsten & Webb, 1978; Verkman & Frosch, 1985).

11.2.2 Slow dyes

Potential-dependent partitioning between the extracellular medium and either the membrane or the cytoplasm is the general process underlying the mechanisms of slow dyes. Unlike fast potentiometric indicators, slow dyes have to be charged, so that the voltage difference can pull the dyes from one compartment to another. The change in environment is what produces the voltage-dependent spectral change. Three chromophore types have yielded useful slow

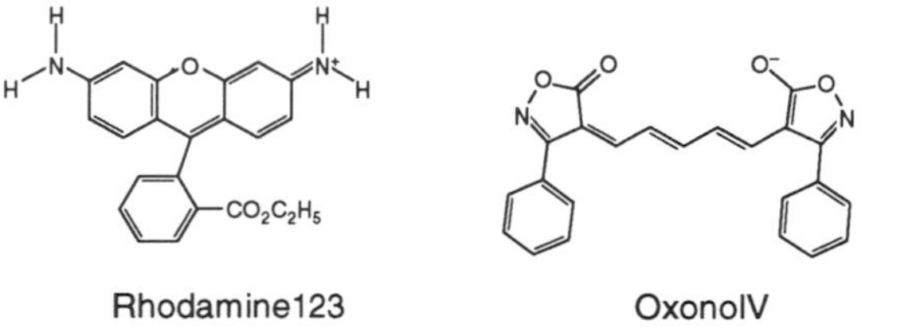

Figure 11.3 A representative sampling of slow dyes. TMRE is a redistribution dye for single cell membrane potential imaging (Farkas *et al.*, 1989). DiSC$_3$(5) is a very sensitive dye for cell suspensions (Sims *et al.*, 1974). Rhodamine 123 is a mitochondrial stain for qualitative single cell measurements (Johnson *et al.*, 1981). Oxonol V is a negatively charged dye used for suspensions of mitochondria and submitochondrial particles (Smith *et al.*, 1976).

dyes: cyanines, oxonols and rhodamines. Each of these have special features that make them applicable to different kinds of experimental requirements. Figure 11.3 shows some of the more important slow dyes.

The cyanine class of symmetrical dyes with delocalized positive charges were originally introduced by Alan Waggoner and have proven to be extraordinarily sensitive probes for potential changes in populations of non-excitable cells (Sims et al., 1974; Waggoner, 1979, 1985). A large number of these dyes with varying hydrocarbon chain lengths, numbers of methine groups in the bridging polyene, and heterocyclic nuclei are available. Depending on the nature of the dye and its concentration, potential-dependent uptake can effect either an increase or a decrease in fluorescence intensity. In general, the fluorescence of these dyes is enhanced upon membrane binding – thus accumulation of dye leads to fluorescence enhancement. However, at high dye:lipid ratios many of the dyes have a tendency to aggregate, resulting in self-quenching of fluorescence. The latter is the case for $diSC_3(5)$ (Figure 11.3), which can lose 98% of its fluorescence when a cell or vesicle preparation is polarized to 100 mV (Sims et al., 1974; Loew et al., 1983, 1985a). $DiOC_6(3)$ has less of a tendency to aggregate and displays an increased fluorescence quantum yield as it binds to the plasma and organelle membranes. Its lipophilicity is responsible for its use as a stain for mitochondria and endoplasmic reticulum (Terasaki et al., 1984). However, dye association with intracellular organelles can lead to cytotoxicity and/or misinterpreted fluorescence changes (Korchak et al., 1982) and appropriate controls must be performed to check for this possibility in experiments on bulk cell suspensions with cyanines.

Anionic oxonols also show enhanced fluorescence upon binding to membranes, but, because of their negative charge, binding is promoted by depolarization. More importantly, the negative charge lessens intracellular uptake of oxonol dyes, solving some of the difficulties encountered with the cyanines. They are, however, less sensitive than the cyanines. Bis-oxonol – and its relatives with barbituric acid nuclei – have been used in fluorescence experiments on cell suspensions (Brauner et al., 1984; Mohr & Fewtrell, 1987; Labrecque et al., 1989). Oxonol V (Fig. 11.3) is among several similar oxonols developed by Chance, Smith and Bashford (Smith et al., 1976; Bashford et al., 1979; Smith & Chance, 1979) for dual-wavelength absorbance measurements on energy-transducing organelle suspensions.

The membrane potential of individual cells can be monitored with a fluorescence microscope and a cationic redistribution dye. Most of these have been qualitative studies; this is because of the difficulty in calibrating the fluorescence intensities determined

from variably sized cells as well as the complexity of the fluorescence perturbations induced by the dynamic intracellular environment. A cyanine dye was used to distinguish between quiescent and cycling cells based on the difference in dye accumulation (Cohen et al., 1981). Rhodamine 123 (Fig. 11.3) was originally introduced as a mitochondrial stain by Chen and co-workers (Johnson et al., 1980, 1981; Chen, 1988). It has been used largely in qualitative studies of mitochondrial membrane potential and has been especially effective in flow cytometry applications. Chen's laboratory recently introduced a cyanine dye which forms J-aggregates when it is concentrated in highly polarized mitochondria; since the aggregate has a red emission and the monomer fluoresces green, the colour of the emitted light provides a direct indication of mitochondrial potential (Reers et al., 1991; Smiley et al., 1991).

In order to develop a quantitative assay of membrane potential in individual cells, it would be desirable to use a permeable redistribution dye with spectral characteristics having minimal environment sensitivity. Thus, the fluorescence intensity will reflect the degree of Nernstian accumulation of dye only and can therefore be readily interpreted. This laboratory (Ehrenberg et al., 1988) has synthesized two rhodamine dyes, TMRE (Fig. 11.3) and TMRM, which are very similar to rhodamine 123 but have the free amino groups substituted with methyl substituents. This makes these dyes more permeable than rhodamine 123 and also blocks any poorly reversible hydrogen bonding interactions with anionic sites in the mitochondrial inner membrane and matrix. This, combined with the general environmental insensitivity of rhodamine fluorescence and the low tendency of these dyes to aggregate, makes them good 'Nernstian' indicators of membrane potential. That is to say, the ratio of fluorescence intensities measured in two compartments separated by a membrane, viz. F^{in}/F^{out}, when properly corrected for background dye binding, can be inserted into the Nernst equation to provide a direct measure of the potential difference between the compartments:

$$\triangle V_m = - \frac{RT}{F} \ln \frac{F^{in}}{F^{out}} \qquad [11.1]$$

Where R is the ideal gas constant, T the absolute temperature and F is Faraday's constant in this equation. This approach was successfully applied to the plasma membrane potential of several different cell types (Ehrenberg et al., 1988). It has also been used to follow changes in mitochondrial membrane potential via digital imaging microscopy (Farkas et al., 1989). These studies will be described in more detail in the following section.

11.3 MAPPING MEMBRANE POTENTIAL BY DIGITAL FLUORESCENCE MICROSCOPY

Appropriate choices of either the fast or slow dyes can make it possible to study phenomena involving membrane potential in individual cells under the microscope. As has been noted above, the primary application of the fast dyes has been in studies of the electrical activity in excitable systems. These are the subject of the chapter by Wu and Cohen, elsewhere in this volume (Chapter 30). In this section, I would like to describe the use of the fast dyes to map membrane potential variations along the surface of single cells. This depends on the ability of these membrane-staining dyes to sense local electric fields, rather than on their speed. Slow dyes have been mainly applied to studies of cell and organelle suspensions (Freedman & Laris, 1981, 1988; Waggoner, 1985). Later in this section, the ability of the 'Nernstian' dyes to determine membrane potential in single cells will be described and their promise as indicators of membrane potential in individual mitochondria will be explored.

11.3.1 Membrane potential maps from dual-wavelength fluorescence ratio images

As detailed in Section 11.2, potentiometric indicators classified as 'fast' stain the outer leaflet of the plasma membrane lipid bilayer. These dye molecules individually sense the electric field in the membrane with a resultant perturbation of their spectral properties. They are therefore capable of reporting rapid changes in membrane potential and can be used to follow the spread of an action potential down a single neuron (Grinvald, H. et al., 1981; Krauthamer & Ross, 1984; Ross & Krauthamer, 1984; Shrager et al., 1987; Shrager & Rubinstein, 1990). Similarly, they can be used to detect temporally constant spatial variations in membrane potential along the surface of cells. This latter application requires some special dye properties; among these is a potential dependent spectral shift which can become the basis for dual-wavelength ratiometric measurements.

The success of dual-wavelength ratiometric methods for measurements with fluorescent cation indicators (Rink et al., 1982; Grynkiewicz et al., 1985; Tsien & Poenie, 1986) is largely due to the ability of this approach to eliminate artifactual variations in total dye fluorescence from the assay. Thus cells in suspension can be assayed without concern for small variations in the level of dye loading from sample to sample. More importantly, optically heterogeneous specimens, such as single cells under the fluorescence microscope, can have their cation distributions mapped via digital radio imaging (Tsien & Poenie, 1986). The availability of several commercial dual-wavelength fluorescence spectrometers and imaging systems has made this technology accessible to many laboratories. While dual-wavelength differential absorbance techniques have been occasionally applied to potentiometric indicators (Bashford et al., 1979; Freedman & Hoffman, 1979 and references cited therein), dual-wavelength ratiometric fluorescence measurements have only recently been implemented for membrane potential (Montana et al., 1989; Ehrenberg et al., 1990; Jesurum & Gross, 1991; Bedlack et al., 1992).

Clearly, this approach requires a dye which undergoes a potential dependent spectral shift, and the ANEP naphthylstyryl chromophore meets this criterion (Fluhler et al., 1985). Using di-4-ANEPPS, we have demonstrated the viability of this idea with both a dual-wavelength spectrofluorometer on lipid vesicle suspensions and a fluorescence microscope equipped with dual-wavelength digital imaging equipment to study individual cells (Montana et al., 1989). The lipid vesicles were prepared in either high $[K^+]$ or high $[Na^+]$ media and diluted into buffers containing varying proportions of these two ions. Addition of the potassium-selective ionophore valinomycin produced a membrane potential defined by the $[K^+]$ gradient across the vesicle membrane. As shown in Fig. 11.4, the dual-excitation ratio does provide a good indicator of membrane potential; more importantly, it is insensitive to potential sources of artifact such as photobleaching and non-potentiometric fluorescence changes due to changes in the level of dye binding. Similarly, the efficacy of the dual-wavelength approach was demonstrated for single cell images by showing that the ratio was linearly dependent on the membrane potential induced by an externally applied electric field on spherical HeLa cells (Montana et al., 1989); the calibration was achieved by solving Laplace's equation for this geometry (Gross et al., 1986; Ehrenberg et al., 1987).

Recently, we initiated a study of electric field effects in cells with complex geometries where the induced membrane potential cannot be simply predicted from analytical solutions of Laplace's equation (Ehrenberg et al., 1990; Bedlack et al., 1992). Specifically, we are interested in the ability of electric fields to direct neurite growth and have studied the initial signal transduction events in differentiated N1E-115 neuroblastoma cells. Since these cells have both complex geometries and voltage-gated channels, determination of the membrane potential distribution is only possible by direct experimental mapping using the dual-wavelength ratiometric approach. Di-4-ANEPPS was rapidly internalized into these cells; therefore, di-8-ANEPPS was used to stain the cells. The data were

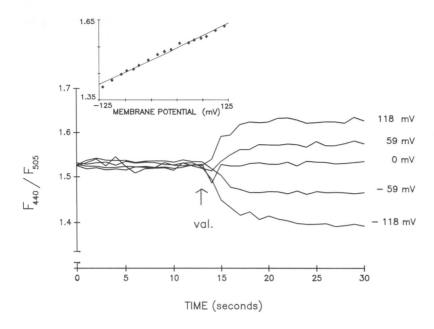

Figure 11.4 Ratio of di-4-ANEPPS fluorescence excited at 440 nm and 505 nm for a series of lipid vesicle suspensions maintained at different membrane potentials via valinomycin-mediated potassium diffusion gradients. Valinomycin is added at the arrow and the resultant membrane potentials are indicated to the right. The inset shows the linearity of the ratio vs. membrane potential for 18 such experiments. (Reprinted with permission from Montana *et al.*, 1989. Copyright 1989 American Chemical Society.)

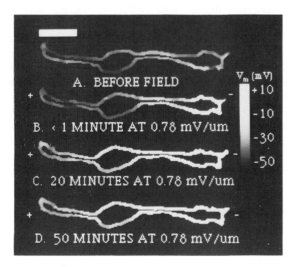

Figure 11.5 Ratio images of the periphery of a N1E-115 neuroblastoma cell stained with di-8-ANEPPS and subjected to a uniform electric field of 0.78 mV μm⁻¹. Fluorescence is excited at 450 and 530 nm in rapid succession with emission collected at >580 nm using a cooled slow-read-out 14 bit CCD camera. The two images are segmented by thresholding to isolate the bright membrane fluorescence along the cell periphery. The 450 nm/530 nm ratio images are then calculated with a 32 bit precision and calibrated via valinomycin mediated K⁺ diffusion potentials. The grey scale bar indicates the range of membrane potentials; the white bar at the top is 50 μm.

collected with a slow read-out cooled CCD camera to provide sufficient linearity, dynamic range, and signal-to-noise. Fluorescence was collected with excitation wavelengths alternately at 450 nm and 530 nm using interference filters; emission was selected with a 580 nm long-pass filter. As Shown in Fig. 11.5 the bright 'ring stain' pattern around the cell periphery was segmented to assure sufficient signal-to-noise for accurate ratios and provide a simplified map of the membrane potential distribution. The ratio image in the absence of the field is fairly uniform; the field applied to the cell in Fig. 11.5, 7.8 V cm⁻¹, induces a marked increase in the ratio on the cathodal side. This represents a physiologically relevant electric field, i.e. a field strength which induces the growth response *in vivo* and *in vitro*; the results show that significant changes in local membrane potential are effected by this magnitude field. The ratio was calibrated according to membrane potential by using valinomycin-mediated K⁺ diffusion potentials. As can be seen, the side of the cell facing the cathode is depolarized while the anodal side is relatively unaffected. Thus, the simple expectation of equal but opposite membrane potential changes on the opposite faces of the cell is not what is found experimentally. The technique of dual-wavelength ratiometric imaging of membrane potential may become an important tool in investigations of localized cellular responses to external stimuli.

11.3.2 Nernstian dyes permit imaging of cell and mitochondrial membrane potential

As noted in Section 11.2, this laboratory has developed a pair of cationic rhodamine derivatives, TMRE (Fig. 11.3) and TMRM, with properties suited for quantitation of membrane potential in individual cells

(Ehrenberg *et al.*, 1988). A microscope photometer was employed in this original study to obtain fluorescence intensities in the cytosol and in the extracellular medium as required by equation [11.1] above. To avoid contaminating signal from the very bright mitochondria, the measuring pinhole aperture was positioned over a mitochondria-free region (often the cell nucleus). The measured fluorescence intensities had to be corrected for a small amount of non-potential-dependent binding of the dye as determined from cells depolarized with K^+/valinomycin. Another important consideration in correctly determining F^{in}/F^{out} is the depth of field of the fluorescence microscope. The problem is that the emission is collected from the entire thickness of medium in the bathing solution while only the thickness of the cell contributes to the cytosolic fluorescence. Several measures were taken by Ehrenberg *et al.* (1988) to deal with this problem:

(1) The experiments were carried out on slides with a minimum volume of bathing medium above the cells.
(2) Small apertures in both the excitation path and the emission path limited the out-of-focus contributions to the fluorescence detected by the photomultiplier tube.
(3) The out-of-focus contribution to the fluorescence was determined by using a slide in which the cells were bathed in medium containing fluoresceinated dextran; because the dextran is excluded from the cell, a correction to F^{in}/F^{out} may be derived from the fluorescence detected while focusing in the centre of the cell relative to fluorescence detected while focusing on a cell-free region of the slide.

These considerations are not so easily dealt with if quantitative imaging microscopy is to be substituted for microphotometry. In particular, the issue of out-of-focus contributions cannot be solved because of the necessarily wide field and measuring apertures in the imaging microscope optical path. On the other hand, imaging permits one to acquire spatial information such as the ability to simultaneously monitor mitochondrial and plasma membrane changes (Farkas *et al.*, 1989); using simple segmentation techniques to divide the cell into primarily mitochondrial and cytosolic fluorescence, it is possible to distinguish and separate the kinetics of the membrane potential changes induced by various interventions.

While ordinary wide-field microscopy is successful in qualitatively monitoring kinetic changes in mitochondria, the actual magnitude of mitochondrial membrane potential cannot be quantitated. This is because it is not possible to obtain an accurate measure of fluorescence intensity emanating from such a small compartment with the broad depth of field set by the microscope optics – the corrections and special measures described above for measuring the plasma membrane potential are inadequate for mitochondria. This limitation may be overcome if confocal microscopy is employed with its very narrow depth of field. The confocal microscope is essentially the same as a microscope photometer with pinhole field and measuring apertures to reduce out-of-focus fluorescence; in the confocal microscope these apertures are scanned in concert over the microscope field to produce an image (White *et al.*, 1987; Brakenhoff *et al.*, 1989; Pawley, 1990; see also Chapter 17). This approach has been demonstrated for Nernstian dye accumulation in mitochondria (Farkas *et al.*, 1989; Loew *et al.*, 1990; Loew, 1991) and is illustrated with a series of confocal images of a NIH 3T3 fibroblast cell equilibrated with 100 nM TMRE in Fig. 11.6.

The intensities measured in the brightest mitochondrial regions of Fig. 11.6(a) are as much as 1000 times higher than the intensity of TMRE fluorescence in the extracellular medium; the intensity in the cytosol is a factor of 15 higher than the extracellular fluorescence. After correction for non-potentiometric binding, which amounts to a factor of 2 for cytosol and 3 for mitochondria (Fig. 11.6(b)), a membrane potential map can be generated through a logarithmic transform. This is displayed in Fig. 11.6(c) where the contrast between compartments is indicative of the potential across the membrane separating them. The potential determined in this manner for the plasma membrane is believed to be accurate (*ca,* -50 mV for the fibroblast). However, several open problems made the determination of individual *mitochondrial* potentials of uncertain validity. A major problem is that the width of the optical slice is limited to about 0.6 µm even with the smallest confocal pinhole and the highest numerical aperture objective; mitochondria are typically ellipsoidal objects with their smallest dimensions in this same range. So even confocal microscopy may not be able to provide sufficiently narrow depth of field for objects as small as mitochondria. Most importantly, for a given mitochondrion in a given image there is currently no way of judging how serious this problem is. A second limitation of confocal microscopy involves the high excitation light intensities necessitated by the technique (at least in the current generation of confocal microscopes), because it is based on the principle of rejecting the fluorescence emission emanating from outside the plane of focus; thus, single measurements on living cells can be performed, but the accumulation of large image sets for either kinetics or three-dimensional reconstructions lead to fading of the dye and/or phototoxic effects.

An alternative to confocal microscopy for accurate 3D imaging is the restoration of out-of-focus contributions

(a)

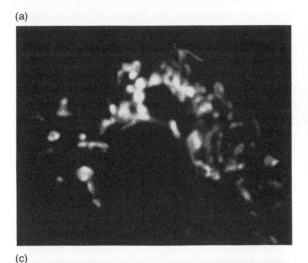

(b)

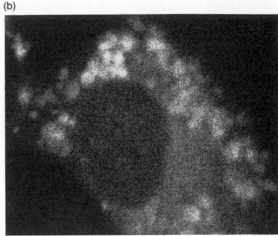

(c)

Figure 11.6 Confocal microscopy of NIH 3T3 fibroblast equilibrated with 100 nM TMRE. (a) A Biorad MRC 600 confocal microscope was set in the accumulate mode to acquire a series of 32 8-bit images into a 16-bit image buffer. This additional dynamic range was required to detect both the high fluorescence from the mitochondria and the relatively low fluorescence signal from the cytosol; the latter, however, is not perceptible in the figure. (b) After equilibration with 0.5 μM valinomycin and 150 mM K^+ to completely depolarize the cell and mitochondrial membranes, the same cell displays very little TMRE binding. Data from this image can be used to correct for non-potentiometric binding. (c) A logarithmic transformation of the image in (a) after correction for non-potentiometric binding provides an image with a grey scale that is proportional to membrane potential.

to wide-field images by computer-based deconvolution techniques (Agard, 1984; Carrington & Fogarty, 1987; Agard *et al.*, 1989; Fay *et al.*, 1989). Since this method restores out-of-focus light to its point of origin, it requires lower overall doses of excitation light than confocal microscopy. Furthermore, since it is based on a rigorous experimentally determined definition of the optical characteristics of the microscope, it is possible to model the behaviour of objects such as mitochondria and validate the measurements. This approach, therefore, shows great promise and is being pursued as an alternative to confocal microscopy for the quantitative imaging of mitochondrial membrane potential with Nernstian dyes.

11.4 CONCLUSION

The determination of membrane potential with potentiometric dyes requires an appropriate match of

the biological preparation, indicator probe, optical detection scheme and the scientific question to be answered. This review has focused on the variety of potentiometric indicators that have been developed over the past 20 years and has attempted to provide sufficient information so that new experiments can be designed. This process was illustrated with applications of slow and fast dyes to digital imaging microscopy of membrane potential distributions in single cells.

This was not intended as an extensive review of the subject. Indeed, several noteworthy applications of fast dyes in excitable systems have employed digital imaging microscopy. Patterns of visual stimuli have been imaged on the monkey cerebral cortex (Blasdel & Salama, 1986). Spiral waves of excitation were visualized in isolated cardiac muscle (Davidenko *et al.*, 1992). Evoked activity in salamander olfactory bulb was spatially and temporally defined with a video detection system (Kauer, 1988). For studies which require millisecond time resolution, arrays of 100–500 discrete detectors can be used to develop low spatial resolution maps of potential changes. These methods are reviewed in Chapter 30. In general, therefore, the current state of technology requires some compromise between optimal temporal and spatial resolution. As the imaging and dye technologies improve, such compromises should become unnecessary.

REFERENCES

Agard D.A. (1984) *Ann. Rev. Biophys. Bioengng* **13**, 191–219.

Agard D.A., Hiraoka Y., Shaw P. & Sedat J.W. (1989) *Methods Cell Biol.* **30**, 353–377.

Bashford C.L., Chance B., Smith J.C. & Yoshida T. (1979) *Biophys. J.* **25**, 63–85.

Bedlack R.S., Wei M.-d & Loew L.M. (1992) *Neuron* **9**, 393–403.

Blasdel G.G. & Salama G. (1986) *Nature* **321**, 579–585.

Brakenhoff G.J., van Spronsen E.A., van der Voort H.T.M. & Nanninga N. (1989) *Methods Cell Biol.* **30**, 379–398.

Brauner T., Hulser D.F. & Strasser R.J. (1984) *Biochim. Biophys. Acta* **771**, 208.

Carrington W. & Fogarty K.E. (1987) *Proceedings of the 13th Annual Northeast Bioengineering Conference*, K. Foster (ed.). IEEE, pp. 108–111.

Chen L.B. (1988) *Ann. Rev. Cell. Biol.* **4**, 155–181.

Chien C.-B. & Pine J. (1991) *Biophys. J.* **60**, 697–711.

Cohen L.B. & Salzberg B.M. (1978) *Rev. Physiol. Biochem. Pharmac.* **83**, 35–88.

Cohen L.B., Salzberg B.M., Davila H.V., Ross W.N., Landowne D., Waggoner A.S. & Wang C.H. (1974) *J. Membr. Biol.* **19**, 1–36.

Cohen R.L., Muirhead K.A., Gill J.E., Waggoner A.S. & Horan P.K. (1981) *Nature* **290**, 593–595.

Davidenko J.M., Pertsov A.V., Salomonsz R., Baxter W. & Jalife J. (1992) *Nature* **355**, 349–351.

Davila H.V., Salzberg B.M., Cohen L.B. & Waggoner A.S. (1973) *Nature New Biol.* **241**, 159–160.

Dragsten P.R. & Webb W.W. (1978) *Biochemistry* **17**, 5228–5240.

Ehrenberg B., Farkas D.L., Fluhler E.N., Lojewska Z. & Loew L.M. (1987) *Biophys. J.* **51**, 833–837.

Ehrenberg B., Montana V., Wei M.-d, Wuskell J.P. & Loew L.M. (1988) *Biophys. J.* **53**, 785–794.

Ehrenberg B., Wei M. & Loew L.M. (1990) *Biophys. J.* **57**, 484a.

Farkas D.L., Wei M., Febbroriello P., Carson J.H. & Loew L.M. (1989) *Biophys. J.* **56**, 1053–1069.

Fay F.S., Carrington W. & Fogarty K.E. (1989) *J. Microsc.* **153**, 133–149.

Fluhler E., Burnham V.G. & Loew L.M. (1985) *Biochemistry* **24**, 5749–5755.

Freedman J.C. & Hoffman J.F. (1979) *J. Gen. Physiol.* **74**, 187–212.

Freedman J.C. & Laris P.C. (1981) In *International Review of Cytology*, Supplement 12. Academic Press, New York, pp. 177–246.

Freedman J.C. & Laris P.C. (1988) In *Spectroscopic Membrane Probes*, Vol. 3, L.M. Loew (ed.). CRC Press, Boca Raton, pp. 1–50.

George E.B., Nyirjesy P., Pratap P.R., Freedman J.C. & Waggoner A.S. (1988) *J. Membr. Biol.* **105**, 55–64.

Grinvald A.S., Hildesheim R., Farber I.C. & Arglister J. (1982) *Biophys. J.* **39**, 301–308.

Grinvald A., Fine A., Farber I.C. & Hildesheim R. (1983) *Biophys. J.* **42**, 195–198.

Grinvald A., Salzberg B.M., Lev-Ram V. & Hildesheim R. (1987) *Biophys. J.* **51**, 643–651.

Grinvald H., Ross W.N. & Farber I. (1981) *Proc. Natl. Acad. Sci. USA* **78**, 3245–3249.

Gross D., Loew L.M. & Webb W.W. (1986) *Biophys. J.* **50**, 339–348.

Grynkiewicz G., Poenie M. & Tsien R.Y. (1985) *J. Biol. Chem.* **260**, 3440–3450.

Gupta R.K., Salzberg B.M., Grinvald A., Cohen L.B., Kamino K., Lesher S., Boyle M.B., Waggoner A.S. & Wang C. (1981) *J. Membr. Biol.* **58**, 123–137.

Hassner A., Birnbaum D. & Loew L.M. (1984) *J. Org. Chem.* **49**, 2546–2551.

Jesurum A. & Gross D.J. (1991) *Biophys. J.* **59**, 526a.

Johnson L.V., Walsh M.L. & Chen L.B. (1980) *Proc. Natl. Acad. Sci. USA* **77**, 990–994.

Johnson L.V., Walsh M.L., Bockus B.J. & Chen L.B. (1981) *J. Cell Biol.* **88**, 526–535.

Kauer J.S. (1988) *Nature* **331**, 166–168.

Korchak H.M., Rich A.M., Wilkenfeld C., Rutherford L.E. & Weissman G. (1982) *Biochem. Biophys. Res. Commun.* **108**, 1495–1501.

Krauthamer V. & Ross W.N. (1984) *J. Neurosci.* **4**, 673.

Labrecque G.F., Holowka D. & Baird B. (1989) *J. Immunol.* **142**, 236–243.

Liptay W. (1969) *Angew. Chem. Internat. Edit.* **8**, 177–188.

Loew L.M. (1988) *Spectroscopic Membrane Probes*. CRC Press, Boca Raton, FL.

Loew L.M. (1991) *New Techniques of Optical Microscopy and Microspectrophotometry*, R.J. Cherry (ed.). The Macmillan Press, London, pp. 255–272.

Loew L.M. & Simpson L. (1981) *Biophys. J.* **34**, 353–365.

Loew L.M., Bonneville G.W. & Surow J. (1978) *Biochemistry* **17**, 4065–4071.

Loew L.M., Simpson L., Hassner A. & Alexanian V. (1979) *J. Am. Chem. Soc.* **101**, 5439–5440.

Loew L.M., Rosenberg I., Bridge M. & Gitler C. (1983) *Biochemistry* **22**, 837–844.

Loew L.M., Benson L., Lazarovici P. & Rosenberg I. (1985a) *Biochemistry* **24**, 2101–2104.

Loew L.M., Cohen L.B., Salzberg B.M., Obaid A.L. & Bezanilla F. (1985b) *Biophys. J.* **47**, 71–77.

Loew L.M., Farkas D.L. & Wei M.-d (1990) In *Optical Microscopy for Biology*, B. Herman & K. Jacobson (eds). Wiley-Liss, New York, pp. 131–142.

Loew L.M., Cohen L.B., Dix J., Fluhler E.N., Montana V., Salama G. & Wu J.-Y. (1992) *J. Membr. Biol.* **130**, 1–10.

Lojewska Z. & Loew L.M. (1987) *Biochim. Biophys. Acta* **899**, 104–112.

Lojewska Z., Farkas D.L., Ehrenberg B. & Loew L.M. (1989) *Biophys. J.* **56**, 121–128.

London J.A., Zecevic D. & Cohen L.B. (1987) *J. Neurosci.* **7**, 649–661.

Mohr C.F. & Fewtrell C. (1987) *J. Immunol.* **138**, 1564–1570.

Montana V., Farkas D.L. & Loew L.M. (1989) *Biochemistry* **28**, 4536–4539.

Müller W., Windisch H. & Tritthart H.A. (1989) *Biophys. J.* **56**, 623–629.

Pawley J.B. (1990) *Handbook of Biological Confocal Microscopy*. Plenum Press, New York.

Platt J.R. (1956) *J. Chem. Phys.* **25**, 80–105.

Reers M., Smith T.W. & Chen L.B. (1991) *Biochemistry* **30**, 4480–4486.

Rink T.J., Tsien R.Y. & Pozzan T. (1982) *J. Cell. Biol.* **95**, 189–196.

Rosenbaum D.S., Kaplan D.T., Kanai A., Jackson L., Garan H., Cohen R.J. & Salama G. (1991) *Circulation* **84**, 1333–1345.

Ross W.N. & Krauthamer V. (1984) *J. Neurosci.* **4**, 659–672.

Ross W.N. & Reichardt L.F. (1979) *J. Membr. Biol.* **48**, 343–356.

Ross W.N., Salzberg B.M., Cohen L.B., Grinvald A., Davila H.V., Waggoner A.S. & Wang C.H. (1977) *J. Membr. Biol.* **33**, 141–183.

Shrager P. & Rubinstein C.T. (1990) *J. Gen. Physiol.* **95**, 867–890.

Shrager P., Chiu S.Y., Ritchie J.M., Zecevic D. & Cohen L.B. (1987) *Biophys. J.* **51**, 351–355.

Sims P.J., Waggoner A.S., Wang C.-H. & Hoffman J.F. (1974) *Biochemistry* **13**, 3315–3330.

Smiley S.T., Reers M., Mottola-Hartshorn C., Lin M., Chen A., Smith T.W., Steele G.D. & Chen L.B. (1991) *Proc. Natl. Acad. Sci. USA* **88**, 3671–3675.

Smith J.C. & Chance B. (1979) *J. Membr. Biol.* **46**, 255.

Smith J.C., Russ P., Cooperman B.S. & Chance B. (1976) *Biochemistry* **15**, 5094–5105.

Terasaki M., Song J., Wong, J.R., Weiss M.J. & Chen L.B. (1984) *Cell* **38**, 101–108.

Tsien R.Y. & Poenie M. (1986) *Trends Biochem. Sci.*, **11**, 450–455.

Verkman A.S. & Frosch M.P. (1985) *Biochemistry* **24**, 7117–7122.

Waggoner A.S. (1979) *Ann. Rev. Biophys. Bioeng.* **8**, 847–868.

Waggoner A.S. (1985) In *The Enzymes of Biological Membranes*, A.N. Martonosi (ed.). Plenum, New York, pp. 313–331.

Waggoner A.S., Wang C.H. & Tolles R.L. (1977) *J. Membr. Biol.* **33**, 109–140.

White J.G., Amos W.B. & Fordham M. (1987) *J. Cell. Biol.* **105**, 41–48.

CHAPTER TWELVE

Quantitative Real-Time Imaging of Optical Probes in Living Cells

W.T. MASON,[1] J. HOYLAND,[1] I. DAVISON,[1] M.A. CAREW,[1]
B. SOMASUNDARAM,[1] R. TREGEAR,[1] R. ZOREC,[2] P.M. LLEDO,[3]
G. SHANKAR,[4] & M. HORTON[4]

[1] Department of Neurobiology, AFRC Institute of Animal Physiology, Babraham, Cambridge, UK
[2] Institute of Pathophysiology, University of Ljubljana, Ljubljana, Slovenia
[3] Institut Alfred Fessard, CNRS, Gif-sur-Yvette, France
[4] Haemopoiesis Research Group, ICRF, St Bartholomew's Hospital, London, UK

12.1 INTRODUCTION

The conventional view of the light microscope as an instrument to study fixed tissue has been transformed in recent years. Today, the microscope can be justifiably viewed as a dynamic instrument at the core of studies of biological activity in real time. The microscope is, however, only a means to an end. Our ability to study events occurring in living cells is largely due to the development of chemical probes capable of specifically targeting biological activity and to the development of computer-controlled instrumentation for acquiring data from living tissue at very high rates.

The ability to interface fast video cameras and computer technology to the conventional microscope has made it possible not only to make qualitative observations, but to derive quantitative image data from single cells, at speeds of up to 30 video frames per second if we use video cameras or confocal laser scanning technology, or many hundreds of samples per second if we use photon-counting technology with photomultipliers. In the future, there is good reason to believe that even with the high data content of imaging and increasing sensitivity of photosensitive detectors, speeds of data acquisition may increase further – well in excess of the video frame rate.

In this chapter, we shall discuss some of the key developments enabling scientists to increase their observational and analytical abilities, and apply these skills to previously intractable problems. Image processing is at the heart of virtually all of these new approaches.

Every cell has ionic charge gradients generated by ions like calcium, sodium and potassium. Gradients in intracellular hydrogen ion concentration also are important to many biological processes. Ionic concentrations inside cells change quickly and dramatically and underlie a wide range of cellular processes including development, growth, secretion and reproduction, so it is important to observe and understand them. As we have been able to apply image processing and low-light-level image capture from signals emitted by optical probes, it has also become clear that large standing ionic gradients occur within cells, and may persist for many seconds during stimulus or suppression of cellular activity. The source of ionic changes may also occur in widely different parts of the same cell, and even with small cells of only 10 μm or so, these ionic pools may be detected.

This article will focus mainly on how image acquisition and analysis is making it possible to study ionic gradients in living cells by using optical probes. Some specific applications from our laboratory using this technology will be used to illustrate the potential for such techniques.

12.2 MULTIDISCIPLINARY ADVANCES

A number of important scientific advances in diverse fields including chemistry, biology, physics, electronics, microscopy and computing have come together to produce an exciting range of new technologies for single cell study of living biological systems.

Image processing underlies these new technologies, and major advances on several fronts have opened up the world of computer-based image acquisition and analysis to a wide range of scientific endeavour. Some of these developments have included:

- the synthesis of chemical probes for cellular function, ranging from chemiluminescent and fluorescent dyes sensitive to ions such as calcium, sodium, chloride, protons and intracellular proteins, to fluorescent probes sensitive to cell surface proteins which give cells sensitivity to the outside world;
- the development of new technology, low-light-level video cameras and ultra-high sensitivity photomultiplier tubes which can detect and image faint fluorescence in real time, at very high speeds;
- fast microelectronic circuitry which can be interfaced to these cameras and controlled by computer, such that video images can rapidly be captured, averaged and stored in digital format in real time for later processing; and
- sophisticated image-analysis software which can perform quantitative measurements on digital image or photon data.

These converging approaches have resulted in new capability to measure in real time the fast changes in intracellular ionic concentration or cell surface protein organization in living cells, with limited disruption of normal cell function.

12.3 OBSERVING BIOLOGICAL ACTIVITY IN 'REAL TIME'

The phrase 'real time' is often applied to some specialized approaches for study of biological systems. Most important events at the cellular level are dynamically changing on the order of seconds or less, and to capture these events in a meaningful way requires hardware and software which can work quickly – in real time – while the events themselves are occurring.

Response times of some of the ion-sensitive probes discussed here occur on the order of about 10 ms or less. The video cameras and photomultiplier tubes used as sensors can respond on a similar time-scale. So the hardware and software used for capturing, processing and storing signals must be about as fast, or ideally even faster.

12.4 CHEMICAL PROBES FOR FUNCTION OF LIVING CELLS

12.4.1 Intracellular ions

Some of the first measurements of ionic activity used photoproteins such as the calcium-sensitive molecule aequorin, which emitted light when the protein combined with calcium ions. Other calcium-sensitive dyes including arsenazo and murexide also provided important advances. However, these dyes had some disadvantages in that they were accessible to only a few scientists, since introducing them into single cells required microinjection and dissociation constants were not always suited to the concentrations found in small cells. Recent isolation of cDNA for apo-aequorin, and isoaequorins, combined with synthetic analogues of coelenterazine has created a new supply of photoproteins (Rizzuto *et al.*, 1992) which expand their detection range and in some cases may even allow ratio imaging. Of course, detection of calcium by aequorin is only one use of this versatile indicator which, in principle, can be used as a reporter gene to examine transcriptional activity of cloned genomic sequences. A new strategy for measuring the activity of protein kinases, applicable to studies on live cells, has been developed by engineering kinase recognition peptides into aequorin and luciferase.

However, from 1982 the development of new fluorescent dyes by Tsien and colleagues was reported which was to revolutionize our ability to investigate ionic activity in single cells (Grynkiewicz *et al.*, 1985). These dyes are sensitive to minute concentrations of intracellular ions such as occur in single cells. One chemical form of the dye (the acetoxymethyl ester) can be loaded into single cells, so virtually all scientists can use the dyes. Most cells contain endogenous esterases which rapidly (*ca.* 5–30 min) hydrolyse the dye to form the free acid, which is trapped in the cell and is ion-sensitive.

A number of such dyes are commercially available (Table 12.1). These dyes have differing wavelengths of light output, but broadly they all emit photons of light in the visible spectrum. Their light output is well-matched to available detectors, including both photomultipliers and intensified video cameras.

For instance, the two most commonly used dyes for

Table 12.1 Various optical probes currently in use. Dyes fall into three categories: those with ion-dependent dual-excitation spectra, those which are dual-emission, and those which show ionic dependence only at a single wavelength.

Ion	Dye	Excitation wavelength (nm)		Emission wavelength (nm)	
		1	2	1	2
Dual-excitation dyes					
Ca^{2+}	Fura-2	340	380	510	
	Fura-5	340	380	510	
	Fura Red	480–500	425–450	660	
H^+(pH)	BCECF	440	490	530	
	SNARF-6	500	560	610	
	SNAFL-1	500	560	600	
Na^+	SBFI	340	380	510	
K^+	PBFI	340	380	510	
Mg^{2+}	Mag-fura-2	340	380	510	
Dual-emission dyes					
Ca^{2+}	Indo-1	350		405	480
H^+	SNAFL-1	500		540	635
	SNARF-2	500		550	640
	DCH	405		435	520
Na^+	FCRYP-2	350		405	480
Single wavelength dyes					
Ca^{2+}	Fluo-3	505		530	
	Calcium Green	505		530	
	Calcium Orange	550		575	
	Calcium Crimson	590		610	
Cl^-	SPQ	350		440	
	Fluorescein	495		535	
	Rhodamine	550		595	

measuring intracellular calcium are fura-2 and indo-1 (Fig. 12.1). The dyes have high quantum efficiency and are sensitive to calcium at concentrations from 30 nM to 5 μM or so. Fura-2 displays a single emission peak at 510 nm, but two calcium-dependent absorption maxima, one at 340 nm which increases with increasing ionized calcium and a second at 380 nm which similarly decreases with a rise in ionized calcium. Indo-1 is generally excited by only a single wavelength of light (340–360 nm), but emits light at two different calcium-sensitive wavelengths (405 and 490 nm). With all such probes, choice of excitation wavelength will influence the wavelengths of light emitted and the specific dynamic range of the dye response.

Considerable work is taking place today in development of new probes for ionic activity in living cells. The field of calcium is attracting most attention and other chapters in this book will focus on such work. New probes for calcium ions include Calcium Green, Calcium Crimson and Calcium Orange, and Fura Red. Of these, Fura Red is perhaps of most interest because it exhibits fluorescence which decreases as Ca^{2+} increases and it has a very large Stokes's shift, of about

175 nm. As such, it may be potentially combined with probes such as fluo-3, a single-wavelength Ca^{2+} probe whose fluorescence increases with Ca^{2+} activity. Because these dyes are excited near convenient visible laser lines, they have the potential to be simultaneously loaded into a single cell for ratiometric confocal imaging, or for dynamic video imaging with CCD cameras in systems not configured with UV-transmitting optics.

Other work for new calcium probes, for example, includes the development of new indicators with higher and lower affinity for Ca^{2+}. Fura-5, for example, has an affinity for Ca^{2+} of 40 nm compared to Fura-2 with an affinity of 135 nm at room temperature. Other probes are also being developed for magnesium ions, with excellent sensitivity and specificity.

One essential problem with many of the fluorochromes used for ion measurement has been that they are taken up into intracellular compartments with time. New probes such as fura-2-dextran (see Chapter 3) have been developed which are reported to minimize such effects. Fura-2 is coupled to the dextran molecule and can be microinjected into cells, as there is no acetoxymethyl ester version available. This may

A.

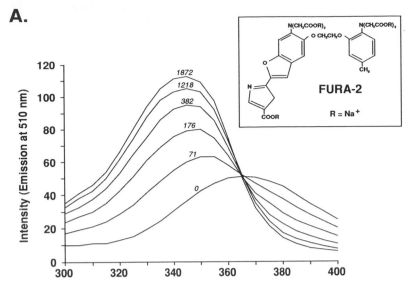

B.

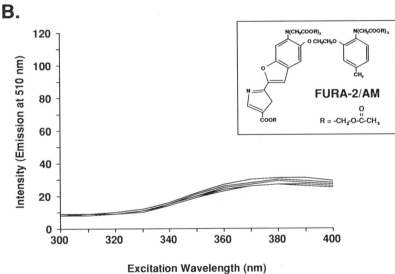

Figure 12.1 (A) Spectra of fura-2 free acid, the calcium-sensitive optical probe. Note that as ionized calcium concentration increases, 340 nm fluorescence (measured at 510 nm) increases, but 380 nm fluorescence decreases. This permits ratio measurements to be made. (B) Similar spectra of fura-2 acetoxymethyl ester, used to load the cells, showing relative calcium-independence of fluorescence. Esterases inside the single cell hydrolyse this version of the dye to form the free acid shown in (A).

well permit much extended measurement times on loaded cells, as the conjugate is only slowly extruded from cells, and thus remains active as a sensor much longer than the free acid version. Fura-2-dextran is as sensitive to Ca^{2+} as is fura-2 free acid.

12.5 REAL-TIME VIDEO IMAGING OF ION-SENSITIVE FLUORESCENT DYES

Fluorescence ratio imaging is at the heart of dynamic video-enhanced light microscopy for optical probes detecting ions. If fluorescent images are obtained as a pair at 340 and 380 nm (with fura-2 for instance),

and the images are ratioed on a point-by-point, or pixel-by-pixel basis (a pixel is the single resolving unit of a video camera, many thousands of which are combined together to give an overall image), the resulting 'ratio image' is proportional to ionized calcium concentration and reduces the chance of possible artifacts due to uneven loading or partitioning of dye within the cell, or varying cell thickness and dye concentration.

Indo-1 is generally used for photometric measurements – it has a slightly faster time response in terms of dissociation time constant (typically estimated at 5–20 ms), and can be used with static optical beam-splitters to separate the emitted light and focus it onto two photomultiplier tubes as a continuous signal.

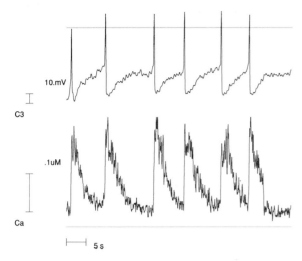

Figure 12.2 Photometric experiment using indo-1 and the PhoCal system to measure simultaneously action potentials (top record) and calcium transients in a spiking GH3 clonal pituitary cell. Two emission wavelengths were recorded using fast photon counting, with about 5 sample points per ms.

Ratio measurements are also employed. This approach has the advantage that no movement need take place in order to change filter position, and so measurements can be fast and vibration free.

Another dye, BCECF, measures intracellular pH as an optical signal, using similar dual-wavelength imaging technology. This dye is excited at 440 and 490 nm and measured at about 510–520 nm. Increases in pH increase the fluorescence of the dye excited at 490, but have little effect on 440 nm fluorescence. This dye is being successfully combined with fura-2 to provide simultaneous imaging of Ca^{2+} and pH. For this type of work, a four-position filter wheel is used to excite in turn the probes at 340, 380, 440 and 490 nm, having loaded both dyes into the single cell simultaneously. Fluorescence is measured with a 515 dichroic long-pass filter and a 535 band-pass filter, and this provides strong signals for both dyes with minimal overlap of interference of the two probes.

Optical probes are also available for sodium and chloride, but frustratingly little progress has been made in development of a good, selective probe for potassium in living cells. PBFI, a probe reported as being of possible utility for measurement of intracellular K^+, is now regarded as of little use, possibly because it is metabolized or compartmentalized inside the cell. However, it may well have utility as a probe for extracellular K^+.

12.5.1 Other probes for cell function

Dynamic video-imaging technology has been used for studies of intracellular ionic concentration, but other

low-light-level applications are also possible given that an appropriate probe and an optical sensor are available. Antibodies to sperm recognition antigens labelled with fluorescein have been used to reveal antigen localization on living sperm. In anterior pituitary cells which control the levels of hormones in the body important for fertility, growth, lactation, stress and so forth, newly released prolactin and growth hormone have been detected using antibodies labelled with fluorescent probes (Fig. 12.3). Cell surface receptors for peptides can be imaged using a biologically active fluorescent analogue of the chemical messengers from the brain.

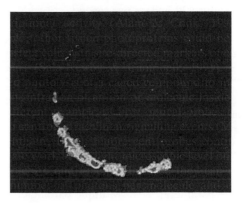

Figure 12.3 Polyclonal antibody to prolactin hormone was used as a rapid label on living anterior pituitary cells stimulated with the prolactin-releasing peptide TRH (100 nM). Visualization after fast labelling of only 10 min duration was enabled by use of low-light-level camera technology, which could effectively detect the small fluorescence levels due to recently released hormone confined to granules on the cell surface.

This area of imaging is currently little exploited, and will be an important area of future development and application for the increasingly sophisticated capabilities of the technology.

12.6 PHOTOMETRIC DETECTION VERSUS LASER SCANNING CONFOCAL VERSUS DYNAMIC VIDEO IMAGING

The technologies for studying optical probes in biological systems can be divided into three categories: (1) photometric technology, (2) video-enhanced light microscopy, and (3) confocal laser scanning microscopy. **Photometric technology** permits measurement of photons on fast time-scales, and gives a measurement of changes with respect to time. It is a *temporal* measurement only.

Video cameras and laser scanning confocal technology

permit a second dimension of observation. **Video-enhanced light microscopy**, or **dynamic video imaging** permits not only *temporal* measurements, but also *spatial* measurements of biological activity. The comparative characteristics of imaging and photometry are shown in Table 12.2. For most of the work with the ion-sensitive fluorescent probes, intensified cameras are used since the working levels of light emitted by these dyes are not detectable by normal video cameras alone. With this approach, a photosensitive array – usually combined with a front-end intensifier to obtain light amplification – is used to image the cell or cells under study. Typically these arrays might provide up to 512×512 pixel resolution, and can be used to capture up to 30 images per second.

Table 12.2 Comparison between imaging versus photometry measurement of optical probes in living cells.

Imaging	Photometry
Spatial information	Limited spatial information
Slow (1–30 samples s^{-1})	Fast (5000 samples s^{-1})
Detectors are less sensitive	Detectors are more sensitive
High data content	Low data content
Multiple parameter acquisition is difficult	Multiple parameter acquisition is easy
Results usually off-line	One-line results display
Higher cost	Lower cost

Confocal laser scanning microscopy (CLSM) is a new imaging technique which again provides spatial and temporal resolution, although the method of imaging may vary widely from manufacturer to manufacturer. Images of 512×512 pixels may be obtained also, but they are constructed by sequential scanning a small point of laser light across the sample, and typically detecting emitted photons with photomultiplier tubes. This technology is discussed elsewhere in this volume. CLSM technology has the advantage of providing a very small depth of field, eliminating out-of-focus photons and thus producing very fine detail. It may also permit optical sectioning through the cell, building up a 3D image from serial optical sections. A disadvantage, however, is that images obtained with the technique tend to be rather noisy, so frame-averaging is required to reduce noise. Although some systems can acquire confocal images at video frame rate and under excitation conditions where minimal bleaching occurs, it is usually necessary to image 20–100 images to obtain high-quality images. The latest CLSM technology also permits work with UV-excited probes, although at a cost. In general, the CLSM microscopes currently available seem to divide

between those best suited for obtaining 3D information, and those developed for high-speed temporal studies.

CLSM technology offers several specific advantages:

(1) *Resolution is enhanced.* With conventional light microscopy and fluorescent ratio imaging, the information sought is collected from a wide area and is thus defocused, limiting resolution. With confocal technology, photons are collected only from a very small plane of interest, and successive fields can be compiled by 3D reconstruction software to give precise localization of calcium waves, for instance.

(2) *Inaccuracy due to z-axis localization of ions or indicator dye is improved.* Because the confocal microscope rejects out-of-focus light, a whole class of geometric artifacts are eliminated.

(3) *Background due to stray light is eliminated.* Point scanning confocal microscopy is intrinsically insensitive to ambient light. Experiments can be performed in a lighted room. In conventional microscopy, care must be taken to prevent ambient light reaching the objectives. This point is of particular importance given the low light level emissions of the probes to be used, where stray light threatens the experiments.

(4) *Background due to medium is reduced.* Confocal microscopy is immune to medium fluorescence arising from, for example:

- indicator dye remaining from the loading process
- indicator dye leaking out of cells
- phenol red
- serum

In conventional microscopy, efforts must be made to minimize all of these background signals. Spatial or temporal variation in any of these background components limit the precision of the experiment.

(5) *Haloing is reduced.* Conventional microscopic images show haloing: the cell boundary is difficult to distinguish clearly because of out-of-focus light from above and below the focal plane. Confocal microscopy provides clear definition of the cell edges, or subcellular organelle edges.

Thus both CLSM and conventional video microscopy potentially provide a quantitative approach to imaging optical probes. Both techniques have the advantage that digital image-analysis techniques permit a wide range of image information to be obtained and can yield both quantitative and qualitative data.

Most importantly, ratio imaging eliminates artifacts due to probe localization and cell geometry. Many of the best ion-sensitive and the new nucleotide-sensitive probes change spectral properties at two wavelengths.

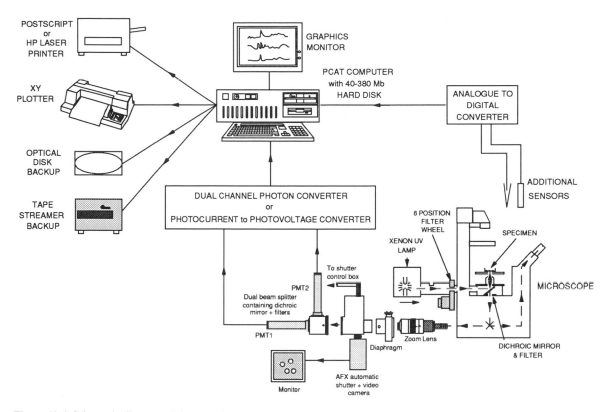

Figure 12.4 Schematic diagram of the experimental arrangement options for the MagiCal system used in our laboratory.

Ratio analysis of the two images produces accurate quantitation and reduces many artifacts associated with dye localization and cell thickness.

12.7   PHOTOMETRIC TECHNOLOGY

One method of measuring fluorescent probes for ions is by use of low-cost photometric technology, where photomultiplier tubes are employed to detect light and produce a signal as a photocurrent or photon counts proportional to light intensity (Fig. 12.4). The system used in our laboratory is called PhoCal (Applied Imaging, UK). This approach is fast, since it is not restricted by the video frame rate of normal cameras, but it has the drawback that it provides only temporal information with no spatial dimensions. Therefore, the absence of spatial information also means that lower specification computers without fast image memory can be used (Figs 12.5 and 12.6). However, to obtain valuable data about the location as well as temporal changes which calcium undergoes, it is necessary to employ the more sophisticated technology of real-time fluorescence ratio imaging.

Detection of light emission by photomultiplier tubes

can be accomplished either by photocurrent to photovoltage conversion or with specialized fast photon counters. The relative advantages of these two approaches are shown in Table 12.3. In brief, photon counting is somewhat more expensive but more sensitive and may be particularly well-suited for work with small cells.

Photometric detection can now also be combined

Table 12.3 Comparison between photocurrent and photon counting measurement of optical probes in living cells.

Photocurrent	*Photon counting*
Lower temporal resolution	High temporal resolution
Time resolution dependent on filter setting	Time resolution dependent on bin width
Integrated current measurement	Bin counting
Integrates photons	Single photon detection
Lower dynamic range	Higher dynamic range
Limited at low light levels	Better for low light levels
Suited to large cells (20 μm plus)	Better for small cells (10–20 μm diameter)
Lower cost	Higher cost

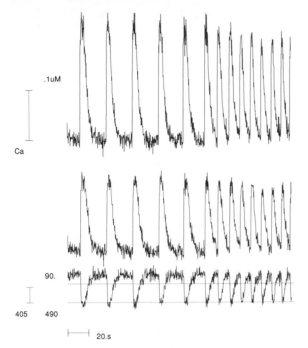

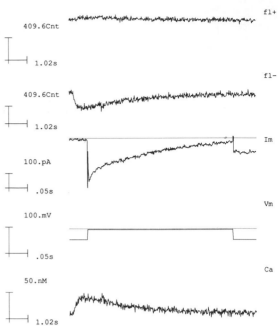

Figure 12.5 Dual wavelength photocurrent measurements of intracellular calcium in an isolated rat muscle cell, loaded with indo-1. Top trace is the calculated intracellular calcium concentration; bottom two traces show the 405 and 490 nm traces of light emission from indo-1 respectively. Note that one wavelength increases in intensity and the other decreases as calcium rises. Muscle cell was stimulated with electrical field stimulation.

Figure 12.7 Measurement of simultaneous intracellular calcium, together with membrane current and voltage using PhoClamp, which permits voltage or current clamp measurements together with two wavelengths of photon information using photon counting. Here a voltage clamp step is imposed which elicits a long-lasting rise in intracellular calcium, the latter persisting for about 10 times longer than the current transient. Note the intracellular calcium trace is on a different time-scale than the voltage and current traces.

with electrophysiology (see Chapter 22). PhoCal and its companion system PhoClamp provide additional and simultaneous analogue-to-digital conversion at high frequencies. PhoClamp, for instance, is ideally suited to voltage and current clamp experiments as it also controls digitally generation of voltage protocols, and can function on two independent time bases. This makes it possible to accumulate fast electrophysiological signals (up to 35 kHz) while at the same time recording the somewhat slower responses of calcium ions, for example (Fig. 12.7).

12.8 DYNAMIC VIDEO RATIO IMAGING OF IONS IN CELLS

Dynamic video imaging is one of the most exciting of the new technologies for observing probes in living cells, due to its power in resolving optical probes within cells in terms of both time and space. The MagiCal system used in our laboratory facilitates real-time imaging experiments. A typical configuration is shown in Fig. 12.8.

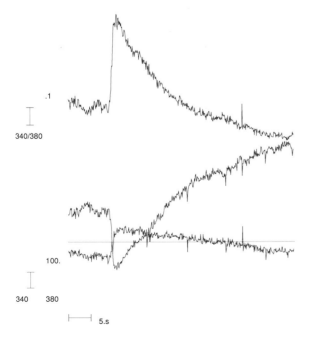

Figure 12.6 Dual-wavelength photon counting measurements of intracellular calcium in a GH3 anterior pituitary cell stimulated with TRH and loaded with fura-2. Excitation wavelengths 340 and 380 nm.

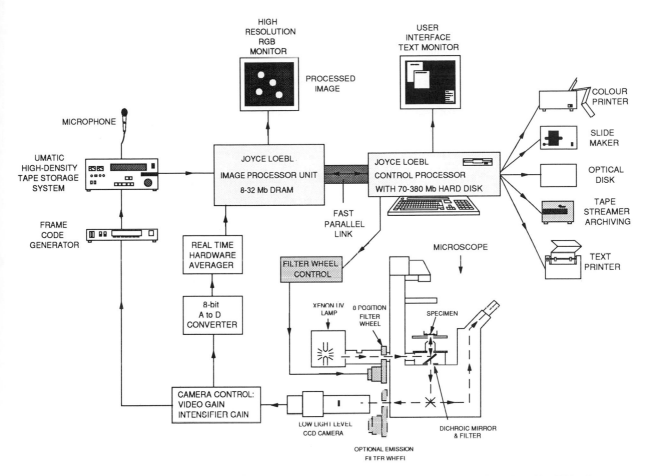

Figure 12.8 Schematic diagram of the MagiCal system used in our laboratory for dynamic video imaging. This shows interaction and control of various components. Images can be stored directly into dynamic random access memory in the image processor unit, or on magnetic video tape together with a unique frame code for off-line playback and analysis. Output of data and archiving of records is available on a number of common peripherals.

Living cells are loaded with fluorescent probe and mounted on a microscope, illuminated by a stable, wide-spectrum light source such as a xenon lamp. An image of the specimen is projected from the microscope onto the face-plate of an intensifying video camera which produces a standard analogue video signal. Fluorescence is detected with programmable resolution from 64 to 512 pixels in both horizontal and vertical axes. An intensifying camera is used because the fluorescent image is very faint. The faintness is due to compromises in the experimental arrangement: the dye concentration has to be low enough to avoid toxic effects to the cells under study and the light source must not be so bright that it bleaches the dye.

12.9 CONFOCAL LASER SCANNING MICROSCOPY

This technique is discussed elsewhere in other chapters of this volume, and a variety of approaches are available (Brakenhoff *et al.*, 1989; Lichtman *et al.*, 1989; Shotton, 1989; Wright *et al.*, 1989). Briefly, an upright or inverted stage microscope can be employed. With point scanning technology, a laser is rapidly moved across the specimen, either mechanically or with acousto-optical deflectors. Emission light from the probe is detected by using photomultiplier technology, and the image reconstructed in either computer memory or on a chip capable of providing a video signal. Thus, the illuminating wavelength is determined by the laser line wavelengths available, and it is not straightforward to perform multiple excitation experiments, although dual-emission wavelength experiments can be performed relatively easily. Recently several manufacturers have developed systems capable of using ultraviolet lasers, although there is a cost penalty. For Ca^{2+} determinations, for instance, fluo-3 – a single-wavelength Ca^{2+}-sensitive probe – has been used extensively, although a single-wavelength probe is difficult to calibrate accurately, and ratio-imaging experiments have not been routine with most commercial confocal laser scanning instruments. This situation is changing rapidly. Several confocal microscopes offer dual photomultiplier detection facilities and it is to be expected that quantitative confocal ratio imaging will become used more widely. Confocal technology can potentially obtain resolutions of 0.5–1.0 µm (probably realistically at the upper end of this range), with substantial elimination of out-of-focus information. Compared to imaging with video-enhanced light microscopy, where resolutions are probably 1–2 µm but with substantial contribution from out-of-focus light emission (outside this focal plane), confocal technology offers interesting prospects for real-time imaging, albeit on somewhat slower usable time-scales than possible with CCD-detector based technology. One drawback of confocal technology compared to traditional CCD-based imaging with intensified CCD cameras is that laser illumination levels for confocal technology are 3–4 times greater than that with epifluorescence, although the dwell time per pixel is very brief and probe bleaching need not necessarily prove a problem.

12.10 RESOLUTION ENHANCEMENT USING DECONVOLUTION

Despite their importance for detection of optical probes, conventional epifluorescence microscopes are less than ideal for imaging cells and do not produce a perfect image of the specimen. Out-of-focus information is obtained, with reduced contrast and enlarged depth of field, and the resulting effect is image blurring. This is because the finite numerical aperture of a microscope objective produces image distortion and reduced contrast. Confocal microscopy goes some way to eliminating this problem, but has other disadvantages. An alternative possibility is the use of digital image processing which can produce sharpened images given knowledge of the point spread function of the microscope optics in use, which can be evaluated easily (Fig. 12.9). Several in- and out-of-focus images from a single specimen are captured (a minimum of three) and these are subjected to deconvolution to remove out-of-focus information. The end-result is a highly sharpened image with increased information content. Several authors have described such an approach (Fay *et al.*, 1989; Monck *et al.*, 1992) and purported to obtain resolution of Ca^{2+} gradients in regions less than 0.5 µm wide. Doubtless commercial systems will become available, although it is unlikely they will be able to provide such imaging capability on fast time-scales. Other tangible benefits may be obtained, however, and the suggestion of Monck *et al.* that the combination of image deconvolution with pulsed laser technology to produce massive probe excitation could well result in thin section images on a time-scale of microseconds.

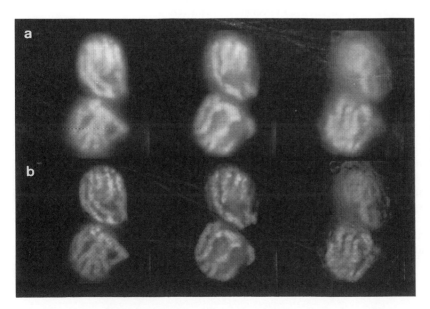

Figure 12.9 Use of image deconvolution to sharpen images obtained with fluorescence light microscopy. Deconvolution of wide-field data using the Jansson method. (a) Sections from original data measured with a vibratome section (approximately 100 µm in thickness) of a root of *Vicia faba* stained with the fluorescent dye DAPI. Cell in late telophase. (b) Equivalent sections from reconstruction. Note removal of out-of-focus flare and the enhancement of the fine spiral substructure of the chromosomes.

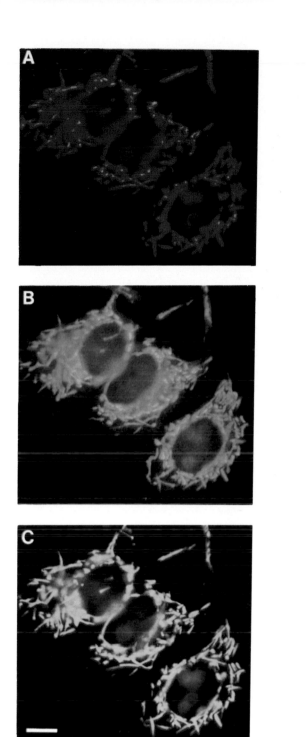

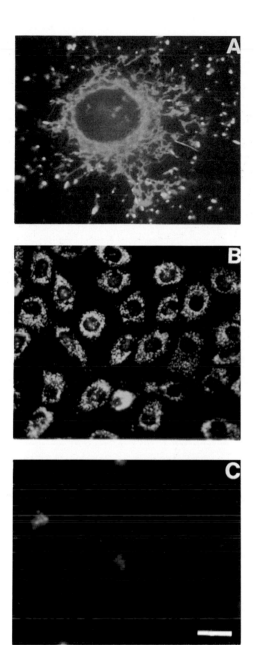

Plate 9.2 Epifluorescence localization of JC-1 in drug-treated cells. (A) Untreated control of CCL22 bovine kidney cells in high K^+ buffer; (B) treated with nigericin (5 μg ml^{-1}) plus ouabain (5 μg ml^{-1}) for 30 min in high K^+ buffer; (C) treated with FCCP (5 μM for 30 min) in high K^+ buffer.

Plate 9.1 Localization of JC-1 by epifluorescence microscopy. (A) Red fluorescence under green excitation corresponding to J-aggregates fluorescence described above; (B) green fluorescence under blue excitation with short pass filter corresponding to monomer fluorescence described above; (C) orange fluorescence under a long pass filter that allows both red fluorescence from J-aggregate and green fluorescence from monomer to be detected simultaneously.

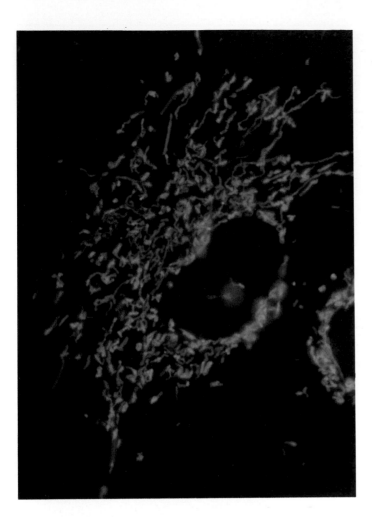

Plate 9.3 Epifluorescence localization of monomer and J-aggregates of JC-1 in an African green monkey kidney CV-1 cell. As seen, there are two types of mitochondria, one with green fluorescence, and, the other, orange.

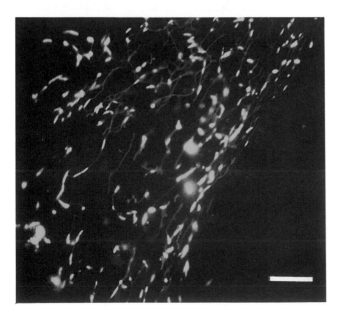

Plate 9.4 Epifluorescence localization of monomer and J-aggregates of JC-1 in a human Forskin fibroblast FS-2. As seen, both green and orange fluorescence may be detected within one contiguous mitochondrion.

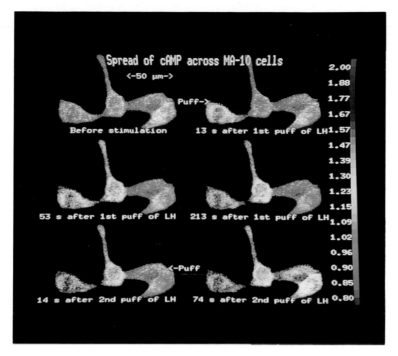

Plate 10.1 Imaging of cAMP spreading from local sites of hormonal stimulation of MA-10 cells. The MA-10 line, derived from a Leydig cell tumour (Ascoli, 1981; Podesta *et al.*, 1991) was grown on glass coverslips and maintained in cell culture in a humidified incubator at 37°C, supplied with 5% CO_2 in Waymouth's medium supplemented with 15% horse serum. These cells are stimulated by gonadotropic hormones such as human chorionic gonadotrophin (hCG) and luteinizing hormone (LH) to elevate cAMP and increase both biosynthesis and secretion of a mixture of steroids, predominantly progesterone (Ascoli, 1981). Cells were washed with HEPES-buffered Hank's balanced salt solution (HBSS) and placed in a recording chamber maintained at 33°C. Four individual cells were microinjected with FlCRhR[1] and imaged with a high-speed, dual-emission confocal microscope (Tsien, 1990). Excitation at 488 nm was accomplished with an argon laser and simultaneous single wavelength emission images (500–535 and >560 nm, respectively) were acquired and stored on an optical memory disk. The single wavelength images were background corrected, and a log ratio image was calculated (short/long wavelength). The log ratio image was corrected for any shading errors by subtraction of the log ratio image obtained from a uniform field of a 1:1 (v/v) mixture of 1 mM fluorescein in aqueous pH 7.6 potassium phosphate with 1 mM rhodamine B in propylene glycol. The corrected image was then pseudocoloured to represent changes in the ratio value. The scale calibrates the colour scale in terms of ratios from 0.8 to 2.0, where 1.0 is defined as the emission ratio obtained from the reference fluorescein–rhodamine B mixture, and low ratios represent low cAMP values and high ratios represent higher levels of cAMP. The resting ratios begin at or slightly below 1.0, and probably correspond to effectively zero free cAMP. A micropipette filled with 20 ng ml[-1] LH was then brought within 5–10 μm of the left-most cell, and at a time = 0 a brief pressure pulse (puff) was manually applied with a syringe to release the hormone locally. The second panel shows the same field of cells 13 s after the puff. In this panel, the left-most cell displays a gradient from orange (ratio about 1.3) near the site of stimulation to blue-green further away. In panel 3, 53 s after the puff, the left-most cell has nearly reached a maximum and its neighbour is beginning to respond. In panel 4, 213 s after the puff, the second cell has increased its cAMP further, either by bath diffusion of the hormone or by diffusion of cAMP through gap-junctions. However, the ratio has still not increased in the two right-most cells, showing that propagation is spatially limited. In panel 5, the pipette was moved to the other side of the field and the two right-most cells were puffed with LH; within 14 s, a small beachhead of cAMP elevation was observable. In panel 6, 74 s after the second puff, the cAMP increase had spread throughout both cells. The dark oval zones within the first and third cells are their nuclei, which partially exclude the FlCRhR. This experiment was performed by Brian Bacskai and Clotilde Randriamampita in collaboration with José Lemos, Ernesto Podesta and Mario Ascoli.

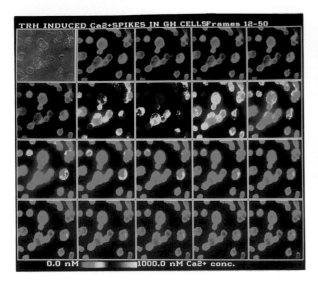

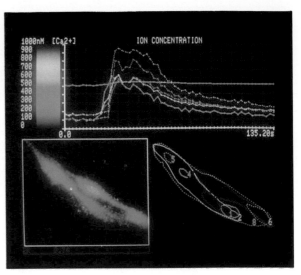

Plate 12.1 Colour image montage of pituitary cell responsiveness to TRH measured using fura-2 by the MiraCal system with an 8 bit cooled slow-scan camera. The large dynamic range of the detector also makes it possible to measure brightfield phase contrast images, here included in the upper left hand corner of the montage as a monochrome image. Other images are ratio images of 340/380 nm fluorescence, where blue is low intracellular calcium concentration and red is high.

Plate 12.2 Measurement of intracellular calcium in different areas of a human smooth muscle cell loaded with fura-2, during stimulation with 1 μM thrombin which causes the generation of IP3. A light pen is used to identify irregular regions of interest which can then be individually graphed simultaneously.

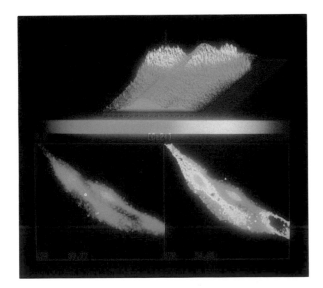

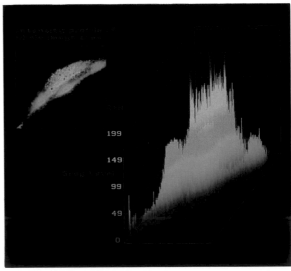

Plate 12.3 Two-dimensional pixel profile of a calcium wave in two human smooth muscle cells, cultured from human female mammary artery. Pixels were plotted along the full field of view as a function of time, with calcium concentration on the z-axis. Zero time is at the rear of the pixel profile.

Plate 12.4 Three-dimensional profile of calcium concentration in the same cell as Plate 12.3 with x and y dimensions as micrometres and $[Ca^{2+}]_i$ as the z-axis.

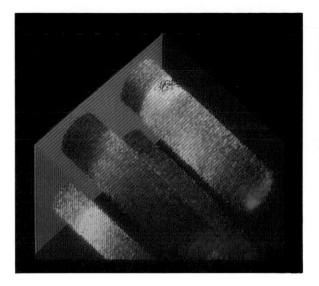

Plate 12.5 Axonometric projection of intracellular calcium in four pituitary cells measured with fura-2 and projected using MagiCal. This allows precise views of calcium rises during a cell stimulation. Images were time-averaged ($n = 8$) and are ratios of 340/380 fluorescence viewed at 510 nm.

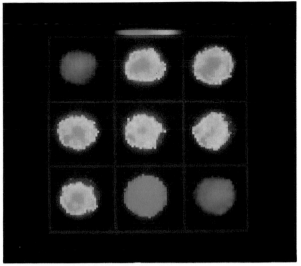

Plate 12.6 Fluorescence ratio imaging was used to study ionic calcium gradients within single growth hormone-secreting cells in response to the hypothalamic peptide growth hormone-releasing factor. Red colours show high ionized calcium concentrations of about 1 μM, and these are noted mainly in the subplasmalemmal region.

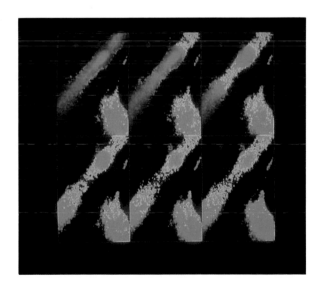

Plate 12.7 Image montage of six sequential images of a calcium wave spreading down the length of a human smooth muscle cell stimulated by thrombin, measured using MagiCal. The images shown here were measured with fura-2.

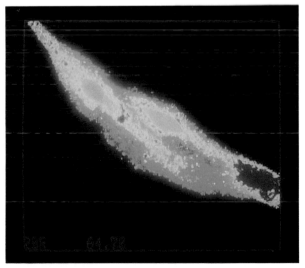

Plate 12.8 High-power view of two fura-2-loaded human smooth muscle cells at the height of thrombin stimulation. Note non-homogeneous localization of calcium in the cell. Cytosolic calcium is high but intranuclear calcium is at a somewhat lower level. This demonstrates the resolving power of fura-2 when used with ratiometric measurements.

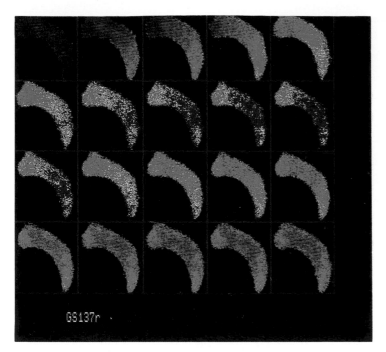

Plate 12.9 Montage of ratio images of intracellular calcium in a rat osteoclast imaged with fura-2 and displayed at approximately 2 s intervals following the control shown in the upper left hand panel. Note spreading wave of calcium originating from one end of the cell. Stimulus was *t*BuBHQ, and is likely to originate from calcium release from endoplasmic reticulum.

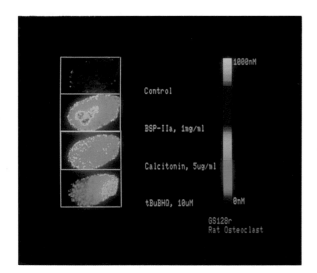

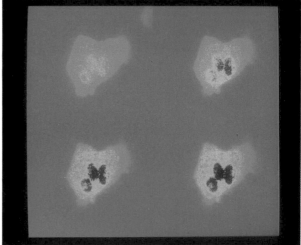

Plate 12.10 Spatial characteristics of the intracellular calcium signal in rat osteoclasts, imaged using conventional fura-2 non-confocal ratiometric imaging. Shown here is a single multinucleated osteoclast imaged with an intensified CCD camera. This figure shows a comparison of the spatial characteristics of the calcium signal in response to bone sialoprotein (BSP-IIA), salmon calcitonin and *t*BuBHQ, all additions being preceded by a buffer wash. BSP-IIA, a candidate for an endogenous ligand for the vitronectin receptor used as the stimulus, appears to elicit a response in the nuclear or perinuclear region of the cell. Calcitonin elicits a generalized rise in intracellular calcium which depends on extracellular calcium. *t*BuBHQ inhibits ATP-dependent sequestration of calcium into the endoplasmic reticulum and releases ER calcium without affecting nuclear calcium uptake.

Plate 12.11 Preliminary data obtained with a Noran confocal laser scanning microscope of osteoclasts loaded with the non-ratiometric calcium probe fluo-3. Images are taken at a single wavelength and run from top left to bottom right in time, being separated in time by about 2 s. BSP-IIA was added after frame one, and this gave rise to an immediate sustained nuclear calcium signal. Note the excellent definition of the putative nuclear calcium localization compared to that obtained in the second panel of Fig. 12.10.

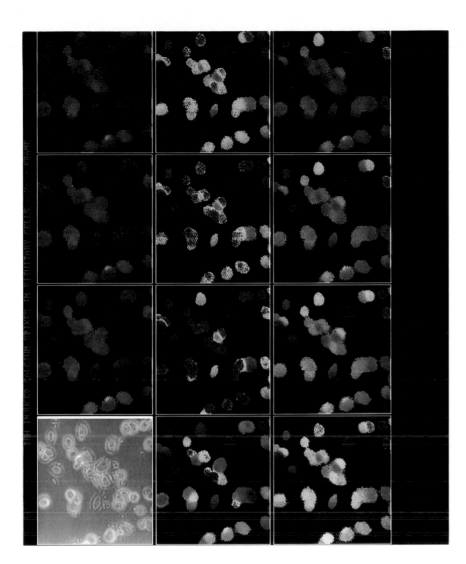

Plate 13.1 Sequential image montage to show which cells respond to stimulation.

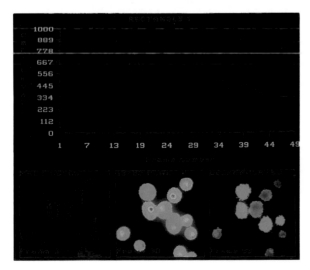

Plate 13.2 Selected example to show temporal and spatial data together: ratio analysis indicates a Ca^{2+} signal in the mast cell nucleus.

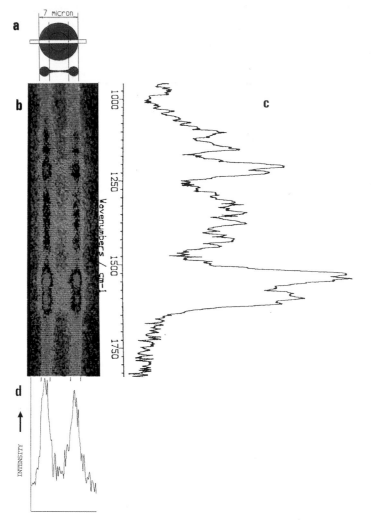

Plate 18.1 Example of the application of line illumination in a Raman microspectroscopic experiment. (a) Line illumination of an erythrocyte (top view and cross-section). (b) Image captured by the CCD chip: measuring time: 30 s, total laser power: 25 mW, microscope objective: see Fig. 18.14. (c) Spectral information (the Raman spectrum of the haemoglobin in the erythrocyte) is found along the vertical axis. (d) Spatial information, in this case the distribution of haemoglobin in the cell, is found along the horizontal axis. This graph is based on the integrated intensity of the Raman signal in the spectral interval 1450–1650 cm^{-1}.

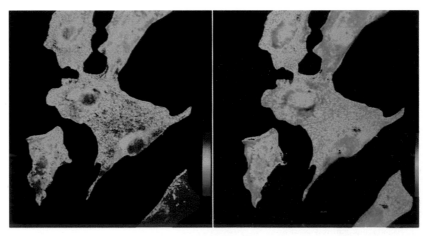

Plate 27.1 MeroCaM 2 fluorescence during serum stimulation of quiescent Swiss 3T3 fibroblasts. Serum-deprived Swiss 3T3 fibroblasts were microinjected with meroCaM 2, an indicator of calcium-calmodulin binding, and then challenged with whole serum medium. In the pseudocolour display here, higher emission values correspond to warmer colours. Emission intensities were normalized relative to prestimulation values to correct for differences in cell thicknesses and instrumental factors. Serum induced a rapid and transient increase in intracellular calcium, as measured with Calcium Green, an intracellular calcium indicator. The calcium transient induced an intial increase in meroCaM 2 excitation (left image: 15 s after stimulation) followed by a decrease (right image: 1 min 45 s after stimulation). Heterogeneous distribution of calcium-calmodulin binding can be seen in some cells.

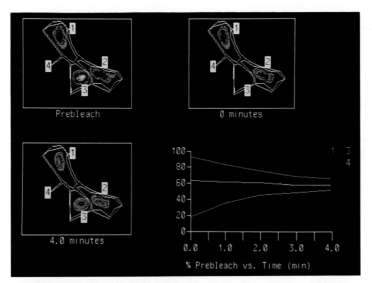

Plate 29.1 Intercellular communication in carboxyfluorescein-labelled HT cells. The centre cell (no. 3) was bleached, and fluorescence in both neighbouring cells (nos 1 and 2) and the entire triplet (no. 4) was monitored. Three pseudo-colour plots show flurorescence before bleaching, immediately after the bleach and after recovery, with the summary of the time dependent recovery also shown for all four areas monitored.

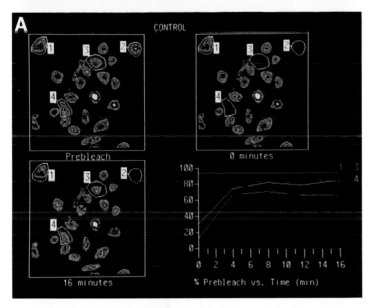

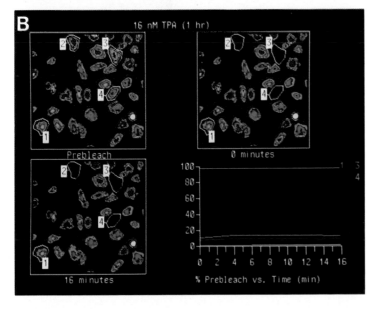

Plate 29.2 Digitized pseudo-colour images of SB-3 cells under control conditions (A), and after treatment with 16 nM TPA for 1 h (B). Images (225 × 225 μm) of cells prior to, immediately after, and 16 min after photobleaching selected cells are shown. A plot with the percentage recovery of fluorescence at 4 min intervals is displayed for each of the cells selected. The plotted data were corrected for background decline of fluorescence using the non-photobleached cell in area no. 1. The single cell in area no. 2 served as a negative control for fluorescence redistribution.

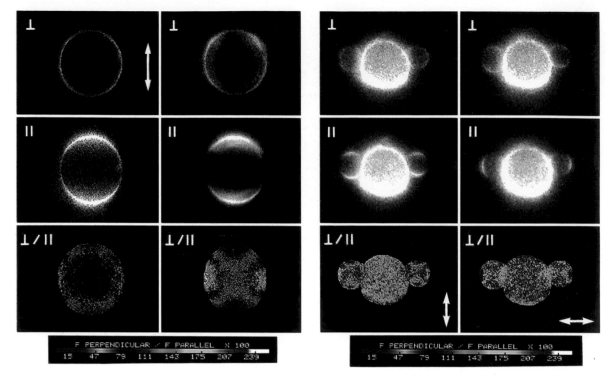

Plate 32.1 Polarized fluorescence of TMA-DPH-labelled liposomes composed of fluid-phase POPC (left) or gel-phase DPPC (right). Top and middle rows: Fluorescence images, after background subtraction, obtained with the emission polarizer oriented perpendicular and parallel, respectively, to the excitation light polarization direction (indicated by the arrow). The top, right side, bottom, and left side of each image corresponds to $\rho = 0°$, $90°$, $180°$, and $270°$, respectively, and $\gamma = 0°$ (see Fig. 32.1). Bottom row: Pseudo-colour polarization ratio images, after mapping. Liposome diameters are 31 μm (left) and 38 μm (right).

Plate 32.2 Fluorescence polarization measurement of a blebbing rat hepatocyte labelled with TMA-DPH. Top and middle rows: Fluorescence images, after background subtraction, obtained with the emission polarizer oriented perpendicular and parallel, respectively, to the excitation light polarization direction for each of two excitation polarizer orientations (indicated by the arrows). The focal plane passes through the centres of the two plasma membrane blebs. Around the bleb perimeters, $\rho = 0°$ and $180°$ corresponds to the top and bottom, respectively, of the bleb images in the left-hand column and the right and left sides, respectively, of the bleb images in the right-hand column. Bottom row: Pseudo-colour polarization ratio images, after mapping. Cell diameter is 25 μm. (Reprinted with permission from *The FASEB Journal*.)

Table 12.4 Comparison between intensified CCD cameras and slow-scan CCD cameras.

Intensified CCD	Slow-scan CCD
Ambient room temperature	Cooled to -50 to $-150°C$
Fast video frame rates ($25\ s^{-1}$)	Slow and variable read-out ($1\ s^{-1}$)
Limited dynamic range ($\sim10^3$)	High dynamic range ($\sim10^6$)
Easy to saturate	Difficult to saturate
Noisy due to intensifier	Ultra low noise – no intensifier
Fast averager circuitry required	Averaging on CCD face
No integration except in hardware	Integration by CCD
Generally low resolution 8 bit (1 part in 256)	Very high resolution 12 bit to 16 bit (1 part in 4096 to 1 part in 65 536)
Blue–green or red spectral response	Red spectral response
Must read-out complete frame	Programmable pixel read-out
No line scanning	Fast line scanning
Not usable for bright-field	Usable for bright-field to low light level

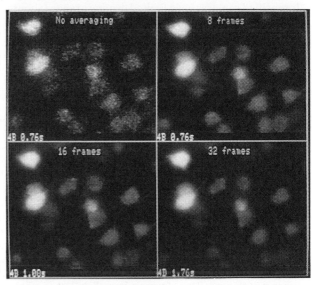

Figure 12.10 Benefits of using real-time signal averaging to reduce noise in an image from an intensified charge-coupled device, video frame rate camera.

12.11 LOW-LIGHT LEVEL CAMERAS FOR FLUORESCENCE RATIO IMAGING

Cameras employed for real-time fluorescence ratio imaging are similar to those used for astrophysics. Several different types of detectors are discussed in this volume. Table 12.4 summarizes the pros and cons of two of the most widely used cameras.

Because the signals emitted from optical probes are very faint and because it is desirable to maintain excitation light at low levels to avoid bleaching of the probe, it is not possible to use standard video cameras for real-time fluorescence ratio imaging. Instead, either intensified video cameras or cooled, slow-scan read-out cameras are typically employed. Both employ charge-coupled device detectors, but configured in different ways, with different electronic outputs and different noise levels.

Intensified video cameras are generally used for very fast applications where video signals are required. They are generally two-stage, with an optically coated front-end intensifier which governs the spectral sensitivity of the camera and this in turn is coupled optically with a lens or with a fibre optic taper to the video camera stage. Coupling with a fibre optic taper is preferable to an optic relay lens as light loss is minimized. Typically, a relay-lens-coupled detector

will be 5–10 times less sensitive than a fibre-optic-coupled system. Most cameras for this work are custom-designed. The first stage of the camera provides intensified input via a microchannel plate. Typically 10^5 or 10^6 lux is the light level required to be detected. A fibre optic taper then reduces the image area onto a CCD image sensor. These devices generally put out a standard video signal which can be displayed on a television monitor and captured using video frame-grabbers. The video signal is composed of odd and even lines (interlaced), generally producing 625 lines per frame and 25 frames per second in Europe or 30 frames per second in North America. Many systems use only the odd or even lines, permitting a filter to be changed in between for very fast applications. For slower applications, both odd and even lines may be acquired. These detectors may be somewhat noisy due to the intensification process, but new generation CCD technology is impressive and single-frame images obtained without signal-averaging to reduce noise can contain very usable data (Fig. 12.10).

Because the data flow is so high (for instance, 512 × 512 pixels times 30 frames per second is 7 864 320, nearly 8 million pieces of data per second), high-quality frame-grabbers with high-speed averaging are required if information flow is not to be lost. This averaging or integration is generally performed after an analogue image has been acquired, and following averaging the image is read out into computer memory through a high-speed analogue-to-digital converter, typically at 8 bit accuracy to yield 256 grey levels.

Intensified cameras have limited dynamic range (about 10^3), but this is quite well-suited to most available optical probes such as fura-2 which have dynamic ranges of about 30.

A slower, but potentially powerful type of detection technology is the cooled slow-scan CCD camera. 'Slow-scan' CCD cameras typically consist of a surface-mounted chip which is subjected to cooling with either liquid nitrogen or using thermoelectric cooling devices. The aim in both cases is to reduce dark current (noise) on the chip, and thus provide the capability to accumulate photon levels on the chip face for long periods of time without elevating the background signal. This results in higher signal-to-noise ratios and very low dark current, a result of long integrating exposures of the camera face to the specimen fluorescence. With these cameras, signal integration is performed directly on the detector chip. They are valuable for studying optical signals which do not vary greatly with time. These cameras produce lower noise images and possess higher dynamic range but can take many seconds to integrate and read out the image. Dynamic ranges of 10^5–10^6 are achievable, and this means that higher resolution analogue-to-digital converters can be employed. Thus 8, 12 or 16 bit conversion is usually possible, with 16 bit conversion providing up to 65 536 grey levels. Integration times

of many seconds to minutes may be used, without appreciable increase in background image noise. One important factor with these detectors is that the line-by-line read-out process can introduce noise itself, due to capacitive coupling between adjacent elements. Also, high bit-rate conversion is slower. Thus, higher bit conversion images may take a number of seconds to read out. It is worth noting that for the applications being discussed in this chapter, an 8 bit cooled CCD is worth strong consideration as they possess low dark current, can be read out quickly and are less expensive than higher bit conversion devices. The 256 grey levels obtained are comparable to that obtained with intensified CCD technology, and usable images can be obtained at the rate of 1–2 s^{-1}.

On a related note, silicon intensified target (SIT) tubes in general are inappropriate, as they have a high time-lag leading to image 'streaking' and tend to be more noisy. They cannot thus be used reliably at video frame rates as they produce carry-over of image from one wavelength to another, which can give rise to artifacts. Although some manufacturers still offer these for real-time fluorescence imaging, their use should be discouraged in favour of CCD technology which has high discharge rates, eliminating image carry-over.

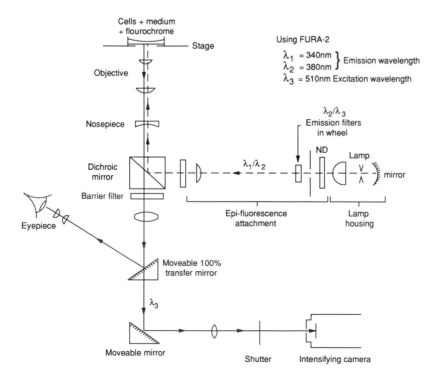

Fig. 12.11 Optical arrangement of the fluorescent microscope using an excitation filter wheel.

12.12 LIGHT WAVELENGTH SELECTION

Three possibilities exist for changing light to illuminate biological specimens: (a) a rotating filter wheel mounted on a stepping motor which is computer controlled and allows selection of wavelengths; (b) a pair of grating monochromators with alternating light selected by a rotating mirror; and (c) use of tunable dye lasers for variation of light wavelength, producing brief pulsed flashes of light on demand of the computer software.

Grating monochromators may be useful for some scientists interested in development of optical probes, but they are expensive and the amount of available light is limited unless the slit widths are opened wide. They allow spectra at variable wavelengths to be taken in a living cell, however.

Rotating filter wheels permit multiple wavelengths by using the wide range of interference filters available, and neutral-density filters can balance the amount of emitted light if necessary. This is important for imaging with a video signal, because to achieve the full range of dynamic

Table 12.5 MagiCal and MiraCal – differentiation of two typical high- and low-cost imaging systems.

Characteristic	MagiCal	MiraCal
Description	Top of the range product	Entry level – upgradable to MagiCal
Cost	Above £70 000 for basic system	Below £35 000 for basic system
Performance	High performance and fast	Mid performance and slower
Capture basis	Video-based capture system; can access most video cameras	Line scan read-out system; dedicated non-video camera system
Recommended detector	Intensified video frame rate CCD	Slow-scan cooled CCD or intensified video frame rate CCD
Interface to video camera	Yes	Yes
Interface to cooled read-out CCD	No	Yes
Image size	Up to 512 × 512 pixels; typically use 256 × 256 pixels	192 × 165 pixels up to 512 × 512 (video)
Sub-array scanning	No	Yes
Exposure time	Video-locked to 40 ms (PAL); 30 ms (NTSC)	Variable from software
Absolute usable capture speed for low light levels	25 images per second	5 images per second
Time between ratio images	80 ms	1000–2000 ms
Image storage	Up to 32 Mbyte of DRAM for fast access and display; thence to hard disk	Storage to computer memory and hard disk
Playback/animation speed	Fast; up to 10 times real time if required and variable from keyboard	Slow; dependent on disk controller speed but typically 10 times slower than MagiCal
Display monitors	Split-screen display; two colour monitors – one for text and one images	Single-colour display for text and images
Software options	About 600; full morphometric analysis available in package	Limited to about 100; offers limited morphometry – only linear spatial measurements
Software relationships	Can network to ORACal and other software packages for Magiscan	Can network to ORACal, but is also stand-alone
Single pixel access and user interface with images	Light pen	Mouse
Tape recorder mass storage	Tape recorder mass storage, for up to 60 min of continuous data	Not available
Software imaging averaging	Available as standard	Available as standard
Hardware image averaging	Available as option	Detector used for integration
Image digitization	8 bit, 256 grey levels	8 bit, 256 grey levels
Spectral sensitivity	Generally good blue sensitivity	Optimal for red sensitivity
Frame integration capacity	256 frames	Unlimited
Tape streamer backup	Yes	Yes
Optical disk backup	Yes	Yes

gain requires that image intensities at both ratioing wavelengths are approximately within an order of magnitude of each other. Otherwise, artifactual results can be obtained due either to saturation of the camera or the analogue-to-digital converter, or to use of the camera in a range below optimal sensitivity. The optical path employed is shown in Fig. 12.11.

The rotating filter wheel has additional advantages In other respects, since more than two wavelengths can be employed. With the Applied Imaging filter wheel used in our laboratory (Applied Imaging, Hylton Park, Wessington Way, Sunderland, Tyne & Wear SR5 3HD and 2340A Walsh Avenue, Santa Clara, CA 95051), eight filter positions can be employed, so that multiple fluorochromes can be used. For instance, simultaneous observations of both fura-2 and BCECF probes loaded in a single cell can be made by quadruple excitation at 340 and 380 nm (for fura-2) and 440 and 490 nm (for BCECF), passing light through a 510 nm dichroic filter and with all observations at a 520 nm centre filter. Calibration indicates negligible interference between the two dyes,

so simultaneous measurements of intracellular calcium and protons can be made.

12.13 COMPUTER HARDWARE FOR FLUORESCENCE RATIO IMAGING

A number of manufacturers offer systems for fluorescence ratio imaging. These variously run on either PCs or Apple computers, although powerful UNIX-based workstations are likely to enter the market in the near future, and are already available for some 3D image reconstruction problems. There is insufficient space to conduct a comprehensive survey of available systems here and such a survey would quickly date as manufacturers upgrade and develop their systems.

12.13.1 Imaging technology

Our laboratory uses two approaches for fluorescence ratio imaging. These systems have been developed in

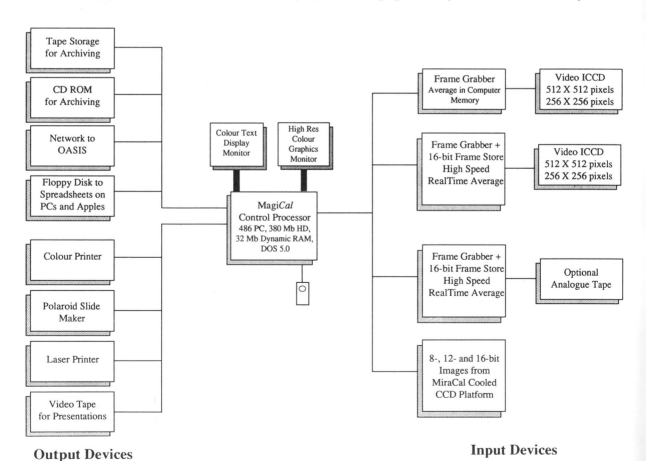

Output Devices

Input Devices

Figure 12.12 Input and output system diagram of MagiCal showing different input and output configurations options, including a wide range of detector options. MagiCal is essentially a video camera-based system.

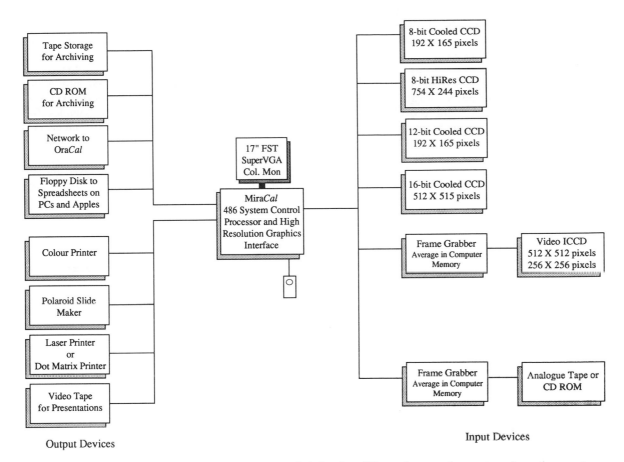

Figure 12.13 Input and output system diagram of MiraCal showing different input and output configurations options, including a wide range of detector options. MiraCal is most efficiently used with slow-scan, line read-out CCD detectors.

conjunction with Applied Imaging. Called MagiCal and MiraCal, their respective features are compared in Table 12.5. Whereas MagiCal is a fast, truly real-time image-acquisition system capable of grabbing up to 25 frames per second from a video-based intensified CCD camera and offering a real-time frame-averager card, MiraCal is based on a cooled slow-scan read-out camera and can produce up to 5 images per second given availability of light emission. There is a cost differential of more than 2:1, with the complexity of faster real-time imaging costing substantially more.

12.13.1.1   MagiCal

The MagiCal system, shown schematically in Fig. 12.12, is based on a 80486 control processor in a PC with an expanded backplane to accommodate a variety of customized interface cards for signal-averaging, digitization and real-time processing.

This is linked by a special fast parallel link to an image-processing unit (IPU) comprising a bit-slice processor with 48 bit instructions at 150 ns per instruction. The IPU contains dedicated hardware for video input to and output from an image memory which can contain up to 4096 × 4096 16-bit pixels. Image memory access is through a special pipe-lined processor running in parallel with a purpose-built central processor unit (CPU). Additional purpose-designed hardware performs real-time averaging, subtraction, ratioing, control of filter changer and video tape recorder, etc.

12.13.1.2   MiraCal

MiraCal can be configured in a variety of options (Fig. 12.13). It is also configured on a 486 system processor, making use of advanced graphics display technology. It uses a compact, cooled read-out camera with 8 bit analogue-to-digital conversion, and 190 × 165 pixels. Other cameras are also supported, including higher resolution 12 and 16 bit detectors up to 512 × 512 pixels, although these are not required for most ratio-imaging work. Image integration and filter wheel control can be achieved from the icon-like interface, such that variable exposure times for each wavelength image can be requested to make full use of the camera dynamic range. Images are written either to hard disk or more rapidly to system memory, which can be up to 32 Mbytes. At maximum speeds, the system can capture up to 5 images per second or alternatively log data from 20 regions of interest to a file available for graphing. The resulting data provide ratiometric images of more than adequate resolution for most biological applications, yielding data files of around 32 kbytes. A sample montage image from pituitary cells imaged with this system during TRH stimulation is shown in Plate 12.1.

12.13.2   Software for image analysis

MagiCal's software component is TARDIS (Time Analysis and Ratioing of Digital Image Sequences), written in Microsoft Pascal and bit-slice microcode, and running in an MSDOS environment. MiraCal is described elsewhere in this volume (see Chapter 13). It is written in Microsoft C and also runs in an MSDOS environment. Both systems provide a wide variety of manual and automated functions for image display and quantitative analysis.

MagiCal and MiraCal have been designed for compatibility with other MSDOS-based applications. In MagiCal, menus of options (Fig. 12.14) are presented on a separate screen from the images so that users are guided through appropriate operations at each stage. MiraCal is a single screen with a high-resolution user interface, allowing presentation of images, options, text and other facilities on a single-colour screen. Options are selected either by mouse or by single keys from the keyboard and it is not necessary to know special commands to drive the systems. Operations performed on sequences or subsequences of images range from protocols for control of image capture and real-time processing through to image analysis. Up to 20 different kinds of images can be captured, to include samples for background subtraction, shade correction and experimental images. Multiple calibration tables can be applied to selected image frames. Sequences can be edited within the image memory and saved on, or reloaded from, disks or video tape. A montage facility enables selected frames to be juxtaposed, with annotation and graphical enhancement for presenting results in a concise way.

12.14   CAPTURING AND REAL-TIME PROCESSING OF VIDEO SIGNALS

In the digitization process employed with MagiCal, each video line is quantized into discrete time intervals and in each interval the signal intensity is read as one of 256 discrete grey levels in the case of 8 bit digitization (Fig. 12.15(a)). Digitization is performed in order to read an approximation of the video signal as an array of digital values into the image processor unit (IPU) for processing and analysis; this is performed at the rate of 25 full video frames (with up to 512 × 512 pixels) per second.

Standard European video frame rates are 25 frames per second (one frame every 40 ms), and at maximal resolution of 512 × 512 pixels, image sizes are 256 kbytes, or about 262 000 pixels. To process these in real time, dedicated hardware has been developed to average and ratio images at this rate.

Having digitized the image, both signal and noise are captured. Noise arising from the intensifying camera can be reduced by processing the array of numbers which represent the image. A simple way to reduce noise is to average several successive frames as they arrive, or to integrate the signal which may be done in software or on the camera face itself in the case of cooled CCD cameras. The MagiCal hardware allows selection of the number of frames to be averaged, and by selecting the number of bits to be shifted out, the signal can be averaged or integrated (Fig. 12.15(b)). Averaging and integrating can also be done later in the system by software but this is slower. It will also require more memory to store the raw data until averaging, and thus limit the possible number of images in a given experiment. MiraCal on the other hand stores images either directly to a RAM drive in

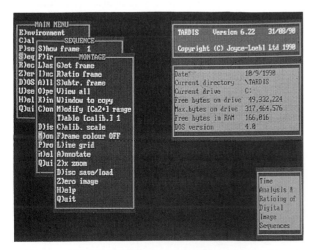

Figure 12.14 Control of MagiCal is by use of a number of easy access windows which permit control of data acquisition and subsequent image analysis for quantitative and qualitative purposes.

(a)

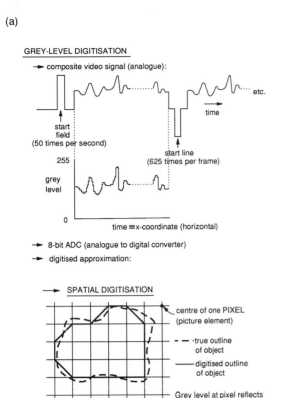

(b)

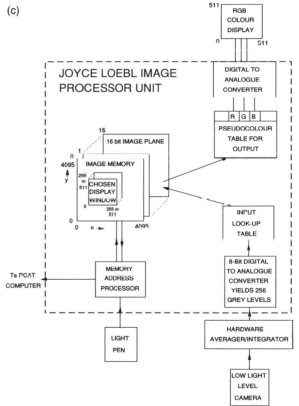

(c)

Figure 12.15 (a) Schematic explanation of grey level and spatial digitization of video signal. (b) Schematic diagram showing the mechanism of record accumulation with the MagiCal averager/integrator. The 8 bit video signal is accumulated at resolutions up to 512 × 512 pixels in a 16 bit memory buffer. This value is divided by 1 for integration purposes, or by the number of frames accumulated for an average. Finally, the averaged or integrated signal is transferred to the image-processor unit. (c) Diagram of the image processor unit for ultimate display on the graphics monitor. The light pen is interfaced through the memory address processor which permits direct identification of defined pixels sets for analysis or display. The grey levels are converted into pseudocolour and finally displayed as an analogue RGB (composite of red, green, blue) signal.

system memory or to hard disk as a single step.

The images are stored in dynamic random access memory (DRAM) configured typically as a frame store device with up to 4095 × 4095 pixel image planes (Fig. 12.15(c)). Input and output look-up tables provide the high-resolution system image storage, as well as the ability to output the values at each pixel as a colour. This is called 'pseudo-colour' or 'false' colour because the user chooses the colour to be displayed to suit his own preferences. Hence, in many of the examples in this article, blue represents low calcium concentration and red is high calcium concentration.

Dynamic video imaging of optical probes for ions is routinely carried out at 256 × 256 pixel resolution. The MagiCal system from Applied Imaging is supplied with 8 or 32 Mbytes of DRAM memory. With a 32 Mbyte system, some 512 images (at 256 × 256 pixels) can be collected. Image memory can be reconfigured, however, to provide storage of many more images at lower pixel dimensions. Smaller pixel windows can be defined prior to capture to increase the allowed use of memory.

MagiCal and MiraCal systems are fitted with hard disks ranging from 120 to 500 Mbytes capacity for permanent data storage, and image-compression routines (such as the public domain PKARC program) can be employed before permanent archiving on tape streamers or the new generation of rewritable optical disk storage subsystems.

12.15   INPUT AND OUTPUT

The digitized image enters an image analyser, which is a digital computer with special hardware to process and analyse images rapidly. The image analyser controls the entire system, ensuring that the filter changer has the required optical filter in place as each image enters the camera, controlling the flow of images through the averager, and providing the control interface between the operator and the machine. It displays images, analyses them to draw graphs, puts resultant data into disk files, prints images and tables of results, and performs other analytical tasks.

Whilst dynamic video imaging allows high-power inspection of, say, ionic concentration within a single cell at high magnification, the use of lower power objectives enables collection of data from many cells at one time. The ability to select the area to be analysed on a pixel-by-pixel basis enables the user either to produce traces of fluorescence from a large number of single cells (which can then be exported as ASCII files of numbers into a spreadsheet for averaging and statistical analysis), or alternatively the entire field containing perhaps 20–100 cells can be selected to produce a graph of the population response for a given experiment.

12.16   IMAGE PROCESSING: AVERAGING, BACKGROUND CORRECTION AND RATIOING

Once images have been captured, they are processed prior to analysis. Background fluorescence is removed by capturing images at each desired wavelength, without cells, but in conditions identical to the experiment. These background images are subtracted pixel-by-pixel from each of the cell images before further processing. Another factor to allow for is uneven illumination of the field of view, i.e. 'shade correction'.

Ratioing to give a calibrated ion concentration involves applying a formula at every pixel, which takes into account an experimentally determined constant appropriate to the dye–ion interactions, dye quantum efficiency and system optics, intensity ratios for the pixel in each of the two images, and calibratable extremes of ratio intensity measured in the experimental arrangement.

Image analysers generally store a whole number at each pixel, having 256 possible values. Tables are used within the software to map the possible range of ion concentrations either linearly or logarithmically onto the range from 0 to 255. Ratioed images contain whole numbers representing ratios or ion concentrations and measurements made on these images use the tables to look up the true values. The tables also speed the computation of ratioed images: rather than performing a time-consuming division at every pixel the result is looked up in a table addressed by the two intensity values being ratioed. A pseudo-colour look-up table is then mapped onto the range of grey values or ion concentrations.

12.17   ANALYSING DIGITIZED VIDEO IMAGES

Image analysis extracts measurements, thereby reducing the information to manageable numbers. The general procedure is to identify regions of interest within images, to measure these regions and to plot the measurements as a function of time or distance in a graph or a table.

The simplest way to segment the image is to trace round areas of interest directly on the image display, using a light pen. The system can enlarge the images while this is done, to make delineation more accurate. Another method is to select all pixels which lie within a given range of intensity or ion concentration and then to detect automatically contiguous groups of such pixels for measurement. This method is called *thresholding*: selecting pixels whose intensities lie between a pair of values.

Fluorescent images may not show important structures in cells, and will probably not show the true boundaries of the cells. It is therefore important to be able to capture a bright-field image along with the fluorescent image and to use that for defining regions of interest. This can be done by image superposition of a phase-contrast image with a fluorescent ratio image, by image addition.

12.18   PRESENTING DATA

Software in the MagiCal and MiraCal image analyser can present results in a wide variety of ways, which

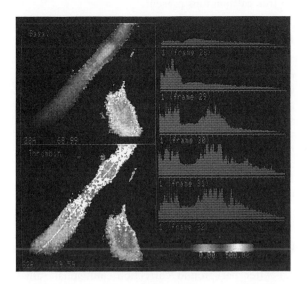

Figure 12.16 Example of pixel intensity profiling which can be used to analyse spatial changes in ion distribution. This shows a fura-2 loaded smooth muscle cell before (top left hand image) and during stimulation (bottom left hand image) with 1 μM thrombin, which triggers a calcium wave spreading down the cytoplasm. A freehand line was defined along the cell axis with the light pen, and sequential pixel profiles developed at various time intervals during stimulation. Reading the profiles from top to bottom shows clearly the spread of the calcium wave along the cell.

contribute to gaining insight into what really happened during an experiment. Examples include

- superimposing graphs of different regions, either of the same or of different cells, to compare their behaviour (Plate 12.2);
- histograms of the frequency of occurrence of ion concentrations at all of the pixels in a region;
- profiling pixel intensity along lines defined through cells and comparing these with profiles of the same lines from other images in the sequence (Fig. 12.16), or plotting pixel profiles as a function of time (Plate 12.3);
- three-dimensional views of ion concentration profiles across a region (plate 12.4);
- plotting images in a stack, using time as a third dimension, to follow where changes occur;
- animating sequences to compare the changing ion profiles as a function of time (Plate 12.5); and
- allowing tables of results to be analysed statistically, studied in spreadsheets, graphed in different ways, etc.

12.19 CELL CULTURE AND LOADING OF FLUORESCENT PROBES

Routine cell preparation and culturing techniques are employed. The cells intended for study are plated onto thin glass coverslips, usually no. 1.5. This permits focusing from below using an inverted microscope and allows passage of the lower wavelengths of ultraviolet light required to excite many of the available fluorescent probes (<360 nm). Plastic media cannot be used as it absorbs ultraviolet excitation light.

Cells may be loaded with ion-sensitive fluorescent probe in one of two ways. The free acid form of the probe may be directly loaded through the cell membrane by micropipette or patch pipette. For fura-2, a concentration of around 50–100 μM has been found to be satisfactory. This method, however, requires techniques which are often unavailable in many laboratories. The great attraction of many of these probes is that they are available in the acetoxy-methyl ester form. These non-polar ester derivatives may be added to the extracellular medium where they will diffuse across the membrane to be hydrolysed by non-specific cytoplasmic esterases, resulting in the membrane-impermeable free acid form. Generally, the ester form of the dyes is insensitive to changes in ion concentration. Unfortunately most plant cells studied so far appear to have low esterase activity, and this has made experiments difficult unless micro-injection is used.

For experiments measuring intracellular free calcium using fura-2 or measuring pH using BCECF or other probes for other ions, the probe can be initially made up as a stock solution of 2 mM in dimethyl-sulphoxide (DMSO). Cells are normally incubated in a 4 μM solution of the acetoxymethyl ester form (fura-2-AM, BCECF-AM or other) made up in a standard extracellular medium such as one containing (in mM): 125 NaCl, 5 KCl, 1.8 CaCl$_2$, 2 MgCl$_2$, 0.5 NaH$_2$PO$_4$, 5 NaHCO$_3$, 10 HEPES, 10 glucose, 0.1 BSA, pH 7.2) for 30 min at 37°C. Washing at least three times with fresh medium removes any excess probe and subsequent microscopic examination reveals a clear signal from the intracellular free acid with minimal compartment-alization within the intracellular organelles. Pluronic acid, a weak detergent, can also help with loading dye into some cells which are resistant to such procedures.

Data collection from cells should start within approximately 2 h of loading as they have been seen to lose responsiveness when loaded for extended periods. Maintaining them at room temperature rather than 37°C does, however, appear to extend this time. The cause is not fully understood but may be due to the probe partitioning into compartments which do not normally undergo changes in ionic concentration, or simply the normal biological function of the cells may be affected.

12.20   CALIBRATION OF ION-SENSITIVE DYES IN LIVING CELLS AND IN SOLUTION

Calibration for free ion concentration may be performed by two distinct methods, either managing free acid solutions of the probe with various known ion concentrations or imaging the cells of interest loaded with the ester derivative of the probe under permeabilized conditions.

Although the fluorescent dyes developed for ion-sensitive measurements can also be utilized as single-wave length excitation or emission probes, their power is only fully utilized using dual-excitation or dual-emission measurements with probes such as fura-2 and BCECF. These eliminate many of the artifacts associated with simple fluorescence intensity measurements. Fluo-3, a calcium-sensitive optical probe which changes spectral properties at only a single wavelength when calcium activity changes, is commonly used with single wavelength confocal microscopy, but is prone to artifacts and difficult to calibrate reliably.

In all the following methods, the probe is excited with either a single or dual wavelength and measured respectively at a dual or single wavelength. All calibration is performed in similar medium and at the same pH as used for subsequent experiments.

The following methods use, as an example, intracellular free calcium measurements using the dual-excitation fluorescent probe fura-2 which is available in both the free acid and acetoxymethyl ester forms (Molecular Probes, Eugene, OR). Many other probes for different ions are now available but the general methods are valid for any dual-excitation probe. For pH-sensitive dyes such as BCECF, permeabilization with nigericin could be used in place of ionomycin as in the following example.

Data required for calibration may be entered into the TARDIS software by entering the maximum (R_{max}) and minimum (R_{min}) ratios obtained into the following equation which is pre-programmed into the software.

$$\text{Ion concentration} = K_d * \beta * [(R - R_{min}) / (R_{max} - R)]$$

where K_d is the dissociation constant of the probe; β is (intensity at the upper wavelength at R_{min}) / (intensity of the upper wavelength at R_{max}); R is the measured ratio; R_{max} is the ratio when the probe is saturated with calcium ions; and R_{min} is the ratio with no free calcium ions present.

12.20.1   Free acid solution method

The simplest method for initial calibration is to prepare two solutions of medium, one containing 5 mM $CaCl_2$ which will saturate the probe with free calcium ions and result in the maximum ratio obtainable. The other should have the $CaCl_2$ substituted by 1–10 mM EGTA which will bind all free calcium ions to result in the minimum ratio obtainable. Both must contain the free acid form of the probe at a concentration of about 50–100 μM which is the approximate concentration found in cells loaded with the ester derivative.

Readings of these solutions are taken and the ratios and constants obtained by the following method:

(1) A background frame is captured at each wavelength to correct for any background noise.
(2) After background correction, individual values are measured graphically using the standard computer software. Raw values at each wavelength and ratio values are determined in high and low calcium. This procedure provides an estimate of R_{min} and R_{max} which can then be entered into the software.
(3) The raw 380 nm value at minimum calcium is divided by the raw 380 nm value at maximum calcium. This is the β value which is required to be entered into the equation. Note that this value will differ with changes in the optical components of the system.

Our software also has facility for entering an upper and lower calcium value, which permits the full range of pseudo-colour to be utilized. This is adjusted experiment-to-experiment, but of course does not affect the absolute value of ionic concentration determined.

12.21   INTRACELLULAR METHOD

This is the preferred method of calibration. The cells are loaded with the acetoxymethyl ester derivative of the probe as previously described and the cell membrane permeabilized to ions with a specific ionophore. For calcium, ionomycin at a concentration of about 2 μM has been found to be particularly suitable for calcium. Another popular calcium ionophore, the antibiotic A23187 is only suitable as its halogenated analogue 4-bromo-A23187, as native A23187 exhibits autofluorescence at the upper fura-2 wavelength. Some laboratories have also used low concentrations of digitonin or saponin to permeabilize cells for calibration; effective concentrations will vary from cell to cell.

The R_{max} measurement should be made in elevated extracellular $CaCl_2$. A concentration of 10 mM has been found to be sufficient to saturate the probe within the cell. Addition of the ionomycin causes a rapid, sustained rise in intracellular calcium which reaches maximum within tens of seconds. This whole process should be captured and time allowed for equilibration to measure the R_{min} value. This may now be determined on the same field by washing the cells two or three times with calcium-free medium containing 1–10 mM EGTA. Transport across the cell membrane – resulting in binding of free calcium ions – may require at least 15 min.

12.22 BIOLOGICAL APPLICATIONS OF REAL-TIME QUANTITATIVE MICROSCOPY

The research of our laboratory is concerned with the intracellular signals which control biological function. Calcium is clearly an important messenger, and other intracellular molecules such as nucleotides, G-proteins, phosphoinositides, calcium-binding proteins and indeed the cytoskeletal matrix itself are also important modulators. Many of these molecules act through release of calcium from intracellular pools or regulation of calcium entry from the extracellular pool, or they may be regulated by calcium ions themselves. Our work with fluorescence ratio imaging has employed mainly the dual-excitation dye fura-2, and more recent work has utilized this dye in combination with BCECF to gain information about simultaneous changes in calcium and pH.

The following sections represent case studies of different experimental approaches which have utilized optical probes for study of ionic homeostasis in living cells, using different physico-chemical features of the probes to provide different classes of information.

12.22.1 Case study 1: intracellular calcium measurements

Pituitary hormone secretion is controlled by hypothalamic peptides and transmitters. Thyrotropin releasing hormone (TRH), the releasing factor for prolactin, causes generation of inositol trisphosphate (IP_3) in these cells and this in turn releases calcium ions from an intracellular pool. In addition, cyclic AMP (cAMP) also stimulates rises in intracellular calcium, although this varies considerably from cell to cell (Fig. 12.17). Figure 12.18 shows the spatial nature of the TRH response in fura-2-loaded cells, revealing that responses

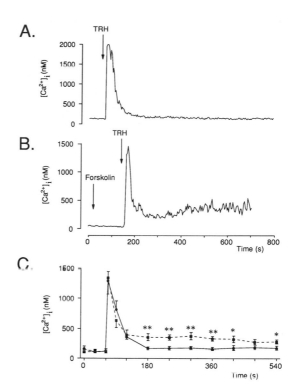

Figure 12.17 This illustrates typical findings on normal pituitary cells cultured from the bovine anterior pituitary. Initial application of forskolin, which promotes a rise in intracellular cAMP, causes modest but sustained rises in intracellular calcium activity, imaged in fura-2-loaded cells. Subsequent application of TRH at 100 nM promotes a rapid but transient rise in calcium activity, with a sustained plateau phase. (A) shows the normal response to TRH; (B) shows the TRH response in the presence of forskolin – note plateau phase following initial calcium transient; and (C) shows the average response of six cells for each treatment. Data were extracted from MagiCal as ASCII files and plotted in Lotus 1–2–3 and Freehand graphics and labelling packages.

are not spatially heterogeneous. From other experiments we believe this plateau phase may be the result of cAMP action, since it is not observed in these cells when cAMP has not been elevated: TRH application alone promotes only the transient rise but no sustained plateau phase (Plate 12.2).

Figure 12.19 also shows that imaging assists us in evaluating anisotropy – differences in behaviour amongst cells. Virtually no two cells ever behave identically, and population responses as measured in a spectrofluorometer cuvette can be misleading. In cell populations where temporal asynchrony is prominent, such as in the response of lymphocytes to the mitogen concanavalin A (Con A), the apparent population response is very small. Individual cell responses to Con A may, on the other hand, be large, exhibiting transient rises to calcium concentrations in excess of

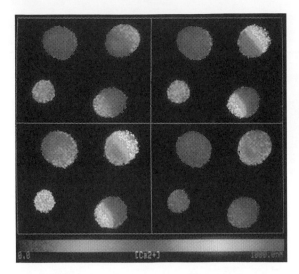

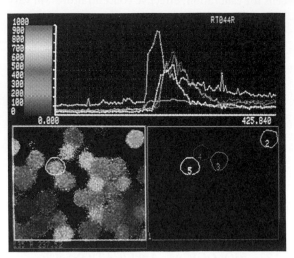

Figure 12.18 Four ratio images of $[Ca^{2+}]_i$ in bovine anterior pituitary cells imaged with fura-2 using MagiCal, and stimulated with TRH. Frames are displayed at approximately 3 s intervals following the control shown in the upper left hand panel. Note calcium rise occurs in only a portion of two of the cells, more homogeneously in a third and to only a small extent in the fourth.

Figure 12.19 Anistotropy amongst cell populations can be addressed with dynamic video imaging. Lymphocytes were loaded with fura-2 and stimulated with mitogenic concentrations of concanavalin A. This caused large but transient rises in $[Ca^{2+}]_i$. These increases were asynchronous with time, and their temporal properties and amplitude vary widely from cell to cell.

1000 nM, but they occur with latencies varying from 1 s to 1 min after Con A application. The average response is, therefore, misleadingly small.

12.22.2   Case study 2: localizing calcium changes within cells

Figure 12.20 shows another potentially valuable use of imaging technology for looking at the spatial localization of intracellular calcium within the cytoplasm. Here we used isolated rat somatotrophs – growth hormone secreting cells – and stimulated hormone secretion with the 44 amino acid peptide growth hormone releasing factor (GHRF). In contrast to the prolactin secreting cells, GHRF modulates intracellular calcium rises primarily through the plasma membrane, by stimulating cAMP levels inside the cell, since both basal oscillations and the rise in frequency and extent of calcium inside the somatotroph are not observed when extracellular calcium is removed. In Plate 12.6, a typical rise in intracellular calcium when the cell is GHRF-stimulated is shown. Prominent elevation of intracellular calcium takes place in the sub-plasmalemmal area, indicating that this region experiences the highest elevation of calcium during stimulation. Similar behaviour is noted regardless of the stimulus used – GHRF or cAMP – and is also observed to be qualitatively identical during spontaneous basal oscillations. Prolactin cells exhibit different behaviour, namely that in many cells stimulation with

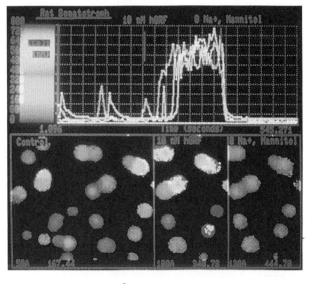

Figure 12.20 Basal $[Ca^{2+}]_i$ oscillations and growth hormone-releasing factor induced rises in $[Ca^{2+}]_i$ and ionic oscillation frequency in single growth hormone-secreting cells. In the last part of the record, removal of extracellular sodium diminished intracellular calcium concentration.

TRH causes a calcium rise in the cytoplasm which is localized to one pole of the cell only, and may be related to localized membrane traffic taking place during the secretory event.

We have also examined smooth muscle cells from the human female internal mammary artery, in which thrombin causes a prominent but transient increase in intracellular calcium due to IP$_3$ generation. Fluorescence ratio imaging allows us to understand the local changes in calcium which trigger gross intracellular changes. Plate 12.7 shows a typical experiment where thrombin was applied at 1 μM concentration, with 2 mM external calcium. Although total intracellular calcium rises and falls, the local nature of these changes is complex. Typically, following thrombin a large rise in calcium ion concentration is detected which is initiated in one pole of the cell, and then spreads within 2–5 s as a wave along the cell. Repeated waves of calcium will occur if thrombin application is maintained at a low concentration, but the wave always originates from the same intracellular region. Again, these waves are independent of extracellular calcium. The nuclei of these smooth muscle cells appears to have a moderate resting level of calcium which does not change when cytoplasmic calcium rises and falls with this stimulus, suggesting the compartment is 'tight' to cytoplasmic calcium. However, the nuclear compartment will show elevated calcium changes when the cell is permeabilized with ionomycin, so fura-2 localized there does apparently act as a valid ionized calcium indicator.

12.22.3 Case study 3: isosbestic point imaging – simultaneous measurement of ionic activity and ionic influx

Measurements of intracellular calcium activity with fura-2 reflect the ionic concentration in the cell cytoplasm, and this can provide useful information about modulation of cellular activity. As mentioned above, these measurements are derived from ratio images of light measured at 510 nm and excited at 340 and 380 nM respectively. However, if fura-2 is excited at 360 nm, the emission at 510 nm is virtually independent of calcium ion concentration. Furthermore, the 360 nm fluorescence is strongly quenched if Mn^{2+} ions bind to the probe. By assuming that manganese can act as a surrogate for calcium and enter through calcium channels, the rate of calcium entry can be monitored (Fig. 12.21) as well as intracellular calcium concentration.

This technique has been used successfully in a number of cell types and is a valuable addition to the information which can be obtained purely from ratio fluorescence measurements with fura-2.

12.22.4 Case study 4: imaging intracellular calcium store refilling

The nature and control of the plasma membrane calcium channels that mediate entry of extracellular calcium in non-excitable cells are only partially understood, and the method of isosbestic point sampling described above is one possible approach which can be used with optical probes. Normally calcium influx follows the agonist-induced discharge of intracellular calcium stores. The stores then refill as a consequence of the increase in cytosolic calcium concentration. Models for refilling of stores include control by the fullness of the store *per se* (e.g. the capacitance model (Putney, 1990)) as well as control by various second messengers produced by the agonist (Irvine, 1990).

Figure 12.21 The MagiCal system was used with 3-wavelength excitation. Extracellular Mn at 100 μM had little effect on 360 nm fluorescence, indicating little or no calcium entry under resting conditions, consistent with the low ionic permeability of these cells. TRH induced a rapid but transient [Ca^{2+}]$_i$ rise (monitored simultaneously with 340 and 380 nm excitation) as usually observed, with little initial change in the 360 nm signal. This figure also shows the experimentally observed increase and decrease in 340 and 380 signal during an intracellular calcium transient, which represented a transient increase from about 70 nm to somewhat over 1200 nm. However, as the calcium transient induced by TRH declines, the 360 nm signal also begins to decline, indicating that calcium entry has been activated, presumably to assist in refilling of the discharged intracellular stores.

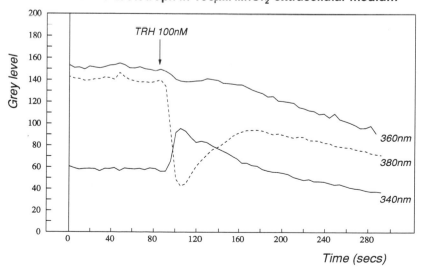

Bovine Lactotroph in 100μM MnCl$_2$ extracellular medium

TRH 100nM

Grey level / Time (secs)

360nm
380nm
340nm

CELL A

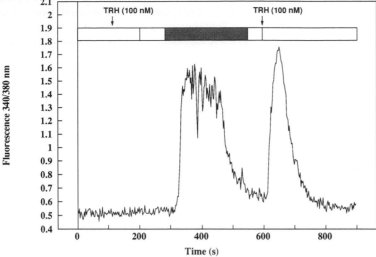

CELL B

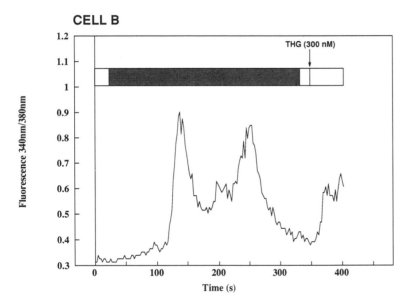

Figure 12.22 Normal rat pituitary cells loaded with fura-2 using protocols to examine calcium store reloading. Both cells were initially Ca-depleted using low Ca medium for 50 min, having been loaded with fura-2. In (A), TRH treatment has no effect on $[Ca^{2+}]_i$ after this treatment, but calcium restoration allows the stores to recharge in the absence of TRH (during shaded bar). Subsequent TRH application causes a large rise in $[Ca^{2+}]_i$ once stores have been recharged. (B) Similar experiment showing oscillations in $[Ca^{2+}]_i$ during re-loading, perhaps due to uptake into more than one compartment. Application of thapsigargin – a blocker of endoplasmic reticulum Ca-ATPase, also discharges this calcium store.

Activation of calcium influx by store depletion can be studied by allowing the intracellular calcium stores to empty (if only partially) by incubation of the cells in nominally calcium-free medium. Addition of calcium to the medium causes an increase in $[Ca^{2+}]_i$, which may enter the cell through the channels involved in a normal agonist-mediated response.

An example of this approach is shown in Fig. 12.22. Single rat adenohypophyseal cells were loaded with fura-2 (4 μM, 30 min at 37°C) and $[Ca^{2+}]_i$ measured by ratiometric video imaging. Cells were incubated for 50 min in medium to which no $CaCl_2$ had been added, in order to deplete calcium stores. The concentration of calcium in the low-calcium medium was less than

1 μM, measured with fura-2 free acid. In these experiments, the ratio of 340/380 nm fluorescence excitation is displayed, a measure of $[Ca^{2+}]_i$.

Recordings of $[Ca^{2+}]_i$ in two cells in separate experiments (Fig. 12.22(A) and (B)) are shown. Cell A was challenged with TRH after incubation in low-Ca medium in order to determine the time required for depletion of stores. Cell B was incubated for the same time, but the first TRH addition was omitted, and thapsigargin (THG, a blocker of the endoplasmic Ca-ATPase) was added instead of the second TRH challenge. In both cases, a large increase in $[Ca^{2+}]_i$ was seen only when calcium was returned to the cell. TRH (cell A) and THG (cell B) both elicited calcium responses, presumably from refilled stores. The refilling phase is large and prolonged and can be characterized in terms of modulators of calcium channels and signalling pathways.

In summary, ratio imaging can be used to study the depletion and refilling time courses and characteristics of intracellular calcium stores, important sites for cellular calcium homeostasis. The main advantage of this type of approach is that store-depletion-activated calcium influx can be separated from that activated by an agonist.

12.22.5 Case study 5: manipulation of intracellular calcium-responsive compartments – ratiometric imaging to detect nuclear calcium changes

Use of imaging technology can play a valuable role in detecting intracellular and intranuclear changes in calcium concentration. In many cells, the ratiometric calcium probes can be loaded into the cytoplasm and appear also to be taken up into nuclear and perhaps other compartments. We have been interested in whether the fura probe, for instance, can effectively image nuclear calcium changes.

In human smooth muscle cells (Plate 12.8) it is clear that the nuclear compartment loads with calcium indicator and calibration of the signal revealed a resting ionized calcium level of about 300–500 nM, although large changes in cytosolic calcium do not perturb the level of calcium to any extent. However, application of ionomycin to the cells showed convincingly that fura in the nuclear compartment was accurately reporting calcium concentration changes, as a large rise in calcium could be induced. Thus, the nucleus was concluded to be relatively impermeant to small ions and presumably to other small molecules.

Recent work on multinucleated osteoclast cells has provided a totally different picture (Plate 12.9). In these cells, resting nuclear ionized calcium is low, probably about 50–100 nM. We have used agonists for the vitronectin receptor, a member of the integrin family highly expressed in osteoclasts. Peptides bearing the RGD sequence have been shown to bind to the vitronectin receptor, and promote osteoclast retraction. Addition of RGD-sequence-containing peptides caused a large rise in ionized calcium localized to the nucleus (Plate 12.10), and mainly in the absence of any cytosolic changes.

We sought to determine whether such changes could be artifactual, by using other putative agents which might stimulate intracellular calcium changes by acting either on the surface membrane or on intranuclear compartments. Calcitonin, which causes plasma membrane entry of calcium, elicited a generalized cytosolic rise not localized to any specific compartment of the cell. On the other hand, tBuBHQ, an inhibitor of the Ca^{2+}-ATPase in endoplasmic reticulum, caused in the same cell a markedly different rise in cytosolic calcium, presumably due to localized release of intracellular calcium from the endoplasmic reticulum. This study raises the fascinating possibility that specific intracellular signalling molecules exist which can evoke intranuclear calcium modulation in the absence of cytosolic changes *per se*, and that the nucleus may possess intranuclear stores of calcium ions. The role for such changes is, however, unclear, but could be linked to early gene activation or modulation of gene transcription.

The possibility that these changes are artifactual has been further eliminated by preliminary use of a real-time confocal microscope (Noran Instruments, Madison WI). Using fluo-3 as a single wavelength calcium probe under conditions where a focal plane of 0.7–1.0 μm was achieved, discrete nuclear calcium changes evidenced by increases in fluo-3 emission were observed after addition of several RGD-sequence-containing peptides (Plate 12.11).

12.22.6 Case study 6: four-wavelength imaging of intracellular calcium and pH

Virtually all intracellular processes are sensitive to cytosolic calcium ions. Often, changes in intracellular calcium activity are associated with changes in intracellular pH (pH_i), which is also believed to play an important role in cell signalling (reviewed in Frelin *et al.*, 1988). However, the interrelationship between cytosolic mechanisms controlling levels of calcium and hydrogen ions in anterior pituitary cells is still poorly understood. This is mainly because of the difficulty in accurate measurement of the respective cytosolic activities. The use of ion-sensitive microelectrodes is not feasible in small cells and fluorescent imaging of dyes which are ion-specific, easily introduced into the cytosol and sensitive to physiological concentrations of calcium and hydrogen ions has made it possible to undertake such studies in single living cells (Rink *et al.*, 1982; Grynkiewicz *et al.*, 1985).

This case study reports in detail a method of simultaneous microfluoriometric measurements of calcium using fura-2 and H^+ using the pH-sensitive dye BCECF simultaneously loaded in a single cell to investigate cytosolic pH and calcium activity in single bovine lactotrophs.

Optical cross-talk between the fluorochromes under our experimental conditions was determined to be insignificant. We found that resting pH_i and $[Ca^{2+}]_i$ are correlated and that alkalinization (and sometimes acidification) induces partial release of Ca^{2+} from agonist-sensitive intracellular stores. A role for extracellular Na^+ in the recovery of pH from acid load was also identified, which may be important for ionic homeostasis. Further, removal of extracellular Na$^+$ resulted in a decrease in resting [Ca^{2+}]$_i$ and pH$_i$. This suggests that Na$^+$/H$^+$ may play a role in determining resting [Ca^{2+}]$_i$.

Lactotroph cell cultures were incubated either with 1–8 µM fura-2 acetoxymethyl ester (Molecular Probes) or with 0.5–4 µM BCECF acetoxymethyl ester (fura-2-AM, BCECF-AM; Novabiochem, UK) or for dual-loading experiments with both. Loading was at 37°C for 30 min. Stock solutions of 1 µM TRH and 150 mM NH$_4$Cl were prepared in bathing media. Solution changes were introduced by superfusion (1–5 ml min^{-1}). In some experiments using the weak base NH$_4$Cl, we were able to estimate the intracellular buffer capacity for H$^+$ using the method of Szatkowsky and Thomas (1989) by extrapolating the initial change in pH upon the addition of NH$_4$Cl.

Dye loading resulted in an apparently uniform distribution of fluorescence throughout the cell cytoplasm, consistent with previous reports (Malgaroli *et al.*, 1987; Eisner *et al.*, 1989; Akerman *et al.*, 1991). When changes in pH and [Ca^{2+}]$_i$ were measured simultaneously, cells were loaded with both dyes. All experiments were performed using the MagiCal imaging system at 37°C.

12.22.6.1 Fura-2 measurements

Fura-2 was excited alternately with 340 nm and 380 nm light (10 and 13 nm half-bandwidth respectively) by means of a xenon arc lamp and a computer controlled rotating filter wheel (Zorec *et al.*, 1990; Akerman *et al.*, 1991). Emitted light was collected via a 510 nm dichroic mirror. The calibration approach was as described previously (Grynkiewicz *et al.*, 1985) where the apparent dissociation constant, K_d, for fura-2 was taken to be 240 nM, and 1–2 µM ionomycin was employed to equilibrate internal and external calcium.

12.22.6.2 BCECF measurements

BCECF was alternately excited with 440 and 490 nm light, and light collected similarly to the experiments with fura-2, but through a 520 nm long-pass filter. Excitation filters for BCECF and fura-2 were mounted in the same filter wheel (Applied Imaging, Sunderland, UK). The sequence of excitation could be selected by computer control. Calibration of the pH measurements (Rink *et al.*, 1982) was performed with solutions: (in mM) 130 KCl, 10 NaCl, 1 MgSO$_4$, buffered with either 10 mM HEPES, PIPES (piperazine-*N,N'*-bis-2-[ethanesulphonic acid]) or MES (2[*N*-morpholino]ethanesulphonic acid) adjusted to various pH values (KOH, HCl) from 5.5 to 8. Equalization of pH$_i$ with external pH (pH$_o$) in these solutions was obtained

by extracellular application of K$^+$–H$^+$ exchanger nigericin (2–5 µM). pH was calculated from the ratio of fluorescence intensity at 490/440 nm, as described previously (Eisner *et al.*, 1989). The apparent pK for BCECF was calculated (non-linear regression) from the nigericin calibration to be 6.97 ± 0.05 (mean ± SEM, n = 38), similar to previous reports.

12.22.6.3 Simultaneous measurements of pH$_i$ and [Ca^{2+}]$_i$

Simultaneous changes in [Ca^{2+}]$_i$ and pH$_i$ can be measured by loading both fura-2 and BCECF in the same cell (Simpson & Rink, 1987; Zavoico & Cragoe, 1988; Miyata *et al.*, 1989; Tsunoda *et al.*, 1991), providing that any cross-talk between the dyes is taken into account (Zavoico & Cragoe, 1988). We mounted four excitation filters in the rotating filter wheel, to excite BCECF and fura-2, where the emission (barrier) filters were as for BCECF measurements. Calibrations for both dyes were performed on cells loaded simultaneously with fura-2 and BCECF. We found that equivalent loading concentrations of fura-2 and BCECF resulted in the BCECF fluorescence emission being considerably brighter (2- to 3-fold) than that of fura-2, in agreement with previous reports (Zavoico & Cragoe, 1988; Miyata *et al.*, 1989). To enable simultaneous four-wavelength measurements with a single output determination, fura-2 concentration was generally increased to about four-fold more than that for BCECF. In addition, fluorescence excitation intensity was balanced by addition of quartz neutral-density filters (0.1–0.5) into 440 and 490 nm excitation pathways. Relative emission intensities of fura-2 were kept smaller than those of BCECF.

12.22.6.4 Recording and analysis

Emission signals were collected with a video frame rate intensified charge-coupled device (CCD) camera (Photonic Sciences, Robertsbridge, UK), digitized at 8 bit resolution, with 4–36 frames being averaged at a single wavelength to reduce noise arising from the low-light-level camera. This gave a ratio image about once every 2–3 s. The ratio image was calculated with a look-up table using parameters obtained from the aforementioned calibration approaches for BCECF, fura-2 or BCECF/fura-2 measurements. Individual ratio frames were produced using pixel-by-pixel division of either paired 340/380 frames or 490/440 frames, using average images captured close in time for the calculation of pH$_i$ and [Ca^{2+}]$_i$. Continuous traces of average pH$_i$ and [Ca^{2+}]$_i$ were obtained by defining a region around a cell image using the light pen, thresholding at an intensity value equivalent to approximately pH = 5.8 or 10 nM [Ca^{2+}]$_i$, to exclude background noise. These values were chosen after inspection of the pixel grey level amplitude histogram.

12.22.6.5   Are simultaneous measurements of pH_i and $[Ca^{2+}]_i$ prone to artifacts?

Optical cross-talk between dyes (Zavoico & Crageo, 1988) and pH-sensitivity of the fura-2 apparent K_d (Negulescu & Machen, 1990) represent the two main potential artifacts in work of this type. Some optical spillover was expected because, although we used narrow-band excitation filters, the absorption spectrum of fura-2 is broad and potentially overlaps with the BCECF absorption. BCECF also absorbs at 340 nm excitation. At the 440 nm excitation wavelength of BCECF, cross-excitation of fura-2 might be expected to give rise typically to about 5–10% of the fluorescence intensity compared to that emitted due to 380 nm excitation (Haugland, 1989), and at 340 nm excitation BCECF absorbs about 5–8% relative to the absorption at 500 nm excitation (Zavoico & Cragoe, 1988). In order to reduce cross-talk between the two dyes, relative amounts of both dyes can be varied. By lowering fura-2 fluorescence, the contribution to the BCECF signal is reduced and contamination in the estimation of pH_i by variations of $[Ca^{2+}]_i$ is decreased. The contamination of emission signal at 440 nm excitation, due to fura-2, would result in a logarithmically proportional change in the pH_i estimation.

The contribution of fura-2 to the BCECF signal at 440 nm excitation was measured as a function of the amount of fura-2 cells previously loaded for a standard 30 min with BCECF. After the cells were loaded with BCECF-AM, they were exposed to further loading with fura-2-AM for intervals of 10 min to yield various final loading concentrations. In between, the relative fluorescence intensities were measured as an average of 50 frames. Sequential loading of fura-2 resulted in an increased amount of fura-2 in cells, whereas the amount of BCECF was unchanged (judged from the 490 nm excitation signal; Fig. 12.23). A 4- to 5-fold increase in emission fluorescence excited at 380 nm resulted in a 15% increase in fluorescence intensity recorded at 440 nm excitation, whereas 490 nm there was no apparent effect of fura-2 loading (Fig. 12.23(A)). Estimated pH_i (Fig. 12.23(B)) determined from the 490/440 nm ratio fluorescence was therefore declining as a function of sequential loading of fura-2. At the end of the experiment when dye loading was 10 min longer, as used in all other experiments here, the net effect was that estimated resting pH_i decreased by about 4% (change in pH of 0.3). Thus a systematic error in pH estimation is expected due to relatively high amounts of fura-2 in the cytosol. In agreement

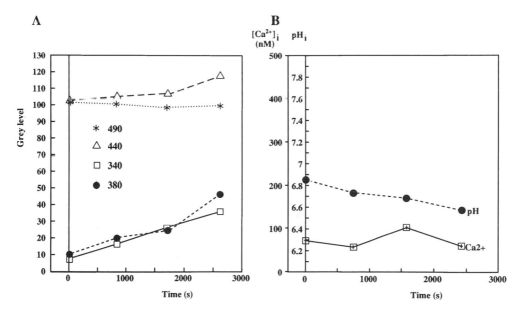

Figure 12.23 The effect of sequential loading of fura-2-AM on BCECF-AM-loaded bovine lactotrophs. Cells were loaded with 1 μM BCECF-AM for 30 min at 37°C in the bathing medium containing 5 mM CaCl₂, then they were exposed sequentially to loading with 4 μM fura-2-AM for 10 min. Signals were acquired (50 frames) between sessions of loading. Relative fluorescence intensities excited at 490 nm (*), 440 nm (open triangle), 340 nm (open square) and 380 nm (filled circle) were measured at 520 nm and displayed on an arbitrary scale (A). Note that relative intensities at 440 nm increased with the relative increase in fluorescence intensities at 340 and 380 nm excitation. (B) pH_i (filled circles) and $[Ca^{2+}]_i$ (open squares) levels estimated from the experiment in A. Note that there is no correlation between the two measurements. However, the pH_i is declining which is correlated to the relative increase in the fluorescence intensities of fura-2-AM, excited at 340 and 380 nm (see panel A).

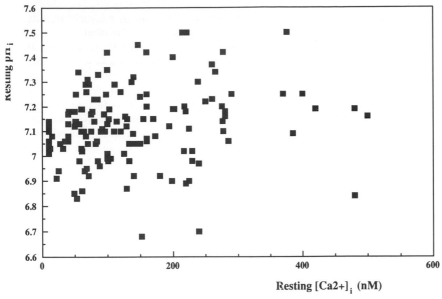

Figure 12.24 Relationship between resting pH_i and resting $[Ca^{2+}]_i$ in single bovine lactotrophs. Correlation coefficient of 0.18 is not significantly different from zero ($P<0.1$). Average resting pH_i was 7.12 $\pm$ 0.15 and average $[Ca^{2+}]_i$ was 136 $\pm$ 105 nM ($n = 145$, mean $\pm$ standard deviation).

with this, if cells were loaded with an excess of fura-2 and fluoresced highly at 340 nm (79 $\pm$ 9 a.u. (arbitrary units)) and 380 nm (76 $\pm$ 3 a.u.) in comparison to fluorescence at 440 nm (24 $\pm$ 1 a.u.) and 490 nm (12 $\pm$ 1 a.u.), estimated resting pH_i was 6.54 $\pm$ 0.06, whereas resting $[Ca^{2+}]_i$ was 112 $\pm$ 12 nM (mean $\pm$ SEM, $n = 6$; extracellular $CaCl_2$ 5 mM). Thus, relative fura-2 emission fluorescence levels were kept smaller than those of BCECF. The former were typically a third to a half of those recorded at 440 and 490 nm excitation (see Fig. 12.23). The systematic error was not significantly different from the variability of pH measurements in the absence of fura-2 (see Eisner *et al.*, 1989; and Fig. 12.24). Therefore, measurements of pH_i and $[Ca^{2+}]_i$ from cells simultaneously loaded with BCECF and fura-2 are feasible, especially as changes in $[Ca^{2+}]_i$ were often more than three orders of magnitude and changes in pH more than 1 unit. Changes in $[Ca^{2+}]_i$ were not significantly different from those recorded in the absence of BCECF (compare results of Akerman *et al.*, 1991).

Another possible source of artifact is the sensitivity to pH of the fura-2/Ca^{2+} apparent dissociation constant (K_d). From measurements of Negulescu and Machen (1990) a 2- to 3-fold increase in K_d is expected upon acidification (from resting to pH around 6), but K_d is insensitive to alkalinization (from resting to pH around 7.5). Therefore, $[Ca^{2+}]_i$ should be underestimated at acidic pH.

12.22.6.6   Resting pH_i and $[Ca^{2+}]_i$

Measurements with only BCECF loaded into cells revealed resting pH_i to be 7.14 $\pm$ 0.10 ($n = 12$), in agreement with previous reports on pituitary (Shorte *et al.*, 1991) and other cell types (Frelin *et al.*, 1988).

With only fura-2 loading, resting $[Ca^{2+}]_i$ was around 100 nM (Akerman *et al.*, 1991). Similar results were obtained when both dyes were in the cytosol (Fig. 12.24). Under these conditions, resting pH_i was 7.12 $\pm$ 0.15, whereas resting $[Ca^{2+}]_i$ was 136 $\pm$ 105 nM ($n = 145$). Comparison between the pH_i and $[Ca^{2+}]_i$ data (Fig. 12.24) revealed no significant correlation ($P>0.1$).

12.22.6.7   Alkalinization induces release of Ca^{2+} from agonist-sensitive cytosolic stores

Thyrotropin releasing hormone (TRH), a physiological regulator of prolactin secretion, was applied to cells to evoke a transient release of calcium from intracellular stores (Akerman *et al.*, 1991), and the weak base NH_4Cl to evoke alkaline pH changes (Frelin *et al.*, 1988). Repetitive applications of TRH to lactotrophs can cause repetitive release of calcium from an intracellular pool, but only if 3–10 min recovery time is allowed between applications. The effects of TRH and NH_4Cl applications on $[Ca^{2+}]_i$ and pH_i are shown in Fig. 12.25. TRH (100 nM) application induced a rapid rise in $[Ca^{2+}]_i$, peaking at around 1000 nM.

A concomitant but slight increase in pH of about 0.05 units in some cells (Fig. 12.25(A)) was not reliably recorded in all experiments. If this alkalinization was due to an artifact, then such a pH response should not be recorded in the absence of fura-2. This was confirmed in separate experiments where cells were loaded with BCECF only and the application of TRH failed to produce a detectable pH rise in any of 22 cells examined, consistent with previous reports (Törnquist & Tashjian, 1991). Therefore, the small apparent alkalinization sometimes measured during

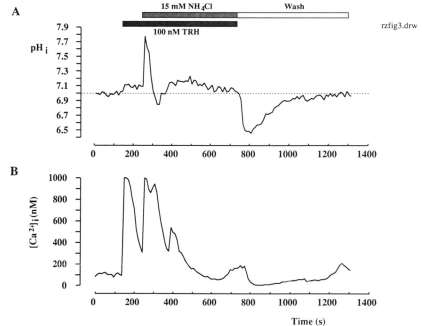

Figure 12.25 The effect of TRH (100 nM) and NH$_4$Cl (15 mM) application on cytosolic pH (A) and calcium activity (B) in a single bovine lactotroph loaded with fura-2-AM (4 μM) and BCECF-AM (1 μM). Both agents have been applied as boluses and were washed from the chamber by the superfusion. Small increase in pH$_i$ during the TRH application is most likely an artifact, due to the optical cross-talk between the dyes. Bathing medium contained 5 mM CaCl$_2$.

TRH application is most likely due to optical cross-talk, although changes smaller than 0.05 units, measured after the application of TRH (Hallam & Tashjian, 1987) could be detected with this approach in a single cell.

In some experiments, immediately after the TRH application, NH$_4$Cl (15 mM) was applied which resulted in a rapid alkalinization of about 0.5–0.7 units. This was followed by a rapid decline in pH towards resting levels. In 38 out of 64 cells the recovery from alkalinization was biphasic (Fig. 12.25(A)), whereas in others, pH recovered monotonically towards resting pH$_i$ (see Fig. 12.28(B)). After the first transient rise of [Ca^{2+}]$_i$ due to alkalinization (Fig. 12.25(A)), a short period of acidification followed. This may be due to the competition of Ca^{2+} and H$^+$ for common binding sites (Ives & Daniel, 1987). More likely, this may also be due to the bolus application of NH$_4$Cl into the bathing solution and the subsequent equilibration of NH$_4$Cl concentration.

Upon the removal of NH$_4$Cl from the bathing medium, the cytosol became acidic by about 0.3–0.7 units, followed by recovery towards resting pH$_i$ (Fig. 12.25(A)). A similar response in pH and [Ca^{2+}]$_i$ to an application of NH$_4$Cl was recorded, if TRH application was omitted from the experimental protocol (Fig. 12.27).

Alkalinization induced a fast, transient increase in [Ca^{2+}]$_i$, peaking at values typically around 600–1800 nM (1006 ± 467, $n = 31$), similar to those induced by TRH. The recovery towards resting [Ca^{2+}]$_i$ pro-

ceeded with a similar rate to the TRH-induced Ca^{2+} rise, but was not usually correlated with the change in pH, and was not affected by the presence of TRH, since a similar time-course was recorded in the absence of TRH (Fig. 12.27). Moreover, there was no correlation between the resting pH$_i$, measured prior to the application of NH$_4$Cl and the transient rise in [Ca^{2+}]$_i$.

Figure 12.25 shows an example of multiple [Ca^{2+}]$_i$ transients following alkalinization. Such a response was recorded in 23% of cells, whereas in 49% of cells alkalinization produced only one transient. In the latter group two cells responded with a transient [Ca^{2+}]$_i$ rise which was delayed from onset of alkalinization by 100 and 240 s. In the remaining cells, [Ca^{2+}]$_i$ increased by only about 2 to 3 times the resting value (such as that seen at the end of the NH$_4$Cl pulse in Fig. 12.25(B), see also Fig. 12.28; total number of cells = 42; data with TRH and without pooled together). Similar responses in [Ca^{2+}]$_i$ and pH$_i$ to NH$_4$Cl applications were recorded in cells loaded only by fura-2 or BCECF respectively.

In Fig. 12.26, the extracellular Ca^{2+} was removed before the application of TRH and NH$_4$Cl, to test whether the transient rise in Ca^{2+} is from intracellular stores. In the absence of extracellular calcium [Ca^{2+}]$_i$ declined to a lower resting value (Fig. 12.26(A) and (B)) consistent with a role of a tonic influx of Ca^{2+} determining [Ca^{2+}]$_i$ at rest (Akerman et al., 1991; Zorec et al., 1991). During the application of NH$_4$Cl the transient rise in [Ca^{2+}]$_i$ persisted (Fig. 12.26(A)),

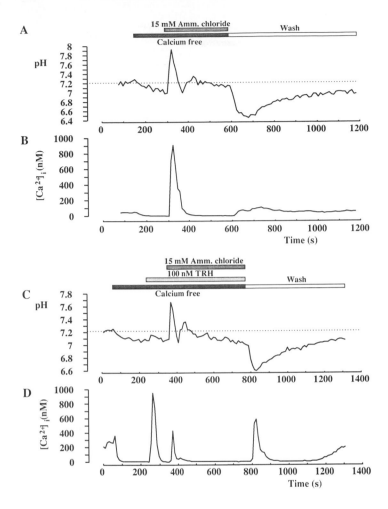

Figure 12.26 (A) The effect of NH$_4$Cl on [Ca^{2+}]$_i$ in the presence of low extracellular Ca^{2+}. Calcium-free medium was prepared by omission of CaCl$_2$ in the bathing medium. NH$_4$Cl was applied as in Fig. 12.25, and it induced similar changes in pH$_i$ as in Fig. 12.25. Cell was loaded with 4 µM fura-2-AM and 1 µM BCECF-AM. Bathing medium contained 5 mM CaCl$_2$. Panel B shows the effect of the application of TRH and NH$_4$Cl on [Ca^{2+}]$_i$ in the absence of extracellular calcium. Note that the increase in [Ca^{2+}]$_i$ is smaller than evoked by the application of TRH. Cell was loaded with 2 µM BCECF-AM and 4 µM fura-2-AM; bathing medium contained 5 mM CaCl$_2$.

which indicates that the source of Ca^{2+} is mainly from intracellular stores. The average increase in [Ca^{2+}]$_i$ in this experiment was 1555 ± 568 nM ($n = 10$). Note that the peak value of the [Ca^{2+}]$_i$ transient is as high as that evoked in the presence of extracellular calcium (Fig. 12.25). When TRH application preceded NH$_4$Cl-induced alkalinization in the absence of extracellular calcium, the amplitude of the transient rise in [Ca^{2+}]$_i$ due to alkalinization was diminished. The application of TRH elicited an increase in [Ca^{2+}]$_i$ of 772 ± 309 nM, whereas the following application of NH$_4$Cl increased [Ca^{2+}]$_i$ to 377 ± 302 ($n = 20$). The addition of NH$_4$Cl in Fig. 12.26 induced a similar alkalinization as in Figure 12.25.

These results suggest that alkalinization induces release of calcium, at least in part, from agonist-sensitive intracellular stores (Danthuluri *et al.*, 1990). This is further supported by experiments where TRH was applied after NH$_4$Cl (Fig. 12.27). In this experiment only 5 out of 21 cells responded with a transient increase in [Ca^{2+}]$_i$ (550–1000 nM). In 30% of cells

there was a sustained decrease in [Ca^{2+}]$_i$, in the rest there was a small delayed transient rise in [Ca^{2+}]$_i$ of about 50–100 nM (Fig. 12.27), or there was no change. A representative response to TRH application during the NH$_4$Cl alkalinization is shown in Fig. 12.26. This contrasts with the responsiveness of lactotroph cells exposed to TRH; more than 90% of cells are responsive (Akerman *et al.*, 1991). These results are consistent with the view that NH$_4$Cl application depletes calcium from agonist-sensitive calcium stores.

12.22.6.8   *Correlation between resting [Ca^{2+}]$_i$ and pH$_i$*

To observe a fast increase in [Ca^{2+}]$_i$, a bolus application of NH$_4$Cl was required. If NH$_4$Cl was applied slowly by superfusion, the fast increase in [Ca^{2+}]$_i$ was difficult to observe, or absent. This suggests that the rate and/or amplitude of alkalinization determines the response of [Ca^{2+}]$_i$ (Fig. 12.28(A) and (B)). In Fig. 12.29 cells were superfused by progressively increasing concentrations of NH$_4$Cl (from 0.5 to 15 mM). This

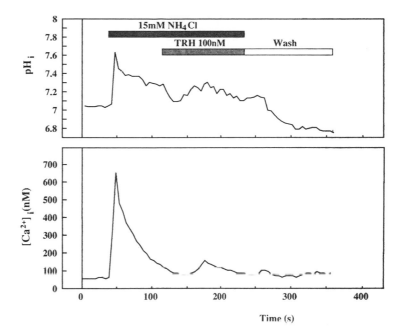

Figure 12.27 The effect of NH₄Cl (15 mM) application followed by the application of TRH (100 nM) on pHᵢ and [Ca²⁺]ᵢ. The methods of drug application are as in Fig. 12.25. Note that the response in [Ca²⁺]ᵢ to TRH application is delayed, whereas the small TRH-induced acidification seems to be not. Cells loaded as in Fig. 12.25. Bathing medium contained 1.8 mM CaCl₂.

resulted in a gradual increase in pH_i as well as in $[Ca^{2+}]_i$, but transient increases in $[Ca^{2+}]_i$ were not observed in 11 out of 12 cells. The relationship between alkalinization and $[Ca^{2+}]_i$ is shown in Fig. 12.29, where a small increase in $[Ca^{2+}]_i$ may be due to increased influx of Ca^{2+} through channels (Shorte *et al.*, 1991), the permeability of which may be affected by pH_i (Prod'hom *et al.*, 1987). NH₄Cl-evoked depolarization (Kaila & Voipio, 1990) may also contribute to the increase in $[Ca^{2+}]_i$.

12.22.6.9 *Acidification and transient increase in [Ca²⁺]ᵢ*

Upon the removal of NH₄Cl intracellular acidification induced a transient increase in $[Ca^{2+}]_i$ with a temporal profile similar to that induced by TRH or intracellular alkalinization but in many cells there was no change in $[Ca^{2+}]_i$ (see Figure 12.25). The increase in $[Ca^{2+}]_i$ is very likely due to release from intracellular stores, since the $[Ca^{2+}]_i$ transient was observed in the absence of extracellular calcium. Representative recordings on two cells are shown (Fig. 12.28(A)) to highlight the difference in the onset of the transient rise in $[Ca^{2+}]_i$.

If nigericin (10 μM) was applied to cells in normal bathing medium, an acidification was imposed due to the induced exchange of K^+ for external H^+ (Simpson & Rink, 1987), resulting in a transient increase in $[Ca^{2+}]_i$ (Fig. 12.28(B)). The amplitude and time-course were similar to the transient rise in $[Ca^{2+}]_i$ induced by removal of NH₄Cl in the absence of extracellular calcium (Fig. 12.28(A)). These results

imply that in some cells acidification induces a transient release of calcium from intracellular stores.

Estimated buffering capacity for H^+ in the experiment of Fig. 12.28 was 19 ± 5 mM (pH unit $\times$ 1)$^{-1}$, $n = 4$ [min. 11 and max. 29 mM (pH unit $\times$ 1)$^{-1}$], which compares favourably with measurements on other cell types (Szatkowsky & Thomas, 1989).

12.22.6.10 *Feasibility of simultaneous measurements of cytosolic pH and [Ca²⁺]ᵢ*

The aim of this work (Case Study 6) was to validate the use of simultaneous fluorescence measurements to cells loaded both with fura-2 and BCECF (Simpson & Rink, 1987) to study interactions between pH_i and $[Ca^{2+}]_i$. Despite a slight cross-talk between the two optical probes, meaningful results on the relationship between pH_i and Ca^{2+} homeostasis were obtained. The main source of artifact is the limited spectral separation between the excitation protocols for the two dyes. If the relative fluorescence of fura-2 was kept smaller than that of BCECF, the cross-talk between the two measurements was low, and the systematic error in the pH estimation was smaller than the variability of the pH measurements with only BCECF in the cytosol (see Eisner *et al.*, 1989). Zavoico and Cragoe (1988) have shown that *in vitro* conditions (in a cuvette) BCECF amounts should be kept smaller than fura-2 to avoid contribution of BCECF at 340 nm excitation to the emission signal at 500 nm due to the small absorption of BCECF at 340 nm excitation. A similar effect, although much smaller (Miyata *et al.*,

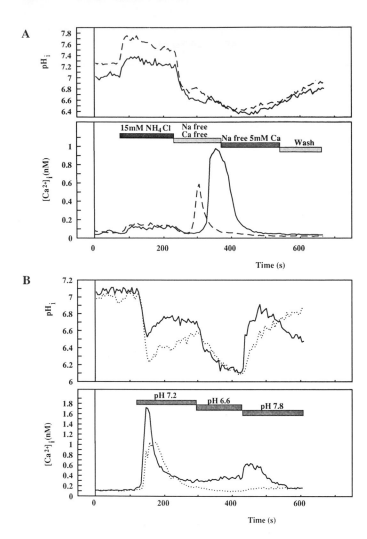

Figure 12.28 (A) The effect of NH$_4$Cl (15 mM) applied by the superfusion system and the subsequent effect of the washout by Na$^+$- and Ca^{2+}-free medium on [Ca^{2+}]$_i$ and pH$_i$. These media were prepared by replacing NaCl by choline chloride with no added CaCl$_2$. Note the absence of a fast rise in [Ca^{2+}]$_i$ following the exchange of solutions containing 15 mM NH$_4$Cl. Recordings from two cells are shown to highlight the difference in the onset in the acid-evoked increase in [Ca^{2+}]$_i$. These are not due to the delays in the exchange of solutions, but rather to variability in cell responsiveness. (B) The effect of nigericin (5 µM) on [Ca^{2+}]$_i$ and pH$_i$ superfused into the chamber at various pH levels. Bathing medium as in Fig. 12.25, but contained 1.8 mM CaCl$_2$. It was adjusted to stated pH levels by the use of HCl or NaOH prior to the experiment. Note that levels of [Ca^{2+}]$_i$ indicated by the ordinates should be multiplied by 1000.

1989), can be expected in our *in situ* conditions (in a single cell), where the ratio fura-2/BCECF was 0.3–0.5. Fluctuations in pH (0.5–0.7) are expected to minimally affect the contribution of BCECF to the emission signal due to the absorption of BCECF at 340 nm. For instance, it can be estimated that fluorescence emission of BCECF excited at 340 nm would change by about 0.5% with an alkalinization from 6.5 to 7.0 pH (assuming linear relationship between absorption of BCECF at 340 nm and pH). Thus BCECF signal contamination due to absorption at 340 nm light can be assumed to be constant, and the estimation of [Ca^{2+}]$_i$ is possible, since the calibration was carried out in the presence of both dyes in the cytosol.

Close inspection of figure 1B of Zavoico and Cragoe (1988) reveals that the mixture of fura-2 (200 nM) and BCECF (20 nM), when excited at 340 nm, fluoresces more at low Ca^{2+} than in the presence of high Ca^{2+}.

Considering that this mixture should behave approximately as a plain fura-2 solution, the aforementioned finding is unexpected (see Grynkiewicz *et al.*, 1985). An improvement in the spectral separation between the calcium and pH determination protocols can be expected with the new generation of dyes, such as fura Red in combination with BCECF, although this will require more complex optical systems.

12.23 SUMMARY

The development of optical probes for biological activity combined with powerful image-acquisition and analysis technology is having a major effect on the study of physiological parameters of living cells. The best of these probes appear not to interfere with normal cellular processes, although they do require

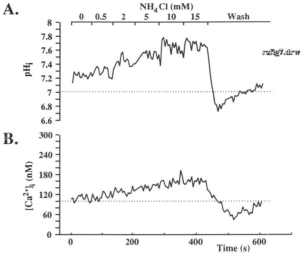

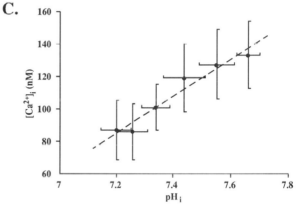

Figure 12.29 The effect of progressive increase in NH$_4$Cl (0.5–15 mM) on pH$_i$ (A) and [Ca^{2+}]$_i$, obtained by the superfusion system. Traces represent a typical response in a cell. (C) Relationship between measurements of average [Ca^{2+}]$_i$ and peak changes in pH$_i$ (in six cells) during superfusion of NH$_4$Cl-containing solutions (as above). Cells were loaded with 4 μM fura-2-AM and 2 μM BCECF-AM. Bathing solution contained 1.8 mM CaCl$_2$. Line drawn through the points is of the form $y = 112.3x - 723.4$, $n = 6$; correlation coefficient is 0.97.

caution in interpretation of data as the probes may localize within cells, rendering them insensitive or modifying their properties.

New probes for ions, cyclic nucleotides, cellular enzymes and genetic material are under development and will further extend this capability. Photometric and imaging technology allow the application of these probes to yield both temporal and spatial data. The scientist is now presented with a variety of means for imaging these low-light-level probes, including use of a variety of different detectors and imaging systems. Video frame rate systems offer high spatial and temporal resolution generally at a high cost; often the high acquisition speeds may not be warranted by the biological characteristics of the system. Lower cost systems like the MiraCal system used in our laboratory provide a reasonable speed of acquisition well-suited to most biological events, and at a low cost. Confocal imaging technology provides somehat slower image acquisition, but with improved spatial resolution and less out-of-focus information. New techniques for

image deconvolution will permit resolution enhancement for both confocal and video frame rate systems, but at a cost of both time and money, given the extensive computational power required.

New and even more powerful image-processing systems will be developed, and cost/benefit will probably improve in coming years. Prototype imaging systems with industrial cameras and direct computer memory access can now produce up to about 200 images per second at high resolution, permitting fast cellular processes to be revealed. In addition, novel probes such as for genetic material and more rapid observational and analytical capabilities will probably emerge in the coming decade.

Technologies for fluorescence imaging will become more accessible to a wider number of laboratories. With this, the power of quantitative image-processing and optical probe technology will be increasingly realized by scientists in academia, industry and government, and the authorities who provide funding for these establishments.

ACKNOWLEDGEMENTS

We thank the Agricultural and Food Research Council, Medical Research Council, Applied Imaging Ltd, Kabi Pharmacia, Wellcome Trust, Nuffield Foundation, Guggenheim Foundation and British Heart Foundation for valuable funding which has supported various aspects of the work discussed here.

SELECTED BIBLIOGRAPHY

A comprehensive bibliography of the intracellular ion measurement literature would be too extensive to detail here, but the references below will provide some worthwhile reading. In addition, the February/March 1990 issue of *Cell Calcium* (Volume 11) provides a very comprehensive recent treatment of additional methods and applications for work in this area.

Akerman S.N., Zorec R., Cheek T.R., Moreton R.B., Berridge M.J. & Mason W.T. (1991) *Endocrinology* **128**, 475–488.

Ben-Jonathan N. (1985) *Endocrine Rev.* **6**, 564–589.

Brakenhoff G.J., van Spronsen E.A., van de Voort H.T.M. & Nanninga N. (1989) *Methods Cell Biol.* **30**, 379–398.

Cheek T.R., Jackson T.R., O'Sullivan A.J., Moreton R.B., Berridge M.J. & Burgoyne R.D. (1989) *Cell Biol.* **109**, 1219–1227.

Cobbold P.H. & Rink T.J. (1987) *Biochem. J.* **248**, 313–328.

Danthuluri N.R., Kim D. & Brock T.A. (1990) *J. Biol. Chem.* **265**, 19071–19076.

Eisner D.A., Nichols C.G., O'Neill S.C., Smith G.L. & Valdeolmillos M. (1989) *J. Physiol.* **411**, 393–418.

Fay F.S., Carrington W. & Fogarty K.E. (1989) *J. Microsc.* **153**, 133–149.

Frelin C., Vigne P., Ladoux A. & Lazdunski M. (1988) *Eur. J. Biochem.* **174**, 3–14.

Grinstein S. & Rothstein A. (1986) *J. Membr. Biol.* **90**, 1–12.

Grinstein S., Rotin D. & Mason M.J. (1989) *Biochim. Biophys. Acta* **988**, 73–97.

Grynkiewicz G., Poenie M. & Tsien R.Y. (1985) *J. Biol. Chem.* **260**, 3440–3450.

Hallam T.J. & Tashjian Jr. A.H. (1987) *Biochem. J.* **242**, 411–416.

Haugland R.P. (1989) *Handbook of Fluorescent Probes and Research Chemicals.* Molecular Probes Inc., Eugene, OR, USA.

Hertelendy F., Todd, H., Peake G.T., Machlin J., Johnston G. & Pouns G. (1971) *Endocrinology* **89**, 1256–1262.

Ingram C.D., Keefe P.D. & Wooding P.B. (1988) *Cell Tissue Res.* **252**, 655–659.

Inoue S. (1986) *Video Microscopy.* Plenum Press, New York & London, 584 pp.

Irvine R.F. (1990) *FEBS Lett.* **263**, 5–9.

Ives E. & Daniel T.O. (1987) *Proc. Natl. Acad. Sci. USA* **84**, 1950–1954.

Kaila K. & Voipio J. (1990) *Pflügers Arch.* **416**, 501–511.

Kato M., Hoyland J., Sikdar S.K. & Mason W.T. (1992) *J. Physiol.* **447**, 171–189.

Lichtman, W., Sunderland S.J. & Wilkinson R.S. (1989) *The New Biology* **1**, 75–82.

Malgaroli A., Vallar L., Elahi F.R., Pozzan T., Spada A. & Meldolesi J. (1987) *J. Biol. Chem.* **262**, 13920–13927.

Mason W.T., Rawlings S.R., Cobbett P., Sikdar S.K., Zorec R., Akerman S.N., Benham C.D., Berridge M.J., Cheek T. & Moreton R.B. (1988) *J. Exp. Biol.* **139**, 287–316.

Mason W.T., Sikdar S.K., Zorec R., Akerman S., Rawlings S.R., Cheek T., Moreton R. & Berridge M. (1989) In *Secretion and Its Control*, M. Conn (ed.). Rockefeller University Press, New York, pp. 225–238.

Mason W.T., Hoyland J., Rawlings S.R. & Relf G.T. (1990) *Methods Neurosci.* **3**, 109–135.

Miyata H., Hayashi H., Suzuki S., Noda N., Kobayashi A., Fujiwake H., Hirano M. & Yamazaki N. (1989) *Biochem. Biophys. Res. Commun.* **163**, 500–505.

Monck J.R., Reynolds E.E., Thomas A.P. & Williamson J.R. (1988) *J. Biol. Chem.* **263**, 4569–4575.

Monck J.R., Oberhauser A.F., Keating T.J. & Fernandez J.M. (1992) *J. Cell Biol.* **116**, 745–759.

Negulescu P.A. & Machen T.E. (1990) *J. Membr. Biol.* **116**, 239–248.

Neylon C.B., Hoyland J., Mason W.T. & Irvine R. (1990) *Am. J. Physiol.* **259**, C675–C68.

O'Sullivan A.J. & Burgoyne R.D. (1988) *Biochim. Biophys. Acta* **969**, 211–216.

O'Sullivan A.J., Cheek T.R., Moreton R.B., Berridge M.J. & Burgoyne R.D. (1989) *EMBO J.* **8**, 401–411.

Prod'hom B., Pietrobon D. & Hess P. (1987) *Nature* **329**, 243–246.

Putney J. (1990) *Cell Calcium* **11**, 611–624.

Rink T.J., Tsien R.Y. & Pozzan T. (1982) *J. Cell. Biol.* **95**, 189–196.

Rizzuto R., Simpson A.W.M., Brini M. & Pozzan T. (1992) *Nature* **358**, 325–327.

Schlegel W., Winiger B.P., Mollard P., Vacher P., Wuarin F., Zahnd G.R., Wollheim C.B. & Dufy B. (1987) *Nature* **329**, 719–721.

Shorte S.L., Collingridge G.L., Randall A.D., Chappell J.B. & Schofield J.G. (1991) *Cell Calcium* **12**, 301–312.

Shotton D.M. (1989) *J. Cell Sci.* **94**, 175–206.

Sikdar S.K., Zorec R., Brown D. & Mason W.T. (1989) *FEBS Lett.* **253**, 88–92.

Sikdar S.K., Zorec R. & Mason W.T. (1990) *FEBS Lett.* **273**, 150–154.

Simpson A.W.M. & Rink T.J. (1987) *FEBS Lett.* **222**, 144–148.

Szatkowsky M.S. & Thomas R.C. (1989) *J. Physiol.* **409**, 89–101.

Taylor D.L. & Wang Y.L. (eds) (1989) *Methods Cell Biol.* **29 & 30.**

Törnquist K. & Tashjian A.H. (1991) *Endocrinology* **128**, 242–250.

Tsunoda Y., Matsuno K. & Tashiro Y. (1991) *Exp. Cell Res.* **193**, 356–363.

Williams D.A. & Fay F.S. (1990) *Cell Calcium* **11**, 75–83.

Winiger B.P. & Schlegel W. (1988) *Biochem. J.* **255**, 161–167.

Winiger B.P., Wuarin, F., Zahnd G.R., Wollheim C.B. & Schlegel W. (1987) *Endocrinology* **121**, 2222–2228.

Wright S.J., Walker J.S., Schatten H., Simerly C., McCarthy J.J. & Schatten G. (1989) *J. Cell Sci.* **94**, 617–624.

Zavoico G.B. & Cragoe E.J. Jr. (1988) *J. Biol. Chem.* **263**, 9635–9639.

Zorec R., Tester M., Macek P. & Mason W.T. (1990) *J. Membr. Biol.* **118**, 243–249.

Zorec R., Sikdar, S.K. & Mason W.T. (1991) *J. Gen. Physiol.* **97**, 473–497.

Zorec R., Hoyland J., Relf G., Bunting R. & Mason W.T. (1993) *J. Physiol.* (in press).

CHAPTER THIRTEEN

A Small-Area Cooled CCD Camera and Software for Fluorescence Ratio Imaging

G.J. LAW & W. O'BRIEN

Department of Neurobiology, AFRC Institute of Animal Physiology and Genetics Research, Babraham, Cambridge, UK

13.1 INTRODUCTION

Fluorescence is a powerful tool to probe biological structure and function. Detection of a single molecule tagged with a fluorophore is testimony to the extreme sensitivity of fluorescence and photon-counting techniques (Hirschfield, 1977). Rapid advances in the number of fluorescent probes and in low-light-level imaging has generated much interest in the field of quantitative fluorescence microscopy (Tsien, 1989; Mason *et al.*, 1990). Dynamic ratio imaging of fluorescent probes inside living cells has required the use of image intensifiers and large amounts of fast-access computer memory to capture a series of images at video rate. Image intensifiers are excellent devices to detect low light levels at video rate but they can suffer from electronic noise and substantial frame-averaging is common for quantitative analysis. A typical elapsed time might be close to 1 s per image. The purpose of this chapter is to demonstrate that integration of a similar low-light-level source directly on a cooled charged-coupled device (CCD) before rapid transfer of this image to a hard disk on a PC-compatible computer which supports super video graphics adapter (sVGA) graphics has produced comparable data at a substantial financial saving. As each image is only 32 kbyte even a small 40 Mbyte disk allows a sequence of approximately 1200 frames to be captured. Conversion of your microscope into a tool for quantitative analysis of a single cell fluorescence may now cost much less than you think and our proven approach offers a practical guide in this book.

13.2 CHOOSING THE RIGHT EQUIPMENT

This book along with volumes 29 and 30 of *Methods in Cell Biology* entitled *Fluorescence Microscopy of Living Cells in Culture*, provides much background information on this subject. What kind of image-processing and analysis system is best, if at all, will depend on your application. I say, 'if it all', because certain applications with optical probes might be best done with a photomultiplier tube (PMT), especially if your data acquisition needs to be much faster than video-rate (see Chapter 22 for fast photometric applications), or depends only on the average cell response where a PMT-based spectrofluorometer is more than adequate. Immunofluorescence has been the most common application of fluorescence in cell biology to detect cellular antigens. A 35 mm camera and film may suffice to record the result you want here. Traditionally, autoradiography has used X-ray film but reusable imaging-plate technology offers many benefits, as discussed in Johnston *et al.* (1990). Digital-imaging cameras can act as densitometers to quantitate image information from 2D gels, blots and microtitre dishes (Urwin & Jackson, 1991); and are certainly not restricted to microscope use.

A main aim of this chapter is to discuss the practical benefits of working with a small-area imaging chip and on-chip integration. The performance of the camera and the features of the software to control it will be described with experimental data. As this book represents a practical guide to the use of optical probes in the biological sciences we have tried to present our chapter in this context.

13.3 OPTICAL PROBES

In an ideal situation the choice of optical probe would be dictated by the need of the scientist but in practice the range of probes available limits what can be done (see Haugland (1989) which contains a reference list of some 5000 titled entries).

Small changes in absorption of a large input signal are difficult to measure accurately and in general this kind of optical probe is less sensitive than one which emits photons over a low background signal (but see Chapter 30). For this reason we shall briefly discuss only the more sensitive optical probes which generate luminescence and can be subdivided into two major types: (1) chemiluminescence (or bioluminescence); and (2) fluorescence.

Chemiluminescence is different from fluorescence because a chemical–oxidation reaction is required to trigger production of a photon from a chemiluminescent substrate, whereas absorption of energy from incident photons is necessary for fluorescence. One advantage of chemiluminescent probes then is that they can operate under conditions of an extremely low background signal because they do not suffer from problems of scattered light, or autofluorescence, caused by excitation of the biological specimen. In contrast, however, a chemiluminescent molecule can only generate one photon flash before becoming extinct, whereas a good fluorophore may generate as many as 10^6 photons over a long (or short) period of time. Ultrasensitive detection down to 10^{-20} mol, however, has been reported with a chemiluminescent substrate (phenylphospate-substituted dioxetane) for alkaline phosphate, which takes advantage of a powerful enzyme-driven amplification of photon flux in the presence of excess susbtrate (Beck & Koster, 1990). For comparison, the detection limit is about 10^{-14} mol for conventional use of a fluorescent dye

which has a high quantum yield like that of fluorescein (Beck & Koster, 1990). Cellular autofluorescence is clearly visible and can limit probe detection but is much less of a problem at wavelengths >600 nm (Aubin, 1979; Benson *et al.*, 1979).

Numerous approaches can be implemented to reduce the interference of background signal to achieve better sensitivity with fluorescence. One approach is to excite and collect light from a reduced sample volume; and this is what is effectively done with a high numerical aperture lens on a microscope which operates on a larger fraction of the total emission than a traditional cuvette-based system. Confocal microscopy can reduce the reaction volume to about 1 μm^3 and minimize the amount of background interference. Other improvements in sensitivity are possible with time-resolved fluorescence measurements using long-lived fluorophores. Here the excitation source is pulsed and the instantaneous scattered light and short-lived autofluorescence is allowed to decay to zero before measurements of the remaining long-lived fluorescence is carried out. Time-gated fluorescence can increase sensitivity by several orders of magnitude to compare with that of chemiluminescence (Diamandis & Christopoulos, 1990).

Although most of the work with chemiluminescent substrates has led to the development of extremely sensitive biochemical assays, a common need to physically separate free from bound molecules has precluded a wide application to living cells. Bioluminescence provides a unique opportunity to express and target photoproteins inside living cells. Expression vectors for firefly, luciferase, fused to a promoter sequence has become a powerful tool to examine transcriptional activity (Alam & Cook, 1990). In principle, other fusion photoproteins could be made to transfect cells with site-directed markers of cellular functions (Sala-Newby & Campbell, 1991). Light-induced photolysis of a caged compound to increase the concentration of an ion or molecule inside a cell is a different example of how optical probes can be used to control intracellular signalling events (McCray & Trentham, 1989). Fluorescent probes continue to dominate work at the light microscopic level on single cell structure and function because they can emit a large number of photons per molecule and can be made specific enough to work in a complex chemical environment.

13.3.1 Detection of radioisotopes

Scintillation is widely used in biochemistry to detect molecules labelled with radioisotopes. In an aqueous environment relatively weak β-emitters (i.e. tritium) need to be close to scintillant molecules in order to produce light, otherwise the energy is dissipated and lost to the solvent. Energy transfer between ^3H and a fluorophore, however, can be enhanced by trapping the acceptor in close proximity to the radiation energy, as observed in radioluminescence. Amersham have developed this concept to provide a range of scintilation proximity assays which do not require separation of free from bound probe. Others have reported 240 cps for interactions between micromolar levels of [^3H]palmitate and an acceptor fluorophore in biological membranes. It is tempting to speculate that other radiolabelled molecules could be detected inside cells by this approach.

13.3.2   Quantitative analysis

Measurement of changes in dye intensity at a single wavelength is not sufficient for quantitative analysis of fluorescence in living cells because this approach does not distinguish between a change in the amount of bound dye from that of free dye. Time-resolve spectroscopy can sometimes differentiate free from bound probe but for image analysis the usual solution is to ratio intensities at two different wavelengths. Fluorescent probes which yield a spectral shift upon target recognition are therefore best suited for ratio imaging. To date the main application of fluorescence ratio imaging has been to monitor changes in ion concentrations in living cells (see special issue of *Cell Calcium*, February/March 1990).

At present, however, only a few optical probes are available to quantitate biochemical changes in a single living cell. New probes (e.g. cyclic AMP) are emerging from fluorescence resonance energy transfer (FRET) studies (see Chapter 9). FRET requires two fluorophores with overlapping absorption and emission spectra to be brought into close proximity (a few angstroms apart); this results in a spectral shift as energy is transferred from the donor to acceptor fluorophore. A distinct advantage of FRET is that there is a similar change in fluorescence ratio emission for any given pair of dyes in close proximity and this physical property can overcome the difficulty of engineering chemical relay to couple binding to detection into a single probe molecule.

No longer is the field of optical probe design restricted to the photo-chemist; in fact, you might already have all the necessary starting ingredients to make a new FRET probe. Optical probes which have their complete sensor function built into a single molecule are in some ways preferential to ones that work in tandem, but few exist. A calcium-sensitive fluorescent analogue of calmodulin is a good example of how a fluorescent protein might be used to monitor activation of a protein inside living cells (Hahn *et al.*, 1990). Covalent attachment of an inert second dye molecule whose spectrum differs from that of a single-wavelength probe might provide a way to generate more ratiometric probes in the future. More detailed information about optical probe technology can be found throughout this book; and in other references provided.

13.3.3   Introduction into cells

Optical probes can access target molecules in solution, on the cell surface and inside fixed cells but, in general, cannot freely diffuse across the plasma membrane of an intact cell. Acetoxymethyl ester groups on some of the low-molecular-weight probes have enabled them to permeate the cell membrane barrier and be trapped inside cells by the activity of intracellular esterases, but many probes cannot be loaded this way. Chemical or electrical, permeabilisation is often used to introduce probes into cells but these approaches work best when shown to be reversible to minimize loss of cell function.

Microinjection under pressure offers a means to selectively deliver probes into cells. We recently evaluated a Zeiss automated microinjection system to deliver probes into cells. A cell-'impermeant' optical probe for Ca^{2+} ions was mircroinjected into several cells and their subsequent receptor-mediated Ca^{2+} signals were taken as strong indication of cell viability (data not shown). Vesicle delivery of probes constitutes the main alternative to punching holes into cells and has involved the use of endocytosis, red blood cells, reconstituted viral envelopes and other liposomes. A large amount of probe can be loaded into red blood cells by osmotic shock and their content is delivered by fusion to recipient cells expressing haemagglutinin (Doxsey *et al.*, 1985). Unfortunately it has proven difficult to express this protein in many cell types which restricts the use of this technique (A. Helenius, pers. commun.).

13.4   PHOTON DETECTION

13.4.1   Low-light levels

Attenuation of the excitation beam is necessary to minimize photobleach of a fluorophore and this results in low light levels. Calculations based on illumination levels deemed necessary for quantitative fluorescence microscopy of living cells concluded that for a 512 × 512 array, each element will detect a signal due to a photon once every 32.8 s or once in 1000 video frames (Spring & Lowy, 1990). Worse still, the likelihood of the observed signal being real must be weighed against the possibility of it being due to noise in the detector. In confocal microscopy the emission beam typically dwells on each detector element for 1–10 μs and the above calculation would negate any chance of this working; but the reality is that many photons are available if the excitation intensity is increased and confocal images are possible. Even so, at 4 μs per point a 512 × 512 array would take about 1 s to scan and 'photon-capture' time is effectively reduced by a quarter of a million-fold compared to continuous integration for 1 s. Large-area illumination and detection is inherently faster than using a single-point raster scan to build up an image and multiple-point scanning is now available on some confocal microscopes. At present, two types of detectors are used for image analysis of cellular fluorescence projected from a conventional (non-confocal) microscope: intensified CCD (Inoue, 1986; Art, 1990) and the slow-scan cooled CCD (see Chapter 21).

13.4.2   Image-intensifiers

In a typical image intensifier photons strike a photocathode to generate electrons which are amplified and targeted onto a phosphor screen to cause a large burst of photons to the CCD. This gain overcomes the electronic noise of the CCD and enables detection. Unfortunately, the light-collecting efficiency of a photocathode is substantially less than that of a CCD so many photons are now never detected. A CCD has about a 5–50-fold better chance of capturing a photon (500–600 nm) than a photocathode. Intensified CCDs have a low intrascene dynamic range for one video frame. This is often set by the recovery time of the microchannel plate (MCP) which acts as a gain stage. Special high conductivity MCPs can achieve local count rates of up to 4–5 events per channel per video frame but with an individual channel size of 6–8 μm demagnification is required to maximize their density on each pixel element of a CCD (20 μm^2). Spring and Smith (1987) calculated a signal-to-noise of 11.7 for their intensified camera. An 8 bit (256 grey levels) analogue-to-digital converter is adequate to cover this dynamic range and is often used for video-rate imaging (see Mason et al., 1990). Image intensifiers are the best suited to capture fast image data.

13.4.3   Slow-scan CCD

Slow-scan CCDs are currently the most sensitive photometric detectors available for low-light-level scientific imaging applications. State-of-the-art CCDs can capture more than 90% of incident photons and each photon-generated electron can be represented by one digital unit. Each pixel can store thousands of electrons and a wide intrascene dynamic range of 65 536 grey levels is possible. Real objects can extend over several pixels and so even higher dynamic ranges

are possible. In pursuit of photometric excellence the electronic circuitry and sampling techniques employed to reduce noise during read-out has resulted in slow read-out times of several seconds and this time penalty appears to preclude their application to measuring rapid changes in signals with living cells. High-speed frame rates (100 per second), however, are possible by using faster analogue-to-digital converters (ADC), fewer pixels and multiple serial registers. A fast ADC usually means less precision and one has to decide what digital accuracy is adequate to cover the dynamic range of your signal response. An 8 bit ADC only has 16 meaningful levels because of a square-root error function. Fortunately, this has proven adequate for many changes in dye-loaded cells. This subject is extensively covered in Chapter 21 and will not be repeated here. Suffice it to say that slow-scan cameras can operate at the sub-second frame rates needed for biological data.

What we hope to show in the following data section is that these kind of cameras, when adequately cooled, do have enough sensitivity and precision to quantify changes in dye signals at the single cell level. The main difference between this work and that previously published by others (Conner et al., 1987) is that we have summated light by optical reduction rather than employing the conventional method of on-chip binning on a full-size CCD. A small-area CCD acts as a fast subarray without the need to discard data and the extra cost of large-area CCDs.

13.5   SMALL-AREA IMAGING SYSTEM

13.5.1   Equipment

All experiments were carried out on an epifluorescence, inverted stage microscope (Nikon Diaphot), fitted with quartz optics for UV transmission and a 100% side-port to camera. Special consideration was given to filter and lenses to maximize intensity of light reaching the camera. To excite the calcium ion indicator fura-2 at 340 nm and 380 nm band-pass filters (Omega, Brattleboro, VT, USA) were mounted in a computer-controlled filter wheel (Applied Imaging, Santa Clara, CA, USA). Excitation light was provided by a 100 W xenon arc lamp and neutral density filters (typical value of 1% transmission) were placed in the light path to minimize photobleach. A 430 nm dichroic mirror was mounted in the epicube to cleanly separate excitation from emission light sources.

Fura-2 has a broad emission spectra in solution and a 480 nm long-pass filter was chosen because transmission was about 2.5× more than a commonly used 510/40 band-pass filter. A 40×/1.3 NA, oil-immersion objective lens and a 0.4× projection lens was found to be the best combination from Nikon stock to maximize image brightness. Images were captured on a Texas Instruments TC211 CCD (192×165 pixels) operating at −20°C and after exposure were digitized to 8 bits (256 grey levels).

This image sensor was interfaced to a 386/33 MHz PC supplied with 8 Mbytes of memory, a maths coprocessor, sVGA, internal tape back-up drive; and a 500 Mbyte hard disk. Data were compressed with a public domain PK-ARC program and archived on tape.

A modified and updated version of this system, including extended software and additional camera interfaces, is now available from Applied Imaging. If a fluorescence microscope is not available, most microscopes can be converted for fluorescence observation by addition of a low-cost makler objective lens and excitation light source (Davies, 1991).

13.5.2 Spatial resolution

The main purpose of this imaging system is to record fluorescence intensity changes at the level of a single cell within a cell population and some sacrifice on spatial information was introduced to maximize light intensity and cell number. With the 40×/0.4 lens combination each pixel samples a $1.05 \times 1.20\ \mu m$ rectangle in the object plane ($0.42 \times 0.48\ \mu m$ for a 100× lens). This spatial sampling frequency is less than the diffraction limit of light (0.2–$0.3\ \mu m$). Undersampling can lead to aliasing and patterns not in the original object can occur because of a reduced periodic sampling. The Nyquist theory states that one needs to sample at least twice the highest spatial frequency in the object to resolve it and avoid possible artifacts (Inoue, 1986).

Fluorescence emission acts like a point source and diffuses out in all directions and this out-of-focus light can reduce axial spatial resolution to less than 5 µm in conventional microscopy (Hiraoka *et al.*, 1987). In view of the contamination of the in-focus image by out-of-focus light, the above pixel size would appear to be adequate to resolve changes in fluorescence intensity as seen under conditions of conventional microscopy. In the presence of confocal optics, fluorescence imaging appears to be restricted to a depth of focus of 0.7 µm for a NA 1.4 objective lens. For comparison, polarized light microscopy without confocal optics can make 0.15 µm slices (Inoue, 1990). Part of this difference may be due to fluorescence blurring which happens when a cone of light illuminates a sample, whereas phase-dependent imaging does not suffer from this effect. Deblurring algorithms can correct for out-of-focus information to improve optical sectioning with fluorescent probes (Hiraoka *et al.*,

1987). Early schemes operated on a stack of images collected at different points of focus to calculate and remove out-of-focus information; this unfortunately precluded real-time analysis. Recent work, however, has reported a method to deblur individual images to obtain apparent thin-section and ratiometric Ca^{2+} images (Monck *et al.*, 1992). Deblurring might enhance confocal images as well.

13.5.3 Timing details

To demonstrate the performance of the system cells were loaded with fura-2AM (4 µM/30 min) and alternate 340 nm then 380 nm images were captured and written to hard disk. A typical integration time was 0.5–1.0 s for each image and 0.5 s was required to save to disk and change the filter wheel ready for the next image. For faster data acquisition single frames could be saved at a rate of 5–10 s^{-1} on a ram drive and ratioed against a fixed image obtained at the isosbestic wavelength which is at 360 nm for fura-2. To take full advantage of this method a high-intensity and shuttered light source is recommended to reduce exposure time (no data available at present). We find, however, that one ratio every 1.5–3.0 s has been adequate to monitor changes in intracellular Ca^{2+} ion concentration ($[Ca^{2+}]_i$) in many types of cells.

13.5.4 Calibration

All images were background subtracted and appropriately ratioed on a pixel-by-pixel basis. $[Ca^{2+}]_i$ levels were calculated according to the equation of Grynkiewicz *et al.*, (1985): $[Ca^{2+}]_i = K_d\beta(R-R_{min})\ /\ (R_{max}-R)$ with the aid of a look-up table to speed up the process (see Mason *et al.*, 1990 for more details). The constants R_{min}, R_{max} and β were determined empirically using fura-2-loaded cells exposed to the calcium ionophore ionomycin (5 µM) and either 5 mM $[Ca^{2+}]$ for R_{max} or zero added Ca^{2+} ions and 10 mM EGTA for R_{min}. The value of β was calculated as the ratio of fluorescence at 380 nm for dye in minimal and maximal calcium ion concentration. The values obtained were: R_{min} 0.2, R_{max} 1.8, β 3.0; and K_d was taken as 224 nM as reported by Grynkiewicz *et al.*, (1985).

13.6   DATA SECTION

13.6.1   Raw data and graphical analysis

To verify that the small-area and cooled CCD had enough sensitivity to detect changes in intracellular Ca^{2+} ion concentration as reported by fura-2 we examined the individual changes in fluorescence

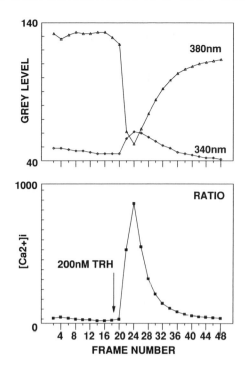

Figure 13.1 Typical changes in fluorescence intensity at 340 and 380 nm for fura-2-loaded cells.

average plots can be made along with changes at individual wavelength. Selected frames can be played (i.e. 40–86 etc.) and full control over the y-axis is possible on screen. Data can be exported to spreadsheets like Lotus 1-2-3 and Excel for further off-line analysis. Storage of raw data allows a careful check on suspect Ca^{2+} responses and subsequent re-analysis with new calibration constants whenever necessary. Data compression typically reduces image-size to 10 kbyte and 50 or more ratio-pairs may be stored on a floppy disk whilst thousands can be stored on hard disk or archived to tape.

13.6.2 Showing which cells respond

Electrophysiologists would often like to know which cells will respond to a test substance before they patch-clamp. Biochemists who measure a small change in ion concentration in a cell population quite often want to know whether every cell displays this small response or if only a subpopulation is responsive, producing a much larger signal than the average. Cell population work tends to average out the individual ionic responses and this process can mask out unique signalling patterns of certain cells. On the other hand, a common question for a single cell biochemist is whether an observation is meaningful at the cell population level. In all these cases an ability to switch between single cell and population would be a clear advantage. A sequential image montage of 30–50 cells is a rapid way of solving this issue. Plate 13.1 shows a sequence of images composed of individual frames taken once every 2 s. Each image represents a pseudocolour map of $[Ca^{2+}]_i$ in purified bovine lactotrophs from anterior pituitary cells. The colour table goes from blue to green to yellow to red $[Ca^{2+}]_i$ increases. The first frame shows the phase-contrast image at the focal plane used for fluorescence imaging. TRH was added at the end of the last frame on the top row (right hand side) and the next frame shows some response. A peak effect was apparent one frame later but some cells clearly remained blue and unresponsive throughout the whole sequence. A delayed but sustained response is noted in one cell. Graphical analysis would be necessary to more clearly define temporal changes. Just a few coverslips would be enough to build up a good idea of the population response whilst retaining individual cell data as well.

intensity at 340 and 380 nm under conditions of minimal bleach. Fig. 13.1 shows typical changes for rat pituitary cells loaded with fura-2 and challenged with 200 nM thyrotropin-releasing hormone (TRH) as indicated by the arrow. TRH has been previously shown to increase IP3 in these cells and cause a transient increase in cytosolic Ca^{2+} concentrations. Cells were excited at each wavelength for 0.8 s in this example and their pre-stimulatory intensity levels remained constant (<5% change). This result shows that photobleach is not a serious problem.

Individual intensity values of approximately 45 (340 nm) and 132 (380 nm) indicate good use of the camera's dynamic range. The software can control exposure time independently for each wavelength to match the signals to the dynamic range of the camera, if necessary. Addition of TRH induced a rapid increase in the 340 signal and a concomitant decrease in the 380 signal. A peak response typically represented a 4–5-fold increase in the initial 340/380 ratio before declining back towards basal levels. These kind of changes are as expected for a true Ca^{2+} response as recorded by fura-2. Calculated changes in intracellular calcium ion concentration against time are shown in the bottom panel of Fig. 13.1.

To analyse the response of many different cells a user-defined box is first placed around each cell and the image sequence played back. Up to 20 regions of interest can be defined in one pass. Single cell and

13.6.3 Nuclear hot spots – real or artifact?

The following experiment was chosen as an example of the importance of recording raw data for inspection. In addition, it also shows that light levels are high enough to use a 100/1.3 lens and gives a good example of the current type of screen layout where graphical analysis has been mixed with selected images. Mast cells represent a good model to study the mechanism by which cells trigger release of hormones stored in dense-core granules in the cytoplasm. In this experiment rat peritoneal mast cells were loaded with fura-2 and challenged with 5 µg ml^{-1} of compound 48/80 to degranulate them. Addition of 48/80 apparently induced a highly localized Ca^{2+} spike appearing as a red hot spot in cells shown in the middle image of Figure 13.2 with temporal changes above. Phase-contrast microscopy indicated that these spots were localized to a nuclear region. On closer inspection of this effect it was apparent that the 380 nm signal did not return to basal and that granules outside the cell appeared to be stained with fura-2. These observations suggest a loss of fura-2 from degranulated mast cells. Ratio imaging is supposed to some extent to compensate for problems of differences in dye concentration. Degranulation and dye loss might unmask a localized Ca^{2+} response in the nucleus, but a possible artifact cannot be dismissed with these data alone. Our recent data have, however, indicated that true Ca^{2+} signals can be localized to mast cell nuclei (data not shown). The reader is referred to Chapter 16, which extensively covers the pitfalls of ratio imaging.

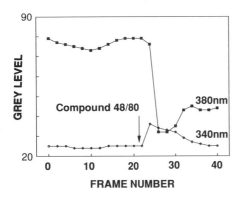

Figure 13.2 Raw data shows no recovery in the signal at 380 nm which was used to calculate [Ca^{2+}]$_i$ in Plate 13.2.

13.7 FINAL COMMENTS

At present there is no one universal imaging system to meet every possible application of optical probes in the biological sciences. An ideal detector would be limited by photon output of the source and provide enough sensitivity and dynamic range to accurately measure discrete changes as they occur. At extremely low light levels signal becomes limited by the statistics of photon flux and the time allowed to sample data becomes important. Low-noise read-out often takes

several seconds for a full-frame CCD and this represents a long time during which no signal is being detected. Image intensifiers might be more appropriate for full-frame imaging at video-rate. On-chip binning and faster read-out can dramatically speed up the image refresh-rate of a cooled CCD. A low-cost option is to summate light in the optical path before integration of this condensed image on a small-area and cooled CCD camera. Reduction in pixel number and digital precision enables quite fast processing and extensive storage on a rather inexpensive PC.

ACKNOWLEDGEMENTS

We would like to acknowledge the generous financial support of the Agricultural and Food Research Council and the laboratory of Dr W.T. Mason for continuous support and enthusiasm for this project. G.J. Law would like to thank Dr P.S. Dannies for supporting his interest in fluorescence microscopy and to make it known that all credit must go to W. O'Brien who came in as a summer student unfamiliar with image processing and 'C' language and compiled this program.

REFERENCES

Alam J. & Cook J.L. (1990) *Anal. Biochem.* **188**, 245–254.
Art J. (1990) *Confocal Microscopy Handbook*. Plenum Press, New York, pp. 127–139.
Aubin J.E. (1979) *J. Histochem. Cytochem.* **27**, 36–43.
Benson R.C., Meyer R.A., Zaruba, M.E. & McKhann G.M. (1979) *J. Histochem. Cytochem.* **27**, 44–48.
Beck S. & Koster H. (1990) *Anal. Chem.* **62**, 2258–2270.
Connor J.A., Cornwall M.C. & Williams G.H. (1987) *J. Biol. Chem.* **262**, 2919–2927.
Davies P.A. (1991) *Microscopy & Analysis* November, p. 37.
Diamandis E.P. & Christopoulos T.K. (1990) *Anal. Chem.* **62**, 1149A.
Doxsey, S.J., Sambrook J., Helenius A. & White J (1985) *J. Cell Biol.* **101**, 19–27.
Grynkiewicz G., Poenie M. & Tsien R.Y. (1985) *J. Biol. Chem.* **260**, 3440–3450.
Han K.M., Waggoner A.S. & Taylor D. (1990) *J. Biol. Chem.* **265**, 20335–20345.
Haugland C. (1989) [Conti RP] *Handbook of Fluorescent Probes and Research Chemicals*. Molecular Probes Inc, Eugene OR.
Hiraoka Y., Sedat J.W. & Agard D.A. (1987) *Science* **238**, 36–41.
Hirshfield T. (1977) *Appl. Spectrosc.* **31**, 245.
Inoue S. (1986) *Video Microscopy*. Plenum Press, New York.
Inoue S. (1990) *Confocal Microscopy Handbook*. Plenum Press, New York, pp. 1–13.

Johnston R.F., Pickett S.C. & Barker D.L. (1990) *Electrophoresis* **11**, 255–260.

McCray J.A. & Trentham D.R. (1989) *Ann. Rev. Biophys. Biophys. Chem.* **18**, 239–270.

Mason, W.T., Hoyland J., Rawlings S.R. & Relf G.T. (1990) *Methods Neurosci.* **3**, 109–135.

Monck, J.R., Oberhauser A.F., Keating T.J. & Fernandez J.M. (1992) *J. Cell Biol.* **116**, 745–759.

Sala-Newby G.B. & Campbell A.K. (1991) *Biochem. J.* **279**, 727–732.

Spring K.R. & Lowy R.J. (1990) *Methods Cell Biol.* **29a**, 15, 270–289.

Spring K.R. & Smith P.D. (1987) *J. Microsc.* **147**, 265–278.

Tsien R. (1989) *Ann. Rev. Neurosci.* **12**, 227–253.

Urwin V.E. & Jackson P.G. (1991) *Anal. Biochemistry* **195**, 30–37.

CHAPTER FOURTEEN

Multiparameter Imaging of Cellular Function

GARY R. BRIGHT

Department of Physiology and Biophysics, School of Medicine, Case Western Reserve University, Cleveland, OH, USA

14.1 INTRODUCTION

The cytoplasm of living cells is a highly organized and dynamic biochemical system. The plasma membrane does not just contain a collection of enzymes that carry out their respective functions, but maintains a complex dynamic environment of precise, temporally and spatially orchestrated reactions. Cellular regulation of stimulus–response coupling involves a delicate inter-relationship of second messenger systems, including changes in concentrations of ions and cyclic nucleotides, phosphorylation events, lipid metabolism, and gene expression. An understanding of cell regulation requires an understanding of how this delicate balance is maintained and regulated. Because of this organization it is important to study specific events within the context of the cellular environment.

Fluorescence spectroscopic imaging provides a unique tool for studying cellular biochemistry at the single cell level. The combination of sensitive fluorescent probes, selective for specific physiological parameters and the ability to map these parameters in 2, 3 and 4 dimensions, has resulted in new insights in cell biology. The most prominent applications have been fluorescence ratio imaging for quantitation of concentrations of ions (see Tsien, 1989; Bright *et al.*, 1989;

Bright 1993) and the redistribution kinetics of cellular proteins (see Taylor *et al.*, 1989).

The ability to measure responses in single cells has led to the discovery of events such as regular oscillations in cytoplasmic calcium (see Jacob *et al.*, 1988). Most of these changes have been average changes over the whole cell. Some cells have been large enough to see subcellular oscillations in $[Ca^{2+}]$ that vary in position within the cells. These waves may be unidirectional as in muscle cells (Wier *et al.*, 1987; Takamatsu & Wier, 1990) or complex patterns, such as spirals, as found in *Xenopus* oocytes (Lechleiter *et al.*, 1991). In addition, the ionic responses of cells within a population can be very heterogeneous (see Millard *et al.*, 1988; Bright *et al.*, 1989; Gylfe *et al.*, 1991).

Fluorescent analogues of cytoskeletal elements have been used to study aspects of cytoskeletal organization and dynamics. Fundamental aspects of both actin (McKenna *et al.*, 1985; Amato & Taylor, 1986) and tubulin (Mitchison, 1989) dynamics have been approached using fluorescently labelled proteins in conjunction with photobleaching and photoactivation techniques.

A particularly exciting direction is the study of cellular interactions. Poenie *et al.* (1987) followed the changes in calcium concentration upon interaction of cytotoxic cells with target cells. Further studies like this will no doubt open up many aspects of cellular interactions that are important in developmental biology.

Although a large amount of information has been gained by the application of various fluorescence spectroscopic imaging techniques, there is much more information potentially available. Knowing that a given biochemical event occurs at a given point in time may not be enough to understand the biological process. Time of occurrence post-stimulus may not be the important factor. It may be the changes in a given parameter relative to those of another parameter that are important. If this is the case then being able to measure one parameter at a time is inadequate, particularly given the heterogeneity of responses that has been observed (see Bright *et al.*, 1989). The ability

to measure multiple parameters within single cells will allow these interrelationships to be studied. The focus of this review is on the practical issues and technology necessary for basic multiparameter measurement of cellular biochemistry.

The basic issues involved in multiparameter experiments (DeBiasio *et al.*, 1987; Waggoner *et al.*, 1989) include appropriate selection of fluorescent probes with contrasting wavelengths, ability to tune the excitation, dichroic reflector, and emission paths of the fluorescent microscope to appropriate wavelengths, and the ability to acquire and analyse the subsequent data. The integration and automation of many of these capabilities becomes essential for routine experimentation. Indeed, this lack of integration has limited current applications.

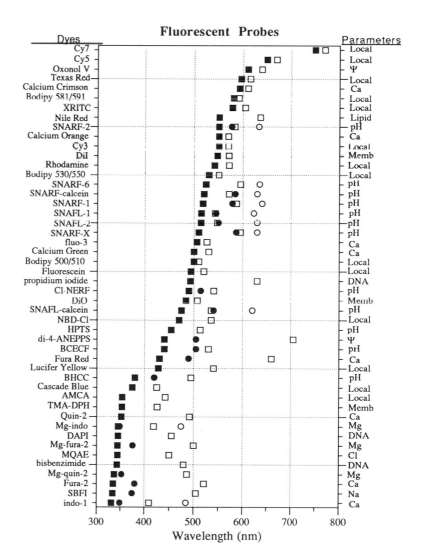

Figure 14.1 Compilation of a variety of fluorescent probes. Wavelengths were compiled from a number of sources (Molecular Probes, Inc; Lambda Probes & Diagnostics; DeBiasio *et al.*, 1987; Gross & Loew, 1989; Bright, unpublished observations). The actual wavelengths used in an experiment may vary from these, since many of the data were derived from catalogues of peak wavelengths. The closed and open symbols represent excitation and emission wavelengths, respectively. Where multiple symbols exist, more than one wavelength was reported.

14.2 FLUORESCENT PROBES

14.2.1 Biochemical parameters

The basis of the selectivity and sensitivity of fluorescence spectroscopic imaging is, of course, the fluorescent probe. A variety of fluorescent probes exist as (1) fluorescent analogues of cellular enzymes and other proteins for localization; (2) reporter groups of various physiological parameters, such as ion concentration; (3) membrane potential, (4) lipid membrane structure, and (5) enzyme activity. The types of probes, and the parameters of each, is outside the coverage of this review but has been extensively covered by others (Haugland, 1983; Kohen *et al.*, 1989; Tsien & Waggoner, 1990; see also Chapters 1 and 3 in this volume and several chapters in *Methods in Cell Biology*, Vols 29 and 30).

14.2.2 Wavelength selection

Knowing the parameters to be measured, the goal is to select an appropriate probe. If only one probe is involved the selection can be very easy. Since the subject here is the measurement of multiparameter, the goal is to select two or more probes with contrasting wavelengths. In this instance, contrasting wavelength means a combination of excitation and emission wavelengths which allows discrimination of each probe in the presence of the others. For example, two probes would be considered to have contrasting wavelengths if they had different excitation but the same emission wavelength.

These combinations become particularly difficult when the dye is a ratio dye, where two or more excitation or emission wavelengths are needed for one dye. In practice, this will tend to limit the number and combination of probes because the ratio dye requires more of the available spectrum. Further developments in dyes will hopefully provide greater choices of wavelengths for a given parameter. Molecular Probes Inc. has succeeded in expanding the range of dyes (variety of wavelengths) available for measuring both pH (SNAFL and SNARF series) and Ca^{2+} (Calcium Green, Orange and Crimson series). Fig. 14.1 illustrates a compilation of dyes and wavelengths. The wavelengths are concentrated between 300 and 600 nm. The reason for this is that most dyes are for visual inspection, thus, wavelengths visible by the human eye are stressed. Future developments will hopefully expand the spectral range of useful probes since electronic detection systems typically possess a much wider wavelength sensitivity than the human eye. Current detection systems exhibit spectral ranges of 300–900 nm, thus doubling the usable spectrum.

Fig. 14.2 illustrates three common dyes that are easily used in combination. Included are two dyes potentially useful for following enzyme dynamics and one for monitoring cytoplasmic pH. Figure 14.3 illustrates two ratio dyes, fura-2 (Ca^{2+}) and SNARF-1 (pH), that can be used in combination. This combination is more complex than the first in that four excitation and two emission wavelengths are involved. There is also significant excitation of both dyes when using 340 and 380 nm light needed specifically for fura-2. Since only the fura-2 emission is analysed, it will reflect Ca, although SNARF is being excited. This combination of dyes, although with different wavelengths, has recently been used in a non-imaging application to measure both Ca^{2+} and pH in living cells (Martinez-Zaguilan *et al.*, 1991). A similar study using fura-2 and BCECF has also been reported (Simpson & Rink, 1987).

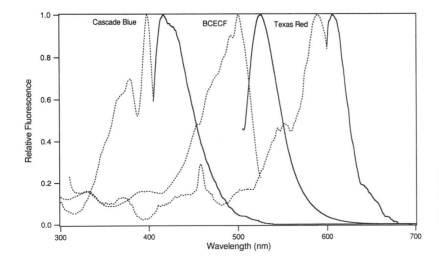

Figure 14.2 Spectra of dyes with contrasting wavelengths. This example includes two dyes potentially useful for localization studies (Cascade Blue, Texas Red) and one dye for a physiological parameter (pH, BCECF).

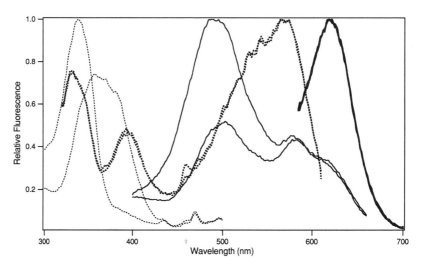

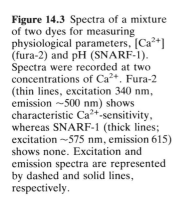

Figure 14.3 Spectra of a mixture of two dyes for measuring physiological parameters, $[Ca^{2+}]$ (fura-2) and pH (SNARF-1). Spectra were recorded at two concentrations of Ca^{2+}. Fura-2 (thin lines, excitation 340 nm, emission ~500 nm) shows characteristic Ca^{2+}-sensitivity, whereas SNARF-1 (thick lines; excitation ~575 nm, emission 615) shows none. Excitation and emission spectra are represented by dashed and solid lines, respectively.

Another issue in selecting a probe is the shape of the spectrum. Some dyes, for example the cyanines (see DeBiasio *et al.*, 1987), possess significant absorbance outside the major peak, that is, a rather large shoulder exists. Second, absorption may exist at other wavelengths. For example, since fluorescent dyes invariably involve ring structures, there is usually a significant absorbance in the near-ultraviolet, even though the major absorption wavelength occurs at a much longer wavelength. If such a dye is mixed with a dye with a peak absorbance in the ultraviolet, both dyes will be excited. Discrimination would require obviously different emission wavelengths. Another key aspect is that since absorption occurs, photobleaching of the dye not being measured will occur. Thus, selection involves choosing minimal overlap of excitation and emission wavelengths. It may be necessary to excite or collect fluorescence away from the peak excitation or emission to achieve maximal discrimination. Thus, selection of appropriate wavelengths means more than simply choosing the optimum for a given dye.

14.2.3 Delivery into cells

A significant aspect of using multiple probes involves delivery of each to the cell. There are a variety of means of introducing dyes into cells including: (1) physical introduction via perturbation of cell membrane integrity, such as needle microinjection or a physical shock; (2) metabolic incorporation, such as can occur with some lipid probes; and (3) the more popular diffusional loading of acetoxymethyl ester (AM) forms of dyes followed by enzymatic removal of the AM by intracellular esterases. Delivery techniques have been reviewed recently by McNeil (1989).

Basic difficulties occur with the need to load different dyes by different methods. The basic issue is how many insults is a cell able to survive? Even if all dyes are of the AM form, there are problems. The AM forms of dyes are often either toxic or cause significant short-term perturbation of the cells. For example, BCECF-AM, a pH-sensitive dye, is very well-tolerated by cells (see Bright *et al.*, 1987). Cells can be easily labelled. However, if monitored immediately after loading, the cells exhibit large numbers of membrane blebs on the surface. These blebs go away and the cells look fine after a recovery period of ~15–30 min. A basic question remains as to whether this perturbation has altered the cell in some way. For our own work, we can detect no significant change in the growth characteristics of BCECF-labelled cells compared to unlabelled cells. When several dyes are loaded into cells this problem can be exaggerated. Thus, the conditions for delivery of the probe may be an issue when it comes to making the best selection.

14.3 DATA ACQUISITION

In order to acquire multiparameter data, each dye must be excited and the emission recorded. For ratio dyes this may require multiple excitation and/or emission wavelengths. Thus, the acquisition of these data requires an instrument capable of varying the wavelength of the input and output light. For modern microscopes based on reflected (epi-)fluorescence this involves (1) a source with all the required wavelengths, (2) a means of selecting the appropriate wavelengths from the source, (3) a means of selecting the appropriate emitted light, (4) a detector to quantitatively record the emitted light, and (5) the ability to save the data recorded from the camera for analysis.

14.3.1   Light source

The light source must possess all the necessary wavelengths needed for the probes selected. In some cases, this may involve the need for multiple sources. The three most popular sources are the tungsten–halogen filament lamp, xenon arc lamp, and mercury arc lamp. The tungsten–halogen lamp is the most common since standard microscopes are delivered with this type of lamp. Although usable at wavelengths >400 nm, most of the spectral energy is concentrated at wavelengths >600 nm (see Inoue, 1986), which limits its usefulness for most dyes (see Fig. 14.1). These lamps are unsuitable for dyes with ultraviolet wavelengths, such as fura-2 and bisbenzimide. We have successfully used this lamp for ratio imaging of BCECF (450 nm, 495 nm excitation; Bright et al., 1987). However, we often ran into problems of insufficient excitation energy and eventually switched to a xenon arc lamp (Bright et al., 1989)

The standard power supplies provided with these lamps are usually inadequate. Experience has shown that these power supplies are unable to provide suitably stable illumination levels with typical building supply voltages. Also as the lamp ages, the shape of the output spectrum, the colour temperature, changes.

Arc lamps, such as the xenon and mercury, have more often been the lamps of choice because of the increased ultraviolent wavelength output. Xenon lamps have a relatively even distribution of wavelengths <800 nm, but some very large spectral lines >800 nm. Mercury lamps have a number of spectral lines in the range 300–600 nm. In some cases, the presence of these spectral lines may be undesirable. In the case of the xenon lamp, the near-infrared lines put an increased burden on interference filters which must block these wavelengths. A desirable aspect is that the increased intensity of spectral lines in the ultraviolet, such as present in the mercury lamp, can more than compensate for inefficiencies in transmission found in some microscopes. In addition, for specific purposes, such as photoactivation (see Chapters 26 and 31), the availability of the 365 nm spectral line makes the mercury lamp the one of choice. The disadvantages of arc lamps include a tendency to flicker due to arc wander, particuarly in mercury lamps. With proper burn-in, this can be almost eliminated in modern lamps.

For high-power requirements, often needed for fluorescence photobleaching recovery (Jacobson et al., 1982), and optical trapping (Block, 1990), lasers are often needed. The major disadvantage of lasers is the restricted availability of wavelengths as can be seen in Fig. 14.4. Wavelengths <400 nm are often only available in the most expensive laser systems. Increased advances in lasers are providing more systems which are available as air-cooled and are therefore cheaper and easier to maintain. In the longer term, tunable diode lasers will provide a wide selection of wavelengths. A discussion of lasers and related issues can be found in Gratton and vanderVen (1990).

14.3.2   Wavelength selection

One of the most important functionalities required for multiparameter fluoresence imaging is the ability to tune the excitation and emission light paths with respect to wavelength on demand. There are at least three components: the excitation path, dichroic reflector, and emission path. Each of these can be independently tuned for maxmimal flexibility. The wavelength tuning devices used may vary depending upon the light source and path. Several criteria are important in selecting a tuning device, including (1) wavelengths available, (2) band-pass, (3) tuning speed, (4) available aperture, and (5) whether the device is suitable for an imaging light path.

The two fundamental approaches to wavelength tuning are either to use a series of static devices, such as interference filters, and mechanically place the appropriate element in place or to use an acoustic-optic device. Many of the wavelength selection devices in use today were added for fluorescence ratio imaging.

14.3.2.1   Excitation

The excitation path requires a device that can choose the appropriate wavelength from the light source. A variety of devices exist for tuning wavelengths from the source. The type of device depends upon the design of the excitation path and the position within the system at which the device is placed.

The simplest means of selecting a narrow wavelength range from a source is to use an interference filter. These filters transmit a selected wavelength range by the generation of multiple constructive and destructive interferences. Interference filters have secondary transmission peaks at wavelengths which are harmonics of the principal wavelength and must be blocked by additional coatings. A discussion of interference filters and associated issues can be found in Marcus (1988). Fig. 14.4 presents a compilation of many of the interference filters that are available for several common probes. The specifications of a filter include a centre wavelength and band-pass. The wider the band-pass, the greater the amount of energy that is transmitted. Interference filters are fairly easily

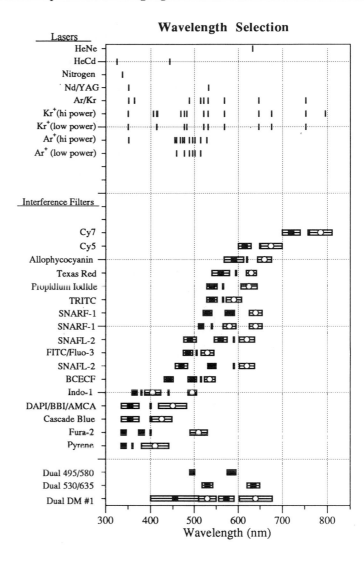

Figure 14.4 Compilation of filter sets and laser lines usuable for wavelength selection. For lasers, each vertical line represents a laser line available in the output. For interference filters, the symbol represents the centre wavelength and the bars represent the band-pass. Vertical bars represent the dichroic reflectors. (Sources of information include Omega Optical, Inc. (Brattleboro, VT, product literature); DeBiasio *et al.*, 1987; Gratton & vanderVen, 1990; Bright, unpublished observations; and miscellaneous product literature.)

designed these days for almost any wavelength and band-pass. Filters are available in large apertures and are well-suited for imaging.

For tuning, several discrete filters with appropriate centre wavelengths and band-passes are necessary. The cheapest and slowest method of wavelength tuning is manual placement of a filter into the excitation path. This may be by placement of a filter into a slot on the microscope or by movement of a filter holder which holds two or more filter sets. The disadvantages of manual tuning are that the vibrations can be very large and a settling time must be introduced before image acquisition. This severely limits the speed of tuning. This was one of the major limitations in our previous study (DeBiasio *et al.*, 1987). Most standard fluorescence filter holders supplied only carry 2–4 filter sets. These often hold complete sets of excitation, dichroic reflector and

emission filters and thus require disassembly and reassembly to alter one component. Thus, trying different combinations can be tedious.

The simplest means of automatically selecting the excitation wavelength is to place a filter wheel between the lamp housing condenser and the microscope. A filter wheel is a multi-position wheel with a different interference filter at each position. Most condenser lens systems supplied with microscopes have a sufficient adjustment range to accommodate the introduction of a filter wheel. The goal of the filter wheel is to position the appropriate filter in the light path upon command. The wavelength tuning speed is related to the speed of the wheel. Filter wheels are of moderate speeds. The fastest we are aware of is about ~50 ms for switching between adjacent filters (Sutter Instruments, Novato, CA). A more typical system (Ludl Electronic Products, Inc., Hawthorne, NY) is on the order of ~100 ms. A characteristic of filter wheels is that different times are required to select a wavelength depending upon which position the filter occupies. Thus, when a quick change must be performed, the next filter should be in an adjacent position in the wheel.

Dual monochromators have been popular excitation sources since they were introduced by Tsien *et al.*, (1985) for fluorescence ratio imaging. The modification added by Tsien *et al.* was the addition of a chopper mirror which alternately directed light from two lamp-monochromator systems into the microscope. Each monochromator is tuned to a given wavelength and then the object alternately excited. This type of system provides easy selection of a wide range of wavelengths. It is capable of switching between the two wavelengths very quickly. These systems are usually designed to switch at video-rates (30 frames per second). The major disadvantage for multiparameter imaging is that only two wavelengths can be switched quickly. Both monochromators must be changed for any additional wavelengths. Filters give spatially more uniform illumination than monochromator slits. Monochromators are not suitable for imaging light paths. The physical large size, mechanical complexity and expense limit the applicability to multiparameter imaging.

Monochromators also have some unique advantages. Monochromators are freely and independently adjustable in wavelength and band-pass. It is easy to balance the excitation energy by varying the slit width (band-pass). For filters, which have fixed band-pass, intensity must be adjusted with neutral-density filters. Monochromators also permit full spectral scanning which can be helpful in diagnosing problems with the system and for investigating the environment of the probe within the cell cytoplasm.

A variation on the Tsien system is to use two monochromators, one lamp and a mirror, either on a solenoid-based transport device (Linderman *et al.*, 1990) or a galvanometer (Ryan *et al.*, 1990), to select which monochromator to direct the lamp. Newer versions of this approach use a single monochromator and lamp and provide fast wavelength changes based upon a faster means for scanning the diffraction grating within the monochromator. In this device only one monochromator and lamp is necessary. This approach is not as fast for switching between the two wavelengths, but it is much more flexible since it is not limited to just two wavelengths.

A second basic mechanism for wavelength tuning is to use an acousto-optic device. An acousto-optic device is effectively a tunable diffraction grating. A standing wave is established within a birefringent crystal using acoustic transducers. By adjusting the frequency and power of the acoustic energy both wavelength and intensity can be varied on a millisecond time-scale without any moving parts.

Acousto-optical devices have existed for many years for use with lasers. Since they were designed for lasers the aperture is very small. It is not easy to enlarge the aperture due to the physics involved. Thus, they are limited to being used with lasers and the wavelengths available are limited to those of the multiline laser. These laser-dependent systems are not useful in an imaging light path. These sytems work by physically deflecting the diffracted light beam. Tuning occurs by placing a slit at the entrance to the microscope and adjusting the appropriate laser line on the slit, thus illuminating the specimen. The temperature of these devices must be well-controlled otherwise significant drift of the beam can occur.

A second type of acousto-optical device has recently been described (Kurtz *et al.*, 1987). The acousto-optical tunable filter provides a large aperture, is suitable for broad-band non-coherent light, and provides fast switching time and intensity control. These systems provide a spatially stable first-order beam, thus making it suitable for an imaging light path. They are capable of rapid spectral scans. On the surface they appear to provide an ideal mechanism for wavelength tuning. However, in practice, these devices have an extremely narrow band-pass (~0.5–1 nm in visible range) which varies with frequency, and does not always provide a stable image. With this narrow band-pass, very powerful excitation sources are necessary. Although commercially available, they are currently very expensive. The potentially more useful ones are the small-aperture devices for use with fibre optics (Infrared Fiber Systems, Inc. Silver Spring, MD).

More recently, a new device has become available which combines a birefringent filter and liquid crystals (VariSpec; Cambridge Research and Instrumentation, Inc., Cambridge, MA). This device is currently available with a tunable range of 400–740 nm and 15 mm aperture. The band-pass is set at time of manufacture from 2 to 15 nm. Wavelength switching is fast. This device may provide the promise of the AOTF.

14.3.2.2 *Dichroic reflector*

The dichroic reflector (DR) is an integral part of the reflected fluorescence microscope. It reflects the excitation light onto the specimen and transmits the emitted light to the detector. The DR is typically a long- or short-pass interference filter designed to operate at 45° in an imaging light path. In a multiparameter microscope a means of tuning this filter is necessary. Since this filter must be mounted at 45°, a mechanical switching is about the only approach that can be used. Of course, manual switching is possible but not very useful as discussed above. One approach is to mount a filter wheel such that it places the DR at 45°.

The speed of wavelength switching is relatively slow since it is based on a filter wheel. The major disadvantage is the movement of an element in the imaging light path. The switching between two DRs can lead to a lateral shift in the images. These translations must be corrected in order to correlate the resulting images. In our device (Eastern Microscope, Raleigh, NC) this translation has been constant, thus the same translation corrections can be used throughout the experiment. An alternative is to include enough adjustments in the design of the filter wheel to align each of the DRs in the wheel. Usually it is more cost-effective to make corrections with the image processor after acquisition.

An alternative to physically moving the DR is to use a multiple-pass DR (Fig. 14.4; Omega Optical, Brattleboro, NH). This DR reflects multiple wavelengths and transmits multiple wavelengths (Fig. 14.4). Thus, no filter change is necessary if the appropriate reflection and transmission bands are available. The advantages are speed, since no changes are needed, and no misalignment of resulting images. The disadvantages are that the combination of wavelengths available is significantly limited compared to the filter wheel system. This necessarily limits the combination of probes that can be used together.

14.3.2.3 Emission

The emission component of the light path further selects the emission wavelength of the probe. Careful selection of emission wavelengths is often necessary for discrimination of multiple probes. All of the methods of tuning the excitation light path that are suitable for imaging, such as a filter wheel, and acousto-optic tunable filters are appropriate for the emission light path. Since this path is an imaging path, the alignment and positioning of the filter are more important than in the excitation path. Also, translation corrections are typically needed. In addition, just as multipass DRs are available, multipass transmission filters are also available. Thus, by selective excitation, images of different probes can be recorded without the need to tune this path.

14.3.3 Detection and acquisition

The quality of the image data recorded is determined by the detector. Important properties of the detector which control the quality of the signal include sensitivity, dynamic range, resolution, geometric distortion and noise characteristics (Bright & Taylor,

1986). The basic requirements of a detector for multiparameter imaging include (1) the ability to detect all the wavelengths of interest, (2) sufficient sensitivity to accurately record images at low light levels, (3) the ability to respond fast enough to acquire each of the images of a multiparameter set on the necessary time-scale, and (4) a large intrascene dynamic range.

The benefits of a large dynamic range are particularly important in multiparameter imaging. It is often a complex task to balance the concentration of the different probes such that the output intensities are in the same range. If the detector has a large dynamic range then getting all the intensities to fall within the range of the camera is much simpler.

The selection of detectors today often comes down to choosing between an intensified video-rate CCD camera and a slow-scan, cooled CCD. Cameras, such as the SIT and ISIT, although still used, suffer from the characteristics of the electrostatic intensifiers used, relatively long lags and limited regions of linear radiometric response, thus limiting the intrascene dynamic range (Bright & Taylor, 1986). The video-rate CCD exhibits a linear radiometric response over a wider range of input intensities. The lag is almost non-existent (Bookman, 1990). Coupling this camera with a microchannel plate intensifier results in a very sensitive, fast-responding camera with few distortions (Spring & Lowy, 1989; Bookman, 1990; see also Chapter 20). The intensifier usually limits the dynamic range to a level similar to conventional cameras. The wavelength sensitivity is determined by the particular photocathode material used on the intensifier. Although many types of material exist, the more common ones easily cover the wavelength range of most common dyes currently in use.

Image acquisition from a video camera requires an image digitizer. These systems provide digitization of the image, typically with 8 bit grey scale resolution. Real-time image-averaging is necessary to improve the signal-to-noise ratio of the low-light-level images (Bright & Taylor, 1986).

The slow-scan, cooled CCD camera is almost the perfect camera. By cooling the CCD, the sensitivity at 520 nm is comparable to an intensified video CCD. It exhibits significantly greater sensitivity to wavelengths >600 nm. The wavelength sensitivity range can be as large as 200–1000 nm (Aikens et al., 1989; see also Chapter 21). Thus, as new probes with wavelengths >600 nm are developed (see DeBiasio et al., 1987) this camera will be the one of choice. The radiometric response is also linear over the full range from preamplifier noise level to full capacity.

A significant benefit of cooled CCD cameras is the large dynamic range. Although it is possible to digitize the image at 8 bits, the real benefits come when using

12, 14 or 16 bit digitizers. This results in ranges of 4096, 16 384 and 65 536 gray levels, respectively compared to 256 for 8 bits. This also leads to the biggest disadvantage of these cameras, read-out speed. Cooled CCD cameras do not generally hook up to standard video digitizers. They are usually supplied with boxes that acquire the image from the CCD chip using one of the above digitizers into computer memory. The speeds of the 12, 14 and 16 bit digitizers are significantly less than the 8 bit video-rate digitizers, generally ~50–500 kHz, compared to ~8.5 MHz. The read-out rate will also be related to the size of the CCD chip. Thus, chips with 512×512 pixels, similar in resolution to video-rate cameras, will be read out much faster than a 2048×2048 pixel chip. More extensive information regarding cooled CCD cameras can be found in Aikens et al., (1989; see also Chapter 21). There are applications where the unique capabilities of cooled CCD cameras, i.e. frame transfer, can be used to great advantage (see Aikens et al., 1989; Linderman et al., 1990).

Image acquisition from cooled CCD cameras is usually by digital transfer from the camera controller to a computer workstation. The size of the image can have a dramatic impact on the choice of computer. An 8 bit, 512×512 image (0.25 Mbytes) is considerably smaller than the 16 bit, 1024×1024 image (2.0 Mbytes). This is an important consideration when determining what kind of computer is used.

14.4 DATA ANALYSIS

Once the data are recorded, it is important to restore the images from any distortion that may have been introduced during the acquisition. This involves restoration from radiometric non-linearities and geometric distortions (Jeričevic et al., 1989) and any misregistrations due to filter changes in the imaging light path (see above), and possibly magnification corrections (see Waggoner et al., 1989; Keller, 1990). In some cases, spatially dependent fluorescence photobleaching must be taken into account (Benson et al., 1985).

Coordinated display of the multiple parameters was the extent of the data analysis in our previous study (DeBiasio et al., 1987). The basic approach was to map each of the parameters to a different colour in a composite image. By displaying the composite images of the temporal sequence the interrelationships became obvious. Other kinds of experiments will no doubt require more sophisticated types of analyses. A similar evolution is already occurring in the field of flow cytometry, where multidimensional analyses are being explored.

14.5 CORRELATION WITH OTHER RECORDING MODALITIES

There are other types of information that can be recorded along with fluorescence images. It is possible to view a video-enhanced transmitted light image of the field simultaneously with the fluorescence (Foskett, 1988; Tsien & Harootunian, 1990; Spring, 1990). Thus, field selection and focusing can occur without exposure to the excitation light, consequently limiting photobleaching. Mobilities of cellular components can also be measured by integrating a fluorescence photobleaching recovery (FPR or FRAP) instrument into the multiparameter imaging system (Jacobson et al., 1982; Kapitza et al., 1985; Thomas & Webb, 1990). Availability of so-called 'caged' compounds provides a means of optically manipulating the cellular environment (see Chapters 26 and 31). These compounds can be used to perform concentration jump experiments within the cytoplasm. For cells and organelles that are moving it is possible to catch them in an optical trap (Ashkin et al., 1987; Block, 1990). It is also possible to correlate electrophysiological data with changes in the cytoplasm (see Chapter 22). In the future, it will be possible to include information about fluorescence lifetimes of probes (Lackowicz & Berndt, 1991) which may open another dimension.

14.6 COMPUTER ASPECTS

The computer is responsible for coordinating the various peripheral devices attached to the microscope, acquiring the images and subsequent analyses. The system must be able to accomplish this on the timescale of the biological process being monitored. The computer bottleneck is typically at the image acquisition and storage stage. Since each image consists of typically 0.25 Mbytes of data, the rate at which the machine can save this data is important. Conventional computer hard disks are relatively slow, particularly those found in personal computers. For workstations there are a variety of disks available able to save full $512 \times 512 \times 8$ bit images in real time. For real-time recording, optical video disks are typically the most cost-effective approach.

The choice of computer hardware is generally not as important as the software. For routine multiparameter imaging measurements, flexibility becomes the dominant factor. Most commercial software for fluorescence imaging has been developed for a specific purpose, such as ratio imaging. The difficulty comes when trying to adapt this turnkey software to other

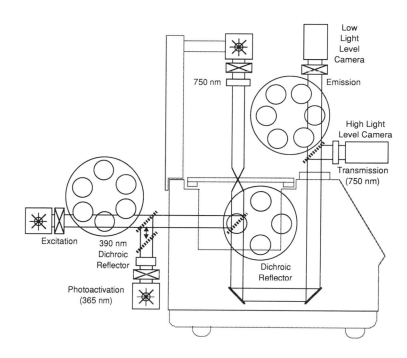

Figure 14.5 Example of multiparameter microscope.

purposes. Most software packages lack the flexibility of integrating the control of several more filter wheels, shutters and cameras, in addition to new analysis alogrithms necessary for multiparameter imaging. Most often, contracted custom programming is proposed as the solution by these vendors.

Maximum flexibility comes from an object-oriented approach to software design. The goal is a software system where the user can add new modules to an existing system to extend functionality. For example, if a new filter wheel is added, then the addition of another filter wheel module would allow immediate accessibility by the user. These new modules may be either purchased from a vendor or programmed by the user. Along with this approach is the ability of the user to fully define how the system operates. Since commercial vendors do not supply code for their software, minor changes are usually impossible.

Examples of this approach to software design include ISee (Inovision Corp., Durham, NC), Visualization Workbench (Paragon Imaging, Lowell, MA), AVS (Stardent, Boston, MA), Khoros (University of New Mexico), and LabView (National Instruments, Austin, TX). These systems provide an icon-driven programming interface which makes software development and modifications much simpler than conventional programming. ISee is unique in that it integrates hardware control, image acquisition and image analysis into one package. Visualization Workbench and AVS currently focus on image analysis and display. These systems run on UNIX workstations and are designed for inte-

gration into a networked environment. LabView is a system for data acquisition and analysis that has image-processing functions available from other vendors.

14.7   SYSTEM DESCRIPTION

A fairly simple multiparameter imaging system is illustrated in Fig. 14.5. This system takes into account most of the simpler components discussed above. The functionality provided includes multiparameter fluorescence imaging with simultaneous transmitted light imaging (either phase, DIC, or asymmetric illumination) with the option of photo-release of caged compounds. Using the x, y, z, scanning stage (not shown), the system is capable of systematically acquiring data from many fields.

Most of the components of this system are available off-the-shelf. The three filter wheels (excitation and emission, Ludl Electronic Products, Hawthorne, NY; dichroic, Eastern Microscope, Raleigh, NC), four shutters (Vincent Associates), and scanning stage (Ludl Electronic Products) are controlled by an integrated controller (Ludl Electronic Products). All filters are from Omega Optical, Inc. and Chroma Technology Corp. (Brattleboro, VT). The cameras include an intensified video CCD (Videoscope International, Washington, DC) and a standard video CCD (Ikagami). The system is controlled by a Sun Microsystems workstation (Mountain View, CA) with Datacube

(Peabody, MA) image-processing hardware, 1 Gbyte of disk space and an erasable optical disk system (Pinnacle Micro, Irvine, CA) for archival storage. Software for both hardware control and analysis is based on ISee (Inovision Corp, Durham, NC) and our own custom programs.

14.8   CONCLUSION

The ability to correlate multiple biochemical parameters at the single cell level will allow the exploration of interrelationships that are currently not visible. It is hoped that these interrelationships will clear our cloudy understanding of many aspects of cellular biochemistry. The instrumentation to perform these experiments is a simple evolution of current technology.

ACKNOWLEDGEMENTS

I would like to thank Dr Jean Welter for his comments and suggestions on the manuscript. This work has been supported, in part, by grants from the Diabetes Association of Greater Cleveland (no. 322–89) and The Council for Tobacco Research, USA (2683).

REFERENCES

Aikens R.S., Agard D.A. & Sedat J.W. (1989) *Methods Cell Biol.* **29**, 292–314.

Amato P.A. & Taylor D.L. (1986) *J. Cell Biol.* **102**, 1074–1084.

Ashkin A., Dziedzic J.M. & Yamane T. (1987) *Nature* **330**, 769–771.

Benson D.M., Bryan, J., Plant A.L., Gotto Jr. A.M. & Smith L.C. (1985) *J. Cell Biol.* **100**, 1309–1323.

Block S. (1990) In *Non-invasive Techniques in Cell Biology*, J.K. Foskett & S. Grinstein (eds). Wiley-Liss, New York, pp. 352–384.

Bookman R.J. (1990) In *Optical Methods in Biology*, B. Herman & D. Jacobsen (eds). Wiley-Liss, New York, pp. 235–250.

Brakenhoff G.J., van Spronsen E.A., van der Voort H.T.M. & Nanninga N. (1989) *Methods Cell Biol.* **30**, 379–398.

Bright G.R. (1993) In *Optical Microscopy: Emerging Methods and Applications*, B. Herman & J.J. Lemasters (eds). Academic Press, New York, pp. 87–114.

Bright G.R. & Taylor D.L. (1986) In *Applications of Fluorescence in the Biomedical Sciences*, D.L. Taylor, F. Lanni, A.S. Waggoner, R.F. Murphy & R.R. Birge (eds). Alan R. Liss, New York, pp. 269–288.

Bright G.R., Fisher G.W., Rogowska J. & Taylor D.L. (1987) *J. Cell Biol.* **104**, 1019–1033.

Bright G.R., Fisher G.W., Rogowska J. & Taylor D.L. (1989) *Methods Cell Biol.* **30**, 157–192.

DeBiasio R., Bright G.R., Ernst L.A., Waggoner A.S. & Taylor D.L. (1987) *J. Cell Biol.* **105**, 1613–1622.

Foskett J.K. (1988) *Am. J. Physiol.* **255**, C566–C571.

Gilkey J.C., Jaffe, L.F., Ridgway E.B. & Reynolds G.T. (1978) *J. Cell Biol.* **76**, 448–466.

Gratton E. & vander Ven M.J. (1990) In *Handbook of Biological Confocal Microscopy*, J.B. Pawley (ed.). Plenum Press, New York, pp. 53–67.

Gross D. & Loew L.M. (1989) *Methods Cell Biol.* **30**, 193–219.

Gylfe E., Grapengiesser E. & Hellman B. (1991) *Cell Calcium* **12**, 229–240.

Harootunian A.T., Kao J.P.Y., Paranjape S. & Tsien R.Y. (1991) *Science* **251** 75–78.

Haugland R.P. (1983) In *Excited States of Biopolymers*, R.F. Steiner (ed.). Plenum Press, New York, pp. 29–58.

Herman B. (1989) *Methods Cell Biol.* **30**, 220–245.

Inoue S. (1986) *Video Microscopy*. Plenum Press, New York.

Jacob R., Merritt J.E., Hallam, T.J. & Rink, T.J. (1988) *Nature* **335**, 40–45.

Jacobson K., Elson E., Koppel D. & Webb W.W. (1982) *Nature* **330**, 769–771.

Jeričević Z., Wiese B., Bryan J. & Smith L.C. (1989) *Methods Cell Biol.* **30**, 48–84.

Jovin T.M. & Arndt-Jovin D. (1989) In *Cell Structure and Function by Microspectrofluorometry*, E. Kohen & J.G. Hirschberg (eds). Academic Press, New York, pp. 99–118.

Kapitza H.G., McGregor G. & Jacobson K.A. (1985) *Proc. Natl. Acad. Sci. USA* **82**, 4122–4126.

Keller H.E. (1990) In *Handbook of Biological Confocal Microscopy*, J.B. Pawley, (ed.). Plenum Press, New York, pp. 77–86.

Kohen E., Kohen C., Hirschberg J.G., Sanus R., Schachtschabel D.O. & Nestor J. (1989) In *Cell Structure and Function by Microspectrofluorometry*, E. Kohen & J.G. Hirschberg (eds). Academic Press, New York, pp. 199–228.

Kurtz I., Dwelle R. & Katzka P. (1987) *Rev. Sci. Instrum.* **58**, 1996–2003.

Lakowicz J.R. & Berndt K.W. (1991) *Rev. Sci. Instrum.* **62**, 1727–1734.

Lechleiter J.S., Girard S., Peralta E. & Clapham D. (1991) *Science* **252**, 123–126.

Linderman J.J., Harris L.J., Slakey L.L. & Gross D.J. (1990) *Cell Calcium* **11**, 131–144.

McKenna N.M., Meigs, J.B. & Wang Y.-L. (1985) *J. Cell Biol.* **100**, 292–296.

McNeil P.L. (1989) *Methods Cell Biol.* **29**, 153–174.

Marcus D.A. (1988) *Cell Motil. Cytoskeleton* **10**, 62–70.

Marks P.W., Kruskal B.A. & Maxfield F.R. (1988) *J. Cell. Physiol.* **136**, 519–525.

Martinez-Zaguilan R., Martinez G.M., Lattanzio F. & Gillies R.J. (1991) *Am. J. Physiol.* **260**, C297–C307.

Millard P.J., Gross D., Webb W.W. & Fewtrell C. (1988) *Proc. Natl. Acad. Sci. USA* **85**, 1854–1858.

Mitchison T.J. (1989) *J. Cell Biol.* **109**, 637–652.

Pawley J.B. (1990) *Handbook of Biological Confocal Microscopy*. Plenum Press, New York.

Poenie M., Tsien R.T. & Schmitt-Verhulst A.M. (1987) *EMBO J.* **6**, 2223–2232.

Ryan T.A., Millard P.J. & Webb W.W. (1990) *Cell Calcium* **11**, 145–156.

Simpson A.W.M. & Rink T.J. (1987) *FEBS Lett.* **222**, 144–148.

Spring K.R. (1990) In *Optical Methods in Biology*, B. Herman & D. Jacobson (eds). Wiley-Liss, New York.

Spring K.R. & Lowy R.J. (1989) *Methods Cell Biol.* **29**, 270–291.

Spring K.R. & Smith P.D. (1987) *J. Microsc.* **147**, 265–278.

Takamatsu T. & Wier W.G. (1990) *FASEB J.* **4**, 1519–1525.

Taylor D.L., Amato P.A., Luby-Phelps K. & McNeil P. (1984) *Trends Biochem Sci.* **9**, 88–91.

Taylor D.L., Bright G.R., DeBiasio R., Fisher G.W., Luby-Phelps K. & Wang L.-L. (1989) In *Cell Structure and Function by Microspectrofluorometry*, E. Kohen & J.G. Hirschberg (eds). Academic Press, New York, pp. 297–313.

Thomas J. & Webb W.W. (1990) In *Noninvasive Techniques in Cell Biology*, J.K. Foskett & S. Grinstein (eds). Wiley-Liss, New York, pp. 129–152.

Tsien R.Y. (1989) *Methods Cell Biol.* **30**, 127–156.

Tsien R.T. & Harootunian A.T. (1990) *Cell Calcium* **11**, 93–110.

Tsien R.Y. & Waggoner A. (1990) In *Handbook of Biological Confocal Microscopy*, J.B. Pawley (ed.). Plenum Press, New York, pp. 169–178.

Tsien R.T., Rink T.J. & Poenie M. (1985) *Cell Calcium* **6**, 145–157.

Waggoner A., DeBiasio R., Conrad, P., Bright G.R., Ernst L., Ryan K., Nederlof M. & Taylor D.L. (1989) *Methods Cell Biol.* **30**, 449–478.

Wier W.G., Cannell M.B., Berlin J.R., Marban E. & Lederer W.J. (1987) *Science* **235**, 325–328.

CHAPTER FIFTEEN

Digital Image Analysis: Software Approaches and Applications

GRAHAM T. RELF

Regional Medical Physics Department, Freeman Hospital, Newcastle upon Tyne, UK

15.1 INTRODUCTION

Computing is a relatively new tool in experimental science, and its power is increasing at a rapid rate while its cost plummets. In dual-wavelength fluorescence experiments the computer becomes a vital component of the instrumentation in a special way, by providing a means of ratioing images, ratioing being an arithmetical operation which has to be performed at every point in the image and which cannot reasonably be carried out by purely optical means.

The computers which have now become everyday machines are fundamentally digital. That is, they deal with discrete numbers: integers or fractions. They cannot handle real numbers in the mathematical sense, whereas our belief that the physical world is accurately modelled by real numbers implies that we should measure continually varying, or analogue, quantities. The first compromise in applying computers to images acquired from a microscope is therefore that they must be digitized. In practice this means that the image is approximated by a regular matrix of picture elements, called pixels, rather like the squares of a canvas for tapestry. Each pixel has a value representing the grey level (intensity) or colour of

that part of the scene. The pixel has a finite area on the image sensor, or camera, and integrates all of the light falling on that small area during the period of time used for acquiring the image.

The maturity and low cost of television technology means that it offers suitable camera designs which incorporate some of the digitization requirements. The well-known thermionic tube camera scans the scene as a raster pattern of horizontal lines, thereby digitizing the image vertically. The resulting video signal is still an analogue voltage varying with time along each line, but by sampling that voltage at a fixed frequency we can digitize it in the horizontal sense also. The computer can therefore sample intensity values at regular time intervals which represent all of the pixels in the image. There are two scanning rates currently used worldwide for video: 25 Hz, with 625 lines per frame (image) in the PAL and SECAM systems, or 30 Hz, with 525 lines per frame in the NTSC system. Alternatives are emerging with the introduction of high-definition television.

In recent years the tube camera has been increasingly replaced by charge-coupled device (CCD) cameras, using light-sensitive silicon elements arranged in a rectangular array. Cameras built from these devices can be very small. They usually incorporate circuitry to produce a standard analogue video signal but the discrete nature of the silicon array can also be exploited directly if the camera has suitable control circuitry. Note that whereas in tube cameras the pixel areas can be adjacent, in solid-state cameras there may well be gaps between adjacent light-sensitive areas of silicon.

The intensities sampled from the video signal are also analogue values and a further step is required to convert these to discrete numbers for computation. An analogue-to-digital converter (ADC) is an electronic device for doing this. The ADC is driven at the required frequency for the horizontal sampling and produces a number in a range which is a power of 2. The power of 2 arises from the binary nature of electronic switches (on or off). A convenient unit of computer data storage is the byte, which consists of 8 bits (binary digits), and which can represent a number in the range from 0 to 255 (i.e. 2^8). The voltage range from a video camera can be represented accurately using 256 grey levels. Any further levels would simply be lost in electronic noise fluctuations; indeed ordinary cameras never provide as many as 8 bits of reproducible data.

15.2 IMAGE PROCESSING AND ANALYSIS

By introducing some computing terms we have seen how an image from a camera is digitized to produce grey level data as an array of pixels. The resulting data are stored in the computer's memory and this complete process is called acquisition, image capture, or frame-grabbing. The stored image, typically comprising 8-bit pixels, is called a grey image.

Image processing consists of transforming the data by some algorithm (i.e. a numerical procedure), usually in order to enhance it in some way. For example, an automatic procedure can be used to identify all the pixels which lie on or near sudden changes of brightness, which might correspond to the edges of objects which are subsequently to be measured. The result of this image processing is still an image, containing the same amount of data.

Image analysis, on the other hand, goes further by identifying pixels which lie within the objects to be measured and distinguishing these pixels from those in the general background. The general term for making this distinction between pixels is segmentation. One simple method of segmenting an image is to select those pixels which lie within a given range of grey levels. This method is thresholding. Figure 15.1 shows an image which has first been processed to find the edges of objects, so that edge pixels are those with the highest grey levels in the resulting image, and then thresholded. The result of segmentation

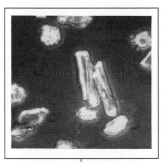

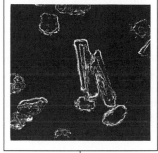

Image processing edge filter

Threshold brightest range

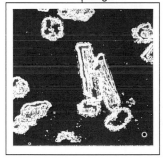

Figure 15.1 Edge detection and thresholding.

is a binary image, so called because its pixels have only two grey levels, representing selected pixels and background pixels. The binary image can also be processed as an image and then each contiguous group of pixels can be detected as an object. The geometrical description of the object then enables various measurements to be made, such as the object's area, perimeter or centre of gravity, or, by referring back to the original grey image and looking at just those pixels which lie in the detected object, we might measure the total intensity or, through a calibration curve, the optical density of the object.

It is important to remember at the measuring stage that the original was digitized. There is consequently a limit to the possible accuracy of the shape detected to represent the object and hence of any measurements made. For small objects, just a few pixels across, different algorithms for making a particular measurement can produce significantly different answers. A notoriously difficult measurement is that of perimeter. The difficulty can be envisaged by considering maps of a coastline which have been drawn at different scales; the lengths of estuaries in particular may appear to be wildly different. The researcher must therefore know exactly how each measurement has been made in order to be able to assess experimental accuracy.

Having found the objects in an image and measured them, the final result of image analysis will typically be tables and graphs made from the measurements. Figure 15.2 summarizes the stages of image analysis we have been considering. Note that during this process there has been a huge reduction in the amount of data. A digitized image typically comprises 10^5 or 10^6 pixels. A sequence of images from an experiment lasting a few minutes could easily consist of 10^9 bytes. The measuring process reduces this to some 10^2 values in a table or graph which the experimenter can assimilate. The equipment required to do this effectively is not simply a general-purpose computer.

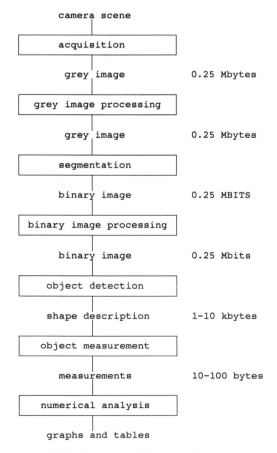

Figure 15.2 Principal stages of image analysis.

15.3   REQUIREMENTS FOR DUAL-WAVELENGTH FLUORESCENCE STUDIES

15.3.1   Real-time acquisition?

In acquiring images there are some fundamental decisions to be made because at the present state of the art, cameras which digitize in real time (25 or 30 frames s^{-1}) produce a result which is electronically noisy, this effect being accentuated by the intensification required for working with faint fluorescent sources. This means that several successive video frames need to be averaged before a useful result is obtained. Typically 8 frames are averaged, which brings the acquisition rate down to 3 or 4 frames s^{-1}. The averaging has to be done for each wavelength required before ratioing, so the best practical time resolution is even worse. In some experiments it is possible to capture most images at one wavelength, only occasionally capturing one at another wavelength for reference: if the second wavelength is at the isosbestic point in the spectrum the intensities within the image should vary very little (see e.g. Mason *et al.*, 1990).

On the other hand, cooled CCD cameras can produce images with negligible noise. The drawback with these is that they cannot be read at video rates; typically several seconds or even minutes are required for reading the data out for each image. However, some cooled CCD cameras now have built-in processing abilities enabling only selected portions of the image to be read, thus saving time. These portions could be rectangles covering individual cells. Software could be written for the controlling computer to determine from an initial full-view image where these rectangles should be, perhaps with interactive assistance from the user, and set the camera accordingly. A factor in favour of cooled CCD cameras is that with longer light integration times and low noise the number of significant bits of intensity data can be greater than 8. Even if the computer system analyses only 8 bit pixels, these 8 bits can be extracted from a wider dynamic range (perhaps 16 bits) without having to alter gain settings (and consequently recalibrating). In this case the hardware should permit any contiguous 8 bits from the pixel data to be selected for transfer to the computer, and this selection should be under the user's control through software.

Another consideration at the acquisition step is shading correction. It is impossible to illuminate the specimen uniformly in a microscope and the intensifying stages of cameras produce far from uniform sensitivity across the field of view. The image of a nominally uniform field will, therefore, generally have significantly varying intensity levels. However, to a good approximation the intensities at different wavelengths vary across the field in proportion to each other, so if the images are to be combined by ratioing then the shading effects cancel. This is, of course, why the ratioing technique is so useful. Be aware of digitization effects though: if intensity levels are very low, a change by a single grey level due to shading might have a large effect on the ratio.

Shading can be corrected by software if a uniform reference image can be presented to the system first, for it to store. This is a standard technique in image-processing software. With fluorescence it may be difficult to achieve a uniform reference image though.

15.3.2 Combining images

The next step in image acquisition might be some hardware for ratioing pairs of images as they enter the computer. It is now practical to have a single electronic card in a PC which will do this in real time, by using look-up tables (LUTs) held in memory chips. These tables need to be alterable to allow recalibration for each experiment and must therefore be held on the card in read–write memory (RAM). Again, software must provide the user with simple means of creating LUTs for a given experiment. The action of the LUT is indicated in Fig. 15.3. For a given position (x, y) in each of two images A and B the pixels have grey levels gA and gB. These grey levels, each on a scale from 0 to 255, are used as indexes for the LUT. Held at the coordinate (gA, gB) in the table is the grey level which will represent the ratio gA/gB in the resulting image. Even the ratioed image will usually be an array of integers because (a) most image-processing, analysis and display software works on such images and it would make considerably more work in developing software if non-standard techniques were used and (b) fractional numbers require more than 1 byte per pixel in order to maintain similar accuracy: there would at least have to be a place-keeper for a decimal point for each pixel. Therefore, we use integer grey levels on a scale from 0 to 255 even in the ratioed image, and a simple one-dimensional table (effectively a calibration curve) relates these grey levels to ratios, or indeed directly to ion concentrations.

The look-up table can be set to combine two images in any way, not just by ratios. But remember that shading will need to be corrected for non-ratio methods.

In designing a practical system, ratioing by hardware may not be a good idea because data are being lost when the pairs of images are combined. Experimenters often wish to keep all of the raw data to give more flexibility in analysing them later, perhaps to test new hypotheses. The look-up table technique

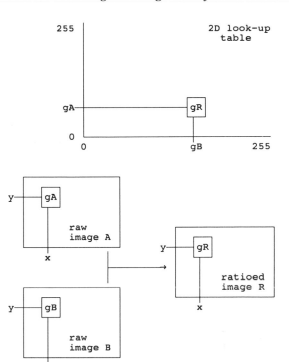

Figure 15.3 Ratioing images by look-up tables.

discussed above can also be carried out very quickly in software, but not in real time to keep up with the video signal; in this case images are usually captured first and ratioed later. There is no doubt, however, that seeing ratios on a display while the experiment is in progress can be useful. So, while the budget may only permit the software method to begin with, the ideal system shows ratios by real-time hardware during the experiment but also saves the raw data.

15.3.3 Archiving images

Remember that each image typically consists of 512×512 pixels = 0.25 Mbytes. Data archiving is therefore an important aspect of system design.

Raw data can be archived in analogue form on video tape or video disk, the latter being more permanent but also more expensive. In either case the equipment must have sufficient bandwidth to be able to recall the full information. Domestic VTRs do not have adequate bandwidth: the more expensive professional video recorders must be used. If, instead of storing the analogue signal, the data are saved after digitization, then perfect recall from the storage medium is guaranteed (barring faults). We are then in the realms of digital magnetic tape and magnetic or optical disks. The cheapest of these per byte is tape: streamer tapes for PCs are readily available and easy to use. The drawback of tape can be speed of access but some types now search and find data very quickly. For example, digital data storage (DDS) tapes, a similar technology to hi-fi DAT, have a capacity of 2 Gbytes at a cost of £20 per tape; the drives for these take about 20–30 s on average to find a file but the data transfer rate is then about 200 kbytes s^{-1}. DDS drives cost about £2500. The rewritable optical disks now becoming available are probably the best medium for storing sequences of images because they have a high capacity (of the order of gigabytes; 1 gigabyte = 4000 images each 512 × 512 × 8 bits) and they are erasable and removable. Their long-term storage characteristics are not yet proven, however. There are two main types of rewritable optical disks: phase change disks and magneto-optical disks. The former are faster because they erase and rewrite in a single operation whereas the magneto-optical disk takes three passes over the data area to rewrite it. The magneto-optical disks are claimed to be more reliable for frequently rewritten data (up to 1 million correct rewrites are now specified by manufacturers). The previous generation of optical disks, the so-called WORMs (write once, read many times) are probably the best for long-term archiving, particularly because they do not have to be protected from stray magnetic fields.

A drawback of any digital storage method when compared with analogue video recording is speed. Special hardware would be needed if digital data transfer were to run at the same rate as video recording, because each 512 × 512 frame has to be stored in 1/25th (PAL) or 1/30th (NTSC) of a second. The data rate is therefore 6.5 Mbytes s^{-1}, which no PC processor could handle by itself. To use digital storage is only really practical either to capture images less frequently or to store the images temporarily in a large amount of main memory (RAM) during the experiment and transfer them to disks afterwards. Very large RAMs ('solid-state disks') are available but they are expensive and volatile (power must be maintained to keep the data). Figure 15.4 compares prices and capacities for the principal mass-storage options.

A recent development which promises to alleviate this speed problem is in the field of data compression. Chips are now available for real-time image compression. Be aware, however, that the main market for these is in communicating document images, where the appearance of the result is all-important. The compression and decompression algorithms do not preserve every grey level exactly, so ratioing the results might well produce wrong answers. If considering the use of such devices it is necessary to check in detail how their algorithms work and to test their results.

Having said that, considerable data reduction in

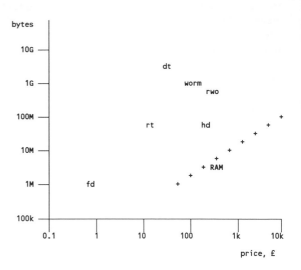

		Mbyte/£	Drive cost(£)
dt	= digital tape (DDS) or ExaByte VHS	120	2500
fd	= floppy disk	1.5	50
hd	= hard disk (magnetic)	0.5	200
rt	= streamer tape cartridge	10	300
rwo	= read/write optical disk	5	2000
worm	= write-once optical disk	10	
++	= solid-state disk (RAM)	0.01	—

Figure 15.4 Capacities and prices of available mass-storage media. The drive cost is only included for hard disks in the graph.

fluorescence images can easily be obtained if background areas, away from regions of interest, can be thrown away, e.g. by setting all pixels in those areas to grey level zero. It depends whether the experimenter really wishes to keep everything, just in case! Public domain software, such as the well-known PKZIP, can achieve good compression factors on images in which there are large uniform areas but has very little effect on raw 8-bit images in which background grey levels are fluctuating noisily.

15.3.4   Image display

Displaying the images from the computer presents no difficulty with present technology. Even the humble PC is now sufficient. If it is fitted with a video graphics adapter (VGA) it can display images of up to 320 × 200 pixels with 256 grey levels or up to 640 × 480 pixels with 16 grey levels. The former may well suffice for fluorescence work but to obtain 256 grey levels at the higher resolution it is necessary to fit a Super VGA card to the PC. While VGA is standard as far as software is concerned, no standard exists for Super VGA and there are several competing low-cost cards available. Most commercial imaging software is able to use all the common Super VGA cards, however. Other high-resolution formats such as TIGA and XGA are available but not widely enough used yet to be able to tell which will become a real standard. These are also more expensive.

Images can be displayed by the VGA (etc.) either as monochrome shades of grey or by using colour. The graphics adapter card uses look-up tables, usually held in RAM (read–write memory) to map the grey levels stored in the image arrays into colours as the image is displayed. There are three ways of colouring an image: pseudo-colour, real colour and false colour.

For a pseudo-colour display each grey level, representing monochrome intensity, is arbitrarily assigned to a (usually different) colour. This not only looks pretty but it often gives the contrast necessary to enable fine details to be seen which were not apparent in the monochrome image. It also makes it easier to read against a scale of, for example, ion concentration: red can represent one particular concentration, yellow another, and so on.

Real colour displays require real colour input from the camera which really captures three component images, in red, green and blue (RGB). In the computer these are normally held as three images, each effectively monochrome, and transformations can be performed to different basis vectors; for example, a 3 × 3 matrix transformation can be used to change each pixel from RGB to hue, luminosity and saturation (HLS) to extract different information. This transformation can often be a useful image-processing step before segmentation. For fluorescence work, real colour is unlikely to be useful. The spectral characteristics of the three sensors in the camera are unlikely to match or even significantly overlap the principal wavebands of any given biologically useful fluorochrome.

False colour, the third technique, might well be useful though. Suppose we had captured three monochrome images at different wavelengths by using filters. These could then be treated as if they were the components of an RGB image and displayed as such. This enables a meaningful display of three superimposed images to be shown to the experimenter in a way which can be interpreted straightforwardly. This is also very useful in the unfortunately quite common cases where the real wavelengths of fluorescence are too close to each other to be distinguished if viewed in real colour. The idea can be extended to false colour images with more than three components.

Real colour display boards are readily available for PCs. A typical specification uses 8 bits for each colour, therefore storing and displaying 24 bits of information for each pixel, that is 16.7 million colours! This is quite

unnecessary, because the eye cannot distinguish anything like as many colours. It is always possible to convert to an 8 bit pseudo-colour image, as explained by Preston (1991).

15.3.5 System software

No computer is usable until it has its first level of software running on it: its operating system. Programs for specific applications, such as fluorescence imaging, are run from the operating system which provides fundamental services to the application: accessing files on disks, communicating with the keyboard and display, getting time from a clock, etc.

'Turn-key systems' are ones which have been set up so that as soon as the operating system has started it runs the required application program automatically, so the user is never directly aware of the operating system. Most laboratory computers today are more general-purpose; the typical PC or Macintosh machine presents a user interface from which the experimenter chooses various programs to run. The current fashion is for graphical user interfaces (GUIs) in which icons are manipulated on the display screen with a pointing device. The Apple Macintosh is based on such a system and more recently the PC has become endowed with Microsoft Windows which has very similar properties. Such systems are multi-tasking. That is, they can run several programs 'simultaneously' as far as the user is aware, by assigning short time slices to each program in succession. They are message-passing systems in which the various programs can communicate with each other. This gives them great advantages in terms of common facilities for screen-handling and printer output and for generally presenting a standard face to the user who can therefore learn to use new programs quickly. So far so good, but an important aspect for our purposes is that these systems are not real-time systems: they do not ensure response to outside events in any particular minimum time. (This is particularly because they are not interrupt-driven.) Their use in laboratory instruments measuring event times accurately, therefore, has to be treated with caution unless special-purpose data acquisition hardware is responding to events and measuring time independently of the host computer. It is also necessary to warn against using machines which are multi-user or working on a network, for similar reasons: the desired action may not occur at the required time and overall performance is reduced by extraneous activities in the system.

Unfortunately, the PC running old-fashioned DOS is really more suited to the task in hand. This is doubly unfortunate because there are now several attractive imaging software packages available in the GUI environments from which otherwise suitable programs for specific applications might be developed.

15.3.6   Application software

It is easy to underestimate the amount of work involved in developing a complete program for use in fluorescence imaging experiments. The best course is to obtain ready-made software from a reputable source who can provide support in the event of any difficulties. This is, of course, easier said than done. Software of this nature usually forms part of a complete system, including all the hardware from microscope to colour printer. This is for a good reason: the system is complex and the organization which designed it can only reasonably support it in the field if they have control over all the components.

For those who nevertheless wish to design such software the following recommendations will help.

(a) Use a well-standardized, readily available, structured high-level language. First choice would be Pascal. Almost as good, but easier to make time-consuming mistakes in, is C.

(b) Document the design before writing any code. Specify exactly what the software is required to do and design the means by which it will achieve it. A good specification will be easy to turn into a users' guide at the end. Not only will this approach save time in the development by minimizing wrong turns but it will make the software remain useful for much longer, because it will be maintainable into the future as needs change. It is all too easy to jump enthusiastically into programming: at first, while the program is small, it is easy to comprehend. So the programmer says it will be fully documented afterwards. But that never happens: there is always something more pressing to do then.

(c) Spend plenty of time designing the user interface and make it as consistent and simple to use as possible. This is an art, however. There are many books on the subject but really easy-to-use programs are rare. Even popular user interfaces such as the GUIs have many cumbersome aspects. Furthermore, writing software of any size to run in the current GUI environments is very complicated.

(d) Break the design into clear, preferably small, well-defined modules. The object-oriented programming (OOP) techniques which are now becoming popular are worth mastering because they aid this process very effectively. The most recent versions of the Turbo languages from Borland provide OOP libraries. Other manufacturers provide versions of C++, which is an OOP extension of C.

(e) If it all possible obtain a well-supported library of image-processing routines on which to build, because re-inventing these would be very time-consuming. Try to find one which has been developed with all the above points taken into consideration. There are several professional image processing magazines in which these products are advertised.

A suggested list of modules required for such a program follows. Each should be designed as a separately compiled unit containing the relevant data structures and all of the routines necessary to access those data. This kind of encapsulation is the essence of OOP: other modules cannot access the data directly but only through the routines provided in the module containing the data definitions.

- The user interface: standard routines for each type of interaction, e.g. a menu, a data-entry box, displaying pages of help text, etc.
- Image memory layout and access.
- Camera control.
- Filter control.
- Any other microscope control.
- Hardware ratioer control and alternative software ratioing.
- Hardware averager control and software averaging.
- Image acquisition: timing, overall management of the above modules.
- Input to image memory from the final stage of digitization.
- Output to the display.
- Image sequence storage on disc and retrieval.
- Output to printers.
- Calibration: both spatial and fluorescence (ratioing etc.).
- Image-processing operations: processing individual images.
- Processing and editing sequences of images.
- Segmenting objects from image sequences.
- Measuring objects.
- Graphing.
- Tabulating.
- A module for each special facility, e.g. producing a montage of images, profiling, histogramming, plotting 3D views, etc.

Although this is only a summary it is hoped that the reader will appreciate that when the task is broken down into topics like this, the amount of software required is considerable. It is not something to be undertaken lightly.

REFERENCES

Mason W.T., Hoyland J., Rawlings S.R. & Relf G.T. (1990) In *Methods in Neurosciences*, Vol. 3, *Quantitative and Qualitative Microscopy*, P.M. Conn (ed.). Academic Press, London, pp. 109–135.

Preston K. (1991) *Photonics Spectra* April, pp. 119–121.

CHAPTER SIXTEEN

Fluorescent Probes in Practice – Potential Artifacts

J. HOYLAND

Department of Neurobiology, AFRC Institute of Animal Physiology and Genetics Research, Babraham, Cambridge, UK

16.1 INTRODUCTION

In reviewing the potential artifacts and pitfalls in the use of fluorescent probes which may befall the unwary, it is important to look at the general application modes in which fluorescent probes may be used. For example, fura-2, the calcium-sensitive probe, may be used either as a single-wavelength qualitative probe simply to image the spatial distribution of calcium. Alternatively, it may be used in dual-wavelength ratio mode in a calibrated system to quantify calcium concentrations and fluxes. The potential artifacts will be very different in the two systems, indeed, simply utilizing fura-2 in its ratio mode corrects most of the problems associated with single-wavelength measurements.

Most of the following chapter refers to the potential artifacts which may affect the *measurement* of ion concentrations. It therefore generally relates to imaging and photometry for *ratio* measurements rather than simple fluorescence. The variety of difficulties experienced in dynamic video imaging are as diverse as the measurements we make. This contribution, therefore, partly re-states the obvious, establishing an understanding of the basics, but more importantly hopes to stimulate constructive thought on recognizing and solving the inevitable problems which arise when systems and probes are pushed toward their limits.

16.2 PHOTOBLEACHING

All fluorescent probes will photobleach to a greater or lesser extent when excited with a suitable wavelength, at a rate proportional to the intensity of the incident light. While this may not present a problem for some applications such as the simple spatial mapping of hormone distribution, it does seriously affect any attempt to quantify ion concentrations using single-wavelength probes.

The most obvious practical way to reduce photobleaching is to minimize the light reaching the probe. Unfortunately, reducing the incident light will also reduce the emitted light intensity, so optimum conditions for image analysis must include the following.

16.2.1 Optimal loading

Optimal loading of a particular probe in a particular cell type should be assessed by experiment. A maximal signal will only result from maximal loading, but care must be taken to ensure the probe is not significantly buffering the ion of interest. A compromise must therefore be sought between maximizing the signal and minimizing probe concentration. In practice, intracellular probe concentrations of 30–100 μM are usually suitable. Loading may be achieved either by direct microinjection or by acetoxymethyl ester loading across the cell membrane (Tsein, 1981).

16.2.2   Maximum sensitivity of data collection

The availability of suitably sensitive detectors has played a crucial role in the development of technology to measure dynamic ion changes in living cells. While adequately sensitive photomultiplier tubes have been available for some years, more recent developments in intensified charge-coupled device (CCD) video cameras have allowed direct imaging of probes for measurement of ion concentrations and fluxes. In addition, fast computer access and processing now allows ratiometric techniques to be applied to video imaging, eliminating many of the possible artifacts and errors associated with single-wavelength techniques. Other factors which will improve sensitivity are the use of excitation and emission filters which are well-matched to the probe. The emission filter should have a bandwidth covering at least 90% of the emission spectrum. The use of an objective of the highest numerical aperture available will also improve sensitivity.

16.2.3   Ratiometric measurements

Utilizing ratiometric measurments will correct for uneven probe loading between cells and across each cell. It will correct for differences in cell membrane thickness, where the membrane unevenly absorbs some of the emitted light, and it will correct for part of the photobleaching problem. Correction of these artifacts is, however, limited to the dynamic range of the detection system and any combination which takes either of the images used for the ratio outside its dynamic range will cause problems as described in the section below. Furthermore, significant photo-bleaching has been shown in fura-2 to not only reduce the light output of the probe, but to also shift the emission spectrum (Becker & Fay, 1987). At just 8% photobleaching, using a standard ratio calibration, this results in an underestimation of up to 20% of 200 nM $[Ca^{2+}]_i$.

16.2.4   Minimum illumination time

Minimize the time actually illuminating the probe by exciting it only while capturing data. Extraneous room light must be kept to a minimum as it will contribute to photobleaching not only during an experiment but also while loading and storing loaded cells before an experiment.

16.2.5   Minimum oxygen

Minimize the oxygen concentration as it has been shown to play a major role in photobleaching of the fura-2/calcium system (Becker & Fay, 1987). The lowest concentration concomitant with good cell viability will greatly reduce the rate of bleaching.

16.3   DYNAMIC RANGE

Whatever type of system is employed to detect the emission from fluorescent probes, it will be limited to a specific dynamic range. That is, there will be a lower level at which detection is not possible and there will be an upper level at which the system saturates. The value at the maximum divided by that at the minimum is known as the dynamic range of the system. In addition to the values for the system as a whole, due regard must be paid to saturation in any single part of the system.

For example, an imaging system will have an overall dynamic range which is governed by the range of the probe, the range of the camera (which may have two

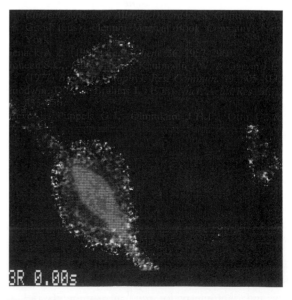

Figure 16.1 A typical ratio image of a cell which is saturating at one or both excitation wavelengths. The edges of the cell are usually ill-defined and it may appear larger than its real size due to scattered light. The ratio image from a properly illuminated cell will usually appear relatively uniform across its surface but show considerable pixel-to-pixel noise consistent with the use of image intensifiers. The saturated cell exhibits a 'flat' centre containing unusually low pixel-to-pixel noise.

intensifier stages, both capable of saturation and a CCD detector whch may saturate) and the range of the analogue-to-digital converter (ADC) used to digitize the image. There will also be a maximum ratio set by the ratioing range in the calibration table. In an ideal system the effective 'gain' of each component is set so they all saturate at the same level. This, however, is not always possible and great care must be taken to ensure that no part of the system will either saturate or fall below its minimum detection threshold during an experiment. (Fig. 16.1)

The most usual saturation artifact in ratio imaging is perhaps concerned with the setting of the maximum ratio level when ratioing. This is necessary to optimize the display of changes in ion concentration. If, for example, a cell undergoes a change of 200 nM during an experiment, it would be unwise to display it on an axis with a span of 5000 nM. In this case the maximum level set when calculating the ratios would be set to about 500 nM to make the changes more obvious. If, however, this level is set, and ion concentrations exceed this value, then the system will simply show the maximum value for all higher ratios, resulting in a serious under-reading error. The better image-analysis systems software do, however, allow 'test ratioing' and inform the user when this condition is approached. Care must be taken, however, to 'test ratio' the *highest* ratio images of a sequence.

Another frequent situation resulting in under-reading of ion concentration occurs when a component of the image-collection system saturates, this may be the ADC, the CCD chip or the intensifier stages. The safest way to eliminate this type of potential artifact is to collect and inspect the original images before they are ratioed, as the changes in fluorescence during an experiment cannot always be predicted. Some systems which perform the ratioing operation 'on line', and do not allow access to the original images, leave themselves open to this artifact.

The problems that occur when images fall below the minimum detectable value are less serious and more easily overcome. With the better software, a minimum threshold level is set, below which all values are treated as zero and therefore do not show up on the image. In addition, any pixels which then have a zero value automatically result in a zero value ratio and are therefore not displayed on the ratio image. This problem may develop during an experiment due to a combination of changing ion concentration, photobleaching and probe leakage or exocytosis. It causes a deterioration in the ratio image quality as pixels disappear but does not indicate erroneous values.

16.4 PROBE LOADING

Probes may be loaded into the cytoplasm either by direct microinjection of the free acid probe through the plasma membrane or by a membrane permeable chemical form which is hydrolysed and trapped inside the cell. Microinjection may be regarded, from a chemical point of view, as the simplest option but requires delicate and expensive facilities. The most usual method of loading in imaging laboratories is therefore by the chemical method.

Most probes in general use are now available in an acetoxymethyl ester (AM) form which passes readily across most cell membranes. Once inside, non-specific esterases hydrolyse it to its free acid form. As the cell membrane is not permeable to this form of the probe it is trapped and concentrated inside. For example, loading with a solution of only 4 µM outside the cell for 30 min at 37°C may result in intracellular concentrations of fura-2 free acid of 50–100 µM.

As with all experimental systems, care must be taken to ensure that the probe is not significantly affecting the measurement of the parameter of interest. In this case there is a risk of the probe buffering the ion of interest and experimental controls must be devised to ensure this is not the case. A simple 'dose–response' curve for probe loading should reveal the maximum safe loading concentration for a particular probe in a cell type. In addition, it should also be noted that probes and/or the carrier (often dimethylsulphoxide, DMSO) are generally toxic to many cell types. They should, therefore, be used within 2–3 h of loading, preferably as soon as hydrolysation is judged to be complete, which may be 10–20 min after washing the extracellular probe away. A further wash at this point is advisable to remove any unhydrolysed probe which has passed out of the cells across the plasma membrane.

It should also be noted that the use of AM probes for loading can result in loading of organelles within the cells (Steinberg *et al*, 1987; DiVirgilio *et al.*, 1990). While these may give a good imaging signal, the probe inside may not be available to cytosolic free ions. Muted responses in some cell types may therefore be caused by the presence of organelles rather than small changes in ion concentration as indicated. Loading the cells with free acid probe by microinjection would not load the organelles and would reveal the real extent of the cytosolic free ion concentration.

16.5 ION CALIBRATION

Calibration of a ratiometric imaging system may be performed either by measuring the ratios of solutions of known ion concentration or by loading the cells with probe and forcing maximum and minimum ratios. First the loaded cells are made permeable with a suitable ionophore, allowing free ions from the extracellular medium to saturate the probe. This will achieve the maximum ratio possible. A chelator of the ion of interest is added to preferentially bind all the free ions to result in the minimum ratio obtainable. Maximum ratios obtainable by saturating the probe in *solutions* can be up to double that of saturated probe inside *cells*. It is therefore most desirable to calibrate a system using the cells to be used in experiments rather than any other method. The reasons for this difference are not entirely clear but evidence suggests that intracellular viscosity (Poenie, 1990) and compartmentalization of the probe are the main causes. The choice of ionophore will be governed by the ion of interest but it should be noted that some have been shown to be fluorescent at the wavelengths used to excite some probes.

Difficulties in calibration may be encountered if the cells of interest contain secretory granules. Permeabilizing the cells with ionophore has been shown also to permeabilize secretory granules, especially to H^+ ions. (Almers & Neher, 1985). Internal compartments of secretory granules may be quite acidic so this can result in a significant acidification in intracellular pH during calibration. A large variation in the values obtained for maximum (R_{max}) and minimum (R_{min}) ratios have been reported even in supposedly homogeneous cell populations. While the ion readily saturates the probe for the R_{max} value, chelating the ion to measure R_{min} may be more difficult and can take 15–30 min to reach a plateau. An alternative method to acquire the (R_{min}) value is to use $MnCl_2$ to quench the fura-2 signal (see Hesketh *et al.*, 1983).

16.6   CELL MOVEMENT AND FAST ION FLUXES

Cells used for image analysis of fluorescent probes are usually plated on thin glass coverslips to allow transmission of ultraviolet wavelengths and enable viewing on an inverted stage microscope. Cells used in ratiometric image analysis must remain stationary during the experiment. Usually, experimental protocol requires changing the extracellular medium either for the addition of an agonist or perfusion of the cells. It is therefore imperative that they adhere firmly to the coverslip. Even a small movement of the cells while adding an agonist is likely to result in a crescent-shaped artifact of apparently high ion concentration caused by shifting of the denominator image when

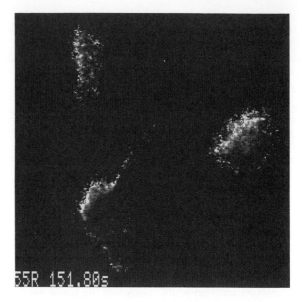

Figure 16.2 Movement of these cells during the addition of agonist results in a high ratio on the top left edge which may be interpreted as a fast polar response. This example of the GH3 cell line loaded with fura-2 is relatively obvious as all the cells in the field exhibit the effect. If, however, just one cell from a field of 20 moves, the artifact may be much more believable.

ratioing (Fig. 16.2). In the worst case cells are completely lost from the field of view. Some cell types in culture on the glass coverslip adhere quite well whereas others require it to be coated with a substrate such as poly-L-lysine to aid adhesion. The choice of substrate will depend on the cell type and application; for example, fibrinogen has been used to bind human platelets (Heemskert *et al.*, 1992) but care must be taken to ensure that it does not interfere with the normal function of the cell.

In a dual-emission ratiometric system, images are captured sequentially and then ratioed. The resultant images are therefore not true ratios as the individual images are not captured at the same time. The shorter the time between capturing image pairs, the better the approximation to a true ratio measurement. The better image analysis systems have capture rates up to 25 frames per second or 40 ms between images used to calculate a ratio. While the time between pairs may be set to be very much longer, thus enabling slower, longer experiments to be performed while ensuring the ratio image is a close approximation to a true ratio.

Very fast changes in ion concentration will also be subject to a similar error. This will result in under-read values during fast transient increases in ion concentration and over-read values as the concentration decreases. The only solution in this case is to increase the data capture rate by moving to a faster system. While they have other limitations, ratio photometric systems may be more appropriate with capture rates of up to 200 Hz.

16.7 AUTOFLUORESCENCE

Many of the probes employed to monitor intracellular ion concentrations require excitation wavelengths well into the ultraviolet region. Unfortunately, a number of natural peptides are also fluorescent in this region. This problem is especially prevalent in plant tissue and mammalian pancreatic cells. If the component from autofluorescence is small compared to the contribution from the probe it may be disregarded. Otherwise, levels of autofluorescence must be recorded before loading so they may be subtracted from the loaded images. In practice, this is rather difficult as the excitation intensity must be set, to measure autofluorescence, before loading with probe. Preliminary experiments must, therefore, establish a loading protocol which gives a consistent level of fluorescence. Plated cells must also be either kept on the microscope stage while loading or accurately repositioned afterwards.

16.8 INTERACTIONS BETWEEN MULTIPLE PROBES

Currently, the better imaging systems allow simultaneous monitoring of multiple probes. Intracellular pH and calcium, for example, are known to interact and play pivotal roles in cell signalling and secretion. BCECF and fura-2 may be used to monitor pH and calcium, their excitation wavelengths are 440, 490 nm and 340, 380 nm respectively. Emission spectra from both probes overlap at about 520 nm so they may both be monitored using a single emission filter set. There will, however, be significant optical cross-talk between the probes both in the combined emission spectrum and as emission from the lower wavelength probe exciting the other. Fura-2 in this case emits a broad spectrum, peaking at 510 nm, so there is a significant proportion at 490 nm, the upper excitation wavelength of BCECF. Experimental protocols must, therefore, be devised to measure both types of cross-talk in the cells of interest so corrections may be made (Fig. 16.3). It is also advantageous to keep the lower wavelength probe concentration as low as possible consistent with adequate signal for good signal-to-noise. This may usually be achieved by weighting the loading concentration of each probe (Zorec et al., 1993).

Care must also be taken when loading the ester form of multiple probes. It has been found essential to load probes simultaneously as the probes appear to compete for the intracellular esterase activity. Great difficulty has been experienced with sequential loading of multiple probes.

16.9 AVERAGING AND INTENSIFIER NOISE

Some applications, such as monitoring fast ion fluxes, call for maximal capture rates. Others, such as

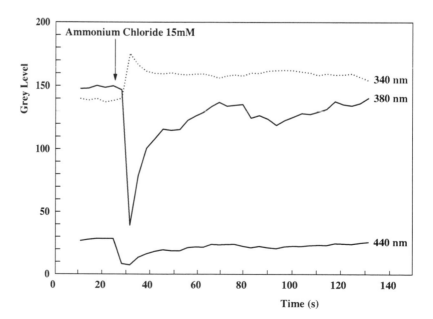

Figure 16.3 The optical cross-talk of fura-2 loaded bovine lactotrophs from excitation at 380 nm and 440 nm. Analysis here reveals cross-talk of 14.4 ± 3.6% which must be corrected for when using multiple probes.

mapping the spatial distribution of ion concentrations within cells, may call for slower rates with higher spatial resolution. For the reasons mentioned above, most video imaging systems currently use intensified CCD cameras to capture images. They do, however, have an inherent problem in that microchannel plate intensifiers introduce a degree of electronic noise into the final image. The most effective way to improve the signal-to-noise ratio is to average a number of video frames (Mason *et al.*, 1990). This may either be achieved in software after storing the frames or, in the better systems, on-line prior to storage and subsequent analysis. There is, of course, a trade-off between temporal resolution and signal-to-noise ratio so the experimental objectives must dictate the balance. Generally, averaging four or eight video frames gives adequate signal-to-noise ratio for many purposes which gives a temporal resolution of 0.3 to 0.6 s between ratio image pairs.

16.10   PROBE LEAKAGE AND EXOCYTOSIS

The cell membrane is not completely impermeable to the free acid form of fluorescent probes. There is a small leakage of probe to the extracellular medium (DiVirgilio *et al.*, 1990). In a constantly perfused system this causes no problems other than a very small drift of intensity which is easily accommodated by ratiometric image analysis. In static systems with no perfusion, long-term leakage may increase the background fluorescence slightly but the effect is minimal.

A potentially serious effect has, however, been reported in mast cells (Almers & Neher, 1985). Acetoxymethyl ester loading here appears to load secretory granules with probe. On stimulation, the granules are released from the cell membrane to deposit their high concentrations of probe to the immediately surrounding area. This appears as a ring of high ion concentration around the edge of the cell and may be misinterpreted as an ingress of ions around the periphery. Examination of the images prior to ratioing would reveal the difference, a loss of probe would result in a sharp reduction of emission at both ratio wavelengths.

16.11   PROBE KINETICS

Most of the work on probe kinetics has been performed in solutions rather than cells. Measurements of association and dissociation constants have been performed for fura-2 and azo-1 (Kao & Tsien, 1988) by the temperature jump relaxation method and for fura-2 and indo-1 (Jackson *et al.*, 1987) by stopped flow measurements. The practical implications of this work are that fura-2 requires 5–10 ms to reach equilibrium at 20°C in solution of ionic strength 140 mM. While the response will be faster at 37°C, other factors such as viscosity and spatial microheterogeneity may slow intracellular measurements. Some fast calcium fluxes may, therefore, be misinterpreted as the probe kinetics may be the limiting factor. However, the main rate-limiting factor for a system will be the maximum capture rate of the system. For imaging systems this will generally be video frame-rate or 40 ms per image. Photometric systems can run much faster and, in some cases, exceed the response time of the probe.

REFERENCES

Almers W. & Neher E. (1985) *FEBS Lett.* **192**, 13–18.
Becker P.L. & Fay F.S. (1987) *Am. Phys. Soc. Special Comm.* C613–C618.
DiVirgilio F., Steinberg T.H & Silverstein S.C. (1990) *Cell Calcium* **11**, 57–62.
Grynkiewicz G., Poenie M. & Tsien R.Y. (1985) *J. Biol. Chem.* **260**, 3440–3450.
Heemskerk J.W.M., Hoyland J., Mason W.T. & Sage S.O. (1992) *Biochem. J.* **283**, 379–383.
Hesketh T.R., Smith G.A., Moore J.P., Taylor M.V. & Metcalfe J.C. (1983) *J. Biol. Chem.* **258**, 4876–4882.
Jackson A.P., Timmereman M.P., Bagshaw C.R. & Ashley C.C. (1987) *FEBS Lett.* **216**, 35–39.
Kao J.P.Y. & Tsien R.Y. (1988) *Biophys. J.* **53**, 635–639.
Mason W.T., Hoyland J., Rawlings S.R. & Relf G.T. (1990) *Methods Neurosci.* **3**, 109–135.
Poenie M. (1990) *Cell Calcium* **11**, 85–91.
Steinberg S.F., Bilezikian J.P. & Al-Awquti Q. (1987) *Am. Phys. Soc. Special Comm.* C744–C747.
Tsien R.Y. (1981) *Nature* **290**, 527–528.
Zorec R., Hoyland J. & Mason W.T. (1993) *Eur. J. Physiol.* (in press).

CHAPTER SEVENTEEN

Confocal Microscopy – Principles, Practice and Options

C.J.R. SHEPPARD

Physical Optics Department, School of Physics, University of Sydney, NSW, Australia

17.1 INTRODUCTION

Confocal microscopy is particularly advantageous for 3D imaging of thick objects as a result of its optical sectioning property (Sheppard, 1987). It is widely used in the fluorescence mode for imaging biological objects of various types, but is also used in the bright-field reflection mode for imaging objects of different forms.

Light from a laser is focused by an objective lens onto the object, and the reflected, or fluorescent, light focused onto a photodetector via a beam-splitter. In the confocal microscope, a confocal aperture, or pinhole, is placed in front of the photomultiplier tube (or other) detector (Fig. 17.1). An image is built up by scanning of the focused spot relative to the object, and is usually stored in a computer imaging system.

Confocal microscopy is one particular imaging mode of the scanning optical microscope, which exhibits various advantages over conventional instruments as a result of the image being formed by a scanning technique.

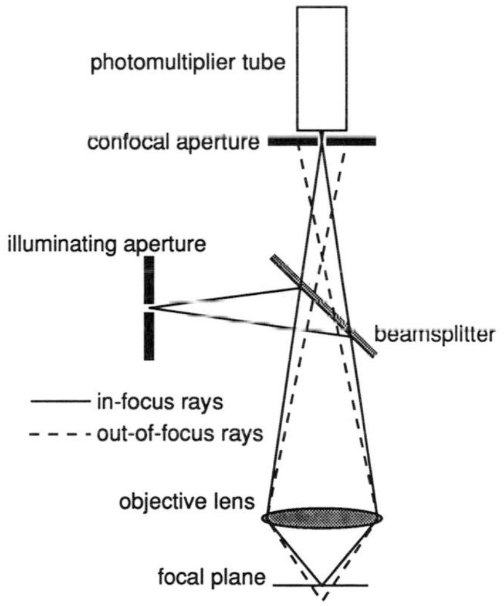

17.2 ADVANTAGES OF SCANNING

Broadly, the advantages of scanning optical microscopy stem from two main properties. First is the fact

Figure 17.1 Schematic diagram of a confocal microscope. Light originating from points away from the focal plane is defocused at the confocal aperture and thus detected weakly.

that the image is measured in the form of an electronic signal, which allows a whole range of electronic image-processing techniques, both analogue and digital, to be employed. These include image-enhancement techniques such as frame-averaging, contrast enhancement, edge enhancement, and image subtraction to show changes or movement; image-restoration techniques for resolution enhancement and noise reduction, and image-analysis techniques such as feature recognition and cell sizing and counting.

Second is the property that imaging in a scanning microscope is achieved by illuminating the object with a finely focused light spot. This allows a number of novel optical imaging modes to be employed such as confocal imaging or differential phase contrast, but also introduces the possibility of imaging modes in which the incident light spot produces some related effect in the specimen which can be monitored to produce an image.

A particularly important class of such methods occurs when the wavelength of the detected radiation differs from that of the illumination. An advantage of using scanning techniques for such spectroscopic imaging is that imaging methods, performed with the incident radiation, are separated from wavelength selection and analysis of the emitted radiation, thus simplifying system design and resulting in superior performance. The detection system may also have greater sensitivity because it does not have to image. This is of great advantage in fluorescence microscopy, which also results in the further advantage that the resolution is determined by the shorter incident wavelength, rather than the longer fluorescence wavelength.

Fluorescence, or luminescence, microscopy can give information concerning spatial variations in excitation states, binding energies, band structure, molecular configuration, structural defects, and the concentration of different atomic and molecular species.

Use of a pulsed laser allows investigation of transient effects such as the lifetime of excited states, and capture and emission cross-sections. Other examples of spectroscopy which may be performed using scanning techniques include absorption spectroscopy, Raman spectroscopy, resonance Raman spectroscopy, coherent anti-Stokes Raman spectroscopy (CARS), two-photon fluorescence, photoelectron spectroscopy, and photo-acoustic spectroscopy.

17.3   CONFOCAL MICROSCOPY

In a conventional microscope the object is illuminated using a large-area incoherent source via the condenser, and each point of the object imaged by the objective lens. It is the objective which is responsible for determining the resolution of the system. If the image is now measured point by point by a detector of small aperture, the image is unchanged providing the detector is small enough. In a scanning microscope, however, we have a point source and a large area detector, rather than a large area source and a point detector. Reciprocity argues that image formation is identical in a scanning microscope and a conventional microscope. However, now it is the first lens (also termed the objective, but sometimes the projector) which determines the resolution.

In the confocal microscope (Fig. 17.1) we use both a point source and a point detector, achieved by placing a pinhole in front of the detector, so that in this case both lenses take part equally in the image-formation process and the resolution is improved by a factor of about 1.4 times. Figure 17.2 shows the theoretical image of a single point object, illustrating the sharper image produced by the confocal microscope. Here the coordinate v is a dimensionless optical coordinate, related to the true radius r in the image by

$$v = (2\pi r \sin\alpha)/\lambda \qquad [17.1]$$

where $\sin\alpha$ is the numerical aperture of the objective and λ the wavelength of the illumination. The curves apply both for bright-field and also for fluorescence if the wavelength of the fluorescent light is approximately equal to that of the incident light.

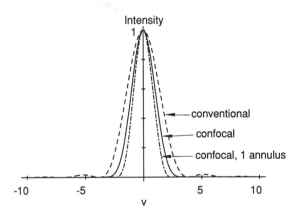

Figure 17.2 The image of a single point in conventional and confocal microscopes with circular pupils, and also in a confocal microscope with one circular and one thin annular pupil. The confocal image is a factor 1.4 sharper than the conventional.

Also shown is the confocal image for a system when an annular mask is placed in the pupil of the objective for any one of the two passes through the objective, thus resulting in an even sharper image. Other advantages of confocal microscopy are, first, that out-of-focus information is rejected by the pinhole so that an optical section is imaged and, second, that unwanted scattered light is also rejected by the pinhole. Confocal systems can operate in confocal transmission, but commercial systems usually operate in the reflection geometry. This makes the operation of the system much easier as the point detector can be arranged to coincide with the image in the beamsplitter of the point source, resulting in coincidence of the illuminating and detection spots. Often confocal microscopes are used in a fluorescence mode, rather than detecting the light reflected by the object.

Light emanating from regions of the specimen separated from the focal plane are defocused at the pinhole plane and hence rejected, thus resulting in an optical sectioning effect. By scanning in the depth direction sequential sections can be studied and a complete three-dimensional image built up. The 3D data can, for example, be stored in a computer for subsequent processing and display.

Of course, 3D images are difficult to display directly, so instead we can extract sections oriented in some arbitrary direction. For example, xz images are sections parallel to the system axis. We can also produce projections in an arbitrary direction, in which the depth information is suppressed. Projections can be produced either by summation of sections resulting in an extended focus image, or alternatively by detecting the peak signal in depth, giving an autofocus image. Either approach can be achieved using digital or analogue methods, and an extended focus image can also be generated directly by photographic integration on a cathode ray tube. Two projections produced at slightly differing angles, either by the extended focus or autofocus method, give a stereoscopic image pair.

As well as recording the peak signal in depth, as in the autofocus method, we can record the depth position in order to locate its position in depth. This depth information can be displayed as a surface profile image in grey levels or colour, or can be combined with the peak information to produce colour-coded images or reconstructed views. The sensitivity of the depth measurement is about a few nanometres, or can be better than 0.1 nm using confocal interference techniques.

Projections in different directions can be computed and stored for animated display, but at present information content of a full 3D image is too great to permit real-time rotation.

By using a flat reflecting surface, such as a mirror, as specimen the imaging performance in the depth direction can be investigated by axial scanning and observation of the resulting defocus signal (Sheppard & Cogswell, 1990). This can also be achieved either by tilting the mirror slightly, scanning in a transverse plane and examining the line-scan signal, or by producing an xz image and again extracting a line-scan. The defocus signal gives much information about the performance of the optical system. Ideally, it should be a smooth, narrow response with weak sidelobes. In practice it is made broader as a result of the finite size of the pinhole and often degraded by lack of alignment or the presence of aberrations. In particular, the presence of strong sidelobes or auxiliary peaks is very detrimental to three-dimensional imaging performance. Confocal microscopy is very sensitive to the presence of small amounts of aberration such as spherical aberration or astigmatism which can be introduced by the use of incorrect coverglass thickness, or the effects of the mounting medium when focusing deep into the specimen. In these cases, observation of the defocus signal allows optimization of the imaging performance by altering the tube length at which the objective is used or by insertion of correction lenses.

The transverse resolution of the microscope is proportional to the numerical aperture of the objective. However, the axial imaging performance depends more strongly on objective aperture. The axial imaging performance can be quantified by observing the defocus signal. The width of the peak of the axial response varies considerably with aperture, so that in order to get good axial imaging it is necessary to use the largest possible aperture. Figure 17.3 shows the width of the axial response in wavelengths at which the intensity has dropped to one-half (the full-width half-maximum, or FWHM). The variation in width for both dry and oil-immersion objectives are shown. It should be noted that the response is sharper for a dry, rather than an oil-immersion, objective of the same numerical aperture, although of course higher numerical apertures are available for immersion objectives. It should also be remembered that immersion objectives also have better aberration correction than dry objectives.

At present, most commercial confocal microscopes are designed for fluorescence operation, so that to many people confocal and fluorescence are almost synonymous. However, it should be stressed that there are many useful non-fluorescent applications using confocal brightfield techniques. The main advantages

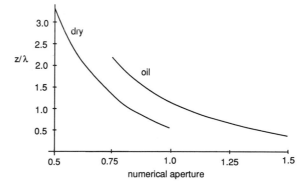

Figure 17.3 The width of the axial response in a confocal microscope from a plane reflector at which the intensity has dropped to one-half (FWHM).

of the fluorescence mode of confocal microscopy are that specific stains can be introduced, and also that high-quality 3D images can be produced without problems arising from coherent optical noise (speckle).

As described earlier, resolution of a scanning fluorescence microscope is superior to that in a conventional fluorescence microscope. There is a further improvement for confocal fluorescence microscopy, so that, for example, if the fluorescence wavelength is 1.5-times the primary wavelength the cut-off in the spatial frequency response is 1.5-times as great in a scanning compared with a conventional fluorescence microscope, but 2.5-times as great in a confocal fluorescence system. The price one pays for this dramatic improvement in resolution is the decreased signal strength of the confocal arrangement, although we argue later that the reduction in signal is caused by the rejection of unwanted, out-of-focus light.

Most commercial confocal fluorescence systems use an argon ion laser to provide a main line at 488 nm which is close to the absorption peak of FITC. The other strong line of the argon ion laser at 514 nm can be used to excite rhodamine or Texas Red. An alternative laser for fluorescence work is the helium–cadmium laser which gives lines at 442 nm and 325 nm. The former can be used with acridine orange, fluorescein and Feulgen–Schiff staining, the latter with Hoeschst or DAPI. Either helium–neon (633 nm) or krypton ion (647 nm) can be used to excite chlorophyll *b*. The krypton ion laser can give a large number of lines throughout the visible and ultraviolet. A further laser of interest is the frequency doubled neodymium YAG laser at 532 nm.

Confocal microscopy in the bright-field (non-fluorescence) reflection mode has many applications in the materials science and industrial areas, and can also be exploited in biological studies (Cogswell & Sheppard, 1990). The tandem scanning microscope, which uses a rotating aperture disk for scanning, is often used in a bright-field mode as signal levels are frequently low in the fluorescence mode. Reflected bright-field has the advantage of giving extremely sharp depth imaging, sharper than in confocal fluorescence, but in some cases there can be difficulties in interpretation as a result of coherent noise (speckle). This is not usually a problem when the specimen consists of reflecting surfaces, but can be when it is necessary to investigate refractive index variations within a semitransparent object. The coherent noise is reduced by using the autofocus, or more effectively by the extended focus, technique and in some cases the visibility of the image features can be improved by filtering.

A further important technique of confocal reflection microscopy is the use of immunogold probes (Cogswell & Sheppard, 1990). In this case the gold particles are spatially well-separated so that coherent interactions are not important, and the gold particles can be located in three-dimensional space. Scanning microscopes can be very sensitive for detecting scattering from these gold particles, and preparations with particles as small as 5 nm in diameter without silver enhancement have been imaged. In many cases, however, the lower limit to the size of the particles is set by the requirement that the scattering, which varies with the fourth power of the diameter, is strong compared with scattering by the refractive index variations in the specimen itself.

Although much of the early work on confocal microscopy was performed in the transmission mode, this is now not nearly as frequently used as the confocal reflection geometry. The reason for this is that there is extreme difficulty in aligning, and maintaining alignment during scanning, due to refractive effects in the specimen. This can be to some degree alleviated by using a double-pass method, where the transmitted light is reflected back through the specimen and detected using the usual reflected light detector (Cogswell & Sheppard, 1990). This method can also be used to increase the signal in confocal fluorescence microscopy.

Confocal transmission retains the resolution improvement of confocal imaging, and also results in an improvement in depth imaging, but not to such a marked degree as in confocal reflection. Confocal transmission microscopy can also be performed in a differential phase-contrast mode and by mixing with a coherent reflection signal can result in further improvement in depth imaging.

Scanning techniques are well-suited to interference methods, and in particular confocal interference microscopy is a powerful technique (Sheppard *et al.*, 1988). This is possible because confocal imaging is a coherent process, and is further simplified experimentally by the long coherence length of lasers. Confocal methods have the major advantage for interference microscopy that the shape of the reference beam wavefront is immaterial, only its phase and amplitude at the detector pinhole being of importance, removing the requirement for matched optics and making alignment much less critical. Furthermore, it is possible to combine the system with multiple detectors and real-time processing to extract image information. Confocal interference microscopy can be used for investigation of refractive index or surface height variations, or for obtaining the signal phase for restoration of images or measurement of system aberrations.

Confocal interference microscopy can be performed in transmission, using a Mach-Zehnder arrangement, or in reflection using a Michelson geometry. High sensitivity can be achieved using either phase-shifting or heterodyne technique. The heterodyne technique itself also exhibits an imaging property, which allows non-confocal images to be formed without an objective lens being necessary. This may prove useful in focusing deep into a specimen which would be impossible with a real lens. Alternatively combining illumination of the object by a focused spot, and heterodyne detection, can result in confocal imaging without the use of a physical pinhole.

As described earlier, the axial imaging performance of a confocal microscope can be investigated by observation of the variation with defocus of the signal from a reflecting plane as specimen (the defocus signal). Confocal imaging is a coherent technique so that if only the intensity of the defocus signal is measured the phase information is lost. However, if interference methods are used to extract the phase and amplitude, the aberrations of the imaging system can be determined by a simple Fourier transformation of the defocus signal (Sheppard et al., 1988). This is useful for optimization of the imaging system, or to provide information for image reconstruction.

17.4 PRACTICAL ASPECTS

In a scanning microscope the object is illuminated with a focused light spot which is scanned relative to the object (Fig. 17.1). This can be achieved either by scanning the light spot, or by scanning the object itself. Most commercial instruments at present scan the light spot. The two most widely used methods employ galvo-mirror scanners for both the x- and y-scan, or a Nipkow disk as is used in the tandem scanning microscope. Galvo mirrors are either of the resonant variety which can oscillate at high speed but are fixed-frequency devices, or more usually of the feedback-stabilized type, which can scan at a line frequency of about 1 kHz. Nipkow disk scanners, in which a disk with an array of holes is rotated, because they have multiple apertures, allow the use of a white light source. Their main disadvantages are that signal level can be low, and that a television camera must be used in order to produce an electronic image signal. Alternative beam-scanning systems include polygon mirror scanners, which achieve high scanning speeds but are difficult to sychronize, and acousto-optic scanners, which allow TV scanning rates but suffer from chromatic variations which rule them out for fluorescence applications without use of special imaging geometries.

The main advantage of object scanning is that the optical system is then completely unchanging during scanning so that the imaging properties are unvarying across the field. Beam-scanning systems, on the other hand, can exhibit brightness variations across the field,

a fall-off resolution at the edges of the field, and also noticeable curvature of field. Object scanning is thus preferable for quantitative work or in image reconstruction. The disadvantages of object scanning are that imaging can be rather slow, taking perhaps a few seconds to record a single frame, and that use of electrical probes is more difficult. The speed consideration is perhaps not so important when it is realized that often long scan times are necessary in order to collect sufficient light to produce a low-noise image.

The final method of scanning which should be considered is mechanical scanning of the objective lens. This again has the advantage that the light travels on-axis through the optical system, thereby giving good quantitative imaging. On the other hand, again speed is rather low.

A number of alternative arrangements are used for beam-scanning microscopes. Usually a microscope eyepiece, or lens of similar design, is used to produce a focus which subsequently illuminates the objective lens. In order that the beam fills the objective lens aperture, the axis of rotation of the scanning mirrors must be situated close to the plane of the entrance pupil of the eyepiece. A single mirror placed here and scanned in both the x and y directions is the simplest design. Alternatively, two separated galvo mirrors can be used, coupled by a telecentric system. Finally, two close-coupled galvo mirrors can be used. Sometimes the axis of rotation of one is offset so that it both translates and rotates.

In an object scanning system, the optical system can be simplified as the system is operated in an on-axis condition. It is found that it is possible to optimize performance by small corrections to the effective tube length of the objective. Mechanical object scanning is achieved using electromechanical devices, stepper motors or piezoelectric actuators. One feature which must receive some attention is to ensure that the plane of scanning is accurately located. This can be achieved by mounting the specimen stage on leaf springs or stretched wires.

An optical system only gives its best performance if it is accurately aligned. In particular a confocal microscope must be aligned correctly or it will exhibit artifacts, especially with a small pinhole size. First of all the beam expanding system must be adjusted and the beam collimated so that the objective is used at its correct tube-length. The easiest way of aligning the system is to examine a planar object with a pinhole of large diameter. Without refocusing the object the pinhole size is reduced and its axial and transverse position adjusted for maximum signal. Alternatively, the three-dimensional image of a point-like object can be observed.

Many types of detector can be used for scanning

optical microscopy. If signal level is high, as, for example, in transmission or in reflection from a surface such as bone or tooth, a simple photodiode can be employed. For confocal fluorescence applications signal level is low, and is reduced further as the thickness of the optical section is reduced. Then highly sensitive detectors such as photomultiplier tubes, perhaps cooled, and sometimes with photon counting are advantageous.

The relevant properties of a detector are quantum efficiency, sensitivity, dynamic range and linearity. Conventional photocathodes have a quantum efficiency of only about 20%, but this figure can be raised to above 30% by using negative electron affinity photocathodes. Silicon detectors of either the photodiode, avalanche photodiode or CCD variety can have quantum efficiencies greater than 80%. Avalanche photodiodes in principle combine higher quantum efficiency with the ability to photon count with pulse height analysis for rejection of dark current. However, at present there are difficult practical problems in their use. Charge-coupled devices combine high quantum efficiency with the facility for integration within the device and the possibility of cooling to improve noise performance. CCDs are usually of the area or linear array variety, and in some cases additional information can be extracted from this spatial information.

At present there are no objective lenses specially designed for scanning microscopy, and so objectives must be chosen from commercially available types. For laser microscopy the objective only need be corrected for one wavelength (or two, including a fluorescent wavelength) which should allow lenses of increased numerical aperture or working distance to be developed. Similarly, in microscopes with on-axis scanning, off-axis aberrations are unimportant, giving further flexibility in the design. For example, an immersion lens, designed for one wavelength and with a limited field of view for on-axis scanning, with numerical aperture of 1.4 and working distance of 1 mm seems feasible.

Achromat lenses have some advantages over apochromat lenses for some applications as they have lower loss, as well as being considerably cheaper. We have found fluorite lenses to be a good compromise for on-axis scanning. However, for beam scanning the off-axis aberrations and field curvature are too large. Again, with beam-scanning confocal microscopes, there is a need for high aperture lenses (to give good collection efficiency) of low magnification. These are not necessary for use with on-axis microscopes as then one objective can be used to cover the whole range of magnifications, simply by altering the amplitude of scan. Confocal microscopes are more sensitive to aberrations than conventional instruments, particularly for imaging in the depth direction. So incorporation of

a correction collar is useful. Finally, for many biological studies water-immersion objectives are preferable, because with uncovered specimens this removes both the reflection and aberrations produced by the coverglass–specimen interface.

It is worth pointing out that many of the early scanning laser microscopes did not use a computer for image manipulation, but rather used analogue electronics together with a long persistence display. The main impetus for employing digital techniques was to provide image storage to improve the real-time observations of images, but in fact very high quality images with several thousands of lines resolution can be produced by photographing directly the cathode-ray tube display. Similarly, many forms of image processing such as contrast enhancement, filtering, image addition and subtraction, and so on can be achieved using analogue methods. The use of digital methods of course greatly extends the flexibility of the system, but nevertheless analogue processing facilities are worth retaining in order to improve the dynamic range of the recorded data.

A single 512 × 512 image contains a quarter of a megabyte of information. This means that a 3D image consisting of many sections requires large memory and also processing time for 3D manipulation. For this reason images which consist mainly of 'empty space' can be stored and processed alternatively using a vector scan method. Actually, good stereoscopic effects can be obtained from few sections, and it is also possible to store projections directly, rather than sections, which can also greatly reduce the quantity of data.

Most scanning microscopes generate a single image in a time of the order of a second, which means that with a commercial imaging system a slow-scan input is necessary. Alternatively personal computers nowadays have large enough memories to permit direct storage of the images in their RAM. Personal computers can provide most of the functions necessary for scanning microscopes, but are not fast enough for real-time manipulation of 3D images.

A range of standard image enhancement methods can be used to advantage in scanning optical microscopy. These include contrast enhancement by linear stretching or histogram equalization, edge enhancement by filtering in either the spatial or Fourier domain, low-pass or median filtering to reduce noise in images, image averaging and so on. Most of these methods can be employed with 3D as well as 2D data sets.

Once a 3D image has been stored, projections in arbitrary directions can be produced by image rotation. This is a computationally intensive process, and an alternative is to stack sections with an appropriate pixel offset between adjacent sections. The sections may be stacked by summation, corresponding to the extended focus method or by selecting the peak signal, corresponding to the autofocus method, as described earlier.

17.5   SYSTEM PERFORMANCE

For a small pinhole size the system behaves as a true confocal microscope, but for large pinhole sizes imaging performance is identical to that in a conventional microscope. Thus, between these two limits, the size of the pinhole affects the imaging performance greatly. In practice, microscope users often open up the size of the pinhole in order to get more signal from a weakly fluorescent or scattering object, so it is important to consider what effect this will have on the various imaging properties.

Figure 17.4 shows the effect of pinhole normalized radius v_d, defined by equation [1] for the *image* space, on the strength of the signal from a planar in-focus object, and the axial resolution of a planar object, assuming the system is shot noise limited (Sheppard *et al.*, 1991). The curves apply for a confocal fluorescence system. As the pinhole size is increased the signal increases, but the axial resolution decreases. It is seen that the signal rises quickly with increasing pinhole size, reaching about 70% for a value of v_d of 4. Thereafter, there is little increase in signal level from a planar object. One might ask why then microscope users sometimes open up the pinhole much larger than this? The answer is that by using a larger pinhole the axial resolution is decreased and hence for a thick object we collect a larger signal, simply because the optical sectioning effect is weaker.

The behaviour of the confocal fluorescence system is shown for both pinhole and slit forms of confocal aperture. It has been claimed that slit apertures can give improved signal strength (Sheppard & Mao, 1988), as indeed is apparent from Figure 17.4, but in fact this occurs only because a thicker section is imaged, i.e. it is directly a result of the decreased axial resolution. Overall, therefore, there seems little to be gained from using a slit aperture from the point of view of imaging performance.

Assuming the system to be shot noise limited the variation in the signal-to-noise ratio with pinhole size has been investigated. The signal-to-noise ratio is found to increase monotonically with pinhole size (Sheppard *et al.*, 1991). In practice, there will be stray light present, which will also contribute to the noise observed, and the detector can also cause additional noise. These effects can greatly influence the noise behaviour of the system.

We have made measurements of signal and strength of stray light using an object scanning microscope (Sheppard, 1991). Very good agreement was found between the variation in the strength of the signal with pinhole size measured and predicted by theory. These measurements were used to calibrate the optical coordinate v_d. If the stray light is assumed to be uniformly distributed over the pinhole plane, its strength is proportional to the pinhole areas, which was found to give a good fit to the experimental data. The flare varied between 10^{-9} and 10^{-6} of the total signal for the pinhole sizes used. For small pinhole sizes the ratio signal/stray light became constant at about 2.5×10^7. This means that, in this particular microscope, the signal can be as low as 4×10^{-8} relative to the incident power before it is equal in strength to the stray light.

As the stray light is constant in strength throughout the image, it can be subtracted from the signal electronically to improve contrast at very low signal levels. However, its presence does affect the signal-to-noise ratio, which rises from zero for small values of v_d, and eventually drops for large values of v_d. Interestingly, as the signal strength decreases, there is less latitude in the choice of pinhole size. The optimum pinhole size for maximizing signal-to-noise ratio also decreases. For a relative signal strength of 4×10^{-8}, for example, the optimum pinhole size has reduced to $v_d = 2.3$. This argument suggests that opening up the pinhole to detect more light may not be desirable as more stray light will be detected also.

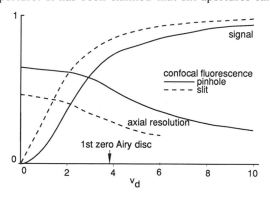

Figure 17.4 Comparison of the axial resolution and signal strength for a confocal fluorescence system with pinholes and slits of varying sizes. The first zero of the Airy disk in the pinhole plane is shown for comparison.

REFERENCES

Cogswell C.J. & Sheppard C.J.R. (1990) In *Confocal Microscopy*, T. Wilson (ed.). Academic Press, London, pp. 213–243.

Sheppard C.J.R. (1987) In *Advances in Optical and Electron*

Microscopy, Vol. 10, R Barer & V.E. Cosslett (eds). Academic Press, London, pp. 1–98.

Sheppard C.J.R. (1991) *Micron* **22**, 239–243.

Sheppard C.J.R. & Cogswell C.J. (1990) *J. Microsc.* **159**, 179–194.

Sheppard C.J.R. & Mao X.Q. (1988) *J. Mod. Opt.* **35**, 1169–1185.

Sheppard C.J.R., Hamilton D.K. & Matthews H.J. (1988) *Proc. SPIE* **1028**, 92–95.

Sheppard C.J.R., Cogswell C.J. & Gu M. (1991) *Scanning* **13**, 233–240.

CHAPTER EIGHTEEN

Confocal Raman Microspectroscopy

G.J. PUPPELS, M. VAN ROOIJEN, C. OTTO & J. GREVE

Biophysical Technology Group, Department of Applied Physics, University of Twente, 7500 AE Enschede, The Netherlands

Fluorescence methods are at this moment the most successful research tools in cell biology. Their fundamental disadvantage is, however, that with the exception of those cases where intrinsic chromophores are investigated, probes have to be introduced into the system under study. The fact that the manner in which system and probe influence each other is often unknown, can hamper the interpretation of experimental results. For this reason the development of direct analysis techniques is important.

Raman spectroscopy is a technique that provides molecular structural and compositional information without using probes. Recent advances in instrumentation have improved the sensitivity of the technique, now enabling the study of chromosomes and single living cells.

18.1   INTRODUCTION

Many of the questions posed by modern cell biology are tackled by means of fluorescence techniques. As amply illustrated in other chapters of this volume, the key to solving these questions often lies in the development of suitable fluorescent probes. However, if the aim is to obtain information about a biological system it should always be kept in mind that fluorescence methods work indirectly. The introduction of a probe means that the information obtained is about the system with probe. Unless it is precisely known what influence the probe has on the system and, vice versa, how the system affects probe characteristics, the interpretation of experimental results has to be made with great caution.

This point is illustrated by the difficulties that were encountered in the interpretation of some well-known discoveries made by fluorescence microscopy.

In 1968 Caspersson and co-workers found that staining of fixed metaphase chromosomes with quinacrine-mustard gave rise to a fluorescent banding pattern on the chromosomes (Caspersson et al., 1968). The question was, of course, what caused these banding patterns. The initial idea was that the mustard moiety would preferentially interact with DNA rich in guanine (G)–cytosine (C) basepairs. Experiments with quinacrine, however, led to the same results so that this explanation could not be upheld. Instead it was found that in vitro adenine (A)–cytosine (C)-rich DNA enhanced quinacrine fluorescence, while GC-rich DNA quenched quinacrine fluorescence, (Weisblum & de Haseth 1972; Pachman & Rigter,

1972). The Q-banding pattern, as it was called, therefore appeared to reflect local variations in chromosomal DNA base composition, AT-rich regions fluorescing brightly. But that turned out to be only a partial explanation, for it was found that, for example, the centromeric heterochromatin of mouse chromosomes, containing AT-rich satellite DNA, showed dimmer fluorescence than the less AT-rich chromosome arms (Rowley & Bodmer, 1971). During interphase the centromeric heterochromatin fluoresced brightly (Natarajan & Gropp, 1972). It became clear that the presence of proteins exerted a strong influence on the quinacrine fluorescence of chromosomes, necessitating further studies of the binding mechanisms and fluorescence characteristics of the dye (Sumner, 1982). Even now, after more than 20 years of intensive research and the discovery of many other types of metaphase chromosome banding patterns, our understanding of this phenomenon is far from complete.

A more recent discovery was the binding of fluorescently labelled anti-Z-DNA antibodies to polytene chromosomes (Nordheim *et al.*, 1981; Arndt-Jovin *et al.*, 1983). This attracted a great deal of attention because it could indicate that this left-handed double helical DNA form is present and possibly has a function in nature. A clear positive staining of the chromosomes was only observed, however, after fixation of the chromosomes in (3:1) alcohol–acetic acid or after incubation of unfixed chromosomes at low pH, prior to their incubation with the labelled anti-Z-DNA antibodies (Robert-Nicoud *et al.*, 1984). The question that remains is, therefore, whether the absence of or only marginal antibody binding to unfixed chromosomes means that there is no Z-DNA and that fixation or low pH treatment induces this DNA structure. An alternative explanation could be that the Z-DNA in the unfixed chromosomes is not accessible to the antibodies, e.g. because it is already complexed with a Z-DNA binding protein, or because the antibodies cannot penetrate the compact chromosomal chromatin. Another question that can be asked with respect to the use of anti-Z-DNA antibodies in this case is how well the antibodies can be screened against other (unknown) DNA structures that may be induced by the chromosome fixation (Puppels, 1991).

Ideally, the introduction of a probe in a system should not lead to any perturbation of the system nor of the probe. In mathematical terms:

$$(\text{system}) + (\text{probe}) = (\text{system} + \text{probe}) \quad [18.1]$$

The two examples above underline, however, that this is not usually the case and that despite the versatility of fluorescence methods a definite need remains for direct analysis techniques, i.e. techniques that do not require the use of probes. This chapter deals with one such technique: Raman spectroscopy. Continuing progress in instrumentation has brought this technique to a level of sensitivity where it can now be applied in studies of single living cells and chromosomes (Puppels *et al.*, 1990b).

18.2   RAMAN SPECTROSCOPY

Raman spectroscopy provides information about molecular composition and structure and about interactions between molecules. The technique is based on an inelastic light scattering process, predicted in 1923 by Smekal (Smekal, 1923) and first observed by the Indian scientists Raman and Krishnan in 1928 (Raman & Krishnan, 1928). A theoretical treatise of Raman scattering has been given in a number of textbooks (e.g. Koningstein 1971; Long, 1977) and will be omitted here.

18.2.1   Raman scattering

The processes of Stokes, anti-Stokes and resonance Raman scattering and of Rayleigh and fluorescence scattering are schematically depicted in the energy level diagrams of Fig. 18.1. Rayleigh scattering and other elastic light scattering processes result from an interaction between a radiation field and a molecule characterized by the absence of energy exchange. Stokes Raman scattering occurs when the interaction promotes the molecule to a higher vibrational energy level. In that case the scattered photon possesses a lower energy than the incident photon, and therefore a longer wavelength. In an anti-Stokes Raman process the opposite occurs and the scattered photon has a shorter wavelength than the incident photon. If the radiation field is monochromatic – nowadays only lasers are used of course – Raman scattering thus gives rise to a spectrum of discrete lines, the positions of which correspond to the energies needed to excite vibrational modes. The shift of a Raman line relative to the wavelength of the exciting radiation is expressed in relative wavenumbers:

$$\triangle \; \text{cm}^{-1} = (\frac{1}{\lambda_0} - \frac{1}{\lambda_s}) \, 10^{-2} \quad [18.2]$$

where $[\lambda] = m$, λ_0 = excitation wavelength, λ_s = wavelength scattered light.

A molecule of N atoms possesses $3N$ degrees of freedom. After subtraction of translational and rotational modes $3N-6$ independent vibrational modes remain ($3N-5$ for a linear molecule). Those modes that can be coupled to the electric field component of the incident radiation via a change in molecular

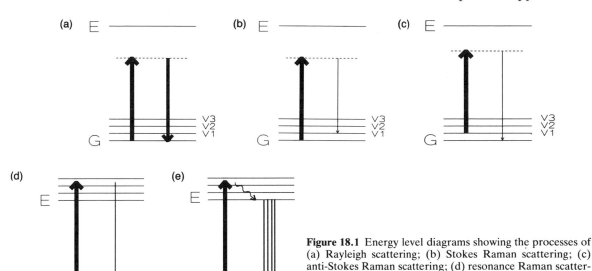

Figure 18.1 Energy level diagrams showing the processes of (a) Rayleigh scattering; (b) Stokes Raman scattering; (c) anti-Stokes Raman scattering; (d) resonance Raman scattering; (e) fluorescence scattering. G, E_i, Ground and excited electronic states; V_i, excited vibrational states. The width of the downward arrows illustrates the different signal intensities for the different processes (not to scale).

polarizability are Raman 'active', i.e. will give rise to a line in the Raman spectrum.

The intensity of Raman scattered light is many orders of magnitude lower than that of Rayleigh scattered light. An enhancement of the Raman signal can be obtained by resonantly exciting molecules. An additional advantage of resonance Raman spectroscopy is that different molecular subgroups can be separately resonantly excited and investigated.

The molecular structural information in a Raman spectrum is contained in the *position*, *intensity*, *polarization* and *width* of the lines.

(1) The *position* of a Raman line corresponds to the frequency of a vibration, which depends on molecular structure, the masses of the atoms involved and on their chemical bonds.

(2) The *intensity* of a Raman line depends linearly on the number of scatterers and on the laser light intensity. In resonance Raman experiments it depends on both ground state and excited state properties.

(3) Raman scattering may be anisotropic due to the fact that the molecular polarizability is a tensor. The *polarization* properties of Raman lines depend on the symmetry of the molecules. The polarization of a line can be a valuable aid in assigning that line to a particular vibration.

(4) The *width* of a Raman line depends on molecular structural heterogeneity and is influenced by dynamical processes and molecular re-orientational motions taking place within a time-frame of about 0.1–10 ps (Bartoli & Litovitz, 1972).

The parameters mentioned often depend in a subtle way on the precise chemical microenvironment of the molecules under study.

In thermal equilibrium conditions the relative number of molecules in ground state and higher vibrational states is given by the Boltzmann distribution function.

$$\frac{N_1}{N_0} = \exp\left(-\, h\nu_{\text{vib}}/kT\right) \qquad [18.3]$$

where h = Planck's constant, k = Boltzmann's constant, T = absolute temperature, N_1 = number of molecules in higher vibrational state, and N_0 = number of molecules in ground state.

The intensity ratio of the anti-Stokes Raman and Stokes Raman signals therefore decreases almost exponentially with frequency shift. At 1000 cm^{-1} the anti-Stokes Raman signal is about two orders of magnitude weaker than the corresponding Stokes Raman signal. For this reason usually only the Stokes's signal is measured.

Fluorescently scattered light also undergoes a Stokes shift and often the two processes interfere. Because fluorescence scattering is a process with a much higher quantum yield than Raman scattering it is usually the Raman spectroscopist whose efforts are frustrated. Even trace amounts of luminescing sample impurities can make it impossible to obtain a Raman spectrum. In flow cytometry, on the other hand, the lower level of fluorescence that can be detected is determined by water Raman scattering (Steen, 1990; Jett *et al.*, 1990). The simplest solution to these

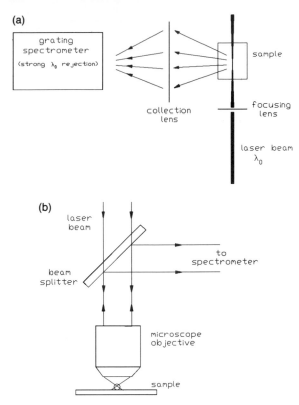

Figure 18.2 (a) Schematic of a Raman spectroscopy set-up; (b) Raman microspectroscopy set-up.

that are used are either photomultiplier tubes, operated in photon counting mode (single-channel detection) or array detectors such as intensified photodiode arrays or unintensified slow-scan charge-coupled device (CCD) cameras (multichannel detection). Especially when multichannel detection is employed, where a relatively low wavelength dispersion is desired in order to capture a large part of the Raman spectrum simultaneously, efficient Rayleigh line suppression demands some creativity (Mathies & Yu, 1978; Flaugh *et al.*, 1984; Baek *et al.*, 1988; Puppels *et al.*, 1990a; Rich & Cook, 1991).

In 1975 two groups independently reported the development of Raman microspectrometers (Rosasco *et al.*, 1975; Delhaye & Dhamelincourt, 1975). The design of Delhaye and Dhamelincourt, in which a microscope and a spectrometer are optically coupled, has found widespread acceptance (Fig. 18.2(b)). In a Raman microspectrometer laser light is focused on a sample by means of a microscope objective. The light that is scattered by the sample is collected by the same objective and focused onto the entrance of the spectrometer. In this way it is possible to study very small (micro-sized) samples, or to probe larger objects at specific sites and so combine spatial (morphological) information with spectroscopical information. Early applications included environmental particulate pollution studies and the identification of foreign body inclusions in tissue (Etz & Blaha, 1980).

problems is for a Raman spectroscopist to choose an excitation wavelength away from the excitation spectrum of the fluorophore (see, for example, Yu *et al.*, 1987, p. 56, and also Section 18.4.2.2) and for the flow cytometrist to choose a fluorophore with an emission spectrum away from the strong water bands in the Raman spectrum. Other more technically involved methods use phase-sensitive detection (to separate Raman from fluorescence signal), based on the fact that Raman scattering is an instantaneous process whereas in fluorescence, excited state lifetimes are often in the order of nanoseconds (Genack, 1984).

18.2.2   Instrumentation

The instrumentation needed for a Raman experiment is conceptually simple. A schematic is shown in Fig. 18.2(a). The Raman effect is very weak and the technique became practicable and popular only after the invention of the laser, because of the possibility of tightly focusing the laser light, thus producing a high photon density in a sample. Scattered light is collected and its spectrum analysed by means of a grating spectrometer. A prerequisite for a successful Raman experiment is a strong suppression of the Rayleigh line at zero wavenumber shift. The detectors

18.2.3   Raman spectroscopy and molecular biology

Since the late 1960s and early 1970s Raman spectroscopy has been extensively used to study biological (macro)molecules such as nucleic acids, proteins and protein-bound chromophores, lipids and polyenes. For reviews and textbooks see, for example, Carey (1982), Parker (1983), Clark and Hester (1986), Spiro (1987–88).

The complexity of the spectra of these large molecules is high, due to the large number of lines, which often overlap each other (see, for example, Fig. 18.10, 18.12–18.17). It is nevertheless possible to extract a great deal of information from such spectra. The assignment of lines to particular vibrations and the interpretation of spectra in terms of molecular structure and composition of a sample is achieved by normal mode calculations (e.g. Abe *et al.*, 1978; Krimm, 1987), combined X-ray diffraction and Raman spectroscopic experiments (e.g. Erfurth *et al.*, 1975; Goodwin & Brahms, 1978; Benevides & Thomas, 1983), isotope substitution experiments, which lead to shifts in Raman line frequencies (e.g. Lord & Thomas, 1967; Benevides & Thomas, 1985; Oertling *et al.*, 1988) and, in the case of large molecules or complexes, by obtaining spectra of molecular subgroups or

Table 18.1 Raman spectroscopic information about biological molecules and characteristic applications.

	Information	Applications
Nucleic acids	Sugar–phosphate backbone conformation Glycosidic bond orientation Nucleoside sugar pucker Base composition Base stacking Hydrogen bonding interactions	Study of DNA secondary structure and transitions (e.g. B to Z) Characterization of DNA melting and premelting phenomena Monitoring of DNA protonation and denaturation at low pH Characterization of natural and synthetic DNAs/RNAs Influence DNA–protein interactions on DNA structure
Protein	Protein main chain secondary structure Presence & conformation of disulphide, SH and CS groups Presence & microenvironment of aromatic amino acids: tyrosine – residue acting as H-bond donor or donor & acceptor tryptophan – residue buried or exposed	Study of protein heat denaturation Investigation of eye lenses Muscle fibre protein conformation with muscle in relaxed and contracted state
Protein-bound chromophores, e.g. metallo-porphyrins	Oxidation state & spin state of the core metal Type of axial ligands and peripheral substituents Ground and excited state geometry of the molecule	Determination structure–function relations Monitoring of *in vitro* and *in vivo* enzymatic processes Identification of reaction intermediates
Lipids	Conformation of hydrocarbon chains Microenvironment of head and tail group molecules	Monitoring of phase transitions in membranes Characterization of membrane fluidity

See Carey (1982), Parker (1983), Clark and Hester (1986), Spiro (1987–88).

molecules separately. A brief overview of the type of information about biological molecules that can be obtained by means of Raman spectroscopy and some characteristic applications are given in Table 18.1.

18.2.4 Raman spectroscopy of single cells

Although Raman spectroscopy can provide a wealth of information and microspectrometers have, in principle, a diffraction-limited spatial resolution, the technique has until now been little used in studies at the level of a single cell. The reason for this is the small Raman scattering cross-section of many biological molecules, including nucleic acids, proteins and lipids. The applicability of Raman spectroscopy in *in situ* studies of molecules in a single cell thus far depended on favourable circumstances such as the possibility of resonance enhancement of the Raman signal. In this manner, for example, Barry and Mathies (1982) obtained spectra of visual pigments in single photoreceptor cells of toad and goldfish. Also the spectra of haemoglobin in a red blood cell (Jeannesson *et al.*, 1986) and spectra of single bacterial cells

(Dalterio *et al.*, 1986) were measured in this way. The very high concentration of DNA in sperm cells allowed a spectrum of DNA inside a single salmon sperm head to be obtained (Kubasek *et al.*, 1986). Single cell spectra were furthermore recorded of very large cells such as a muscle fibre (Pézolet *et al.*, 1980) and a giant lymph node cell (Abraham & Etz, 1979).

In order to obtain resonance enhancement for the study of nucleic acids and proteins, UV-laser excitation has to be employed. A UV-Raman microspectrometer has been developed that enables the recording of single cell spectra with submicrometre spatial resolution in this way (Sureau *et al.*, 1990). Using resonance excitation in single cell studies is not without risk, however, because of the danger of photodegradation. In studies of isolated purified compounds or model systems in solution this risk can be decreased by constantly refreshing the sample in the laser beam, e.g. by employing a flow chamber or a spinning cell (Carey, 1982). These measures cannot be taken in single cell studies of course.

Recently surface enhanced Raman spectroscopy was used to detect very low concentrations of antitumour drugs in the nucleus and cytoplasm of single living cells (Nabiev *et al.*, 1991). This technique relies on the strong enhancement of the Raman signal of molecules close to a metal surface. In this study small silver hydrosols were introduced into the cell, by means of endocytosis. Drug concentrations down to 10^{-10} M can be detected in this way. Although this method solves the problem of low Raman signal intensities, similar questions as in fluorescence studies arise. What is the influence of the metal surface on the system studied? What determines the accessability of molecules to the metal surface?

The only other way to make Raman spectroscopy more widely applicable in single cell studies is then to improve the sensitivity of the instrumentation. This has been the approach in our laboratory, which has led to the realization of the confocal Raman microspectrometer described in the next section.

18.3 THE CONFOCAL RAMAN MICROSPECTROMETER (CRM)

The CRM, which enables the recording of high-quality Raman spectra of single cells and chromosomes, is shown in Fig. 18.3. In brief, this section deals with three important aspects of its design: the *sensitivity* of the instrument, its *spatial resolution*, and laser light-induced *damage* to cells and chromosomes.

18.3.1 Sensitivity

By optimizing the design of the spectrometer for the use of one fixed laser excitation wavelength it was possible to raise the signal detection efficiency of the CRM to an extent where up to 15% of the signal collected by the microscope objective is actually detected. All components of the set-up are optimized for Raman experiments with 660 nm laser excitation. A narrow-band filter is used to optically couple microscope and spectrometer; 80% of the incoming laser light is transmitted, whereas Raman scattered light between 300 and 3000 cm^{-1} is reflected with an efficiency $\geq 99\%$. The chevron-type Raman notch filter set suppresses the intensity of the scattered laser light. The beam of light entering the spectrometer is reflected back and forth between two band-pass filters. At each reflection >80% of the laser light is transmitted, and in that way effectively separated from the Raman scattered light, which is efficiently reflected ($R \geq 99\%$ 600–2600 cm^{-1}). After 12 such reflections laser light intensity is suppressed by a factor 10^{8}, at the cost of only 10–20% of the Raman scattered light (Puppels *et al.*, 1990a). This efficient blocking of laser light allows the use of a ruled blazed grating for wavelength dispersion, despite the fact that this type of grating has a much higher stray light level than holographic gratings. The advantages of ruled gratings are their much higher efficiency and the fact that their performance is much less polarization-dependent than that of holographic gratings. The Raman spectrum is

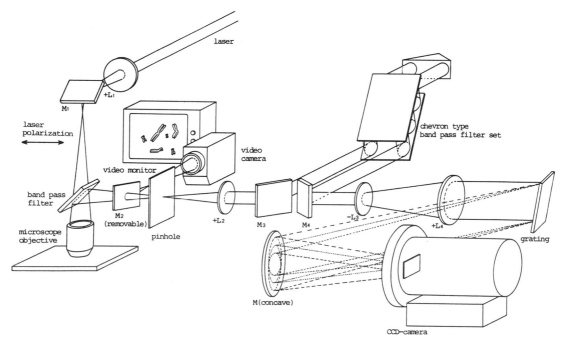

Figure 18.3 The confocal Raman microspectrometer. (From Puppels *et al.*, 1991a, reproduced by permission of John Wiley & Sons Ltd.)

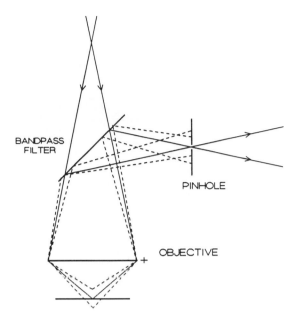

Figure 18.4 Suppression of out-of-focus signal contributions by means of confocal signal detection.

imaged onto a liquid-nitrogen-cooled slow-scan CCD camera (Wright Instruments Ltd, EEV P8603 chip). These two-dimensional detectors are ideal for Raman spectroscopic purposes. They combine a high quantum

efficiency (~ 40% at 700 nm for this type) with noiseless operation. The cooling of the chip to 140 K eliminates dark current and read-out noise is limited to about 10 electrons per channel. This means that at virtually any signal intensity level the signal-to-noise ratio is photon-noise-limited.

18.3.2 Spatial resolution

High numerical aperture microscope objectives are used. This does not only ensure a submicrometre lateral spatial resolution but also signal collection under a wide solid angle.

Raman signal is emitted not only by the object under investigation but also by its surroundings, e.g. the substrate on which a chromosome is deposited and the buffer in which it is immersed. This effect is comparable to the blurring of an image in a fluorescence microscope due to out-of-focus fluorescence. The background Raman signal adds photon noise to the Raman signal of interest. These signal contributions can be suppressed by means of confocal detection of the Raman signal (Fig. 18.4) (Brakenhoff *et al.*, 1979, see also Chapter 17). In the CRM in general a pinhole with a diameter of 100 μm is used because it was found that in this way a significant reduction of background signal was obtained without losing Raman signal of the object in focus. The effect of the pinhole

Figure 18.5 The influence of the size of the pinhole in the image plane of the objective on the Raman signal obtained from a 0.5 μm polystyrene bead on a fused silica substrate in water. (A) Measurement without pinhole; (B) background signal without pinhole; (C) 100 μm pinhole (diameter); (D) 50 μm pinhole; (E) 25 μm pinhole. It can be seen that the 100 μm pinhole strongly reduces background signal contributions without affecting the Raman signal from the object in focus. Smaller pinholes further reduce background signal intensity but also the signal contributions from the polystyrene bead. (From Puppels *et al.*, 1991a, reproduced by permission of John Wiley & Sons Ltd.).

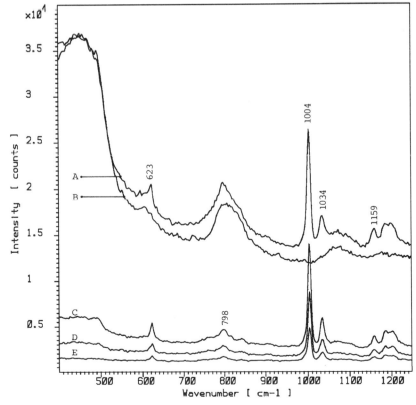

is illustrated by the measurements in Fig. 18.5. A pinhole of this size does not improve lateral spatial resolution, so that this is determined by the size of the laser focus. The dimensions of the actual measuring volume of the CRM, equipped with a ×63 Zeiss plan neofluar water-immersion objective (NA 1.2) were determined by scanning a small polystyrene bead (0.22 μm) through the laser focus and measuring the intensity of the polystyrene Raman signal as a function of the position of the bead (Puppels *et al.*, 1991a). The FWHM-values of the resulting curves showed a lateral spatial resolution of 0.45 μm (both directions) and an axial resolution of 1.3 μm. In theory FWHM-values of 0.31 μm (lateral) and 0.8 μm (axial) should be possible for this system (Van der Voort & Brakenhoff, 1990).

18.3.3 Laser light-induced damage

The CRM was originally designed to work with laser light of 514.5 nm (Puppels *et al.*, 1989; Greve *et al.*, 1989). It turned out, however, that all samples investigated – cells, metaphase and polytene chromosomes – were invariably damaged by the laser light (see Fig. 18.6), even at laser powers below 1 mW. This damage became visible under a light microscope in the form of a 'paling' of the sample at the site of the laser focus. Simultaneously with the paling the intensity of the Raman signal decreased. Interestingly, a line due to a guanine ring vibration at 1487 cm^{-1} was often found to diminish faster than the other lines of the spectrum (Fig. 18.7). Experiments with concentrated DNA and histone solutions showed that the 514.5 nm laser light did not affect the main chromatin constituents in their purified uncomplexed form (Puppels *et al.*, 1991b). It could also be shown that excessive heating of the samples due to laser light absorption, or indirectly due to substrate heating did not play a role, and possible effects due to multiphoton absorption could also be excluded.

The choice of the laser wavelength used for excitation was found to be crucial. Whereas at wavelengths below 514.5 nm the same paling of samples was observed, this was not the case upon irradiation with laser light of 632.8 nm and above. Moreover, the Raman spectra of chromosomes and cells remained constant upon prolonged irradiation with 660 nm laser light of the same site (Puppels *et al.*, 1991b). Also cells were found to be able to survive irradiation with much higher laser powers at 632.8 nm or 660 nm than at 514.5 nm (Fig. 18.8). Of course the 'safety' of 660 nm laser light is not a general rule, but it holds true for 'chromatin samples' such as chromosomes and cell nuclei. When (strongly) absorbing chromophores, such as haem proteins, are present in a cell, laser light intensity and signal acquisition time

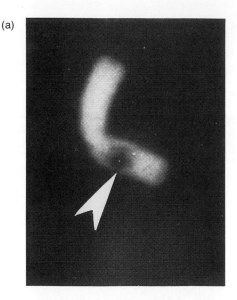

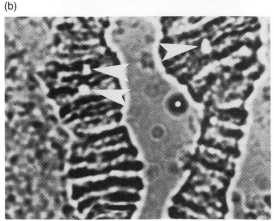

Figure 18.6 Paling of chromosomes by 514.5 nm laser light. (a) Metaphase chromosome (unfixed, magnification ×2000); (b) polytene chromosome (fixed, magnification ×1500). (From Puppels *et al.*, 1991b, reproduced by permission of Academic Press Inc.).

have to be carefully balanced in order to avoid artifacts due to photochemical reactions.

Photochemical effects appear to be the most plausible cause of the observed sample degradation at 514.5 nm. Since purified DNA and histones are not susceptible to this radiation damage, other compounds that act as photosensitizers must be present in chromosomes and cells. DNA bases (especially guanine) and a number of amino acids (methionine, tyrosine, tryptophan, histidine and cysteine) can be degraded in photodynamical processes requiring the presence of light, a sensitizer and oxygen (Foote, 1976). In such a process laser excitation brings the sensitizer via the singlet state in a long-lived triplet state. It can then react with oxygen to form reactive oxygen species that could cause the observed radiation

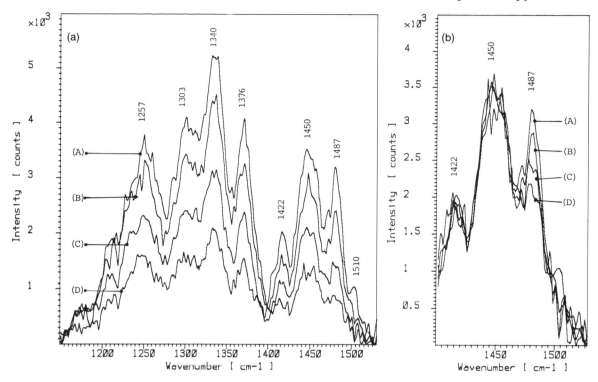

Figure 18.7 Effect of laser light (514.5 nm)-induced sample degradation on the Raman spectrum of metaphase chromosomes. Raman line assignments are given in Table 18.2. (a) Decrease of the intensity of the Raman signal, as recorded during measurements on isolated (unfixed) single (Chinese hamster lung cell) metaphase chromosomes. Shown are spectra obtained during (A) 1st minute of irradiation; (B) 3rd minute; (C) 5th minute; and (D) 9th minute of irradiation. (Averages of 10 measurements, for more details see Puppels *et al.* 1991b.) (b) Illustration of the sensitivity of the 1487 cm^{-1} guanine line to 514.5 nm light irradiation. Detail of the spectra shown in Fig. 18.7(a). The spectra were scaled to have equal intensity in the spectral interval 1150–1460 cm^{-1}. (From Puppels *et al.*, 1991b, reproduced by permission of Academic Press Inc.)

damage through oxidation of the DNA bases or amino acids leading to lesions in these molecules. It appears that these, as yet unidentified, sensitizers are excited by blue or green laser light but not by the 660 nm laser light used in the CRM.

The evidence for laser light-induced sample degradation at 514.5 nm may offer an alternative explanation for the results of Goodwin and Brahms (1978). In that work Raman spectra of a number of chromatin samples isolated and differing in the relative amount of non-histone protein (NHP), were recorded. It was found that the intensity of the guanine line at 1487 cm^{-1}, which results mainly from an N(7)=C(8) stretching vibration, depended on the amount of NHP present in the chromatin. The more NHP was present, the lower the intensity of this line. Since the N7 group of the guanine molecule involved in this vibration is projected into the large groove of the DNA molecule, this result was interpreted as evidence that NHP molecules interact with DNA via the large groove. However, in view of the fact that in that work 514.5 nm laser light was used, these results could also be the result of laser light-induced sample degradation, as evidenced by Fig. 18.7(b).

18.4 APPLICATIONS

The CRM offers the possibility of investigating microscopically small biological objects. It can be used to determine which molecules are located where and in what form, conformation and quantity inside a living cell or a chromosome. Experiments can take the form of 'probing' an object at a specific site, e.g. to learn more about the chemical composition. 'Mapping' is another possibility. Here an object is scanned through the laser focus. In this way variations in chemical composition and/or molecular conformation can be analysed and linked with, for example, morphological features. In the future we hope to be able to monitor changes in local molecular composition and conformation due to cellular processes, i.e. add the time domain to our experiments.

A marked difference between working with intact biological structures and experiments with model systems is that, especially in the case of cells, the molecular composition of the sample is not precisely known. The first step in analysing the spectra is, therefore, to identify the molecules most prominently

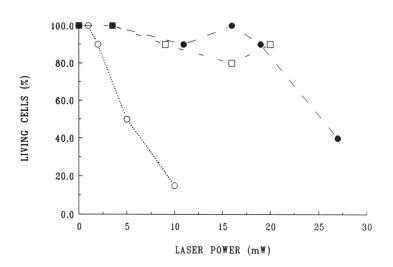

Figure 18.8 Percentage of cells (human lymphocytes) surviving a 5 min laser light irradiation as a function of laser light power for 514.5 nm, 632.8 nm and 660 nm. ○····○ 514.5 nm, ×63 Zeiss plan neofluar water-immersion objective (NA 1.2); focus diameter: ~ 0.5 μm. □····□ 632.8 nm, ×40 Nikon E Plan objective; focus diameter: ~ 1 μm. ●····● 660 nm, same objective as for 514.5 nm. The laser beam was focused in the centre of the cells. Each point in the graph represents 10 irradiated cells (20 for 660 nm, 25 mW). Control samples contained >90% living cells. (From Puppels *et al.*, 1991b, reproduced by permission of Academic Press Inc.)

contributing to the Raman signal. After that an analysis of the structure of these molecules in their natural environment can be made and compared with that of the isolated compound or a model system under controlled conditions.

Raman spectroscopy as a direct technique lacks the selectivity of, for example, immunofluorescence techniques. All molecules present in the measuring volume add to the Raman signal. This can lead to very complex spectra that are difficult to interpret. But on the other hand it can also put immunofluorescence results in a different perspective. In a Raman spectroscopic study of the effects of low pH treatments of polytene chromosomes, needed to evoke the binding of anti-Z-DNA antibodies, it was found that these treatments not only led to Z-DNA immunoreactivity but also to large-scale structural changes in chromosomal DNA structure. Under such conditions, which cannot be recognized by the immunofluorescence studies, the biological significance of the anti-Z-DNA antibody binding to chromosomes becomes very doubtful (Puppels, 1991). This example illustrates that rather than competitive techniques, Raman and (immuno) fluorescence methods should be considered as complementary techniques. In fact, it is possible to combine the two, because most fluorophores cannot be excited at 660 nm, enabling fluoresence-free Raman measurements of labelled objects. As shown below (Section 18.4.2.2) specific cell subsets can be made available for Raman spectroscopic investigation, by labelling them with a specific antibody conjugated to a fluorescent label and sorting them on a flow cytometer on the basis of their fluorescence signal. Such a labelling could also be used to facilitate the localization of specific structures, e.g. sites of transcription or replica-

tion in a cell by means of fluorescence microscopy, prior to a Raman measurement.

18.4.1   Chromatin in chromosomes and cells

DNA in cells is complexed to proteins. Proteins fold and package the DNA (histones) and are essential in such fundamental processes as replication and transcription of the DNA (non-histone proteins). In non-dividing, transcriptionally inactive cells chromatin non-histone protein (NHP) content is usually low (Bradbury *et al.*, 1981; Mathews & Van Holde, 1990). The way chromatin is organized strongly influences the transcriptional activity of the DNA. It is a well-known fact that the (inactive) heterochromatin in cells is much more condensed than (active) euchromatin. In R-banded metaphase chromosomes the stained bands are relatively rich in transcribed genes (Bickmore & Sumner, 1989). The fact that metaphase chromosome regions react differently to the R-banding treatment in a manner that correlates with transcriptional activity during interphase, is another reflection of the fact that active and inactive chromatin are organized in a different way. In polytene chromosomes intense transcription of a gene leads to localized puffing, i.e. swelling of the chromosome at the position where the gene is located.

The CRM is employed to characterize (local) chromatin composition in a number of systems. Shown below are results of measurements of polytene chromosomes, metaphase chromosomes and nuclei in intact human white blood cells. Table 18.2 gives a listing of DNA and protein vibrations and their assignments.

Table 18.2 Raman line assignments for (B-)DNA–protein samples.

Line position (cm⁻¹)	Assignment		
622	Phenylalanine		
645	Tyrosine		
669	Thymine		
681	Guanine ring breathing		
729	Adenine ring breathing		
749	Thymine ring breathing		
782 ⎱ 787	Cytosine ring breathing		
790 ⎰	DNA:PO₂ symmetric stretching		
833	DNA:PO₂ asymmetric stretching, tyrosine		
853	Tyrosine		
896	DNA:backbone		
925	DNA:backbone		
932	Protein:α-helix C–C skeletal mode		
1004	Phenylalanine	⎱	
1017	DNA:backbone C–O stretching		Protein: C–C and
1032	Phenylalanine		C–N stretching modes
1057	DNA:backbone C–O stretching		
1094	DNA:PO₂ symmetric stretching	⎰	
1126	Protein: C–N stretching		
1176	Tyrosine, phenylalanine		
1211	Thymine, tyrosine, phenylalanine	⎱	Protein: Amide III
1240	Thymine		β-Sheet: ~1230–1240
1255	Cytosine		Random coil: ~1240–1250
		⎰	α-Helix: ~1260–1300
1304	Adenine, cytosine	⎱	Protein: CH₂/CH₃
1340	Adenine	⎰	deformations
1376	Thymine, adenine, guanine		
1422	Adenine		
1449	Protein:CH₂/CH₃ deformations		
1487	Guanine, adenine		
1511	Adenine		
1578	Guanine, adenine		
1606	Phenylalanine, tyrosine		
1617	Tyrosine, phenylalanine	⎱	α-Helix: ~1645–1655
1640–1700	Protein: amide I		Random coil: ~1660–1670
1670	Thymine 30	⎰	β-Sheet: ~1665–1680

From Erfurth and Peticolas (1975), Thomas *et al.* (1977, 1986), Tu (1982), Prescott *et al.* (1984), Yu *et al.* (1987).

18.4.1.1 Polytene chromosomes*

Polytene chromosomes are a strongly amplified form of interphase chromosomes, found, for example, in the salivary gland cells of *Drosophila* and *Chironomus*. They arise through many rounds of DNA replication without subsequent separation of daughter chromatids. The chromatids run through the entire length of a chromosome. Degrees of polyteny (the number of chromatids in a chromosome) of up to 8000 can be encountered.

These giant chromosomes possess characteristic patterns of alternating dark bands and light interbands, which are readily observable under a light microscope (see Fig. 18.9). The fact that an order-of-magnitude correspondence exists between the number of genes and the number of bands and interbands has given rise to the concept that the banding pattern reflects functionally different chromatin domains. Beermann (1952) found that puffing of a chromosome region was the result of intense transcription. Based on that work others hypothesized that genes reside in bands. Evidence for transcriptional activity in up to 70% of all bands of *Drosophila melanogaster* was found (Ananiev & Barsky, 1978). However, evidence to the contrary was also reported showing transcriptional activity primarily in interbands, diffuse bands and puffs (Fujita & Takamoto 1963; Semeshin *et al.*,

* Work in cooperation with Drs M. Robert-Nicoud, D.J. Arndt-Jovin and T.M. Jovin of the Max Planck Institute for Biophysical Chemistry, Göttingen, Germany.

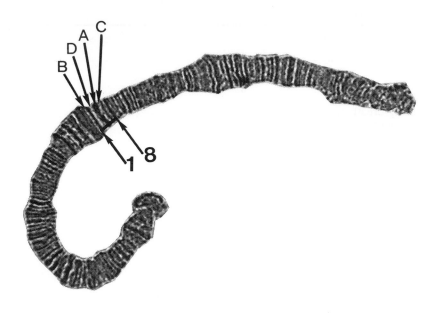

Figure 18.9 *Chironomus thummi thummi* chromosome II. Indicated are the positions of the measurements shown in Fig. 18.10 (arrows A, B, C and D) and the eight bands and interbands (between arrows 1 and 8) for which relative protein and DNA concentrations, protein–DNA ratio and relative chromatin concentration were determined (Fig. 18.11). Bar denotes 10 μm. (From Puppels *et al.*, 1991d, copyright 1991 by San Francisco Press, Inc., with permission.)

1979; Jamrich *et al.*, 1977). A more recently forwarded hypothesis states that the band–interband pattern is a reflection of local transcriptional activity rather than of the distribution of genic and non-genic material (Zhimulev *et al.*, 1981; Hill & Rudkin, 1987).

Only little is known about the organization of band and interband chromatin in terms of DNA and protein concentration, although it is known that in general the compaction of chromatin does influence transcriptional activity. The first studies of the band–interband pattern of *Chironomus thummi thummi* chromosomes with the CRM indicate that their characterization by means of Raman spectroscopy is a very promising approach. Spectra were recorded of 3:1 ethanol:acetic acid) fixed polytene chromosomes from squash preparations and of unfixed chromosomes, isolated under physiological conditions (Robert-Nicoud *et al.*, 1984). In general, squash preparations are used in many experiments because of the very distinct and stable banding pattern of the chromosomes. The isolation of unfixed chromosomes is more time-consuming and the contrast between bands and interbands, as observed under a microscope is lower. But, of course, the unfixed chromosomes provide much more information about the native state of the chromosomes.

Raman spectra was recorded of eight neighbouring bands and interbands of the fixed chromosome shown in Fig. 18.9. Spectra of different bands were virtually indistinguishable. Spectra of different interbands showed large variations both in absolute signal intensity and in relative peak intensities. Figure 18.10 shows four of the spectra that were obtained. On the basis of the intensitites of 1093 cm^{-1} DNA backbone line

and the 1448 cm^{-1} line due to protein CH deformations, DNA and protein concentrations can be determined (method described in Puppels *et al.*, 1991c). Shown in Fig. 18.11 are relative protein and DNA concentrations for the eight bands and interbands as well as the protein-DNA ratio (mass/mass) and total relative chromatin concentration. It shows clearly that DNA and protein concentrations are highest and nearly invariable in the bands. Both concentrations are very variable in the interbands. Parts (D) and (E) of Fig. 18.11 illustrate a very intriguing aspect: the local chromatin concentration in polytene chromosomes appears to be inversely related to the protein–DNA ratio.

In Fig. 18.12 the results of measurements on bands and interbands of unfixed chromosomes are shown. These confirm the fixed chromosome results. Total chromatin concentration is lower in interbands (lower Raman signal intensity) and protein–DNA ratio is higher in interbands (Table 18.3). The position of the amide I and III lines in the spectra provide information about protein secondary structure. The maximum of the amide I line for bands is at 1662 cm^{-1} and for interbands at 1666 cm^{-1}. In the difference spectrum the amide I maximum is found at 1669 cm^{-1}. This shift of the amide I vibrations to higher wavenumbers indicates that on average the interband proteins contain a higher relative amount of β-sheet or random coil domains than the band proteins. The absence of a strong line below 1240 cm^{-1} in the amide III region in Fig. 18.12(C) which would be indicative of β-sheet structure, makes clear that it is especially the relative amount of random coil domains that is higher in

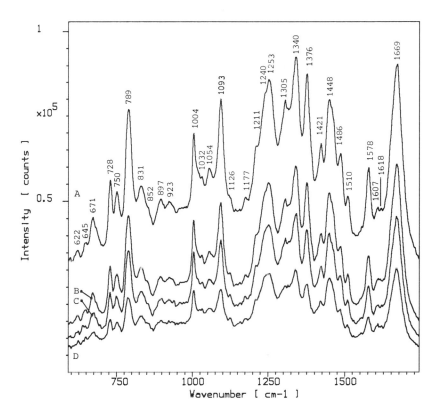

Figure 18.10 Raman spectra obtained from the fixed chromosome shown in Fig. 18.9; (A) band, (B,C,D) interbands. Laser power: 15 mW (660 nm); signed integration time: 10 min; ×63 Zeiss Plan Neofluar water-immersion objective (NA 1.2, 0.12–0.22). The spectra have been shifted along the ordinate for clarity. Background (phosphate-buffered saline and fused silica) signal has been subtracted. (From Puppels *et al.*, 1991d, copyright 1991 by San Francisco Press, Inc., with permission.)

interbands than in bands. This can be due either to actual local variations in the secondary structure of the chromosomal proteins, or to local variations in chromosomal protein composition. Electron microscopic (Ananiev & Barsky, 1985) and immunofluorescence (Bustin *et al.*, 1977; Kurth *et al.*, 1978) studies have shown that the DNA in both bands and interbands is organized in a nucleosomal form. The protein–DNA mass ratio in nucleosomes is 1:1. As indicated in Table 18.3 this means that bands contain much less non-histone proteins than interbands. Therefore it would seem likely that the differences in protein secondary structure are due to differences in protein composition. The high intensity of the phenylalanine line at 1004 cm^{-1} in Fig. 18.12(C) could be a further indication of this, as it is most likely explained by a higher relative amount of phenylalanine residues in the interband proteins.

18.4.1.2 Metaphase chromosomes

During mitosis the chromatin in a cell condenses into metaphase chromosomes, and transcription stops. The chromosomes consist of two identical chromatids connected in the centromere (see Fig. 18.6(a)). Metaphase chromosome banding was mentioned in the introduction of this chapter. Unlike the case of

polytene chromosomes, here physico-chemical treatments or endonuclease digestion of the chromosomes followed by staining are needed to make the banding pattern visible. Since the discovery of quinacrine-banding (Caspersson *et al.*, 1968) many other types of banding patterns have been discovered (reviewed in, for example, Babu & Verma, 1989). Although the banding pattern cannot be observed on untreated unstained chromosomes it is clear that they must reflect one or more properties of the chromosomes in their native state.

Raman microspectroscopy may play an important role in elucidating the origin of the banding patterns. The experiments with polytene chromosomes show that it is a sensitive monitor of local variations in chromatin composition and structure. Figure 18.13 shows that good Raman spectra can also be obtained of the much smaller metaphase chromosomes. Spectrum (a) was obtained of a metaphase chromosome isolated under near physiological conditions, spectrum (b) of a 3:1 methanol:acetic acid-fixed chromosome from a squash preparation dried in air. Comparison of the two spectra shows that Raman spectroscopy is also a sensitive monitor of the molecular structural changes that take place as a result of the treatments a chromosome undergoes.

The shift of the amide I maximum to 1668 cm^{-1} in

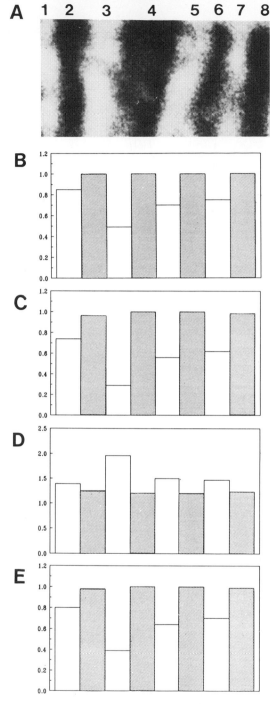

Figure 18.11 Variation in protein and DNA concentrations in eight adjoining bands and interbands of the chromosome of Fig. 11.9. (A) Magnification of Fig. 11.9; (B) protein concentration (normalized to band 4); (C) DNA concentration (normalized to band 4); (D) protein–DNA ratio (absolute values); (E) total chromatin concentration (protein + DNA; normalized to band 4). (B,C,E) error ± 5%, (D) error ± 19% (for absolute values), ± 10% (for normalized values). (From Puppels *et al.*, 1991d, copyright 1991 by San Francisco Press, Inc., with permission.)

spectrum (b), indicates a change in protein secondary structure, which is propably due to protein denaturation by the fixative. The intensity increase of the phenylalanine line at 1004 cm^{-1} could have the same cause, indicating a change in the microenvironment of the phenyl rings. However, at this time it cannot be excluded that during the fixation and squashing of the cells the chromosomes are contaminated with, for example, cytoplasmic proteins.

The fixation and drying of the chromosomes also alters the DNA structure. Whereas in the unfixed chromosome the DNA is in a double helical right-handed B-form as evidenced by the characteristic vibrations at 1094 cm^{-1} and 835 cm^{-1} (backbone) and 784, 750, 731 and 682 cm^{-1} (ring breathing vibrations of C, T, A and G respectively (see Table 18.2), characteristic of C$_2$-endo sugar pucker), in the spectrum of the fixed chromosome small lines at 708, 690, 660 and 645 cm^{-1} could indicate the presence of a small amount of A-DNA, a double helical right-handed DNA form found under low humidity conditions (Erfurth *et al.*, 1975). The high intensity of 1245 cm^{-1} in the fixed chromosome spectrum may be due to a partially denatured state of the DNA leading to a hypochromic effect due to unstacking of T-rings. Further evidence of this is the increased intensity of the A-line at 731 cm^{-1} which may have the same cause, i.e. unstacking of A-rings. We assume that the strong line at 928 cm^{-1} in spectrum (b) is due to acetic acid bound to the chromosome. Its intensity varies widely between different preparations.

We expect that a systematic Raman spectroscopic study of the local chromatin composition of metaphase chromosomes and the changes therein caused by physico-chemical banding treatments will provide valuable information concerning the origin of the banding patterns.

18.4.1.3   Cell nuclei

The spatial resolution of the CRM makes it possible to obtain spectra of chromatin in the nucleus of an intact cell free from signal contributions of the cytoplasm (Puppels *et al.*, 1991c). Figure 18.14 shows a human neutrophilic granulocyte nucleus spectrum, averaged over five measurements. As in the chromosome spectra all lines except the 1048 cm^{-1} line can be assigned to DNA and protein vibrations (see Table 18.2). Possible RNA or phospholipid signal contributions are not discernible. The spectra are very reproducible. Only slight vibrations (~ 10%) in the intensity ratio of the 1449 cm^{-1} and 1094 cm^{-1} lines are found (indicating small variations in protein-DNA ratio). A conspicuous aspect is that the protein–DNA ratio is high (~ 2.3:1). This indicates the presence of a large amount of non-histone proteins which would

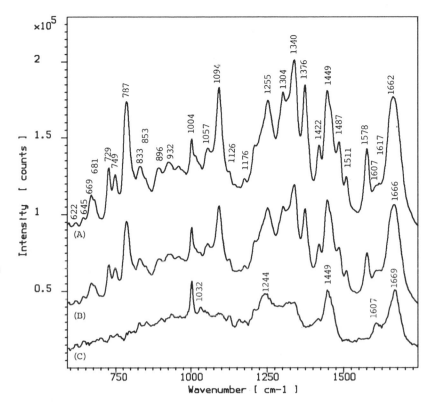

Figure 18.12 Raman spectra of unfixed *Chironomus thummi thummi* polytene chromosomes. (A) Average for four different bands; (B) average of four different interbands; (C) difference spectrum B–A after normalizing. (A) and (B) with respect to the 1094 cm⁻¹ DNA line (shown 3 times enlarged). Laser power: 15 mW (660 nm); measuring time: 10 min; microscope objective ×63 Zeiss Plan Neofluar water-immersion (NA 1.2, 0.12–0.22). Spectra shifted along ordinate for clarity. (From Puppels *et al.*, 1991e, reproduced by permission of The Royal Society of Chemistry.)

Table 18.3 Protein and DNA concentrations of bands and interbands of unfixed *Chironomus thummi thummi* polytene chromosomes.

	Protein[a] (± 5%)	DNA[a] (± 5%)	Protein:DNA[b] (mass/mass) (± 19%)	Total chromatin[a] (± 5%)	NHP:histone[b,c] (mass/mass)
Bands	1.00	1.00	1.3	1.00	0.3 (± 0.3)
Interbands	0.79	0.65	1.6	0.76	0.6 (± 0.3)

[a] Normalized values.
[b] Absolute values.
[c] Assumption: DNA:histones = 1:1 (mass/mass).

not be expected in a cell with low transcriptional activity, such as a mature neutrophil (Klebanoff & Clark, 1978). A second unexpected result is the presence of a line at 1048 cm⁻¹. This line is not present in spectra of free DNA, nucleosomes, isolated chromatin, polytene or metaphase chromosomes (Erfurth & Peticolas, 1975; Thomas *et al.*, 1977; Goodwin & Brahms, 1978; Thomas *et al.*, 1980, 1986; Savoie *et al.*, 1985; Hayashi *et al.*, 1986) and is also not found in cytoplasmic spectra. Moreover, it is the only line that strongly varies in intensity in the nuclear spectra, sometimes being absent altogether. This may indicate that it originates from a compound or molecular complex unevenly distributed in the nucleus. The 1048 cm⁻¹ line has thus far been found in

spectra of lymphocytes, neutrophils, eosinophils and basophils, but an assignment is still lacking.

18.4.2 The cytoplasm

The molecular composition of the cytoplasm of a cell is very complex and the distribution of molecules inhomogeneous. Because all molecules in the measuring volume contribute Raman signals this could easily give rise to unreproducible and uninterpretable spectra. Raman microspectroscopy can therefore be most successfully applied when one or a few types of molecules are present in much higher quantities than all other molecules either throughout the cytoplasm or in specific, identifiable cytoplasmic organelles, or

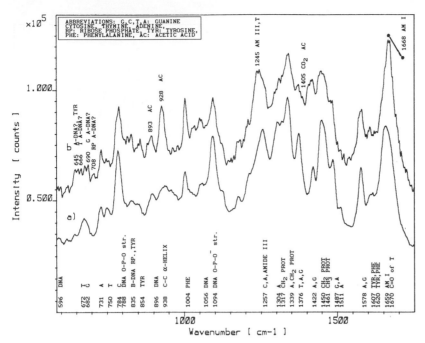

Figure 18.13 Raman spectra of single Chinese hamster lung cell chromosomes, obtained on chromatid arm midway between centromere and telomere. (a) Unfixed chromosomes in hypotonic buffer; (b) fixed chromosomes, dry in air (enlarged by a factor of 2.5). Laser power: 10 mW (660 nm); measuring time: (a) 15 min, (b) 40 min; microscope objective ×63 Zeiss Plan Neofluar water-immersion (NA 1.2). (From Puppels *et al.*, 1990c, reproduced by permission of John Wiley & Sons Ltd.)

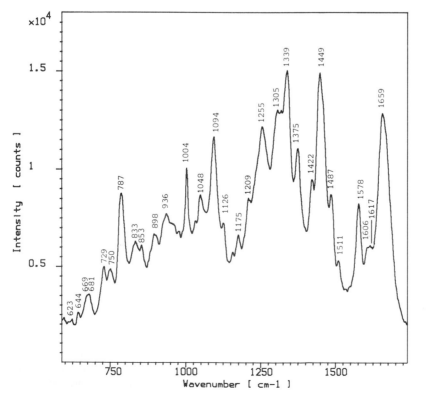

Figure 18.14 Raman spectrum of the nucleus of intact human neutrophils. The spectrum is averaged over five measurements on different cells. Laser power: 6 mW (660 nm); measuring time: 150 s per measurement; microscope objective: ×63 Zeiss Plan Neofluar water-immersion (NA 1.2). (From Puppels *et al.*, 1991c, reproduced from the *Biophysical Journal* by copyright permission of the Biophysical Society.)

when one or a few types of molecules give a much stronger Raman signal than all others.

Our studies of human white blood cells indicate that these restrictions still leave plenty of room for Raman microspectroscopical investigations of their functioning in the immune system.

18.4.2.1 Myeloperoxidase in neutrophilic granulocytes

Neutrophils are very motile cells that respond to chemotactic stimuli. Their main fuctions are phagocytosis, killing and digestion of bacteria and other microorganisms. They possess some 200 cytoplasmic granules of about 0.2 μm in diameter which contain a wide variety of oxidative metabolites and digestive enzymes. Two types of granules are distinguished: azurophilic and specific granules. A prominent enzyme of the azurophilic granules is myeloperoxidase (MPO), a haem-protein (Klebanoff & Clark, 1978). A neutrophil contains about 6 pg of MPO (Schultz & Kaminker, 1962). This enzyme catalyses the oxidation of chloride to hypochlorous acid, which is cytotoxic to bacteria (Parker, 1984). After phagocytosis, cytoplasmic granules fuse with the phagosome and degradation of the ingested object commences. The active site haem groups of MPO are strong Raman

scatterers, even when exited off-resonance. A comparison of spectra (A) and (B) of Fig. 18.15 makes clear that the cytoplasmic Raman spectrum of neutrophils is dominated by signal contributions of the MPO-haem groups (Puppels et al., 1991c). This means that this enzyme can be studied in vivo, and that it may be possible to study its activation and inactivation during the course of phagocytosis and digestion of a foreign object.

18.4.2.2 Basophilic granulocytes: immunofluorescence and Raman spectroscopy combined

The frequency of basophils in peripheral blood is very low (< 1% of the total white blood cell fraction). They possess cytoplasmic granules up to 1.2 μm in diameter which contain large amounts of heparin and histamine (Galli & Dvorak, 1978). Basophils play an important role in hypersensitivity responses (Ho et al., 1978). In order to obtain enriched basophil fractions they were labelled with fluorescein-conjugated human IgE antiserum (kept at 4°C from then on, in order to avoid degranulation) and sorted by means of fluorescence-activated cell sorting (Puppels et al., 1991c). The fluorescent labelling did not interfere with the Raman measurement because the fluorescein absorption band

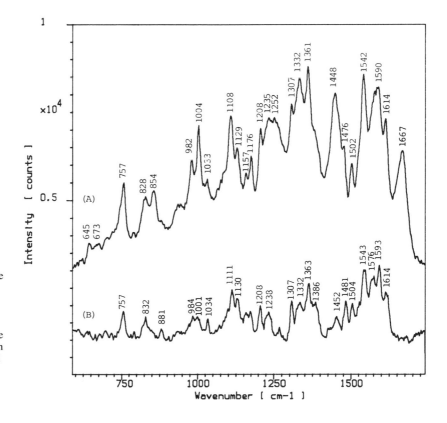

Figure 18.15 (A) Human neutrophil cytoplasmic Raman spectrum, averaged over five measurements on different cells. Laser power: 6 mW (660 nm); measuring time: 30 s per measurement; microscope objective: see Fig. 18.14. (B) Raman spectrum of isolated native (oxidized) human MPO. Sample concentration: 230 μM litre⁻¹ in 200 mM phosphate buffer; laser power: 10 mW (660 nm); measuring time 18 × 100 s; sample measured in a flow system. (From Puppels et al., 1991c, reproduced from the *Biophysical Journal* by copyright permission of the Biophysical Society.)

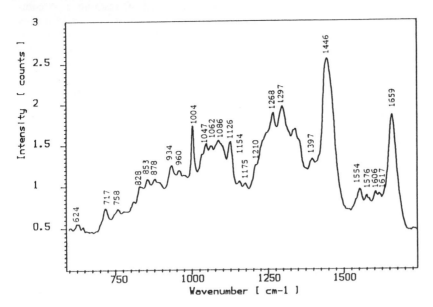

Figure 18.16 Human basophilic granulocyte cytoplasmic Raman spectrum. Average of four measurements on different cells. Conditions: see Fig. 18.14.

is located around 490 nm, far away from the 660 nm laser light used in the Raman experiment.

The cytoplasmic spectra of the basophils (Fig. 18.16) showed large qualitative variations and showed little similarity with published spectra of heparin or histamine (Cabassi *et al.*, 1978; Itabashi *et al.*, 1982). This may be due to the fact that these molecules are stored in the granules in a form or environment different from that of the isolated molecules used in the heparin and histamine studies. The lines at 1268 and 1297 cm^{-1} suggest phospholipid signal contributions (Wallach & Verma, 1975; Lippert *et al.*, 1975; Goheen *et al.*, 1977). The presence of phospholipids could also account for the position of the CH$_2$ deformations line at 1446 cm^{-1}, which is lower than that usually found in proteins ($\sim$ 1449–1451) and for part of the signal found between 1040 and 1130 cm^{-1}. Phospholipid signal contributions could originate from the dense multiple concentric membranous arrays found in the cytoplasmic granules (Dvorak & Ackermann, 1989). The most important result of the basophil experiments is that they show that it is possible to combine fluorescence and Raman techniques.

18.4.2.3 *Lymphocytes*

The task of the lymphocytes, which can be divided in many different subgroups each with their own function, is to establish the immune response of the body against virus-infected cells and cancer cells, i.e. to kill these cells.

Carotenoids appear to play a beneficial role in this part of the immune system. Epidemiological studies have shown that a high dietary intake of carotenoids

decreases the risk of many types of cancer (Peto *et al.*, 1981), and that many cancer patients have a lower blood plasma level of carotenoids than healthy individuals (Wald *et al.*, 1980). Experimental studies have shown that, for example, tumour development is slower in mice fed a β-carotene-supplemented diet than in mice given the same diet without β-carotene (Tomita *et al.*, 1987). In humans it was found that increased intake of β-carotene increased the number of OKT4$^+$ lymphocytes (helper/inducer T-lymphocytes) in peripheral blood (Alexander *et al.*, 1985). Carotenoids could exert their influence on the immune system via their provitamin A activity, via their strong antioxidant capacities, or via other, as yet unknown, mechanisms. Recent review papers on this subject are Bendich (1990) and De Vet (1990).

In order to investigate the mode of action of the carotenoids it is necessary to know their whereabouts. In view of their role in the immune system lymphocytes are obvious candidates. Raman spectra of pellets of lymphocytes indeed show strong carotenoid lines (Del Priore *et al.*, 1984). Because the CRM enables single cell Raman measurements with a high spatial resolution it is now possible precisely to locate the carotenoids inside the lymphocytes.

Figure 18.17 illustrates this. It shows a Raman spectrum obtained in the cytoplasm of a lymphocyte, at a position with a high concentration of carotenoids. Moreover, by means of immunofluorescence-activated cell sorting, different lymphocyte subsets can be made available for Raman measurements.

Knowing which subsets do, and which do not contain carotenoids and knowing where in the cells carotenoids are located means that a much more

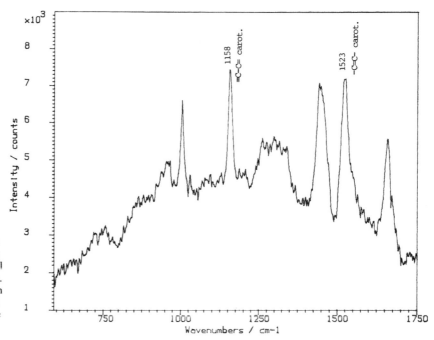

Figure 18.17 Human lymphocyte cytoplasmic spectrum with strong carotenoid lines, indicating a local high concentration of carotenoids. Conditions: see Fig. 18.15. (From Puppels *et al.*, 1991f, reproduced by permission of Elsevier Science Publishers.)

direct approach to investigating their claimed role in the immune system is possible.

18.5   FUTURE DEVELOPMENTS

Venturing into the field of cell biology an exciting future lies ahead for Raman microspectroscopy. The enormous body of data, compiled in nearly 25 years of Raman studies of biological molecules, makes it possible to fruitfully apply this technique in the investigation of complex biological structures such as chromosomes and single living cells.

Further development in instrumentation in our laboratory is aimed at obtaining images of objects in Raman scattered light. The Raman equivalent of fluorescence microscopy will doubtless further enhance the potential of Raman spectroscopy in cell biology.

Raman images can be obtained in a number of ways. In principle it should be possible to scan the whole object through the laser focus and record a Raman spectrum at each spatially resolvable position within the object. An image of the object could then be reconstructed, e.g. on the basis of the intensity of a selectable line of the Raman spectrum. It would, however, take a very long time to collect all the spectra.

A two-dimensional detector (such as a CCD camera) makes it possible to use line illumination of a sample (Bowden *et al.*, 1990) (Fig. 18.18), one dimension of the detector being used for spectral

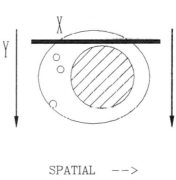

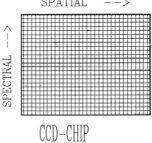

Figure 18.18 Raman microspectroscopy employing line illumination, using one dimension of a CCD camera for spectral information and one dimension for spatial information.

information, the other for spatial information. This form of illumination will be especially useful in studies of chromosomal banding patterns since these can, in fact, be looked upon as one-dimensional objects. A line illumination and detection option has recently been implemented in the CRM. The result of a test of this option is shown in Plate 18.1. An erythrocyte on a fused silica substrate in phosphate-buffered saline was illuminated as depicted in (a). The haemoglobin Raman signal was detected by the CCD camera and the image it captured is shown in (b). Different colours signify different signal intensities. Spectral information is found along the vertical axis (c), spatial information along the horizontal axis. The spatial information is in this case the distribution of haemoglobin in the cell (d), which is determined by the doughnut-like shape of the erythrocyte.

A third possibility to achieve Raman imaging is to use global illumination of an object and directly image the object in Raman light, scattered at a selected wavenumber. This means that one Raman line has to be filtered out from the light scattered by the object. The feasibility of such direct imaging has already been demonstrated with instruments based on a grating monochromator (Delhaye & Dhamelincourt, 1975) or a filter monochromator (Batchelder *et al.*, 1991). Such Raman microscopes could, for example, be used to determine the distribution of a certain type of molecule in a cell on the basis of a marker Raman line and monitor changes in this distribution in the course of a cellular process, in much the same way as fluorescent probes are used in fluorescence microscopy.

ACKNOWLEDGEMENT

The authors would like to thank Renate Maagdenberg for preparation of the manuscript.

REFERENCES

Abe M., Kitagawa T. & Kyogoku Y. (1978) *J. Chem. Phys.* **69**, 4526–4534.

Abraham J.L. & Etz E.S. (1979) *Science* **206**, 716–718.

Alexander M., Newmark H. & Miller R.G. (1985) *Immunol. Lett.* **9**, 211–214.

Ananiev E.V. & Barsky V.E. (1978) *Chromosoma* **65**, 359–371.

Ananiev E.V. & Barsky V.E. (1985) *Chromosoma* **95**, 104–112.

Arndt-Jovin D.J., Robert-Nicoud M., Zarling D.A., Greider C., Weimer E. & Jovin T.M. (1983) *Proc. Natl. Acad. Sci. USA* **80**, 4344–4348.

Babu A. & Verma R.S. (1989) *Int. Rev. Cytol.* **108**, 1–59.

Baek M., Nelson W.H., Britt D. & Sperry J.F. (1988) *Appl. Spectrosc.* **42**, 1312–1314.

Barry B. & Mathies R. (1982) *J. Cell Biol.* **94**, 479–482.

Bartoli F.J. & Litovitz T.A. (1972) *J. Chem. Phys.* **56**, 401–411.

Batchelder, D.N., Cheng C., Müller W. & Smith B.J.E.

(1991) *Makromol. Chem. Macromol. Symp.* **46**, 171–179.

Beermann W. (1952) *Chromosoma* **5**, 139–152.

Benevides J.M. & Thomas Jr. G.J. (1983) *Nucl. Acids Res.* **11**, 5746–5761.

Benevides J.M. & Thomas Jr. G.J. (1985) *Biopolymers* **24**, 667–682.

Bendich A. (1990) In *Carotenoids: Chemistry and Biology*, N.I. Krinsky, M.M. Mathews-Roth & R.F. Taylor (eds). Plenum Press, New York, pp. 323–336.

Bendich A. & Shapiro S.S. (1986) *J. Nutr.* **116**, 2254–2262.

Bickmore W.A. & Sumner A.T. (1989) *Trends Genet.* **5**(5), 144–148.

Bowden M., Gardiner D.J., Rice G. & Gerrard D.L. (1990) *J. Raman Spectrosc.* **21**, 37–41.

Bradbury E.M., Maclean N. & Matthews H.R. (1981) *DNA, Chromatin and Chromosomes*. Blackwell, Oxford.

Brakenhoff G.J., Blom P. & Barends P. (1979) *J. Microsc.* **117**(20), 219–232.

Bustin M., Kurth P.D., Moundrianakis E.N., Goldblatt D., Sperking R. & Rizzo W.B. (1977) *Cold Spring Habor Symp. Quant. Biol.* **XLII**, 379–388.

Cabassi F., Casu B. & Perlin A.S. (1978) *Carbohydr. Res.* **63**, 1–11.

Carey P.R. (1982) *Biochemical Applications of Raman and Resonance Raman Spectroscopies*. Academic Press, New York.

Caspersson T., Farber S., Foley G.E., Kudynowski J., Modest E.J., Simonsson E., Wagh U. & Zech L. (1968) *Exp. Cell Res.* **49**, 219–222.

Clark R.J.H. & Hester R.E. (eds) (1986) *Spectroscopy of Biological Systems*. John Wiley & Sons, Chichester.

Dalterio R.A., Nelson W.H., Britt D., Sperry J.F. & Purcell F.J. (1986) *Appl. Spectrosc.* **40**, 271–272.

Delhaye M. & Dhamelincourt P. (1975) *J. Raman Spectrosc.* **3**, 33–43.

Del Priore L.V., Lewis A. & Schat K.A. (1984) *Membr. Biochem.* **5**, 97–108.

De Vet H.C.W. (1990) PhD thesis, University of Maastricht, Maastricht, The Netherlands.

Dvorak A.M. & Ackermann S.J. (1989) *Lab. Invest.* **60**, 557–567.

Erfurth S. & Peticolas W.L. (1975) *Biopolymers* **14**, 247–264.

Erfurth S.C., Bond P.J. & Peticolas W.L. (1975) *Biopolymers* **14**, 1245–1257.

Etz E.S. & Blaha J.J. (1980) NBS special publication 533, 153–197.

Flaugh P.L., O'Donnell S.E. & Asher S.A. (1984) *Appl. Spectrosc.* **38**, 847–850.

Foote C.S. (1976) In *Free Radicals in Biology*, Vol. II, W.A. Pryor (ed.). Academic Press, New York, pp. 85–133.

Fujita S. & Takamoto K. (1963) *Nature* **200**, 494–495.

Galli S.J. & Dvorak H.F. (1978) In *Cellular, Molecular and Clinical Aspects of Allergic Disorders*, S. Gupta & R.A. Good (eds). Plenum Medical Book Company, New York.

Genack A.Z. (1984) *Anal. Chem.* **56**, 2957–2960.

Goheen S.C., Gilman T.H., Kaufmann J.W. & Garvin J.E. (1977) *Biochem. Biophys. Res. Commun.* **79**, 805–814.

Goodwin, D.C. & Brahms J. (1978) *Nucl. Acids Res.* **5**(3), 835–850.

Greve J., Puppels G.J., Olminkhof J.H.F., Otto C. &

De Mul F.F.M. (1989) In *Spectroscopy of Biological Molecules*, A. Bertoluzza, C. Fagnano & P. Monti (eds). Societa Editrice Esculapio s.r.l., Bologna, pp. 401–404.

Hayashi H., Nishimura Y., Katahira M. & Tsuboi M. (1986) *Nucl. Acids Res.* **14**(6), 2583–2596.

Hill R.J. & Rudkin G.T. (1987) *BioEssays* **7**(1) 35–40.

Ho P.C., Lewis R.A., Austen F. & Orage R.P. (1978) In *Cellular, Molecular and Clinical Aspects of Allergic Disorders*, S. Gupta & R.A. Good (eds). Plenum Medical Book Company, New York.

Itabashi M., Shoji K. & Itoh K. (1982) *Inorg. Chem.* **21**, 3485–3489.

Jamrich M., Greenleaf A.L. & Bautz E.K.F. (1977) *Proc. Natl. Acad. Sci. USA* **74**, 2079–2083.

Jeannesson P., Angiboust J.F., Jardillier J.C. & Manfait M. (1986) In *Proceedings IEEE/8th Annual Conference of the Engineering in Medicine and Biology Society*, 1404–1406.

Jett J.H., Keller R.A., Martin J.C., Nguyen D.C. & Saunders G.C. (1990) In *Flowcytometry and Sorting* (2nd edn), M.R. Melamed, T. Lindmo & M.L. Mendelsohn (eds). Wiley-Liss, New York, pp. 381–396.

Klebanoff S.J. & Clark R.A. (1978) *The Neutrophil: Function and Disorders*. North-Holland, Amsterdam.

Koningstein J.A. (1971) *Introduction to the Theory of the Raman Effect*. D. Reidel Publishing Company, Dordrecht.

Krimm S. (1987) In *Biological Applications of Raman Spectroscopy*, Vol. I, T. G. Spiro (ed.). John Wiley & Sons, New York, pp. 1–45.

Kubasek W.L, Wang Y., Thomas G.A., Patapoff T.W., Schoenwaelder K.-H., Van der Sande J.H. & Peticolas W.L. (1986) *Biochemistry* **25**, 7440–7445.

Kurth P.D., Moundrianakis E.N. & Bustin M. (1978) *J. Cell Biol.* **78**, 910–918.

Lippert J.L., Gorczyca L.E. & Meiklejohn G. (1975) *Biochim. Biophys. Acta* **382**, 51–57.

Long D.A. (1977) *Raman Spectroscopy*. McGraw-Hill, New York.

Lord R.C. & Thomas Jr. G.J. (1967) *Spectrochim Acta* **23A**, 2551–2591.

Mathews C.K. & Van Holde K.E. (1990) *Biochemistry*. Benjamin/Cummings, Redwood.

Mathies R. & Yu N.-T. (1978) *J. Raman Spectrosc.* **7**, 349–352.

Nabiev I.R., Morjani H. & Manfait, M. (1991) *Eur. Biophys. J.* **19**, 311–316.

Natarajan A.T. & Gropp A. (1972) *Exp. Cell Res.* **74**, 245–250.

Nordheim A., Pardue M.L., Lafer E.M., Möller A., Stollar B.D. & Rick A. (1981) *Nature* **294**, 417–422.

Oertling W.A., Hoogland H., Babcock G.T. & Wever R. (1988) *Biochemistry* **27**, 5395–5400.

Pachman U. & Rigter R. (1972) *Exp. Cell Res.* **72**, 602–608.

Parker C.W. (1984) In *Fundamental Immunology*, W.E. Paul (ed.). Raven Press, New York, pp. 697–747.

Parker F.S. (1983) *Application of Infrared, Raman and Resonance Raman Spectroscopy in Biochemistry*. Plenum Press, New York.

Peto R., Doll R., Buckley J.D. & Sporn M.B. (1981) *Nature (Lond.)* **290**, 201–208.

Pézolet M., Pigeon-Gosselin M., Nadeau J. & Caillé J.-P. (1980) *Biophys. J.* **31**, 1–8.

Prescott B., Steinmetz W. & Thomas Jr. G.J. (1984) *Biopolymers* **23**, 235–256.

Puppels G.J. (1991) PhD thesis, University of Twente, Enschede.

Puppels G.J., Olminkhof, J.H.F., Otto C., De Mul F.F.M. & Greve J. (1989) In *Spectroscopy of Biological Molecules*, A. Bertoluzza, C. Fagnano & P. Monti (eds). Societa Editrice Esculapio s.r.l., Bologna, pp. 357–358.

Puppels G.J., Huizinga A., Krabbe H.W., De Boer H.A., Gijsbers G. & De Mul F.F.M. (1990a) *Rev. Sci. Instrum.* **61**(12), 3709–3712

Puppels G.J., De Mul F.F.M., Otto C., Greve J., Robert-Nicoud M., Arndt-Jovin D.J. & Jovin T.M. (1990b) *Nature* **347**, 301–303.

Puppels G.J., Otto C., De Mul F.F.M. & Greve J. (1990c) In *Proceedings of the 12th International Conference on Raman Spectroscopy*, J.R. Durig & J.F. Sullivan (eds). John Wiley & Sons, Chichester.

Puppels G.J., Colier W., Olminkhof J.H.F., Otto C., De Mul F.F.M. & Greve J. (1991a) *J. Raman Spectrosc.* **22**, 217–225.

Puppels G.J., Olminkhof, J.H.F., Segers-Nolten G.M.J., Otto C., De Mul F.F.M. & Greve J. (1991b) *Exp. Cell Res.* **195**, 361–367.

Puppels G.J., Garritsen H.S.P., Segers-Nolten G.M.J., De Mul F.F.M. & Greve J. (1991c) *Biophys. J.* **60**, 1046–1056.

Puppels G.J., Otto C. & Greve J. (1991d) In *Microbeam Analysis*, D.G. Howitt (ed.). San Francisco Press, San Francisco, pp. 85–87.

Puppels G.J., Otto C., Greve J., Robert-Nicoud M., Arndt-Jovin D.J. & Jovin T.M. (1991e) In *Spectroscopy of Biological Molecules*, R.E. Hester & R.B. Girling (eds). The Royal Society of Chemistry, Cambridge, pp. 301–302.

Puppels G.J., Otto C. & Greve J. (1991f) *Trends Anal. Chem.* **10**(8), 249–253.

Raman C.V. & Krishnan K.S. (1928) *Nature* **121**, 501.

Rich C. & Cook D. (1991) *SPIE Proceedings*, 1461.

Robert-Nicoud M., Arndt-Jovin D.J., Zarling D.A. & Jovin T.M. (1984) *EMBO J.* **3**(4), 721–731.

Rosasco G.J., Etz E.S. & Cassatt W.A. (1975) *Appl. Spectrosc.* **29**, 396–404.

Rowley J.D. & Bodmer W.F. (1971) *Nature* **231**, 503–506.

Savoie R., Jutier J.-J., Alex S., Nadeau P. & Lewis P.N. (1985) *Biophys. J.* **47**, 451–459.

Schultz J. & Kaminker K. (1962) *Arch. Biochem. Biophys.* **96**, 465–467.

Semeshin V.F., Zhimulev I.V. & Belyaeva E.S. (1979) *Chromosoma* **73**, 163–177.

Smekal C.V. (1923) *Naturwissenschaften* **11**, 873.

Spiro T.G. (ed.) (1987–88) *Biological Applications of Raman Spectroscopy I–III*, John Wiley & Sons, New York.

Steen H.B. (1990) In *Flowcytometry and Sorting* (2nd edn), M.R. Melamed, T. Lindmo & M.L. Mendelsohn (eds). Wiley-Liss, New York, pp. 11–25.

Sumner A.T. (1982) *Cancer Genet. Cytogenet.* **6**, 59–87.

Sureau F., Chinsky L., Amirand C., Ballini J.P., Duquesne M., Laigle A., Turpin P.Y. & Vigny P. (1990) *Appl. Spectrosc.* **44**, 1047–1051.

Thomas Jr. G.J., Prescott B. & Olins D.E. (1977) *Science* **197**, 385–388.

Thomas Jr. G.J., Prescott B. & Hamilton M.G. (1980) *Biochemistry* **19**, 3604–3613.

Thomas Jr. G.J., Benevides J.M. & Prescott B. (1986) In *Biomolecular Stereodynamics*, R.H. Sarma & M.H. Sarma (eds). Adenine Press, Guilderland, New York, pp. 227–253.

Tomita Y., Himeno K., Nomoto K., Endo H. & Hirohata T. (1987) *J. Natl. Cancer Inst.* **78**, 679–680.

Tu A.T. (1982) *Raman Spectroscopy in Biology: Principles and Applications*. John Wiley & Sons, New York.

Van der Voort H.T.M. & Brakenhoff G.J. (1990) *J. Microsc.* **158**(1), 43–54.

Wald N., Idle M., Borcham J. & Baily A. (1980) *Lancet* **2**, 813–815.

Wallach D.F.H. & Verma S.P. (1975) *Biochim. Biophys. Acta* **382**, 542–551.

Weisblum B. & de Haseth P.L. (1972) *Proc. Natl. Acad. Sci. USA* **69**, 629–632.

Yu N.-T., DeNagel D.C., Jui-Yuan Ho D. & Kuck J.F.R. (1987) In *Biological Applications of Raman Spectroscopy*, Vol. 1, T.G. Spiro (ed.). John Wiley & Sons, New York, pp. 47–80.

Zhimulev I.F., Belyaeva E.S. & Semeshin V.F. (1981) *Crit. Rev. Biochem.* **11**, 303–340.

CHAPTER NINETEEN

Dual-Excitation Confocal Fluorescence Microscopy

IRA KURTZ

Division of Nephrology, Department of Medicine, UCLA School of Medicine, Los Angeles, CA, USA

19.1 INTRODUCTION

The biological problems which are studied with confocal fluorescence microscopy can be categorized into two groups: (1) the study of living or fixed preparations whose fluorescence properties do not change with time and (2) the study of living preparations whose fluorescence changes temporally and/or spatially in up to three dimensions. A number of investigations have focused on the former category, but there are few reports which fall into the latter category. An example of studies which involve the change of fluorescence temporally and spatially in a living cell is the measurement of ion activity such as pH using fluorescent probes. Although these probes have been used to measure ion activity in single cells using regular epifluorescence microscopy, only recently have measurements of ion activity been performed using confocal microscopy.

19.2 MEASUREMENT OF pH_i WITH BCECF

Since the introduction of carboxyfluorescein (CF) by Thomas *et al.* in 1979, fluorescent techniques for measuring intracellular pH (pH_i) have virtually replaced microelectrode techniques (Thomas *et al.*, 1979). Optical measurements of pH_i have a rapid response time, high sensitivity and provide spatial information. Because of the excess leakage rate of CF, Tsien and colleagues developed BCECF in 1982 (Rink *et al.*, 1982). This probe is presently the most widely used pH probe. BCECF has a pK of approximately 6.9 and a very slow leakage rate. To load cells with the dye, the acetoxymethyl ester derivative is used which is lipid-permeable and therefore crosses cell membranes easily. Inside the cell, esterases cleave the probe, leaving the charged form of the dye trapped in the cytoplasm.

BCECF is excited at two wavelengths: 490 nm (pH-sensitive wavelength) and at 440 nm (isosbestic wavelength). The 490/440 nm excitation ratio varies with pH_i and is linear between pH 6.5 and 7.4 in most cells. The dye is calibrated *in vivo* using the high K^+ nigericin technique (Thomas *et al.*, 1979). Using a microfluorometer coupled to either a photomultiplier tube or a two-dimensional detector, pH_i can be measured in single cells (Tanasugarn *et al.*, 1984; Wang & Kurtz, 1990). However, if the epithelium is heterogeneous in the depth dimension, e.g. kidney tubule, gastric glands, measurements of pH_i in single cells become less accurate because of the potential acquisition of

Figure 19.1 Pseudo-Nomarski image of an isolated perfused rabbit cortical collecting duct. (40× objective, 8× zoom). The small round cells are intercalated cells and the larger polygonal cells are principal cells. (Adapted from Wang & Kurtz, 1990.)

out-of-focus fluorescence information. This occurs because of the poor resolution in the z dimension of regular epifluorescence microscopes. The use of regular epifluorescence microscopy also makes it difficult to study spatial pH differences within different compartments in a single cell, i.e. cytoplasm versus nucleus, endocytotic vesicles. The cortical collecting duct, a portion of the kidney tubule studied in our laboratory, is cylindrical in shape and possesses at least two cell types: intercalated and principal cells (Fig. 19.1). The cylindrical shape and heterogeneous nature of this preparation necessitated the development of a new optical approach for monitoring pH_i in this preparation.

The confocal microscope has improved resolution in the z dimension (approximately 0.5 μm) Zimmer *et al.*, 1988; Wells *et al.*, 1989). Confocal microscopes are excited with either a laser or a white light source (xenon, mercury arc lamp). These two types of confocal microscopes differ in their sensitivity, speed of data acquisition and wavelength selection. White-light-based systems offer a wide range of excitation wavelength choice; have rapid rate of image acquisition (30 per second) and are less sensitive than the laser-based systems. The latter have limited excitation wavelength capability and slower image acquisition rates. Most laser-based systems utilize a low-powered air-cooled argon laser which emits at two wavelengths: 488 nm and 514 nm.

In our initial efforts to measure pH_i confocally, a laser-based system was chosen because of its greater sensitivity. The argon 488 nm line excites the pH-sensitive wavelength of BCECF. However, argon lasers do not emit at the isosbestic wavelength of BCECF. The newer pH dyes carboxy-SNARF (semi-

naphthorhodafluor) and carboxy-SNAFL (seminaphthofluorescein) are excited at a single laser line, 514 nm, and emit at two peak wavelengths whose intensity varies inversely with pH (*Bioprobes*, 1991). Given the ability to measure two emission wavelengths using two photomultiplier tubes, the initial logical approach in measuring pH_i with a laser scanning confocal microscope appeared to be to use carboxy-SNARF-1. However, studies with the dye revealed an excessive leakage rate. In addition, the pK of the dye is approximately 7.6, making it difficult to measure pH_i up to 6.5. Therefore, we chose to use BCECF. In order to excite BCECF at its isosbestic wavelength a 15 mW helium–cadmium laser was also coupled to the confocal microscope. By alternating the two excitation sources pH_i could be measured in real time.

19.3 DESIGN OF A DUAL-EXCITATION LASER SCANNING CONFOCAL FLUORESCENCE MICROSCOPE FOR MEASURING pH_i WITH BCECF

In the first design of the instrument, an MRC-500 scanning unit (Bio-Rad) was interfaced with an inverted Nikon Diaphot microscope (Fig. 19.2) (Wang & Kurtz, 1990). More recently an MRC-600 scanning unit has replaced the original unit. The emission port was used to excite and collect emitted fluorescence from the kidney tubule on the microscope stage. A polarized 25 mW argon laser (model 5425A, Ion Laser Technology) and a polarized 15 mW helium–cadmium laser (He–Cd) (model 4214B, Liconix) were coupled to the MRC-600 scanning unit. The two laser beams

Figure 19.2 (a) View of the coupling of a 25 mW argon laser (right) and a 15 mW He–Cd laser (left) to the MRC-600 scanning unit. (Adapted from Kurtz & Emmons, 1993.) (b) Close-up view of the optics in front of each laser (He-Cd (right), argon laser (left)).

were combined by reflecting the 442 nm He–Cd laser with a mirror (Oriel) onto a dichroic mirror (Omega Optics) inserted at an angle of 45° in front of the argon laser. The dichroic mirror reflected the 442 nm light and transmitted 488 and 514 nm light to the scanning unit. An electronic shutter (Vincent Assoc.) was placed at an angle in front of each laser. The shutters were opened and closed alternately under computer control. The duration of shutter opening and time between the opening of one shutter and closing of the second shutter were software-selectable. In all experiments a 40× fluorite objective was used. This objective has a high optical throughput and minimal longitudinal chromatic aberration (Keller, 1989). An excitation filter cube in the MRC-600 scanner reflected 442 nm and 488 nm light to the microscope.

Photomultiplier tube no. 1 was used for fluorescence imaging and photomultiplier no. 2 was used for bright-field imaging. A fibre bundle was used to transfer transmitted light to photomultiplier tube no. 2 in the MRC-600 scanner. Software was written (1) to control the timing parameters of the shutters and (2) to digitize fluorescence images sequentially at 442 nm and 488 nm excitation. A zoom factor of 3 or 4 times was used. To monitor pH_i from more than one spatial location in the xy-plane, up to eight regions at each excitation wavelength were measured as a function of time. Up to eight excitation ratios were displayed in real time with or without background subtraction. The data

were stored in an ASCII file and could be imported into a spreadsheet program for plotting and analysis.

The optical properties of the initial MRC-500 based system were characterized. To minimize curvature of field and radial chromatic aberration the data were acquired as close to the optical axis of the objective as possible (Wells *et al.*, 1989). A fluorite objective was used to minimize chromatic aberration (Keller, 1989). The z-axis resolution in the reflected light mode at 488 nm with the pinhole at 0.96 nm was approximately 1 μm. In the fluorescence mode, the excitation and emitted wavelength are not the same. Chromatic aberration prevents the excitation and emission wavelengths from following the same optical path (Wells *et al.*, 1989). The emitted light will not be imaged at the detector pinhole which decreases the z-axis resolution. At 488 nm excitation with the detector pinhole closed maximally, the signal-to-noise ratio from cells loaded with BCECF was too low. Therefore, the pinhole was increased to 1.68 nm. The z-axis resolution at this pinhole size was 1.8 μm at 488 nm excitation and 1.1 μm at 442 nm excitation.

19.4 MEASUREMENT OF pH_i IN THE CORTICAL COLLECTING TUBULE

As depicted in Fig. 19.1 the cortical collecting tubule possesses more than one cell type, i.e. principal and intercalated cells. The cylindrical shape of the epithe-lium and its heterogeneous nature make this prepara-tion difficult to study using optical methodologies. In studies of this preparation, the tubule is cannulated with micropipettes which permits the luminal as well as the basolateral solutions to be changed rapidly. Movement of the preparation is particularly proble-matic when using a confocal microscope. However, a laminar flow chamber and floating table have ameliorated this problem. We measure pH_i in single principal and intercalated cells with BCECF. An example of an experiment measuring pH_i in a single principal cell is depicted in Fig. 19.3. Principal cells were found to have a basolateral Na^+/H^+ antiporter, a Na^+/base cotransporter and a Na^+-independent Cl^-/base exchange (Wang & Kurtz, 1990). In separate studies, cells were discovered to have a basolateral Na^+-dependent organic transporter which transported BCECF and was stilbene-inhibitable (Emmons & Kurtz, 1990). More recently, studies from our labora-tory have demonstrated that the majority of inter-calated cells have both apical and basolateral Cl^-/base exchangers (Emmons & Kurtz, 1991). This finding suggests that the current model of two types of intercalated cells, i.e. α-cells (basolateral Cl^-/base exchange) and β-cells (apical Cl^-/base exchange), needs to be modified.

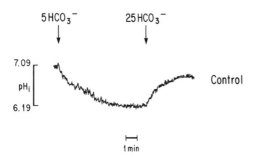

Figure 19.3 Measurement of pH_i in a single principal cell. Basolateral HCO_3^- was decreased from 25 to 5 mM and increased after several minutes to 25 mM. By measuring the rate of change of pH_i under different experimental condi-tions, the mechanisms of H^+/base transport across the apical and basolateral membrane can be determined. (Adapted from Wang & Kurtz 1990.)

19.5   MEASUREMENT OF ENDOCYTIC VESICLE pH

To measure the pH of the endocytic vesicles *in vivo*, it is necessary to target dyes with the appropriate pK, to these acidic compartments. Recently Dunn and colleagues have designed fluorescent conjugates of transferrin and dextran (Dunn *et al*., 1991). The probes

are based on new rhodol dyes Cl-NERF and DM-NERF. The probes are excited at 485 nm (isosbestic wavelength) and 514 nm (pH-sensitive wavelength). Unlike the two excitation wavelengths used to measure pH with BCECF, the rhodol dyes can be excited with the two wavelengths (488 nm and 514 nm) emitted by an argon ion laser. The pK of both dyes is low; Cl-NERF approximately 3.0–3.5 and DM-NERF 5.0–5.5. With these dyes it should be possible to generate three-dimensional images of the endocytic pathway *in vivo*.

19.6   FUTURE DEVELOPMENTS

In the typical laser-based confocal microscope, a single argon laser is coupled to the laser scanning unit. In future, in order to increase the number of excitation wavelengths, several lasers will be utilized coupled to the scanning unit with fibre optics. This will allow the lasers to be placed in a separate area rather than on the optical table.

A second limitation of present-day confocal micro-scopes is the speed of data acquisition (time for scanning the laser across the preparation and for writing data to disk). One approach is to store images on an optical disk for digitization at a later time. It is probable that the speed of laser scanning will increase so that laser-based systems will be able to acquire 30 images per second. White-light-based systems will probably improve in sensitivity and Z-axis resolution. Once the sensitivity and resolution are comparable to laser-based systems, white light confocal microscopes will be the system of choice because they offer a wide range of excitation wavelengths and the necessary speed of data acquisition.

ACKNOWLEDGEMENT

This work is supported by an NIH Grant no. 851 IG-4. Dr Kurtz is an Established Investigator of the American Heart Association.

REFERENCES

Bioprobes (1991) Molecular Probes Inc., Eugene, OR.

Dunn K.W., Maxfield F.R., Whitaker J.E. & Haugland, R.P. (1991) *Biophys. J.* **59**, 345a (abstract).

Emmons C. & Kurtz I. (1990) *J. Am. Soc. Nephrology* **1**, 697 (abstract).

Emmons C. & Kurtz I. (1991) *J. Am. Soc. Nephrology* **2**, 699 (abstract).

Keller H.E. (1989) In *The Handbook of Biological Confocal Microscopy*, J. Pawley (ed.). IMR, Madison, WI, pp. 69–77.

Kurtz I. & Emmons C. (1993) *Meth. Cell Biol.*, in press.

Rink T.J., Tsien R.Y. & Pozzan, T. (1982) *J. Cell Biol.* **95**, 189–196.

Tanasugarn L., McNeil P., Reynolds G.T. & Taylor D.L. (1984) *J. Cell Biol.* **98**, 717–724.

Thomas J.A., Buchsbaum R.N., Zimniak A. & Racker E. (1979) *Biochemistry* **18**, 2210–2218.

Wang X. & Kurtz I. (1990) *Am. J. Physiol.* **259**, C365–C373.

Wells K.S., Sandison D.R., Strickler J. & Webb W.W. (1989) In *The Handbook of Biological Confocal Microscopy*, J. Pawley (ed.). IMR, Madison, WI, pp. 23–35.

Zimmer F.J., Dreyer C. & Hausen P. (1988) *J. Cell Biol.* **106**, 1435–1444.

CHAPTER TWENTY

Properties of Low-Light-Level Video Frame Rate Cameras

P. TOMKINS & A. LYONS

Photonic Sciences, Robertsbridge, East Sussex, UK

20.1 APPROACHES TO LOW-LIGHT-LEVEL IMAGING

There are many advantages in adopting standard video signals for use in scientific imaging systems. One of the tremendous advantages is the wide range of image-acquisition products and software written around data presented as a video signal. With this, of course, comes cost advantages. A further benefit is that by using standard video you can consider your imaging system as modular, e.g. you can change the camera with great ease, allowing wide applications and flexibility. One of the conveniences of video-rate imaging is the high time resolution it affords. Further advantages come from using video recorders, where the TV image is recorded, and can be later integrated in a computer at leisure, editing out unwanted data, and replaying the experiments as many times as necessary, all using the original data set, stored once on a relatively cheap, affordable machine. Dual- or multiple-wavelength excitation, multiple-wavelength imaging using fast filter wheels, and other dynamic or multiplexed techniques become possible when one is freed of the destructive read-out and long cycle times of slow-scan devices. This said, there are some applications where none of these features are useful, and slow-scan

systems compete. But for many, the very ability of video systems to show your image building up in the frame store along with a live display actually during the acquisition, will always feel more comfortable than the hopeful 'shot in the dark' technique of integrating CCDs, especially where valuable or short-lived material is being used and you do not have time for too much trial and error.

Video image-acquisition products are inevitably prolific and some really sophisticated systems can be put together quite economically. Highly refined application-specific software packages are now becoming established that support these products.

However, the high-speed read-out and real-time capabilities of video systems are designed for high light levels and as such most video cameras just will not produce a useful image at the low light levels that are often of interest to the scientist. The old-fashioned approach to this was to build some kind of integration into the camera tube, and this was the principle behind 'SIT' and 'ISIT' cameras. This technique produced lag, and to some extent offset the response-time advantages of video. Modern techniques, where high-speed, high-sensitivity cameras feed real-time data to computers, provide an ideal solution, so the researcher can put in just the integration wanted (with just as much lag as the experiment will allow) by the use of a computer controlled frame-grabber and image-processing software.

There are two main types of noise that limit the low level at which an image can be usefully obtained. First is thermal noise. This is reduced in slow-scan systems by cooling. Second is 'read-out' noise which is overcome both by good system design and only reading out once at the end of an integration. At video rates, both noise sources contribute to limit the lower light level at which a CCD can produce a useful image.

Basically, at room temperature, the noise of most normal TV detectors is simply too high to enable low light images to be realized. The solution to this is to provide significant amplification of the image before the imaging element, so that commercial TV rate detectors are actually working at the light levels for which they are designed. This is done by means of image intensifiers. These devices produce an image at the output many times (several thousand) brighter than the original input. By doing this, the signal level is so great at the image sensor that its intrinsic noise can be ignored.

Clearly, if the light level is increased by a factor of say 1000 at the CCD, a good picture will result. To do this, it is only necessary to interpose (with suitable optical coupling) an image intensifier between the incoming image and the CCD. In this example, the CCD working in TV (uncooled) mode, has been made 1000 times more sensitive by use of the image intensifier.

This technique works very well indeed. There is, however, a fundamental law of physics that limits us, due to the fact that light comes in discrete packets or 'quanta' called photons, and when you get down to low level signals, providing you see individual photons clearly, you will not get more sensitivity with more gain. In these cases, the system is said to be 'quantum limited'.

20.2 IMAGE INTENSIFIERS, MAXIMIZING SIGNAL, MINIMIZING NOISE

To detect really low-light-level images, high sensitivity is absolutely necessary. More than this, it is important to make sure that your weak signal is not buried in noise, so low noise characteristics are necessary. Ideally one wants to maximize the 'signal-to-noise' ratio. Regrettably, this is measured differently by just about everyone, and no real standard allowing intercomparison is conveniently available for the sytems we are discussing. But the principle is valid; the best camera for any application is one that produces more signal and less noise, or a higher ratio of the two, for a given situation.

Fundamental to sensitivity is the efficient use of the precious few photons that are available. Image intensifiers use a light-sensitive input coating called a 'photocathode' to convert photons to electrons. These electrons are later amplified inside the intensifier. The percentage of photons that give rise to photoelectrons is called the quantum efficiency of the photocathode. The higher the quantum efficiency at the wavelength you are interested in, the higher the sensitivity of the overall detector is likely to be. Note that photocathode quantum efficiency is not the overall efficiency of the detector – just an important factor. There are other parameters (such as how efficiently the photoelectrons are themselves detected within the intensifier) that go together to make the overall efficiency always less than the photocathode quantum efficiency. Since the cathode quantum efficiency always limits the theoretical maximum possible overall efficiency, this should always be maximized wherever possible.

Two important definitions are the quantum efficiency (QE) of the photocathode and the detective quantum efficiency (DQE) of the overall detector system. The DQE is the fraction (expressed as a percentage) of incident photons that are detected by the system. In an ideal electro-optical detector, the DQE equals the QE. In reality, internal losses cause the DQE to be less than the QE sometimes in systems not designed for scientific applications by a surprisingly large factor. Quantum efficiency, when applied to photcathodes, is sometimes called responsive quantum efficiency (RQE). DQE can be related to RQE by a constant of proportionality, which could be called 'internal efficiency' of the detector. RQE is a strong function of the wavelength at which the measurement is made. The internal efficiency, which is a result of internal electron optical design, is wavelength-independent. This means that a knowledge of the RQE of the photocathode of a device is sufficient to indicate how the DQE varies as a function of wavelength of input light. It is not easy to find out the absolute DQE without sophisticated measurement. It is often the difference in this value that distinguishes a 'good' camera from a 'bad' one.

20.3 THERMAL NOISE

Both photocathodes and semiconductor detectors are subject to random emission or production of electrons in the absence of light. This is a result of thermal effects within the sensitive material, and is given the name thermal noise. Without analysing the origins, it is sufficient for us to know that thermal noise increases rapidly with temperature. Thermal noise doubles for (approximately) every 7–8°C rise in temperature.

Similarly, this noise reduces by a factor of two for every 7–8°C. The cooling of the detector is the principle of noise reduction for slow-scan CCDs. So much for the rate of change of thermal noise with temperature, but what of its absolute value? For photoemissive materials (photocathodes), there can be a tremendous range in the amount of thermal noise produced. As a general rule, photocathodes that respond well in the red and near-infrared will be noisier than photocathodes where the peak response is in the blue. This is because long-wavelength photons have less energy than short-wavelength ones ($e = h$), therefore, electrons are released more easily in the red cathode than in the blue one, so thermal effects give rise to more emission. Typical photocathodes at room temperature can give rise to anywhere between 100 and 10 000 electrons $cm^{-2} s^{-1}$. The actual value depends very much on the specific manufacturing process, and is not only a function of spectral response, but a combination of this with material formulation, manufacturing process and some, still not fully understood, 'seemingly random' variables. For this reason, it is important that devices be not only manufactured for, but carefully selected for demanding scientific applications.

20.4 READ-OUT NOISE

Read-out noise is the noise associated with the process of converting the optical signal into a voltage wave form. Again, it is a function primarily of design. The read-out noise parameters that affect slow-scan CCDs are much less important in intensified CCD cameras. However, it is necessary to apply good design practices to avoid fixed pattern noise, such as clock feed-through from generating 'vertical' and possibly 'horizontal' stripes or bands across the image.

20.5 NON-THERMAL NOISE

If a detector is cooled and its thermal noise plotted, it will be seen to decrease, as described above, but at a certain point the rate of decrease will virtually cease, and an almost constant noise level almost indistinguishable from thermal noise or true photon events will be found. The origin of these events is internal to the detector structure and in a photoemissive device originates from two main sources. First, from internal micro-discharges which emit very low light levels and are seen by the photocathode thus producing electrons. Secondly, from direct electron emission from microscopic high points on the photoemissive surface,

this being called field emission. The two can be distinguished since internal discharges produce a non-thermal noise component that covers an area, or possibly all of the image, whereas field emission appears as bright fixed spots. A third source of noise that can be eliminated by design is faint corona discharge around high-voltage structures external to the vacuum whose light can fall on to the photcathode.

20.6 TEMPORAL CHARACTERISTICS OF PHOTOCATHODE NOISE

If a detector based on a photoemissive material sensitive to visible light, such as a photomultiplier or image intensifier, is exposed (when not operating) to a high level of illumination, then placed in complete darkness, the noise measured will be much higher than had the detector been kept in darkness prior to the measurement. This effect may produce increases in noise of several orders of magnitude. The noise so induced decays with a time constant of seconds to tens of seconds depending on the particular photocathode. The magnitude of the effect is dependent upon the intensity of illumination, to some extent duration of illumination and particularly to the illuminating wavelength. The effect is far more pronounced when UV radiation is allowed to fall on the photocathode.

This has important practical consequences as regards experimental technique when low-light-level images are to be detected, particularly when low and stable background noise is required. The guidelines, therefore, are:

(1) Wherever possible, keep detectors in a dark environment prior to use.
(2) When changing samples, do not simply switch off the detector and expose it to room light, or other strong light sources. Protect the photocathode from such exposure.
(3) Never, if using fluorescence techniques, allow the exciting UV radiation to illuminate or to fall on the photocathode.
(4) If you need to expose the photocathode to high light levels (for installation or cleaning purposes), do this in incandescent lighting. Daylight contains much UV and fluorescent lighting emits a surprisingly large quantity of UV radiation.

It should be noted that this 'charging effect' is quite reversible and in no way damages the detector, it simply raises the noise. If this effect is noted, the best solution is to leave the detector running in total darkness with power applied until the noise has fallen to a reasonable level.

20.7 GAIN USE AND ABUSE

If an ordinary light level video camera such as a CCD or vidicon is operated at normal light levels a perfectly acceptable image results. At 1% of this light level a perfectly acceptable image can be obtained from the detector by putting an image intensifier with a gain of 100 in front of it. Indeed, this arrangement is 100 times as sensitive as the original. It is tempting to believe that by increasing the gain more and more, ever greater sensitivity will be achieved. This is not the case, as more and more gain is employed at fainter and fainter light levels the image will be seen to become progressively more grainy. These grains are, in fact, the amplified result of individual detected photons. At these light levels the quantum nature of light becomes apparent and adding further gain will simply make the speckle resulting from a photon appear brighter. In fact, providing that each detected photon speckle is clearly visible at the output of the camera, then there is no advantage in increasing the gain further. In these low light regimes extra sensitivity can only be achieved with detectors of higher QE, i.e. detectors that detect a larger fraction of the few photons in the input image. The quantum nature of light, therefore, imposes a useful limit to the maximum gain that can be employed in an intensified camera with good effect. One should therefore beware of claims for cameras with incredibly high gains, unless there is a design reason for this, e.g. (operation in the photon counting mode) since very high gains can actually limit rather than enhance performance.

For the following discussion of gain, we shall consider how the term may be applied to the image intensifier of an intensified CCD (ICCD) camera. There are several ways of defining gain. The most common is 'luminous gain'; this parameter, which is largely a hangover from military technology, is not particularly useful for scientific purposes. It is the ratio of output signal to input signal when measured with an eye-corrected photometer using a tungsten lamp as the light source. Typical image intensifiers have luminous gains in the order of 10 000–30 000 (for a second-generation device incorporating a micro-channel plate electron multiplier). This figure is misleadingly high since most intensifiers respond to the near-infrared of the tungsten lamp to which the eye-corrected photometer is blind. Thus the input signal is undersampled and a large figure obtained for the ratio.

A far more useful quantity is the radiant power gain or photon gain. The former, measured in units of watts per watt is the ratio of optical power out of the intensifier to optical power in. Expressed as photon gain it is a measure of the number of photons out per photon in. Clearly this is a function of wavelength and is best defined at the peak emission wavelength of the output phosphor, which is usually in the green.

Typical second-generation intensifiers have radiant power gains or photon gains in the order of 1000–2000. Yet a third kind of gain is the photoelectron gain. This is the number of photons emitted at the screen per detected photon, i.e. per emitted photoelectron since only a small fraction (RQE) of the incident photons give rise to photoelectrons (i.e. are detected) then the photoelectron gain is clearly the photon gain divided by the DQE.

The latter two types of gain interrelate in a way that is most important as regards image quality. The radiant power gain is the product of the photoelectron gain and DQE. Consider the following three image intensifiers.

DQE	Photoelectron gain	Radiant power gain
10%	1 000	100
1%	10 000	100
0.1%	100 000	100

These three intensifers would all have the same radiant power gain, and could have exactly the same luminous gain – in other words the published specifications for the three devices could be identical. Now consider how the three devices respond if given a signal of 1000 photons. The first device will give an output image of 100 speckles, each of 1000 photons. The second device will produce ten output speckles each of 10 000 photons, and the third simply one bright spot of 100 000 photons.

The first intensifier would produce a basic image, the second limited spatial information and the third no image information at all. Therefore, it is vital to understand how photon gain is distributed between DQE and photoelectron gain in order to assess the potential imaging performance of a device – and none of this information is usually available in intensifier manufacturers' information. As we will discuss later, the only way you can tell if a camera designer has done a good job is to test the end product in your application.

This said, a given camera will have a constant DQE at a given wavelength, and if the gain is controllable then all the better. For higher light levels a low gain is required, and at the lowest light levels the maximum gain compatible with clear detection of individual photon events is needed. For this reason, a wide user-controllable gain range makes a camera far more versatile as regards the light level over which it can do meaningful science. In intensified CCD cameras this is not just an electronic gain but, as described above, is the factor by which electron showers are actively multiplied (truly, in number, within the tube) before reaching the image read-out element.

20.8 SPECTRAL RESPONSE

We have described previously how quantum efficiency of photocathodes varies with the wavelength of the incoming radiation. If the response is plotted as a function of wavelength, a 'spectral response curve', whose form is roughly an upturned 'U' is obtained. The peak will correspond to a particular part of the spectrum, and it is important to try to match this to the science, or at least ensure that there is reasonable response at the wavelength of interest.

The spectral response can also allow comparison of response at different wavelengths, important, for example, if ratioing images of orange and blue fluorescent dyes.

There is a very limited choice of response curves available in imaging devices, unlike photomultipliers, in which a wider choice has long been available.

The most common photocathodes in image intensifiers as used in ICCDs are 'S25' and 'S20'. Typical curves are shown in Fig. 20.1. These curves are not absolute and there is tremendous variation between devices and this again highlights the importance of careful selection for particular applications. The curve represents the average. The difference between S25 and S20 is basically one of photocathode formulation and process, details that do not need to concern us here, since we are more interested in the performance.

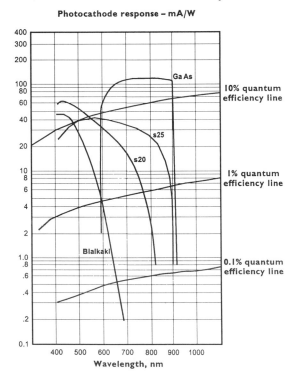

Photocathode response – mA/W

Figure 20.1 Typical photocathode spectral sensitivities.

Terms you will encounter when reading manufacurers literature are:

(1) 'Luminous sensitivity' – expressed in microamps per lumen. In this mode the intensifier or other device is run as a photodiode and the current that can be drawn per unit of incident light is recorded. Unfortunately this is done with a 2854 K tungsten lamp, which emits predominantly in the infrared, so photocathodes that have responses longer than 700 nm will have higher luminous sensitivity figures than ones that do not.

This figure is only of use for intercomparison of devices with the same spectral response curve. In absolute terms, it is only truly a measure of how good the photocathode is at detecting a tungsten lamp at 2854 K. For example, a 400 µA lm^{-1} S25 photocathode might have a quantum efficiency at 500 nm of 5%, whereas a 200 µA lm^{-1} S20 might achieve 10%. If you are working at 500 nm, the 'lower sensitivity' device will produce twice the signal! As such, beware of luminous sensitivity figures, these are best left where they originated – in military specifications for gun sights! Scientists need more meaningful parameters.

(2) '*Radiant sensitivity*' – This is measured as luminous sensitivity, but the light source is a monochromatic one of known optical power. The wavelength is varied and the photocathode current drawn is plotted as a function of wavelength. Radiant sensitivity is given in milliamps per watt and does at least provide real information about the response of the device.

(3) '*Quantum efficiency*' – A much more meaningful measurement for scientific purposes, the quantum efficiency as a percentage is plotted as a function of wavelength. This is related directly to the radiant sensitivity, but allows for the difference in energies between photons of different wavelength.

Often, spectral response curves will be shown in milliamps per watt with lines of equal quantum efficiency shown on them. Again, it should be noted, that a higher response (shown in milliamps per watt or %QE) at the wavelength you are working at gives better detection efficiency and more signal.

20.9 ICCD CAMERA RESOLUTION CONSIDERATIONS

In digitizing the video output the data are mapped into an array of picture elements called pixels. Systems are commonly 512 × 512, 768 × 512 and now programmable up to 1024 × 1024 pixels.

Resolution is quoted in two ways: either 'TV lines

per picture height' which is the number of separate vertical lines that can be resolved across the screen in a width equal to the picture height, at limiting resolution, which is $\frac{3}{4}$ of the picture width; or 'TV lines per picture width' which is simply the number that can be resolved across the full picture width.

Since the aspect ratio of the image (in CCIR) is 3:4, then the resolution in TV terms per picture height is simply $\frac{3}{4}$ of the number per picture width. Simple enough, but often used to confuse (and not always innocently). Again, when reading data, make sure you compare like with like. The 'per picture height' concept is rather old-fashioned and stems from early TV days. 'Per picture width' is more meaningful since you normally want to acquire the whole picture into the computer anyway.

Resolution is one of those areas which (like 'gain') is often chased to its limit with very little benefit. There is no advantage in having a very high-resolution camera if you are going to digitize to 256 × 256 or even 512 × 512. However, it is better to 'oversample' a little and have somewhat more resolution (in TV lines per picture width) than you have pixels in your frame store per horizontal line.

There is little advantage in going for a camera with 1000 + TV lines resolution with a 512 × 512 frame store. Of course, if you have a 1024 × 1024 frame store then you might consider a really high-resolution camera for your science. These are few and far between at present and since camera design is balancing a variety of performance characteristics, if you want such resolution, it is likely that you will have to compromise on some of the features available in present, moderate-resolution systems.

Another problem that arises is the confusion between resolution and image clarity. What really matters is the modulation transfer function (MTF) rather than limiting resolution. This function is a plot of contrast (as percentage) as a function of spatial frequency when looking at a black and white (100% contrast) target. The contast seen through the camera will fall as the spatial frequency increases, down to 4% at the 'limiting' resolution. The contrast at low spatial frequencies tells you how clearly you will see what *is* 'resolved', or has 'black and white' degenerated to shades of misty grey.

In Fig. 20.2 we show the MTF curves of two cameras which have the same resolution: camera A will give clear, crisp images, whilst camera B will look like it was always seeing through fog. For instance at 250 TV lines per picture width, camera A would give 80% contrast between black and white, while camera B would give only 30%. This is an important consideration for some (but only some) applications, yet it is information that is almost never published. Yet again, the only way to test the suitability of a camera for a particular application is to test it. The MTF curve may at least give some insight into why some cameras work better than others even though the 'resolution' is the same.

20.10 THE CHARGE-COUPLED DEVICE

The CCD operating principle is called 'charge-coupling'. Finite amounts of electrical charge, often called 'packets' are created in specific locations in the silicon semiconductor material. Each specific location, a 'pixel' is created by the field of a pair of gate

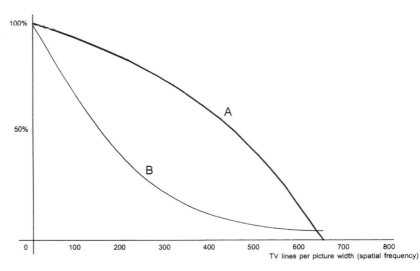

Figure 20.2 MTF curves of two cameras with the same resolution.

100%

50%

0   100   200   300   400   500   600   700   800

TV lines per picture width (spatial frequency)

electrodes very close to the surface of the silicon at that location. By placing the storage elements in a line, for instance, voltages on the adjacent gate electrodes can be alternately raised and lowered and cause the individual charge packets beneath them to be passed from one storage element to the next.

A photoelectric effect by which free electrons are created in a region of silicon results when illuminated by photons in the approximate spectral range of 300 (ultraviolet) to 1100 (near infrared) nm wavelength. Absorption of such incident radiation in the silicon generates a linearly proportional number of free electrons in the specific area illuminated. In a CCD the number of free electrons generated in each site will be directly proportional to the incident radiation on that specific site. A focused image from an optical system creates charge-packets in the finite photosite array that will be a faithful reproduction of the scene projected on its surface.

After an appropriate exposure time, charge-packets are simultaneously transferred by charge-coupling under an adjacent single long gate-electrode, to a parallel CCD analogue transport shift register.

Each charge-packet corresponds to a picture element (pixel) and continues to faithfully represent the total sensed radiant energy which was absorbed in the specific photosite. The transfer gate is immediately returned to the non-transfer clock level (LOW) so photosites can begin integrating the next line of incident image information. At the same time, the CCD analogue transport register, now loaded with a line of picture information in the form of charge-packets from a line of sensor sites, is rapidly clocked to deliver the picture information, in serial format, to the device output circuitry.

20.11 FRAME AND LINE TRANSFER SENSORS

CCD cameras, as used in ICCD cameras, have two main forms of architecture, known as frame transfer and line transfer. Both types are just 2D arrays of photosites, but functionally they are divided into image zones and storage zones. During 'exposure' the image falls in the image zone and builds up charge. At the end of the field time, this pattern is transferred to the storage zone, and is read out during the next field time whilst the subsequent field is integrated in the image zone. A frame transfer sensor has (for example) an active area of 288 vertical by 600 horizontal photosites or pixels. At the end of integra- tion the whole image is moved 288 pixels vertically into the storage zone. This storage zone is protected from charge generated by light by covering it with an

optically opaque mask such as an aluminium coating. The required resolution in the vertical direction is achieved by changing the electrode bias between odd and even frames to effectively displace the potential wells such that twice the number of vertical elements are achieved in effect.

A line transfer sensor truly has the correct number of pixels, but each vertical column has an adjacent storage column. This means that only half the image area is actually light-sensitive. This does not change the resolution, but it does reduce the sensor's quantum efficiency. In an ICCD, where the camera design can compensate easily, the choice of line or frame transfer device is unimportant to the user of intensified, video-rate cameras.

The only difficulty arises when time-resolved work is done. The frame transfer sensor actually integrates one field then the next, each with only 288 elements' resolution. In photon counting type applications, where a photon 'speckle' may last less than a field time, it may end up being imaged only in one field. So a grabbed frame may show lots of speckles which are crossed with a blank line, since the events are imaged in one field only and when the subsequent interlacing field arrives at that spot, the image has gone. If this might cause a problem, use a line transfer sensor.

Line transfer sensors can either integrate each field for a field duration, or integrate each field for a frame duration, reading out 'out of phase'. Higher temporal resolution is achieved with field integration at the cost of a factor of 2 in effective sensitivity.

Unless carrying out sophisticated signal processing of the photon counting type, the CCD architecture will, for the most part, be 'transparent' to the camera user. When integrating many frames in a computer, the type of CCD sensor architecture is virtually irrelevant, and other, performance-related, parameters should guide your selection.

20.12 INTEGRATION OF AN ICCD INTO AN IMAGE-PROCESSING SYSTEM

20.12.1 Line impedance/termination

Most cameras are designed to run into a standard 75 Ω load, and before turning on a camera it is advisable to consult the camera's manual and look up 'video output'. It will probably say '1 V peak-to-peak into 75 Ω'. This is much more than just a load-driving capability. The video output is a 7–8 MHz signal passing along a transmission line, not just a wire. If the load is not 75 Ω, then reflections may occur, producing ghosting, multiple images and loss of resolution and contrast. Also, some cameras may only produce '1 V peak-to-peak' when actually feeding the correct load, and may provide larger voltages with a high impedance load.

Having determined the designed load impedance for the camera, turn to the specification of the frame-grabber. There may well be a switch '75 Ω IN/OUT' on the board, or a link to be made. The manual should tell you, but if it does not it may be advisable to connect a multimeter on 'resistance' range between the pin and earth on the video input, to check if the correct impedance is there. If not, look for a switch, or a solder link, or if still no luck, contact the board manufacturer.

It is *vital* to set up the frame-grabber input impedance to suit the designed load impedance for the camera.

20.12.2 Synchronization

Virtually all video cameras provide 'composite video', which contains sync information. Most frame grabbers are able to separate the sync from the video, and then use this to synchronize digitization and data storage. Some frame-grabbers cannot do this or cannot do it very well and in this case it is necessary to feed a separate sync pulse from the camera to the frame-grabber. Most cameras (or their controllers) provide a 'compound' or 'mixed' sync output and this can be fed to the 'sync in' input on the frame-grabber. This signal will probably be at TTL levels and is best fed by a coaxial cable.

In this configuration, the camera is the 'master clock' and controls the timing of the frame-grabber. In some cases, particularly with fussy computer systems, the frame-grabber is given a disproportionately high level of importance and is configured to drive the camera. In this case the frame-grabbers' 'mixed sync out' should be fed to the 'external sync input' of the camera.

Now the camera's timing is slaved to the computer. A vital point is that if you have a computer and frame-grabber that needs to drive the camera, *do not* buy a camera that will not accept an external sync signal. On the other hand, if you have a friendly image-acquisition system that will take the sync off the video, then it will accept virtually any camera and is far more flexible.

'Mixed' sync is just that: vertical and horizontal pulses both on the same output. It is relatively trivial to sort the pulses out internal to the camera or computer due to their widely differing frequencies and timings. There is, unfortunately, another way of doing it. Here, vertical and horizontal syncs are provided separately, under the names Vd and Hd (originally from 'vertical drive' and 'horizontal drive' from the days of magnetically scanned tube cameras).

These pulses are usually TTL and it is necessary to provide both, in the same direction such that the camera can provide Hd and Vd to the computer via two wires or vice versa (but not one way and one the other, or just one and not the other). This type of synchronization is justly unpopular and many cameras and acquisition products do not support it.

One final note on synchronization, should none of this work, check the polarity of the sync signal. This does not mean is it +ve or −ve with respect to earth (if it is TTL, it will always be 'positive') but whether digital '1' is represented by a logical high or low.

Most mixed syncs are described as 'negative going'. This means that a line sync pulse is represented by a low level on the mixed sync output. Ensure that the camera and the frame-grabber are talking the same language!

20.12.3 Black level

Many cameras produce the standard video signal with an arbitrary DC offset, so the bottom of the sync pulse is, for example, at +1.5 V DC. In most cases, sophisticated frame-grabbers are happy with this, since they will AC couple the signal and then change it to a DC level suitable to their internal circuitry. There are, however, a few otherwise very good products around that are not so accommodating. Such devices tend to take DC values as significant so a black level at 0.5 V DC will result in the computer assigning a mid-grey image to its interpretation of black! Some cameras will let you adjust the DC offset. Almost all frame-grabbers that are not AC coupled will have an adjustment potentiometer on the board to set the black level. A few, difficult products, will allow you to do this only via software.

Ideally check if your frame-grabber produces a black image (i.e. near 0) for a dark input. Do not pursue a precise 'zero' too hard as there may be a slight 'pedestal' when the active video line is slightly above the black level.

If not, and the manual says the input is DC coupled, then find out the accepted range of the frame-grabber. This will probably be levels in the range '+0.3 to 0.8 V DC' or something similar. Now, with the camera connected, feed dark images to the frame-grabber and digitize, adjust the potentiometer, or software setting until you get reducing values and eventually zeros. If this cannot be achieved, then the camera DC offset is probably too great. Check this on an oscilloscope with the input DC coupled. Look in the camera manual on how to change the DC offset, or call the camera manufacturer. Set the camera's DC output level to within the accepted range of the frame-grabber, then make fine adjustments to either or both, until black gives a digital value of near zero.

20.12.4 Setting peak white

Having set black = 0 (roughly) it is now necessary to determine that saturated white = 255 (with 8 bit digitization) or near enough. Again, check that the camera output is specified as '1 V peak-to-peak into 75 Ω' and that this is what the frame-grabber is looking for, and also that the 75 Ω load is connected. With an oscilloscope, adjust camera gain and/or image brightness so that a peak signal level of just over 0.7 V above the black level is obtained.

Now grab a frame and check the pixel values corresponding to the peak part of the signal. These values should be around 255. If not, then there is a problem. Check the manual for the frame-grabber, and if there is a potentiometer to adjust, then follow the procedures described, or adjust the analogue-to-digital converter (ADC) gain in software.

These four steps can be summarized as:

(1) Impedance: Matching, camera to frame-grabber.
(2) Synchronization.
(3) Black level: Setting up so black = 0.
(4) Peak white: Setting up so peak white = 255.

When you have achieved these steps, the camera should be integrated successfully into an image-acquisition system.

20.13 CONCEPTS OF VIDEO

It is useful to understand some of the terms associated with video technology, and to have a grasp of the basic concepts. Video-rate cameras provide serial, analogue data outputs. The image accumulated in the camera is read out as a varying voltage or current (depending on the exposure history of the image point) in a sequence that was originally described by the scanning pattern of the electron beam in tube cameras. The image is read out horizontally line by line, progressing vertically across the image until the whole area has been 'scanned'.

In Europe (and other places where the mains frequency is 50 Hz) 'standard' video (known as CCIR) consists of 25 'pictures' or frames per second. This is chosen to be half the mains frequency so that synchronization can overcome the annoying moving patterns associated with mains pick-up. In North America, the TV standard (EIA) is 30 frames per second, and the mains frequency is 60 Hz.

If, however, the picture was simply to be scanned every 1/25 s, there would be a perceivable flicker. This is overcome by scanning through the picture twice in every 1/25 s, the first time displaying the 'even' lines,

i.e. 0, 2, 4, 6, etc. down the screen, then the second time by displaying the odd lines 1, 3, 5, 7, etc. In CCIR there are 625 horizontal lines counting vertically up the image, and these consist of two 'fields' called 'odd' and 'even', each of 312.5 lines. This means that:

> A frame consists of an odd field and an even field
> A frame takes 40 ms and a field takes 20 ms
> Alternatively the frame frequency is 25 Hz and the field frequency is 50 Hz.

When looking at a TV image, the top complete line is line no. 1 (in terms of display order) the next is line 314, the next is 2, then 315 and so on. This technique, which is invariably used to overcome flicker, is called interlace. An odd field is interlaced with an even field to produce a complete frame.

With 25 frames per second, and 625 lines, the time taken per line is 64 μs, i.e. the line frequency is 15.625 kHz. (This is the high pitched 'whistle' some people can hear from televisions and monitors.)

These numbers are important when considering digitizing video. Digitizing the image to 512 × 512 requires doing the analogue-to-digital conversion at 512 × 15.625 kHz = 8 MHz. This also shows the bandwidth necessary in any video circuitry in order not to degrade the resolution. 'Minimum bandwidth required = resolution × line frequency' is a very simple guide. It is not strictly true, since the image information does not occupy the full line period, as some of this time is taken with timing pulses and other signals – the actual time taken for a scan of the displayed line is 52 μs, the other 12 μs being called the blanking, since there is no picture being 'written' during this time.

It is instructive to consider how a complete video line would appear if displayed on an oscilloscope. Figure 20.3 shows a line of video, where the scene being viewed is black on the left and white on the right, gently grading between the two.

Timing is quite important (especially when integrating a system) so we list these:

For CCIR:
Line period	64 μs
Front porch	1.55 ± 0.25 μs
Sync pulse	4.7 ± 0.1 μs
Horizontal blanking	12.5 ± 0.25 μs

So much for 'horizontal' timings. The image is clearly made of a sequence of these lines, and they are strictly ordered. With a 20 ms field time only 18.4 ms is actually taken up with 'picture', the other 1.6 ms is used for field sync pulses, this corresponding to 25 lines plus one horizontal blanking interval. This 1.6 ms is called the vertical blanking interval. Actually a 625 line picture displays less than 600 horizontal lines, the rest being blanked.

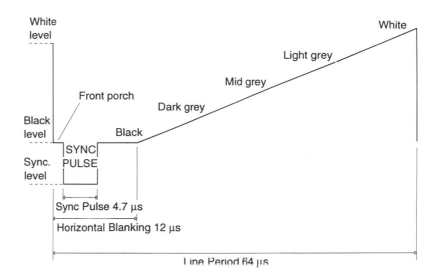

Figure 20.3 A line of video: the scene being viewed is black on the left and white on the right, gently grading between the two.

In summary, each frame comprises 625 lines made up of an odd field and an even field each of 312.5 lines which are interlaced together. Over 25 of these lines are not displayed (blanked) by the vertical blanking interval. At the beginning of each line there is a line sync pulse, embedded in the horizontal blanking interval. During this interval the camera sends out a signal corresponding to 'black' so this is then called the black level.

As regards signal levels, the depth of the sync pulse should be 30% of the overall signal. This means that '1 V peak-to-peak' signal should have a 300 mV difference between sync level and black level, and 700 mV difference between black level and peak white.

The tolerance on these is given as ± 15 mV, but this is seldom adhered to. In practice, sync depths between 200 and 400 mV are quite common and most modern computer systems and video equipment are quite insensitive to such differences. (It is, of course, vital to set up equipment so that both camera and frame-grabber agree what is 'black' and what is 'peak white'.)

American (EIA) TV standards are very similar, except for the difference in timing, with a field frequency of 60 Hz and a line frequency of 15.734 kHz.

It is very important to remember that the output of a standard rate camera is analogue. Though people sometimes request it, there is no point in producing a digital output video camera. It makes much more sense to carry the analogue signal to a high-speed video converter inside the computer, especially since these are widely available at suitably affordable prices – commonly known as 'frame-grabbers'.

A word of advice here. There are 'true' frame-grabbers which grab all of the required frame, and much less expensive devices that grab parts of a series of fields over a few frames to fill up the memory, thus digitizing at a slow rate. The latter units are often useless for the kind of science this book addresses, so beware.

When connecting a video camera to a frame-grabber there are five important questions that should be asked:

(1) Does the camera need to run in to a 75 Ω load and if so, is the frame-grabber input configured this way?
(2) Will the frame-grabber synchronize from the video signal, or does it need another separate sync input?
(3) Does the computer/frame-grabber need to input its sync for the camera to follow?
(4) What signal level does the frame-grabber call 'black' and is the camera set up so that its black level corresponds?
(5) What signal level does the camera call peak white, and is the frame-grabber set up to accept this?

20.14 PHOTON COUNTING IMAGING

When the light levels become really low, and gain is as high as is useful, the image will simply consist of a succession of bright 'speckles' on the screen. This kind of image, rather like looking into a snowstorm, results from the quantum nature of light. The speckles each result from the detection of an individual 'photon' or light quantum. Increasing gain further will only make the individual speckles brighter. The only improvement that can be made in these situations is to ensure maximized DQE in order to obtain the maximum number of speckles for a given number of photons.

This kind of image, which can be summed, integrated or averaged in a conventional frame store to significantly enhance signal-to-noise, can be treated in quite another way. Rather than thinking of the quantized nature of the image as a difficulty, it can be turned to advantage. Most simply, one takes a frame of video and looks at each pixel and decides simply whether there is a bright speckle in it or not. In the computer memory, each pixel has an address, and if, for the frame in question, there was a speckle present in several pixels, the number 'one' is added to the content of each corresponding memory location. This is repeated frame after frame, for each (memory location) pixel adding one when a bright speckle is seen in the corresponding place, zero otherwise. This can be likened to a rather large array of photon-counting photomulipliers, and indeed, the technique has been used very successfully in astronomy and nuclear physics where a huge 2D array of PMT would be quite impractical.

The method is very simple in principle, but rather more difficult to implement in practice. The main problems occur because the 'speckle' sizes are not all the same in terms of intensity even though they each represent precisely one photon. To gain advantage from this technique, it is necessary to reject the smaller speckles that represent noise arising 'later' in the camera's gain mechanism. By so doing, signal-to-noise ratios can be improved, even at these ultra-weak signal levels. This will only work if the 'spread' of speckle brightness somehow enables a distinction to be made between real photon-induced events and noise; otherwise rejecting the small events will throw out as large a fraction of the signal as it does of the noise, and no improvement will result. For this reason, a camera must be designed rather carefully to work in this mode, and normal ICCDs just will not perform.

This technique will not enable you to overcome photocathode dark (or 'thermal') noise, since a photoelectron emitted by the photocathode as a result of dark noise is indistinguishable from one released by a photon. Other problems arise since speckles rarely fall only in one pixel. Even if speckles were exactly the same brightness, then there would be quite a range in pixel value between the occasions where a speckle fell directly on a pixel and the cases where the event occurred at the junction of four pixels (simplistically, a 4:1 ratio. Sophisticated software is needed to prevent this latter case being recorded as four events.

There are, in most intensifier-based cameras, a few bright events that are significantly more intense than the average. These arise from ions, and can be discriminated against by rejecting very bright events.

In essence all this is accomplished by selecting a 'window' function for what qualifies as a photon event, a lower threshold to reject low amplitude noise, and an upper threshold to reject ion events. This is virtually identical to the techniques long established for photomultipliers, except that here we are only working with a 2D array of between a quarter and one million channels! There are clearly trade-offs to be made between maximizing DQE and minimizing noise, as a result of window width and position.

To do this intelligently it is vital to understand the amplitude spread of the photon events, called the 'pulse-height distribution'. Ordinary intensified cameras have a negative-exponential form to their pulse-height distributions, and it is clear that where to set discriminator thresholds in this case is quite arbitrary. To work well, it is necessary to have a distinct 'photon peak' in the distribution, and it is to this end that specially designed cameras become essential.

Attempts have been made to ascribe significance to the speckle size other than 'one photon or zero', by saying 'one, two or three?' etc. with each frame. It is very doubtful whether such narrow photon peaks can be generated by present cameras. Here we see a compromise between photon counting and analogue 'total charge' integration, but such techniques can still contribute a little to noise reduction.

The technique has been called 'pseudo photon counting'. At its best, it is a very powerful technique, and at its worst, is truly 'pseudo' in as much as it will perform the same as analogue techniques at very low light levels, but will probably cost you far more for the equipment.

This kind of imaging is just making its way into biomedical applications and should be given a chance to prove itself in this area as it has already done in the physical sciences. As biologists, readers should take great care, however, to see that they are buying a genuine technical advantage for their science, and not just paying heavily for the latest gimmick. Yet again, the way to do this is by 'hands-on' evaluation in the specific application.

20.15 ICCDS, SITS AND ISITS

The most modern form of low-light-level video-rate imager is, of course, the intensified CCD or 'ICCD'. This technology, without doubt, represents the future of such imaging, and has been described previously. The fundamental configuration is an image intensifier optically coupled by some means to a video-rate CCD camera. These instruments are becoming more prolific, and range from the product designed specifically for demanding scientific applications down to

ICCDs thrown together for rather less discerning customers, usually in the security market. An intermediate class, which can seriously mislead the budget-driven scientist, is the low-quality security camera in a higher quality case with a flashy controller, incorporating a bit of logic and a lot of front panel. The solution to this dilemma is to carry out side by side comparisons on your actual experiment and see which camera delivers the results.

Some very valuable work has been done with the old technology of SIT (silicon intensified target) and ISIT (intensified SIT) cameras. The SIT camera is basically a vidicon TV pick-up tube in which a silicon diode array is used as the target. The photocathode of a SIT tube is the conventional S20, and the front stage is very similar to a single-stage 'first generation' image intensifier. The electrons from the photocathode are imaged onto the silicon target of the vidicon stage by an inverting electrostatic lens, which utilizes a high acceleration voltage (between 6 and 12 kV). When the energetic electrons impact on the silicon target, many electron-hole pairs are produced for each incident electron and this charge is stored at (or near) its site of formation until the scanning electron beam of the vidicon 'reads out' the charge, once per frame. Often over 1000 electrons are generated in the target for each electron emitted by the photocathode. This leads to a sensitivity much higher than that achieved with a vidicon alone.

The ISIT is simply a SIT with an extra stage of image intensification glued to the front. It is typically some 20 times more sensitive than a SIT. Regrettably, both the SIT and ISIT suffer from serious problems that often lead to them being replaced by the ICCD. These problems are:

(1) Target damage: The tubes are prone to 'image burn-in', if exposed to fixed scenes with bright areas for prolonged periods.
(2) Lag: Lag is unfortunately one of the most characteristic problems of SIT-based products.

It is a regrettable fact that lag gets progressively worse as input light level is reduced. This originates from the fundamental physics on which the tube operation depends.

Lag is usually defined as the residual signal (expressed as a percentage of the original) after a signal is removed and three dark fields have been read out. This is often about 5% at 0.01 lux, but at 1 millilux, it can become as high as 30%! Clearly lag seriously complicates any time-resolved work in the order of frame-times at low light levels, and since it is signal-dependent, can really frustrate quantitative work.

This makes the SIT very insensitive to moving images against a fixed background. This is easily demonstrated in a 'microscopy' application; a SIT camera allowed to look at a wet sample will, after a few seconds, show a very fine picture. If, however, a cell moves across the scene being viewed the movement will often not be detected by someone watching the monitor. This has been demonstrated many times. It is a logical consequence of the fact that the SIT uses its lag to improve the signal-to-noise ratio in any frame by averaging over many frames.

20.16 DEFECTS AND THEIR MINIMIZATION

When slow-scan CCDs were first introduced into astronomy, the cosmetic quality was very poor. Though these problems have largely gone away, any sophisticated camera is likely to have some defects. The astronomers worked really hard to develop procedures to clean up images and hide defects, and these are just as applicable to biology and medicine as they are to their original purpose. Also, in intensified CCD cameras, there are quite a few problems that can arise from intensifiers. Some of these problems can be corrected and others can actually limit the performance of the camera. Below we list some of the common camera defects and some comments on how to handle them.

(1) *Defect in the image intensifier*
   (a) *Bright spots.* Generally, a bright spot in the image arising from the intensifier is only to be allowed at the very highest gains. At lower gains such a bright spot should lead to the manufacturer rejecting the intensifier. If the spot is small, 'pixel fixing' can be used. In this technique, a 'mask' is made up in memory which sets the bright pixel(s) to the average value of the adjacent pixels or to a set value, at the user's discretion. Such 'fixes' are very hard to see, unless the fixed area is large, which would only be necessitated by defects that should not be present anyway.
   (b) *Dark spots.* Dark spots are far more acceptable (and frequent) than bright spots. They are usually areas of reduced sensitivity rather than totally black areas. Small black spots can be dealt with by pixel-fixing as with bright spots. Areas of reduced sensitivity can be compensated for rather than masked out, and so information in these areas need not be lost. A section below describes the process.
   (c) *Fixed patterns.* Often, hexagonal patterns, known as 'chicken wire' are seen on the image. This results from the physical structure of the microchannel plate and/or fibre optic

components. These are areas of modified gain and can be almost completely eliminated from final images.

    (d) *Shading*. Many intensifiers have lower gain at the edges than at the centre. This produces a bright patch in the centre of the image. Again, providing this is with reasonable limits (not more than say 4:1), it can be eliminated by subsequent processing.

(2) *Defects in coupling*. Fibre optic coupling can produce fixed patterns (chicken wire) usually hexagonal or square, and dark spots, some of which might be black. Black spots are dealt with by fixing, and the chicken wire by the techniques outlined later.

(3) *Defects in the CCD*

    (a) *Bright/dark colums or rows*. These really should not be allowed in an intensified CCD camera. If you have such defects the remedy is to replace the sensor – certainly do not accept such a device.

    (b) *Bright spots*. These are often encountered, especially in long integrations or at very high video gains. They are usually single pixel, and can be 'fixed' to an average local value.

    (c) *Dark spots*. Again usually single pixel and 'fixable'.

20.16.1 Flat field normalization – the magic touch

In a real camera you will have some of the defects described above. There is a technique, developed by astronomers for simultaneously removing the effects of shading, chicken wire and grey spots. The image is not sensitive to where in the camera these problems arise, and all the various contributions from different components can be treated at one time.

Precise details depend on the configuration of the computer system but the principle is the same: Project onto the camera an image of a uniform featureless object, weakly illuminated, whose brightness is about the same as a typical real object. Integrate this with a typical integration, and look at the image. It might be awful, showing all the shading, chicken wire, dark defects, etc., but this is precisely what you need to see. Make sure that the integration is long enough to iron out quantum statistics, and thus giving a really good image of all the defects. Store this image. Now run some statistics on the image and find the average pixel value. It might vary considerably over the image, and, providing the featureless, uniform object truly was featureless and uniformly illuminated, this represents the variation of response of the camera, but results in obtaining one (and of course only one) average for the whole image. Then divide the whole image, on a pixel-by-pixel basis, by this scaler quantity (the average). Having obtained an 'image' where everything is close to unity (so in practice some care will be needed to avoid loss of accuracy), a pixel in the new image that was at the average value in the original will have a value of 1.00. A pixel in the new image corresponding to a grey pixel in the original of half average brightness will have a value of 0.5. This new image can be called the 'correction image'.

The next step is to divide any real image pixel-by-pixel by the correction image and all the defects seen earlier will virtually disappear.

This technique, 'old hat' to astronomers, can do more than correct for the camera defects. Put your featureless, uniformly illuminated object under the microscope and perform the technique with the camera viewing in this way and you take out shading in the microscope too!

Extensions of the technique can correct for uneven illumination, optical vignetting and a wide range of other shading problems. The technique will not handle true black and white spots, which it can often just reverse. Pixel fixing is still required. Neither of course, can this method correct for distortion. It is, however, when properly applied, a virtually universal approach to shading problems.

CHAPTER TWENTY-ONE

Properties of Low-Light-Level Slow-Scan Detectors

R. AIKENS

Photometrics, Ruscon, Arizona, USA

21.1 INTRODUCTION

Electronic image acquisition has undergone pheno-
menal advances over the last 20 years. The major
imaging technology innovation of the last few years
has been the introduction of a detector called the
charge-coupled device (CCD) which replaces bulky,
less sensitive scanning camera tubes such as vidicons.
This discussion focuses on the theory of CCD imagers,
slow-scan CCD camera implementations and the
utility of CCDs in light microscopy.

21.2 CONTEMPORARY IMAGE-ACQUISITION TECHNOLOGY

21.2.1 Video cameras

The most familiar form of an image-acquisition system
is the hand-held video camera. A modern video
camera employs a CCD and generates a standard
NTSC video signal which is compatible with a
multitude of recording and display devices. A video
camera operates at 30 frames per second and with a
fixed format. Typically, a video image is composed of
525 lines, each with 250–600 elements of horizontal
resolution, depending on the quality of associated
electronics.

Video cameras have been used as a tool in light
microscopy for several years (Inoue, 1986). An
impressive array of video-based equipment is available
which permits the user to digitize, process, display and
archive images obtained with video cameras. However, video cameras have their limitations. Since they
must operate within rigid established constraints, their
performance is limited. Modern video cameras employ
CCDs which are specifically designed to meet the
requirements of video imaging by trading-off features
which the CCD might otherwise offer. There is a new
generation of video components emerging based on
high-definition television (HDTV) which will offer
higher resolution and movie-like image quality.
The components which come out of HDTV will
undoubtedly find utility in scientific imaging; however,
price and performance will be tuned for the high-
volume consumer which will limit the usefulness of
these devices for quantitative imaging.

21.2.2 Intensified video cameras

There is a class of low-light-level video cameras which
employ image-intensifier stages in front of the basic

sensor. The most common of these, the silicon intensified target vidicon (SIT), employs a traditional vidicon target camera tube with an integral image-intensification section. An image is focused on a photocathode and photoelectrons are accelerated toward a standard silicon target in the vidicon section of the tube. Conventional video electronics are used to render a standard NTSC video signal. The intensifier–vidicon combination allows individual photoelectrons to be detected but at the expense of dynamic range. The quantum efficiency of photo-cathodes is low, especially in the red region of the spectrum, which limits the useful range of these cameras. SIT cameras find utility where light levels are low and when it is required to observe a scene in real time. SITs and intensified SITs (ISIT) are expensive as well as fragile and are being replaced in many applications by high-performance cooled CCD cameras. Low-light-level video cameras will continue to find utility in many situations for years to come (Spring *et al.*, 1989).

21.2.3   Cooled slow-scan CCD cameras

The CCD is capable of remarkable performance when operated in conjunction with ideal electronics. In the 1980s, astronomers and scientists at several national laboratories and science centres made enormous progress on the development of essentially ideal CCDs for use in research and industrial environments. Most of that CCD development (Janesick, 1980–91) was not commercialized until the mid-1980s. Innovative CCD cameras were designed to capitalize on the perfor-mance of the CCD instead of conforming to video or television standards. A modern cooled CCD camera is capable of spatial resolution up to 16 million pixels, ultra high intrascene dynamic range in excess of 50 000 to 1, and exceptional sensitivity with the ability to integrate for hours in low-light-level applications. Several manufacturers now produce premium-quality slow-scan cooled CCD cameras which find utility in light microscopy and other disciplines.

21.3   HOW CCDs WORK

21.3.1   Characteristics of silicon

CCDs are processed in silicon much like other MOS integrated circuits. They range from a few square millimetres up to several square centimetres in area. It is useful to examine the properties of silicon to better understand CCD principles.

Silicon is a semiconducting element found between phosphorus and boron in the periodic table of elements. It can be purified and formed into large single crystals and sliced into very thin sheets for fabrication of integrated circuits. Photons of energy greater than about 1 eV, which penetrate silicon, will break the covalent bonds that bind atoms together in the lattice, thus creating hole–electron pairs. Energy in the form of heat, even at ambient room temperature, can also break bonds, creating charge pairs. High–energy particles and cosmic events which are always present, produce multiple hole-electron pairs. Silicon is opaque to light of wavelength shorter than 400 nm and becomes transparent at 1200 nm.

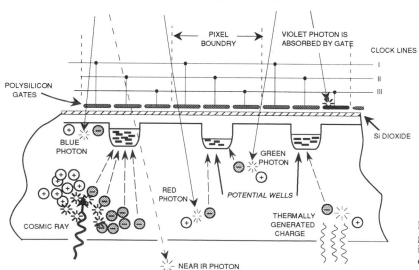

Figure 21.1 Photonic charge generation and collection process in a CCD. Thermally generated charge is also collected in potential wells.

21.3.2 The potential well concept

It is possible to capture the electronic charge generated by photon interaction in silicon before it recombines and to store it in potential wells. Potential wells can be produced near the surface of the silicon through the application of appropriate voltages to a series of insulated gate electrodes. A device capable of integrating photon-generated charge can be configured, as illustrated in Fig. 21.1. In this instance, the gates are positioned to create adjacent, but uncoupled potential wells. Note that light must pass through the electrically conductive polysilicon gates which limits the wavelength response of this device.

After photonic charge has accumulated in the discrete potential wells, it may be transferred through the silicon along channels as illustrated in Fig. 21.2. Photonic charge is swept along by drift fields as the gate potentials are changed in sequence to propagate the well to the right. This process is continued until charge from individual well sites is transferred to an output amplifier for measurement. This charge transfer mechanism is the fundamental principle underlying CCD operation.

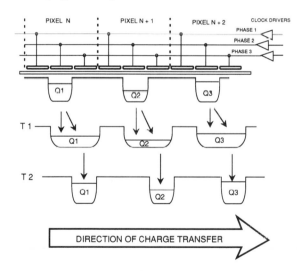

Figure 21.2 Integration of photonic charge and charge transfer concept underlying CCD operation.

21.3.3 The two-dimensional CCD

Figure 12.3 illustrates how the CCD concept can be extended to a two-dimensional structure. The imaging matrix is made up of light-sensitive CCD elements called picture elements or pixels. Note that the exposure of the CCD to light is a separate and distinct operation which precedes the application of the CCD principle to the read-out of integrated charge. After the termination of an exposure, entire rows of accumulated charge may be transported toward a

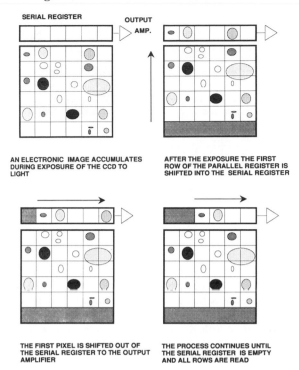

Figure 21.3 A two-dimensional CCD imaging device. The exposure and read-out of integrated charge are two distinctly separate processes.

serial CCD register. In the serial CCD register, individual charge packets are transported to a read-out amplifer where a measurable signal proportional to the amount of charge from each pixel is produced. The read-out process progresses row after row until the entire CCD has been cleared of charge.

Clearly, unless the charge transport process is extremely efficient, electronic charge will be lost or will corrupt adjacent pixels. In fact, modern CCDs exhibit outstanding transfer efficiencies as high as 0.999 999, where 1.0 is ideal. The CCD output amplifier has been developed to yield ultra low noise and high photometric linearity over a wide range of operating levels.

21.3.4 CCD architecture

There are three basic CCD architectures, two of which are illustrated in Fig. 12.4. A full-frame CCD imager with four output registers allowing simultaneous read-out of four imaging quadrants is shown in Fig. 21.4(a). This configuration is used when a short read-out time is required. The frame transfer CCD in Fig. 21.4(b) is really two CCDs in the same package. One is used for imaging, while the other is used for temporary storage and read-out. Photonic charge is allowed to integrate in the active imaging area. After the integration is over, the resulting electronic image is

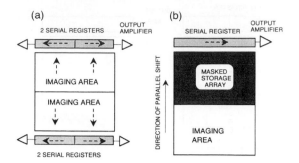

Figure 21.4 Two popular CCD architectures. (a) Full-frame CCD with four read-out registers. This permits high speed read-out, an especially useful feature in large CCD arrays. (b) Frame transfer CCD.

rapidly shifted up into the masked storage register. This shift operation typically takes 2 μs per row, or about 1 ms for a 512 × 512 element CCD. While the storage area is being read out in conventional CCD fashion, the next exposure is made in the image area, thus there is little dead time between frames. A frame transfer CCD may be operated at 30 frames per second to produce a television image. The frame transfer CCD suffers from image smearing which occurs during the image transfer in the presence of light. A synchronous rotary shutter can be employed to eliminate this problem.

A third style of imager, the interline transfer CCD, employs masked transport registers interleaved with the imaging area to allow continuous imaging in a television mode. This CCD design suffers from a loss in sensitivity because light striking the covered transport registers does not contribute any signal. The advantage of this architecture is that it is inexpensive to produce since the total required area is that of only one CCD. Many variations of the interline transfer CCD have been used in television cameras; however, this device is rarely used for scientific imaging because of its poor sensitivity and venetian blind spatial sampling properties.

21.4 THE HIGH-PERFORMANCE SLOW-SCAN CCD CAMERA

When the CCD is removed from the video realm, it can be operated at any speed and in any read-out manner which fits the application. The CCD may be cooled to prevent the formation of thermally generated charge, which is indistinguishable from that created by light. Charge so generated, called dark current, can significantly degrade CCD performance by using up well capacity and creating undesirable noise. Since a slow-scan CCD camera need not

conform to video standards, read-out speed can be reduced to realize ultra-low-noise performance. Novel masking and read-out modes permit very high frame rates through the use of a portion of the CCD for short-term electronic image storage. A well-designed cooled slow-scan CCD camera offers the user a variety of read-out speeds, cooling options, digitizing levels and CCD array sizes.

21.4.1 Subarray read-out

A slow-scan cooled CCD camera may be programmed during read-out so that only pixels in a selected region of interest (ROI) are processed. The CCD exposure is made as usual, but after it is terminated the parallel and serial registers are programmed to slew through regions outside the ROI and process only those pixels within it. Several ROIs may be read out after a single exposure. Since the time to process and digitize each pixel is fixed, the smaller ROIs allow higher frame rates. A 100 × 100 pixel ROI in a 512 × 512 pixel CCD may be read in 10 ms at a pixel read frequency of 1 MHz.

21.4.2 Charge grouping

During the CCD read-out, a slow-scan camera may be programmed to combine rows of charge into the serial register, or to combine two or more charge packets at the output amplifier. This process is commonly referred to as charge grouping or binning. During the binning process, the CCD is operated at reduced resolution in exchange for increased dynamic range and higher frame rate. A 512 × 512 pixel CCD binned 2 × 2 yields an image which has 256 × 256 super-pixels, each with four times the charge capacity of the individual CCD pixels. When programmed as in the example, the entire CCD may be read out in approximately one-quarter the time required for a full resolution image.

21.4.3 Camera implementations

Figure 21.5 is a block diagram of a contemporary slow-scan cooled CCD camera system. The term 'slow-scan' as used here means that the read-out speed of this camera is significantly lower than that of a video camera. The pixel-read frequencies for slow-scan cameras range from 50 kHz up to 2 MHz. The slow-scan read-out allows sufficient time for charge to be effectively transferred in the CCD registers and for the resulting output signals to be digitized to a high degree of accuracy. A host computer is shown in the illustration, because it is an important integral component of a slow-scan CCD camera system. The host computer affords the user the option of setting up a

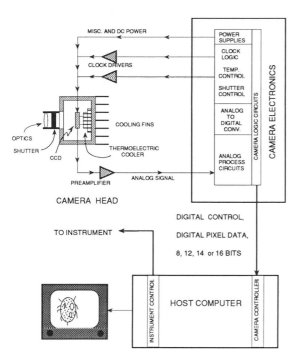

Figure 21.5 A typical slow-scan CCD camera and host computer configuration. An important system element is the software which must efficiently acquire, process and archive large amounts of data.

variety of operating modes under program control. Data processing and archiving are also under control of the host computer. The host is used to operate instrument accessories such as shutters, filter wheels and light sources. The slow-scan CCD camera components shown are the camera controller, camera electronics unit and camera head.

The camera controller contains the logic which causes camera action based on input from the host computer. It also passes digitized pixel data to the host computer, usually over a DMA channel. The camera controller generates all the sequences required for clocking the CCD phases and timing for the analogue processing circuits.

The camera electronics unit transforms digital commands and signals into active CCD clocking levels and sequences. It also performs analogue processing on the CCD output signal prior to digitization. The analogue-to-digital converter is usually contained in the camera electronics unit, although it may, in some cases, be located in the camera head in close proximity to the CCD. A 16 bit digitizer requires very high-performance analogue electronics and low-noise circuit design to effectively utilize the available CCD dynamic range. Digitizers from 8 to 16 bits are commonly used in slow-scan cameras depending on speed and dynamic range requirements.

The camera head contains the CCD and associated cooler. The cooler may be a Peltier device, or in extreme cases where exposure times are long, liquid nitrogen may be used. The CCD is usually enclosed in a hermetically sealed chamber to prevent the formation of frost or condensation. A window is provided to allow the entry of light.

A local preamplifier is employed to provide a strong signal which can drive a low-impedance line to the analogue electronics. The analogue electronics, including the analogue-to-digital converter (ADC), may be located in the camera head, especially if the read frequency is over 2×10^6 pixels per second. A shutter is employed with full-frame CCDs to prevent light from corrupting the CCD image during the read-out process.

CCDs are extremely sensitive to over voltage conditions or electrical transients, therefore, the camera system must be designed to protect the CCD from outside electrical disturbances.

21.5 SLOW-SCAN CCD CAMERA PERFORMANCE

21.5.1 Noise considerations

Noise is composed of undesirable signal components which arise from a variety of sources. There are four significant noise sources in CCD cameras which merit discussion. These are:

(1) KTC noise;
(2) noise from thermally generated charge (dark current);
(3) noise from the CCD amplifier; and
(4) photonic noise.

21.5.1.1 KTC noise

KTC noise is generated during the CCD reset process and can be eliminated with carefully designed analogue processing circuits.

21.5.1.2 Dark current noise

Thermally generated charge is indistinguishable from that generated by photons, hence it is a corrupting factor in a CCD signal. Dark current has several components (Janesick, 1989) and it is beyond the scope of this text to discuss them in detail. A dark current image has a spatial pattern which is often punctuated with 'hot' pixels that are many times brighter than the average background. Dark current should be well-behaved and reproducible in a high-quality scientific grade CCD. At temperatures between ambient and

−60°C, dark current decreases by a factor of two for every 6°C decrease in temperature, so a modest amount of cooling can effect an appreciable dark current reduction.

Since the dark current image is reproducible, it can be 'subtracted off' the object image. This is done by acquiring a 'dark frame' with the same exposure time as the object image and then performing a pixel by pixel dark current subtraction. There is, however, a random noise component in the dark current which may degrade image quality. The noise component of the dark signal is equal to the square root of that signal in electrons. As an example, if a CCD has a dark current of 100 electrons per pixel per second, then in a 1 s exposure the dark signal will be 100 electrons, and the RMS dark noise will be 10 electrons. The noise will increase with integration time and may limit the usefulness of the CCD, especially at longer exposure times. The only recourse when long exposures are necessary is further cooling, which may require the use of a cryogen like liquid nitrogen. In general, for most light microscopy applications a Peltier cooler is satisfactory and it is not necessary to cool the CCD below −40°C.

A recent innovation has resulted in a 40-fold reduction in dark current in CCDs. Multi-pinned phase (MPP) operation (Janesick *et al.*, 1989) has been implemented in CCD products by several CCD manufacturers. When using an MPP CCD, under moderate light level conditions where exposure times are short, a cooling system may not be necessary at all.

21.5.1.3   *CCD amplifier noise*

When a charge packet from a pixel reaches the output of the CCD serial shift register, it is placed on a capacitive node producing a voltage proportional to the quantity of delivered charge. Scientific CCDs are designed to produce between 1 and 5 µV per output electron. The charge-sensitive capacitive node is connected to the gate of an 'on chip' source follower which provides a relatively low output impedance. The noise from the source follower and associated circuitry combine to create a 'noise floor'. The noise floor is invariant in a given camera configuration. A high-quality CCD, coupled with sound camera design, should result in a noise floor of only a few electrons.

21.5.1.4   *Photonic noise*

This noise component arises because of the quantum nature of light. As with the dark current, photonic noise, often called shot noise, is the square root of the photonic signal in electrons. Photon statistics can limit CCD performance when light levels are low or exposure times are too short to capture a significant

number of photons. When a CCD camera is photon noise limited in a given exposure time, there is no way to improve the situation except by increasing the light level or the CCD photon-to-electron conversion efficiency (quantum efficiency). It is important to make the photonic noise much larger than both the dark noise and the noise floor so that their effects are minimized. A useful figure of merit by which the cumulative effect of these noise sources can be quantified, is the signal-to-noise ratio (SNR).

Given all three noise sources, the total noise for a CCD may be written as their quadrature sum:

$$\text{Total noise (NT)} = \sqrt{NA^2 + ND^2 + NP^2}$$

where NA is the amplifer noise, ND is the dark noise, and NP is the photon noise.

The SNR may be written in terms of all the signal and noise components.

$$SNR = \frac{QE \times IP}{\sqrt{NA^2 + ID + IP}}$$

where IP is the total number of captured photons, QE is the quantum efficiency of the CCD, ID is the total dark in electrons, and NA is the amplifier noise.

It is always desirable to maximize the SNR. Since the SNR is a function of several variables, there are always trade-offs which will yield the best CCD performance in a specific application. For example, if light levels are high, then the noise floor and dark current noise may not be relevant and the CCD can be operated at elevated temperatures in a photon noise limited mode. Figure 21.6 is a graphical representation of the major noise sources in a CCD as a function of light level. Note that as soon as the signal exceeds the square of the noise floor, the total noise is dominated by photon statistics.

The sensitivity of a CCD is determined by the properties of silicon and the CCD configuration. The polysilicon gates covering the parallel register are opaque to all wavelengths from 120 to 400 nm. In order to achieve higher sensitivity in the blue and violet portion of the spectrum, CCD manufacturers developed methods for thinning CCDs so that illumination could be brought to the backside of the device. Figure 21.7 shows typical quantum efficiency curves for both front and backside illuminated CCDs. Phosphors which absorb ultraviolet wavelengths and emit in a sensitive region of the CCD response have been very effectively utilized to enhance short wavelength response.

21.5.2   Photometric linearity

The charge generation mechanism in a CCD is intrinsically linear, so that the output signal should be

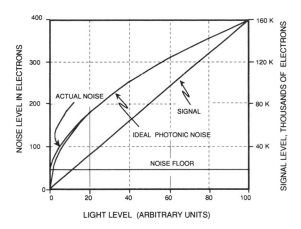

Figure 21.6 Graphical representation of signal and noise in a CCD. Noise floors as low as 5 electrons are possible.

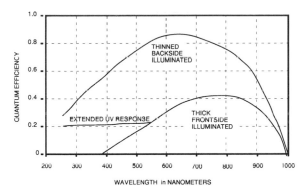

Figure 21.7 Quantum efficiencies of front and back illuminated CCDs. The back illuminated CCD offers nearly ideal response over a large wavelength range.

precisely proportional to the integrated charge at each pixel. The output amplifier is the first place where a departure from linearity may occur. The correct choice of operating potentials can reduce the non-linear effects of the output amplifier to negligible proportions. A high-quality CCD will exhibit linear photometric behaviour to better than 0.1% of full scale from the noise floor up to several hundreds of thousands of electrons.

Photometric linearity is of concern when linear operators are applied to CCD data during image processing. Simple arithmetic operators such as an image ratio, or more complex image reconstruction algorithms which operate in linear transform domains, will produce spurious and erroneous results if the raw input data are produced by a non-linear detector system.

21.5.3 Spatial resolution

The resolution of a CCD camera is determined by the geometry of the CCD pixels, which range in dimensions from 6 to 30 μm on a side. The pixels in a scientific-grade CCD are square and contiguous with no dead space between them. Since pixel size and format are fixed, a CCD image has no geometrical distortion. Square CCD pixels exhibit nearly ideal behaviour in terms of their spatial sampling properties, and it is possible to operate near the Nyquist sampling limit.

To avoid aliasing, an insidious phenomenon which occurs when an image is undersampled, the highest spatial frequencies in the image must be less than half the CCD pixel spatial frequency. Serious image degradation will occur if this rule is not observed. When an image with a periodic spatial pattern is undersampled, aliased components may easily be detected by the presence of a Moiré pattern. The same distortion, however, will occur when imaging an amorphous form, although it may go undetected. When imaging with a light microscope, the diffraction spot should cover several CCD pixels in order to completely avoid aliasing problems. Once an image is sampled with false aliased components, valuable information has been irrevocably lost.

The choice of CCD resolution will depend on the application. Large-format CCDs with 4×10^6 pixels are becoming commonplace, but there are systems implications when considering large CCDs. Each pixel may be digitized to between 8 and 16 bits so that a single image could require 8 Mbytes of storage. With the decreasing cost of memory, optical disk drives and other storage media, image storage is becoming less of a problem. The use of data compression algorithms can lead to further reductions in storage requirements. Another important consideration when choosing CCD resolution is image-processing time. A shading correction requires a pixel-by-pixel subtraction and a division which could require up to 16 million arithmetic operations. The time to process images must be appraised when doing an overall system design. Fortunately, very high-performance computer workstations are becoming available at reasonable prices which also eases this problem

21.5.4 Temporal resolution

A slow-scan camera is a still imaging device similar in nature to a snapshot film camera. Exposure times may range from nanoseconds to hours, depending on the application. The time to read the CCD and digitize its output signal determines maximum pixel read rate. Slow-scan pixel read-out times vary from about 20 μs to 500 ns. CCD performance drops off at pixel read frequencies above 2 MHz. Charge transfer efficiency in the serial register begins to degrade at the higher speeds, especially when the CCD is cooled.

Analogue-to-digital converters impose speed limitations. A 16 bit slow-scan CCD camera cannot operate faster than about 500 kHz per read-out register using current technology. The use of multiple serial register devices can yield a very short overall frame time at the expense of additional digitizers and interface electronics. The analogue processing circuits must settle out to 16 bit precision and a true 16 bit analogue-to-digital conversion takes a significant amount of time. Since CCD images may contain up to 8 Mbytes of digital data, frame rates may be also limited by the ability to move and archive large blocks of data.

21.5.4.1   Digital resolution versus speed

Slow-scan CCD cameras digitize pixels to between 8 and 16 bits depending on the application. Generally, precision must be traded-off against increasing frame rates. The user should decide on the minimum acceptable dynamic range and choose the digitizing level accordingly. Table 21.1 gives typical digital dynamic range as a function of pixel read time for single-channel contemporary slow-scan CCD cameras.

Table 21.1 Digital resolution as a function of pixel read frequency.

Number of bits per pixel	Pixel read frequency
16	20 kHz–300 kHz
14	50 kHz–500 kHz
12	100 kHz–2 MHz
10	500 kHz–8MHz
8	1 MHz–10 MHz

Analogue-to-digital converters are available which operate considerably faster than the rates given in the table. The read frequencies given in Table 21.1 take into account the time for the CCD and associated analogue electronics to settle to the specified digital accuracy prior to digitizing. Multiple serial register devices will offer a speed increase proportional to the number of available output channels. A 1 k × 1 k CCD with four output registers operating at 2 MHz can operate at 8 frames per second.

In summary, temporal resolution is affected by a number of factors, and the user must decide how to trade-off frame rate and dynamic range against SNR to arrive at the 'best' combination of those factors to achieve the desired result.

21.6   APPLICATIONS OF SLOW-SCAN CCD CAMERAS

21.6.1   High-speed framing

Slow-scan cameras by their nature might appear to be poorly suited for high time resolution applications.

There are trade-offs (Ross *et al.*, 1991) which permit frame rates greater than 100 frames per second. As an example, a large format CCD may be programmed to be read out over an ROI of 100 × 100 pixels. If the camera pixel rate is 1 MHz, then the image-acquisition time for the subarray is 10 ms. In addition to the read-out time, the exposure time plus the time required to discard pixels outside the ROI must be taken into account. Figure 21.8 illustrates the subarray read-out principle. the rapid scan overhead, in addition to the image read-out time, is typically 20 μs per row and 0.5 μs per serial shift in a large CCD. If the CCD is a full-frame imager, some method for gating the input image must be employed so that light does not strike the CCD during read-out.

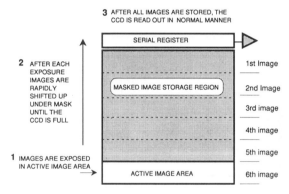

Figure 21.8 Subarray read-out. High frame rates over small regions of interest can be achieved by limiting the field of view. The regions outside the image area are rapidly scanned, and only the pixels in the image are digitized.

It is possible to obtain a series of images very rapidly by using a portion of the CCD as a storage device. Figure 21.9 illustrates this concept. A mask prevents light from striking all but the lower portion of a full-frame CCD. The mask need not be placed directly on the CCD but may be located at another location in the system by using appropriate transfer optics. It is possible to acquire several images in rapid succession by integrating in the open area of the CCD and then shifting the resulting electronic image under the mask. This may be done until the CCD is filled up with images. The CCD is then read out in normal slow-scan fashion. It is feasible to acquire a limited number of images at an effective frame rate of several thousand frames per second using this technique. When only two images in rapid succession are required, the frame transfer CCD is a good choice. This device is useful in ratio imaging for measuring pH levels and calcium concentrations in cells. The frame transfer CCD is particularly useful for ratio imaging since two separate images may be taken within a millisecond.

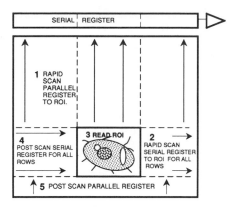

Figure 21.9 High-speed frame transfer imaging. The effective frame rate is determined by the number of unmasked CCD rows.

21.6.2 Shading corrections

When an object is imaged with a CCD and a microscope, CCD background and responsivity variations, system vignetting and scattered background light may mask important information which is present in the image. Because the CCD exhibits near perfect linearity over its entire dynamic range, it is practical, through the use of precise calibration images, to computationally remove shading and responsivity errors and render an image which is radiometrically precise and free of spurious intensity errors. Figure 21.10 shows the sequence of steps which must be taken to perform a shading correction. A dark frame containing CCD structure and dark current background is taken at the same exposure time as the image of interest.

A uniform source flat field image which contains shading errors is then acquired. The most critical step in the shading error correction process is the creation of the uniform source. Not only must it be uniform, it must also be imaged in precisely the same fashion as the object image to be corrected. This is difficult in the case of a fluorescent image, because the microscope stage must be moved away from the object of interest to a region of uniform fluorescence which may be above or below the focal plane. If the microscope is refocused, the illumination pattern will change and the flat field correction will introduce shading errors instead of correcting them. The shading correction process is wavelength-dependent which compounds the problem.

After a flat field shading correction is performed, each pixel becomes an individual radiometrically precise photometer.

21.6.3 Low-light-level imaging

The slow-scan CCD camera is a nearly perfect integrator, and in low-light-level circumstances exposures

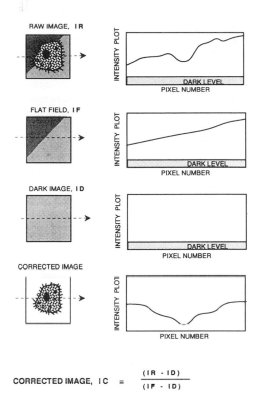

$$\text{CORRECTED IMAGE, } I_C = \frac{(I_R - I_D)}{(I_F - I_D)}$$

Figure 21.10 Shading correction. This process can be accomplished with a single flat field image. Since the CCD is linear over the entire dynamic range the correction works at all intensities.

up to several hours may be taken. During exposures longer than a few minutes, high energy particles and cosmic rays will begin to degrade image quality. Long exposures should be broken up into a sequence of several shorter exposures so that uncorrelated events may be detected and removed from the images. Dark current, system noise and light level should all be considered, and the exposure time must be selected to yield an acceptable SNR when using the CCD at low light levels. In the final analysis, it comes down to the total number of captured photons, which ultimately determines performance. Clean and efficient optics coupled with high quantum efficiency are necessary in low-light-level situations.

21.6.4 Image reconstruction

Digital images produced by a CCD and a light microscope can be enhanced through the application of image-reconstruction techniques. These methods have been successfully employed in spectroscopy, astronomy and microscopy (Agard *et al.*, 1989). An object image is first spatially sampled by the CCD and digitized. The combined effects of the system, including the microscope, CCD and camera electronics are then characterized by imaging a point source which simulates a delta function. The point source image is captured in precisely the same manner as the object image. The recorded object image is the convolution of the system transfer function with the original input image. The corrupting effects introduced by the system may be removed through the application of powerful mathematical transforms. By individually transforming the object image and system transfer function into Fourier space, it is possible to deconvolve the system effects through a simple algebraic operation. Fast Fourier transform software and hardware can deconvolve a one million pixel image in a few seconds. Just as in the flat field problem, the most difficult task is obtaining a calibration image which is a faithful representation of the system transfer function. Through the careful application of Fourier-based image-processing techniques, out-of-focus images may be de-fuzzed and resolution approaching the diffraction limit of the microscope can be achieved.

The successful application of transform techniques places severe demands on camera performance. The camera used to acquire both the object image and the system transfer function must be linear over several orders of intensity level, and the image must be spatially oversampled to ensure that all the information is captured. If the entire process is not carried out very carefully, the results will contain false, spurious image components which can easily be misinterpreted. Attempts to perform image reconstruction with conventional video cameras have never been totally successful because of the poor quality of video image data.

21.7 SUMMARY

The slow-scan CCD camera is a powerful tool which permits light microscopy to be taken to new levels of excellence. Images required with a well-designed slow-scan CCD camera 'stand up' to the severe requirements of image-processing algorithms, which fail when applied to conventional video images. The large dynamic range, superb linearity and photometric integrity of the data produced by slow-scan CCD cameras give the imaging scientist the ability to make measurements which were heretofore impossible. While this technology opens up new opportunities for discovery, video cameras and image intensifiers are still needed to provide total imaging capability. No one type of camera can do everything and the user must decide where the attributes of a slow-scan CCD camera are best utilized. While this discussion has been slanted toward microscopy, there are other equally exciting life sciences applications for slow-scan CCD cameras. Electrophoresis, chromatography, X-ray diffraction, transmission electron microscopy (TEM) and radiology are a few of the other disciplines where slow-scan CCD cameras are being successfully applied.

REFERENCES

Agard D.A. *et al.* (1989) *Methods Cell Biol.* **30**, 353–377.
Inoue S. (1986) *Video Microscopy*. Plenum, New York.
Janesick J. (1980–91) Informal Notes, Jet Propulsion Laboratory, California Institute of Technology, Pasadena, California.
Janesick J. (1989) *Proc. SPIE*, 1071–1015.
Ross N.L. *et al.* (1991) *J. Neurosci. Methods* **36**, 253–261.
Spring Kenneth R. *et al.* (1989) *Methods Cell Biol.* **29**, 269–289.

CHAPTER TWENTY-TWO

Fast Photometric Measurements of Cell Function Combined with Electrophysiology

J. DEMPSTER

Department of Physiology & Pharmacology, Strathclyde University, Glasgow, UK

The development of highly fluorescent ion-sensitive dyes has revolutionized techniques for the measurement of intracellular ion concentrations. With these dyes, it has become possible to make quantitative measurements of intracellular Ca^{2+}, Na^+ and H^+ ion concentrations from a wide variety of cell types in a much less invasive fashion than had been possible with earlier techniques such as ion-sensitive microelectrodes, or photoproteins such as aequorin.

A particular feature of these dyes, which has contributed to the widespread interest in them, is their potential for extracting absolute measurements of intracellular ion concentrations, independent of dye concentration, cell thickness, etc. The binding of inorganic ions to dyes, forming ion–dye complexes, changes the basic light emission or absorbance properties of the dye. For instance, the binding of Ca^{2+} to the dye quin-2 increases its fluorescent emission by up to seven-fold. However, it is difficult to relate such changes directly to Ca^{2+} concentration since the overall magnitude of the signal is dependent on dye concentration which may itself be changing with time (often due to photobleaching). These problems can be avoided using dual-wavelength dyes such as the Ca^{2+} sensitive fura-2 or indo-1 dyes, Na^+-sensitive SBFI and H^+-sensitive BCECF. In these dyes the binding of the inorganic ion has a differential effect at separate parts of the absorbance or emission spectrum. For instance, the absorbance spectrum of fura-2 is modulated by Ca^{2+} binding. In the absence of Ca^{2+}, the dye absorbs ultraviolet light strongest at wavelengths around 335 nm and hence produces the strongest emission when illuminated with light of that wavelength. When Ca^{2+} ions are bound to the molecule the absorbance peak shifts to 362 nm, with the position of the emission spectrum peak (around 510 nm) little affected (Grynkiewicz *et al.*, 1985). Although the absolute amplitude of the fluorescent emission from the dye is dependent on a multitude of factors, including dye concentration, the 335/362 emission ratio is dependent only on the Ca^{2+}-fura/fura ratio and can thus be used as a quantitative measure of Ca^{2+} concentration. Fura-2, and other dyes with ion-sensitive absorbance spectra, are described as **dual-excitation** dyes.

Ca^{2+} has a similar dual-wavelength effect on indo-1, except that it is the emission light spectrum which is sensitive to Ca^{2+} binding. The peak emission wavelength of indo-1 is 482 nm, while with a Ca^{2+} molecule bound it is 398 nm. In this case the 398/482 emission ratio is proportional to Ca^{2+} concentration. Indo-1 is described as a **dual-emission** dye. Further details of the properties of fura-2 and indo-1 can be found in Grynkiewicz *et al.* (1985).

22.1 FLUORESCENT LIGHT MEASUREMENT

At least three different methodological approaches, each with their own advantages, can be taken to the measurement of fluorescent emissions from dye-loaded cells: spectrophotometers, imaging systems, and microscope-based photometric systems.

Spectrophotometric techniques are used in a variety of fields and general-purpose spectrophotometers (e.g. Perkin-Elmer LS-5) for measuring the light emission and absorbance properties of samples in liquid suspension are widely available. Such devices are equipped with light sources capable of producing controllable monochromatic light over a range of wavelengths. This capability makes them particularly suitable for use in studies with dual-excitation dyes such as fura-2. Their main limitation, as research tools in cell physiology, is that measurements are made from large populations of cells in suspension. Consequently, the properties of single cells cannot be studied. Also the rate at which ratio measurments can be made is limited to less than 1 per second due to the time taken to alter the wavelength of the excitation light source.

In general, it is preferable to study the properties of single cells, in order to avoid problems with inhomogeneities in the response of cells within populations. This requires the use of a microscope capable of visualizing individual cells and a system for analysing the image. Within microscope-based systems either a photometric or an imaging approach can be taken. Imaging is discussed in detail elsewhere in this volume, but essentially involves the use of a low-light image intensified or cooled CCD camera to capture the fluorescence images of the cell, and an image-analysis system to process the resulting pictures. The technique permits the spatial distribution of ions such as Ca^{2+} with the cell to be observed and quantified (e.g. Williams *et al.*, 1985). Inhomogeneities in the response of cells within small populations to a given stimulus can also be observed. Again, it is difficult to achieve more than 1–2 Hz (measurements per second) and the large amount of data accumulated per image makes the storage and manipulation of series of images difficult.

In the microscope-based photometric system, fluorescent light emission from the cell is channelled, via the microscope optics, to a photodetector instead of a camera. Thus, in terms of properties, it lies between the cuvette spectrophotometer and the imaging system. Single cells can be studied but detailed spatial information is lost, with only the integrated light emission from the whole cell being collected. The distinct advantage of the system, however, is that significantly faster measurement rates (in the order of 10–1000 Hz are possible, compared to the other methods.

High-speed measurements of intracellular ion concentration changes are of interest in themselves, but the technique is further enhanced when combined with the simultaneous measurement of additional cell parameters: muscle length and tension, and, in particular, electrophysiological measurements such as membrane potential and current. This method has been used extensively in the study of muscle contraction (e.g. Baylor & Hollingsworth, 1988; Eisner *et al.*, 1989) and neurosecretion (Mollard *et al.*, 1989).

The desire to capture and analyse such a varied range of signals from a single cell has placed new demands on the instrumentation for recording experimental signals and the production of stimuli; an area in which the computer has come to play a central role. This chapter discusses some of the issues involved in the development of such computer-based fluorescence/electrophysiological measurement systems.

22.2 A FLUORESCENCE/ ELECTROPHYSIOLOGICAL RECORDING SYSTEM

A diagram of a typical measurement system is shown in Fig. 22.1. Cells or tissues are mounted in an experimental chamber on a microscope adapted for UV fluorescence work (most commonly used dyes have absorbance and emission spectra in the UV range), such as the widely used Nikon Diaphot system (Nikon Corp., Tokyo, Japan). The cells are epi-illuminated with narrow-band UV light via the objective lens, as shown. A rotating filter wheel/changer may exist within the excitation light path to allow excitation at different UV wavelengths, as required when dual-excitation dyes are in use. The resulting fluorescent light emissions from the cells are projected via the microscope optics, through the camera port, on to photomultiplier tube (PMT) light measurement devices. For experiments with dual-emission dyes, two PMTs are used, combined with a wavelength-sensitive 45° dichroic mirror and band-pass filters to allow fluorescent emissions to be measured at two different light wavelengths.

Electrophysiological measurements are made via glass micropipette electrodes attached to the cells. A variety of recording techniques are possible (e.g. see Standen *et al.*, 1987) but the most common one, in the context of fluorescence analysis, is the whole-cell patch-clamp method. A clean fire-polished glass micropipette with a tip diameter of around 1 μm can be made to form a tight seal when pressed against the cell membrane. If a suction pulse is applied to the pipette, the interior membrane under the pipette can be sucked away, providing low resistance access to the

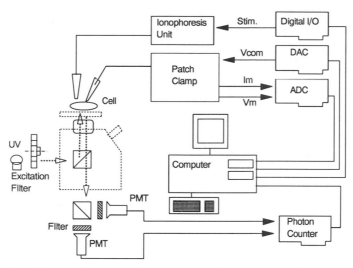

Figure 22.1 Diagram of a system for the combined recording of cell fluorescence and membrane current and potential. A cell is epi-illuminated with UV light. Fluorescent light emissions are captured via the microscope camera port and measured with PMTs. For dual-emission dyes, the emitted light at the characteristic peak emission wavelength for the dye is measured by a pair of PMTs, after being split into two streams using a combination of 45° dichroic mirror and band-pass filters. A filter wheel placed into the excitation filter path can be used to provide alternating UV excitation wavelengths for dual-excitation dyes. Cell membrane current and potential are measured via a micropipette attached to the cell and drugs applied via an ionophoresis unit. Overall timing, application of stimuli, and digital recording of signals are performed by a laboratory computer, equipped with A/D, D/A converter and photon counting interface cards.

inside of the cell. When a patch-clamp amplifier (e.g. EPC-7, List Electronic, Darmstadt, Germany; Axopatch, Axon Instruments, Foster City, CA, USA) is attached to the electrode, the current flowing across the cell membrane can be measured and the cell membrane potential set or 'clamped', to precisely controlled values. A detailed introduction to the patch-clamp method can be found in Sakmann and Neher (1983).

In general, some means is required to apply a a variety of controlled stimuli to the cell under study. Many ion channels (e.g. Na^+, K^+, Ca^{2+}) are voltage-sensitive and can be activated by abrupt depolarizing changes in cell membrane potential. The patch clamp has a command voltage input for this purpose, allowing externally generated voltage patterns to be applied to the cell. Other ion channels may be linked to receptors on the cell surface and activated by the binding of a specific agonist. Agonist may be applied to the cell either ionophoretically or by pressure ejection from a second micropipette.

In summary, a combined fluorescence/electro-physiological recording system produces at least four output signals; two fluorescent light emission channels from the PMTs, cell membrane potential and current from the patch clamp. Stimulus patterns must also be provided to the patch-clamp command voltage input and/or an agonist application system. Overall control of the experiment is usually handled by a computer system which generates the stimulus voltage patterns and may also record PMT and patch-clamp signals.

22.3 THE PHOTOMULTIPLIER TUBE

The fluorescent light emissions from ion-sensitive dyes, although having improved tremendously over the past decade are nevertheless of a low level by normal standards, and require a light measurement device of high sensitivity. This requires the use of the photomultiplier tube rather than simpler solid-state devices such as the photodiode. A PMT is an evacuated glass tube containing an array of electrodes called the dynode chain, with a light-sensitive phosphor-coated surface at one end. A high voltage is applied to the dynode chain. When a photon strikes the sensitive surface, some electrons are knocked out of the phosphor. These electrons are accelerated by the electric field and when they strike the first electrode in the chain, they knock out more electrons which are swept down to the next electrode. A cascade of electrons occurs from one dynode to the next, each contributing more electrons until they finally appear at the anode as a measurable current. A PMT produces a current pulse for each photon striking the tube surface. PMTs with appropriate housings and high-voltage power supplies can be obtained from a number of suppliers, in particular Thorn EMI (Ruislip, Middlesex, UK) and Oriel (Stratford, CT, USA). (It is worth noting that PMT tubes can vary in terms of sensitivity and other operating factors. If work with dual-emission dyes is considered, the PMTs should be purchased as pairs with matching characteristics.)

Unlike most transducers used in the physiological sciences, which produce analogue voltages proportional to the quantity being measured (e.g. pressure, temperature), the PMT produces a series of random short-duration current pulses (50–100 ns) whose average **frequency**, rather than amplitude, is a measure of the incident light level. One consequence of this is that the PMT cannot give an instantaneous measurement of light level. Rather, photon pulses must be accumulated over a finite period of time. Two approaches can be taken to converting the PMT pulse signals into a stable light level signal, each with its own advantages and disadvantages.

In the **photon counting** method, the current pulses are applied to a high-speed digital counter with each pulse incrementing the counter by one unit. The counter is allowed to accumulate counts for a fixed time interval and the average light level expressed as the number of counts acquired in that interval. Photon counting requires a high-speed digital counter capable of responding to short-duration pulses. A pulse discriminator is also required to detect the presence of the PMT current pulse.

Figure 22.2(A) shows a typical photon counting system using the Thorn EMI C660 counter, a high-speed digital counter expansion board for IBM PC compatible computers. It provides two separate counters capable of accumulating over 16 million counts each and responding to pulses occurring randomly at rates up to 10 MHz. Under the control of a program running on the PC, the counters can be made to accumulate counts for preset periods, ranging from 0.5 ms to 20 s. The current pulses produced by the PMT are too small to be measurable by the C660 board directly. A C604 amplifier/discriminator is therefore used to convert the current pulse into a standard ECL (emitter coupled logic) digital pulse. The discriminator also provides a detection threshold which separates the photon pulses from background noise.

The alternative to photon counting is to directly integrate the current output of the PMT to effect a pulse frequency to voltage conversion, as shown in Fig. 22.2(B). The photon current pulses from the PMT charge the capacitor C of the integrator. As the charge from each current pulse accumulates, the integrator output voltage increases. A resistor R, in parallel with the capacitor, causes the stored charge to leak away with a characteristic time constant of $\tau = RC$. The integrator, therefore, produces an average output voltage which is proportional to the balance between the rate at which the PMT supplies current and the integrator time constant. In broad terms, this time constant is equivalent to the counting interval of the photon counter. (A similar effect can be achieved by feeding the PMT output signal through a low-pass filter.)

Photon counting has a number of technical advantages in terms of sensitivity, dynamic range and immunity to noise. In addition to the intermittent current pulses produced by photons, PMTs produce a background noise current independent of any light input, and hence called the **dark current**. This dark current contributes to the signal measured using analogue integration. At very low light levels, where only a few photons are being collected, the integrated dark current overwhelms the light signal. The photon counting system, however, is largely insensitive to dark current noise since it can discriminate between the transient photon-produced current pulses and the

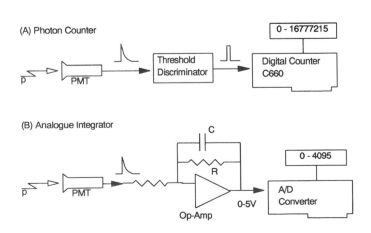

Figure 22.2 Photomultiplier tube light measurement methods. (A) Photon counting. Direct counting of PMT output pulses produced by photons striking the tube surface. Pulses with amplitudes greater than 1 mV are detected using a discriminator, producing a transistor–transistor logic (TTL) or emitter coupled logic (ECL) digital pulse which increments a high-speed digital counter board in a PC. (B) Analogue integration. PMT output is fed into a leaky integrator circuit which produces an analogue output voltage proportional to the photon pulse rate. The voltage level is then digitized by an A/D converter.

background current. Using photon counting the light sensitivity can be extended down to single photons per counting period.

The photon counter can also handle a much wider range of light intensity than the analogue method. The 24 bit counter in the Thorn EMI C660, for instance, can handle a range of 0–16 777 216 photons per counting period. On the other hand, the integrator output is an analogue voltage in the range 0–10 V. When this voltage is digitized, for measurement on a computer, using a 12 bit analogue-to-digital converter (as is commonly used for electrophysiological work) a resolution of only 0–4095 points would be obtained.

Nevertheless, the performance of the analogue integrator is quite adequate for general applications. The range of fluorescence levels usually produced by current dyes do not require the full sensitivity of the photon counter. An integrator is also a much simpler device to construct. Also, unlike the photon counter, it provides a direct analogue voltage which can be stored on a standard instrumentation tape recorder (e.g. Racal Store 4, Racal Electronics, Southampton, UK; Biologic DTR1800, Bio-Logic, Echirolles, France) in the same way as any other analogue signal. In particular, a computer system is not required.

Conversely, the photon counting method lends itself to computer-based approaches. The wide dynamic range of the photon counter and its programmability make it easy to define counting intervals and sensitivity, from within a computer program. At present, both photon counting and analogue integration are widely used in fluorescence measurement systems.

22.4 DUAL-EMISSION DYE MEASUREMENT SYSTEMS

The detailed configuration of a fluorescence recording system depends on whether a dual-emission or excitation dye is being used. Experiments using dual-emission dyes are perhaps the simplest to perform, at least in terms of the recording instrumentation. A configuration such as in Fig. 22.3(A) is used. The fluorescent light, emitted from the dye-loaded cell, is split into two components using a 45° wavelength-sensitive dichroic mirror and passed, via narrow band-pass filters to separate PMTs. The filter passbands are chosen to make each PMT sensitive to light of wavelengths close to the characteristic emission peaks for the free and Ca^{2+}-complexed forms of the dye.

Taking indo-1 as an example, 405 nm and 490 nm filters might be used. An estimate of intracellular Ca^{2+} concentration can be obtained from the ratio

$$R = \frac{F_{405} - F_{back}}{F_{490} - F_{back}} \qquad [22.1]$$

where F_{back} is the background light emission from sources other than the dye within the cell, such as the inherent fluorescence of the cell itself and/or light from external sources. A measurement of F_{back} may be obtained by recording the fluorescent emission before the application of the dye to the cell. R is not linearly related to Ca^{2+} concentration but varies sigmoidally between a minimum value in the absence of Ca^{2+} and a maximum in high Ca^{2+} concentrations. The range of Ca^{2+} concentration measurement is thus somewhat limited.

A quantitative estimate of the Ca^{2+} concentration can be obtained using the binding equation.

$$[Ca^{2+}] = K_d \frac{R - R_{min}}{R_{max} - R} \qquad [22.2]$$

where R_{min} and R_{max} are the minimum and maximum observed ratios and K_d is the effective binding coefficient for Ca^{2+} and the dye. In practice, these three parameters must be determined experimentally for the recording system before equation [22.2] can be used. R_{min} and R_{max} are influenced by the UV transmission properties of the microscope optics, and in particular the bandwidth of the PMT emission filters. Even the K_d may be influenced by environmental factors within the cell such as binding to intracellular proteins. A variety of calibration methods are possible, one of the simplest being to measure values of R for a series of known Ca^{2+} concentrations using mixtures of Ca^{2+}, dye, and a buffer such as EGTA (Fabiato, 1991). R_{min}, R_{max} and K_d may be obtained by fitting equation [22.2] to the R vs. $[Ca^{2+}]$ curve using iterative non-linear curve fitting (as implemented in programs such as Biosoft Fig.P, Biosoft, Cambridge, UK). Such techniques however, do not necessarily account for the effects of dye binding to intracellular proteins (Highsmith et al., 1986), and more sophisticated approaches such as the selective permeabilization of cells bathed in buffered solutions of known Ca^{2+} concentration may be preferable (Williams & Fay, 1990).

22.5 DUAL-EXCITATION DYE MEASUREMENT SYSTEMS

An essentially similar ratiometric approach can be applied to the dual-excitation dyes, by using a UV excitation light source which alternates between the peak absorbance wavelengths for the bound and

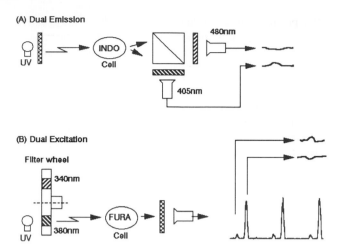

Figure 22.3 Fluroescence measurement system configurations. (A) Dual-emission. A cell loaded with indo-1 is illuminated with UV light at the peak absorption wavelength for the dye (350 nm). The spectral components of the fluorescent emissions, corresponding to the peak emission wavelengths for the Ca^{2+}-bound (405 nm) and free (490 nm) dye forms, are using a 45° dichroic mirror and filters, and measured using a pair of PMTs. $[Ca^{2+}]$ is determined from the ratio of the PMT output signals. (B) Dual-excitation. A cell loaded with Fura-2 is alternately illuminated with filtered UV light at the peak absorption wavelengths for the Ca^{2+}-bound (340 nm) and free (360 nm) dye forms, using a rotating filter wheel. Fluorescent emissions at the peak emission wavelength are measured using a PMT. A computer (or appropriate hardware), synchronized to the rotation of the filter wheel, is used to store measurements from each filter and perform ratio and $[Ca^{2+}]$ calculations.

unbound forms of the dye. This is generally achieved by placing a mechanical filter changer in the excitation light path, as shown in Fig. 22.3(B); either a rotating filter wheel or an alternating filter changer. Since the peak emission wavelength does not shift, only a single-emission filter and PMT is required. Most of the commonly available wheels have at least four filter positions, and some have as many as ten. This is particularly useful since it allows multiple ion measurements to be made when more than one dye is applied to the cell.

The computation of the bound/unbound ratio is more complicated than for dual-emission dyes since the PMT signal is split into a series of discrete pulses as the filters rotate into and out of the light path (see Fig. 22.3(B)). To compute R, as in equation [22.1], it is necessary to determine when each filter is in place in the light path and store a snapshot of the peak emission level. After a filter change cycle has been completed the peak level for each filter episode is retrieved and used to compute the ratio. This process may be software- or hardware-based. For instance, some systems use analogue hardware modules to detect and store the emission level for each filter (e.g. Cairn Research, Sittingbourne, UK) whereas the PhoCal system (Applied Imaging, Gateshead, UK) performs this function digitally using software within the host PC.

High-performance excitation filter changers are difficult to produce and no single design is entirely satisfactory for all purposes. For instance, one approach is to drive the filter with a synchronous motor which can be made to rotate at variable speeds locked to a control frequency. Cairn Instruments used this approach to produce a wheel which can rotate at frequencies as high as 300 Hz. However, such wheels have rather limited capabilites of stepping from one filter position to another, and cannot rotate at less than 3 Hz. Another approach, such as that used in the Sutter Instruments wheel (supplied by Axon Instruments), is to make use of a digitally controllable stepper motor which can rapidly move in any direction in precisely controlled steps. This filter can step between adjacent filters within 50 ms, but cannot easily be made to rotate at constant speeds. The Applied Imaging wheel is intermediate in design between these two; it is based upon a stepper motor, but with a digital phase lock loop circuitry which allows it to constantly rotate at a fixed frequency of 6.25 Hz (for synchronization with a video camera).

If high-speed sampling is required (e.g. >2 ratios per second) a constantly rotating filter wheel is preferable. In general, it is easier to accelerate a wheel up to a high constant speed, than to rapidly start a wheel moving and bring it to a halt again at a new position. The Applied Imaging wheel, for instance, requires

250 ms to step between filter positions, whereas in a constantly rotating mode it sweeps through a filter in 40 ms.

The use of dual-emission or -excitation dyes has a profound effect on the rate at which ratio measurements can be made. Dual-emission dyes allow simultaneous measurement at the characteristic emission peak wavelengths. Consequently, the rate at which ratio measurements can be made is limited only by the response time of the dye itself to changes in ion concentration which are typically in the order of a few milliseconds. On the other hand, since dual-excitation dyes require a mechanically alternated light source, measurements are discontinuous, with the maximum rate determined by time taken to execute a complete filter change cycle. Currently, only the 300 Hz Cairn instruments wheel allows dual-excitation dye measurements at rates approaching those of dual-emission dyes.

Dual-emission dyes are often used in combination with electrophysiological measurements. The fast response time of the method proves valuable in the

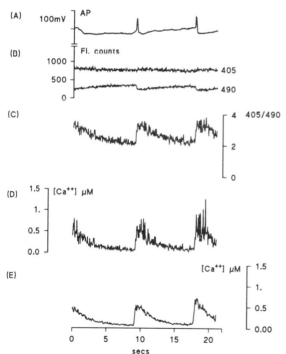

Figure 22.4 Membrane potential and ratiometric Ca^{2+} recordings made using a 'chart recorder'-type fluorescence measurement program. (A) Cell membrane potential from an indo-1 dye-loaded, whole-cell patch-clamped, pituitary cell. (B) Fluorescence emissions at 490 nm and 405 nm, using photon counting with a 2 ms count period. (C) Ratio 405/490 using equation [22.1]. (D) Computed Ca^{2+} concentration using equation [22.2] with $R_{min} = 1.8$, $R_{max} = 4.5$, $K_d = 0.426$ м. (E) $[Ca^{2+}]$ after smoothing of 405 nm, 490 nm signals using a 10 point running averaged. Data courtesy of Dr W.T. Mason, AFRC Babraham.

study of rapid Ca^{2+} transients evoked by action potentials in neurons. Figure 22.4 shows some results from such an experiment: the 405 nm and 490 nm fluorescence emissions were obtained using photon counting over a 2 ms period, the 405/490 ratio was computed using equation [22.1] and $[Ca^{2+}]$ using equation [22.2]. The cell is electrically active, producing a series of spontaneous action potentials, as can be seen from the membrane potential recording. Each action potential produces a transient influx of Ca^{2+} into the cell, increasing intracellular Ca^{2+} concentration.

Figure 22.4 also illustrates some of the difficulties associated with the ratiometric computation of $[Ca^{2+}]$. The fluorescence signals are relatively small and quite noisy, as a consequence of the low light levels and small number of counts accumulated during the short counting period. In general, the standard deviation of the measured light signal is proportional to the square root of the average number of photon counts. An average count of 1000 will exhibit fluctuations with a standard deviation of 31. As can be seen from Fig. 22.4(C), taking the ratio of two such signals increases the noise yet again.

However, compared to the ratio, the computed $[Ca^{2+}]$ is enormously noisy. In addition the variance of the $[Ca^{2+}]$ signal is not constant, but increases dramatically with $[Ca^{2+}]$. This is a consequence of the mathematical form of equation [22.2]. As R approaches R_{max} with increasing Ca^{2+}, the denominator in equation [22.2] becomes very small and any noise on the R signal becomes increasingly magnified. In essence, the $[Ca^{2+}]$ calculation becomes progressively unstable as R approaches R_{max}.

The variance of the fluorescence signals can be reduced by using a longer photon counting period to accumulate larger numbers of counts which are subject to less statistical variations. An alternative approach which can be applied after recording, is to average adjacent count periods. Figure 22.4(D) shows the $[Ca^{2+}]$ computed after a series of 10 adjacent samples in the 405 and 490 signal have been averaged using a 10 point running mean algorithm.

22.6 ANALOGUE SIGNAL DIGITIZATION

If electrophysiological or other signals (e.g. muscle tension) are to be recorded simultaneously with the fluorescence signals, an additional computer interface card may be required for analogue signal digitization. An analogue-to-digital (A/D) converter is a computer controlled voltmeter which can be made to measure a voltage signal and return a number which can be stored in computer memory. Analogue signals are digitized by making a series of such measurements (or samples) at fixed intervals and storing the stream of numbers in memory as a digital signal record.

A complete system providing the hardware for A/D conversion of 8–16 input channels, sample timing, trigger synchronization, and often digital-to-analogue (D/A) conversion, is usually described as a **laboratory interface**. Laboratory interface expansion cards are widely available for the IBM PC family, the most commonly used for electrophysiological work being the Labmaster (available from Axon Instruments), CED 1401 (Cambridge Electronic Design, Cambridge, UK), DT2801A (Data Translation, Marlboro, MA, USA), LAB-PC (National Instruments, Austin, TX, USA). A more detailed discussion of laboratory interfaces can be found in Dempster (1993).

22.7   SOFTWARE FOR RECORDING FLUORESCENCE SIGNALS

Computer-based measurement systems are very dependent upon the availability of appropriate data acquisition and analysis software. Such software can be developed within the laboratory but it is a difficult and time-consuming process. A program for acquiring fluorescence signals must attend to one or more of the following activities.

Photon counting
Excitation filter control
A/D conversion of 1–4 analogue channels
Fluorescence ratio binding calculations
Display & measurement of recorded signals
Plotting of hard copies

It is difficult, in practice, to produce a single program which is ideal for all types of cell fluorescence experiments. This is not due simply to computer memory limitations or other hardware factors, although these are issues which must be considered. Rather, it is that a variety of experimental paradigms are in use and the program features appropriate for one approach may conflict with those required for another.

22.8   THE 'CHART RECORDER' PARADIGM

Many experimental protocols require a continuous unbroken record to be made of the fluorescence, electrical and/or other signals, in much the same way as might be obtained with a chart recorder. Figure 22.4 is an extract from such a recording, in which six data channels were acquired over a period of 95 s. A key feature in such programs is the ability to handle

data files of large size. For instance, the data in Fig. 22.4 was collected using a 300 Hz sampling rate which resulted in a file containing 1 Mbyte of data.

Continuous **recording-to-disk** methods are often required to collect such large records without gaps since many laboratory computers do not have sufficient RAM (random access memory) space to accommodate the whole data file. The incoming signals must therefore be copied to disk while digitization is in progress, using techniques such as the double buffer method. Samples, as they are acquired from the A/D converter and photon counter, are stored in a small (e.g. 16 Kbyte) temporary buffer in RAM which is continuously filled in a cyclic fashion. The buffer is split into two halves, and as each half becomes full its contents are written on to disk while samples continue to be stored in the other half. Continuous sampling rates of around 30 kHz can be readily achieved using this technique on typical IBM PC compatible computers.

Similarly, the program's analysis features must be able to handle such large records efficiently, to display magnified sections of the record, and provide cursor measurements. The chart recorder paradigm proves to be satisfactory for a wide range of experimental protocols where a continuous relatively slow record is required, and a small number of stimuli are applied at infrequent intervals. Much of the work to which intracellular ion measurements have so far been applied falls into this category; and most commercially available measurement systems (e.g. Applied Imaging, Cairn Instruments, Newcastle Photometrics) supply programs which adhere to this paradigm.

It is also worth noting that a number of general-purpose chart recorder programs also exist; such as AxoTape from Axon Instruments, the MacLab Chart program running on the Apple Macintosh (World Precision Instruments USA), and CED's Chart program for the CED 1401 interface. Although unable to directly couple to photon counters, these programs can be used with systems, like that supplied by Cairn instruments, which produce analogue outputs of the fluorescence signals.

22.9   THE 'OSCILLOSCOPE' PARADIGM

While chart recorder programs are, probably, the most generally useful, there are categories of experiment where they have distinct limitations. This is particularly so where the prime focus of interest is the properties of voltage-activated Ca^{2+} currents, and their effect on intracellular Ca^{2+} concentration. Such currents are normally studied using the whole-cell patch-clamp technique, and Ca^{2+} currents are evoked by depolarizing

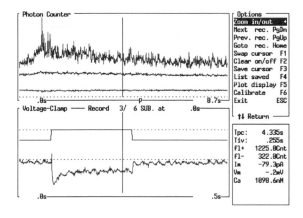

Figure 22.5 Display screen from the PhoClamp fluorescence/voltage clamp program using an 'oscilloscope' paradigm. The upper display panel shows an 8.7 s fluorescence sweep, 490 nm and 405 nm indo-1 fluorescence signals and [Ca^{2+}]. The lower panel shows the associated 500 ms voltage clamp sweep with the Ca^{2+} current which was evoked by a 200 ms depolarizing voltage pulse. The voltage clamp sweep occurred 1 s after the start of the fluorescence sweep.

the cell membrane under the control of a voltage pulse applied to the voltage clamp command input. Such current signals are transient and of brief duration (100–500 ms). The flow of Ca^{2+} ions into the cell through these channels results in changes in intracellular Ca^{2+} concentration which are similarly transient but with a slower time course (1–10 s). In order to study the voltage sensitivity of the Ca^{2+} current and/or concentration transients, it is usually necessary to apply a range of voltage steps of varying amplitude and duration.

These voltage clamp studies are distinctive in that the experimental protocol is split into a series of relatively short, discrete, recording sweeps, rather than a single long continuous record. Also, instead of a relatively small number of stimuli, separated by intervals of minutes, hundreds of precise voltage pulse stimuli are applied, at 10–20 s intervals. This style of experiment suggests the use of an 'oscilloscope' paradigm in the design of the program. Returning to the analogy with conventional recording devices, the oscilloscope is an instrument designed for recording repeatable high-speed transients, synchronized with some external trigger event. In general, electrophysiology software for recording voltage-activated currents, such as Axon Instruments' pCLAMP, follow this oscilloscope paradigm. The generation of repeated pulse stimuli, with a variety of heights and/or widths is itself a distinct task. It is now usual to generate pulse waveforms within the computer and apply them using a D/A converter.

A program designed for recording and analysing voltage-activated currents and fluorescence, therefore, has to perform the following functions.

Record fluorescence sweep
Record current, voltage sweep
Generate voltage clamp command pulses
Apply leak current subtraction
Display and analyse recordings

A display screen from PhoClamp (Applied Imaging), a combined voltage-clamp/fluorescence program, can be seen in Fig. 22.5. The screen is split into two panels, the upper panel showing the fluorescence recording sweep of indo-1 dye (405 nm, 490 nm) fluorescence and computed [Ca^{2+}]. The lower panel shows the electrophysiological sweep with membrane current and potential. Due to the marked differences in the time-course of the Ca^{2+} current and concentration signals. It is convenient to use different sweep durations: 500 ms for the current and 5 s for the fluorescence.

22.10 LEAK CURRENT SUBTRACTION

As an added complication, the current signals often require the process of digital leak current subtraction. One of the difficulties inherent in the voltage clamp technique is ensuring that the current signal being measured is being mediated only by the ion channels under study. Cell membranes contain a wide variety of channel types, selective for most of the ions present in normal physiological salt solutions (Na$^+$, K$^+$, Ca^{2+}, Cl$^-$). Channel types may also have different voltage sensitivities and kinetics. During an experiment, it is normal practice to attempt to isolate a single channel type by blocking other channels using a variety of pharmacological agents. Such blocking strategies are rarely completely effective and a residual leak often remains in addition to the Ca^{2+} current. In many cells the leak current can be significant in comparison with the Ca^{2+} current under study.

In many circumstances, however, a digital subtraction process can be used to remove the leak current component from the current signal. Leak currents are often due to current flowing through imperfections in the pipette–membrane seal, or to passive K$^+$ or Cl$^-$ channels in the cell membrane which are not readily blockable by common pharmacological agents. In either case, the conductance of the leak channels is constant, and the current scales linearly with membrane potential. In contrast, the Ca^{2+} current is highly voltage- and time-dependent, absent in response to hyperpolarizing the voltage steps, and requiring steps to potentials more positive than -40 mV before any current appears. Consequently, the current record, in response to a hyperpolarizing (or small depolarizing) step contains only the leak current component. The leak component can be removed from the Ca^{2+} current, evoked by large depolarizations, by scaling the leak record in proportion to the ratio of voltage step sizes and subtracting it from the Ca^{2+} current record. Further details of digital leak subtraction can be found in Dempster (1992).

The protocol for acquiring a leak subtracted Ca^{2+} current and the resulting intracellular Ca^{2+} concentration changes is as follows:

(1) acquire a series of leak current records;
(2) create average leak current and store on file;
(3) start fluorescence recording sweep;
(4) after a delay, start membrane current, potential recording sweep;
(5) generate voltage clamp command step;
(6) on completion of fluorescence sweep, store on file;
(7) scale and subtract leak current from Ca^{2+} current and store leak subtracted current on file.

22.11 COMPUTER SYSTEMS DESIGNS

The combination of fluorescence and electrophysiological signal recording is quite a demanding task for a computer system. The computer must perform a complex task of generating the voltage clamp command waveforms and coordinating these with the recording of current and voltage from the patch clamp and the fluorescence signals. Ideally, incoming signals should also be displayed on screen as they are digitized and also the fluorescence ratio calculated on-line.

The timing requirements of each channel may differ, as seen in Fig 22.5. A typical Ca^{2+} current may last 100–500 ms, and a 1000 sample point record would require a sampling rate of 2 Hz per channel. Fluorescence signals, being of longer duration generally require sampling at rates no more than 0.5 kHz per channel. Fortunately, the modern laboratory computer is more than fast enough to support these tasks, although care has to be taken with the design of the software. An important approach is to devote critical timing tasks to subsystems such as the laboratory interface unit and the digital photon counter.

In the PhoClamp program, fluorescence measurements are made using a Thorn EMI C660 photon counter board and A/D and D/A conversion with a National Instruments LAB-PC laboratory interface. Both of these devices have on-board clocks and are capable of a significant degree of autonomy in their operation. Overall timing and control of recording sweeps is coordinated by the PhoClamp master program, running on the host computer, which displays results on screen as they are acquired and performs ratiometric calculations and other analysis functions.

The key to an effective design is to manage the transfer of data into and out of the computer with the minimum load on the central processing unit (CPU) of the computer. Fortunately, the IBM PC is equipped with a variety of data transfer channels, eight direct memory access (DMA) channels and 16 interrupt lines, which make it well-suited to handling these multiple tasks.

DMA is a technique for automatically transferring data directly between a peripheral device, such as a laboratory interface, and the host computer's RAM memory without involving the CPU. It is very efficient, taking only 1–2 μs to transfer a byte of data, and leaves the CPU free for other functions. Most of the modern IBM PC-based laboratory interfaces used in the electrophysiological laboratory support the use of at least one DMA channel.

Interrupt lines provide a simpler, though less efficient, alternative to the DMA data transfer method. An interrupt line can be used to signal the host computer that data are ready to be transferred (e.g. an A/D conversion has completed). It causes the CPU to interrupt the program which it is currently executing and transfer control to an interrupt service routine, a program which transfers the data from the interface into memory. Once the transfer is complete, control is returned to the original program. Compared to DMA, a considerable overhead is involved in the use of an interrupt, in that the state of the interrupted program (i.e. the contents of the CPU registers) must be saved before and restored after the execution of the interrupt service routine to allow the program to continue correctly. A detailed discussion of the operation of DMA and interrupts can be found in Eggebrecht (1990).

A/D conversion of the current and voltage signals makes the highest demand on the system, requiring samples to be stored in the computer at an aggregate rate of 4 kHz. The LAB-PC interface is programmed to collect a sweep of 256–2048 samples, alternating between two analogue inputs, digitizing the signals at strictly timed intervals under the control of its sampling clock. Samples are transferred into a storage buffer in computer memory using a DMA channel. Once programmed, the complete sequence of operations is totally independent of the CPU, and very efficient.

Ideally, all of the three main input and output streams of data (A/D converter, photon counter, D/A converter) would be carried via their own DMA channel, leaving the CPU free for supervisory operations. However, although there are sufficient free DMA channels on the PC(6), not all of the required peripheral devices, particularly the photon counter, can support DMA. Also most laboratory interfaces support DMA for either A/D or D/A conversion, but not both simultaneously (the Data Translation DT2831 and DT2841 interfaces are an exception).

Fortunately, the sampling and undate rates required for the photon counters and for D/A output are sufficiently low that the overheads incurred using interrupt lines are acceptable. The Thorn EMI C660 photon counter does not support DMA but can be made to initiate an interrupt request whenever a count period has been completed. It requires servicing at rates of no more than 500 Hz and is attached to interrupt IRQ7. D/A waveform generation is timed using the PC system clock and an interrupt service routine attached to IRQ0. This clock normally produces an interrupt event every 55 ms but it can be reprogrammed to higher rates (see Dempster, 1993 for details). Two D/A channels are used; one to provide the command voltage waveform, the other a synchronization pulse used to trigger the A/D sweep.

22.12   CONCLUSION

As is no doubt apparent, the development of a high-speed fluorescence measurement system is a complex task, involving some expertise in a number of quite disparate fields: fluorescence microscopy, properties of dyes, computer hardware and software. Developing such a system in-house can be a time-consuming process, unless a complete system description can be obtained from another laboratory. It is often more practical to purchase a system from a supplier specializing in this area, who will provide a complete integrated system.

Among the best-known of the specialist companies in the UK are Cairn Instruments, Newcastle Photometrics and the PhoCal system marketed by Applied Imaging Ltd (developed jointly by the author and Dr W.T. Mason). Each of these companies can supply both dual-emission and dual-excitation systems, and produce their own excitation filter changers. PhoCal and the Newcastle Photometrics system are based on IBM PC family computers while the Cairn Instruments system is a primarily a hardware-based system providing analogue signal outputs (although computer software in the form of an IBM PC-based chart recorder is also supplied).

Producers of the more traditional spectrophotometers and/or imaging systems, such as Spex or Photon Technology International, and microscope companies such as Nikon or Zeiss, are also becoming interested in this area.

Fluorescent dyes are useful tools in many areas of cell research and, undoubtedly this area will continue to develop. New dyes with different ion sensitivities continue to be produced. The measurement systems, and particularly the computer software, have by no means reached the ultimate in terms of performance and versatility. For example, the potential exists to study excitation–secretion coupling by combining the measurement of intracellular Ca^{2+} with the electrophysiological measurement of cell capacity (a measure of secretory vesicle fusion), using a lock-in amplifer (Lindau, 1991; Zorec et al., 1991). Similarly, experiments using intracellular concentration jump stimuli are possible using UV-activated caged Ca^{2+} buffers such as Nitr-5 and DM-nitrophen (Zucker, 1992).

Even though great advances are being made in fluorescence imaging (including the use of confocal microscopy), bringing improvements in time resolution and reductions in cost, it is likely that the simplicity and sensitivity of the microphotometric measurement system will guarantee it a continued place in the electrophysiological laboratory.

REFERENCES

Baylor S.M. & Hollingsworth (1988) *J. Physiol.* **403**, 151–192.

Dempster J. (1993) *The Computer Analysis of Electrophysiological Signals*. Academic Press, London.

Eggebrecht L.C. (1990) *Interfacing to the IBM Personal Computer* (2nd edn). Howard W. Sams & Co. Indianapolis.

Eisner D.A., Nichols C.G., O'Neill S.C., Smith G.L. & Valdeolmillos M. (1989) *J. Physiol.* **411**, 393–418.

Fabiato A. (1991) In *Cellular Calcium: A Practical Approach*, J.G. McCormack & P.H. Cobbold (eds). IRL Press, Oxford, pp. 159–175.

Grynkiewicz G., Poenie M. & Tsein R.Y. (1985) *J. Biol. Chem.* **260**, 3440–3450.

Highsmith S., Bloebaum P. & Snowdowne K.W. (1986) *Biochem. Biophys. Res. Commun.* **138**, 1153–1162.

Lindau M. (1991) *Q. Rev. Biophys.* **24**, 75–101.

Mollard P., Guerineau N., Audin J. & Dufy B. (1989) *Biochem. Biophys. Res. Commun.* **164**, 1045–1052.

Sakmann B. & Neher E. (1983) *Single-channel Recording*. Plenum Press, New York.

Standen N.B., Gray P.T.A. & Whitaker M.J. (1987) *Microelectrode Techniques – The Plymouth Workshop Handbook*. Company of Biologists, Cambridge.

Williams D.A. & Fay R.S. (1990) *Cell Calcium* **11**, 75–83.

Williams D.A., Fogarty, K.E., Tsein R.Y. & Gray F.S. (1985) *Nature* **18**, 558–561.

Zorec R., Henigman F., Mason W.T. & Kordas M. (1991) In *Methods in Neurosciences*, Vol. 4, P.M. Conn (ed.). Academic Press, London, pp. 194–210.

Zucker R.S. (1992) *Cell Calcium* **13**, 29–40.

CHAPTER TWENTY-THREE

Flow Cytometry: Use of Multiparameter Kinetics to Evaluate Several Activation Parameters Simultaneously in Individual Living Cells

ELIZABETH R. SIMONS

Boston University School of Medicine, Boston, MA, USA

23.1 INTRODUCTION

It has always been important to know whether and how cells function. In that context, it is desirable, wherever possible, to correlate the various expressions of cellular function or stimulus responses temporally with each other and, where applicable, with receptor occupancy. Furthermore, it has become apparent that, in a given sample of cells, all cells do not respond equally or synchronously.

One of the most important current challenges in cell biochemistry is to delineate the mechanisms by which cells transduce information, those by which the resultant signals are transmitted, and those which eventually lead to expressions of the particular cell's function. In order to acquire enough information to resolve these mechanisms, it is essential to correlate these stimulus-induced responses or consequences with the corresponding state of the specific receptors' occupancy in real time. It is also important to determine whether each of the functions observed depends on another earlier one, and whether it is expressed by all cells, by that fraction of cells which has occupied receptors, or by a fraction of cells which is, *a priori*, identical in terms of receptor occupancy to another which does not respond. Information of this kind can lead to temporal resolution of responses through determination of: Which step comes first? Does it depend upon another which starts earlier? Are more receptors occupied on responding than on non-responding cells, implying receptor cooperativity?

23.2 FLUORESCENT TECHNIQUES – GENERAL ADVANTAGES AND DISADVANTAGES

These questions, as well as the need to work with fewer cells as both cells and agonists become more expensive and/or difficult to acquire, cannot be resolved by the use of radioactive probes because the latter applications involve separation of cells from their supernatant so that each can be counted separately, a technique which is (a) too slow for the rapid time points needed (often 1–5 s intervals), (b) too wasteful of cells, as each time point requires a new sample, (c) difficult to use if more than one parameter is to be followed simultaneously, and (d) incapable of distinguishing subpopulations of cells, so that it cannot be determined whether a response which is 50% of the control's reflects a 50% response by each cell or a full response by 50% of the cells.

Routine fluorescence measurements of cell suspensions, or of large numbers of cells attached to a matrix, avoid the problems (a) and (b) above; (c) is a disadvantage which exists for most fluorometers although one (Hitachi 2000) permits scanning of up to four emissions from two separate excitation wavelengths, or of any combination of four parameters, and (d) exists for all measurements which reflect the average over large numbers of cells. Conversely, fluorescence microscopy permits evaluation of each individual cell, therefore avoids (d), but it is laborious to use for large numbers of cells and can usually measure only one parameter at a time (in most microscopes there are two excitation wavelengths but only one emission wavelength).

Another source of errors in fluorescence measurements of cell suspensions is the possibility, indeed the probability, that extra- as well as intracellular fluorescence is being measured. Ideally, 100% of either should be being measured; any combination of extra- and intracellular indications of the same parameter creates problems in the final quantitative evaluation of that parameter. For some distributive probes, such as the thiocyanines used to measure membrane potentials, or 9-aminoacridine used to measure pH_{in}, it is the extracellular probe's fluorescence which gives an indication of how much the positively charged probe has penetrated into the negatively charged cell, for the cyanines, or of how much protonated amine is external, for 9-aminoacridine, since the intracellular probe is quenched in each case (Deamer et al., 1972; Sims et al., 1974). In each case, the fluorescence in a single compartment is being measured, and therefore the overall cell membrane potential or the cytoplasmic pH, respectively, can be calculated. Measurement of extracellular probe is therefore crucial in these cases. Conversely, when measuring intracellular probe, as is true for the in situ probes introduced into the cell as esters, either it must be assumed that there is no de-esterified and therefore indicating probe in the extracellular buffer or this must be corrected for. Unfortunately, most investigators assume there is no contribution from any extracellular probe; however, most cells do leak (some more, some less), most have relatively high permeability to small anions, and some release esterases capable of hydrolysing any residual esters in the buffer. The necessary corrections have been described in the literature (Davies et al., 1987a) and will be explained briefly in Section 23.6. In contrast to suspension studies, when intracellular fluorescence is examined on a flow cytometer, the extracellular milieu is so diluted by the sheath buffer that its contribution to the total fluorescence is negligible. Therefore, flow cytometry obviates the necessity for extracellular probe correction and is the method

of choice for cells which leak cytoplasmic probes readily.

In addition to the need to eliminate artifacts due to probe leakage, some other precautions are necessary. Among these is the question of ratioing of fluorescences, the basis of some of the most useful fluorometric probes of cell functions. In this context, it must be remembered that the use of the so-called 'ratiometric' fluorometric in situ cytoplasmic probes is really only correct when either the ratio system of a flow cytometer, or that of a microscope is being used, each of which evaluates the ratio directly for each cell. That is, the use of ratios to obviate errors dealing with different sizes of cell and/or different probe concentrations within each cell not only depends upon adherence to Beer's law but also assumes that the probe concentration (c) and the light path (l) in the ratioed fluorescences are equal, i.e. assumes that the measurement is made in the same cell at two different wavelengths. This means that $R = F_1/F_2 = f_1 cl/f_2 cl = f_1/f_2$, where f_1 and f_2 are independent molar fluorescences characteristic of the property being measured by the probe. Since mathematically the average of ratios is not equal to the ratio of the averages, the equation holds only when a single cell's c and l (and not the average over all cells, as is true in suspensions) is used. Thus, although ratiometric probes are routinely used in cell suspensions, this procedure involves an implicit assumption that all cells are of equal size and contain equal concentrations of probe.

In flow cytometry, no such simplifying assumption needs to be made if a ratio board is used, since flow cytometers are or can be so equipped. A ratio board takes the ratio of selected photomultiplier outputs for each cell, cell by cell; if, however, the ratio of the average fluorescences (mean fluorescence channel) is taken at the two wavelengths (i.e. of the mean channels in a flow cytometer), this singular advantage of flow cytometry is lost.

23.3 FLOW CYTOMETRY FOR KINETIC STUDIES OF CELLULAR FUNCTIONS

For all of the above reasons, in the past few years we, as well as a number of other investigators, have adapted the flow cytometer, an instrument originally designed to examine cell surface antigens as cell type markers, to perform multiparameter kinetic studies of stimulus responses of suspendable cells, and to correlate these parameters for each cell with the receptor occupancy of that cell (Lazzari et al., 1986, 1990; Davies et al., 1988, 1989, 1990; Ryan et al., 1990; Bernardo et al., 1990; Brunkhorst et al., 1991). The time resolution to date is between 1 and

4 s, achievable by using an injection system which adds the stimulus directly to the thermostated stirred tube containing the already flowing cells (Kelley, 1989).

We have now shown, for platelets as well as for neutrophils, that flow cytometry is particularly useful when one is dealing with small cell samples, low subsaturating doses of stimulus, multiple classes of receptors, and/or heterogeneity in the response and in the proportion of cells which is exhibiting that response (Beard *et al.*, 1986; Lazzari *et al.*, 1986, 1990; Sullivan *et al.*, 1987a; Davies *et al.*, 1988, 1989, 1990; Ryan *et al.*, 1990; Bernardo *et al.*, 1990; Brunkhorst *et al.*, 1991).

23.4 TIME OF ONSET OF INITIAL RESPONSE

Many cellular functions are initiated almost instantaneously upon exposure to agonists; some functions are independent of each other while others are interdependent. It is therefore of interest to measure, if possible, the actual time of onset of each of the functions being studied, as well as the response time for each function relative to all the others.

To date, the ability to measure cells very rapidly in flow cytometers has been limited by the length of time required to mix cells and agonist adequately (considerably shortened in a system equipped with a magnetic stirrer), the time necessary to flow from the sample chamber to the drop-forming tip (dependent on the length and diameter of the sample delivery tube), on the flow rate, and on the desired number of cells per time point. For standard flow cytometers, the earliest post-agonist addition time point at which data could be collected was approximately 10 s in the Becton Dickinson FACS 440 for cells (e.g. platelets) which could not be subjected to the high shear rates arising from flow rates greater than 2000 cells per second. In our hands the 'boost' system designed to push cells through more rapidly caused artifacts as well as broken cells (others may have better luck with this accessory). By moving the stirred thermostated cell chamber nearer the delivery tip, thereby shortening the delivery tube, and installing an injection system, Kelley (1989) was able to achieve initial time observations of 1–2 s. Recently Sklar and his colleagues have announced their intention to build a system capable of observations within approximately 100 ms in a flow system at Los Alamos, NM (NFCR Newsletter, Dec, 1991).

It should be noted that, in this respect, non-flow cytometric measurements have achieved much shorter initial time points. In terms of cell applications, these are the continuous flow, stopped flow and quenched flow systems developed by Gear and by Rink (Gear,

1980; Gear & Burke, 1982; Rink & Sage, 1987; Sage & Rink, 1986, 1987; Jones *et al.*, 1989). However, since these depend on observations of cell suspensions, each yields an average over all the cells, with temporal resolution in the 100 ms range; continuous observation of subpopulations (unless the cells are fixed) is not possible in these rapid kinetics systems, and each time point hence requires a new sample. Nevertheless, until Dr Sklar's system is operative, flow cytometers cannot achieve such early initial time points due largely to the need to mix thoroughly cells and agonists, then flow the mixture through a tube of some finite length into the drop-forming tip without damaging the cells. Some attempts to place the mixing chamber immediately above the tip have been made, and the time when observations can be made within 100–300 ms after addition of agonist is near.

The discussion above refers to initial times of observation, the times which are critical when the sequence of events (e.g. in signal transduction) is of interest. Clearly the temporal resolution, i.e. the time interval between recording of observations, depends strongly upon the flow rate which the cells can tolerate – the faster the flow the higher the number of cells per second and the shorter the time interval between data points.

23.5 PARAMETERS WHICH CAN BE MEASURED

An important limit on the parameters which can be measured by multiparameter flow kinetics (i.e. multiple simultaneous fluorescences measured on a fluorescence-activated cell sorter, FACS) is the availability of non-mutally interfering intracellular or cell surface-linkable fluorescent probes. New probes are constantly being developed, in many cases by Molecular Probes Inc., Eugene, OR, USA, which has become the world's main supplier of fluorescent probes and whose large catalogue lists hundreds of probes and their properties. Not all probes, however, can be excited by wavelengths corresponding to an accessible laser line, so that probes usable on a laser flow cytometer are more limited than those usable for suspension fluorometry or fluorescence microscopy when the accessible wavelengths are only limited by the available filters (if microscopes or filter fluorometers are used) or not limited at all (if fluorometers having both excitation and emission monochromators are used).

Although cell types and classes differ in the specific responses to agonists, it is now clear that most (though not all) mammalian cells respond to a specific stimulus (i.e. one acting through a receptor mediated pathway) by undergoing a series of changes:

(1) A relatively rapid but small (and often missed) hyperpolarization, which is followed by a slower and much larger depolarization. Depending on the cell, the depolarization may be dependent on the $[K^+]$ and $[Na^+]$ or just $[Na^+]$ transmembrane gradient.

(2) A relatively rapid acidification which does not appear to depend upon the $[Na^+]$ gradient, but which is followed by an alkalinization which proceeds via Na^+/H^+ countertransport.

(3) A transient cytoplasmic $[Ca^{2+}]$ increase (within <1 s in some cells, 30 times slower in others) in which the Ca^{2+} generally comes from intracellular stores such as the endoplasmic reticulum, although some cells count on a Ca^{2+} influx from the extracellular milieu for at least part of the transient.

(4) Activation of at least one phospholipase in the plasma membrane (most are Ca^{2+} dependent) and release of the appropriate products, e.g. arachidonic acid if phospholipase A2, inositol trisphosphate and diacylglyceride if phospholipase C is the initially activated enzyme.

(5) For secretory cells, an eventual degranulation, though the times of granule content release vary widely.

(6) For phagocytic cells, a phagocytic vacuole (phagovacuole) will form sometime during steps 1–5, and the oxidative burst will be initiated so that oxidizing products, as well as the specific granules' lytic enzymes are released into the phagovacuole.

(7) If a cell is able to respond and the agonist is a chemotactic agent, chemotaxis will be initiated sometime during the above processes.

For most cells it is not yet known whether any of the above events depends upon the prior occurrence of any of the others, or whether all can be initiated simultaneously by the same event. Almost all of the events except chemotaxis can readily be followed continuously by flow ctyometry kinetics.

23.6 CLASSES OF FLUORESCENT PROBES

There are a number of classes of probes, and only examples of each will be given here, rather than the entire and exhaustive list. Many of the currently available probes are well-described in reviews by Waggoner (1976, 1988, 1990) and in the Molecular Probes Handbook (Haugland, 1989).

The overall principles which govern the use of these probes differ:

23.6.1 Distributive probes

Membrane potential and some pH probes distribute themselves between the cell and the external milieu according to the property being evaluated. In the case of membrane potentials, this is the net negative charge of the cell with respect to the buffer in which it is suspended. In the case of pH probes, it is the difference between the extra- and the intracellular pH. These probes will distribute into every compartment across whose membrane a potential or pH gradient, respectively, exists, and therefore will be distributed within all cellular organelles. Unless an image-enhanced fluorometer is used, the net membrane potential or pH will therefore be an average over the whole cell, its organelles and interior compartments.

23.6.2 Distributive probe concentration limitations

There are two general precautions with respect to distributive probes: (1) Enough probe must be used to ensure that the distribution is dependent upon the property to be evaluated, not upon the quantity of probe available. As Hoffman and Laris (1974) described when these probes were first prepared by Waggoner (Sims et al., 1974), for each new cell type the lowest concentration of probe which will yield a result independent of that concentration must be evaluated. (2) Too much cationic probe must not be used since, being of opposite charge, it will tend to neutralize the net negative charge of the cell and, within it, tend to concentrate in the most negative compartment. Similarly, excess amine pH probe must not be used since it will become protonated in the cytoplasm and will therefore alter the cytoplasmic pH. The balancing of these two mutually contradictory conditions is critical to the correct use of a distributive probe (Freedman & Novac, 1983;, Freedman & Laris, 1988).

It is self-evident that probes which tend to multimerize and self-quench within the cell, such as the dithiocyanine indicators of membrane potentials (Sims et al., 1974) or the aminoacridine indicators of cytoplasmic pH (Deamer et al., 1972), are not suitable for flow cytometry as only the cell and not its surrounding medium register a fluorescence in flow cytometry. There are, however, non-self-associating cyanine probes such as the dioxa and dicarbo series (see below) which do fluoresce in the cell and are therefore usable for flow cytometric measurements of membrane potentials (Seligman & Gallin, 1983). Although some other amines such as acridine and methyl orange have been suggested as distributive pH probes (Nuccitelli & Deamer, 1982), these have been found to be difficult to quantitate due to binding, largely to sites on cell membranes, and they are therefore used qualitatively but not quantitatively.

23.6.3 *In situ* cytoplasmic probes

In situ cytoplasmic probes depend upon the principle that uncharged entities can traverse a cell membrane easily, whereas charged ones are trapped on one side or the other (Thomas *et al.*, 1979; Tsien, 1980; Grynkiewicz *et al.*, 1985). This is rather a broad principle, often honoured in the breach since some cell membranes are relatively permeable to anions, but it appears to hold well for many classes of cells. The probes' designers then reasoned that, if an esterified (and preferably non-fluorescent probe) could diffuse into a mammalian cell's cytoplasm, the ubiquitous and relatively unspecific cytoplasmic esterases would hydrolyse the probe, yielding a fluorescent indicator which is trapped in the cytoplasm because of its newly acquired charge. The assumption was that, because most of the esterases are contained in the cytoplasm, the eventual localization of the de-esterified probe would be in the cytoplasm. We have shown this assumption to be correct for at least three such probes, the Ca^{2+} probes indo-1 and fura-2, and the pH probe BCECF (see below).

23.6.4 Intensity of fluorescence as an indicator

As indicated below, there are *in situ* probes whose fluorescence intensity F at a given wavelength is, over the probe concentration region in which Beer's law is obeyed, (i.e. $F=f[X]$) a linear function of the concentration, $[X]$, of the ion for which it is specific, assuming the probe concentration and the light path through the cell (i.e. the cell size) to be constant. A calibration curve can then be used to evaluate the actual concentration from the fluorescence intensity. The calibration curves, for a probe which changes only in the intensity of its fluorescence as a function of $[X]$, will have a slope dependent upon the internal concentration of probe, i.e. will be parallel but not superimposed. In order to calculate the actual concentration of X, the slope of the calibration curve is determined by measuring the maximal and minimal fluorescences of X under the conditions and for the type and number of cells being studied, in order to locate the linear correlation. These curves should be used only in the region in which they are approximately linear. It should be remembered, in this connection, that such a calibration curve is always S-shaped with an inflection point at the pK, pCa^{2+} or, in the general case, pX.

23.6.5 Ratio of fluorescence as an indicator

If the *in situ* probe being used undergoes a wavelength shift of the absorbance or the emission maximum upon binding X, the ratio technique described above can be used (Thomas *et al.*, 1979; Tsien, 1980; Grynkiewicz *et al.*, 1985). The ratio R of fluorescence emissions at two wavelengths, elicited at a single excitation wavelength, or of fluorescence emissions at a single wavelength excited by light at the two different excitation wavelengths, is then independent of the cell size (i.e. the length of the light path) and of the probe concentration. In some cases, there is an isosbestic point at which the fluorescence is independent of the concentration of X; the ratio technique is then still valid. It is, however, not valid if the two chosen wavelengths for the ratio fall on the same emission or absorbance peak; in such a case, the ratio of two intensities on essentially the same straight line (the side of the absorbance or emission peak) is taken, which will yield a constant and will be independent of the concentration of X.

23.6.6 *In situ* probe concentration limitations

For many of the *in situ* probes, it is preferable to use the lowest concentration at which the probe's fluorescence changes can be detected. That is, unlike the distributive probes, there is no possible error in using a very low concentration of probe. There are several problems that may occur if the probe concentration chosen is too high:

(1) The probe concentration within the cells may exceed adherence to Beer's law, and self-quenching of fluorescence may occur. This is particularly true of fluorescein derivatives.
(2) The probe ester may not be fully hydrolysed, leading to a continuous change in intracellular fluorescence and the possibility that extracellular ester may be present and be hydrolysed in the external buffer. Furthermore, excess ester may penetrate into some of the internal organelles or granules; if these contain no esterases, the problem of excess ester leakage remains, if they do contain esterases the measurement is now not of the cytoplasmic pH but rather of the pH of both cytoplasmic and organellar compartments.
(3) Since leakage is concentration gradient-dependent, the probe as well as its ester will leak more readily to the exterior, causing the artifacts already described above.
(4) Cytoplasmic hydrolysis of any ester leads to acidification of the cytoplasm which may or may not have enough protein content to act as buffer. For example, we have found that loading platelets or neutrophils with a 15 μM probe ester for 15 min at 37°C leads to a sustained reduction of the cytoplasmic pH from approximately 7 to 6.85.

23.7 SPECIFIC PROBES FOR PARAMETERS OF CELL FUNCTION

The following list and brief description of some of the currently most popular and frequently used probes is not meant to be all inclusive; it is merely meant to illustrate the use of certain types of probes, and to present their advantages and disadvantages in specific applications. The examples include those probes we have found most useful for cell function studies by multiparameter flow kinetics.

23.7.1 Intracellular cations – Ca^{2+} and Mg^{2+}

The initial *in situ* Ca^{2+} probe described by Tsien (1980) was quin-2, a probe whose fluorescence intensity increased upon binding of Ca^{2+}, but which exhibited no wavelength shift. It could be introduced as a membrane-permeable acetoxymethyl ester, making it preferable to previously available Ca^{2+} indicators such as aequorin which is not membrane-permeable and must be introduced into the cell cytoplasm by more stressful techniques such as electroporation or 'scrape loading'. Quin-2, however, had several disadvantages, including a relatively low quantum yield, a high Ca^{2+} buffering capacity, and the absence of a wavelength shift upon binding Ca^{2+}. Tsien and his colleagues therefore developed a new series of probes, indo-1 and fura-2 (Grynkiewicz et al., 1985) (later followed by fura-5, a longer wavelength ratiometric analogue). These indicators were introduced into cells as acetoxymethyl esters, and de-esterified in the cytoplasm. However, unlike quin-2 which exhibited no emission or excitation wavelength shift after binding Ca^{2+}, indo-1 undergoes a large emission wavelength change (from 485 nm for Ca^{2+} to 405 nm for Ca^{2+}-bound indo-1, both excited at 357 nm) while fura-2 undergoes a somewhat smaller excitation wavelength shift (380 nm Ca^{2+}-free to 340 nm Ca^{2+}-bound, with both emissions at 510 nm). The esters are almost non-fluorescent, but an excess of unhydrolysed fura-2-AM can lead to artifactual ratios as it also emits fluorescence at 510 nm; there is no problem with indo-1-AM which fluoresces at 455 nm when excited at 357 nm. Comparable Mg^{2+} indicators, such as mag-fura-2 and mag-fura-5, also exist. There are now newer and, for some applications, preferable Ca^{2+} and Mg^{2+} indicators, including fluo-3, rhod-2, fura-5, Calcium Green, Calcium Orange and Calcium Crimson and the comparable Mg^{2+} indicators, which can be excited by an argon laser in the visible range, and exhibit large Stokes's shifts (some with emissions around 600 nm) but not all exhibit wavelength maximum differences between the ligand-bound and ligand-free states.

The current literature tends to describe experiments using indo-1 for flow cytometric studies and fura-2 for fluorescence microscopy because it is more convenient to use a single excitation wavelength in the former, a single emission in the latter, and ratiometric probes are always preferable to the non-ratiometric (i.e. single wavelength) indicators. We have shown by organellar separation (Borregaard et al., 1983; DelBuono et al., 1989), followed by fluorometry, that the two probes, when used with all the aforementioned corrections, at concentrations below 5 μM, give fully comparable results, and that >95% of the probe resides in the cytoplasm under these conditions.

23.7.2 Intracellular cations – H^+

The earliest fluorescent pH_{in} probes for cells were the distributive probe 9-aminoacridine (Deamer et al., 1972) and other acridine derivatives (Nuccitelli & Deamer, 1982). Because these self-associate in cells, they are not recommended for flow cytometric measurements. Methyl orange, an unquantitative pH probe, has been used as a qualitative probe of some of the highly negative compartments of mammalian cells.

The first of the *in situ* probes was fluorescein diacetate, soon replaced by 5(6)-carboxyfluorescein diacetate, whose de-esterified form was more readily retained within the cell (Thomas et al., 1979), and by dimethyl-5(6)-carboxyfluorescein (Simons et al., 1982). A number of fluorescein and rhodamine derivatives have since appeared, the one most frequently used being BCECF (2', 7'-bis-(2-carboxyethyl)-5-(and 6)-carboxyfluorescein) (Siffert et al., 1987b; Davies et al., 1987a, 1990; Sullivan et al., 1987b), which permeates the cell membrane in the acetoxymethyl ester form. Since all of these pH indicators are weak acids, their pK is of prime importance, and the choice of probe must be made on the basis of the range of pH expected to be encountered, and the pK of the probes. Some of these probes also have very short linear regions on the pH calibration curve, and therefore only a short pH span over which they are sensitive enough to be used as pH_{in} indicators.

The choice of cytoplasmic pH probe must be predicated not only on the desired pH range to be observed, but also on the excitation wavelengths available and on the desirability of ratiometric vs. non-ratiometric probes. For example, BCECF has been such a successful pH probe for flow cytometric measurements, in spite of a relatively low quantum yield, because it has a pH-dependent absorbance maximum at 500 nm, easily excitable by the 488 nm line of an argon laser, and a pH-independent absorbance at 450 nm, excitable by the 457 nm argon line. Emission from both excitations is observable at 500 nm. In general, the emission fluorescence is easy to isolate by appropriate band-pass filters – it is the excitation wavelengths which must be deliverable by a laser if a flow cytometer is to be used. For ordinary fluorometry, the latter restriction is, of course, not applicable. Another advantage of BCECF and other similarly fluorescein-based probes is that it can easily be used in conjunction with indo-1 so that pH and Ca^{2+} changes can be observed simultaneously (Davies *et al.*, 1990).

23.7.3   Intracellular cations – Na^+ and K^+

The measurement of cytoplasmic Na^+ changes has been attempted, both in the flow cytometer and in the fluorescence microscope. While a probe for each of these already exists (SBFI and PBFI), improved versions are needed which have a better level of selectivity. The selectivity of SBFI and PBFI *in situ* appears to be only a factor of 2, which makes SBFI, for example, difficult to use in a 120 mM K^+, 25 mM Na^+ environment like the cytoplasm of a cell. A 100-fold higher selectivity for one ion over another (i.e. a difference of two orders of magnitude in the dissociation constant K_d of the selected ion from the probe) is desirable in a 'good' indicator. Such a difference is present, for example, in mag-fura-2 which has a 100-fold lower K_d for Mg^{2+} than for Ca^{2+}. In contrast, the two-fold difference in SBFI affinity and Na^+ and K^+ means that a cytoplasmic change in Na^+ would be difficult to detect in a mammalian cell whose $[K^+]$ is approximately 120 nM. Although PBFI could, in principle, be used to detect a change in that $[K^+]$, the high concentration of probe which would be required is likely, as described above, to abnormally lower the cytoplasmic pH due to the number of protons released as PBFI-AM is lysed by cytoplasmic esterases.

23.7.4   Transmembrane potentials

Any living mammalian cell retains its functional integrity by maintaining an electrical potential, the transmembrane potential, between the interior and the exterior of the cell. If there are organelles, there is generally also a potential across the organellar membranes. The retention of the overall transmembrane potential requires energy, in the form of ATP, since the major portion is attributable to the transmembrane gradients of Na^+ and K^+, usually maintained by an ATP-driven Na^+/K^+ ATPase in the membrane. For that reason, poisoning of the Na^+/K^+ ATPase with compounds such as ouabain, or collapse of the ionic gradients with the appropriate ionophores, profoundly perturbs the transmembrane potential, although, as shown below, the portion of that potential attributable to the K^+ and Na^+ gradients, respectively, appears to differ for each cell.

In most cells, all or part of the membrane potential change observed in response to stimulation is attributable to the opening of either an antiport (e.g. the Na^+/H^+ antiport as described for platelets initially by Horne and Simons (1978) and found to be a counter-transport by Davies *et al.*, (1987b) or a Na^+ or K^+ involving channel (e.g. the hyperpolarization in neutrophils) (Korchak & Weissmann, 1978, Lazzari *et al.*, 1986, 1990). Although, in general, anions permeate the plasma membrane freely and therefore establish a gradient based purely upon the existent Donnan potential, that distribution, especially for Cl^- and HCO_3^- ions, also contributes to the overall potential, the extent of contribution being cell- and cell-environment-dependent (Wright & Diamond, 1977; Korchak *et al.*, 1982; Simchowitz, 1988; Lazzari *et al.*, 1990). Similarly, each cell type's internal organelles vary, as does the extent to which the potential across their membranes contributes to the overall potential of the cell; but these contributions must be considered. For example, it has been estimated (Sims *et al.*, 1974; Waggoner, 1988) that mitochondria can account for >50% of the overall potential, but, unless the organelles can be isolated (DelBuono *et al.*, 1989) or are absent, as in erythrocytes, the individual organelles' contributions cannot usually be evaluated except in cells large enough for microelectrodes to be inserted reliably into the desired location.

The thermodynamic parameters contributing to the overall transmembrane potential include the chemical potentials attributable to each individual ion's concentration gradient across internal as well as plasma membranes, to the osmotic pressure across these membranes, and to the Donnan equilibrium (Hoffman & Laris, 1974; Freedman & Laris, 1988; Simons *et al.*, 1988; Gallin & McKinney, 1988b). It has been shown that, overall, the Goldman constant field equation, also known as the Hodgkin-Katz equation, applies to these cells:

$$Y = (RT/F) \ln \frac{\{P_K K^+_{\ out} + P_{Na}Na^+_{\ out} + P_{Cl}Cl^-_{\ in}\}}{\{P_K K^+_{in} + P_{Na}Na^+_{in} + P_{Cl}Cl^-_{out}\}}$$

where V is the transmembrane potential and P represents the membrane permeability of the specific ion named. In general, all other ionic contributions, including those of protons, can be neglected when dealing with the overall potential of a cell.

Since membrane potentials reflect the overall charge difference between the inside and the outside of a cell, all membrane potential probes involve the measurement of their relative distribution and, of necessity, reflect the overall potential which includes not only the transplasma membrane but also transorganellar membrane potentials.

As discussed earlier in this chapter, two types of distributive fluorescent membrane potential probes exist, those whose contribution to fluorescence is intracellular and those whose fluorescence intracellularly is quenched, so that only the extracellular probe is detectable. Clearly one can use either type in fluorometers, but only the intracellularly fluorescing probe in flow cytometers.

23.7.5 Cyanines

The dithio, dioxa or dicarboxy series of the cyanines are highly conjugated probes whose properties depend upon the number of carbons in the conjugated chain and upon the length of the N-attached C tails. They were first described by Waggoner and his colleagues (Sims *et al.*, 1974; Waggoner, 1988). All are lipophilic quarternary ammonium salts which distribute between the external buffer and the cell according to its net negative charge. Some, like the dithio series, have a tendency to self-associate in the cell membranes, thereby quenching the probe fluorescence; the sole fluorescence which can be observed, therefore, is that in the extracellular buffer, rendering this probe useless for flow cytometry. In contrast, other cyanines such as the dioxa series do not exhibit a self-associating tendency; since they tend to have a very low quantum yield in aqueous buffers, the sole fluorescence which can be observed is that emitted by the intracellular probe. The dioxa cyanines ($diOC_X(Y)$) with a varying number of carbon atoms (Y) in the conjugated link between the rings, and various numbers of carbon atoms (X) attached to the two heterocyclic rings, have been used successfully in flow cytometry by a number of investigators (Seligman *et al.*, 1982, 1984; Lazzari *et al.*, 1986, 1990; Bernardo *et al.*, 1990; Seligman, 1990; Brunkhorst *et al.*, 1991).

23.7.6 Oxonols

Since most oxonols are themselves negatively charged, the reasons why their distribution reflects the overall transmembrane potential remain unclear. They have relatively low Stokes's shifts, but have nevertheless been used for membrane potential measurements in cell suspensions (Bashford *et al.*, 1985), and in flow cytometric measurements (Lazzari *et al.*, 1990).

In published cases in which several different probes have been used to evaluate membrane potentials and their changes (Horne *et al.*, 1981; Seligman & Gallin, 1983; Lazzari *et al.*, 1990), the results have been identical. Thus, when used correctly, any of the above probes yields the same information.

23.7.7 Phospholipase activation

Although it has long been known that stimulation of secretory cells activates the phospholipid metabolic pathways, it has only been possible to follow the kinetics of activation of phospholipases A2, C and D by isolation and identification (by HPLC or TLC) of the phospholytic products, at time intervals from milliseconds to seconds and minutes after exposure of the cells to a stimulus.

Ca^{2+} availability is usually required by these enzymes, but the question of whether the Ca^{2+} comes from extra- or intracellular stores, and how high a concentration of Ca^{2+} must be available before the phospholipase is active, had not been resolved because simultaneous studies had not been possible.

Fluorometric techniques for following phospholipase activation have been tried, but largely unsuccessfully. Although fluorescently tagged phospholipids, lysophospholipids and fatty acids have been available for a number of years, and can be incorporated into cell membranes, a resultant fluorescence-detectable susceptibility to phospholipase hydrolysis had not been achieved. Recently, however, we reported our success with a new probe, bis-Bodipy-phosphatidylcholine, developed in collaboration with Molecular Probes, which can be incorporated into neutrophil (and platelet) plasma membranes (Meshulam *et al.*, 1991). Because the phospholipid carries a Bodipy group (a fluorescein derivative) in both the 1 and 2 positions, there is total quenching in the resting cell membrane and the probe is not fluorescent. Stimulation activates one or more of the aforementioned phospholipases. If the activated enzyme is phospholipase A1 or A2 (PLA1, PLA2), one of the Bodipy-labelled fatty acids will by lysed off, the quenching will be relieved, and both the lysophospholipid remaining in the membrane and the fatty acid (which is usually released into the exterior buffer or into the cytoplasm, depending on the membrane leaflet in which the phospholipidolysis occurs) wil be fluorescent. The appearance of this fluorescence can then be followed continuously.

For our experiments, this was done by flow cytometry since we are interested in responding subpopulations. The results permitted us to resolve the Ca^{2+} vs. activation of PLA question: when neutrophils are stimulated with immune complexes, the maximal Ca^{2+} transient is over and redistribution within the neutrophil is well under way before PLA activation can be detected, at approximately 20 s. Of perhaps equal importance is the ability to distinguish between different mechanisms of cell activation: in contrast to immune complexes, which activate neutrophil PLA, chemotactic peptides do not activate this enzyme at all within the first 2.5 min after exposure to neutrophils.

Thus the activation of at least one of the pathways of phospholipid metabolism is now accessible to flow cytometric measurements. Probes for the other phospholipases of interest are currently under development.

23.7.8 Oxidative burst

Among the consequences of phagocyte stimulation is the liberation of oxidative products, via activation of an NADPH-dependent oxidase some of whose components, in the resting cell, reside in the azurophilic granule membrane (Borregaard *et al.*, 1983), others in the cytoplasm (Curnutte *et al.*, 1987a,b) and the remainder in the plasma membrane. Activation of the phagocyte leads eventually to degranulation, a process of exocytosis in which granule membranes fuse with the plasma membrane. The consequent assembly of the oxidase, as well as release of the granules' superoxide dismutase and myeloperoxidase, leads to an event called the oxidative burst, the release of O_2^-, O_2^{-2}, $O\cdot$, $OH\cdot$ and, if Cl^- is present OCl^-. The release occurs preferentially into the phagocytic vacuole, but also into the extracellular milieu. Quantitation of the oxidative burst, in terms of the rate or extent of release of superoxide, is usually accomplished by following the difference in absorbance when oxidized cytochrome C is reduced by the activated neutrophil released superoxide (Whitin *et al.*, 1980, 1981).

Chemiluminescence has also been used. However, neither technique is adaptable to fluorescence spectroscopy. Bass and his colleagues were the first to adapt oxidizable hydrofluorophores to measurements of the rate of appearance of products of activated phagocytes' oxidative burst in a flow cytometer (Bass *et al.*, 1983, 1986). The membrane-permeant non-fluorescent ester of their indicator, dihydrodichlorofluorescein (originally called dichlorofluorescin) remained in the cytoplasm of the resting neutrophils as a non-fluorescent free acid after esterolysis. Upon stimulation of the phagocytes, and consequent initiation of the oxidative burst, the products are released into the extracellular buffer and into the phagocytic vacuoles. A certain proportion of these released products diffuses back into the cytoplasm of the cells and oxidizes the dihydrochlorofluorescein. The resultant increase in fluorescence can be followed on a fluorometer or on a flow cytometer. However, the sensitivity of the assay was limited by the time required for rediffusion from the extracellular buffer or from the phagovacuole into the cell cytoplasm, by the extent of that diffusion, and by the extent to which catalase and peroxidase destroyed the oxidative burst products before they could react with the dihydrochlorofluorescein.

To overcome this problem, the oxidizable fluorophore has now been attached directly to the agonist (Ryan *et al.*, 1990). Dihydrodichlorofluorescein-labelled immune complexes and similarly labelled chemotactic peptides are now commercially available (Molecular Probes). We have shown that, for immune complex-stimulated neutrophils, a much larger fluorescence increase is detectable much sooner after exposure of the neutrophils to the stimulus than for the cytoplasmic probe (Ryan *et al.*, 1990). This is true largely because most of the oxidative products are released directly into the phagocytic vacuole, where they are not as subject to dilution as in extracellular buffer, and the concentration of oxidants is therefore much higher very near the reduced probe. For chemotactic peptides, which elicit a much smaller oxidative burst, detection is nevertheless possible without resorting to the addition of cytochalasin B, the agent usually added when fMLP-induced oxidative burst is measured by any of the other techniques. Furthermore, since there is always a small amount of already oxidized material in the preparations, it is possible to measure agonist binding as well as oxidative product in the same experiment, with the same photomultiplier.

These probes, as well as their more recently available dihydrorhodamine counterparts, should be applicable to fluorescence microscopy as well as flow cytometry. Furthermore, the same oxidizable groups have been linked to particles such as zymosan or bacterial fragments which, when opsonized, can stimulate phagocytes via the Fc or the C3b receptors; the particles are then phagocytized, and, in response to the oxidative burst, exhibit an increase in fluorescence proportional to the superoxide and/or peroxide release.

23.7.9 Phagocytosis

For flow cytometry, phagocytosis is most readily demonstrated by the use of fluorophore-labelled particles such as beads, latex particles, zymosan, LPS, bacterial fragments, etc. The same fluorophores described above as antibody and ligand labels can be employed, though coupling is sometimes best-achieved by using the succinimidyl ester rather than the thiocyanate or the anhydride of the fluorophore. In order to be ingestible or recognized by phagocytic cells, the labelled particles may need to be opsonized, a technique which generally does not perturb the fluorescence. The fluorescence of agonist which is too well bound to the surface to be washed off, yet is not phagocytosed, can be quenched with an antibody to the fluorophore.

23.8 NON-KINETIC APPLICATIONS OF RELEVANCE TO MULTIPARAMETER FLOW KINETIC CORRELATIONS

It is not possible here to discuss all of the measurements which can be accomplished by flow cytometry. The instrument was originally developed for cell type identification, detection of fluorescently labelled markers on the plasma membrane surface of the cells, and sorting of the cells according to these markers. Similarly, the use of flow cytometers to study cell cycle stages and ploidy will not be covered here. Rather, only static or equilibrium measurements will be discussed here and those that happen very rapidly and that are associated with kinetic functions, such as receptor occupancy.

23.8.1 Receptor occupancy – surface proteins and their ligands

Since the flow cytometer was originally designed to recognize and, eventually, to sort cells according to their surface 'markers', i.e. the receptors and binding proteins which are exposed on the outer surface of the membrane, the number of possible labels is already too large to be reviewed here *in toto* and increases almost daily.

In some cases the fluorophore is bound directly to the ligand, in others to a primary antibody to the protein being recognized, in still other situations, especially where magnification of the fluorescent signal is necessary, the fluorophore is bound to the second antibody. For flow cytometry, the usual fluorophores for this purpose are excited at 488 nm and are derivatives of fluorescein, of rhodamine, or of the phycobiliproteins. In some cases the fluorophore is designed for internal energy transfer so that an emission at long wavelengths can be achieved. The lymphocyte literature is particularly rich in usage of cell surface labels.

23.8.2 Nucleic acids

While there are many nucleic acid probes used routinely to determine the ploidy or the phase of a cell, virtually all require fixation and permeabilization of the cells. Since this makes them incompatible with cell function measurements, the propidium and ethidium derivatives will not be discussed here. In contrast, although others are in the planning stages, the only probe which appears to have been used to measure the DNA content of a living and functioning cell is Hoechst 33342, a probe which binds in the minor groove of DNA, which is excited at 357 nm and fluoresces at 485 nm. The AT-selective DAPI probe is also double-stranded DNA-specific, and exhibits similar fluorescence characteristics to the Hoechst probe, albeit with a higher intensity and a somewhat greater photostability. It is, however, less readily inserted into living mammalian cells, and has been used mainly in fungi and prokaryotes. The Hoechst 33342 excitation and emission wavelengths do not interfere with those of fluorescein, rhodamine or phycoerythrin derivatives, so that simultaneous obser-

vations of cell cycle and existent surface-exposed proteins can be performed.

23.9 CONCLUSION

It should be clear from the above that this chapter presents only the beginnings of the possible applications of flow cytometry to kinetic measurements of cell function. As more probes become available and as the ability to design non-interference between the probes increases, the number of possible applications of multiparameter flow kinetic techniques to not only identify cells but follow their functional responses while monitoring the receptor occupancy, the type of cell, the size, etc., will increase. The limitations arise from our ability as researchers.

ACKNOWLEDGEMENTS

I thank all of the members of my laboratory group who, over the years and now, have done the research and accumulated the data on much of which this review relies. I am also grateful to the National Institute of Health, whose grants HL15335, HL07501, HL19717, HL33565 and AM31056 supported this laboratory's studies.

REFERENCES

Bashford C.L., Alder G.M., Gray M.A., Micklem K.J., Taylor C.C., Turek P.J. & Pasternak C.A. (1985) *J. Cell Physiol.* **123**, 326–336.

Bass D.A., DeChatelet L.R., Szejda P., Seeds M.C. & Thomas M. (1983) *J. Immunol.* **130**, 1910–1917.

Bass D.A., Olbatantz P., Szejda P., Seeds M.C. & McCall C.E. (1986) *J. Immunol.* **136**, 860–866.

Beard C.J., Key L., Newburger P.E., Ezekowitz R.A.B., Arceci R., Miller B., Proto P., Ryan T., Anast C. & Simons E.R. (1986) *J. Lab. Clin. Med.* **198**, 498–505.

Bernardo J., Newburger P.E., Brennan L., Brink H.F., Bresnick S.A., Weil G. & Simons E.R. (1990) *J. Leuk. Biol.* **47**, 265–274.

Borregaard N., Heiple J.M., Simons E.R. & Clark R.A. (1983) *J. Cell Biol.* **97**, 52–61.

Brunkhorst B., Lazzari K.G., Strohmeier G., Weil G. & Simons E.R. (1991) *J. Biol. Chem.* **266**, 13035–13043.

Curnutte J.T., Kuver R. & Babior B.M. (1987a) *J. Biol. Chem.* **262**, 6450–6452.

Curnutte J.T., Scott P.J. & Babior B.M. (1987b) *J. Clin. Invest.* **83**, 1236–1240.

Davies T.A., Dunn J.M. & Simons E.R. (1987a) *Anal. Biochem.* **167**, 118–123.

308   E.R. Simons

Davies T.A., Katona E., Vasilescu V., Cragoe E.J. Jr. & Simons E.R. (1987b) *Biochim. Biophys. Acta* **903**, 381–387.

Davies T.A., Drotts D., Weil G.J. & Simons E.R. (1988) *Cytometry* **9**, 138–142.

Davies T.A., Drotts D.L., Weil G.J. & Simons E.R. (1989) *J. Biol. Chem.* **264**, 19600–19606.

Davies T.A., Zabe N., Weil G.J., Drotts D. & Simons E.R. (1990) *J. Biol. Chem.* **265**, 11522–11526.

Deamer D.W., Prince R.C. & Crofts A.R. (1972) *Biochim. Biophys. Acta* **274**, 323.

DelBuono B.J., Luscinskas F.W. & Simons E.R. (1989) *J. Cell Physiol.* **141**, 636–644.

Freedman J.C. & Laris P.C. (1988) In *Spectroscopic Membrane Probes*, L. Loew (ed.). CRC Press, Boca Raton, FL, pp. 1–49.

Freedman J.C. & Novac T.S. (1983) *J. Membr. Biol.* **72**, 59–74.

Gallin E.K. & McKinney L.C. (1988a) *J. Membr. Biol.* **103**, 55–66.

Gallin E.K. & McKinney L.C. (1988b) In *Cell Physiology of Blood*, R.B. Gunn & J.C. Parker (eds). Rockefeller University Press, New York, pp. 315–332.

Gear A.R.L. (1980) *J. Lab. Clin. Med.* **100**, 866–886.

Gear A.R.L. & Burke D. (1982) *Blood* **60**, 1231–1234.

Greenberg-Sepersky S.M. & Simons E.R. (1984) *J. Biol. Chem.* **259**, 1502–1508.

Grynkiewicz G., Poenie M. & Tsien R.Y. (1985) *J. Biol. Chem.* **260**, 3340–3350.

Haugland C. (1989) *Handbook of Fluorescent Probes and Research Chemicals*. Molecular Probes Inc., Eugene OR.

Hoffman J.F. & Laris P.C. (1974) *J. Physiol.* **239**, 519–552.

Horne W.C. & Simons E.R. (1978) *Blood* **51**, 741–749.

Horne W.C. & Simons E.R. (1979) *Thromb. Res.* **13**, 599–607.

Horne W.C., Norman N.E., Schwartz D.B. & Simons E.R. (1981) *Eur. J. Biochem.* **120**, 295–302.

Jones G.D., Carty D.J., Freas D.L., Spears J.T. & Gear A.R.L. (1989) *Biochem. J.* **262**, 611.

Kelley K. (1989) *Cytometry* **10**, 796.

Korchak H.M. & Weissmann G. (1978) *Proc. Natl. Acad. Sci. USA* **75**, 3818–3822.

Korchak H.M. & Weissmann G. (1980) *Biochim. Biophys. Acta* **601**, 180–194.

Korchak H.M., Eisenstat B.A., Smolen J.E., Rutherford L.E., Dunham P.B. & Weissmann G. (1982) *J. Biol. Chem.* **257**, 6919–6922.

Lazzari K., Proto P.J. & Simons E.R. (1986) *J. Biol. Chem.* **261**, 9710–9713.

Lazzari K.G., Proto P. & Simons E.R. (1990) *J. Biol. Chem.* **265**, 10959–10967.

Luscinskas F.W., Mark D.E., Brunkhorst B., Lionetti F.J., Cragoe E.J. Jr. & Simons E.R. (1988) *J. Cell. Physiol.* **134**, 211–219.

Meshulam T., Herscovitz H., Casavant D., Bernbardo J., Roman R., Haugland R.P., Diamond R.D. & Simons E.R. (1991) *J. Cell Biol.* **115**, 361a.

Nuccitelli R. & Deamer D. (1982) *Intracellular pH: Its Measurement, Regulation and Utilization in Cellular Functions*. A.R. Liss, New York.

Pollock W.R. & Rink T.J. (1986) *Biochem. Biophys. Res. Commun.* **139**, 308–314.

Pollock R.W., Sage S.D. & Rink T.J. (1987) *FEBS Lett.* **10**, 132–136.

Rink T.J. & Sage S.D. (1987) *Physiol. Soc.* **369**, 115.

Ryan T.C., Weil G.J., Newburger P.E., Haugland R.P. & Simons E.R. (1990) *J. Immunol. Methods* **130**, 223–233.

Sage S.D. & Rink T.J. (1986) *Biochem. Biophys. Res. Commun.* **139**, 1124–1129.

Sage S.D. & Rink T.J. (1987) *J. Biol. Chem.* **262**, 16364–16360.

Seligman B.E. (1990) *Curr. Topics Membranes Transport* **35**, 103–125.

Seligman B.E. & Gallin J.I. (1983) *J. Cell. Physiol.* **115**, 105–115.

Seligman B.E., Fletcher M.P. & Gallin J.I. (1982) *J. Biol. Chem.* **257**, 6280–6286.

Seligman B.E., Chused T.M. & Gallin J.I. (1984) *J. Immunol.* **133**, 2641–2646.

Siffert W., Fox G., Muckenhoff K. & Scheid P. (1984) *FEBS Lett.* **172**, 272.

Siffert W., Siffert G. & Scheid P. (1987a) *Biochem. J.* **208**, 9.

Siffert W., Siffert G., Scheid P., Reimens T. & Ackerman J.W.N. (1987b) *FEBS Lett.* **212**, 123–126.

Simchowitz L. (1988) In *Cell Physiology of Blood*, R.B. Gunn & J.C. Parker (eds). The Rockefeller University Press, New York, pp. 194–208.

Simons E.R. (1988) In *Energetics of Secretion Responses* Vol. 1, J.W.N. Akkerman (ed.). CRC Press, Boca Raton, FL, pp. 105–120.

Simons E.R. & Greenberg-Sepersky S.M. (1987) In *Platelet Responses and Metabolism* Vol. III, H. Holmsen (ed.). CRC Press, Boca Raton, FL, pp. 31–49.

Simons E.R., Schwartz D.B. & Norman N.E. (1982) In *Intracellular pH: Its Measurement, Regulation and Utilization in Cellular Functions*, R. Nuccitelli & D. Deamer (eds). A.R. Liss, New York.

Simons E.R., Davies T.A., Greenberg S.M., Dunn J.M. & Horne W.C. (1988) In *Cell Physiology of Blood*, R.B. Gunn & J.C. Parker (eds). Rockefeller University Press, New York, pp. 266–279.

Sims P.J., Waggoner A.S., Wang C.H. & Hoffman J. (1974) *Biochemistry* **13**, 3315–3329.

Sullivan R., Griffin J.D., Fredette J.P., Leavitt J.L. & Simons E.R. (1987b) *J. Immunol.* **139**, 3422–3430.

Sullivan R., Melnick D.A., Malech H., Meshulam T., Simons E.R., Lazzari K.G., Proto P., Gadenne A.-S., Leavitt J.L. & Griffin J.D. (1987a) *J. Biol. Chem.* **262**, 1274–1281.

Sullivan R., Fredette J.P., Leavitt J.L., Gadene A.-S., Griffin J.D. & Simons E.R. (1989) *J. Cell. Physiol.* **139**, 361–369.

Thomas J.A., Buchsbaum R.N., Zimniak A. & Racker E. (1979) *Biochemistry* **18**, 2210–2218.

Tsien R.Y. (1980) *Biochemistry* **19**, 2396–2404.

Waggoner A.S. (1979) *Ann. Rev. Biophys. Bioengng* **9**, 47–68.

Waggoner A.S. (1988) In *Cell Physiology of Blood*, R.B. Gunn & J.C. Parker (eds). The Rockefeller University Press, New York, pp. 210–215.

Waggoner A.S. (1990) In *Flow Cytometry and Sorting*, M.R. Malamed, T. Lindow & M.L. Mendelsohn (eds). Wiley Liss, New York, pp. 209–225.

Whitin J.C., Chapman C.E., Simons E.R., Chovaniec M.E.

& Cohen J. (1980) *J. Biol. Chem.* **255**, 1874–1878.

Whitin J.C., Clark R.A., Simons E.R. & Cohen H.J. (1981) *J. Biol. Chem.* **256**, 8904–8906.

Wright E.M. & Diamond J.M. (1977) *Physiol. Rev.* **57**, 109–156.

Zavoico G.B. & Cragoe E.J. Jr. (1988) *J. Biol. Chem.* **263**, 9635–9639.

Zavoico G.B., Cragoe E.J. Jr. & Feinstein M.B. (1986) *J. Biol. Chem.* **261**, 13160.

CHAPTER TWENTY-FOUR

Flow Cytometric Analysis and Sorting of Viable Cells

JAN F. KEIJ[1] & HANS HERWEIJER[2]

[1] Department of Molecular Pathology, Institute of Applied Radiobiology and Immunology/TNO, Rijswijk, The Netherlands
[2] Department of Medical Oncology, Daniel den Hoed Cancer Center, Rotterdam, The Netherlands

24.1 INTRODUCTION

Fluorescence emitted from specific probes allows for the sensitive analysis of a great variety of cellular molecules and functions. Several fluorescent probes are presented in detail in this volume. In practice, three methods are used to detect the emitted fluorescence: (1) fluorometry, (2) fluorescence microscopy, and (3) flow cytometry.

A flow cytometer enables multiparameter measurements of individual cells at analysis rates of several thousands of cells per second. In general, 3–16 parameters of each cell can be measured simultaneously, depending on the complexity of the instrument. This makes flow cytometry ideally suited for the analysis of heterogeneous cell populations. Also, the high throughput rates allow for the accurate identification of rare cells within reasonable measurement times. An added feature of some flow cytometers is their capability to sort analysed cells into highly purified populations. The flow sorted cells can then be used for further biological or morphological studies.

Compared to fluorometry, which involves bulk measurements, the main advantage of flow cytometry is the possibility to analyse single cells. Also, a flow cytometer is capable of more sensitive measurements, as only the cell-bound fluorescence is measured. Flow cytometers are superior to fluorometers and fluorescence microscopes in their ability to measure several parameters per cell simultaneously. When compared to fluorescence microscopes, flow cytometers are the instrument of choice when it comes to high analysis rates. However, flow cytometers do not allow repeated analysis of a specific cell over time, unlike fluorescence microscopes.

In this chapter, we give a general description of the instrumentation and software needed for the analysis and sorting of viable cells. A selection of hardware modifications for special-purpose analyses is presented. Two example applications from our laboratory illustrate the possibilities of flow cytometry.

24.2 FLOW CYTOMETER HARDWARE

24.2.1 General description

In a flow cytometer, cells flow through a focused light beam in a single file (Fig. 24.1). Light scattered by the cells yields information on cell size and morphology. Fluorescence, from cellular pigments or added probes, can be measured simultaneously, for qualitative and

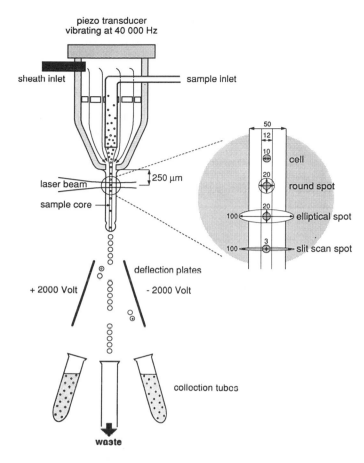

Figure 24.1 Schematic representation of the fluidics of a flow sorter. As the sample flows into the conical nozzle, it is accelerated and aligned in the centre of the fast-flowing sheath fluid. About 250 μm below the nozzle exit, the sample is illuminated by a focused laser light beam. Several spot shapes are depicted (scale in micrometres) that are discussed in the text. Cell sorting is accomplished by deflecting cell-containing droplets by electrically charged plates. Droplets are formed by vibrating the nozzle holder at a fixed frequency.

quantitative analyses. Depending on the complexity of the instrument, 3–16 parameters can be collected per cell. After the analogue signals are digitized, they are transferred to a data storage medium for future analysis.

Flow sorters are flow cytometers that can sort cells into purified populations by deflecting electrically charged droplets containing the cells of choice (Fig. 24.1). Modern flow sorters can reliably sort cells (under sterile conditions) at cell rates of up to 5000 cells s^{-1}, at purities greater than 97%. This provides a means to correlate flow cytometric measurements of specific cell types with other (biological) assays. The main advantages of flow cytometric analysis, speed of analysis, sensitivity and reproducibility, allow for rapid detection of subpopulations and rare cells. Due to these features, flow cytometry is predominantly applied in the fields of haematology, immunology and cancer research. With the availability of new user-friendly instruments, flow cytometry has become a routine technique in the clinical environment.

24.2.2 Flow system and sample delivery

For accurate measurements, it is essential that the sample remains optimally aligned with respect to the illumination source and the light collecting detectors. For this purpose the sample is introduced into a fast flowing sheath fluid in a conical nozzle tip (Fig. 24.1). Laminar flow of the sample and sheath fluids in the emerging fluid jet fixes the sample position to its centre (hydrodynamic focusing). Fluid jet velocities of 10 m s^{-1} are common and are achieved by a slight overpressure ($\approx$ 0.5 atm) on the sheath fluid container. Nozzle tips are available with round orifices ranging from 50 to 400 μm in diameter, allowing the analysis of almost all cell types and unicellular organisms. Some flow cytometers are equipped with Coulter orifices for simultaneous cell volume measurements.

Sheath fluid, usually saline or phosphate-buffered saline, is filtered before use. Filtration through 0.45 or 0.22 μm filters is usually sufficient to eliminate small light-scattering debris. The clean sheath fluid is delivered to the nozzle tip from the storage container by positive pressure from an air or nitrogen cylinder. For cell sorting, it is essential that the flow of the sheath fluid is stable. Stability of the sheath fluid velocity is ensured by flow resistors in the sheath fluid tubing. The entire flow system can be sterilized by autoclaving or by rinsing with diluted bleach, 70% ethanol, or formalin. An in-line sterile 0.45 μm filter is included in the sheath tubing to further minimize the chance of infectious particles and/or light-scattering debris entering the cell sorter fluidics. Considering the openness of the system, these precautions lead to a surprisingly low incidence of infection of the sorted cells, when they are cultured.

Samples are either introduced into the nozzle by a mechanically driven syringe or by positive air pressure. For sorting purposes the positive pressure option is preferred, as it allows stirring and regulation of the temperature of the sample. Facilities for sample mixing and temperature control are available on most instruments.

24.2.3 Sample illumination

Argon ion lasers are used primarily as a light source for probe excitation. They offer a stable output of sufficient power and are easy to focus onto the sample with a single lens. Also, selection of the excitation wavelength is straightforward. The polarized laser light beam is focused onto the fluid jet forming a round spot of 20–50 μm in diameter (Fig. 24.1). Due to the Gaussian intensity distribution of the laser light spot, slight deviations in the position of the sample can result in significant deviations in the measured fluorescence intensities. Applications that require a high degree of accuracy, such as DNA-ploidy analyses, have led to the use of cross-cylindrical lenses. These lenses yield wider (elliptical) spots. Thus, the generated signal will be less sensitive to the position of the cells within the horizontal plane of the sample core (Fig. 24.1).

Addition of a second and third laser to the cytometer allows for multi-wavelength excitation, increasing the number of parameters that can be measured. The second laser is focused onto the fluid jet at a small distance from the first laser. Selective emission collection from each laser is accomplished by a combination of split mirrors and pinholes in the optical path (Fig. 24.2). If one of the lasers is a dye laser, then the selection of the excitation wavelength is only limited by the dye laser substrate used.

Small air-cooled lasers are now available for 325 nm and 441 nm (helium–cadmium He–Cd), 488 nm (argon ion, Ar) and 632 nm (helium–neon, He–Ne). These lasers are much cheaper than the water-cooled Ar lasers and they require no additional power and water-cooling provisions. Bigler (1987) obtained almost identical data quality when comparing a 10 mW He–Cd laser to a 100 mW Ar laser for the analysis of DNA histograms. Shapiro (1983) successfully used a He–Cd laser for ratiometric pH measurements with fluorescein derivatives. Loken et al., (1987) showed that a 20 mW He–Ne laser provided sufficient output power for immunological applications involving allophycocyanine (APC) labels. Air-cooled Ar lasers (488 nm) are now standard in several commercial instruments. These developments have led to cheaper and more user-friendly flow cytometers. Arc lamps, still used in some flow cytometers, have been almost completely abandoned, for they are less stable in several respects such as output power and alignment.

24.2.4 Light scatter and fluorescence collection

When cells enter the excitation beam, light is scattered in all directions. Light scattered at 3–15° with respect to the direction of the excitation beam (forward scatter, FSC) is collected by a photodiode. FSC yields information on cell size (Mullaney et al., 1969). Laser light passing directly through the cells and the fluid jet does not contain size-related information. Therefore, it is blocked by an obscuration bar in front of

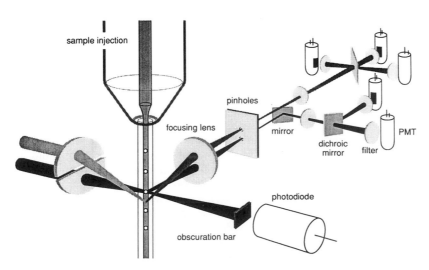

Figure 24.2 Schematic representation of the optics of a two-laser flow cytometer. Cells are illuminated by two laser beams, which are individually focused on the sample core at a small distance to each other. The laser light from the bottom laser is blocked by the forward scatter (FSC) obscuration bar; light passing the bar contains the FSC signal. Fluorescence emitted from the sample is focused by the focusing lens, separated by pinholes and focused onto the photomultipliers (PMTs) after passing through the appropriate mirrors and filters. The side scatter (SSC) obscuration bar was not drawn for clarity.

sample injection

pinholes

focusing lens

mirror

dichroic mirror

filter

PMT

photodiode

obscuration bar

the FSC detector, thus increasing the signal-to-noise ratio. Light scattered at a 90° (80–100°) angle to the direction of the beam (side scatter, SSC) is focused on a photomultiplier tube (PMT). A PMT is used as this signal is much weaker than the FSC. Again, an obscuration bar in front of the collection optics is used to block light scattered by the fluid jet. SSC yields information on cell morphology: granular and ruffled cells scatter more light than clear and round cells. Light hitting the SSC PMT is filtered by band-pass or short-pass filters to avoid co-collection of strong fluorescence signals. In single laser applications, a filter is not required in the FSC detection as this signal is much stronger than any other collected signal.

Bivariate plots of FSC and SSC are routinely used to differentiate between lymphocytes, monocytes, blasts and granulocytes in peripheral blood and bone marrow samples (Salzman et al., 1975; Visser et al., 1978). Dead and/or damaged cells can be distinguished from viable cells as they yield lower FSC and higher SSC signals. A probe for viability, such as propidium iodide (PI), used by Dangl and Herzenberg (1982), or Calcofluor White M2R (Berglund et al., 1987) can be added to distinguish living from dead cells more accurately.

Fluorescent light is focused on a PMT after passing through an appropriate optical filter (Fig. 24.2). When several probes with different emission spectra are excited simultaneously, dichroic mirrors are inserted in the optical path to separate the two emission signals. Dichroic mirrors are preferred to simple beam-splitters, as they allow for better spectral separation and higher light yields. For example, a flow cytometer equipped with an Ar laser emitting 488 nm light can be set up to measure five parameters. In the forward direction: FSC. In the perpendicular direction: SSC using a 488/10-BP filter, fluorescein isothiocyanate (FITC) conjugated to an antibody, through a 530/30-BP filter, phycoerythrin (PE) conjugated to a second antibody, through a 585/45-BP filter and finally PI (for the exclusion of dead cells) through a 620-LP filter. Despite the optical filtering, spectral overlap is a common problem in applications where FITC- and PE-conjugated antibodies are used as probes. Circuits for the correction of this spectral overlap are standard on most instruments.

When more probes are required or when spectral overlap of two probes used cannot be corrected, a second laser can be applied. After adding the essential pinholes and split mirrors to eliminate cross-talk of the emitted signals, it is possible to measure one or two additional probes. If the second laser is a He–Ne laser, it is possible to collect the emission bands of 660/30-BP for APC (conjugated antibody) or of 680/20-BP for the lyrically described membrane potential probe $DiIC_6(5)$ (Shapiro, 1981). A second Ar laser

emitting UV light at 351 nm and 363 nm allows for the collection of either DAPI fluorescence for DNA cell cycle analysis or the co-collection of the two emission bands from the dye indo-1 (intracellular Ca^{2+}) or from the intracellular pH probe DCH (Valet et al., 1981). The optical plankton analyser (OPA), which was developed in our laboratory, is equipped with three lasers. Using a He–Ne, a He–Cd and a water-cooled Ar laser, this instrument can identify algae in heterogeneous samples based on differential (endogenous) pigment distributions.

In addition to colour filters, polarizers have been used in special applications. Eosinophils can readily be identified from bivariate scatterplots of polarized versus depolarized SSC (Terstappen et al., 1988). Polarizers have also been added to the fluorescence collection set-up for membrane polarization studies using DPH (Fox & Delohery, 1987).

24.2.5 Signal processing

Before the electronics system starts processing the incoming pulses, it must be triggered. One parameter is selected to function as the trigger parameter. Usually this is the FSC, as all cells generate a strong FSC signal and it is also a convenient parameter to exclude debris and dead cells. After setting the trigger level, the system will only process events with a signal intensity higher than this threshold value. The signals from the detectors are amplified and processed individually. First, the PMT current is amplified and converted to a voltage signal in a pre-amplifier (pre-amp). A baseline restoration circuit subtracts any DC voltage from the AC signal, thus providing a stable reference value of zero volts.

Following pre-amplification, the converted signals are further amplified by either linear or logarithmic amplifiers. The amplified signals, ranging from 0 to 10 V, are processed through peak-sense-and-hold (PSH) circuits, which retain the peak heights for the analogue-to-digital conversion (ADC) of the data. The ADCs convert the peak values to digital data with 8–28 bit resolution, depending on the instrument. Digital data from each parameter generated from a single cell, are then sent to a computer for storage and future or on-line analysis.

The more sophisticated multi-laser flow sorters, processing up to 16 parameters, feature additional parameters such as (1) peak area integration, (2) peak width, and (3) the ratio of two parameters. Peak area integration is used when the (large) diameter of the measured cells leads to a non-linear response in peak height measurements. Peak width or time-of-flight (TOF) is used to detect doublets and clumps of cells (Fig. 24.3). The ratio circuit is used for probes such as indo-1, which yields cell-size-independent information from the ratio of two emission/excitation bands. The RELACS III (Van den Engh & Stokdijk, 1989) in our laboratory can also collect additional DC parameters. This feature allows the constant monitoring of sample pH and temperature and of time as a parameter for measuring kinetic processes.

24.2.6 Cell sorting

Most sorters employ droplet deflection to sort desired cells. Originally designed for graphic printing by Sweet (1965), electrostatic droplet charging and deflection in a flow cytometer was first used by Fulwyler (1965). In a flow sorter the fluid jet is broken into droplets by vibrating the nozzle tip assembly at a constant frequency. Two deflection plates, charged to around +2000 and −2000 V, are situated downstream (Fig. 24.1). Initially, the entire jet is maintained at 0 V, allowing the droplets to pass down between the charged plates. Two sort windows can be set up for simultaneous sorting of two populations. After each cell is analysed, the sorter electronics check whether the cell belongs to one of these two populations. If so, a sort pulse is programmed to charge the fluid jet at a preset delay time. This delay time is equal to the time it takes for the cell to travel from the laser-jet intersection to the break-off point of the fluid jet. By charging the fluid jet to either +100 or −100 V, the droplet breaking off from the jet will be charged. Travelling down between the two deflection plates, the droplet will move towards the oppositely charged plate. When cells are detected that do not fulfil the pre-set sorting criteria, the droplet is not charged and passes undeflected into a waste container. The sorted cells are collected into tubes, which can be cooled to retain viability.

Due to drift in the position of the break-off point during sorting, the correct timing of the charging pulse may be off significantly. Instead of the desired cell, this results in deflecting an empty droplet. By increasing the charging pulse length to several droplets (two or three) per sorted cell, the probability of deflecting the selected cell is increased. However, sorting one droplet per cell facilitates higher sort rates. Electronic circuits, which monitor the occurrence of coincident arrival of cells in the laser beam and in the deflection envelope, can be set to either increase sorting yield or sorting purity.

Although droplet sorters are most common, electromechanical fluid switch sorters are commercially available. Due to low sorting rates (500 cells s^{-1}), these sorters are of limited interest for general applications. Still, sorting of infectious material can only be done safely in such a closed system. Currently, a fluid switch sorter is under development in our institute for the sorting of very long (0.1–1 mm) and

fragile algae (Dubelaar et al., 1989). Because of their large size, most types of algae could never be sorted in a droplet sorter.

Two basically different sorters have been developed based on killing the undesired cells instead of collecting the desired cells. The first instrument was designed to electroporate undesired cells. Bakker Schut et al. (1990) showed that purities of 95% could be achieved at sort rates of 5000 cells s^{-1}. The second instrument, the ZAPPER (Herweijer et al., 1988), kills unwanted cells by exposing them to a second lethal laser beam. Purities of greater than 99% were achieved at sort rates of 30 000 cells s^{-1}.

24.2.7 Analysis and sorting speed

Speed is an important feature of flow cytometry. Cell analysis rates are limited by several aspects in the flow cytometer design. Dead time of the slowest electronic component determines the maximum achievable analysis rate. Considering a dead time of 50 μs, this would yield a maximum obtainable cell rate of 20 000 cells s^{-1}. Since cells are not evenly distributed throughout the sample core, high cell rates result in the coincident arrival of cells in the laser beam. Peak height analysis will no longer yield relevant information on the measured parameters of the individual cells. More realistic cell rates are therefore 2000–5000 cells s^{-1}. To overcome coincidence limitations on the analysis rate, high-speed flow cytometers have been built. An instrument described by Peters et al. (1985) employs high sheath fluid rates of 50 m s^{-1}, requiring 15 atm on the sheath container. Sorting is restricted to chromosomes, as the survival of sorted cells was shown to be significantly reduced.

When more than one laser is used, the cell rate and sort rate are further reduced by the time it takes for the measured cell to travel from the first to the last laser spot. As most instruments must completely process the measurement of each cell before a new one can be measured, this travel time adds about 40 μs to the dead time. A parallel data processing system was developed (Van den Engh & Stokdijk 1989), reducing the total dead time of the instrument to the time required for analogue-to-digital conversion of the data. The digital data are stored in a temporary first-in-first-out memory (FIFO) in each ADC. After the cell has passed through all laser spots, the collected data from the cell are transferred from the FIFOs to the storage computer. This parallel system, with an effective dead time of 15 μs, has been used in our laboratory for several years and formed the basis for the ZAPPER. A faster version (4 μs dead time) has already been shown to analyse cells at rates of 100 000 cells s^{-1} (Keij et al., 1991). When completed, the new ZAPPER will be equipped with a 257 nm laser for the killing of unwanted cells at this very high cell rate.

24.2.8 Calibration

Before quantitative measurements can be made, the electronics of the whole instrument must be calibrated. Normally, this is done by a skilled engineer. First, the individual components (pre-amp, amp, PSH, ADC) are checked. Second, the components are connected and checked as a whole. Third, the instrument is aligned with preset PMT voltages and amplifier gains and a known standard sample. After inspecting the optical filters for scratches, the PMT performance is checked at several PMT voltages to verify the linear range. For daily use, the instrument's performance can be verified with calibration beads containing known quantities of fluorochrome. These standard beads can also be used for intercomparison of experiments. For reproducible compensation of FITC and PE spectral overlap, standard bead mixtures are commercially available.

Whenever quantitative analysis of an intracellular probe is desired, the response of this probe must be calibrated. Procedures have been described for Ca^{2+} probes that involve establishing 0 and 100% values (Rabinovitch *et al.*, 1986). Dye-loaded cells in media with known Ca^{2+} concentrations, supplemented with detergents or ionophores, are used. Similar procedures have also been described for pH calibration (Cook & Fox, 1988).

Recently, a general, more convenient method was described for probes that yield quantitative information from the ratio of two emission bands (Kachel *et al.*, 1990). Calibrated buffer samples supplemented with probe were passed through the flow cytometer. By inserting a chopper in the excitation beam, the continuous fluorescence signals were made to pulsate. The electronics could now be triggered with these pulses, allowing easy measurement and calibration of the signals. Fluorescent drug accumulation can either be calibrated this way or by HPLC analysis of the extracted fluorescent drug. When drug accumulation may lead to fluorescence quenching, and thus to an underestimation of the concentration, HPLC analysis is preferred (Herweijer *et al.*, 1989).

24.3 SOFTWARE FOR DATA ACQUISITION AND ANALYSIS

24.3.1 Data acquisition

Collection of raw data from a flow cytometer is usually done in list mode, i.e. all measured parameter values from every cell are transferred to the storage computer. Transfer rates depend on the computer used and on the data-processing electronics of the flow cytometer. The relatively inexpensive microcomputers commonly used are the rate-limiting factor. Still, most flow cytometers allow for multiparameter data transfer of thousands of cells per second. Faster electronics are constantly being developed and implemented in dedicated and commercially built instruments.

Instead of storage of list mode data, single histograms and bivariate histograms can be stored from selected parameters. This is often done to reduce the amount of data stored. Debris and dead cells can be excluded from data acquisition by setting acquisition windows/gates containing only the desired events.

24.3.2 Data analysis

Data retrieved from the storage medium can be analysed either manually or automatically. The data analysis procedure is aimed at extracting information from a multiparameter space. It involves the identification and enumeration of subpopulations and the calculation of mean fluorescence intensities. In general, dead cells and clumps are first gated out on the basis of scatter, pulse width and PI or Hoechst 33258 measurements (Fig. 24.3). Then, major subpopulations are identified on the basis of scatter parameters. For example, a FSC versus SSC scatterplot can be used to select lymphocytes in a peripheral blood sample. Finally, the other measured parameters are analysed, e.g. antibody distributions, to identify subpopulations. For Ca^{2+}, pH or membrane fluidity probes, the ratio of the fluorescence values collected at two wavelengths is calculated. Output consists of scatterplots, contourplots, histograms, and statistics. As the number of collected parameters increases, analysis of the data becomes more complex. This has led to programs that display up to six scatterplots simultaneously and employ colours to distinguish the gated populations.

Several commercially available software packages provide extensive data analysis capabilities. Beside the analysis capabilities, most packages feature options for data transfer to other computer systems. Completely automated programs are available for DNA cell cycle analysis, reticulocyte enumeration and lymphocyte immunophenotyping.

24.4 SPECIAL-PURPOSE MODIFICATIONS

24.4.1 Hardware modifications

For real-time and kinetics measurements, timers are required. Timers for flow cytometers were first described by Martin & Swartzendruber (1980). They used a ramp generator, which produced a linear increase in voltage over a period of 10 min. Sacrificing one detection parameter, the output of this ramp generator was processed by the electronics and stored in the data storage computer as a parameter for elapsed time. On most flow cytometers, neither ramp generators nor timers are included. Still, kinetics measurements are possible by adding the internal clock time of the computer to the collected data set of each cell as it is transferred (Dive *et al.*, 1987). During the development of the RELACS III (Van den Engh & Stokdijk, 1989), a different approach was chosen. Besides eight AC parameters, this instrument can process four DC parameters. For real-time analysis, one DC parameter is used for pulses from a timer running at 10 Hz. The timer pulses are transferred to the storage computer at this rate. At a resolution of 24 bits the timer can run for 466 h. If a cell coincides with a pulse from the timer, the cell data are transferred first, followed by the timer data. During analysis, the parameter identification numbers, which are transferred to the computer along with the measured peak heights, are used to distinguish cell and timer data. This option has been used to measure anthracycline drug accumulation in human peripheral blood cells (Nooter *et al.*, 1990). The ZAPPER will be equipped with an identical timer, but in this instrument the timer is only sampled when cell data are transferred to the computer. Thus, each cell is tagged with the exact time of measurement.

Over the years, the optics of flow cytometers have remained fairly untouched, although two modifications are worth mentioning. Watson (1989) has described a modification that allows a 20-fold increase in the amount of collected fluorescence. Some commercial instruments now contain a similar modification. This modification should prove useful in the detection of dull signals. The second modification, slit-scan analysis, involves the measuring of cells passing through a beam focused to a spot much narrower in height than the diameter of the cell (Fig. 24.1). Instead of sampling peak height, 256 measuring points per cell are collected as it passes through the laser beam, creating a fluorescence profile of a cell. The total length of the pulse is a measure for the cell diameter. Such fluorescence profiles have been used to measure the ratio of the cellular and nuclear diameters in cytological specimens (Wheeless *et al.*, 1984).

Besides sorting cells into tubes for subsequent culturing, several other cell collection methods have been developed. A very elegant modification was made by Jongkind & Verkerk (1984), who collected cells into oil droplets that allowed enzyme analysis of single cells. For cell morphology studies, cells have

been sorted onto glass slides. The sheath fluid was made hypotonic to reduce cell shrinkage while drying (Alberti *et al.*, 1984). Cells have also been collected onto Millipore filters for subsequent DNA hybridization. After sorting cord blood leukocytes, Bianchi *et al.*, (1987) could determine the sex from as few as 50 newborn cells, using a Y-chromosome specific probe. Single cell deposition units, which are commercially available, allow sorting of a preset number of cells into 96-well plates. These units have been used by Jongkind and Verkerk (1984) for culturing sorted fibroblasts and by Visser *et al.* (1984) for culturing sorted pluripotent haemopoietic stem cells from mouse bone marrow.

24.4.2 Sample delivery modifications

For long-term real-time drug uptake experiments, the quality of the sample is critical with respect to viability and homogeneity. Therefore, the sample must be gently stirred. This assures (1) a constant flow of cells through the laser beam, (2) mixing of reagents added during the experiment, and (3) minimal clump formation. The temperature of the cells must also be regulated. This can be done by a thermostatted water jacket. If reagents are to be added during a run, an inlet should be available. Such sample delivery modules have been described by Omann *et al.* (1985) and Nooter *et al.* (1990). If it is important to analyse the effect of a reagent rapidly, then it is essential to reduce the sample delivery time to a minimum. Recently, a modification has been described by Kelley (1991) that shortens the delivery time of the sample to the measuring point to less than 1 s.

Automated sample delivery units based on 96-well plates are available for routine applications. The performance of these units can be programmed and they should prove valuable in experiments where reproducibility of sample delivery is of the utmost importance (Pennings *et al.*, 1987). Complete, automatic sample preparation and delivery units have recently become available from a commercial vendor.

24.5 APPLICATIONS

24.5.1 Real-time drug uptake analysis

Real-time cytotoxic drug uptake experiments with cell lines and human leukaemia cells have been performed routinely in our laboratory. These types of experiments make it possible to measure the cellular drug accumulation kinetics of various cell types in complex cell samples (peripheral blood). Comparison of different drugs and effects of additional reagents can be determined. An example of real-time analysis

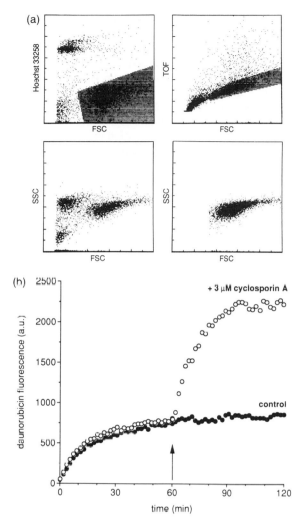

Figure 24.3 Daunorubicin uptake in drug-resistant human A2780 cells as a function of time. Cells were sampled at 2 min intervals for a total of 120 min. Cyclosporin A was added to the sample at $t = 60$ min, as is indicated by the arrow. In the control sample, medium was added at $t = 60$. The uptake curves were generated after the clumps and dead cells were excluded by setting gates using the FSC, TOF and Hoechst 33258 signals. The ungated and gated scatterplots are shown in the bottom left and bottom right panels, respectively. The gates are displayed in the top two panels.

of cytotoxic drug uptake is depicted in Fig. 24.3. In this experiment, human A2780 cells, expressing high levels of the drug-resistance-encoding *mdr1*-gene, were exposed to 2 μM of daunorubicin. This drug fluoresces brightly upon excitation with 488 nm light.

One reaction tube containing 10 ml of cell suspension (2×10^5 cells ml^{-1}) is used to start the measurement. First, a control level of fluorescence is measured, then the drug is injected into the sample ($t = 0$). The subsequent increase in drug fluorescence was continuously monitored for the next 60 min

(Herweijer *et al.*, 1989; Nooter *et al.*, 1990). At preset time intervals (every 2 min), the data of 2000 cells were stored to disk by the controlling computer. In this example, five parameters (FSC, SSC, TOF, Hoechst 33258 fluorescence, daunorubicin fluorescence) were collected for each cell. At $t = 60$, 3 μM cyclosporin A (CsA) was added to the sample, and the measurement was continued for another 60 min. In a second sample, medium was added at $t = 60$ as a control for the action of CsA.

Analysis was done with the kinetics module from the data analysis software. First, gates were set using TOF and Hoechst 33258 signals to exclude clumps and dead cells, respectively (Fig. 24.3). The next step in the analysis procedure involved the calculation of the mean daunorubicin fluorescence of the gated cells at each different time point. After completion of the analysis, these mean fluorescence levels are plotted versus time. From these curves (Fig. 24.3), the uptake kinetics of daunorubicin in A2780 cells can be clearly seen. Steady-state drug accumulation is reached after about 60 min. The addition of CsA led to a rapid increase in daunorubicin accumulation. This is caused by an inhibition of the drug efflux pump, i.e. the cell membrane component known as P-glycoprotein that is encoded by the *mdr1*-gene.

Analyses as described above, will lead to a better insight into the transport of drug molecules into and out of the cell. In the specific case of multidrug resistance, it allowed us to demonstrate active transport of daunorubicin out of fresh human leukaemia cells (Herweijer *et al.*, 1990). This provided a model for the phenomenon of drug resistance during treatment of patients. The method also proved valuable in the study of modulators of P-glycoprotein. The most effective of these modulators (CsA) is now evaluated in clinical trials.

24.5.2 Multiparameter analysis of haemopoietic cells

Flow cytometers have been used extensively in studies of the haemopoietic system. In our laboratory, flow cytometric procedures have been developed to isolate primitive haemopoietic cells (Visser *et al.*, 1984; De Vries, 1988). The current method involves sorting of cells in the lymphoid-blast region that display a combination of cell surface markers (Fig. 24.4). The sorted population is subsequently stained with rhodamine 123, which is routinely used as a mitochondrial marker. When used as the final step in the purification procedure, sorting of cells emitting low rhodamine 123 fluorescence yields highly purified primitive cell populations.

However, in sorting experiments the identification of the bone marrow subpopulations by FSC–SSC scatterplots can be less than optimal. This can be

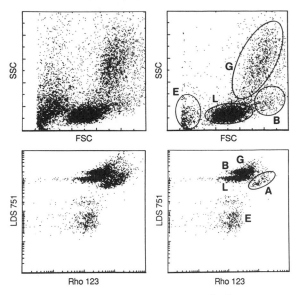

Figure 24.4 Six-parameter analysis of a murine bone marrow sample. Displayed are the FSC–SSC scatterplot of all cells (upper left) and the FSC–SSC scatterplot of gated cells after exclusion of dead and clumped cells (upper right). The lower panels show the corresponding fluorescence distributions. Subpopulations were identified as erythrocytes (E), lymphocytes (L), blasts (B) and granulocytes (G). Cluster (A) possibly represents damaged apoptotic cells as they were found in the lower FSC regions of the leukocyte cluster.

caused by dead cells, clumps, light noise from the oscillating fluid jet and degranulation of an aging sample. Therefore, we have tested dye combinations in an attempt to replace FSC and SSC.

In Fig. 24.4, a murine bone marrow sample is depicted that was stained with 0.1 μg ml⁻¹ Hoechst 33258 (dead cells), 0.2 μg ml⁻¹ Rho 123 (mitochondria) and 2 μg ml⁻¹ LDS 751 (all nucleated cells). Dead cells and clumps were eliminated using Hoechst and TOF signals using gates similar to those shown in Fig. 24.3 for the 3T3 cells. The four major subpopulations can clearly be distinguished in the FSC–SSC scatterplot (Fig. 24.4). Gating the populations in the Rho 123–LDS 751 plot, revealed that the cells expressing low LDS 751 fluorescence were erythrocytes. Intermediate LDS 751 fluorescence was emitted by lymphocytes, while blasts and granulocytes could not be separated completely from each other in the bright LDS 751 cluster.

The small cluster of cells, which display bright Rho 123 and reduced LDS 751 fluorescence are possibly apoptotic cells. This is based on their location in the leukocyte FSC–SSC scatterplot: the subpopulation is located in the lower FSC region, typical for apoptotic cells (Swat *et al.*, 1991). Sorting experiments will be performed to confirm this hypothesis.

24.6 CONCLUDING REMARKS

In this chapter, we have described the principles of the hardware and software for the flow cytometric analysis and sorting of viable cells. Applications involving fixed cells or cellular components, such as chromosomes, do not require significantly different hardware. However, the use of fluorescent probes applied to fixed cells or cell particles is beyond the scope of this volume.

Again, we would like to stress the main features of flow cytometry: speed of analysis and multiparameter analysis of viable single cells. This makes flow cytometry the technique of choice for applications requiring complicated multiparameter analysis and for processing large numbers of cells present in heterogeneous samples. Rare cells can thus be analysed in a statistically reliable way. Also, once a flow cytometer is aligned, large numbers of samples can be run within a short time.

For further reading on flow cytometric hardware and applications, we recommend the recently revised classic *Flow Cytometry and Sorting* (Melamed *et al.*, 1990). *Practical Flow Cytometry* (Shapiro, 1988) contains a treasure of information on the choice and use of fluorescent probes and provides 'how to' information on construction and operation of flow cytometers.

ACKNOWLEDGEMENTS

Funds for the development of the RELACS and ZAPPER flow cytometers were provided by the Dutch research organizations TNO and NWO. The work of J.F.K. is supported by NWO, grant number 900–538–023; the work of H.H. by the Dutch Cancer Society, grant number RRTI 88–8.

REFERENCES

Alberti S., Stovel R. & Herzenberg L.A. (1984) *Cytometry* **6**, 644–647.
Bakker Schut T.C., De Grooth B.G. & Greve J. (1990) *Cytometry* **11**, 659–606.
Berglund D.L., Taffs R.E. & Robertson N.P. (1987) *Cytometry* **8**, 421–426.
Bianchi D.W., Harris P., Flint A. & Latt S.A. (1987) *Cytometry* **8**, 197–202.
Bigler R.D. (1987) *Cytometry* **8**, 441–444.
Cook J.A. & Fox M.H. (1988) *Cytometry* **9**, 441–447.
Dangl J.L. & Herzenberg L.A. (1982) *J. Immunol. Methods* **52**, 1–14.

De Vries P. (1988) PhD thesis, Erasmus University, Rotterdam, The Netherlands.

Dive C., Workman P. & Watson J.V. (1987) *Cytometry* **8**, 552–561.

Dubelaar G.J.B., Groenewegen A.C., Stokdijk W. & Van den Engh G.J. (1989) *Cytometry* **10**, 529–539.

Fox M.H. & Delohery T.M. (1987) *Cytometry* **8**, 20–25.

Fulwyler M.J. (1965) *Science* **156**, 910–911.

Herweijer H., Stokdijk W. & Visser J.W.M. (1988) *Cytometry* **9**, 143–149.

Herweijer H., Van den Engh G. & Nooter K. (1989) *Cytometry* **10**, 463–468.

Herweijer H., Sonneveld P., Baas F. & Nooter K. (1990) *J. Natl. Cancer Inst.* **82**, 1133–1140.

Jongkind J.F. & Verkerk A. (1984) *Cytometry* **5**, 182–187.

Kachel V., Kempski O., Peters J. & Schödel F. (1990) *Cytometry* **11**, 913–915.

Keij J.F., Van Rotterdam A., Groenewegen A.C., Stokdijk W. & Visser J.W.M. (1991) *Cytometry* **12**, 398–404.

Kelley K.A. (1991) *Cytometry* **12**, 464–468.

Loken M.R., Keij J.F. & Kelley K.A. (1987) *Cytometry* **8**, 96–100.

Martin J.C. & Swartzendruber D.E. (1980) *Science* **207**, 199–200.

Melamed M.R., Mullaney P.F. & Mendelsohn M.L. (eds) (1990) *Flow Cytometry and Sorting*. Wiley-Liss, New York.

Mullaney P.F., Van Dilla M., Coulter J.R. & Dean P.N. (1969) *Rev. Sci. Instrum.* **40**, 1029–1032.

Nooter K., Herweijer H., Jonker R. & Van den Engh G. (1990) In *Methods in Cell Biology*, Vol. 33, Z. Darzynkiewicz & H.A. Crissman (eds). Academic Press, San Diego, pp. 631–645.

Omann G.M., Finney D.A. & Sklar L.A. (1985) *Cytometry* **6**, 69–73.

Pennings A., Speth P., Wessels H. & Haanen C. (1987) *Cytometry* **8**, 335–338.

Peters D., Branscomb E., Dean P., Merrill T., Pinkel D., Van Dilla M. & Gray J.W. (1985) *Cytometry* **6**, 290–301,

Rabinovitch P.S., June C.H. & Grossmann A & Ledbetter J.A. (1986) *J. Immunol.* **137**, 952–961.

Salzman G.C., Crowell J.M., Martin J.C., Trujillo T.T., Romero A., Mullaney P.F. & LaBauve P.M. (1975) *Acta Cytol. (Praha)* **19**, 374–377.

Shapiro H.M. (1981) *Cytometry* **1**, 301–312.

Shapiro H.M. (1983) *Cytometry* **3**, 227–243.

Shapiro H.M. (1988) *Practical Flow Cytometry*. Alan R. Liss, New York.

Swat W., Ignatowizc L. & Kiselow P. (1991) *J. Immunol. Methods* **137**, 79–87.

Sweet R.G. (1965) *Rev. Sci. Instrum.* **36**, 131–136.

Terstappen L.W.M.M., De Grooth B.G., Visscher K., Van Kouterik F.A. & Greve J. (1988) *Cytometry* **9**, 39–43.

Valet G., Raffael A., Moroder L., Wunsch E. & Ruhenstroth-Bauer G. (1981) *Naturwissenschaften* **68**, 265–266.

Van den Engh G.J. & Stokdijk W. (1989) *Cytometry* **10**, 282–293.

Visser J.W.M., Cram L.S., Martin J.C., Salzman G.C. & Price B.J. (1978) In *Pulse Cytophotometry*, Part III, D. Lutz (ed.). European Press, Ghent, Belgium, pp. 187–192.

Visser J.W.M., Bauman J.G.J., Mulder A.H., Eliason J.F. & De Leeuw A.M. (1984) *J. Exp. Med.* **59**, 1576–1590.

Watson J.V. (1989) *Cytometry* **10**, 681–688.

Wheeless L.L., Lopez P.A., Berkan T.K., Wood, J.C.S. & Patten S.F. (1984) *Cytometry* **5**, 1–8.

CHAPTER TWENTY-FIVE

Introducing and Calibrating Fluorescent Probes in Cells and Organelles

DAVID A. WILLIAMS, STEPHEN H. CODY & PHILIP N. DUBBIN

Department of Physiology, The University of Melbourne, Australia

25.1 INTRODUCTION

The last decade has seen a rapid escalation in interest in the regulation of intracellular processes by changes in both the levels and distribution of a number of physiologically important cations such as Ca^{2+}, H^+, Na^+, K^+ and Mg^{2+}. Much of this interest is directly related to improvements in both the methodologies of digital imaging microscopy (for review see Moore *et al.*, 1990; Tsien & Harootunian, 1990), and in the design and chemical synthesis of fluorescent probes for detection of specific intracellular cations (Grynkiewicz *et al.*, 1985; Minta *et al.*, 1989).

The ultimate aim in the intracellular use of an ion-sensitive fluorophore is to obtain a measurement of the average concentration (from an image or intensity level) or intracellular distribution of the ion in question, under conditions where the result itself is not influenced by the methodology or through the presence of the fluorophore. The general problems encountered in minimizing the error in such measurements include: (1) reducing cell damage or by-product liberation while introducing readily detected levels of the fluorophore, (2) minimizing the intracellular levels of fluorophore to abolish artifacts resulting from the buffering of ion concentrations following reversible association of ion and probe, (3) limiting the distribution of the probe to the intracellular sites of interest (e.g. cytosol or individual organelles) and (4) determining the exact effects of the intracellular environment (e.g. viscosity, binding) on the behaviour of the fluorophore.

Innumerable research articles dealing with the use of ion-sensitive fluorophores have been published. Recently, a special issue of the journal *Cell Calcium* (February/March 1990) brought together a number of articles which emphasized the intricacies which are inherent in making accurate intracellular ion measurements at high spatial and/or temporal resolution with these fluorophores. The cell types which were described were many and varied and a number of useful experimental paradigms for calibration of fluorescent signals were proposed. It is not the aim of the present chapter to summarize all of this information, but rather to emphasize a number of general principles of loading and calibration, and to detail any major advances that may have occurred since this last synopsis appeared.

One recently advancing area of optical methodology which is gaining increased use in physiological research is that of laser scanning confocal microscopy (for review see Chapters 17 and 19 and Pawley, 1990). The removal of most of the out-of-focus information from two-dimensional images has allowed for improvements in spatial resolution and, in essence, allows for the production of optical sections of living cells in isolation or within tissues. Coupled with specific fluorescent probes, confocal microscopy may provide for highly accurate spatial 2D and 3D distribution maps of ion concentrations in cells. Many of our recent projects have involved this combination and a number of the calibration issues that we will deal with are a result of the specifications of the hardware of the present laser scanning confocal microscopes. The utilization of argon ion lasers in most commercially available systems limits excitation wavelengths to the major bands emitted by such lasers, generally 488 and 514 nm. As such, there has been recent interest in the use of ion-specific fluorophores which are structurally related to fluorescein or rhodamine and are excited in the visible wavelength spectrum. In particular the Ca^{2+} indicators rhod-2, fluo-3 and Fura Red, and the SNARF and SNAFL series of pH indicators (Molecular Probes Inc., Eugene, OR) are already being widely employed for quantitative measurements of cation concentrations with confocal microscopy.

Laser scanning confocal microscopy has much to offer in the study of cellular physiology and pathology. In motile cell systems, such as isolated muscle cells, the confounding effects of cell contraction (or movement) and out-of-focus information on fluorescence levels can be minimized through the ability to confine data acquisition to restricted and well-defined volumes within the cell. These volumes may be represented by complete two-dimensional slices of the cell or by specific areas of interest within the cell, both of which have been demonstrated in some of the studies we describe here.

25.2 GENERAL PRINCIPLES OF THE LOADING PROCESS

The loading protocol utilized to introduce fluorophores into cells should be specifically devised for that cell type. In general, all internalized indicator should ideally be confined to the cytosol rather than allowed to accumulate in intracellular organelles such as mitochondria and nucleus. Fluorophore concentrations should be restricted to levels which, although allowing for acceptable signal-to-noise and signal-to-autofluorescence ratios, minimize the potential for buffering of the target ion. A number of methods have been devised to introduce fluorophores into the internal environment of cells or tissues. The most widely used of these include: (1) incubation of cells (tissues) with esterified derivatives of fluorophores (the most commonly used technique), (2) pressure

injection of cells with microelectrodes containing acidic (charged) dyes, (3) ionophoresis with microelectrodes, and (4) bathing cell suspensions with fluorophores and a chemical agent which transiently and reversibly increases the permeability of the cell membrane to low-molecular-weight substances.

The factors which influence the dye levels and distribution that result from incubation with esterified dye derivatives can be divided into two main types: *fixed* (cell specific) and *variable* (experimentally modifiable). Cell-specific factors include the surface area of the cell membrane and the location and amount of cellular esterases which carry out the de-esterification of internalized dye. It is these factors which may be the major determinants of the inefficient loading which has been reported to occur in certain cell types such as polymorphonuclear leukocytes (neutrophils) (Scanlon et al., 1987).

Incubation time and temperature, agitation of the incubation mixture, and initial dye concentration (external) and cell density are the experimental variables which influence the success of the loading process and should be modified to suit the cell type in use and the experimental requirements. In general, experiments which require kinetic measurements of changes in cytosolic ion concentrations require: short loading times (10–30 min), high (>30°C) temperatures, low dye concentrations and high cell densities. The latter two variables have a complex influence on internal dye concentrations as has been described in detail previously (Moore et al., 1990). In short, the combination of these factors controls, in a probabilistic fashion, the maximum potential number of dye molecules that associate with (and become internalized by) an individual cell. In some situations the ion concentrations of organelles such as the mitochondria, nucleus and endoplasmic reticulum are also of experimental interest, as may be deep-lying cell layers in a multicellular preparation. These cases necessitate long loading times (>60 min), lower temperatures (<30°C), high dye concentrations and/or low cell densities.

Temperature has an obvious influence on the loading process and the distribution of fluorophore, particularly in tissue samples and intact structures such as organs or whole muscles. High temperatures within the physiological range (30–37°C) induce elevated esterase enzyme activity. Internalized dye is likely to be rapidly cleaved in the first compartment (usually the cytosol), or cell it enters (if in a tissue preparation), and as a result the distribution of the dye is restricted. Lowering the temperature of the incubation mixture below this range has dramatic effects on enzymatic processes such as de-esterification (i.e. high Q_{10}) but has much less influence on a physical process (low Q_{10}) such as diffusion. Fluorophores are able to diffuse extensively before cleavage restricts their distribution. In practice, we have found that it is essential that a low-temperature incubation (5–10°C) should be followed by a period of incubation at higher temperature. This ensures that partially cleaved dye does not accumulate in cells or organelles, as these moieties have been shown to introduce significant error into determination of intracellular Ca^{2+} concentrations (Highsmith *et al.*, 1986; Luckoff, 1986; Scanlon *et al.*, 1987). We have found that this loading paradigm has proven to be particularly useful for large multicellular structures such as tumour cell spheroids (see later) and intact skeletal muscles, and also in tissues which possess high levels of extracellular esterase activity.

The ease of use of esterified derivatives for introducing fluorophores into the majority of cell types would seem to make alternative methods unnecessary. However, the often significant fluorescence contribution of dye within intracellular organelles (Williford *et al.*, 1990), and uncleaved dye (Highsmith *et al.*, 1986; Luckoff, 1986; Scanlon *et al.*, 1987) to the measured fluorescent signals, and the possibility of dye binding to intracellular sites within cells (Baylor & Hollingworth, 1988) have been the impetus for the formulation of alternatives which allow introduction of cell-impermeant, charged forms directly into cells.

Several of these methods, such as pressure injection and ionophoresis, require the use of fine-tipped (<1–2 μm diameter) microelectrodes to pierce the plasmalemma of the target cell. Positive pressure or the passage of electric current (generally negative) is then used to expel fluorophore into the cytosol. This process necessarily results in a small degree of cell damage which is often reversible, and where multicellular tissues are used, may necessitate multiple impalements to produce enough loaded cells for fluorescent measurements. However, the use of ionophoresis electrodes for cell loading provides a method for accurately calculating the level of fluorophore introduced into target cells. This method has been recently described (Williams *et al.*, 1990b), and requires knowledge of the pulse amplitude and duration of the ionophoresis protocol, and the volume of the injected cell. The tip diameter of glass microelectrodes varies greatly and as such it was essential to eject fura-2 from a standard microelectrode (e.g. one of a batch from a program-controlled microelectrode puller) into a solution droplet mimicking the ionic composition of the cytosol to calibrate the expulsion rate of electrodes. The fura-2 concentration of the droplets was determined from a standard curve of solution absorption measurements made spectrophotometrically at 360 nm. The fura-2 concentration could then be determined for a range of ionophoresis protocols and allowed formulation of a calibration curve relating total charge passed by an electrode (pulse amplitude × duration) to number of moles of ejected fura-2. Values were then expressed in molar terms by using an estimation of the volume of the target cell, or even more accurately, by using the cytosolic volume fraction of the cell (which was usually the target of the electrode).

Many electrophysiologists use patch microelectrodes to make recordings of whole cell currents while perfusing the intracellular environment with fluorophores contained in the electrode. This coupling of methods has resulted in elegant studies of ionic currents and ion fluxes in many different cell types (Nehr & Almers, 1986; Barcenas-Ruiz *et al.*, 1987; Cannell *et al.*, 1987). The major concern that is raised about measurements of intracellular ion concentrations made under these circumstances is that the cytosolic contents may be modified through dilution by the significantly larger volume of the patch electrode.

Chemical agents have also been used to increase the permeability of cell membranes to small-molecular-weight substances such as the cation-sensitive fluorophores. These methods have similarities in that they induce a state of enhanced cell membrane permeability through incubation of cells (or tissues) with certain chemical agents. These treatments include low pH solutions (acid loading; Bush & Jones, 1988), and solutions with high ATP concentrations.

Recently, the use of the detergent digitonin to enhance the uptake of dyes by plant cells was described (Timmers *et al.*, 1991). Incubation of carrot somatic embryos with fluo-3-AM alone, or with fluo-3 (acidic form) in the presence of either low pH (acid loading) or the dispersant Pluronic F-127 (Molecular Probes) resulted in little appreciable dye internalization. However, if a low concentration (0.025–0.1% (wt/vol.)) of digitonin was added to the incubation medium containing fluo-3, significant levels of intracellular fluorescence were recorded in the cytosol of embryo cells. Importantly, this treatment did not interfere with the normal development of the embryos, indicating that the degree of membrane damage was low or was transient.

Perhaps the least utilized of the chemical methods for introducing low-molecular-weight substances into cells is a method of reversible hyper-permeabilization (Sutherland *et al.*, 1980). This technique utilizes the transient removal of divalent cations to make the external plasmalemma leaky, during which time constituents of the external medium (including introduced species such as fluorophores) become internalized in the treated tissue. Although this procedure was adopted to introduce the photoprotein aequorin into smooth muscle cells and tissue (DeFeo & Morgan, 1980), there are no reports of its use with fluorescent probes. In our preliminary studies the technique was found to be well-suited to loading tissue samples with the de-esterified (acidic) forms of Ca^{2+}-sensitive indicators.

25.3 GENERAL PRINCIPLES OF THE CALIBRATION PROCESS

Calibration of intracellular signals of ion-sensitive fluorophores has been the subject of a number of recent research articles (Borzak *et al.*, 1990; Groden *et al.*, 1991; Owen, 1991; Uto *et al.*, 1991). Some of this information has served to reinforce or supplement that previously presented in *Cell Calcium* (February/ March 1990), but serves to emphasize the importance of this exercise if accurate measurements of intracellular ion concentrations are desired. Most of the problems encountered in the calibration process result from dye that is internalized by cells but which exhibits properties that differ from those of the acidic (ionized) form under defined ionic conditions. The major conditions which have been proposed to alter the intracellular behaviour of fluorescent dyes include viscosity (Poenie, 1990) and intracellular binding of internalized dye (Baylor & Hollingworth, 1988; Blatter & Weir, 1990).

Techniques which attempt to directly define the *intracellular* fluorescence of an indicator have the greatest potential for producing accurate measurements of ionic changes in experimental situations. These techniques are characteristically difficult and require well-defined buffering systems for the specific ion, and solutions containing mixtures of ionophores which allow equilibration of the intracellular and extracellular pools of the required ions. Ionomycin (Calbiochem), a cation ionophore with a high specificity for Ca^{2+} over other ions (i.e. Mg^{2+}, K^+, Na^+), is often employed for equilibration of Ca^{2+} pools, but the Ca^{2+}-transporting ability exhibits strong pH-sensitivity (Lui & Hermann, 1978). As such it is essential to significantly elevate the pH of the medium bathing the cells or tissue (pH 8–10) to facilitate Ca^{2+} equilibration, but this is not always possible given the constraints of many experiments. Alternative *pH-insensitive* ionophores include A23187 (Calbiochem), which exhibits autofluorescence at the wavelength combinations employed with fura-2, and the non-fluorescent, brominated derivative Br-A23187. This latter ionophore has found increased usage with the UV-excitable fluorophores fura-2 and indo-1, whereas A23187 is suitable for Ca^{2+} equilibration of cells loaded with the dyes excitable in the visible spectrum (e.g. fluo-3, rhod-2 and Fura Red).

A new method for the *in vivo* calibration of fura-2, involving the construction of a calibration curve from intracellular fluorescence measurements, has been described recently for sections of isolated mammalian skeletal muscle fibres (Bakker *et al.*, 1993). Fibres were maintained under paraffin oil and exposed to a range of mixtures of strongly Ca^{2+}-buffered solutions of a known free [Ca^{2+}], and concentrations of Mg^{2+}, K^+, H^+ and organic anions that approximated those of the intact fibre. Ca^{2+} concentration was buffered with either of the chelators EGTA and BAPTA, the latter constituent being used because of its pH-insensitivity in the physiological pH range.

Small droplets of the Ca^{2+}-buffered solution along with fura-2 free acid (100 μM) and saponin (100 μg ml^{-1}) were applied (pressure ejection from a patch pipette) under oil directly to the fibre bundle. Saponin permeabilized the sarcolemma allowing equilibration of the bathing solution with the cytosol of the fibres. The bundles were equilibrated in test solutions until the fibres fluoresced uniformly at a stable intensity level (usually 15 min). Paraffin oil restricted the leakage of fura-2 from the bundle even when fibres went into supercontracture at high Ca^{2+} concentrations. The fluorescence was recorded following excitation at both 350 and 380 nm, and calibration curves of the fluorescence ratio (350/380 nm) and Ca^{2+} concentration were derived with a non-linear curve fitting routine employing a least sum of squares procedure. Values for the K_d of fura-2 resulting from the *in vivo* technique were identical to those found *in vitro* in the same Ca^{2+}-buffered solutions. However, the constants that are normally employed in the calibration of fura-2 were all found to be lower in the muscle fibres than the *in vitro* values. R_{max} (14.6$\pm$0.05) and R_{min} (9.4$\pm$0.15) were both lowered by a similar magnitude to that shown previously in smooth muscle cells (Williams *et al.*, 1985), and this emphasizes the importance of employing the most accurate constants for the calibration procedure.

Where it is not possible to implement techniques such as this a comprehensive *in vitro* calibration process is still desirable as this will allow determination of the dissociation constant of the dye:ion complex, and calibration parameters (R_{max} and R_{min}), under the conditions that will be employed experimentally with cell or tissue preparations. Instead, many research groups simply adopt the values presented in the initial description of fura-2 characteristics (Grynkiewicz *et al.*, 1985), and as previously stated (Williams & Fay, 1990), these values were not meant to be used under all experimental conditions in any cell or tissue type.

25.4 PUTTING PRINCIPLES INTO PRACTICE

To cover the essential issues relevant to the introduction and calibration of all the fluorescent probes in living cells and tissues is an onerous task. The list of available fluorophores is immense and is growing all the time (for example, see Chapter 3 and Molecular Probes Catalogue). In an effort to cover the most important issues we have decided to remain within our direct area of expertise and summarize the major projects that we are presently undertaking, or have recently completed, while discussing the major difficulties encountered with the use of fluorescent probes, and detailing the solutions devised to alleviate or minimize these problems.

25.4.1   Spontaneous calcium changes in isolated cardiac myocytes

Image analysis of Ca^{2+} dynamics with high spatial and temporal resolution in isolated spontaneously contracting myocytes has been severely limited by both the methodology and the capabilities of Ca^{2+}-sensitive probes (e.g. aequorin, fura-2) which have been employed to date. We have sought to investigate the properties of progagated spontaneous calcium release (SCR) with laser scanning confocal microscopy by visualizing the intracellular fluorescence fluctuations of a visible wavelength Ca^{2+} indicator, fluo-3 (Minta *et al.*, 1989). The potential advantages of this combination include quantification of Ca^{2+} wave propagation rates in single cardiac myocytes and the potential for lengthy observations of SCR, propagation and interaction in contracting cardiomyocytes.

Suspensions of cardiomyocytes prepared by a standard enzymatic technique from rat left ventricles were loaded with fluo-3 by incubation with the esterified derivative (fluo-3-AM) at 30°C for 30 min, Individual cells were viewed with a laser scanning confocal microscope (Biorad Lasersharp MRC 500, Oxfordshire UK) which was coupled to an inverted Olympus IMT-2 microscope. The details of this optical system have been described previously (Williams *et al.*, 1990a, 1992a).

One major advantage of working with contractile cells is that there is a readily measured physiological parameter, contraction (time and magnitude), that allows direct determination of any deleterious effects that may result from the presence of the fluorophore or its by-products within cells. Analysis of cell shortening following electrical stimulation of fluo-3-containing cells indicated that the isotonic contractile kinetics (maximum cell shortening, contraction time, maximum rate of shortening and lengthening) were indistinguishable from those reported previously for similar cells which did not contain the fluorophore (Delbridge *et al.*, 1989). This strongly suggests that the levels of internalized fluorophore employed in this type of study had little effect, through direct Ca^{2+} buffering or by-product generation, on the physiological

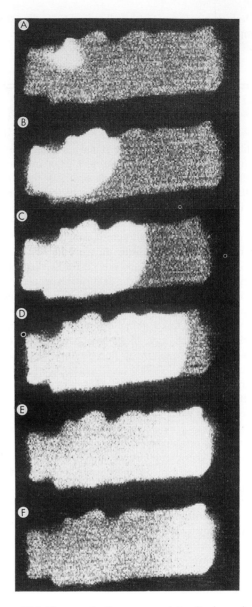

Figure 25.1 Changes in fluorescence intensity in a single spontaneously contracting cardiomyocyte loaded with fluo-3. A broad band of high fluorescence intensity propagated along the cell from upper left to right. Initial cell length (panel A): 100 µm.

parameters (contraction or Ca^{2+} levels) investigated in these cells.

Spontaneously contracting cells exhibited areas of high-intensity fluorescence which propagated throughout the cell in unison with the localized bands of contraction. An example of this type of activity can be seen in the sequence of images displayed in Fig. 25.1. Ca^{2+}-dependent fluorescence originated at a localized site (upper left) of the cell and gradually propagated throughout the cell. The circular front of

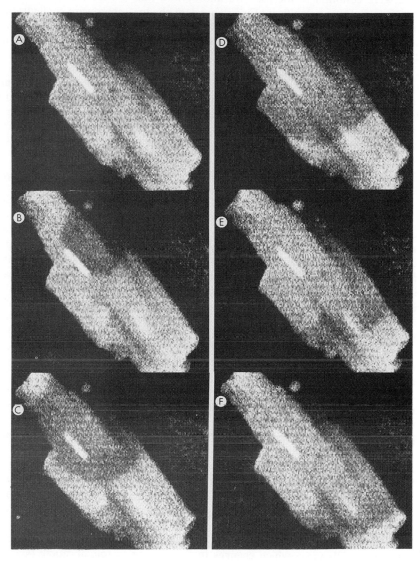

Figure 25.2 The change in intensity of Fura Red fluorescence in an isolated spontaneously contracting cardiac cell. With this fluorophore local increases in Ca^{2+} level are reflected by areas of diminished intensity which clearly propagate throughout the cell. Initial cell length (panel A): 105 μm.

the fluorescence wave indicates that the release and diffusion processes were equal in all directions and generally not affected by the different intracellular structures encountered along and across the cell. As we have described previously in detail (Williams, 1990; Williams *et al.*, 1992a), the fluorescence bands are of limited propagation velocity (50–150 μm), are not subject to contraction and movement artifacts, and are exclusively due to localized Ca^{2+} changes which can be attributed to a spontaneous and propagated Ca^{2+}-release process.

The results of one of the control experiments performed to determine how changes in cell geometry may affect fluorescence intensity is shown in Fig. 25.2. This image series shows the fluorescence of a spontaneously contracting cardiomyocyte loaded with a recently released Ca^{2+}-sensitive fluorophore Fura Red (Molecular Probes). When excited at 488 nm this fluorophore responds to Ca^{2+} increases by exhibiting decreased fluorescence. As a result we would predict that Ca^{2+} waves in contracting cells should be accompanied by bands of reduced fluorescence propagating along the cell, unless the cell contraction resulting from the contraction band is sufficient to cause a large local volume increase, and as a result, a localized increase in fluorescence intensity. As is shown in Fig. 25.2 a dark band, again with a circular wave front, originates locally and passes along this cell. It is evident from this behaviour that the increases in intensity recorded under similar conditions with fluo-3 (see Fig. 25.1) exclusively represent Ca^{2+} fluctuations.

Accurate calibration of the fluctuations in fluorescence intensity of internalized fluo-3 in terms of ionized Ca^{2+} levels presented a number of challenges that were common in the use of a fluorophore that does not undergo a significant Ca^{2+}-dependent change in spectral characteristics. This was previously the case with another Ca^{2+} fluorophore quin-2 (Rink & Pozzan, 1985), which was widely used in the early 1980s (1980–85). Variation of the absolute concentration of intracellular fluorophore within cells or organelles, or as a function of time, could produce intensity variations which could be erroneously interpreted as ion distribution differences. This was particularly evident in cells which exhibited large regional variations in volume (e.g. flattened cells which taper at the edges), cells with large volumes occupied by intracellular organelles, or cells which changed shape during the activity of interest (e.g. contractile cells).

Calibration of fluo-3 fluorescence intensity in cardiac myocytes was performed in several ways. The simplest calibration resulted from the determination of the ratio of maximum (F_{max}) and minimum (F_{min}) fluorescence intensity levels for the fluorophore within individual cells. Experimental fluorescence levels could be scaled relative to these fluorescence limits to allow for the determination of $[Ca^{2+}]$ at any point in time as has previously been described in detail for quin-2 (Rink & Pozzan, 1985). The resulting value for F_{max}/F_{min} of 4.95 ± 0.04 ($n = 14$ individual cells) for cardiac cells varied little from that obtained with fluo-3 acid in Ca^{2+}-containing solutions (5.10 ± 0.17, $n = 3$) under the same experimental conditions, thereby indicating that internalized fluo-3 exhibited similar properties to those of the indicator *in vitro*.

Alternatively, cells were loaded with both fluo-3 and a second Ca^{2+}-sensitive fluorophore fura-2 to allow for implementation of a cross-calibration technique recently described by Williams (1990). The direct Ca^{2+} calibration of fura-2 ratio images, as has been described in detail elsewhere (Moore *et al.*, 1990), provided a baseline *average* $[Ca^{2+}]$ for individual non-stimulated cells upon which changes in fluo-3 fluorescence intensity could be expressed as $[Ca^{2+}]$ changes. This technique also requires knowledge of the fluorescence enhancement of fluo-3 upon Ca^{2+} binding (F_{max}/F_{min}) for the experimental system in use.

These calibration methods have allowed us to make accurate determinations of Ca^{2+} levels in many spontaneously contracting cardiac cells as described in detail in previous reports (Williams, 1990; Williams *et al.*, 1992a). Localized changes in Ca^{2+} concentrations from resting levels of approximately 200 nM to peak values of between 800 nM and 1.5 μM were responsible for the contraction bands evident in the cells we investigated. These procedures have increased the certainty with which quantitative information can be derived from single-wavelength indicators such as fluo-3 and rhod-2, and other indicators (e.g. BCECF) which may be confined to a single-wavelength mode

of data acquisition by the constraints of the standard hardware configurations.

Ca^{2+} oscillations occur in unstimulated cardiac cells from many species, with physiological Ca^{2+} levels in the bathing solutions, and with minimal experimental perturbations of the Ca^{2+} loading state of the cell. Such a powerful and fundamental phenomenon occurring under relatively physiological conditions is clearly relevant to an understanding of the generation of cardiac arrhythmias and the techniques described may allow the necessary further investigation.

25.4.2 Ionized calcium in living plant cells

With the recent realization of the importance of calcium ions in mediation of plant cellular responses (for review see Hepler & Wayne, 1985), there is impetus to improve existing technologies to enable monitoring of intracellular calcium dynamics, especially in response to growth hormones and other physiological stimuli such as gravity and light (Chapter 28). However, plant cells are difficult to load with fluorescent calcium indicators and this is most likely due to extracellular hydrolysis of the esterified derivatives of the fluorophores (Cork, 1985), or the incomplete hydrolysis of the internalized dyes (Brownlee & Wood, 1986). The few studies that have been carried out have resorted to introduction of dyes through microinjection, ionophoresis or acid loading (Bush & Jones, 1988).

Problems of a different kind arise with both intact plant (and animal) tissues, involving uncertainty in analysing signals emanating from individual cells within a given image plane because of contaminating out-of-focus information (Fay *et al.*, 1989: Keith *et al.*, 1985). To circumvent some of the technical difficulties evident with plant cells we have again utilized fluo-3 and a laser scanning confocal microscope to study the role of cations (Ca^{2+} and H^+) in plant cell physiology. This combination enables these processes to be studied in single cells within intact plant tissue preparations, which is an important consideration given that there are not a large number of functional single cell preparations that have been isolated from plant tissues.

To maximize the access of fluorescent esters to cells within plant tissues we found it necessary to remove the barrier provided by the waxy outer cuticle found in intact tissue samples such as coleoptile tips. This was most efficiently achieved by scoring the tissue surface with sharp blades or by applying adhesive tape to, and removing it from, the surface of the coleoptile. The latter alternative enhanced loading in larger areas of the tissue with scoring only resulting in significant internalization of dyes in the localized areas of cuticle removal. Dye access was terminated by rinsing tissue

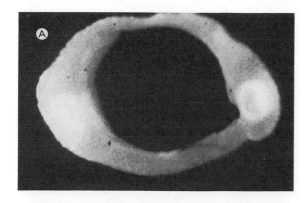

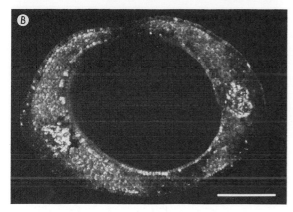

Figure 25.3 (A) Image of the fluorescence of a fluo-3-loaded coleoptile tip as captured by a silicon-intensified-target (SIT) camera following standard epifluorescence illumination. (B) The same coleoptile tip viewed immediately after with laser scanning confocal microscopy. Both images are the average of eight consecutive video frames. Scale bar: 250 μm. (Reproduced from Williams *et al.*, 1990a, by permission of *Cell Calcium*.)

samples in fresh tapwater (40 μM[Ca^{2+}]). When internalized dye was isolated from plant tissues by mechanical homogenization it was invariably found to be fully Ca^{2+} sensitive (Scanlon *et al.*, 1987), indicating that plant cells, once given access to the esterified derivatives of ion-sensitive indicators, are capable of complete cleavage of the molecules.

A low magnification image (Fig. 25.3(A)) of a coleoptile slice following illumination by conventional epifluorescence microscopy, shows the presence of diffuse fluorescence with little indication of cellular detail. The autofluorescence of the two major vascular bundles is the major discernible feature. This image illustrates one of the optical problems which make quantitative evaluation of cellular fluorescence within plant cells difficult, even when successful dye loading has been achieved. Contaminating fluorescence which emanates from focal planes above and below the plane of interest impinges upon the image plane and is a problem in all conventional widefield images,

particularly in large three-dimensional structures such as tissue samples.

An image of the same coleoptile slice acquired with the laser scanning confocal microscope employing the same objectives (see Fig. 25.3(B)), shows a marked improvement in resolution of individual cells. Epidermal layers and the cortical cell mass are clearly visible. Spatial heterogeneity in intracellular fluorescence is clearly evident in higher magnification images (see Fig. 25.4). Vacuoles are discernable but contained little fluorescence, suggesting that the majority of internalized fluo-3 was cleaved and retained by the small volume of cytosol surrounding the large vacuoles before it was able to invade the vacuolar compartment. The slow leakage, or transport of dye into vacuoles via non-specific anion transport mechanisms (Malgaroli *et al.*, 1987), ensures that changes in fluorescence recorded within 60 min of loading exclusively reflect changes in cytosolic [Ca^{2+}].

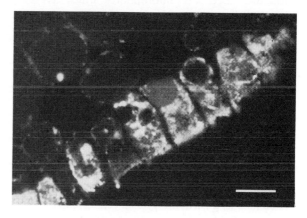

Figure 25.4 High magnification confocal image (eight frames) of a small segment of a fluo-3-loaded coleoptile tip. The intra- and extracellular fluorescence heterogeneity is readily apparent. Scale bar: 25 μm. (Reproduced from Williams *et al.*, 1990a., by permission of *Cell Calcium*.)

It is clear that the many problems associated with the imaging of fluorescent probes to make accurate measurements of intracellular ions in plant tissues can be effectively eliminated by (1) removing the waxy cuticle that presents the major barrier for dye internalization in plant tissues and (2) employing confocal microscopy to reduce the three-dimensional spread of contaminating out-of-focus information in image planes. As a result we have been able to show for the first time with high spatial resolution the elevations in cytosolic Ca^{2+} of maize coleoptiles in response to application of ionophores and plant growth hormones (Williams *et al.*, 1990a). Ca^{2+} changes were calibrated from the fluorescence changes with the same techniques that we have described for cardiomyocytes. Using the techniques and paradigms described here we have also been able to visualize changes in [Ca^{2+}] and pH in epidermal and cortical cells as a result of physiological stimuli such as gravitropism or phototropism (Gehring *et al.*, 1990), and groups are now taking advantage of these advances and pursuing further investigation of plant growth responses (see Chapter 28; Gehring *et al.*, 1991).

25.4.3 Localization of cell nuclei

In the process of analysing Ca^{2+} movements in individual rat cardiomyocytes it has become apparent that the rate of propagation of Ca^{2+} waves through the cell is occasionally influenced by adjacent cell nuclei. It was, therefore, necessary to systematically determine nuclear number and location in these cells. Fluorescent molecules are commonly used to elucidate specific features of cell structure. Apart from their use as labels for numerous monoclonal and polyclonal antibodies there are fluorophores which associate preferentially with specific classes of molecules within cells. We have frequently used ethidium bromide, which intercalates with the double-stranded conformation of DNA and also associates with single-stranded RNA, to investigate the structure, number and position of cell nuclei in both isolated cardiac cells, and single muscle fibres from normal and diseased (dystrophic) skeletal muscles. This fluorophore, although a polar molecule, has significant cell membrane permeability and following a short (10 min) incubation period it quickly enters both cell types where it readily associates with intracellular nuclei acids.

The images shown in Fig. 25.5 give examples of the two most frequently occurring patterns of nuclear distribution. A small number of cells possessed a single centrally placed nucleus (Fig. 25.5(A)), while the majority of individual cells (>90%) had two nuclei placed one at each end of the cell (Fig. 25.5(B)) (Williams, Cody, Delbridge & Harris, unpublished). The numerous strands of mitochondria are also evident in both cell images and are due to the fluorescent staining of the RNA content associated with the structures.

Whereas nuclei are generally distributed in a peripheral spiral in adult skeletal fibres, the presence of centrally located nuclei is thought to be indicative of a muscle fibre which has regenerated following a damage/degeneration sequence (Harris & Johnson, 1978). In muscular dystrophy there is a large turnover of skeletal muscle with continual degeneration and regeneration of individual fibres. In studies of single muscle fibres from the dystrophic mdx mouse we have been able to identify the nuclear distribution pattern in the same fibres from which contractile and morphological properties were measured (Williams *et al.*,

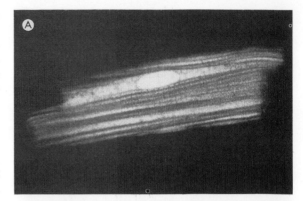

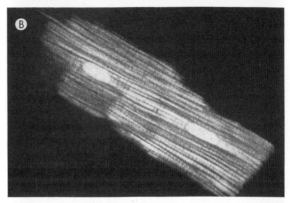

Figure 25.5 Laser scanning confocal microscopic images of individual cardiomyocytes following a 10 min incubation with ethidium bromide. A small percentage (<10%) of all cells possessed a single, centrally placed nucleus (A), while the majority (90%) of cells were binucleated (B). Cell lengths: (A) 95 μm; (B) 150 μm.

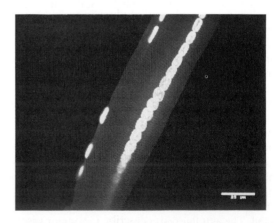

Figure 25.6 A single, enzymatically isolated skeletal muscle fibre from a dystrophic mdx mouse as stained with ethidium bromide. A long chain of centrally located nuclei is evident as are a few peripheral nuclei. Scale bar: 25 μm.

1992b; Head *et al.*, 1992). By doing so we hoped to be able to find correlates between function and the regenerative status (i.e. original or regenerated fibre) of the fibres.

Figure 25.6 illustrates the nuclear distribution pattern of one such fibre and shows a distinctive strand of nuclei placed centrally within the fibre. Rather than finding just the two distribution patterns for nuclei, central strands (as illustrated) and peripheral spirals (not shown), we found a myriad of combinations of these two patterns. These observations suggest that dystrophic fibres do not necessarily undergo complete degeneration following damage but instead are capable of repeated local repair processes, leading to local variations in the nuclear distribution within an individual fibre. This has important implications for interpretation of the disease aetiology and is the subject of further investigation.

25.4.4 Mapping of intracellular pH in isolated cells

The measurement of pH in a variety of different cell types utilizing the unique emission spectra of SNARF-1 in combination with microspectrofluorometry has been reported by a number of groups (Buckler & Vaughan-Jones, 1990; Bassnett, 1990; Seksek et al., 1991; Mariot et al., 1991; Martinez-Zaguilan et al., 1991). The final fluorescence measurement obtained in these studies represented the average intracellular pH with little distinction of the pH of individual intracellular organelles. In contrast, with video imaging techniques, and laser scanning confocal microscopy in particular, we are able to measure the overall contribution of individual organelles to the resting pH distribution of a single living cell. The Lasersharp (Biorad MRC-500) confocal microscope can be readily configured for dual-channel emission detection and with minor modifications the emission spectrum of SNARF-1 can be divided at the isosbestic point with the separate emission bands sent to different detectors. Although the gain and black level of each channel are usually adjusted individually, it is essential if quantitative measurements are desired, to select settings which will cope with the range of fluorescence intensity fluctuations that are expected to occur, and to maintain these settings throughout the experiment and calibration process. The thickness of the optical sections that contributed to the fluorescence emission recorded by each detector was also matched by using equivalent confocal apertures in each emission pathway.

As we have already described, experimentation with concentration and incubation times is necessary for optimization of loading for each cell type used and as a result there have been a wide variation of conditions that have resulted in successful loading of SNARF-1. We have used cultured rat aortic smooth muscle and rabbit proximal tubular cells which were isolated and grown on glass coverslips. The cells were placed in a temperature-controlled perfusion bath

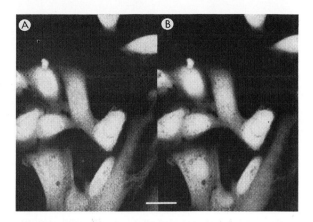

Figure 25.7 Fluorescence images of a field of rabbit proximal tubule cells loaded with the pH-sensitive fluorophore SNARF-1. Images represent wavelengths of the emission spectrum above (A), and below (B) 610 nm. Scale bar: 25 μm.

and perfused with a HEPES-buffered Krebs solution for incubation with 20 μM carboxy-SNARF-1-AM (esterified derivative) for 20 min at 37°C followed by a 10 min period where cells were perfused in fresh, dye-free solutions.

In agreement with previous studies, we have found that the media used for most cell culture procedures often present problems for the accurate measurement of intracellular ion concentrations. Cultured cells should be grown in a medium which is free of pH indicator (the most common of which is Phenol red) as these indicators and their by-products do accumulate in cells during growth, and often contribute to the fluorescence output of the loaded cells. We have found that several cell passages in a medium free of pH indicator is generally sufficient for removal of the contaminating species. In addition, the aliphatic amines which are also found in cell culture media are capable of extracellular cleavage of ester groups and it is essential to replace the culture medium before loading cells.

A typical confocal image of a culture of rabbit proximal tubule cells loaded with SNARF-1 is shown in Fig. 25.7. The fluorescence intensity of each individual image pair (Fig. 25.7(A) and (B)) indicated that the uptake and cleavage of the SNARF-1-AM was consistent between cells and resulted in relatively uniform distribution of fluorophore within each cell. In contrast, there are reports that the majority of the dye becomes associated preferentially with the cytosol (Blank et al., 1990). The intensity of fluorescence within the cells was stable during experiments for up to 200 min with little dye leakage or photobleaching (Dubbin et al., 1991).

A ratio image of the SNARF-1 emission is displayed in Fig. 25.8(A) and represents the pixel-by-pixel division of the image pair (Fig. 25.7(A) and (B)) for these cells. The majority of cells showed a relatively uniform cytosolic pH level, however, in most cases there were pH gradients evident between the cytosol and nucleus of individual cells with the nuclear regions displaying predominantly higher pH than the cytosol. At the end of each experiment the absolute intracellular pH levels were calibrated *in vivo* by treatment of each cell with nigericin (20 μM) in solutions of high [K$^+$] with a range of defined pH levels. A calibration curve constructed from these known pH standards was then used to calculate pH for images acquired during the experiment (Fig. 25.9). The calibration curves of each individual cell within the same field of view were rarely identical. In fact almost all cells showed a unique calibration curve and this may be explained by slight differences in the behaviour of the probe within each cell.

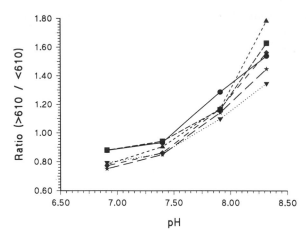

Figure 25.9 Calibration curves relating pH and the fluorescence ratio of SNARF-1 for six individual cells within a single field of rabbit cultured proximal tubule cells. Intracellular pH was equilibrated with that of extracellular buffers following treatment of cells with nigericin.

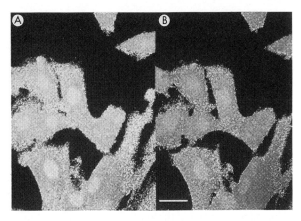

Figure 25.8 Ratio (Fig. 25.7(A)/(B)) images of SNARF-1 emission in the same field of cells, prior to (A) and following (B), the addition of the ionophore nigericin. The distinct differences in pH levels of the nucleus and cytosol of individual cells were virtually abolished by the addition of ionophore. Scale bar: 25 μm.

The preliminary basal cytosolic and nuclear pH levels within the cells that we examined were 7.35 ± 0.08 and 7.44 ± 0.02 (*n* = 9) for rabbit proximal tubule cells, and 7.38 ± 0.01 and 7.54 ± 0.08 (*n* = 5) for rat aortic smooth muscle cells, respectively. There was a large variation in the basal intracellular pH of several different cell cultured types. As also shown by several other groups (Bassnett, 1990; Mariot *et al.*, 1991; Seksek *et al.*, 1991), the intracellular pH in a single population of cultured cells ranged by as much as one unit of pH, so the importance of making measurements in large numbers of cells within a group became apparent. This variation makes it essential that investigation of the effects of physiological agonists be conducted at the level of individual cells.

In order to determine the ability of SNARF-1 to accurately reflect changes of pH within living cells we then experimentally manipulated the internal pH of the cells without damaging their integrity. To do this we utilized a short perfusion with a small quantity of ammonium chloride (NH$_4$Cl). The cells responded with a typical transient rise in pH on addition, and fall after removal, of NH$_4$Cl. By altering the concentration of NH$_4$Cl applied (0.1 μM – 10 mM) we were able to control the initial rise in pH and we found that we could accurately resolve changes as small as 0.03–0.05 pH units. It was also interesting to note that when NH$_4$Cl was added the pH of the nucleus and cytosol both increased and the pH gradient between these compartments was still evident for all concentrations. Using high concentrations of nigericin (10–20 μM) we were able to remove this gradient during controlled experimental conditions (see Fig 25.8(B)) which indicated that the nuclear membrane was actively involved in controlling the intranuclear pH. More importantly, this also showed that the difference was not due to an artifact due to differential uptake or behaviour of the probe in different cellular compartments. The technique of coupling SNARF-1 with a laser scanning confocal microscope system can provide high-resolution spatial information on intracellular pH levels within single living cells during physiological manipulations.

25.4.5   Intracellular pH gradients in mammalian tumour spheroids

Cancer research has employed cell spheroids as models of tumour tissue because they are poorly vascularized, have radial proliferation and distinct pO$_2$ gradients with the tissue. It is also thought that such tissues should exhibit pH gradients and measurements made with pH electrodes have shown a pH of 7.3 at the surface, decreasing to 7.0 at a depth of 400 μm within the spheroid thus supporting this contention (Acker *et al.*, 1987; Carlsson & Acker, 1988). However, the authors of these studies expressed reservations as to whether the microelectrodes used were exclusively reporting levels of intracellular pH because of uncertainty in the location of the tip within the spheroid and because of the tip size of the electrode (2–3 μm). Measurements were therefore conservatively interpreted as representing the pH of the extracellular space within the spheroid. However, this was still an important observation given the known involvement of pH changes in mediating growth response in tumour cells (Carlsson & Acker, 1988).

In a collaborative study with the Department of Surgery, Royal Melbourne Hospital, Australia (Cody, Beischer, Hill, Kaye & Williams, unpublished observations), we have applied confocal microscopy and pH-sensitive fluorophores to unequivocally determine whether these extracellular pH gradients were mimicked in the intracellular environment of the spheroid cells. Rat C6 glioblastoma spheroids (Benda *et al.*, 1971) were cultured with well-established techniques (Yuhas *et al.*, 1977; Sutherland *et al.*, 1971). Loading of spheroids with SNARF-1-AM (ester derivative) was performed in the dark at 10°C for 90 min under constant agitation, to allow for penetration of dye in the spheroids used in these experiments which ranged from 600 to 1500 μm in diameter. However, the limitation to observations of the pH of cells deep within the spheroid was not one of dye distribution, but rather the inability to focus on cells throughout the full depth of the spheroid with the low magnification (×10) and therefore low NA (0.5) objectives required to observe the entire spheroid with the confocal microscope. Therefore, following incubation with SNARF-1-AM spheroids were embedded in agar (2% Sea Plaque; Bioproducts, ME, USA) at 37°C for 15 min. Agar was solidified with brief (2 min) immersions in crushed ice. The embedding step at 37°C also facilitated de-esterification of dye internalized throughout the spheroid. Embedded spheroids were then carefully bisected with a disposable microtome blade (featherblade) to avoid tissue distortion and reveal the necrotic core of the tissue.

A ratio image of the SNARF-1 emission of a spheroid is shown in Fig. 25.10(A). There is evidence of a striking zonation in intracellular pH within the spheroid with a band of cells of uniform pH surrounding a distinctly more acidic core of cells. Analysis of pH in a transect across the tissue (Fig. 25.10(B)) shows that the cells in the outer mass maintained a pH of 7.5–7.6 while the pH of cells in the central cell zone fell progressively to 7.1–7.2 at

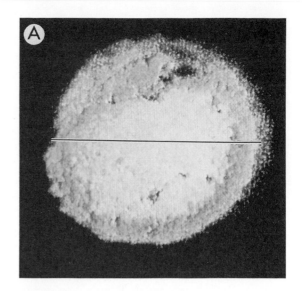

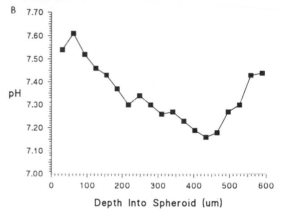

Figure 25.10 (A) A ratio image of the intracellular fluorescence of a SNARF-1-loaded rat C6 glioblastoma spheroid following mechanical bisection. The pH levels of distinct areas of a transect (indicated by the line) through the spheroid are shown graphically in panel B.

the heart of the spheroid. These results are the first clear measurements of cytosolic pH in the cell layers of tumour spheroids. The values are in the same range of values that have been recorded as extracellular values with microelectrodes and confirm a previous hypothesis that both extracellular and intracellular environments will exhibit similar pH gradients (Acker *et al.*, 1987). Further research will determine the relation between these gradients and cell proliferation within spheroids.

25.4.6 Intracellular pH and fatigue in intact mammalian muscles

Peripheral fatigue in skeletal muscle is a complex process which may be mediated at a number of levels of the excitation–contraction coupling process. A

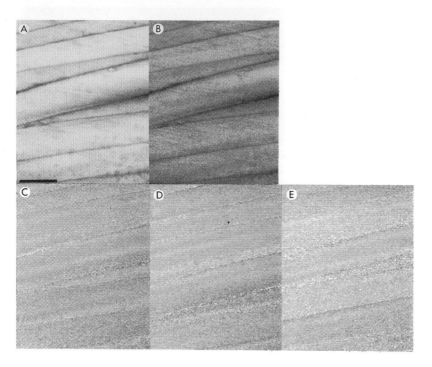

Figure 25.11 Fluorescence images of a small section of peripheral muscle fibres from an intact rat edl skeletal muscle incubated with SNARF-1. (A) and (B) are the component images from which a ratio image (C) was determined which represents the pre-stimulus pH levels of these fibres. Similarly constructed ratio images show the pH distribution of the same fibres immediately following a fatigue-inducing stimulation protocol (D), and following a 15 min period of recovery from fatigue (E).

significant increase in the concentration of intracellular hydrogen ions is often thought to accompany the onset of fatigue but it is still equivocal as to whether this acidification is primarily responsible for peripheral fatigue (for review see Westerblad *et al.*, 1991).

We have investigated the possible involvement of acidification in muscle fatigue by simultaneously determining force generation and intracellular pH in individual fibres of intact rat soleus and extensor digitorum longus (edl) muscles during a stimulation protocol which induced significant fatigue in the tetanic force response of the muscles. Intact muscles were generally incubated with 30 µM SNARF-1-AM in Kreb's Ringer for 30 min at 22°C. Muscles were removed from this mixture, rinsed in fresh Ringer in which they were maintained for a further period (20 min) at 37°C to allow for cleavage of internalized esters.

Stimulation of intact muscles was achieved via platinum field electrodes placed adjacent to each muscle. The fatiguing protocol involved application of 1 s tetanus (100 Hz) every 5 s for 5 min. The muscles were bathed in mammalian Kreb's Ringer and equilibrated at pH 7.4 by constant bubbling with carbogen (95% O_2/5% CO_2) throughout the experimental period.

The maximum isometric tetanic tension of intact muscles was recorded immediately prior to, and following the stimulation protocol, and subsequent to a 15 min recovery period. On average, tetanic force responses decreased reversibly by 48 and 80% in soleus and edl muscles respectively, as a result of the fatigue-inducing protocol.

Images of SNARF-1 emission from both detection channels were collected (as described earlier) and converted to pH ratio images at the same time points of the stimulation protocol. The laser scanning confocal microscope provided us with the ability to look at muscle fibres which remained in the three-dimensional syncytium of the intact muscle. The cytosolic pH of the majority of individual muscle fibres showed no evidence of heterogeneity and this is clearly seen in Fig. 25.11. The initial intracellular cytosolic pH levels for individual fibres of soleus muscles (mean ± SE; 7.30 ± 0.05, $n = 9$ muscles) was not significantly different to that of fibres of the edl muscle (7.21 ± 0.07, $n = 11$ muscles).

As displayed in Fig. 25.11 the fatigue protocol induced little change in cytosolic pH in fibres of the edl muscle and similar observations were made in all muscles of either type subjected to this stimulation procedure. The lack of a detectable change in pH in fatigued skeletal muscle is an interesting observation which has a number of interpretations. It is possible that a significant change in pH is not necessarily involved in the peripheral fatigue response and instead the overproduction or decline in another factor may be more causally related to the force decline.

As all observations were made on peripheral muscle fibres the possibility also remains that fibres towards the core of the muscle may exhibit the significant pH decreases which would cause the large reversible decline in force which is observed. However, *in vivo* all fibres would be expected to receive a similar blood supply and as a result zonal differences in intracellular pH levels throughout the muscle would not be expected. Indeed, in an earlier study by Renaud *et al.*, (1986) tetanic force level was reported to drop by 70% in frog sartorius muscle following an even more severe stimulation protocol, and was accompanied by a pH decrease of 0.4 units measured by microelectrodes in peripheral muscle fibres.

We are presently investigating both of these possibilities by looking at subperipheral fibres with the same techniques, and also by making similar measurements of other ions, particularly Ca^{2+}, with other fluorescent probes such as the combination of fluo-3 and Fura Red that can be quantitatively coupled with confocal microscopy.

ACKNOWLEDGEMENTS

Noel D. Duncan was centrally involved in the unpublished studies of fatigue in skeletal muscle. We are grateful for the technical assistance of Miss Chris Goulter in many of these studies which were supported by The National Heart Foundation (NHF) of Australia, The National Health and Medical Research Council (NH&MRC) of Australia and The Australian Research Council (ARC).

REFERENCES

Acker H., Carlsson J., Holtermann G., Nederman T. & Nylen T. (1987) *Cancer Res.* **47**, 3504–3508.

Bakker A.J., Head S.I., Williams D.A. & Stephenson D.G. (1993) *J. Physiol.* (in press).

Barcenas Ruiz L., Beuckelmann D.J. & Wier W.G. (1987) *Science* **238**, 1720–1722.

Bassnett S. (1990) *J. Physiol.* **431**, 445–464.

Baylor S.M. & Hollingworth S. (1988) *J. Physiol.* **403**, 151–192.

Benda P., Someda K., Messer J. & Sweet W.H. (1971) *J. Neurosurg.* **34**, 310–323.

Blank P.S., Silverman H.S., Chung O.Y., Stern M.D., Hansford R.G., Lakatta E.G. & Capogrossi M.C. (1990) *Biophys. J.* **57**, 137a.

Blatter L.A. & Wier W.G. (1990) *Biophys. J.* **58**, 1491–1499.

Borzak S., Kelly R.A., Kramer B.K., Matoba Y., Marsh J.D. & Reers M. (1990) *Am. J. Physiol.* **29**, H973–981.

Brownlee C. & Wood J.W. (1986) *Nature (Lond.)* **320**, 624–626.

Buckler K.J. & Vaughan-Jones R.D. (1990) *Eur. J. Physiol.* **417**, 234–239.

Bush D.S. & Jones R.L. (1988) *Eur. J. Cell Biol.* **46**, 466–469.

Cannell M.B., Berlin J.R. & Lederer W.J. (1987) *Science* **238**, 1419–1423.

Carlsson J. & Acker H. (1988) *Int. J. Cancer* **42**, 715–720.

Cork R.J. (1985) *Plant Cell Environ* **9**, 157–160.

DeFeo T.T. & Morgan K.G. (1985) *J. Physiol.* **369**, 269–282.

Delbridge L.M., Harris P.J. & Morgan T.O. (1989) *Clin. Exp. Physiol. Pharmacol.* **16**, 179–184.

Dubbin P.N., Cody S.H., Harris P.J. & Williams D.A. (1991) *Proc. Austr. Physiol. Pharmacol. Soc.,* **22**, 83P.

Fay F.S., Carrington W. & Fogarty K.E. (1989) *J. Microsc.* **153**, 133–149.

Gehring C.A., Williams D.A., Cody S.H. & Parish R.W. (1990) *Nature (Lond.)* **345**, 528–530.

Gehring C.A., Irving H.R. & Parish R.W. (1991) *Proc. Natl. Acad. Sci. USA* **87**, 9645–9649.

Groden D.L., Guan Z. & Stokes B.T. (1991) *Cell Calcium* **12**, 279–288.

Grynkiewicz G., Poenie M. & Tsien R.Y. (1985) *J. Biol. Chem.* **260**, 3440–3450.

Harris J.B. & Johnson M.A. (1978) *Clin. Exp. Pharmacol. Physiol.* **5**, 587–600.

Head S.I., Williams D.A. & Stephenson D.G. (1992) *Proc. Royal Soc. Lond. Ser. B,* **248**, 163–169.

Hepler P.K. & Wayne R.O. (1985) *Ann. Rev. Plant Physiol.* **36**, 397–439.

Highsmith S., Bloebaum P. & Snowdowne K.W. (1986) *Biochem. Biophys. Res. Commun.* **138**, 1153–1162.

Keith C.H., Ratan R., Maxfield F.R., Bajer A. & Shelanski M.L. (1985) *Nature* **316**, 848–850.

Luckoff A. (1986) *Cell Calcium* **7**, 233–248.

Lui C. & Hermann T.E. (1978) *J. Biol. Chem.* **253**, 5892.

Malgaroli A., Milani D., Meldolesi J. & Pozzan T. (1987) *J. Cell Biol.* **105**, 2145–2155.

Mariot P., Sartor P., Audin J. & Dufy B. (1991) *Life Sci.* **48**, 245–252.

Martinez-Zaguilan R., Martinez G.M., Lattanzio F. & Gillies R.J. (1991) *Am. J. Physiol.* **260**, C297–C307.

Minta A., Kao J. & Tsien R.Y. (1989) *J. Biol. Chem.* **264**, 8171–8178.

Moore E., Becker P.L., Fogarty F.S., Williams D.A. & Fay F.S. (1990) *Cell Calcium* **11**, 157–179.

Neher E. & Almers W. (1986) *EMBO J.* **5**, 51–53.

Owen C.S. (1991) *Cell Calcium* **12**, 385–393.

Pawley J.B. (1990) *Handbook of Biological Confocal Microscopy*. Plenum Press, New York.

Poenie M. (1990) *Cell Calcium* **11**, 85–91.

Renaud J.M., Allard Y. & Mainwood G.W. (1986) *Can. J. Physiol. Pharmacol.* **64**, 764–767.

Rink T.J. & Pozzan T. (1985) *Cell Calcium* **6**, 133–144.

Roos A. & Boron W.F. (1981) *Physiol. Rev.* **61**, 296–434.

Scanlon M., Williams D.A. & Fay F.S. (1987) *J. Biol. Chem.* **262**, 6308–6312.

Seksek O., Henry-Toulme N., Sureau F. & Bolard J. (1991) *Anal. Biochem.* **193**, 49–54.

Sutherland P.J., Wendt I.R. & Stephenson D.G. (1980) *Proc. Aust. Physiol. Pharmacol. Soc.* **11**, 160P.

Sutherland R.M., McCredie J.A. & Inch W.R. (1971) *J. Natl. Cancer Inst.* **46**, 113–117.

Timmers A.C.J., Reiss H.D. & Schel J.H.N. (1991) *Cell Calcium* **12**, 515–521.

Tsien R.Y. & Harootunian A.T. (1990) *Cell Calcium* **11**, 93–109.

Uto A., Arai H. & Ogawa Y. (1991) *Cell Calcium* **12**, 29–38.

Westerblad H., Lee J.A., Lannergren J. & Allen D.G. (1991) *Am. J. Physiol.* **261**, C195–209.

Williams D.A. (1990) *Cell Calcium* **11**, 589–597.

Williams D.A. & Fay F.S. (1990) *Cell Calcium* **11**, 75–90.

Williams D.A., Fogarty K.E., Tsien R.Y. & Fay F.S. (1985) *Nature* **318**, 558–561.

Williams D.A., Cody S.H., Gehring C.G., Parish R.W. & Harris P.J. (1990a) *Cell Calcium* **11**, 291–298.

Williams D.A., Head S.I., Bakker A.J. & Stephenson D.G. (1990b) *J. Physiol.* **428**, 243–256.

Williams D.A., Delbridge L.M., Cody S.H., Harris P.J. & Morgan T.O. (1992a) *Am. J. Physiol.* **262**, C731–C742.

Williams D.A., Head S.I., Lynch G.S. & Stephenson D.G. (1992b) *J. Physiol.* (in press).

Williamson R.E. & Ashley C.C. (1982) *Nature (Lond.)* **296**, 647–651.

Williford D.J., Sharma V.K., Korth M. & Sheu S.S. (1990) *Circ. Res.* **66**, 234–241.

Yuhas J.M., Li A.P. & Martinez A.O. (1977) *Cancer Res.* **37**, 3639–3643.

CHAPTER TWENTY-SIX
Photolabile Caged Compounds

ALISON M. GURNEY

Department of Pharmacology, United Medical & Dental Schools, St Thomas's Hospital, London, UK

26.1 INTRODUCTION

The application of photolabile caged compounds in biology has been the subject of a number of reviews (Lester & Nerbonne, 1982; Lester *et al.*, 1986; Nerbonne, 1986; Kaplan, 1986; Gurney & Lester, 1987; Kaplan & Somlyo, 1989; McCray & Trentham, 1989; Homsher & Millar, 1990; Kaplan, 1990; Somlyo & Somlyo, 1990). Practical aspects in the use of caged compounds have recently been detailed (Gurney, 1991; Walker, 1991), along with methods for their synthesis and purification (Walker, 1991). It is not the aim of this chapter to provide another review on the subject, but to outline the general properties of caged compounds and to point out some of the problems that have been, or might be, encountered when applying them to study biological pathways. Many caged compounds can now be purchased from Molecular Probes, Oregon, USA or Calbiochem, California, USA.

Photolabile 'caged' compounds are inert precursors of biologically active molecules that can be stimulated with near-UV light to release the active species. When applied extracellularly or incorporated into cells, they provide a means of manipulating physiological or pharmacological pathways on a rapid (milliseconds) time-scale. A flash of light triggers the uncaging of the bioactive molecule, so that it is released at its site of action without a diffusional delay. Some molecules are designed for extracellular application, being particularly useful with intact preparations where restricted diffusion slows the response to the bioactive molecule, or where its effects would be modified by, for example, metabolism, uptake or receptor desensitization. Other 'caged' compounds can be loaded directly into cells to release intracellular second messengers upon photolysis.

Early studies with caged compounds were limited by the availability of only a few molecules. Of those, caged ATP was used to study the energetics of muscle contraction (Goldman *et al.*, 1982; reviewed in Homsher & Millar, 1990) and activation of the Na pump (Kaplan *et al.*, 1978; Kaplan, 1986), while molecules that interact with cholinergic receptors were exploited to probe drug-receptor interactions and the kinetics of nicotinic-receptor channels (Lester & Chang, 1977; Lester & Nerbonne, 1982; Lester *et al.*, 1986). Other possible applications of caged compounds became apparent with the development of caged cyclic nucleotides (Engels & Schlaeger, 1977; Engels & Reidys, 1978; Nerbonne *et al.*, 1984), which allowed cyclic AMP-dependent regulation of the calcium conductance and contraction in cardiac muscle

Table 26.1 Summary of photolabile cage compounds.

Photochemically caged molecules	*References*
Neurotransmitters	
Glutamate	Messenger *et al.*, 1991
Glycine	
γ-Aminobutyric acid (GABA)	McCray & Trentham, 1989
Receptor Ligands	
Carbachol	Milburn *et al.*, 1989
Nicotinic receptor agonists (Bis-Q, QBr)	Lester & Nerbonne, 1982
Cholinergic antagonists	
Phenylephrine	Walker & Trentham, 1988; Somlyo *et al.*, 1988
Nifedipine[a]	
Intracellular second messengers	
inorganic ions: Ca^{2+}	Adams *et al.*, 1988; Kaplan & Ellis-Davies, 1988
Mg^{2+}	Kaplan & Ellis-Davies, 1988
H^+	Nerbonne, 1986; Janko & Reichert, 1987
HPO_4^{2-}	Kaplan *et al.*, 1978; McCray & Trentham, 1989
Zn^{2+}	Blank *et al.*, 1981
Nucleotides: ATP	Kaplan *et al.*, 1978; Walker *et al.*, 1989a
ADP	
GTP	Walker *et al.*, 1989a
GDP	
Cyclic nucleotides: cAMP	
cGMP	Nerbonne *et al.*, 1984; Wootton & Trentham, 1989
Inositol phosphates: IP3	Walker *et al.*, 1987; Walker *et al.*, 1989b
IP2	Walker *et al.*, 1989b
Other intracellular probes	
ATP-γ-S	
ATP-β, γ-NH	
GTP-γ-S	Walker *et al.*, 1989a
GTP-β, γ-NH	
BAPTA	Ferenczi *et al.*, 1989

[a] Nifedipine is inactivated by photolysis.

to be studied directly (Nargeot *et al.*, 1983; Nerbonne *et al.*, 1984; Richard *et al.*, 1985). More recently, a novel chemical approach to caging compounds has enabled a wider variety of probes to be developed (Walker *et al.*, 1988). Around the same time, photolabile calcium chelators were introduced, which release free calcium upon photolysis (Tsien & Zucker, 1985; Adams *et al.*, 1988; Ellis-Davies & Kaplan, 1988). As a result of this increased availability of suitable probes, caged compounds have now become popular for studies on a wide variety of biological responses. Table 26.1 lists some of the currently available molecules.

Optical probes provide a particularly powerful means for manipulating the concentration of molecules inside or outside cells. This is because photochemical reactions tend to be fast and the intensity, duration and area of activating light can be varied with relative ease. In theory, light could be directed at a whole population of cells, or be focused into a spot small enough to irradiate a single cell or part of a cell. The approach should, therefore, allow the possibility of studying how cells respond to perturbations in the concentration of important signalling molecules imposed throughout the cell, or localized to specialized regions of the cell.

26.2 PROPERTIES OF A PHOTOLABILE PROBE

Light is used to initiate a series of reactions. The first phase occurs when the caged compound absorbs a photon of energy, $E = h\nu$, where h is Planck's constant and ν is the frequency of the light. Absorption of a photon promotes the molecule (M) to an excited state (M^*) with higher energy, represented as $M + h\nu \rightarrow M^*$. M^* is a new species with distinct chemical and physical properties. Its extra energy results in chemical reactions known as the dark reactions, which then lead to the formation of stable products. One of these products is the physiologically active moiety, while the other by-products should ideally be biologically inert.

Photochemical reactions take place only under the influence of light, and only light that is absorbed by the molecule can trigger the photochemical reaction.

The amount of light absorbed depends on the extinction coefficient, the intensity and wavelength of the exciting light and the path length of solution through which the light has to pass. Pigmented cells, such as are found in the nervous system of the marine mollusc *Aplysia*, may absorb some of the exciting light and interfere with photolysis (Tsien & Zucker, 1985; Nerbonne & Gurney, 1987). However, most cells are sufficiently transparent that they contribute little to the absorbance of the exciting light, thereby permitting efficient photolysis of caged compounds in isolated cells and in thin tissue preparations. Most of the available caged compounds have an absorbance peak at around 350 nm, a wavelength that is relatively harmless to cells. Absorbance is usually low above 500 nm, so caged compounds are relatively stable under normal laboratory lighting. Stock solutions of most caged compounds can be sufficiently well-protected during use by keeping them in containers wrapped in aluminium foil. Brief exposures to room light while preparing solutions or applying them to experimental preparations do not usually have any detrimental effect. Photolysis typically requires high intensity light in the near *UV* region.

It is not sufficient that a molecule absorbs light for it to be photolabile. Indeed, many molecules absorb light but remain chemically unchanged, e.g. dyes. With other molecules, the extra energy produced by the absorption of a photon results in fluorescence or phosphorescence. Even with a photolabile molecule, the excited intermediate M^* can lose its extra energy and decay back to the ground state M rather than initiate the dark reactions. Clearly an efficient photochemical reaction would be one in which the absorbed light results in product formation a high percentage of the time. The quantum yield of the reaction, which is defined as the ratio of product molecules formed to photons absorbed, provides a measure of the effectiveness of the absorbed light at triggering the photochemical reaction. Thus less light would be required to trigger the release of active molecules from caged compounds with a high extinction coefficient and a high quantum yield. The quantum yields of commercially available probes are all sufficiently high to permit concentration changes in the physiological range to be produced with flash lamps or lasers. In practice, the amount of photolysis produced by a flash of light also depends on the lifetime of the excited intermediate state, relative to the duration of the flash. If the intermediate is short lived, then multiple excitations may occur during the flash, resulting in a higher percentage conversion than would be predicted from the quantum yield. This phenomenon is exemplified by the caged cyclic nucleotides, which have much lower quantum yields than caged ATP, but are photolysed about 50% as well as caged ATP by a

1 ms flash (Wootton & Trentham, 1989). Multiple excitations can apparently occur even with very brief (50 ns) laser pulses.

The ability of photolabile caged compounds to release active molecules rapidly makes them especially useful to biologists. Millisecond time resolution can be achieved with many caged compounds, which is faster than that provided by most other rapid application techniques. Thus photolabile probes have been used to study the kinetics of such rapid events as the activation of receptors (see, for example, Lester & Chang, 1977; Lester & Nerbonne, 1982; Lester *et al.*, 1986; Gurney & Lester, 1987; Milburn *et al.*, 1989; Somlyo & Somlyo, 1990) and ion channels (Gurney *et al.*, 1987; Karpen *et al.*, 1988; Lando & Zucker, 1988; Ogden *et al.*, 1990), muscle contraction (reviewed in Homsher & Millar, 1990; Somlyo & Somlyo, 1990) and neurotransmitter release (Zucker & Hayden, 1988). Nevertheless, although the speed offered by photolysis is one of the main reasons for developing caged compounds, it should not be assumed that photorelease is always fast. The rate of photorelease varies enormously among different caged compounds (see below), and in some cases it can take 100 ms or so for complete photolysis (e.g. Wootton & Trentham, 1989). Thus, when a caged compound is to be used to measure the kinetics of a biological process, it is important that photorelease from the probe proceeds with sufficient speed that it does not limit the time-course of the response being studied.

26.3   THE CHEMISTRY OF PHOTOLABILE COMPOUNDS

Two types of photochemical reaction have been exploited in the development of photolabile caged compounds. These are the *cis* ⇔ *trans* photoisomerization of azobenzenes (Fig. 26.1(a)) and the photochemical cleavage of *o*-nitrobenzyl groups (Fig. 26.1(b)).

26.3.1   Azobenzenes

Azobenzene derivatives have been used most extensively to probe the kinetics of nicotinic receptors (reviewed in Lester & Nerbonne, 1982; Lester *et al.*, 1986; Gurney & Lester, 1987). These studies were made possible because the *cis* and *trans* stereoisomers differ in their pharmacological properties, and interconversion between the isomeric forms can be controlled by irradiating the molecules at different wavelengths. For example, the best-studied azobenzene, Bis-Q (3,3'bis-(α-(trimethylammonium)methyl)-azobenzene), is a potent agonist at nicotinic receptors

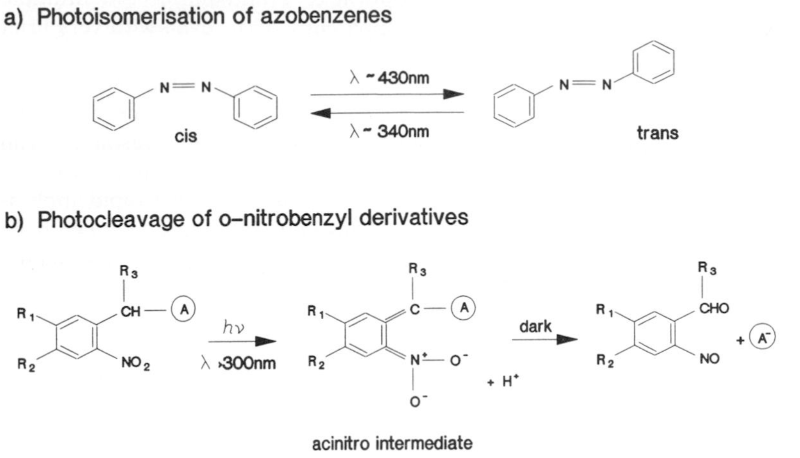

a) Photoisomerisation of azobenzenes

cis $\lambda \sim 430nm$ → ← $\lambda \sim 340nm$ trans

b) Photocleavage of o–nitrobenzyl derivatives

acinitro intermediate

Figure 26.1 (a) Photoisomerization of azobenzene is reversible. The *trans* isomer is produced with light of 410–450 nm, while the *cis* isomer is promoted at 300–350 nm. (b) The reaction scheme proposed to account for the photocleavage of *o*-nitrobenzyl derivatives. Substitutions at R_1, R_2 and R_3 alter the speed and efficiency of the reaction, as well as the biological activity of the precursor and photoproducts.

in the *trans* form, but not in the *cis* form (Lester & Chang, 1977; Lester *et al.*, 1986). Thus preparations bathed in a solution containing mainly the *cis* isomer of Bis-Q show minimal nicotinic receptor activity, but receptors can be activated by exposing the preparation to a flash of light at 420–440 nm, which converts a large proportion of *cis*-Bis-Q to the *trans* form. The reverse reaction is possible with *trans* to *cis* isomerization favoured at wavelengths of around 340 nm. The advantages offered by these azobenzenes (see Lester & Nerbonne, 1982) are the rapidity of photoisomerization, which occurs within 1 µs of absorbing a photon, and the high quantum yield. In addition, the only effect of light is isomerization, with no competing photoreactions or photoproducts to worry about. On the other hand, even complete irradiation does not produce a pure solution of one or other isomer, but a photostationary mixture of the two.

Photoisomerizations are being exploited in other ways which could lead eventually to the development of a general 'cage' for physiologically active molecules. Attempts thus far have involved forming liposomes (Kano *et al.*, 1981) or micelles (Shinkai *et al.*, 1982) from molecules containing azobenzene groups. These approaches (reviewed in Gurney & Lester, 1987) depend on differences in liposome permeability, or in critical micelle concentrations, when the incorporated azobenzene group is in the *cis* or *trans* configuration. Azobenzene chemistry was also exploited in the first approach to caging divalent cations. By synthesizing an azobenzene compound containing two imino-diacetic acid groups, it proved possible to chelate Zn^{2+} in a light-sensitive and reversible manner (Blank *et al.*, 1981). The authors suggested that protons might also be amenable to caging in this manner, but the method has had limited success in caging other divalent cations such as Ca^{2+}.

26.3.2 *o*-Nitrobenzyl derivatives

Currently, the most widely used and available photo-labile caged compounds exploit the light sensitivity of the *o*-nitrobenzyl moiety (Patchornik *et al.*, 1970), illustrated in Fig. 26.1(b). By linking the molecule of biological interest to an *o*-nitrobenzyl group, it is hidden from the active site. Irradiation then cleaves the precursor at the benzyl carbon, freeing the active molecule along with a proton and a nitroso by-product, which may be either a ketone or an aldehyde depending on the substitution at the benzyl carbon (R_3 in Fig. 26.1(b)). Proton loss occurs simultaneously with the formation of an *aci*-nitro intermediate (Nerbonne, 1986), and the rate of breakdown of the intermediate determines the rate of photorelease of the active species from the caged compound (McCray *et al.*, 1980; Goldman *et al.*, 1984a; Nerbonne, 1986). The *o*-nitrobenzyl group was first exploited in the development of caged ATP and photolabile cyclic nucleotide analogues. It has since been used success-fully to cage a wide variety of intracellular messengers, neurotransmitters, receptor ligands and other useful probes (Table 26.1).

Alternative analogues are available for some caged compounds, with different substitutions at positions R_1, R_2 and R_3 in Fig. 26.1(b). For example, caged ATP can be purchased as either the *o*-nitrobenzyl (NB) ester where R_1, R_2 and R_3 are all protons, the *o*-nitrophenylethyl (NPE) ester where $R_3 = CH_3$, or the dimethoxy *o*-nitrophenylethyl (DMNPE) ester where R_1 and $R_2 = CH_3O$ and $R_3 = CH_3$. These modifications influence the rate and efficiency of the photochemical reaction, as well as the biological activity (see later) of the precursor and the photo-product. A number of studies have examined the structural requirements for fast efficient photolysis (McCray *et al.*, 1980; Nerbonne, 1986; Wootton & Trentham, 1989). Unfortunately, there does not seem to be any general rule regarding the effect that each substituent has on photolysis, because the properties of the photolabile molecule are also influenced by the nature of the biologically active molecule being caged. Incorporating methoxy groups at positions R_1 and R_2 has the beneficial effect of causing a red shift of the absorption maximum of caged compounds, thereby improving light absorption in the 300–360 nm range. This modification also improves the speed and efficiency of photorelease from caged cyclic nucleotides (Nerbonne, 1986), but it has the opposite effect on caged ATP and caged phosphate (Wootton & Trentham, 1989). On the other hand, photorelease from caged carbachol (Milburn et al., 1989) and caged ATP (Kaplan et al., 1978) is accelerated when a methyl group is present at R_3, but introducing the same group into dimethoxy o-nitrobenzyl cyclic nucleotides makes them unstable in aqueous solution (Wootton & Trentham, 1989).

The rate of photorelease, which is likely to be an important consideration in selecting a photolabile probe, can vary markedly among analogues. The release of ATP from NPE-caged ATP proceeds with a rate constant of 84 s^{-1}, compared with only 18 s^{-1} from the DMNPE analogue at 20°C and pH 7 (Wootton & Trentham, 1989). In contrast, inorganic phosphate is released rapidly (21 000 s^{-1}) from DMNPE-caged phosphate under similar conditions (Wootton & Trentham, 1989). Photorelease of cyclic nucleotides is more rapid from DMNB analogues than from NB analogues (Nerbonne, 1986), although the precise rates are not yet clear. It has proved difficult to measure the rates of photolysis of the DMNB analogues using chemical techniques (Wootton & Trentham, 1989), but a biological assay has shown that DMNB-caged cyclic GMP probably photolyses at >3000 s^{-1} at pH 7 (Karpen et al., 1988). Even the slowest rates of release may be acceptable in some experiments, for example, when the aim is simply to elevate the concentration of an intracellular messenger. Photolabile probes probably provide the simplest way of doing this. However, if the aim is to gain kinetic information from the response, the probes should photolyse more rapidly, preferably by

at least an order of magnitude, than the response develops.

26.3.3   Photolabile cation chelators

Caged compounds can be used to control the concentration of cations inside or outside cells. The photolabile cation chelators currently available were reviewed recently (Kaplan, 1990), and practical aspects of their application in biology have been described in detail (Gurney, 1991). Zinc ions were the first to be chelated in a photosensitive manner, using the azobenzene chemistry as indicated above (Blank et al., 1981). Several molecules now exist that can act either as caged cations (nitr-5, DM-nitrophen), which release Ca^{2+} and/or Mg^{2+} upon photolysis, or caged chelators (diazo-2, caged BAPTA), which mop up Ca^{2+} upon photolysis. All of these agents use the photochemistry of the o-nitrobenzyl group, the photolysis of which alters the affinity of the molecule toward Ca^{2+} and/or Mg^{2+}. The affinities of some of these agents for Ca^{2+} and Mg^{2+} before and after photolysis are listed in Table 26.2, along with other properties that bear on their use as photolabile cation chelators.

The nitr-5 family of compounds (Fig. 26.2(a)) were designed around BAPTA (Tsien & Zucker, 1985; Adams et al., 1988), a Ca^{2+} chelator with high selectivity for Ca^{2+} over Mg^{2+}. This selectivity is retained in the photolabile probes, both before and after photolysis and, like BAPTA, their Ca^{2+} affinity shows little dependence on pH above pH 7. The affinity of BAPTA for Ca^{2+} is known to be increased or decreased depending on the nature of electron-withdrawing or donating groups present on one of its aromatic rings (Tsien, 1980). With nitr-5 this group becomes more electron withdrawing after photolysis, resulting in reduced Ca^{2+} affinity (Fig. 26.2(a)). In contrast, the opposite change in diazo-2 (Fig. 26.2(b)) results in an increased affinity for Ca^{2+} after photolysis (Adams et al., 1989). Thus nitr-5 or diazo-2 can be used to respectively elevate or reduce the free Ca^{2+} concentration inside cells. Another recently introduced molecule of this type, diazo-3 (Molecular Probes), displays similar photochemical properties to

Table 26.2 Properties of photolabile Ca^{2+} chelators.

	K_D for Ca^{2+} binding (µM)		Photolysis rate (s^{-1})	Quantum yield	K_D for Mg^{2+} binding (mM)	
	Pre-photolysis	Post photolysis			Pre-photolysis	Post photolysis
Nitr-5[a]	0.145	6	4000	0.03–0.1	8.5	8
DM-nitrophen[b]	0.005	2000	3000	0.18	0.005	3
Diazo-2[c]	2.2	0.073	>2000			
Caged BAPTA[d]	160	0.11	>300			

Data from references
[a] Adams et al. (1988); [b] Kaplan & Ellis-Davies (1988); Kaplan (1990); [c] Adams et al. (1989); [d] Ferenczi et al. (1989).

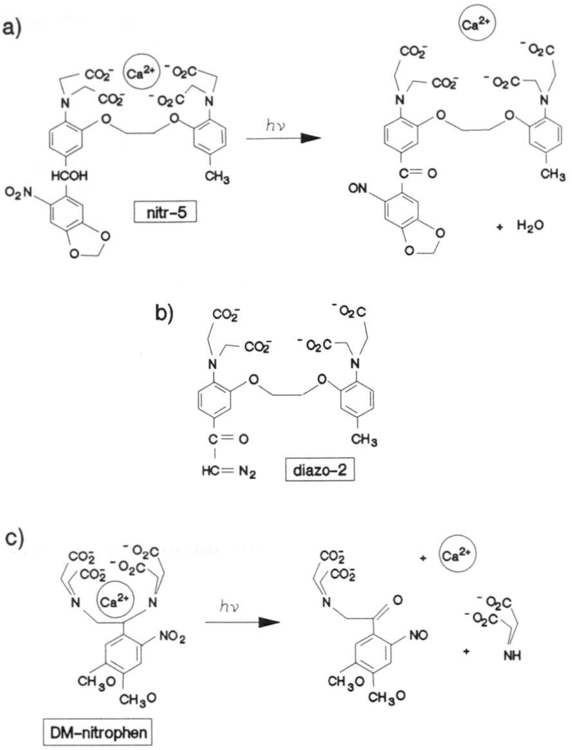

a)

b)

c)

Figure 26.2 Reaction schemes proposed to account for the photolysis of (a) nitr-5 and (c) DM-nitrophen, both of which lose affinity for Ca^{2+} upon photolysis. Diazo-2 (b) is structurally related to nitr-5, but gains affinity for Ca^{2+} on photolysis.

nitr-5 and diazo-2, but it does not bind Ca^{2+} well before or after photolysis. It should therefore prove useful as a control against effects of photolysis produced by the structurally related compounds, that are not due to changes in Ca^{2+} concentration. BAPTA itself has been photochemically caged by derivatizing it with an o-nitrobenzyl group (Ferenczi et al., 1989); photolysis results in the release of BAPTA, which binds up free Ca^{2+}.

DM-nitrophen (Fig. 26.2(c)) represents a different approach to caging divalent cations (Ellis-Davies & Kaplan, 1988). This molecule is based around the chelator EDTA, which binds Ca^{2+} ions tightly, but also has a significant affinity for Mg^{2+}. Thus at the millimolar concentrations of Mg^{2+} present inside cells, a significant fraction of DM-nitrophen would be present as the Mg^{2+} complex, and photolysis would release a mixture of Ca^{2+} and Mg^{2+}. This is a clear disadvantage since it would not permit the effects of a rise in intracellular Ca^{2+} to be studied under physiological conditions (i.e. in the presence of Mg^{2+}). Nevertheless, in preparations where Ca^{2+}-dependent effects can be studied in the absence of Mg^{2+}, it is possible to use DM-nitrophen to selectively release

Ca^{2+}. It is also possible to exploit DM-nitrophen to selectively release Mg^{2+} as, for example, in the regulation of the sodium pump (Klodos & Forbush, 1988).

Nitr-5 and DM-nitrophen, the most easily available and best-studied of the photolabile chelators, both have distinct advantages and disadvantages for use as Ca^{2+} donors (Table 26.2). Both agents release Ca^{2+} rapidly, with a rate of around 3000 s^{-1}. The clearest advantage of nitr-5 is that it is relatively insensitive to pH and the presence of Mg^{2+} ions. Furthermore, unlike DM-nitrophen, it can be loaded into cells non-invasively (see later), so it permits studies in intact cells. DM-nitrophen, on the other hand, binds Ca^{2+} more tightly. Thus at low resting levels of free Ca^{2+} a greater proportion of the probe is bound with Ca^{2+}, enabling more Ca^{2+} to be released by photolysis. DM-nitrophen also displays a greater quantum yield than nitr-5, and its Ca^{2+} affinity undergoes a greater change upon photolysis, both of which add to the Ca^{2+}-releasing ability of the probe.

The two types of chelator also differ in the nature of the reactions triggered by photolysis. Irradiation of nitr-5 induces a structural rearrangement (Fig. 26.2 (a)) to form a molecule with reduced calcium affinity, and photolysis results in the same end products whether it is bound to Ca^{2+} or not. In contrast, DM-nitrophen is thought to undergo cleavage to yield two photoproducts (Fig. 26.2(c)), each with negligible Ca^{2+} affinity. Moreover, the photolysis of DM-nitrophen may yield different products depending on whether or not it is complexed with Ca^{2+} (Ellis-Davies & Kaplan, 1988).

26.4   APPLICATION OF PHOTOLABILE PROBES TO STUDYING BIOLOGICAL PATHWAYS

26.4.1   Caged receptor ligands

The delay between activation of a receptor and a cell's response can provide insight into the mechanisms coupling the receptor to the response. For example, activation of the nicotinic acetylcholine receptor opens an ion channel that is part of the same protein complex. Events mediated by such direct coupling occur on the millisecond time-scale, so to study their activation kinetics requires a very rapid method for applying agonist. At present photochemistry provides the most rapid means of applying nicotinic agonists, submilli-second resolution having been achieved both with the photoisomerizable azobenzene, Bis-Q (Lester & Nerbonne, 1982; Gurney & Lester, 1987), and with caged carbachol (Milburn et al., 1989). Similar time resolution is available with photoisomerizable azobenzene molecules that block either the nicotinic receptor or its channel (Lester & Nerbonne, 1982; Gurney & Lester, 1987). The advantage of these molecules is that they can bypass the diffusional delays normally associated with drug application. Preparations are bathed in the inactive precursor, with light flashes triggering the release of the active ligand on demand. The speed of application allows receptor activation to be temporally separated from receptor desensitization, and from other processes that inactivate the ligand.

Photolysis studies of nicotinic-receptor activation have mostly been performed with Bis-Q and the related molecule, QBr (Lester & Nerbonne, 1982; Gurney & Lester, 1987), because for a long time they were the only caged agonists available. In studying these receptors, caged carbachol may, however, have some advantages. For example, the N-(α-carboxy-2-nitrobenzyl) derivative appears to be more pharmacologically inert than cis Bis-Q before photolysis (Milburn et al., 1989), and carbachol is a well-characterized analogue of the natural transmitter, acetylcholine. Furthermore, caged carbachol should also act as a caged agonist at muscarinic receptors, where Bis-Q is an antagonist (Nargeot et al., 1982). Although a caged acetylcholine molecule would be the most useful for probing the physiological activation of cholinergic receptors, such a molecule has not yet been synthesized.

Caged ligands have also been developed for other receptors (see Table 26.1), although they have not been widely tested in biological experiments. The activation of α-adrenergic receptors in vascular smooth muscle has been studied using caged phenylephrine (NPE ester), which has minimal activity before photolysis (Somlyo et al., 1988). The active agonist is released rather slowly from this caged compound, with a time constant of ~300 ms (Walker & Trentham, 1988), reflecting slow dark reactions. This would be too slow to probe the type of direct excitation–response coupling found at nicotinic receptors. However, following photolysis of caged phenylephrine, vascular muscle contracts with a latency of 1.5 s at 30°C, and this appears to be due entirely to the events linking phenylephrine binding to its receptor and the subsequent response (Somlyo et al., 1988). The contractile response is, however, quite far removed from receptor activation. It has yet to be determined if the rate of phenylephrine release from the caged precursor is fast enough to permit studies of the activation of more immediate events, such as membrane conductances.

26.4.2 Caged intracellular second messengers

There are now a wide variety of caged molecules available for activating, or interfering with, second messenger pathways inside cells (see Table 26.1). For light to photolyse such a molecule at its site of action, the caged compound must be present inside the cell. Methods of incorporating caged compounds into cells are dictated by the preparation under study and by the chemical properties of the probe. For example, the caged cyclic nucleotides are lipid-soluble, and will therefore gain access to the cell interior simply by diffusing from the extracellular medium. However, these molecules may accumulate in lipid components of the cell, and this is likely to influence their concentration at the site of action. Similarly, the photolabile calcium buffer nitr-5 can be incorporated into cells in a non-invasive way, using its acetoxymethyl ester (nitr-5-AM) form. This diffuses across the cell membrane and becomes trapped inside the cell by the action of intracellular enzymes, which cleave the ester to release the free buffer. It can therefore be preloaded using methods that were developed earlier for loading fluorescent Ca indicators (Kao et al., 1989; Valdeolmillos et al., 1989). With this approach, the intracellular concentration of nitr-5 achieved is difficult to predict, and changes in Ca^{2+} concentration induced by photolysis are not known with certainty. On the other hand, nitr-5 can be co-loaded with fluorescent Ca^{2+} indicators, which then report the Ca^{2+} jumps induced by flashes (Kao et al., 1989; Gurney, 1991).

Most other caged second messengers are hydrophilic and do not readily cross the cell membrane. A number of compounds have been used with 'skinned' muscle fibres, in which the cell membranes have been removed (Goldman et al., 1982), or in cells permeabilized with staphylococcal α toxin (Somlyo et al., 1988; Somlyo & Somlyo, 1990), which makes membranes permeable to low-molecular-weight solutes while leaving receptor pathways intact. Most compounds can be incorporated into cells by injection, either through a fine micropipette that impales the cell, or through a fire-polished pipette used in the whole-cell configuration of the tight-seal, patch-clamp technique. Microinjection is the method of choice when second messenger concentrations are to be manipulated while measuring electrical properties of the cell membrane. Flash photolysis is well-suited to electrophysiological studies using the patch-clamp technique, either to study the activation or modulation of ion channels in isolated membrane patches (Chabala et al., 1985; Lester et al., 1986; Karpen et al., 1988), or in whole cells (e.g. Gurney et al., 1985, 1987, 1989; Chabala et al., 1986; Lester et al., 1986; Ogden et al., 1990). The low resistance pathway formed between the recording pipette and the cell interior in the whole-cell configuration, means that the solution filling the pipette equilibrates with, and controls the cell interior. Thus the concentration of the caged compound is known with some degree of certainty. A number of compounds have been successfully incorporated into cells in this way, and millimolar concentrations can be achieved. Provided care is taken to protect the photolabile probe in the pipette from the light, for example by coating the pipette with black Sylgard (Dow Corning, available from BDH), the solution in the pipette provides an essentially unlimited store of unphotolysed chelator. Reproducible responses to photolysis can thus be obtained in a single cell, and it is possible to examine reponses over a wide range of concentrations.

26.4.3 Instrumentation

The only specialized instrumentation required for photolysis experiments with caged compounds is a high-intensity light source, with output in the near *UV*. If time resolution is not important, then an ordinary xenon lamp equipped with a shutter can be used. However, to take advantage of the speed afforded by the photochemistry, the reaction should be started with a very brief flash of light. Flashlamps are available that produce light pulses of less than 1 ms duration with an output of >200 mJ between 300 and 400 nm. This is achieved by discharging a high-voltage capacitor across a short-arc flash tube, which encloses xenon gas. Lamps designed for the photolysis of caged compounds can be obtained from Chadwick Helmuth, El Monte, California (Strobex model 238) or from Hi-Tech Scientific Ltd., Salisbury, the UK distributors for Dr Rapp, Optoelektronik, Hartkrögen 65, D-2000 Hamburg 56, Germany (model JML or JML-E). In each case, the arc light is collected and focused with quartz lenses. The intensity of the flash is determined by the energy discharged through the lamp upon ignition, and is easily varied.

The output of xenon flashlamps has a broad spectrum of 250–1500 nm. The wavelengths most efficient for photolysis (300–400 nm) can be selected by placing filters in the light path. A band-pass filter, such as the Schott UG 11 (Ealing Electro-optics plc, Watford, UK) which shows peak transmission of about 80% at 320 nm, removes short wavelengths that may be damaging to cells and long wavelengths that may warm the cells. We frequently use broader spectrum cut-off filters that only remove wavelengths <300 nm, because this permits greater photolysis. The long wavelengths that remain do not usually present a problem, presumably because they are absorbed as they pass through aqueous solution on their way to the experimental preparation. This can be tested by observing the effects of a flash on the preparation in the absence of any photolabile probe.

It is often desirable to direct flashes through a light guide, which allows the flashlamp to be placed at a distance from the preparation (Fig. 26.3). Light guides are particularly convenient when flash photolysis is to be combined with electrophysiological measurements, because the large trigger pulse (12kV) and the discharge current (2000 A) are not easily shielded from the recording system. With these kinds of experiments it is also important to isolate mechanically the flashlamp from the recording system, because it thumps its housing when it discharges. Fibre optics do not transmit enough UV light to be useful. Liquid light guides have, however, been used successfully for studies with caged compounds. These light guides transmit well from 270 to 720 nm, although the efficiency of transmission depends on the length of the light guide. With a 1 m guide, maximum transmittance is about 80%, but it falls off fairly steeply below 400 nm, such that at 300 nm only 40% is transmitted. The loss of light in this region is unacceptably high if the light guide is used in combination with a UG11 filter. We prefer to remove short wavelengths by placing a borosilicate glass coverslip, about 0.1 mm thick (no. 0 thickness; BDH, Poole, Dorset, UK), at the input to the light guide. The coverslip cuts off sharply below 320 nm, but transmits greater than 90% at longer wavelengths.

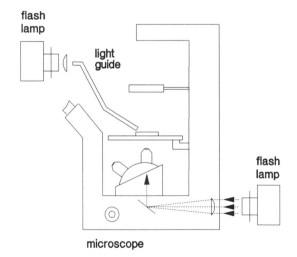

Figure 26.3 Light from a xenon short-arc flashlamp can be directly focused onto a preparation, or directed through a liquid light guide or the optical path of a microscope.

Flashlamps can also be directed onto cells *via* the optical path of a microscope (Fig. 26.3). We have used the epifluorescence port of an Olympus IMT2 inverted microscope. The flashlamp condenser is placed in front of the port and a 200 mm quartz lens focuses the light onto a dichroic mirror, which is selected to reflect maximally below 500 nm. The dichroic mirror is set at a 45° angle to reflect light through the objective lens, which should have as high a numerical aperture

(NA) as possible. With a Nikon 40×, NA 0.85 lens, we can photolyse ~4% of DMNPE-caged ATP with a single flash. Our microscope is limited by a fixed glass lens in the light path. UV-transmitting optics are available with new Olympus microscopes and the comparable Nikon Diaphot. With these it should be possible to achieve greater photolysis, although we have not made any direct comparisons.

Photolysis experiments can also be performed using a pulsed laser (reviewed in McCray & Trentham, 1989), which further improves the time resolution. These lasers have the advantage of emitting high-intensity light at a defined wavelength, with sub-microsecond resolution. Furthermore, the light is emitted as a parallel beam, which can be more easily directed onto small areas. However, lasers capable of producing UV light are expensive. They also tend to be laborious and expensive to maintain and they occupy a lot of bench space. The lasers that have most commonly been used for biological studies with caged compounds include pulsed dye lasers (e.g. Sheridan & Lester, 1982), which are tunable over a range of wavelengths and frequency-doubled ruby lasers (McCray et al., 1980), which produce 50 ns pulses at 347 nm. Others, such as XeF and XeCl excimer lasers, capable of producing near-UV wavelengths can also be used (McCray & Trentham, 1989).

26.4.4 Calibration

For the photosensitive probes to provide useful biological information it is important that the concentration change resulting from photolysis can be determined accurately. With caged calcium, Ca^{2+} release can be simultaneously measured with Ca^{2+} indicators (see Kao et al., 1989), but this is not possible with most other caged compounds. The extent of photolysis can be determined by measuring the concentration of the caged compound and/or its photoproducts in droplets of solution placed at the focal point, before and after flashing. To ensure that all of the drop is irradiated during photolysis, it must be smaller than the light spot that is focused on to it. To prevent evaporation from the droplet, it can be covered with a thin film of mineral oil. When photolysis is directed through a microscope, the area irradiated is very small (~400 μm diameter with 40× objective). To keep the droplet of test solution within this area, small volumes (<0.2 μl) can be micro-injected into a blob (~1 μl) of mineral oil. Flashes are then presented after focusing the objective on the centre of the droplet. Since the volume of the droplet in this case is hard to measure accurately, it is useful to include in the test solution a marker molecule, whose concentration is known and does not change with photolysis. For example, when calibrating the

photolysis of caged ATP in this way, we include a known concentration of adenosine monophosphate, which has a different retention time on high-performance liquid chromatography (HPLC) to ATP and caged ATP.

The concentration of a caged compound may be measured spectrophotometrically or by HPLC analysis. Most HPLC methods use a C18 reversed-phase column or ion-exchange column. Various conditions have been described for separating different caged compounds from their photolysis products using HPLC (Walker, 1991). These compounds include nucleotides (Walker et al., 1989a), cyclic nucleotides (Nerbonne et al., 1984; Wootton & Trentham, 1989), IP3 (Walker et al., 1989b), calcium (Walker, 1991) and carbachol (Milburn et al., 1989). If one flash does not cause detectable photolysis, then separate droplets can be flashed a varying number of times, and the fraction of molecules photolysed by a single flash extrapolated.

26.5 POTENTIAL PROBLEMS ASSOCIATED WITH THE USE OF CAGED COMPOUNDS

There are three possible artifacts in flash photolysis experiments with caged compounds: (1) the caged compound could itself have effects on the tissue even in the dark, (2) the side products of photolysis could be biologically active, and (3) the flash itself might trigger a biological response. Ideally, the caged compound should be inert, and flashes in the 300–400 nm wavelength range should cause no measurable physiological artifact in the absence or presence of the probe. So far, these conditions appear to have been fulfilled with most preparations that are not normally thought of as light-sensitive, and with many of the available caged compounds. However, a few examples of these artifacts have been reported.

26.5.1 Effects of the precursor

Perhaps the most widely studied caged compound is NPE-caged ATP, which in studies of cross-bridge kinetics in skeletal (e.g. Goldman et al., 1984a,b) and smooth (Somlyo et al., 1988) muscle behaved simply as an inert precursor for ATP. Although this analogue has also been successfully exploited to study the kinetics of activation of the sodium pump by ATP, it was found to bind to the same site on the enzyme as free ATP, albeit with lower affinity (Forbush, 1984). It has since been shown to bind to myosin as well (Dantzig et al., 1989). Unphotolysed caged ATP also

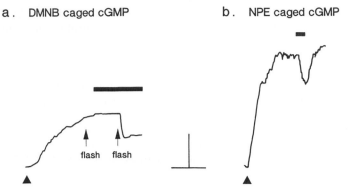

a. DMNB caged cGMP   b. NPE caged cGMP

flash flash

Figure 26.4 Comparison of the effects of DMNB-caged cGMP and NPE-caged cGMP on isometric tension measured in a strip of rabbit main pulmonary artery. Strips were precontracted by changing the perfusing solution to one containing 20 mM K^+ as indicated by the filled triangles. They were then perfused with either DMNB-caged cGMP at 100 µM (a) or NPE-caged cGMP at 10 µM (b), for the times indicated by the horizontal bars. In the presence of the DMNB analogue, the muscle relaxed only when presented with a light flash. In contrast, the NPE analogue caused relaxation on its own. Calibration bars represent 10 min and 50 mg (a) or 40 mg (b). Records a and b from different tissues.

causes partial blockade of ATP-sensitive K^+ channels in cardiac muscle, an effect observed with both the NPE and the DMNPE derivatives (Nichols *et al.*, 1990). In pancreatic β-cells, DMNPE-caged ATP was almost as effective at blocking this type of channel as MgATP, the physiological regulator (Ammälä *et al.*, 1991). Similar blockade of an ATP-sensitive K^+ current occurs in vascular muscle cells with NPE-caged ATP in the absence of light, but in this tissue the DMNPE derivative (up to 20 mM) appears to block the current only after photolysis (Clapp & Gurney, 1992).

The first caged cyclic nucleotides (NB esters) to be synthesized (Engels & Schlaeger, 1977; Engels & Reidys, 1978) were employed originally as lipophilic precursors that could spontaneously hydrolyse inside cardiac muscle cells to generate the free cyclic nucleotides (Korth & Engels, 1979). It proved possible to use NB cAMP as a photolabile precursor in heart muscle, because flashes released sufficient cAMP to enhance the slow inward current at concentrations where pre-flash effects were small (Nargeot *et al.*, 1983). However, the preferred precursor in cardiac tissue is DMNB-caged cAMP, which in the absence of light has no effect on the current or on muscle tension at much higher concentrations than can be tolerated with the NB ester (Nerbonne *et al.*, 1984; Richard *et al.*, 1985). We have similarly compared the effects of the photolabile DMNB and NPE esters of cGMP in vascular muscle (Fig. 26.4), where a rise in intracellular cGMP is thought to mediate relaxation. Without irradiation, concentrations of DMNB-caged cGMP up to 200 µM have no effect on tonic tension developed by exposing the muscle to elevated extracellular K^+, although in the presence of this analogue

flashes do produce relaxation (Fig. 26.4(a)). In contrast, NPE-caged cGMP by itself induces large relaxations even at 50 µM (Fig. 26.4(b)), with light flashes having little further effect (not shown). A similar difference in the activities of these caged cGMP analogues has also been noted in isolated cardiac cells (Charnet & Richard, pers. commun.). It is not yet clear whether the effects of NPE-caged cGMP arise as a result of intracellular hydrolysis to release free cGMP, from a direct interaction with cGMP-dependent kinase or from an interaction with some other cGMP-binding site.

In many tissues, a fall in the intracellular pH decreases the conductance of gap junctions between pairs of cells. The photolabile proton donor, *o*-nitrobenzyl acetate, was used successfully to uncouple cells in the salivary gland of the midge *Chironomus*, where protons were released only upon photolysis (Nerbonne *et al.*, 1982). However, in a number of other cell types the same caged compound reversibly decreased junctional conductance at low concentrations, even without photolysis (Spray *et al.*, 1984). The effect was clearly due to intracellular cleavage of the probe, because it was associated with a fall in the intracellular pH.

Effects of other caged compounds have also been noted in the absence of light. For example, NPE-caged carbachol, an early version of the probe, was found to be active at both nicotinic and muscarinic acetylcholine receptors (Walker *et al.*, 1986). Introducing a carboxylate group into the benzyl carbon to make *N*-(α-carboxy-2-nitrobenzyl) carbachol, subsequently eliminated this activity, while flashes presented in its presence produced rapid activation of nicotinic receptors in BC_3H1 cells (Milburn *et al.*, 1989).

Bis-Q, the photoisomerizable nicotinic receptor agonist suffers from a similar problem. Although the *cis* isomer is much less potent than the *trans* isomer, it is not completely inactive (Lester & Nerbonne, 1982; Nerbonne *et al.*, 1983; Lester *et al.*, 1986). Both isomers also act as antagonists at muscarinic receptors, although the *cis* isomer is less potent (Nargeot *et al.*, 1982). Some photolabile molecules that are active before photolysis can be rapidly inactivated with light, and this property can also be exploited. For example, the calcium antagonist drug nifedipine is rapidly destroyed with a flash, an approach that has been useful in probing the mechanisms by which the drug interacts with voltage-dependent calcium channels (Gurney *et al.*, 1985; Nerbonne *et al.*, 1985).

26.5.2 Effects of the photolysis side products

Photorelease of the 'protected' molecule from *o*-nitrobenzyl derivatives occurs together with the release of a proton and a nitroso by-product. In most cases the pH change resulting from proton loss can be kept to a minimum by heavily buffering the experimental solution. This may not be possible with lipophilic compounds used as intracellular probes, where access of a buffer to the cell interior may be restricted. Nevertheless, in either case the contribution of protons to a response can be evaluated by carrying out control experiments with caged protons. Although, as indicated above, *o*-nitrobenzyl acetate spontaneously hydrolyses in a number of cell types, there are other caged proton compounds that could be used (Janko & Reichert, 1987; Shimada & Berg, 1987). The nitroso photoproducts present a potentially greater problem; they precipitate out of aqueous solution and are highly reactive toward sulphydryl groups on proteins (McCray *et al.*, 1980). Their reactivity can be reduced by derivatizing the benzyl carbon (McCray *et al.*, 1980) with, for example, a methyl group, as in the NPE derivatives ($R_1 = CH_3$ in Fig. 26.1(b)). Despite this, the nitrosoacetophenone produced alongside photolysis of NPE-caged ATP does have effects on cell proteins, such as inactivating the sodium pump of erythrocyte ghosts (Kaplan *et al.*, 1978). The deleterious effects of the nitroso photoproducts can be avoided by including a hydrophilic thiol, such as glutathione or dithiothreitol (DTT), in the solution, usually at millimolar concentrations (Kaplan *et al.*, 1978; Walker *et al.*, 1988; MCray & Trentham, 1989). Fortunately, most receptor ligands and intracellular messengers work in the micromolar concentration range, so effective concentrations of the biologically interesting molecule can be produced with relatively small amounts of the by-product. The photoproducts are only likely to cause a problem when high concentrations of a caged compound are photolysed,

for example, when millimolar concentrations of ATP are needed to activate muscle contraction.

26.5.3 Direct effects of light

Light in the wavelength region of 250 nm is well-known to be damaging to cells. However, there is no evidence that light of 300–400 nm causes any photo-dynamic damage over the time-course of most physiological experiments, even after exposure to many flashes. Other than in preparations that are expected to respond to light (e.g. photoreceptors) there are only a few reports that light of these wavelengths has any measurable effect in the absence of light-sensitive molecules. Flashes on their own transiently activated voltage-clamp currents in intact strips of frog atrial muscle, apparently through a direct effect on the myocardial tissue (Nargeot *et al.*, 1982). However, these effects were small enough and reproducible enough to be subtracted from the responses of interest (Nargeot *et al.*, 1982, 1983). Furthermore, these non-specific effects of light were absent in later studies on isolated cardiac cells (Gurney *et al.*, 1985, 1989; Naebauer *et al.*, 1989).

Light is known to have pronounced effects on blood vessels. Photo-induced relaxation of vascular muscle was first described by Furchgott *et al.* in 1955, and shown to depend on the same wavelengths as those used for flash photolysis experiments. Most studies of the photo-induced relaxation have employed low-level, maintained illumination. However, as illustrated in Fig. 26.5, brief, intense flashes of similar wavelengths and intensity to those that have been widely used in many experiments to photolyse caged compounds, can also relax arterial muscle. The response is quite variable, with some tissues showing pronounced relaxation (Fig. 26.5), while others show

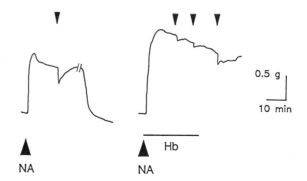

Figure 26.5 Effect of flashes on isometric tension in a strip of rabbit main pulmonary artery. A strip of artery was perfused continuously with 6 μM noradrenaline (NA), and a flash presented during the plateau phase of the contraction. A single flash relaxed the tissue, but this effect was blocked by 5 μM haemoglobin.

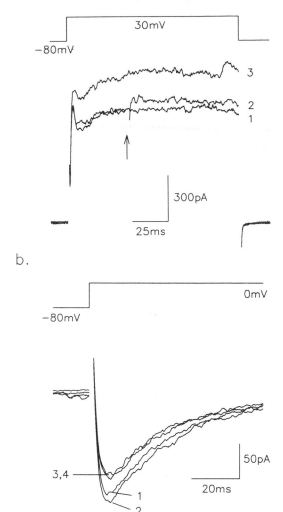

Figure 26.6 Flashes alter outward potassium (a) and inward calcium-channel (b) currents, recorded using the whole-cell patch-clamp technique from isolated smooth muscle cells of rabbit pulmonary artery. Voltage protocols used to activate the currents are illustrated above each set of traces. Currents were activated at 5 s intervals and were recorded in the order indicated by the numbers next to each trace. Experimental details were as described in Clapp and Gurney (1991a, b). The cell in (a) was perfused with physiological solution, and the recording pipette contained an isotonic, HEPES-buffered K^+ solution with 1 mM EGTA. A flash presented to the cell during sweep 2 (arrow) increased current amplitude. The cell in (b) was perfused with a Ba^{2+} (10 mM) containing solution, and the recording pipette contained Cs^+ in place of K^+ to block outward current. A flash presented between sweeps 2 and 3 suppressed the current.

no response (as in Fig. 26.4) or occasionally contraction. The basis of this photo-induced relaxation has been quite widely studied, and evidence is increasing that light stimulates the cytosolic guanylate cyclase

either directly (Karlsson et al., 1984) or more likely indirectly (Furchgott et al., 1985; Wigilius et al., 1990; Wolin et al., 1991). Thus it is possible to block the photo-induced relaxation by exposing the tissue to haemoglobin (Fig. 26.5(b)), which inhibits the activation of guanylate cyclase (Matsunaga & Furchgott, 1989; Furchgott & Jothianandan, 1991). Whatever the mechanism, it is clear that light elevates the intracellular cGMP concentration (Karlsson et al., 1984) and mimics the effect of nitrovasodilator drugs as well as the endothelium-derived relaxing factor (EDRF), which plays an important role in the normal regulation of vascular tone. Flash effects independent of caged compounds can also be observed in isolated vascular muscle cells (Fig. 26.6), where changes in membrane potassium and calcium currents can be observed in the absence of photolabile probes (see also Komori & Bolton, 1991). Again the mechanisms underlying these changes are unclear, but the effects of light resemble those of nitrovasodilator drugs (Clapp et al., 1990; Clapp & Gurney, 1991a). For these reasons, the application of photolabile caged compounds in vascular muscle is not as straightforward as it is in other tissues.

Photo-induced relaxation may only be a problem with intact smooth muscle cells, since a number of caged compounds have been applied successfully to study contraction mechanisms in permeabilized vascular muscle, apparently without interference from the activating light (Somlyo et al., 1988; Somlyo, 1990). Agents that elevate cytoplasmic cGMP only relax tissues that have been precontracted by agonists or elevated K^+ concentrations, and this is also true of light (e.g. Furchgott et al., 1955). Thus, although flashes may not appear to have an effect under basal conditions, in some tissues they may well trigger biochemical reactions that might alter the response to photoactivation of a caged compound. This could be tested by presenting flashes to preparations in which the pathway of interest has already been stimulated. Direct effects of light could have important consequences for preparations other than vascular smooth muscle which contain guanylate cyclase, because EDRF, which is thought to be nitric oxide, activates this enzyme and is increasingly being shown to have widespread biological effects (Moncada et al., 1991).

26.6   CONCLUSION

Photolabile caged compounds have recently become popular probes for the study of biological functions. They provide a means of rapidly changing the concentration of a biologically active molecule at its site of action, either inside a cell or at its extracellular surface. The technique is, however, still in its infancy, and it is likely that many more useful caged compounds will be developed in the near future. In this chapter I have tried to point out some of the problems that might be encountered in using these probes, and ways in which they might be overcome. Nevertheless, in most cases so far, photolabile caged compounds have provided helpful biological information, with relatively few problems encountered.

ACKNOWLEDGEMENTS

I would like to thank Mr G. Allerton-Ross, Ms S. Morgans and Dr L.H. Clapp for allowing me to use the unpublished data in Figs. 26.4–26.6. I also wish to thank Dr S. Bates and Dr Clapp for reading the manuscript and helpful discussions. The work in my laboratory is funded by grants from the Medical Research Council, the British Heart Foundation, the Wellcome Trust and the Royal Society.

REFERENCES

Adams S.R., Kao J.P.Y., Grynkiewicz G., Minta A. & Tsien R.Y. (1988) *J. Am. Chem. Soc.* **110**, 3212–3220.

Adams S.R., Kao J.P.Y. & Tsien R.Y. (1989) *J. Am. Chem. Soc.* **111**, 7957–7968.

Ammälä C., Bokvist K., Galt S. & Rorsman P. (1991) *Biochim. Biophys. Acta.* **1092**, 347–349.

Blank M., Soo L.M., Wassermann N.H. & Erlanger B.F. (1981) *Science* **214**, 70–72.

Chabala L.D., Gurney A.M. & Lester H.A. (1985) *Biophys. J.* **48**, 241–246.

Chabala L.D., Gurney A.M. & Lester H.A. (1986) *J. Physiol.* **371**, 407–433.

Clapp L.H. & Gurney A.M. (1991a) *Pflügers Arch.* **418**, 462–470.

Clapp L.H. & Gurney A.M. (1991b) *Exp. Physiol.* **76**, 677–693.

Clapp L.H. & Gurney A.M. (1992) *Am. J. Physiol. (Heart)* **262**, H916–H920.

Clapp L.H., Allerton-Ross G. & Gurney A.M. (1990) *Biophys. J.* **57**, 158a (abstract).

Dantzig J.A., Goldman Y.E., Luttman M.L., Trentham D.R. & Woodward S.K.A. (1989) *J. Physiol.* **419**, 64P (abstract).

Ellis-Davies G.C.R. & Kaplan J.H. (1988) *J. Org. Chem.* **53**, 1966–1969.

Engels J. & Reidys R. (1978) *Experentia Basel* **34**, 14–15.

Engels J. & Schlaeger E.-J. (1977) *J. Med. Chem.* **20**, 907–911.

Ferenczi M.A., Goldman Y.E. & Trentham D.R. (1989) *J. Physiol.* **418**, 155P (abstract).

Forbush B. III (1984) *Proc. Natl. Acad. Sci. USA* **81**, 5310–5314.

Furchgott R.F. & Jothianandan D. (1991) *Blood Vessels* **28**, 52–61.

Furchgott R.F., Sleator W., McCaman M.W. & Elchlepp I. (1955) *J. Pharmac. Exp. Ther.* **113**, 22–23.

Furchgott R.F., Martin W., Cherry P.D., Jothianandan D. & Villani G.M. (1985) In *Vascular Neuroeffector Mechanisms*, J.A. Bevan, T. Godfraind, R.A. Maxwell, J.C. Stoclet & M. Worcel (eds). Elsevier, Amsterdam, pp. 105–114.

Goldman Y.E., Hibberd M.G., McCray J.A. & Trentham D.R. (1982) *Nature* **300**, 701–705.

Goldman Y.E., Hibberd M.G. & Trentham D.R. (1984a) *J. Physiol.* **354**, 577–604.

Goldman Y.E., Hibberd M.G. & Trentham D.R. (1984b) *J. Physiol.* **354**, 605–624.

Gurney A.M. (1991) In *Cellular Neurobiology: A Practical Approach*, J. Chad & H. Wheal (eds). IRL Press at Oxford University Press, Oxford, pp. 153–177.

Gurney A.M. & Lester H.A. (1987) *Physiol. Rev.* **67**, 583–617.

Gurney A.M., Nerbonne J.M. & Lester H.A. (1985) *J. Gen. Physiol.* **86**, 353–379.

Gurney A.M., Tsien R.Y. & Lester H.A. (1987) *Proc. Natl. Acad. Sci. USA* **84**, 3496–3500.

Gurney A.M., Charnet P., Pye J.M. & Nargeot J. (1989) *Nature* **341**, 65–68.

Homsher E. & Millar N.C. (1990) *Ann. Rev. Physiol.* **52**, 875–896.

Janko K. & Reichert J. (1987) *Biochim. Biophys. Acta* **905**, 409–416.

Kano K., Tanaka Y., Ogawa T., Shimomura M. & Kunitake T. (1981) *Photochem. Photobiol.* **34**, 323–329.

Kao J.P.Y., Harootunian A.T. & Tsien R.Y. (1989) *J. Biol. Chem.* **264**, 8179–8184.

Kaplan J.H. (1986) In *Optical Methods in Cell Physiology (Soc. Gen. Physiol. Ser.)*, P. De Weer & B. Salzberg (eds). Wiley, New York, pp. 385–396.

Kaplan J.H. (1990) *Ann. Rev. Physiol.* **52**, 897–914.

Kaplan J.H. & Ellis-Davies G.C.R. (1988) *Proc. Natl. Acad. Sci. USA* **85**, 6571–6575.

Kaplan J.H. & Somlyo A.P. (1989) *Trends Neurosci.* **12**, 54–59.

Kaplan J.H., Forbush B. III & Hoffman J.F. (1978) *Biochemistry* **17**, 1929–1935.

Karlsson J.O.G., Axelsson K.L. & Andersson R.G.G. (1984) *Life Sci.* **34**, 1555–1563.

Karpen J.W., Zimmerman A.L., Stryer L. & Baylor D.A. (1988) *Proc. Natl. Acad. Sci. USA* **85**, 1287–1291.

Klodos I. & Forbush B. III (1988) *J. Gen. Physiol.* **92**, 46a (abstract).

Komori S. & Bolton T.B. (1991) *Pflügers Arch.* **418**, 437–441.

Korth M. & Engels J. (1979) *Naunyn-Schmiedeberg's Arch. Pharmacol.* **310**, 103–111.

Lando B. & Zucker R.S. (1988) *J. Gen. Physiol.* **93**, 1017–1060.

Lester H.A. & Chang H.W. (1977) *Nature* **266**, 373–374.

Lester H.A. & Nerbonne J.M. (1982) *Ann. Rev. Biophys. Bioengng* **11**, 151–175.

Lester H.A., Chabala L.D., Gurney A.M. & Sheridan R.E. (1986) In *Optical Methods in Cell Physiology (Soc. Gen.*

Physiol. Ser.), P. De Weer & B. Salzberg (eds). Wiley, New York, pp. 447–462.

McCray J.A. & Trentham D.R. (1989) *Ann. Rev. Biophys. Chem.* **18**, 239–270.

McCray J.A., Herbette L., Kihara T. & Trentham D.R. (1980) *Proc. Natl. Acad. Sci. USA* **77**, 7237–7241.

Matsunaga K. & Furchgott R.F. (1989) *J. Pharmac. Exp. Ther.* **248**, 687–695.

Messenger J.B., Katayama Y., Ogden D.C., Corrie J.E.T. & Trentham D.R. (1991) *J. Physiol.* **438**, 293P.

Milburn T., Matsubara N., Billington A.P., Udgaonkar J.B., Walker J.W., Carpenter B.K., Webb W.W., Marque J., Denk W., McCray J.A. & Hess G.P. (1989) *Biochemistry* **28**, 49–55.

Moncada S., Palmer R.M.J. & Higgs E.A. (1991) *Pharmacol. Rev.* **43**, 109–142.

Nargeot J., Lester H.A., Birdsall N.J.M., Stockton J., Wassermann N.H. & Erlanger B.F. (1982) *J. Gen. Physiol.* **79**, 657–678.

Nargeot J., Nerbonne J.M., Engels J. & Lester H.A. (1983) *Proc. Natl. Acad. Sci. USA* **80**, 2395–2399.

Naebauer M., Ellis-Davies G.C.R., Kaplan J.H. & Morad M. (1989) *Am. J. Physiol.* **256**, H916–H920.

Nerbonne J.M. (1986) In *Optical Methods in Cell Physiology (Soc. Gen. Physiol. Ser.)*, P. De Weer & B. Salzberg (eds). Wiley, New York, pp. 417–445.

Nerbonne J.M. & Gurney A.M. (1987) *J. Neurosci.* **7**, 882–893.

Nerbonne J.M., Lester H.A. & Connor J.A. (1982) *Soc. Neurosci. Abstr.* **8**, 945a.

Nerbonne J.M., Sheridan R.E., Chabala L.D. & Lester H.A. (1983) *Molec. Pharmacol.* **23**, 344–349.

Nerbonne J.M., Richard S., Nargeot J. & Lester H.A. (1984) *Nature* **310**, 74–76.

Nerbonne J.M., Richard S. & Nargeot J. (1985) *J. Molec. Cell. Cardiol.* **17**, 511–515.

Nichols C.G., Niggli E. & Lederer W.J. (1990) *Pflügers Arch.* **415**, 510–512.

Ogden D.C., Capiod T., Walker J.W. & Trentham D.R. (1990) *J. Physiol.* **422**, 585–602.

Patchornik A., Amit B. & Woodward R.B. (1970) *J. Am. Chem. Soc.* **92**, 6333–6335.

Richard S., Nerbonne J.M., Nargeot J., Lester H.A. & Garnier D. (1985) *Pflügers Arch.* **403**, 312–317.

Sheridan R.E. and Lester H.A. (1982) *J. Gen. Physiol.* **80**, 499–515.

Shimada K. & Berg H.C. (1987) *J. Molec. Biol.* **193**, 585–589.

Shinkai S., Matsuo K., Harada A. & Manabe O. (1982) *J. Chem. Soc. Perkin Trans.* **2**, 1261–1265.

Somlyo A.P. & Somlyo A.V. (1990) *Ann. Rev. Physiol.* **52**, 857–874.

Somlyo A.P. Walker J.W., Goldman Y.E., Trentham D.R., Kobayashi S., Kitazawa T. & Somlyo A.V. (1988) *Phil. Trans. R. Soc. Lond. B* **320**, 399–414.

Spray D.C., Nerbonne J.M., Campos De Carvalho A., Harris A.L. & Bennet M.V.L. (1984) *J. Cell Biol.* **99**, 174–179.

Tsien R.Y. (1980) *Biochemistry* **19**, 2396–2404.

Tsien R.Y. & Zucker R.S. (1985) *Biophys. J.* **50**, 843–853.

Valdeolmillos M., O'Neill S.C., Smith G.L. & Eisner D.A. (1989) *Pflügers Arch.* **413**, 676–678.

Walker J.W. (1991) In *Cellular Neurobiology: A Practical Approach* J. Chad & H. Wheal (eds). IRL Press at Oxford University Press, Oxford. pp. 179–203.

Walker J.W. & Trentham D.R. (1988) *Biophys. J.* **53**, 596a.

Walker J.W., McCray J.A. & Hess G.P. (1986) *Biochemistry* **25**, 1799–1805.

Walker J.W., Somlyo A.V., Goldman Y.E., Somlyo A.P. & Trentham D.R. (1987) *Nature* **327**, 249–252.

Walker J.W., Reid G.P., McCray J.A. & Trentham D.R. (1988) *J. Am. Chem. Soc.* **110**, 7170–7177.

Walker J.W., Reid G.P. & Trentham D.R. (1989a) *Methods Enzymol.* **172**, 288–301.

Walker J.W., Feeney J. & Trentham D.R. (1989b) *Biochemistry* **28**, 3272–3280.

Wigilius I.M., Axelsson K.L., Andersson R.G.G., Karlsson J.O.G. & Odman S. (1990) *Biochem. Biophys. Res. Commun.* **169**, 129–135.

Wolin M.S., Omar H.A., Mortelliti M.P. & Cherry P.D. (1991) *Am. J. Physiol.* **261**, H1141–H1147.

Wootton J.F. & Trentham D.R. (1989) In *Photochemical Probes in Biochemistry*, P.E. Nielsen (ed.). NATO ASI series C, Vol. 272. Kluwer, Dordrecht, pp. 277–296.

Zucker R.S. & Hayden P.G. (1988) *Nature* **335**, 360–362.

CHAPTER TWENTY-SEVEN

Fluorescent Analogues: Optical Biosensors of the Chemical and Molecular Dynamics of Macromolecules in Living Cells

K. HAHN, J. KOLEGA, J. MONTIBELLER, R. DeBIASIO, P. POST, J. MYERS, & D.L. TAYLOR

Center for Light Microscope Imaging and Biotechnology, Carnegie Mellon University, Pittsburgh, PA, USA

27.1 INTRODUCTION

Fluorescent analogue cytochemistry has grown to produce important insights in a wide variety of fields since 1978, when it was first demonstrated that a protein labelled with a fluorescent dye could function and be investigated within living cells (Taylor & Wang, 1978). The fluorescein-actin used in the first experiments exemplified an approach to analogue production which has been used in the large majority of studies to date. Almost all analogues have been made with dyes whose fluorescence is minimally affected by the intracellular environment, thus focusing on analysis of the analogue's distribution within live cells. Meaningful information has been obtained through adherence to several important principles (Taylor & Wang, 1980; Taylor *et al.*, 1984; Simon & Taylor, 1986; Wang, 1989). Chief among these has been careful *in vitro* characterization of the labelled protein prior to interpretation of live cell data (Wang & Taylor, 1980). Minimal perturbation of the analogues' biological function has been sought during labelling, and the effects of dye attachment have been carefully determined. Furthermore, the smallest possible quantity of analogue has been injected to avoid alteration of normal cell function.

Environmentally insensitive fluorescent analogues of many different proteins have been successfully observed in live cells. Examples include analogues of actin-binding proteins (myosin, vinculin, α-actinin), which have revealed complex changes in the cytoskeleton during a number of physiological processes (Wang *et al.*, 1982; Kreis & Birchmeier, 1982; Simon & Taylor, 1986; Kolega & Taylor, 1991). Labelled tubulin has shown the location and extent of tubulin exchange in the mitotic spindle, thus testing different models of spindle function (Salmon *et al.*, 1984; Wadsworth & Salmon, 1986; Gorbsky *et al.*, 1987). Calmodulin analogues have revealed the changing distribution of calmodulin during mitosis, and indicated transient calmodulin association with various subcellular organelles (Zavortink *et al.*, 1983; Luby-Phelps *et al.*, 1985). A recent addition to this family of probes have been protein analogues labelled with 'caged' fluorophores, whose fluorescence is activated by irradiation (Ware *et al.*, 1986; Mitchison, 1989; Theriot & Mitchison, 1991). This approach has permitted the marking of a subpopulation of analogues for temporal–spatial tracing at high contrast (Mitchison, 1989; Theriot & Mitchison, 1991).

More recently, the first representatives of a new family of fluorescent protein analogues have been developed using environmentally sensitive fluorophores, as proposed earlier (Taylor & Wang, 1980). These analogues have served as indicators of protein activity, in that the attached dye reflected conformational changes or altered ligand binding by the proteins. An environmentally sensitive dye has been used to make an analogue of calmodulin whose fluorescence reflects calcium–calmodulin binding (Hahn *et al.*, 1990). This analogue has been used in conjunction with a fluorescent calcium indicator to correlate the spatial and temporal dynamics of calcium transients and calcium–calmodulin binding in individual, living cells. Details of these studies are described below. Resonance energy transfer between two fluorophores has been applied in an alternate approach to fluorescent sensing of protein activity. Energy transfer is strongly affected by protein changes that alter the distance between donor and acceptor dyes. The approach has been used to monitor actin polymerization *in vitro* (Taylor *et al.*, 1981; Wang & Taylor, 1981). When mixed fluorescein-actin and rhodamine-actin monomers were brought in proximity during polymerization energy transfer increased dramatically. More recently, the catalytic and regulatory subunits of protein kinase A have been labelled with fluorescein and rhodamine (Adams *et al.*, 1991). Dissociation of the subunits by cAMP was observed in live cells as a decrease in energy transfer.

In our laboratory, we are harnessing environmentally sensitive and insensitive protein analogues to decipher the mechanics and regulation of cell motility. We will describe here our efforts to use fluorescent analogue cytochemistry in elucidating the mechanisms by which extracellular stimuli induce cell contraction and initiation of motility. Our aim is to define the often rapid changes in second messengers involved in this process, and their effects on the dynamics of contractile proteins in time and space during cell function.

In the currently established paradigm of stimulus–contraction coupling in non-muscle cells (for reviews see McNeil & Taylor, 1987; Sellers and Adelstein, 1987), binding of ligands to external receptors leads to intracellular production of lipid metabolites and alterations in the concentration and distribution of calcium. We are initially focusing on subsequent regulatory steps involving modulation of the calcium signal through calmodulin and myosin light chain kinase. Calcium binding to calmodulin enables calmodulin to activate myosin light chain kinase, which phosphorylates myosin II regulatory light chains. This phosphorylation activates myosin for contraction.

We will describe protein-based fluorescent indicators (optical biosensors) for observation of individual signalling steps in this process. The production and application of fluorescent analogues used as biosensors of calcium–calmodulin binding will be described, as

will progress towards indicators of myosin light chain phosphorylation. The use of fluorescent analogues of actin and myosin to study the dynamics of contractility will also be described.

We study serum stimulation of quiescent fibroblasts (McNeil & Taylor, 1987) and a 'wound healing' model in which a gap is introduced in a monolayer of fibroblasts using a blunt razor blade (DeBiasio *et al.*, 1988; Fisher *et al.*, 1988). Fibroblasts at the edge of this 'wound' migrate into the gap. In these systems, the timing of stimulation can be controlled, and during wound healing, polarized movement occurs with a predetermined orientation. Cells are observed with a multimode microscope imaging system that permits rapid switching between transmitted light microscopy and various forms of fluorescence microscopy using different detectors and excitation and emission filters (Giuliano *et al.*, 1990).

27.2   MeroCaM 1 AND 2; FLUORESCENT INDICATORS OF CALCIUM–CALMODULIN BINDING

Fluorescent calcium indicators have been invaluable in elucidating the kinetics and intracellular distribution of calcium transients during a wide range of cellular processes (Tsien, 1989). However, subsequent signalling steps have to date remained inaccessible to observation *in vivo*. We have designed a fluorescent indicator of calcium–calmodulin binding which we are now using to correlate intracellular calcium changes with calmodulin activation during growth factor stimulation of quiescent fibroblasts (Hahn *et al.*, 1990). We are correlating the spatial and temporal dynamics of calcium changes and calcium-calmodulin binding. Biochemical evidence indicates that calmodulin's calcium response will depend on its intracellular environment. Calmodulin's calcium affinity is affected by target proteins (Cohen & Klee, 1988), and calmodulin binds to different targets at different calcium levels (Andreason *et al.*, 1983; Cohen & Klee, 1988). Changing subcellular distribution of calmodulin seen in previous studies using environmentally insensitive fluorescent analogues supports regulation of calmodulin by factors in addition to calcium (Pardue *et al.*, 1981; Zavortink *et al.*, 1983; Luby-Phelps *et al.*, 1985).

We have designed our indicators of calcium-calmodulin binding on the basis of previous biochemical studies which show that calmodulin–calcium binding produces a hydrophobic site on calmodulin with affinity for specific small molecules (LaPorte *et al.*, 1980, 1981; Malencik *et al.*, 1981; Manalan & Klee, 1984; Cohen & Klee, 1988). Structure-activity studies enabled synthesis of a novel dye with both highly solvent-sensitive fluorescence and strong affinity for calmodulin's calcium-induced hydrophobic binding site. Through affinity labelling, this dye was covalently attached to calmodulin in a position where it would have access to the calcium-induced hydrophobic pocket. Whenever calcium bound calmodulin, the solvent-sensitive dye moved into the hydrophobic pocket, with a consequent shift in fluorescence. The fluorescence change was shown to be fully reversible on removal of calcium. We named the calcium–calmodulin binding indicator meroCaM 1, for merocyanine dye + calmodulin.

Spectral characterization of meroCaM 1 showed that its excitation spectrum changed in both intensity and peak shape in a calcium-dependent manner. The spectral changes were suitable for ratio imaging, an important technique used to normalize fluorescence of analogues in the cell for differences in cell thickness, excitation intensity, and other factors (Tanasugarn et al., 1984; Bright et al., 1989). The excitation ratio varied sigmoidally with calcium, and showed an overall 3.4-fold change. Unlike other solvent-sensitive fluorophores, the dye used in meroCaM 1 was designed specifically for use in live cells. Optimum excitation ratios were obtained at 532 and 608 nm (emission 623 nm), long wavelengths which do not cause photodamage of cells and are far from the autofluorescence of mammalian cells. The dyes had high absorbance and quantum yield values, decreasing the amount of intracellular protein analogue required to obtain acceptable fluorescence signal.

In order to apply meroCaM 1 to study regulation of calmodulin activity, it was important to characterize the analogue's target protein binding, and to show that the effects of target proteins on calcium affinity remained intact. MeroCaM 1 was shown to retain calmodulin's ability to activate myosin light chain kinase, a calmodulin target protein. It also showed calcium-dependent mobility on native and SDS gels. Known effects of melittin and cAMP-phosphodiesterase on calcium affinity were also observed in meroCaM 1.

The calcium-induced change in meroCaM 1 fluorescence was half maximal at 0.3–0.4 μM, below the 5–10 μM apparent dissociation constant reported for overall calcium–calmodulin binding (Klee, 1988). This could have resulted from the dye stabilizing calmodulin's calcium-bound form, thus affecting the calcium binding constant. Alternately, the dye could have been positioned on the protein where it would reflect primarily binding to the high-affinity calcium sites. These possibilities were distinguished by comparing the calcium dependence of meroCaM 1 and calmodulin cAMP-phosphodiesterase activation. The similar behaviour of the two proteins in this assay indicated that meroCaM 1 fluorescence reflected

binding to high-affinity calcium sites. The effect of cAMP-phosphodiesterase binding on calmodulin's calcium affinity remained intact in meroCaM. The assay did show that meroCaM's maximal activation or affinity for the phosphodiesterase was somewhat reduced from that of calmodulin.

In order to assay the effect of target protein binding on meroCaM 1 fluorescence, the calcium dependence of the excitation ratio was determined in the presence and absence of melittin, a peptide mimic of target proteins' calmodulin binding sites (Comte et al., 1983; Maulet & Cox, 1983; Cox et al., 1985; Seeholzer et al., 1986). These experiments showed that the high and low excitation ratio values, produced at extremes of calcium concentration, were altered by melittin. Although this will complicate quantitation of the extent of calcium–calmodulin binding, quantitative analysis of binding kinetics should not be affected.

We have begun testing meroCaM 1 in live cells. The analogue has shown sufficient brightness and photostability to obtain greater than 10 ratio pairs, and has shown no apparent toxic effects or alteration of cell physiology. In our preliminary studies, meroCaM 1 has been injected in Swiss 3T3 fibroblasts made quiescent by 48 h incubation in low serum medium. Published studies and our own work have shown that these cells respond to whole serum with a rapid and transient upsurge in intracellular calcium (Byron & Villereal, 1989; Tucker & Fay, 1990; Takuwa et al., 1991; McNeil et al., 1985; Taylor, DeBiaiso & Hahn, (pers. commun.)). When the cells containing meroCaM 1 were challenged with whole serum, the intracellular meroCaM 1 excitation ratio increased 10–15% within 1 min and then returned to baseline over a period of several minutes. The filters used to excite and monitor the analogue's fluorescence have not been optimized, so it is anticipated that greater fluorescence responses will ultimately be obtained. Mapping of meroCaM 1 response within individual cells indicated spatial heterogeneity of calmodulin activation.

We have also developed new dyes in an effort to improve meroCaM 1, and have found that substitution of a benzoxazole moiety for benzothiazole in the meroCaM 1 dye results in a strong enhancement of fluorescence response. The new dye has been attached to calmodulin to produce a protein analogue which we call meroCaM 2. The excitation ratio of this indicator shows a 7-fold calcium-dependent change in excitation ratio (excitation 457, 569 nm; emission 587 nm). In preliminary studies, the indicator has been injected in quiescent Swiss 3T3 fibroblasts together with the fluorescent calcium indicator, Calcium Green (plate 27.1) (Molecular Probes, Eugene, OR). Stimulation of the cells with whole serum caused a rapid increase in calcium to maximal levels within 1 min, followed by a decrease which sometimes showed smaller additional maxima. The response of MeroCaM 2 to these calcium changes was quite complex. MeroCaM response generally also reached its maximum (up to 80% increase) within the first minute after stimulation, but afterwards sometimes paralleled calcium changes and sometimes underwent independent oscillations.

The correlation of calcium-calmodulin binding with calcium changes in living cells now appears to be within reach. We hope to extend the approach we have developed to produce indicators of calmodulin binding to target proteins. Published studies indicate that sensitivity of fluorescent calmodulin derivatives to calcium vs. protein binding is strongly dependent on dye structure and the position of dye attachment (LaPorte *et al.*, 1981; Mills *et al.*, 1988). We will test new dyes and explore the use of site-specific mutagenesis to introduce cysteines for attachment of dye at precise positions.

27.3   FLUORESCENT ANALOGUE OF MYOSIN II

We have produced and used a fluorescent analogue of myosin II to directly visualize the dynamics of myosin II during cell movement in single, living Swiss 3T3 fibroblasts. Smooth muscle myosin isolated from chicken gizzard was labelled with tetramethyl rhodamine iodoacetamide (DeBiasio *et al.*, 1988). The dye was covalently bound to both the myosin heavy chain and the 17 kDa light chain, leading to incorporation of 4–6 mol dye per mol protein. Before performing *in vivo* studies, it was essential to establish the biological activity of the analogue through careful biochemical characterization. The analogue was shown to retain native myosin's ability to assemble into filaments as measured by right-angle light scattering (Fig. 27.1), and both labelled and unlabelled filaments were depolymerized by 5 mM ATP. Fluorescent labelling did not affect the analogue's K-EDTA ATPase activity. Labelled and native myosin were phosphorylated to the same extent and at the same rate by chicken gizzard myosin light chain kinase (Fig. 27.2).

The positions of dye attachment on the heavy chain were determined using the papain digestion procedure of Nath *et al.* (1986). One fluorophore was located in the 70 kDa N-terminal portion of myosin's S1 region, and another in the S2 region. No fluorophore was detected in the 25 kDa C-terminal portion of the S1 head. Modification at the latter site has been shown to cause elevation of actin-activated MgATPase activity in unphosphorylated myosin, and to hinder myosin's ability to adopt a folded conformation (Chandra *et al.*, 1985). A gel filtration assay (Trybus

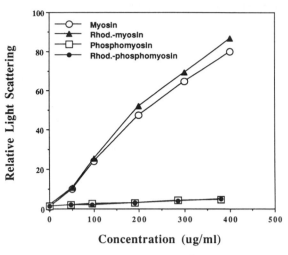

Figure 27.1 Myosin II and rhodamine-myosin II polymerization. The abilities of myosin II and its rhodamine analogue to polymerize were measured by right-angle light scattering using the procedure of McKenna *et al.* (1989). Unlabelled myosin and rhodamine-myosin II were dialysed into 500 mM KCl, 0.1 mM EDTA, 0.1 mM EGTA, 10 mM HEPES, pH 7.5. Dialysates were clarified by centrifugation and the concentrations measured spectrophotometrically using an extinction coefficient of E280 = 0.53 cm^{-1} (0.1%), correcting for 280 nm absorbance due to dye in the case of rhodamine-myosin II. The myosins were diluted 12-fold into an assembly buffer containing 150 mM KCl, 10 mM MgCl$_2$, 1 mM EGTA, 0.1 mM DTT, 10 mM HEPES, pH 7.5 at room temperature. Scattered light intensities were measured at 340 nm using a Perkin-Elmer MPF-3 fluorescence spectrophotometer. To test the sensitivity of the filaments to ATP, the assembly mixtures were made 5 mM in ATP and the measurements were repeated.

et al., 1982) indicated that folding was unaffected by labelling (Fig. 27.3). Actin-activated MgATPase activity was decreased by 0–40% (Table 27.1). In the absence of actin, the labelling had no effect on this activity.

After biochemical characterization, the analogue was injected in living Swiss 3T3 fibroblasts (DeBiasio *et al.*, 1988), where it permitted direct visualization of myosin II dynamics. Fluorescent myosin II became distributed throughout the cell within 1 h after injection, incorporating into the same structures and displaying the same periodic, 'pseudosarcomeric' distribution revealed by immunofluorescence staining of endogenous myosin II (Fig. 27.4) (DeBiasio *et al.*, 1988; McKenna *et al.*, 1989). Unlike immunofluorescence experiments, use of the analogue permitted observation of myosin dynamics over extended periods of time, and time-lapse studies revealed that myosin II was very dynamic. In fibroblasts made quiescent by serum-deprivation, myosin II assembled into stress fibres at the cells margins, moved centripetally toward the nucleus, and then disassembled in the perinuclear

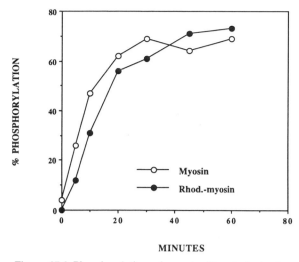

Figure 27.2 Phosphorylation of myosin II and rhodamine-myosin II. Myosin II or rhodamine-myosin II (*ca.* 4 mg ml^{-1}) were phosphorylated at room temperature in a mixture containing 25 mM Tris–HCl, 4 mM MgCl$_2$, 1 mM ATP, 15 μg ml^{-1} bovine brain calmodulin, and CaCl$_2$ 0.2 mM in excess of EGTA, pH 7.5. The reaction was started by the addition of 0.6 μg ml^{-1} smooth muscle MLCK in 1 mg ml^{-1} BSA in Tris. Aliquots of 20 μl were removed at intervals, and the reaction was terminated by dilution. Samples were run on glycerol–urea polyacrylamide gels (Perrie & Perry, 1970). The phosphorylated and dephosphorylated 20 kDa light chain bands were stained with Coomassie blue dye and the relative amounts of each were quantified by densitometry. The rate of phosphorylation was slightly higher for the unlabelled myosin, but this was due to the presence of endogenous kinase which had been eliminated from the rhodamine-myosin II during postlabelling purification.

region, presumably recycling to reassemble at the periphery (Giuliano & Taylor, 1990).

In addition to revealing constitutive cycling through the cytoplasm, the fluorescent analogue helped elucidate the nature of actin–myosin contractility. Stimulation of fibroblasts by serum or growth factors caused shortening of stress fibres with concomitant shortening of the pseudosarcomeric spacing (Giuliano & Taylor, 1990). These observations supported the sliding filament model of fibre shortening, as did quantitative analysis of the fibre shortening induced by cytochalasin, an actin-solating agent (Kolega *et al.*, 1991). The amount of fluorescence associated with stress fibres during cytochalasin-induced shortening was measured by integrating the intensity of digital images. The density of the analogue along fibres increased as the fibre shortened, indicating that myosin II became more concentrated during shortening. The *total* analogue associated with the fibre *decreased* during shortening, indicating a significant loss of the myosin II from the shortening fibre. This partial 'self-destruct' process is an important component of the solation-contraction hypothesis (Taylor & Fechheimer, 1982; Kolega *et al.*, 1991). Thus, the use of the myosin

II fluorescent analogue enabled demonstration of one of the regulatory mechanism of cell contractility.

The ability to follow myosin dynamics in living cells has also illuminated the regulation of cytoplasmic contractility. It has been shown *in vitro* and in extracted cell models that myosin assembly and motor activity is regulated by phosphorylation of its 20 kDa regulatory light chain (Sellers & Adelstein, 1987; Lamb *et al.*, 1988). The fluorescent analogue of myosin II revealed changes in myosin II organization *in vivo* when levels of phosphorylation were raised or lowered by inhibitors of intracellular phosphatases and kinases. When serum-deprived 3T3 fibroblasts were treated with okadaic acid, a phosphatase inhibitor, stress fibres contracted slowly, but incompletely (Fig. 27.5). In contrast, staurosporine, a kinase inhibitor, caused dissolution of fibres without shortening (Fig. 27.6). These observations are consistent with activation of contraction through myosin phosphorylation.

Coinjection of fluorescent actin and myosin analogues in the same cell provided information about these proteins' relative distribution during early stages of locomotion (DeBiasio *et al.*, 1988). In order to correct for subcellular differences in accessible volume and path length, actin and myosin fluorescence was normalized relative to inert fluorescent volume indicators (dextran labelled with fluorescein). Initial cellular protrusions contained an elevated concentration of actin and very little or no myosin II. No well-defined structures were detected in these protrusions by fluorescence or video-enhanced contrast microscopy. Within 1–2 min, well-defined actin- and myosin-containing structures were detected in the same protrusion, together with diffuse actin and myosin. Thus, initial protrusive activity involved a local, relative increase in actin concentration, and did not appear to require structures containing myosin II.

Actin and myosin II were colocalized along thick fibres at the base of protrusions and along dense fibre networks in the leading portion of cells migrating into a wound area. Studies examining fluorescence recovery after photobleaching (FRAP) indicated that the myosin II analogue had essentially no translational mobility in regions near the leading edge, in contrast to a 79% mobile fraction in the perinuclear region. Thus myosin was not readily available for diffusion into a protrusion even though it was present at the protrusion's base and in areas within the leading edge. Active transport may be required to mobilize myosin II into established protrusions, possibly by molecular interactions with actin in the protrusion. The changing distribution of the myosin II analogue among actin fibres suggested possible sites for the action of contractile forces. The increase in myosin II mobility in the perinuclear region is consistent with the disassembly of myosin II.

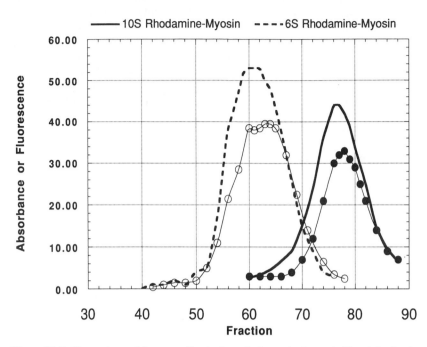

Figure 27.3 Chromatographic assay of ionic strength-dependent myosin II and rhodamine-myosin II conformational change. A 1.5 × 90 cm Sepharose CL-4B column was equilibrated in low salt buffer (150 mM KCl, 5 mM MgCl$_2$, 1 mM EGTA, 10 mM K$_n$PO4, 1 mM MgATP, pH 7.5) at a flow rate of 6 ml h^{-1} maintained using a peristaltic pump. The column was calibrated with proteins IgG, IgM, and BSA. The void and included volumes were determined using Blue dextran 2000 and bromophenol blue, respectively. After dissolving in buffer, myosins were clarified by centrifugation. Fractions of 27 drops (160 fractions total) were collected. The column was then equilibrated in high salt buffer (600 mM KCl, 5 mM MgCl$_2$, 1 mM EGTA, 10 mM K$_n$PO4, pH 7.5) and the procedure repeated. Proteins in low salt were detected by fluorescence, and in high salt by absorbance. The elution volumes of the IgG, IgM, BSA, Blue dextran 2000, and bromophenol blue did not change significantly with ionic strength. Thus, both labelled and unlabelled myosins exhibited phosphate-dependent folding and unfolding.

Table 27.1 Myosin II and rhodamine-myosin II ATPase activity

		Mg^{2+} ATPase	
	K-EDTA	−Actin	+Actin
Myosin	381	<0.1	0.4
Myosin-PO4	–	3.1	27
Rhodamine-myosin	372	0.3	0.5
Rhodamine-myosin-PO4	–	3.7	22

Activities were measured at 25°C using the procedure of Sellers *et al.* (1981). K-EDTA activity was measured in a mixture containing 500 mM KCl, 1 mM ATP, 2 mM EDTA, 15 mM Tris–HCl, and 45–90 nM myosin, pH 7.5. MgATPase activity was measured in 30 mM KCl, 5 mM MgCl$_2$, 1 mM ATP, 0.1 mM EGTA, and 15 mM Tris, pH 7.5, using 540 nM myosin. For actin activation, actin was added at a concentration of 0.3 mg ml^{-1}. Myosins were phosphorylated using the procedure described in Fig. 27.3, except that ATP was replaced with adenosine 5'-O-(3-thiotriphosphate). ATP hydrolysis rates are given as nmol min^{-1} mg^{-1}.

In the future, myosin analogues can be applied in unravelling the functions of myosin II and myosin I, a truncated, separately expressed form of the myosin molecule (for review see Pollard *et al.*, 1991). Locomotion is not wholly dependent on myosin II, as has been demonstrated using cells which do not express myosin II (De Lozanne & Spudich, 1987; Knecht & Loomis, 1987), and in cells microinjected with myosin II antibodies (Honer *et al.*, 1988). These cells move, but with severely altered phenotypes. Fukui *et al.* (1989) reported very different distributions for myosin I and II in *Dictyostelium*, with myosin I predominating in protrusions and myosin II confined to the contractile tail. Such distinct compartmentalization was not apparent in higher cells. However, subtle differences between the localization of myosin I and myosin II in Swiss 3T3 fibroblasts have recently been observed (Taylor & Conrad, pers. commun.).

One would like to know how myosin II mobility

changes during cytochalasin-induced fibre contraction. The solation/contraction hypothesis predicts a myosin II subfraction with radically increased mobility. During a staurosporine-induced fibre dissolution a near complete mobilization of myosin II would be expected. In contrast, okadaic acid-induced contraction would be expected to involve no release of myosin.

The continuous centripetal motion of stress fibres and of other cytoskeletal elements is a widespread phenomenon believed to involve the essential machinery of cell locomotion. However, it is not understood how this motion is driven or polarized. One possibility is that a continuously moving framework is formed and directed through ordered assembly and disassembly of cytoskeletal components. Perhaps certain components or combinations of components serve as template for the addition of others. Multiple-parameter imaging using analogues of myosin II and various actin-binding proteins could establish the order and location of these processes (DeBiasio *et al.*, 1987).

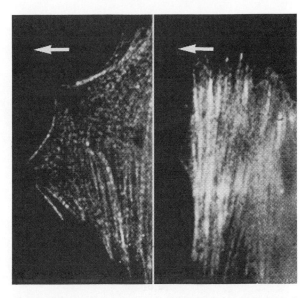

Figure 27.4 Fluorescent images of rhodamine-myosin II and rhodamine-myosin 20 kDa light chain analogues in Swiss 3T3 fibroblasts. Left: light chain analogue. Right: myosin II analogue. These cells are representative of a motile population 3–6 h after wounding. The arrows point in the direction of migration. Both analogues are distributed in a periodic, 'semisarcomeric' pattern.

27.4 FLUORESCENT ANALOGUE OF MYOSIN II REGULATORY LIGHT CHAINS; INDICATOR OF REGULATORY LIGHT CHAIN PHOSPHORYLATION

Smooth muscle and non-muscle cells express a myosin molecule composed of two heavy chains and two pairs of light chains the 17 kDa, or essential, light chains and the 20 kDa, or regulatory, light chains. The regulatory light chains undergo covalent, reversible phosphorylation. As discussed above, phosphorylation of the regulatory light chains is thought to play a key regulatory role, resulting in activation of myosin for contraction. Phosphorylation has been studied to date using dilute solution biochemistry or extracted cell models. To better understand the temporal and spatial dynamics of phosphorylation as they occur in live cells, we have undertaken development of an intracellular indicator of myosin light chain phosphorylation. Building on the success of MeroCaM, our indicator of calcium–calmodulin binding (see above), we are approaching this problem through derivatization of light chains with environmentally sensitive dyes.

Model studies have been carried out using a turkey gizzard light chain labelled with an environmentally insensitive dye, tetramethylrhodamine-5,(6) iodoacetamide. The analogue could be fully phosphorylated *in vitro* with MLCK isolated from smooth muscle, and the phosphorylation rate was 70% that of unlabelled myosin II. When injected in live cells, the analogue exchanged with native light chains. It was incorporated into actin/myosin-based fibres in both non-motile and actively migrating cells, showing the same punctate, periodic fluorescence pattern observed with the whole myosin II analogue (Fig. 27.4). Similar results have been obtained with light chains from striated muscle (Mittal *et al.*, 1987). This *in vivo* exchange provides a versatile method for the introduction of light chains into myosin within live cells. Biochemical studies have shown that light chains are functionally interchangeable among species.

We are now labelling light chains with environmentally sensitive dyes to produce an analogue that will exhibit spectroscopic changes in the form of a wavelength shift or an intensity change upon phosphorylation *in vivo*. Several approaches are being explored. The first of these is labelling of the 20 kDa chicken gizzard light chain with a dye that will 'sense' the charge of the phosphate group. Phosphorylation of the regulatory light chain of chicken gizzard smooth muscle myosin occurs on serine 19. The only cysteine on this light chain is being targeted for specific labelling using cysteine-reactive dyes, and amine-reactive dyes are also being screened. Mutant myosin light chains with the single cysteine moved closer to the phosphorylation site are also being utilized. In a different approach, whole myosin II is being labelled to sense a conformational change that occurs upon phosphorylation.

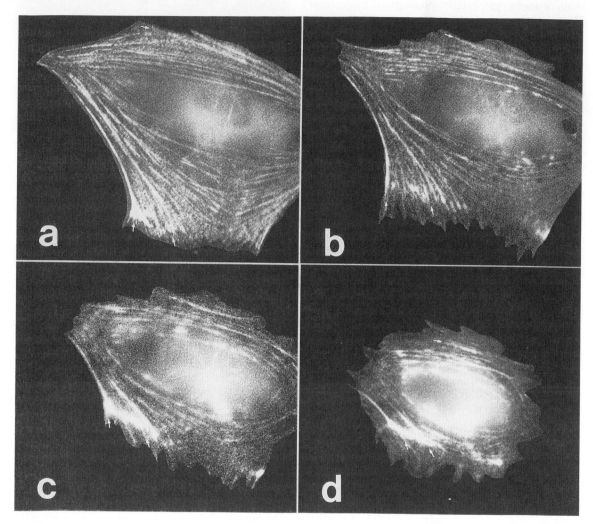

Figure 27.5 Myosin contraction in fibroblasts after treatment with phosphatase inhibitor. Quiescent, serum-deprived Swiss 3T3 fibroblasts were treated with a 400 nM solution of okadaic acid, a phosphatase inhibitor. Images were acquired with a cooled CCD camera at 2.5 min intervals. Small arrows identify selected myosin labelled fibres. (a) Myosin labelled fibres were distributed throughout the quiescent cell, displaying a relatively uniform punctate distribution, (b) 25 min, (c) 40 min, (d) 55 min after okadaic acid treatment. Myosin-labelled fibres contracted from the cell periphery towards the nucleus over time. The cell edges detached from the substrate and retracted.

Analogues labelled with various environmentally sensitive fluorescent dyes are being screened for phosphate sensitivity. Preliminary results demonstrate that these analogues can be phosphorylated with the same efficiency as unlabelled light chains. When the labelled light chains are ultimately observed in live cells, they will be bound to whole myosin. Therefore, the procedure of Katoh and Lowey (1989) has been used to exchange labelled light chains into myosin prior to characterization of fluorescence and biochemical activity. The whole myosin is being fully characterized *in vitro* to ensure that the probe maintains the same properties as native myosin II.

Phosphorylation-sensitive light chains will be micro-injected into cells so that VEC-DIC and fluorescence microscopy can be used to verify normal localization and contractile behaviour. The ability of the analogue to act as a phosphorylation indicator *in vivo* will be tested by comparing spectroscopic changes in live cells with *in vitro* biochemical assays of the state of phosphorylation in these same cells.

The ability to monitor phosphorylation with great temporal and spatial resolution within single cells should provide valuable insights into the rapid events comprising onset of cell movement.

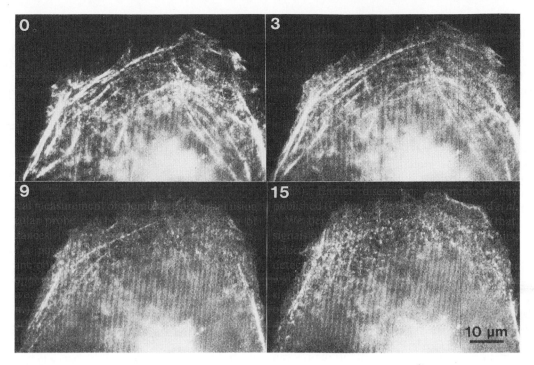

Figure 27.6 Dissolution of stress fibres in the presence of kinase inhibitor. Serum-deprived Swiss 3T3 fibroblasts were microinjected with a rhodamine analogue of myosin II. After 60 min, during which the analogue became distributed through the cytoplasm, the cell was perfused with 50 μM staurosporine while fluorescence images were acquired at 1 min intervals using a cooled CCD camera. Selected images from the sequence are shown with the time in minutes after perfusion indicated in the upper left. Prior to treatment, myosin II was located predominantly in stress fibres. Upon perfusion of inhibitor, stress fibres appeared to dissolve: fluorescence decreased uniformly along the length of the fibres while diffuse cytoplasmic fluorescence increased until little or no fibre structure remained.

27.5   FUTURE STUDIES

Recent developments in design of protein-based indicators hold promise for the observation of a wide range of protein activities *in vivo*, including protein conformational changes, post-translational modification, and ligand binding. Proteins have evolved to recognize physiologically relevant molecules with great specificity, and can be modified to become fluorescent indicators of small molecule concentration. Production of protein-based indicators is currently being pursued through two routes, using environmentally sensitive dyes or fluorescence energy transfer. These approaches have complementary strengths and shortcomings which will determine their applicability in particular situations. Genetic engineering of analogues will be a key tool, enabling precise positioning of attached fluorophores.

The ability to use a single environmentally sensitive fluorophore, rather than the two dyes required for energy transfer, will have obvious advantages in some situations. Maintenance of biological activity in a small protein (such as calmodulin) will be much more difficult after attachment of two dyes. Site-specific attachment of even a single dye is difficult. Intact biological activity will be important in indicators where biological regulation of ligand affinity is being studied, as in indicators revealing the regulation of signalling proteins.

When a protein is modified to become an indicator of small molecule concentration, intact biological activity may, in fact, be undesirable. The indicator should strictly reflect the concentration and distribution of the target ligand. The indicator's distribution and target affinity should not be affected by binding of other ligands, post-translational modification, or other regulatory effects. In such cases, energy transfer would be more applicable because disruption of certain protein activities will be desirable. The primary obstacle to designing useful indicators of small signalling molecules is the unavoidable buffering of these molecules by the indicator itself. Buffering must be minimized by loading very small quantities of indicator. This will require development of bright fluorescent proteins which show strong fluorescence changes when binding their targets. Access to practical small molecule indicators may ultimately depend on the development of very bright dyes suitable for use in living cells.

Indicators of protein–protein binding may be more accessible using currently available methodology. The high concentrations of some intracellular proteins permit loading of relatively large indicator concentrations. Indicators of protein-protein binding could be made by modifying one of the two interacting proteins. Either solvent-sensitive dyes or energy transfer pairs could be used to indicate a protein conformational change induced specifically by the targeted ligand. It may be possible to make indicators with great specificity for a single protein–protein interaction by placing a different dye on each of the interacting proteins, and monitoring energy transfer as an indicator of protein binding. This approach will likely require covalent attachment of the two proteins via a flexible tether. Without such a tether, the labelled proteins could bind to unlabelled, endogenous proteins rather than to each other, necessitating the loading of unacceptably large indicator concentrations to produce detectable energy transfer.

Both energy transfer dyes and environmentally sensitive fluorophores may potentially interact with cellular components to produce artifactual fluorescence changes. Solvent-sensitive fluorophores can interact with lipids and hydrophobic proteins, and energy transfer will be sensitive to microviscosity and factors influencing dye orientation. Environmentally sensitive dyes can be designed with a strong affinity for the tagged protein to avoid artifactual binding to intracellular components. Meaningful interpretation of indicator data will require careful standardization of fluorescence response, both *in vitro* and when possible *in vivo*. Fluorescent calcium indicators have been calibrated within living cells using calcium ionophores and external calcium buffers to control intracellular calcium (Spurgeon *et al.*, 1990).

Intracellular indicators of protein activity promise to yield otherwise inaccessible information about transient protein activity and its regulation in individual, living cells. The cell-to-cell heterogeneity observed in calcium signalling studies suggests that observation of single cells will be critical in untangling the complex interactions of signalling proteins (Byron & Villereal, 1989; Tucker & Fay, 1990). In this and other areas, fluorescent protein indicators should elucidate protein function *in vivo*, thus enabling integration of biochemical data into models of global function. The regulation of many proteins has to date been approachable only through observation of interactions with isolated cellular components. It may soon be possible to observe proteins regulated by the full complement of cellular controls, including those not yet characterized or even postulated.

The most exciting experiments will combine multiple reagents in the same cell or tissue to permit correlations between various parameters in time and space.

Ultimately, we would like to monitor changes in free calcium, calcium binding to calmodulin, phosphorylation of myosin II regulatory light chains and myosin II dynamics in the same experiment. Therefore, the optimal combination of reagents and instrumentation continues to be a great opportunity and challenge.

REFERENCES

Adams S.R., Harootunian A.T., Buechler Y.J., Taylor S.S. & Tsien R.Y. (1991) *Nature* **349**, 694–697.

Andreason T.J., Luetje C.W., Heideman W. & Storm D.R. (1983) *Biochemistry* **22**, 4615–4618.

Bright G.R., Fisher G.W., Rogowska J. & Taylor D.L. (1989) *Methods Cell Biol.* **30**, 15–190

Byron K.L. & Villereal M.L. (1989) *J. Biol. Chem.* **264**, 18234–18239.

Chandra T.S., Nath N., Suzuki H. & Seidel J.C. (1985) *J. Biol. Chem.* **260**, 202–207.

Cohen P. & Klee C. (eds) (1988) *Calmodulin.* Elsevier, New York.

Comte M., Maulet Y. & Cox J. (1983) *Biochem. J.* **209**, 269–272.

Cox J.A., Comte M., Fitton J.E. & DeGrado W.F. (1985) *J. Biol. Chem.* **260**, 2527–2534.

DeBiasio R.L., Bright G.R., Ernst L.A., Waggoner A.S. & Taylor D.L. (1987) *J. Cell Biol.* **105**, 1613–1622.

DeBiasio R.L., Wang L., Fisher G.W. & Taylor D.L. (1988) *J. Cell Biol.* **107**, 2631–2645.

De Lozanne A. & Spudich J.A. (1987) *Science* **236**, 1086–1091.

Fisher G.W., Conrad P.A., DeBiasio R.L. & Taylor D.L. (1988) *Cell Motil. Cytoskeleton* **11**, 235–247.

Fukui Y., Lynch T.J. & Korn E.D. (1989) *Nature* **341**, 328–331.

Giuliano K.A. & Taylor D.L. (1990) *Cell Motil. Cytoskeleton* **16**, 14–21.

Giuliano K.A., Nederlof M.A., DeBiasio R., Lanni F., Waggoner A.S. & Taylor D.L. (1990) In *Optical Microscopy for Biology*, B. Herman & K. Jacobson (eds). Wiley-Liss, New York, pp. 543–537.

Gorbsky G.J., Sammak P.J. & Borisy G.G. (1987) *J. Cell Biol.* **104**, 9–18.

Hahn K.M., Waggoner A.W. & Taylor D.L. (1990) *J. Biol. Chem.* **265**, 20335–20345.

Honer B., Citi S., Kendrick-Jones J. & Jockusch B.M. (1988) *J. Cell Biol.* **107**, 2181–2189.

Katoh T. & Lowey S. (1989) *J. Cell Biol.* **109(4.1)**, 1549–1560.

Klee C.B. (1988) In *Calmodulin*, P. Cohen & C.B. Klee (eds). Elsevier, New York, pp. 35–53.

Knecht D.A. & Loomis W.F. (1987) *Science* **236**, 1081–1086.

Kolega J. & Taylor D.L. (1991) *Cur. Topics Membr.* **38**, 187–206.

Kolega J., Janson L.W. & Taylor D.L. (1991) *J. Cell Biol.* **114**, 993–1003.

Kreis T.E. & Birchmeier W. (1982) *Int. Rev. Cytol.* **75**, 209–227.

Lamb N.J.C., Fernandez A., Conti M.A., Adelstein R., Glass D.B., Welch W.J. & Feramisco J.R. (1988) *J. Biol. Chem.* **106**, 1955–1971.

LaPorte D.C., Wierman B.M. & Storm D.R. (1980) *Biochemistry* **19**, 3814–3819.

LaPorte D.C., Keller C.H., Olwin B.B. & Storm D.R. (1981) *Biochemistry* **20**, 3965–3972.

Luby-Phelps K., Lanni F. & Taylor D.L. (1985) *J. Cell Biol.* **101**, 1245–1256.

McKenna N.M., Wang Y.-L. & Konkel M.E. (1989) *J. Cell Biol.* **109**, 1163–1172.

McNeil P.L. & Taylor D.L. (1987) In *Cell Membranes,* **3**, 365–405.

McNeil P.L., McKenna M.P. & Taylor D.L. (1985) *J. Cell Biol.* **101**, 372–379.

Malencik D.A., Anderson S.R., Shalitin Y. & Schimerlik M.I. (1981) *Biochem. Biophys. Res. Commun.* **101**, 390–395.

Manalan A.S. & Klee C.B. (1984) *Adv. Cyclic Nucleotide Protein Phosphorylation Res.* **18**, 227–277.

Maulet Y. & Cox J.A. (1983) *Biochemistry* **22**, 5680–5686.

Mills J.S., Walsh M.P., Nemcek K. & Johnson J.D. (1988) *Biochemistry* **27**, 991–996.

Mitchison T.J. (1989) *J. Cell Biol.* **109**, 637–652.

Mittal B., Sanger J.M. & Sanger J.W. (1987) *J. Cell Biol.* **105**, 1753–1760.

Nath N., Nag S. & Seidel J.C. (1986) *Biochemistry* **25**, 6169–6176.

Pardue R.L., Kaetzel M.A., Hahn S.H., Brinkley B.R. & Dedman J.R. (1981) *Cell* **23**, 533–542.

Perrie W.T. & Perry S.V. (1970) *Biochem. J.* **119**, 31–38.

Pollard T.D., Doberstein S.K. & Zot H.G. (1991) *Ann. Rev. Physiol.* **53**, 653–691.

Salmon E.D., Leslie R.J., Saxton W.M., Karow M.L. & McIntosh J.R. (1984) *J. Cell Biol.* **99**, 2165–2174.

Sccholzer S.H., Cohn M., Putkey J.A., Means A.R. & Crespi H.L. (1986) *Proc. Natl. Acad. Sci. USA* **83**, 3634–3683.

Sellers J.R. & Adelstein R.S. (1987) *The Enzymes* **XVIII**, 381–418.

Sellers J.R., Pato M.D. & Adelstein R.S. (1981) *J. Biol. Chem.* **256**, 13137–13142.

Simon J.R. & Taylor D.L. (1986) *Methods Enzymol.* **134**, 487–507.

Spurgeon H.A., Stern M.D., Baartz G., Raffaeli S., Hausford R.G., Talo A., Lakatta E.G. & Capogrossi M.C. (1990) *Am. J. Physiol.* **258**, H574–H586.

Takuwa N., Iwamoto A., Kumada M., Yamashita K. & Takuwa Y. (1991) *J. Biol. Chem.* **266**, 1403–1409.

Tanasugarn L., McNeil P., Reynolds G.T. & Taylor D.L. (1984) *J. Cell Biol.* **98(2)**, 717–724.

Taylor D.L. & Fechheimer M. (1982) *Phil. Trans. Roy. Soc. Lond. B* **299**, 185–197.

Taylor D.L. & Wang Y. (1978) *Proc. Natl. Acad. Sci. USA* **75**, 857–861.

Taylor D.L. & Wang Y. (1980) *Nature* **284**, 405–410.

Taylor D.L., Reidler J., Spudich J.A. & Lubert Stryer (1981) *J. Cell Biol.* **89**, 362–367.

Taylor D.L., Amato P.A., Luby-Phelps K. & McNeil P. (1984) *Trends Biochem. Sci.* **9**, 88–91.

Theriot J.A. & Mitchison T.J. (1991) *Nature* **352**, 126–131.

Trybus K.M., Huiatt T.W. & Lowey S. (1982) *Proc. Natl. Acad. Sci. USA* **79**, 6151–6155.

Tsien R.Y. (1989) In *Methods Cell Biol.* **30**, 127–153.

Tucker R.W. & Fay F.S. (1990) *Eur. J. Cell Biol.* **51**, 120–127.

Wadsworth P. & Salmon E.D. (1986) *J. Cell Biol.* **102**, 1032–1038.

Wang Y. (1989) *Methods Cell Biol.* **29**, 1–12.

Wang Y. & Taylor D.L. (1980) *J. Histochem. Cytochem.* **28**, 1198–1206.

Wang Y. & Taylor D.L. (1981) *Cell* **27**, 429–436.

Wang Y., Heiple J.M. & Taylor D.L. (1982) *Methods Cell Biol.* **24B**, 1–11.

Ware B., Brvenik L.J., Cummings R.T., Furukawa R.H. & Krafft G.A. (1986) In *Applications of Fluorescence in the Biomedical Sciences*, D.L. Taylor, A.S. Waggoner, R.F. Murphy, F. Lanni & R.R. Birge (eds). Alan R. Liss, New York, pp. 141–157.

Zavortink M., Welsh M.J. & McIntosh J.R. (1983) *Exp. Cell Res.* **149**, 375–385.

CHAPTER TWENTY-EIGHT

Fluorescence and Luminescence Techniques to Probe Ion Activities in Living Plant Cells

MARK FRICKER[1], MARK TESTER[2] & SIMON GILROY[3]

[1] Department of Plant Sciences, University of Oxford, UK
[2] Department of Botany, University of Adelaide, Australia
[3] Department of Plant Biology, University of California, Berkeley, CA, USA

28.1 INTRODUCTION

Fluorescent and luminescent probes offer unparalleled opportunities to visualize dynamic events within living cells with a minimum of perturbation. Combined with ultra-sensitive imaging systems and powerful computer processing, quantitative measurements of ephemeral gradients, transients and oscillations are possible for the first time. A wealth of dramatic new insights into plant physiology are anticipated in the next few years as more groups apply these powerful techniques to solve old problems and open up exciting new areas of research. The limited work of a small number of groups that are currently equipped to perform these experiments has already highlighted unexpected levels of complexity and dynamic behaviour within plant cells. However, the rate of progress in the plant world has been discouragingly slow by comparison with results from animal systems and most papers are littered with caveats and cautions. The wide diversity of biological problems addressed by plant scientists and the extended range of tissues and species examined means few protocols have been standardized. A significant amount of trial and error is therefore necessary to optimize measurements for each cell type. The main problem seems to lie in the unpredictable behaviour of dyes designed to operate in the respectably well-defined cytosol of animal cells, when confronted with the glorious heterogeneity of plant tissues. The following sections provide a step-by-step guide to measurement of ion activities in plant cells, with particular emphasis on identifying and circumventing plant-specific problem areas.

28.2 TISSUE PREPARATION

The experimental systems used by plant biologists are diverse so the direct application of technology developed for animal systems has not been straightforward. Optical techniques place certain constraints on the type of tissue that can be examined and the extent of processing that may be required. Unicellular, filamentous or relatively flat organisms with restricted 3D growth are readily observed in the microscope, particularly if they are aquatic (e.g. zygotes of *Fucus serratus* – Brownlee & Pulsford, 1988; *Coscindiscus concinnus* and *Guinardia flaccida* – Brownlee *et al.*, 1987; *Mougeotia scalaris* – Russ *et al.*, 1991; *Chara fragilis* – Hodick *et al.*, 1991). To expose the cells of interest in more bulky higher plant tissue may involve peeling, excision, dissection or the formation of protoplasts. It is not surprising that most work on higher plants has concentrated on readily accessible cells such as pollen tubes (Nobiling & Reiss, 1987) or those derived from epidermal layers, such as guard cells (McAinsh et al., 1990; Gilroy et al., 1990; Schroeder & Hagiwara, 1990; Gilroy et al., 1991), root hairs (Clarkson et al., 1988) or stamen hairs (Zhang et al., 1990). Use of protoplasts has allowed measurements on heterogeneous populations derived from roots (Gilroy et al., 1986; Lynch et al., 1989), aleurone layers (Bush & Jones, 1987), cotyledons/hypocotyl (Elliott & Petkoff, 1990) and cells grown in suspension culture (Gilroy et al., 1987, 1989). A few measurements have been made on cells deep within tissues using confocal optical sectioning after physical sectioning (Gehring et al., 1990a,b).

28.3 KEEPING CELLS HAPPY – PERFUSION

Microscope-based visualization is generally preferred over cuvette-based measurements, particularly if a significant contribution to the fluorescence signal is anticipated from dead or damaged cells. Continuous observation also provides information on the state of the tissue and is essential in many cases to monitor responses such as stomatal closure or cytoplasmic streaming. Observation of the tissue at the cell and subcellular level typically requires magnification 100–600-fold. The best optical transfer efficiency is achieved with high numerical aperture (NA) microscope objectives and immersion of the specimen. Submergence reduces light scattering from highly reflective surfaces and is also required for efficient dye loading, rapid application of stimuli or calibration solutions and maintenance of tissue viability during the experimental period.

Ideally the composition of the bathing medium should mimic the environment around the cells in vivo, particularly with respect to ionic composition, water potential and gaseous environment. Continuous perfusion is preferred to simple immersion to reduce boundary layers and allow rapid solution changes. For aquatic organisms, cultured cells or tissues grown in agarose, continuation of the growth conditions is usually sufficient, though the concentration of certain ions may need careful consideration, e.g. Mn^{2+} can enter through Ca^{2+} channels and quench fluorescence from intracellular Ca^{2+} dyes. Other heavy metals may also interfere with some dye loading protocols (Gilroy & Jones, unpublished observations; Gilroy & Fricker, unpublished observations).

Perfusion conditions are more difficult to define for aerial tissues as the turgor relations, apoplastic ion composition and prevailing hormonal status are usually unknown. The potassium activity has a major influence on membrane potential, but the appropriate level is uncertain (Grignon & Sentenac, 1991). Similarly, apoplastic Ca^{2+} levels are not well-defined, but will have a profound effect on cell wall and membrane structure and may modulate cellular responses (Gilroy et al., 1991). Anticipated consequences of immersion include: build up of regions depleted in O_2 and enriched in CO_2 next to non-photosynthetic cells and vice versa for actively photosynthesizing tissue; dilution of apoplastic ions and equilibration of local ion gradients; and increased turgor. There have been no reports examining the significance of these changes.

The specimen needs to be fixed down securely to prevent movement during observation. Various approaches may be appropriate depending on the tissue. Epidermal peels have been immobilized with silicone grease (Gilroy et al., 1990, 1991) or silicone adhesive (e.g. Corning 355 – Blatt, 1991), leaving a clear window so the optics are not compromised. Protoplasts and tissue cultured cells adhere to clean coverslips or more strongly after coating with poly-lysine. However, flow from the perfusion solution and Brownian motion still result in some motion. It is critical to ensure that the immobilization protocol does not affect tissue responses. For example, poly-lysine may induce K^+ channel activity (Reuveni et al., 1985). Certain tissues can be embedded or grown in agarose or gelatin (Clarkson et al., 1988, Zhang et al., 1990).

Cells can be loaded prior to observation and mounted in a closed perfusion system, though microinjection is easiest in situ using an open system. Inverted microscopes are an ideal platform for microinjection of thin, relatively transparent tissue, whilst upright systems with objective rather than stage focusing, offer significant advantages for thick, opaque specimens. The arrival of long-working-distance, water-immersion lenses have reduced the steric problems of microinjection on upright systems, except when a steep angle of penetration is needed.

The temperature is normally set near ambient (20–25°C) through temperature control of the perfusion medium, without additional recourse to a temperature-regulated stage. In a non-perfused system, the sample temperature is likely to increase due to prolonged illumination and lead to temperature-related artifacts. In addition, the composition of the air above open perfusion systems must be regulated and screened to prevent interference from CO_2 exhaled by the microscope operator.

The previous treatment of the tissue and the perfusion regime set the baseline conditions onto which other stimuli are added. The extent of the trauma induced during preparation is difficult to define, however. Plant cells normally alter their metabolic poise continuously in response to changes in their surroundings. Depending on the system, the extent of behavioural modification should be assessed from measurement of parameters such as membrane potential (Brownlee & Pulsford, 1988; Hodick *et al.*, 1991), cytoplasmic streaming (Clarkson *et al.*, 1988), response kinetics (Gilroy *et al.*, 1991), progression through division (Zhang *et al.*, 1990) or growth rate (Hodick *et al.*, 1991). In several cases such measurements have highlighted aberrant behaviour of varying severity, e.g. *Chara* rhizoids stop growing completely when injected with Ca^{2+}-indicating dyes, whilst stamen hair cells only take longer to divide. The contribution of such perturbance may be limited by minimizing manipulation and allowing a recovery time of at least 15–20 min.

28.4 THE NATURE OF THE EXPERIMENT – HOW BIG A SPANNER IN THE WORKS?

Two categories of experiment on signal transduction can be conducted. First, monitoring changes in the ion activity may indicate a role for that ion in transduction of a particular external stimulus or integration of endogenous processes, e.g. progression through the cell cycle. Secondly, changes in the ion activity can be used as an *in vivo* assay for a particular component of a signalling pathway, such as IP_3-stimulated Ca^{2+} release channels (Gilroy *et al.*, 1990).

Many investigations, including our own, have employed very aggressive stimuli designed to elicit the maximum effect. Stepwise increments in stimulus levels from zero to saturating concentrations over a few seconds are rarely encountered *in vivo*, but routinely applied in the laboratory. The oscillations and wobbles diligently recorded by the experimenter as the system recovers from such shocks may reflect many interesting aspects of ion homeostasis but may have little to do with genuine signal transduction. The envelope of the stimulus applied (rise time, magnitude, duration and frequency) requires careful definition. It would be of considerable interest to test whether the same responses result when the stimulus is gradually increased over a prolonged time period or gently modulated about a mean level. It is not inconceivable that many stimuli are normally presented as waves of pulses, with information coded in the frequency as well as magnitude (Due, 1989).

Table 28.1 Fluorescent probes used to report cytosolic Ca^{2+} activities in plant cells.

Dye	Excitation (nm)	Emission (nm)	K_d (nM)	Advantages/ disadvantages
CTC	380	520	440–2600	Membrane-permeant and loads easily. Measures 'membrane-associated' calcium. Responds to Mg^+, H^+, lipid vesicle numbers. May act as a calcium ionophore
Quin-2	340	490	115	Low quantum efficiency, single-wavelength dye, must be loaded at high (0.1–1 mM) concentrations
Indo-1	350	405/485	250	High quantum efficiency. Dual-emission dye. Requires UV excitation. Blue autofluorescence from wall and vacuole significant. Photobleaches more rapidly than Fura-2
Fura-2	340/380	510	220	High quantum efficiency. Dual-excitation dye. Requires UV excitation
Fluo-3	505	526	450	Single-wavelength dye. Can be excited with argon ion lasers for use in confocal microscopy and in combination with UV photolysis of caged probes
Calcium Green	505	526	450	Single-wavelength dye, similar to Fluo-3 with a higher quantum yield. Available linked to high-molecular-weight dextrans
Fura Red	490/440	660		Dual-excitation dye with visible wavelengths. Can be excited with argon ion/He–Cd lasers for use in confocal microscopy and in combination with UV photolysis of caged probes. Emission overlaps chlorophyll autofluorescence

28.5 SELECTION AND USE OF FLUORESCENT PROBES FOR Ca^{2+} AND H$^+$

The basic principle of intracellular ion measurement involves the introduction of a chelating agent whose fluorescent properties alter with the activity of a particular ion. This may be a simple quantitative change in intensity or a shift in either the excitation or emission spectrum. Dyes with spectral shifts are preferable as they permit ratio measurements that distinguish fluorescence changes due to ion binding from those due to dye leakage, bleaching or uneven distribution (Grynkiewicz *et al.*, 1985). There are several features that influence selection of a particular dye:

(1) The ease of loading the dye into a defined compartment, usually the cytosol.
(2) The behaviour of the dye within the cell, including compartmentalization, metabolism and physiological perturbation.
(3) The excitation/emission wavelengths in relation to the spectral sensitivity of the tissue (e.g. blue light responses).
(4) The level of autofluorescence of the tissue.
(5) Combination with other optical techniques such as UV photolysis of caged compounds.

The dyes that have been used so far to measure cytosolic calcium ([Ca^{2+}]$_i$) in plant cells are summarized in Table 28.1. BCECF is the only dye that has been used to measure pH ([H$^+$]$_i$) in plants.

28.5.1 Loading strategy

The objective is to introduce the probe into the cytoplasm of as many cells as possible in sufficient concentration to give good signal-to-noise without significantly increasing the cellular ion buffering or causing toxic effects. Loading the dye into the cytoplasm of plant cells has proved difficult (Cork, 1986; Bush & Jones, 1990). A variety of strategies have emerged, but there are no simple rules as to which will be most effective with a particular tissue. There are essentially six approaches (summarized in Table 28.2). Non-invasive population loading techniques would be preferred but unfortunately do not work with many plant cells.

28.5.1.1 *Permeant dyes: chlorotetracycline*

Chlorotetracycline (also known as chlortetracycline or CTC) is a fluorescent, lipophilic antibiotic isolated from *Streptomyces aureofaciens* that binds divalent and trivalent cations. CTC is readily membrane-permeant

Table 28.2 Loading techniques used to introduce ion-indicating probes into the cytoplasm of plant cells.

Loading technique	Advantages and disadvantages
Permeant dyes	Population of cells loaded with minimum perturbation
Digitonin permeabilization	Population of cells loaded. Technically simple. Considerable perturbation of cell function may be anticipated
AM ester	Population of cells loaded. Technically simple. Problems arise with external hydrolysis, incomplete internal hydrolysis, toxic products and subcellular compartmentalization of esterases and hence dye
Low pH	Population of cells loaded. Technically simple. Many cells stressed or killed by pH load. Dye sticks in wall
Electroporation	Many cells in population loaded, but not to constant level. Requires protoplasts. Severe perturbation, lowers ATP levels. Needs extensive optimization
Microinjection	Applicable to most cell types. Only single cell loaded. Invasive, time-consuming. Technically demanding

and cells are simply loaded by incubation with 10–100 μM dye. This dye is included for completeness, but users should be aware of the potential problems in interpretation of results from plant tissues (see below).

28.5.1.2 *Acetoxymethyl esters*

Most of the ion-selective dyes are impermeant due to one or more carboxyl groups in the molecule which are charged in the physiological pH range. Esterification of the carboxyl groups with acetoxymethyl groups masks their charge and renders the dye membrane permeant. Hydrolysis by intracellular esterases releases the free dye in an active form in the cytoplasm. Problems may arise if:

(1) External hydrolysis occurs (Cork, 1986).
(2) No hydrolysis or incomplete hydrolysis releases only partially activated probe (Brownlee & Wood, 1986; Bush & Jones, 1987, 1990; Elliott & Petkoff, 1990).
(3) Hydrolysis occurs in compartments other than the cytosol.
(4) The products of hydrolysis are toxic (Cork, 1986).

Acetoxymethyl (AM) ester loading works with some plant cells, such as lily endosperm (Keith *et al.*, 1985), aquatic organisms (Brownlee *et al.*, 1987; Dixon *et al.*, 1989) and some developing tissues (Reiss & Nobiling, 1986; Nobiling & Reiss, 1987; Gehring *et al.*, 1990a,b). However, many mature tissues appear to load poorly (Gilroy *et al.*, 1986; McAinsh *et al.*, 1990; Gilroy *et al.*, 1991; Hodick *et al.*, 1993). pH dyes, e.g. BCECF, with a single ester group requiring cleavage give more consistent results than the Ca^{2+} dyes with five ester groups (Bush & Jones, 1990), although level of loading may be variable between cells (Dixon *et al.*, 1989). Mild detergents (e.g. 0.02–0.2% Pluronic F-127 – Gehring *et al.*, 1990b), reduced temperature, ATP permeabilization, increased external pH (Elliott & Petkoff, 1990) and varying ionic conditions may facilitate AM ester loading.

28.5.1.3   Incubation at low pH

Reversible protonation of the carboxyl groups at low external pH (*ca.* 4.5) can be used to mask their charge. Indo-1 probably crosses the plasma membrane as a zwitterion, with 2 amino groups and 2–3 carboxyl groups at pH 4.5 (Bush & Jones, 1987). The protons dissociate at the higher pH of the cytoplasm and the dye is effectively trapped in its anionic (ion-sensitive) form. Low pH loading has been successful with some protoplast types (Bush & Jones, 1987; Lynch *et al.*, 1989; Russ *et al.*, 1991) but not others (Gilroy *et al.*, 1986; Elliott & Petkoff, 1990). In intact cells the dye appears to stick in the wall, possibly through precipitation in localized regions of low pH (Gilroy *et al.*, 1991; Hodick *et al.*, 1991). Charge masking with high levels of other ions, such as K^+, might reduce this problem. Certain cells do not survive low pH treatment (Elliott & Petkoff, 1990; Hodick *et al.*, 1991) and the physiological consequences of pH stress on pH regulation and signalling need careful scrutiny.

28.5.1.4   Electroporation

Pores of variable size can be selectively induced in the plasma membrane of protoplasts by short, high-voltage pulses. Resealing is spontaneous, but can be slowed sufficiently at low temperature to allow free diffusion of dye or other macromolecules into the cytoplasm. A cocktail of low-molecular-weight factors is normally included to replace cytoplasmic components diffusing out of the permeabilized protoplasts. The precise conditions for successful and reversible electropermeabilization of the plasma membrane require careful optimization. Many cells do not survive and the remainder are loaded with variable concentrations of dye. Although respectable measurements of cytoplasmic calcium have been recorded in several cell types using this approach (Gilroy *et al.*, 1986, 1987, 1989; Scheuerlein *et al.*, 1991), the method is not straightforward and is likely to perturb sensitive signalling systems.

28.5.1.5   Microinjection

The cell is impaled with a fine micropipette (*ca.* 0.1–0.3 µm tip) containing 0.1–1 mM dye. Loading has been achieved by diffusion from the tip (*ca.* 20–30 min, Gilroy & Fricker, unpublished observations), iontophoretic injection with continuous current (0.1 nA, 10 min – Gilroy *et al.*, 1991; 10 nA 2–4 min – Hodick *et al.*, 1991) or current pulses (*ca.* 1 nA, 2 Hz, 0.5–10 min – Brownlee & Pulsford, 1988; Clarkson *et al.*, 1988; McAinsh *et al.*, 1990; Zhang *et al.*, 1990). Pressure injection has not yet been examined for ion probes, but is routinely used for other dyes such as Lucifer yellow (Goodwin & Erwee, 1985). There is one report of loading protoplasts via a patch electrode (Schroeder & Hagiwara, 1990).

With patience, most cells can be loaded, but the numbers are limited. Between 60–80% of injections appear to load the cytoplasm rather than the vacuole and the difference in dye distribution between the two compartments is readily observed. The impalement pipette can also function as a microelectrode for measurement of the plasma membrane potential to follow the extent of cell disruption and recovery during the injection procedure (Brownlee & Pulsford, 1988; Hodick *et al.*, 1991).

28.5.1.6   Intracellular dye concentrations

The amount of dye loaded represents a compromise between increasing the signal-to-noise ratio versus buffering effects of the probe. Concentrations have been estimated by comparison with the signal from known concentrations of dye confined to cell-sized volumes, either as droplets in immersion oil (Gilroy *et al.*, 1991) or enclosed in micro-cuvettes (Zhang *et al.*, 1990). Values range from 10 to 300 µM, but are highly dependent on the volume of cytoplasm assumed. In a guard cell, 10 µM dye concentration equates to about 10^7 molecules of fluorophore.

It is not clear at what point these dyes will start to buffer $[Ca^{2+}]_i$ significantly in the plant cytosol and the consequences of different dye loadings have not been examined yet. Direct injection of BAPTA buffers suggests that concentrations in excess of 100 µM are needed to disrupt formation of Ca^{2+} gradients in developing *Fucus* zygotes (Speksnijder *et al.*, 1989).

28.5.2   Maintaining the dye in the cytoplasm

The cytoplasm in a mature plant cell typically occurs as a thin layer less than 1 µm thick sandwiched

Table 28.3 Comparison of measurement techniques used to monitor ion activities in plant cells.

Measurement technique	Loading required	Spatial and temporal resolution	Comments
Fluorometry	Population of single cells in suspension	No spatial resolution. Sampling interval *ca.* 0.05 s minimum	Spectra easy to measure. Heterogeneous responses cannot be distinguished. Autofluorescence easy to correct. Signals from dead and dying cells also included
Flow cytometry	Population of single cells in suspension	No spatial resolution. Sampling interval *ca.* 0.05 s minimum	Need robust cells. Heterogeneous responses appear as an increase in variance
Microscope photometry	Single cells	Whole cells or subcellular regions in large cells measured. Sampling interval 0.05 s minimum typically 1 s	Requires microscope. Measurement of a defined cell region. Autofluorescence correction straightforward. Prone to errors from heterogeneous dye distribution
Camera imaging	Single cell or population of cells	Subcellular regions down to 0.25 μm in x and y; z poorly defined. Sampling interval typically 1–2 s, often intermittent to reduce photobleaching	Allows mapping of spatial transients. Expensive, autofluorescence subtraction difficult
Confocal imaging	Single cell or population in intact tissue	x, y and z resolution well-defined. Maximum $0.2 \times 0.2 \times 0.7$ μm. Temporal resolution dependent on volume sampled from milliseconds for line scan, seconds for 2D section and minutes for 3D data stack	Measurements possible within intact tissue, subcellular resolution excellent, removes out-of-focus blur, simultaneous dual-emission imaging possible. Few ratioable visible dyes for Ca^{2+}, conversely few UV systems for ratio dyes

between the vacuole and the wall, with larger accumulations localized around the nucleus and chloroplasts. In young, rapidly growing cells and some specialized cells, such as tips of root hairs and rhizoids, a greater contiguous volume of cytoplasm may occur, uninterrupted by vacuoles. The distribution of dye follows the distribution of cytoplasm and often appears uneven. The nuclear region appears to accumulate high concentrations of dyes. Confocal imaging indicates that this signal is genuinely located within the nucleus, not in the perinuclear cytoplasm. The majority of reliable measurements are derived from regions rich in cytoplasm where fluorescence signals are strongest such as near the nucleus or oriented parallel to the microscope axis.

Compartmentalization of the dyes into other organelles is a major problem with plant tissues and often limits the time window when cytosolic ion activities are faithfully reported. Dye is transported to the vacuole in many tissues with a variable time-course, from 20 to 30 min (Zhang *et al.*, 1990; Gehring *et al.*, 1990b) to greater than 48 h (Bush & Jones, 1987). The signal from the vacuole is complicated by

the low pH and ionic conditions which will alter both the K_d of the dye and the properties of the fluorochrome moiety. Choosing cells without large vacuoles or regions of large cells where the vacuole is absent avoids this difficulty.

Dye may also accumulate in the endoplasmic reticulum which has a high (>10 μM) intralumenal Ca^{2+} activity (Bush *et al.*, 1989). However, the error introduced into cytosolic measurements is likely to be slight due to the limited amount of dye in the lumen (2% of the total in barley aleurone – Bush *et al.*, 1989; Bush & Jones, 1990), though it may present a problem in the interpretation of 'hot-spots' of Ca^{2+} seen in imaging experiments. Dye may be lost across the plasma membrane by an unknown mechanism, particularly during permeabilization treatments for *in situ* calibration.

Techniques to prevent compartmentalization and extend the available time window for measurement have not been examined in detail in plant cells. Anion channel blockers such as probenecid are known to prevent compartmentalization of other negatively charged dyes such as Lucifer yellow and FITC

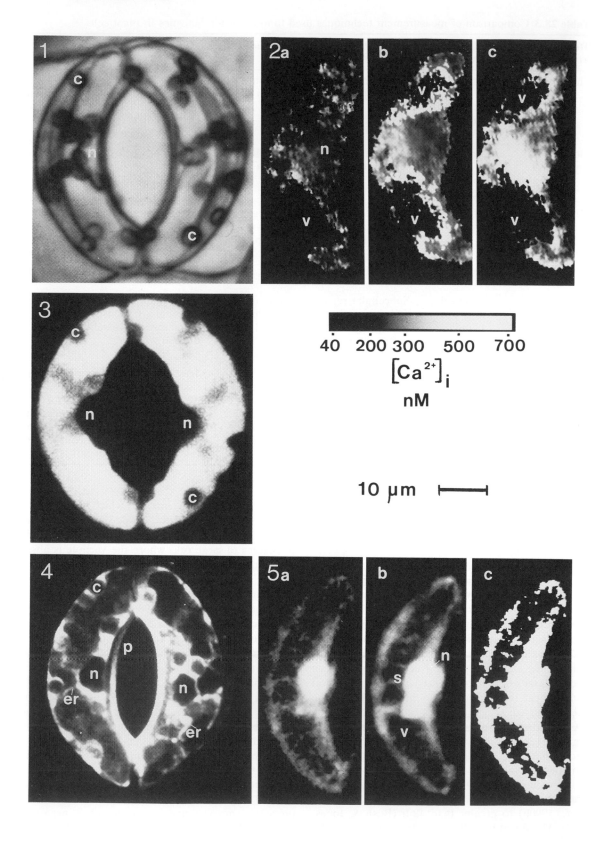

(Oparka, 1991), but have not yet been employed during ion measurements. Dextran-linked dyes appear to prevent compartmentalization and offer a promising route forward, particularly in systems already using microinjection (Hepler & Miller, pers. commun.). Confocal microscopy can separate signals from the cytoplasm and vacuole in situations where compartmentalization does occur and may be used to measure cytoplasmic and vacuolar ion activities simultaneously (Fricker & White, unpublished observations).

28.6 OBSERVATION AND MEASUREMENT

A variety of measurement systems are currently available, offering a balance between cost, sensitivity, spatial resolution and sampling rate. The details are described elsewhere in this book. The main principle for botanical work is to match the hardware to the biology. Some plant processes occur over long time periods (minutes, hours and days) and involve co-ordinated interaction of many cells in the tissue (e.g. Gehring *et al.*, 1990a,b), whilst other questions demand subcellular resolution of rapid transients (Williamson & Ashley, 1982; Schroeder & Hagiwara, 1990; Knight *et al.*, 1991). The principal advantages and disadvantages of the main systems are summarized in Table 28.3. Access to several techniques provides complementary sets of data characterizing both spatial and temporal components of the signal.

28.6.1 Temporal resolution

Most phenomena so far studied in plants take place in the seconds to minutes range and do not require the sophistication of subsecond temporal resolution.

Integration over a 1–10 s sampling interval gives respectable signal-to-noise for most photometry and imaging applications, although intermittent sampling and shuttering of the illumination may be required to minimize photobleaching and photodamage during extended periods of collection (Pheasant & Hepler, 1987; Gilroy *et al.*, 1991). Recent data using aequorin suggest certain processes in higher plants do occur within the second range which will require faster systems for their resolution (Williamson & Ashley, 1982; Knight *et al.*, 1991). Dual-emission dyes enable simultaneous collection of both wavelengths for more rapid sampling and can be implemented easily, except on camera systems. Dual-excitation dyes rely on sequential presentation of two wavelengths, with the interval between exposure defined by the speed of the monochromator, chopper or filter wheel that switches between the two wavelengths.

Techniques measuring large populations of cells record the average response of the population and transients will appear with dramatically reduced amplitude unless all the cells respond synchronously.

28.6.2 Spatial resolution

Ion gradients in plants may exist over a wide range of scales from millimetres in whole coleoptiles (Gehring *et al.*, 1990a,b) to micrometres in tip growing cells (Nobiling & Reiss, 1987; Brownlee & Wood, 1986; Brownlee & Pulsford, 1988). Systems are now available to visualize ions at these widely differing levels of resolution.

In microscope-based photometry systems a physical aperture is used to define an area covering all or part of a single cell over which measurements are averaged. This samples a volume extending above and below the focal plane, defined primarily by the NA of the lens.

Figure 28.1 Bright field image of an open stoma in an epidermal peel from *Commelina communis*. The position of chloroplasts (c) and the nucleus (n) are readily visible.

Figure 28.2 Dynamic changes of $[Ca^{2+}]_i$ in stomatal guard cells of *Commelina communis* induced to close by elevation of extracellular $CaCl_2$ from 20 μM to 1 mM. Guard cells were iontophoretically microinjected with the dual-emission dye Indo-1, and imaged at 0 min (a), 5 min (b) and 10 min (c) after elevation of the extracellular Ca^{2+}. The intensity of the ratio images has been coded according to the inset scale. Regions of low signal strength were masked from the ratio image. Note the localization of increases in $[Ca^{2+}]_i$ to labile 'hot-spots' located near to the vacuole (v) and cytoplasm surrounding the nucleus (n). Further details can be found in Gilroy *et al.* (1991).

Figure 28.3 Single confocal optical section of the stomatal guard cells shown in Fig. 28.1 stained with Acridine Orange (1 μM, 10 min). Dye accumulates predominantly in the vacuole in a pH-dependent fashion. Chloroplasts (c) and the nucleus (n) are visible from their exclusion volume.

Figure 28.4 Linear summation of five confocal optical sections collected at 1 μm intervals from the mid *z*-plane of stomatal guard cells of *Commelina communis* stained with $DiOC_6$ (1 μM, 10 min). The distribution is thought to represent primarily the localization of endoplasmic reticulum (er). The endoplasmic reticulum is concentrated around the nucleus (n) and in the cytoplasm next to the chloroplasts (c). The cuticle around the pore lip (p) is also stained.

Figure 28.5 Dual-excitation confocal fluorescence ratio imaging of cytosolic pH in a stomatal guard cell of *Commelina communis*. Single optical sections near the mid *z*-plane of the cell are shown for excitation of iontophoretically microinjected BCECF at 442 nm (a) and 488 nm (b). The uncalibrated 488 nm/442 nm ratio image is shown in (c). Cytoplasmic strands (s) are clearly visible traversing the vacuole (v). The signal from the nucleus (n) is saturated and gives an erroneous ratio.

The specimen may be moved to allow sampling of other regions allowing some spatial discrimination (e.g. Nobiling & Reiss, 1987).

In camera imaging a real image of the specimen is focused onto the camera face plate. The optics are usually configured such that each pixel typically samples an area equivalent to 0.4–1.2 μm diameter at the focal plane in the specimen (e.g. Fig. 28.2). The volume actually sampled is defined by two cones of light extending on either side of the focal plane, in a similar fashion to photometry systems, thus including information from out-of-focus parts of the specimen. It is possible to correct the image to remove out-of-focus information using software deconvolution algorithms, provided the point spread function of the system can be measured or estimated with sufficient accuracy. Confocal microscopy performs the same task as deconvolution through physical exclusion of out-of-focus information (Fig. 28.5). The volume sampled is well-defined and should improve quantitation, even deep within intact tissues (Williams *et al.*, 1990).

28.6.3 Optimizing the optics

The microscope optics contribute greatly to the quality of the final image. Excitation intensity is rarely a problem, although some components may need to be optimized for UV transmission. Normally additional neutral-density filters are required to attenuate the excitation beam. Recovery of the fluorescent signal is more critical and demands lenses with the highest NA possible at the lowest usable magnification for the resolution needed. Water-immersion lenses probably represent the best optical solution for these imaging applications for a variety of reasons. The NA of the system is limited to the maximum for the glass/aqueous boundary (about 0.9), removing much of the potential increase in light gathering with oil- or glycerin-immersion lenses. The relative path lengths through material of different refractive index do not vary with focus into aqueous media, giving better correction for chromatic and spherical aberration. In addition, focus drift as the coverslip slowly settles on the immersion medium is reduced.

28.6.4 Spectral considerations

Irradiation with light is intrinsic to fluorescence measurements, but also represents a physiologically relevant input to many plant systems and may even trigger Ca^{2+} release from internal stores in some species (Russ *et al.*, 1991). Light drives photosynthesis which alters the energy balance, redox state and Ca^{2+} level (Miller & Sanders, 1987) in the cytoplasm. Many plant cells respond to light quality and intensity with receptors sensitive to blue and red/far-red light.

Other pigments may affect quantitative measurements through selective absorbance of either dye excitation or emission wavelengths or additional autofluorescence. In particular, phenolic compounds in the cell wall and vacuole fluoresce blue when excited with UV wavelengths and chloroplasts fluoresce red when excited with many wavelengths.

Careful selection of the wavelength and bandwidth of the excitation and detection systems can dramatically improve the signal-to-noise ratio. Dyes and filter combinations should be selected to minimize autofluorescence and avoid wavelengths that may trigger physiological responses. Increasing emission filter bandwidth increases the signal but with some loss in the sensitivity of the ratio to ion concentration. Ideally, the signals from both wavelengths for ratio dyes should be comparable and within the linear response range of the detector. This requires judicious use of neutral-density filters.

Fluorometers are excellent for spectral measurements to characterize the behaviour of the dye *in vivo*, such as the extent of ester hydrolysis or tissue autofluorescence (Bush & Jones, 1987; Dixon *et al.*, 1989; Elliott & Petkoff, 1990). Photometry and camera imaging techniques usually rely on wavelength selection *via* interference filters. Excitation in confocal systems is normally limited by the laser wavelengths available, typically from argon ion or krypton/argon ion mixed gas lasers, which are appropriate for single-wavelength excitation (Gehring *et al.*, 1990a,b; Williams *et al.*, 1990). More recently, confocal systems have been built to allow dual excitation of visible wavelength dyes such as BCECF and Fura Red (Fricker & White, 1992), (see Fig. 28.5).

The low emission intensities mean the system must be operated in the dark or using safelights distinct from the measured wavelengths.

28.7   QUANTITATION

A large number of factors affect the sensitivity and precision of measurements. Some estimate of the major sources of error and means of compensating for these are needed. The signal is composed of contributions from dark current, autofluorescence and dye. Dark currents increase dramatically with temperature and should be substracted before digitization.

With photometry techniques, autofluorescence is estimated from the signal measured either prior to loading dye or after quenching the dye at the end of the experiment. However, with imaging techniques, correction is almost impossible as cytoplasmic streaming constantly moves organelles, potentially causing spatial reorganization of autofluorescence. Partial correction can be made in some cases by sampling autofluorescence in corresponding regions of a similar non-loaded cell (Gilroy *et al.*, 1991).

Errors can be introduced into photometry measurements from uneven dye distribution. For example, changes in localized regions of the cytoplasm may be swamped by the large signal derived from the nucleus, which may comprise 30–50% of the total.

28.7.1 Calibration – calcium dyes

The accuracy of measurements depends to a large extent on calibration and becomes increasingly important when small quantitative differences rather than large qualitative changes in $[Ca^{2+}]_i$ between cells are significant. Dye may be calibrated *in vitro* using Ca^{2+}-EGTA buffers to set $[Ca^{2+}]$ in a complex solution designed to mimic the plant cytosol. A typical buffer contains 100 mM KCl, 20 mM NaCl, 1 mM MgSO$_4$, 10 mM HEPES pH 7.2, 1–10 μM dye and 10 mM EGTA plus appropriate amounts of CaCl$_2$ to give the required free $[Ca^{2+}]$. The Ca^{2+}-EGTA buffer stocks should be titrated according to Miller and Smith (1964). The actual free Ca^{2+} is usually calculated using a reiterative computer program that accounts for all the ionic interactions in the calibration buffer. Measurements can be made using volumes of calibration solution similar in size to loaded cells in the experimental apparatus. Calibration is simple to perform and compensates for the optical performance of the measuring system. The accuracy of the calibration depends primarily on the degree of correlation with conditions experienced by the dye in the cytosol. Ionic strength, viscosity and hydrophobicity of the medium have all been identified as potential factors that influence the response of these dyes (Poenie, 1990; Roe *et al.*, 1990; Uto *et al.*, 1991). Viscosity may be increased by addition of 60% sucrose (Zhang *et al.*, 1990) and hydrophobicity altered with 25% ethanol (Russ *et al.*, 1991). Judicious selection of wavelengths where the dye spectrum is less susceptible to potential interference has been suggested. For example, Fura-2 can be measured using the ratio of 340 nm and 365 nm rather than the normal 340 nm and 380 nm (Roe *et al.*, 1990).

Definition of the appropriate composition of the calibration solution can be best estimated from comparison of the dye spectra *in vivo* with that determined *in vitro* under a variety of hydrophobicity, viscosity and ionic composition regimes (Poenie, 1990; Owen, 1991).

For calibration *in situ* dye-loaded cells are permeabilized with a Ca^{2+} ionophore or by detergent treatment at the end of the experiment. Cytosolic $[Ca^{2+}]$ is set by extracellular Ca^{2+}-EGTA buffer solutions. In principle, cytosolic conditions experienced

by the dye should not have changed significantly between the calibration and *in vivo* measurements of $[Ca^{2+}]_i$ during an experiment. One to 10 μM ionomycin or Br-A23187 (a non-fluorescent analogue of A23187) have been used as Ca^{2+} ionophores *in vivo*, however ionomycin has been reported to be much less effective than A23187 in plant cells (Bush & Jones, 1987; Gilroy *et al.*, 1991). This may reflect its requirement for alkaline (pH 9) rather than acidic (< pH 7) conditions normally encountered in perfusion solutions (Liu & Hermann, 1978).

There are two main disadvantages of *in vivo* calibrations. First, whether the homeostatic mechanisms within the cell have really been overcome and secondly, dye tends to leak out rapidly of the cell during permeabilization. There is some confidence if both *in vivo* and *in vitro* calibrations match (Brownlee & Pulsford, 1988; Gilroy *et al.*, 1991). However, this is often not the case (Gilroy *et al.*, 1989) and it is not obvious which calibration is more appropriate.

Calibration of single-wavelength dyes has additional problems as there is no inherent correction for dye leakage or bleaching. *In vivo* calibration is performed to assess the intensity of Ca^{2+}-saturated (F_{max}) and Ca^{2+}-free (F_{min}) fluorescence and $[Ca^{2+}]_i$ usually calculated using published K_d values as (Kao *et al.*, 1989):

$$[Ca^{2+}] = K_d(F - F_{min})/(F_{max} - F)$$

An alternative to determining F_{min} using EGTA buffers is to use Mn^{2+} to quench the fluorescence from the dye. Conveniently, Mn^{2+} is efficiently transported into cells from external concentrations of 0.1–1 mM MnCl$_2$ during incubation with 10 μM ionophore for 10 min. Fluo-3 fluorescence is quenched to $8 \times F_{min}$ whilst F_{max} is known to be $40 \times F_{min}$. These set points can then be used in a modified form of the above equation (Kao *et al.*, 1989; Minta *et al.*, 1989). Our experience in calibrating Fluo-3 in plant cells (Gilroy *et al.*, 1990) is that the Mn^{2+}-quench procedure is more consistent than determining F_{min} with EGTA. Even so, calibration is extremely difficult to perform accurately and in general data from single-wavelength dyes is more qualitatively useful than quantitatively accurate.

28.7.2 Calibration – pH dyes

The approach for calibration of BCECF is essentially similar to those outlined for calcium dyes. Nigericin has been used as an ionophore in some studies, but tends to stress cells rapidly (Pheasant & Hepler, 1987). Permeant weak acids and bases have been used to shift internal pH (Pheasant & Hepler, 1987; Gehring *et al.*, 1990a,b), but also tend to kill cells. *In vivo* calibrations give very poor agreement with *in vitro* measurements

(Dixon *et al.*, 1989) unless performed in pseudo-cytoplasm such as deproteinized coconut water supplemented with 1% ovalbumin (Pheasant & Hepler, 1987). Fortunately the pH response of BCECF is linear over the range 6–8, so pH intervals can be monitored fairly accurately even if the absolute level cannot be determined. The relatively small shift in ratio values means the limits of reliable detection lie between 0.05 and 0.15 pH units (Pheasant & Hepler, 1987).

28.8   DATA PRESENTATION

Effective communication of results is becoming more difficult as the complexity of the data increases. The format for presentation is defined partly by the system generating the data and partly by the publishing costs.

Photometry, fluorometry and flow cytometry systems generate single values that change with time and can be displayed as a simple line trace. The *y*-axis may retain the original fluorescence or ratio value, or represent the ion activity after appropriate calibration. Ideally both scales are presented to indicate the sensitivity of the system to the changes reported. Information on the level of photobleaching and dye leakage is contained in the original traces at each wavelength. The rate of leakage is particularly interesting with some calibration techniques which induce rapid dye loss. On ratio traces this is only apparent as an increase in the variance.

Camera systems generate 2D images. The most effective and visually appealing presentation is pseudocolour representation of the ratio image, as subtle changes in ratio are more readily visible as colour differences rather than changes in grey level. Masking is needed to suppress regions where low, noisy signals create high variation in the ratio value. Typically the ratio is set to zero when one primary signal is below a nominated threshold (Figs 28.2 and 28.5). Choice of this threshold is arbitrary, although some attempts to set a standardized procedure have been made (Gilroy *et al.*, 1991). An *in vitro* calibration is performed at high and low $[Ca^{2+}]$ as the signal strength is progressively reduced by either lowering the dye concentration or introducing more neutral-density filters in the light path. The threshold is then determined as the fluorescence intensity for each wavelength where the ratio differs by more than 10% of the value found at high dye concentrations. An elegant alternative approach is to code the ratio by colour and the average signal strength of both wavelengths as brightness (Tsien & Harootunian, 1990) or saturation (Fricker & White, unpublished data). Ratios derived from areas of weak fluorescence

are still present in the final image but are correspondingly dimmer or more 'washed out'.

The cost of colour reproduction is often prohibitive; representation of data as Y-mod plots with contouring offers one solution, but can be confusing as it is difficult to distinguish areas of low ion activity from those where the ratio has been masked and often impossible to read quantitative data back from the figure.

28.9   PHOTOPROTEINS AS INDICATORS OF $[Ca^{2+}]_i$ IN PLANTS

Aequorin is a Ca^{2+}-dependent photoprotein isolated from the coelenterate *Aequoria victoria* that emits light at 470 nm on binding Ca^{2+} (Blinks, 1979; 1986; Cobbold & Rink, 1987). Aequorin is highly selective for Ca^{2+}, for example Mg^{2+} and K^+ do not trigger luminescence, though these ions may depress the Ca^{2+} sensitivity (Thomas, 1982).

Aequorin has several potential advantages over fluorescent dyes as an indicator for $[Ca^{2+}]_i$. Luminescence measurements have an intrinsically high signal-to-noise ratio, although endogenous chemiluminescence can comprise up to 30% of the signal in plant cells (Gilroy *et al.*, 1989). As a natural protein, aequorin is expected to be non-toxic and remain in the cytoplasm. Light emission is unaffected by pH values greater than 7. Photo-damage associated with excitation illumination for fluorescence is also avoided.

The major limitations to the use of aequorin in plants have been availability of the purified protein and the need to introduce the dye into the cytoplasm by microinjection (Williamson & Ashley, 1982; Okazaki *et al.*, 1987) or electroporation (Gilroy *et al.*, 1989). Supplies of the apoprotein have increased since the gene for aequorin has been cloned (Prasher *et al.*, 1985; Inouye *et al.*, 1985), though obtaining the essential co-factor coelentrazine still presents a problem. Pressure injection is used in preference to iontophoresis to overcome the low iontophoretic mobility of the 20 kDa apo-protein. Premature discharge of the aequorin in the micropipette by Ca^{2+} in the perfusion medium leads to a reduction in the amount of active probe and hence signal introduced into the cell. This can be avoided by either forcing a slow constant stream of aequorin out of the pipette when approaching the cell or plugging the end of the pipette with a small amount of vegetable oil that can be expelled just prior to injection.

More recently the problems of introducing aequorin into the cytoplasm have been elegantly solved using recombinant DNA technology to transform plant cells with the cDNA for the apoprotein. Active aequorin

Table 28.4 Measurement of $[Ca^{2+}]_i$ in plant cells using fluorescent probes.

Species (cell type)	[Dye] and loading strategy	Resting level (nm)± SEM (n)	Treatment and response	Reference
(1) Resting levels				
Phaseolus mungo (root tip protoplasts)	Quin–2 Electroporation 190 μM Fluorometry	171 ±41 (15)	EGTA decreased $[Ca^{2+}]_i$ to 17% in 10 min. Wide variety of loading strategies did not work	Gilroy *et al.*, 1986
Coscindiscus concinnus Guinardia flaccida (entire diatom cell)	Fura–2 AM ester Dual-excitation imaging	110 115	Low ratio in nucleus, increasing towards cell periphery	Brownlee *et al.*, 1987
Hordeum vulgare (aleurone protoplasts)	Indo–1 Low pH 6–85 μM Fluorometry	221	Ionomycin not effective, Fura–2 localized in vacuole. External dye concentrations above 60 μM killed protoplasts	Bush & Jones, 1987
Daucus carota (cultured 6–8 day protoplasts)	Quin–2 Electroporation 100 μM Fluorometry	361±47 (30)	W7, TFP and tetracaine increased $[Ca^{2+}]_i$ to greater than 1 μM in 15 min. Only 10–40% total fluorescence from Quin–2	Gilroy *et al.*, 1987
Hordeum vulgare (mesophyll cell protoplasts) *Daucus carota* (cultured 6–8 day protoplasts)	Quin–2 Electroporation 100 μM Fluorometry	120±62 (3) 361±47 (18)	Quin–2 depleted ATP levels to 30%. Cells failed to divide	Gilroy *et al.*, 1989
Amaranthus tricolour (seedling protoplasts)	Indo–1 AM ester pH 6.8 Fluorometry	113 ±12 (4)	Single-wavelength measurements as AM ester not completely hydrolysed. Acid loading did not work	Elliot & Petkoff, 1990
(2) Standing gradients				
Fucus serratus (rhizoid)	Quin–2 AM ester Imaging		$[Ca^{2+}]_i$ greater than 1 μM in tip and less than 1 μM at base. Parallel Ca^{2+} electrode measurements	Brownlee & Wood, 1986
Lilium longiflorum (pollen tube)	Quin–2 AM ester Ratio photometry	tip=90 base=20 (22)	Cell growth not inhibited. Measured using quin isosbestic point	Nobiling & Reiss, 1987
Fucus serratus (rhizoid)	Fura–2 microinjection 100 μM Dual-excitation imaging	tip=450 ±30 base=10 5±15 (10)	Verapamil decreased $[Ca^{2+}]_i$ in tip from 460 nm to 195 ± 45 (n=5). Nifedipine had no effect. Gradients in 60% of rhizoids	Brownlee & Pulsford, 1988
Lycopersicum esculentum (root hair) *Brassica napus* (root hair)	Fura–2 microinjection Dual-excitation imaging	30–90 in tip	Verapamil decreased $[Ca^{2+}]_i$ initially but levels recovered. Streaming stopped. ABA had no effect. No gradient in some cells. Membrane potential measured	Clarkson *et al.*, 1988

Table 28.4 Continued

Species (cell type)	[Dye] and loading strategy	Resting level (nm)± SEM (n)	Treatment and response	Reference
Chara fragilis (rhizoid)	Indo–1 microinjection Dual-emission imaging	425 ±80 (7)	Increasing $[Ca^{2+}]_e$ and $[K^+]_e$ increased $[Ca^{2+}]_i$ in 30–50% of cells. Gradient only observed in 1 out of 16 cells	Hodick *et al.*, 1991
Daucus carota (whole embryos)	Fluo–3 Digitonin permeabilization Confocal imaging	ND	Elevated $[Ca^{2+}]_i$ in the periphery of embryo and in the cotyledons	Timmers *et al.*, 1991

(3) Transients and signal transduction

Species (cell type)	[Dye] and loading strategy	Resting level (nm)± SEM (n)	Treatment and response	Reference
Commelina communis (guard cell)	Fura–2 microinjection 100 µM	115 ±26 (10)	ABA caused 2–10-fold increase in 80% cells. Peak after 10 min. Possible oscillation reported	McAinsh *et al.*, 1990
Commelina communis (guard cell)	Fluo–3 microinjection 10–50 µM Dual-emission photometry		Photoactivation of caged Ca^{2+} and caged IP_3 increased $[Ca^{2+}]_i$ and caused closure if $[Ca^{2+}]_i$ >600 nM ($n=27$)	Gilroy *et al.*, 1990
Vicia faba (guard cell protoplast)	Fura–2 via patch-electrode 100 µM Photometry	190 ±90 (19)	ABA increased $[Ca^{2+}]_i$ by 2.5–25-fold in 37% cells. Repetitive spikes of Ca^{2+} influx through non-selective channels	Schroeder & Hagiwara, 1990
Commelina communis (guard cell)	*Indo–1* microinjection 10 µM Dual-emission photometry	173±43 (50) 117±38 (25)	Increasing $[Ca^{2+}]_e$, decreasing $[K^+]_e$ and Br-A23187 all increased $[Ca^{2+}]_i$ ABA had variable effect	Gilroy *et al.*, 1991
Haemanthus Katerinae Baker (wall-less endosperm cells)	Quin–2 AM ester Dual-excitation imaging	47–52 (15)	2–3-fold higher $[Ca^{2+}]_i$ in spindle poles versus cytoplasm during mitosis ($n=20$). Cytoplasm 1.6-fold higher than nucleus during interphase. Measured using quin isosbestic point	Keith *et al.*, 1985
Tradescantia virginiana (Stamen hair)	Indo–1 microinjection 50–100 µM Dual-emission photometry	*ca.* 200	Large number of treatments applied Levels of $[Ca^{2+}]_i$ around 1 µM increased rate of chromosome movement. GTPγS had similar effect but not through sustained elevation of $[Ca^{2+}]_i$	Zhang *et al.*, 1990
Zea mays (coleoptile)	Fluo–3 AM ester Single-wavelength confocal	225	Unilateral light – increase to 415 nM in 15 min on shaded side. Gravity – increase to 370 nM on lower side	Gehring *et al.*, 1990a

Table 28.4 Continued

Species (cell type)	[Dye] and loading strategy	Resting level (nm)± SEM (n)	Treatment and response	Reference
Zea mays (coleoptile and root) Petroselinum hortense (hypocotyl and root)	Fluo–3 AM ester Single-wavelength confocal	240–280 85–300	2,4D increased $[Ca^{2+}]_i$ to 380 nM and 230 nm in 4 min, respectively ($n=8$). ABA increased $[Ca^{2+}]_i$ to 320 nm and 590 nm in 6 min, respectively ($n=8$)	Gehring et al., 1990b
Zea mays (root protoplasts)	Indo–1 Low pH Fluorometry	93 (8–11)	Greater than 100 mM NaCl increased $[Ca^{2+}]_i$ to 1260 nM. Li^+ reduced increase. Inositol abolished Li^+ effect	Lynch et al., 1989
Mougeotia scalaris (mature, photosynthetic cell)	Indo–1 pH 4.0 Dual-emission photometry	920±29	External EGTA decreased $[Ca^{2+}]_i$. Illumination increased Ca^{2+} release from internal stores	Russ et al., 1991
Pisum sativum (mesophyll protoplasts)	Indo–1 Fura–2	170±33 (7)	Increasing $[H^+]_i$ and $[Ca^{2+}]_e$ increased $[Ca^{2+}]_i$	Biyasheva & Molotkovskii, 1991
Dryopteris paleacea (spores)	Fura–2 Electroporation Dual-excitation imaging	50	Elevated $[Ca^{2+}]_e$ increased $[Ca^{2+}]_i$ to > 500 nM. Pre-illumination with red light illumination caused a decrease in $[Ca^{2+}]_i$	Scheuerlein et al., 1991

was reconstituted from the expressed apoprotein by incubation with coelentrazine for several hours (Knight et al., 1991). Expression of apo-aequorin is potentially applicable to any transformed plant system and thus may allow measurement of $[Ca^{2+}]_i$ in whole tissues. Exciting future possibilities include the use of imaging combined with tissue-specific promotors or constructs that target the apo-aequorin to specific organelles to monitor $[Ca^{2+}]_i$ in particular cells or compartments. However, the estimated levels of photon emission (0.7 photons s^{-1} per 5 μm^3 – Caswell, 1979) are at the limits of detection for imaging systems.

The aequorin signal is not easy to calibrate. Cytosolic Ca^{2+} activities can be calculated from aequorin luminescence by comparison between the rate of Ca^{2+}-triggered luminescence from the aequorin in the cell to the peak rate of light emission at saturating $[Ca^{2+}]$ (Cobbold & Rink, 1987; Gilroy et al., 1989). There are significant differences, however, between in vivo and in vitro calibrations (Gilroy et al., 1989).

28.10 CALCIUM MEASUREMENTS

CTC is a membrane-permeant 'Ca^{2+}-indicating' dye. It has provided a useful qualitative first step in identifying cells where changes in 'membrane-associated' Ca^{2+} levels occur which might be worth investigating further with the more technically demanding ratio dyes (Reiss & Herth, 1978; Reiss et al., 1985; Timmers et al., 1991). Quantitative measurements are virtually impossible, however, as the dye follows the distribution of intracellular membranes, interacts with membrane potential (Tang & Beeler, 1990), can respond to Mg^{2+} with higher affinity and greater fluorescence enhancement than Ca^{2+}, and may even act as a Ca^{2+} channel (Foissner, 1988).

Results where quantitative estimation of $[Ca^{2+}]_i$ have been attempted using fluorescent probes are collated in Table 28.4 in three groups, roughly corresponding to:

(1) Measurements of resting $[Ca^{2+}]_i$ and factors affecting Ca^{2+} homeostasis.

(2) Detection of standing gradients of Ca^{2+}.
(3) The role of Ca^{2+} in signal transduction.

Resting levels vary from 20 nM to 920 nM in different systems. It is not clear whether these reflect real differences between particular species, e.g. generally higher levels in the algae, or errors introduced by the complex calibration required. For comparison, resting levels measured using Ca^{2+}-selective microelectrodes range between 100 and 400 nM (Miller & Sanders, 1987; Brownlee & Wood, 1986; Felle, 1988, 1989), whilst aequorin studies also indicate resting levels in the micromolar range for algal cells (Williamson & Ashley, 1982; Okazaki *et al.*, 1987).

Standing gradients have been reported in tip-growing cells, ranging from about 100 nM in the sub-tip region to about 500–1000 nM in the tip. The gradient only appears to be sustained during elongation, however, and may disappear if the cell is disturbed (Hodick *et al.*, 1991). the distance over which a gradient is maintained also varies from 1–2 µm (Hepler & Miller, pers. commun.) to 10–20 µm (Brownlee & Pulsford, 1988).

The role of Ca^{2+} in signal transduction has only been examined in a limited number of systems for a small number of stimuli, primarily guard cells, coleoptiles and during phytochrome responses. The results so far obtained vary enormously, even when measured on the same cell type by ostensibly the same technique. It is not clear at the moment how much of this variation resides in genuine biological heterogeneity versus equipment or experimentally induced artifacts.

28.11   pH MEASUREMENTS

Results for cytosolic pH measurements are summarized in Table 28.5. The data are too limited to draw any general conclusions.

28.12   PROBES FOR OTHER COMPARTMENTS

A description of events in the cytoplasm gives a limited viewpoint of dynamic metabolism in plant cells. The massive transport events occurring across the plasma membrane and tonoplast and the ionic balance in the apoplast and vacuole are all potential areas needing research. Many of the probes, particularly Ca^{2+} dyes, have been specifically tailored to respond to cytosolic ion concentrations within a limited range of pH and ionic composition. Alternative probes or modification

of the dyes (e.g. higher K_d values) is required if they are to be useful in other compartments.

28.12.1   Apoplast

The ionic composition of the apoplast is poorly defined, mainly as the high ion exchange capacity of the wall creates a host of polarized microenvironments for selective ion binding. Local ion activities are dependent to a large extent on the interaction between the supply of ions from the neighbouring apoplast or bathing medium, the wall polymer composition and the selective ion transport phenomena occurring in adjacent cells (Grignon & Sentenac, 1991). Membrane potentials, activity of extracellular enzymes, wall structure and binding of ligands to receptors are all likely to be affected by the apoplastic environment. Microelectrode measurements have given highly variable estimates of $[K^+]_e$ ranging from 50 µM (Blatt, 1985), to 3–100 mM (Bowling, 1987) and demonstrated that $[H^+]_e$ may transiently differ by 1–1.5 pH units around neighbouring cells (Edwards *et al.*, 1988). It is not clear what the physiological consequences of these changes are *in vivo*, to what extent they are regulated or whether they also act as signals in their own right. *In vivo* apoplastic measurements probably cannot be performed in perfusion tissue preparations used for determination of cytoplasmic ion levels. Direct imaging techniques for intact tissues need to be developed.

The dye technology for apoplastic measurements is in its infancy and there have been few quantitative reports. Primulin is a textile dye excited at violet wavelengths that exhibits pH-dependent spectral shifts with no apparent toxic effects (Edwards *et al.*, 1988). There is a wide range of soluble pH indicators that may be used, such as umbelliferones (Pfanz & Dietz, 1987). Coupling to dextrans may be desirable to prevent permeation or uptake by non-specific transporters in the membrane.

28.12.2   The vacuole

Dynamic changes in vacuolar morphology and ion transport are of increasing interest. pH measurements rely on uptake and compartmentalization of appropriate dyes in the vacuole. Precise optical sectioning using confocal microscopy allows signals from vacuole, cytoplasm and wall to be distinguished readily and may permit simultaneous measurement of ion activities in all three compartments (Fig. 28.3). An alternative approach is based on the ability of intact leaf cells to take up a variety of fluorescent pH indicators with differing K_d values. Changes in fluorescence can be attributed to pH variation in particular compartments on the basis of (1) the determined

Table 28.5 Measurement of $[H^+]_i$ in plant cells using fluorescent probes.

Species (cell type)	Dye, loading and concentration	pH ± SEM (n)	Treatment and responses	Reference
Tradescantia virginiana (stamen hairs)	BCECF microinjection Dual-excitation photometry	5.7–6.0	Ratio drops for 20–30 min after loading No change through mitosis Calibration in coconut water	Pheasant & Hepler, 1987
Emiliania huxleyi (whole coccolitho-phores)	BCECF AM ester Dual-excitation imaging	7.29 ± 0.11	pH decreased in the absence of HCO_3^- by 0.8 pH units Hydrolysis checked spectrally using fluorometry	Dixon *et al.*, 1989
Zea mays (coleoptiles)	BCECF AM ester single-wavelength confocal	ND	2,4D – pH decreased by 0.3 units ($n=8$) ABA – pH increased by 0.05–0.2 units ($n=11$) FC – pH decreased by 0.1–0.2 units ($n=8$)	Gehring *et al.*, 1990b
Zea mays (coleoptiles)	BCECF AM ester Single-wavelength confocal		pH decreased on shaded side and on lower side by 0.2 units	Gehring *et al.*, 1990a

distribution of the dye and (2) the pH range over which the dye is responsive (Yin *et al.*, 1990). Certain cells contain autofluorescent compounds in the vacuole that respond to pH, which have been imaged to follow changes in vacuolar morphology during guard cell development (Palevitz *et al.*, 1981), but not yet changes of pH in mature guard cells.

Membrane-permeant weak amines accumulate in acidic compartments in response to the pH gradient across the intervening membrane. Fluorescent acridine derivatives that fall into this category can be readily imaged to provide some indication of relative vacuolar pH (Fig. 28.3), though rigorous quantitation has not yet been attempted. Acridine derivatives also exhibit complex and diverse staining behaviour and can be highly phototoxic (Gupta & De, 1988).

28.13 FUTURE PROSPECTS

The number of ions that can be imaged is increasing all the time. Dyes for potassium, sodium, magnesium and chloride exist, but have not yet been used in living plant cells.

It is anticipated that more studies will combine fluorescent measurements with electrophysiological investigations (e.g. Schroeder & Hagiwara, 1990), subtle perturbation of signalling systems with caged compounds (e.g. Gilroy *et al.*, 1990) and detailed comparative morphological description (e.g. Fricker *et al.*, 1991; Figs 28.1, 28.3 and 28.4). Confocal microscopy permits quantitative imaging in larger chunks of intact tissue and provides better quantitation at the single cell level, particularly in combination with ratio techniques (Fricker & White, 1992). More studies on whole organs are likely to appear.

The power of molecular genetics has barely been tapped yet, but offers staggering potential for precisely targeted functional analysis of key signalling and regulatory components, using optical techniques as *in vivo* assay systems. The aequorin transformation studies have elegantly demonstrated the benefits of a molecular biological solution to calcium measurements.

The subcellular compartmentalization of the highly negatively charged ion-indicating dyes introduces some additional questions on the mechanism of their transport in parallel to similar observations on transport of dyes used to trace fluid-phase endocytosis (Oparka, 1991).

Perhaps the greatest advance this new technology offers, however, is simply the development of systems where biochemistry, physiology and cell biology can be directly correlated *in vivo*. Being able to watch dynamic oscillations and heterogeneous responses for the first time represents a critical advance in our ability to describe and understand the molecular world of a cell.

ACKNOWLEDGEMENTS

M.D.F. would like to thank N. White and R. Errington for many helpful discussions and the Nuffield Foundation, Royal Society and the Agricultural and Food Research Council for financial support. M.T. would like to thank the Australian Research Council for financial support and DITAC for a Research Travel Grant.

REFERENCES

Biyashera A.E. & Molotkovoskii V.N. (1991) *Sov. Plant Physiol.* **38**, 184–192.

Blatt M.R. (1985) *J. Exp. Bot.* **36**, 240–251.

Blatt M.R. (1991) *Methods Plant Biochem.* **6**, 281–321.

Blinks J.R. (1979) *Methods Enzymol.* **172**, 164–203.

Blinks J.R. (1986) In *The Heart and Cardiovascular System*, H.A. Fozzard, E. Haber & R.B. Jennings (eds). Scientific Foundation, Raven Pess, New York, pp. 671–701.

Bowling D.J.F. (1987) *J. Exp. Bot.* **38**, 1351–1355.

Brownlee C. & Pulsford A.L. (1988) *J. Cell Sci.* **91**, 249–256.

Brownlee C. & Wood J.W. (1986) *Nature* **320**, 624–626.

Brownlee C., Wood J.W. & Briton D. (1987) *Protoplasma* **140**, 118–122.

Bush D.S. & Jones R.L. (1987) *Cell Calcium* **8**, 455–472.

Bush D.S. & Jones R.L. (1988) *Eur. J. Cell Biol.* **46**, 466–469.

Bush D.S. & Jones R.L. (1990) *Plant Physiol.* **93**, 841–845.

Bush D.S., Biswas A.K. & Jones R.L. (1989) *Planta* **178**, 411–420.

Callaham D.A. & Hepler P.K. (1991) In *Cellular Calcium – A Practical Approach*, J.G. McCormack & P.H. Cobbold (eds). Oxford University Press, Oxford, pp. 383–412.

Caswell A.H. (1979) *Int. Rev. Cytol.* **56**, 145–181.

Clarkson D.T., Brownlee C. & Ayling S.M. (1988) *J. Cell Sci.* **91**, 71–80.

Cobbold P.H. & Rink T.J. (1987) *Biochem. J.* **248**, 313–328.

Cork R.J. (1986) *Plant Cell Environ.* **9**, 157–161.

Dixon G.K., Brownlee C. & Merrett M.J. (1989) *Planta* **178**, 443–449.

Due G. (1989) *Plant, Cell Environ.* **12**, 145–149.

Edwards M.C., Smith G.N. & Bowling D.J.F. (1988) *J. Exp. Bot.* **39**, 1541–1547.

Elliott D.C. & Petkoff H.S. (1990) *Plant Sci.* **67**, 125–131.

Felle H. (1988) *Planta* **176**, 248–255.

Felle H. (1989) *Plant Physiol.* **91**, 1239–1242.

Foissner I. (1988) *J. Phycol.* **24**, 458–467.

Fricker M.D. & White N.S. (1992) *J. Microsc.* **166**, 29–42.

Fricker M.D., Gilroy S.G., Read N.D. & Trewavas A.J. (1991) In *Molecular Biology of Plant Development*, G.I. Jenkins & W. Schuch (eds). *Soc. Exp. Biol. Symp. Series* **44**, pp. 177–190.

Gehring C.A., Williams D.A., Cody S.H. & Parish R.W. (1990a) *Nature* **345**, 528–530.

Gehring C.A., Irving H.R. & Parish R.W. (1990b) *Proc. Natl. Acad. Sci. USA* **87**, 9645–9649.

Gilroy S.G., Hughes W.A. & Trewavas A.J. (1986) *FEBS Lett.* **199**, 217–221.

Gilroy S., Hughes W.A. & Trewavas A.J. (1987) *FEBS Lett.* **212**, 133–137.

Gilroy S.G., Hughes W.A. & Trewavas A.J. (1989) *Plant Physiol.* **90**, 482–491.

Gilroy S.G., Read N.D. & Trewavas A.J. (1990) *Nature* **346**, 769–771.

Gilroy S.G., Fricker M.D., Read N.D. & Trewavas A.J. (1991) *Plant Cell* **3**, 333–444.

Goodwin P.B. & Erwee M.G. (1985) In *Botanical Microscopy*, A.W. Robards (ed.). Oxford University Press, Oxford, pp. 335–358.

Grignon C. & Sentenac H. (1991) *Ann. Rev. Plant Physiol.* **42**, 103–128.

Grynkiewicz G., Poenie M. & Tsien R.Y. (1985) *J. Biol. Chem.* **260**, 3440–3450.

Gupta H.S. & De D.N. (1988) *J. Plant Physiol.* **132**, 254–256.

Hepler P.K. & Callaham D.A. (1987) *J. Cell Biol.* **105**, 2137–2143.

Hodick D., Gilroy S., Fricker M.D. & Trewavas A.J. (1991) *Bot. Acta* **104**, 222–228.

Inouye S., Noguchi M., Sakaki Y., Takagi Y., Miyata T., Iwanaga S., Miyata T. & Tsuji F.I. (1985) *Proc. Natl. Acad. Sci. USA* **82**, 3154–3158.

Inouye S., Sakaki Y., Goto T. & Tsuji F.I. (1986) *Biochemistry* **25**, 8425–8429.

Kao J.P.Y., Harootunian A.C. & Tsien R.Y. (1989) *J. Biol. Chem.* **264**, 8179–8184.

Keith C.H., Ratan R., Maxfield F.R., Bajer A. & Shelanski M.C. (1985) *Nature* **316**, 848–850.

Kikuyama M. & Tazawa M. (1983) *Protoplasma* **117**, 62–67.

Knight M.R., Campbell A.K., Smith S.M. & Trewavas A.J. (1991) *Nature* **352**, 524–526.

Lynch J., Polito V. & Lauchli A. (1989) *Plant Physiol.* **90**, 1271–1274.

Liu C. & Hermann T.E. (1978) *J. Biol. Chem.* **253**, 5892–5895.

McAinsh M.R., Brownlee C. & Hetherington A.M. (1990) *Nature* **343**, 186–188.

Miller A.J. & Sanders D. (1987) *Nature* **326**, 397–400.

Miller D.J. & Smith G.L. (1964) *Am. J. Physiol.* **246**, C160–166.

Minta A., Kao J.P.Y. & Tsien R.Y. (1989) *J. Biol. Chem.* **264**, 8171–8178.

Nobiling R. & Reiss H.-D. (1987) *Protoplasma* **139**, 20–24.

Okazaki Y., Yoshimuto Y., Hiramuto Y. & Tazawa M. (1987) *Protoplasma* **140**, 67–72.

Oparka K.J. (1991) *J. Exp. Bot.* **42**, 565–579.

Owen C.S. (1991) *Cell Calcium* **12**, 385–393.

Palevitz B.A., O'Kane D.J., Kobres R.E. & Raikhel N.V. (1981) *Protoplasma* **109**, 23–55.

Pfanz H. & Dietz K.-J. (1987) *J. Plant Physiol.* **129**, 41–48.

Pheasant D.J. & Hepler P.K. (1987) *Eur. J. Cell Biol.* **43**, 10–13.

Poenie M. (1990) *Cell Calcium* **11**, 85–91.

Prasher D., McCann R.O. & Cormier M.J. (1985) *Biochem. Biophys. Res. Commun.* **126**, 1259–1268.

Reiss H.-D. & Herth W. (1978) *Protoplasma* **97**, 373–377.

Reiss H.-D. & Nobiling R. (1986) *Protoplasma* **131**, 244–246.

Reiss H.-D., Herth W. & Nobiling R. (1985) *Planta* **163**, 84–90.

Reuveni M., Lerner H.R. & Poljakoff-Mayber A. (1985) *Plant Physiol.* **79**, 406–410.

Roe M.W., Lemesters J.J. & Herman B. (1990) *Cell Calcium* **11**, 63–73.

Russ U., Grolig F. & Wagner G. (1991) *Planta* **184**, 105–112.

Scheuerlein R., Schmidt K., Poenie M. & Roux S.J. (1991) *Planta* **184**, 166–174.

Schroeder J.I. & Hagiwara S. (1990) *Proc. Natl. Acad. Sci. USA* **87**, 9305–9309.

Speksnijder J.E., Weisenseel M.H., Chen T.-H. & Jaffe L.F. (1989) *Biol. Bull.* **176**(S), 9–13.

Stern M.D., Spurgeon H.A., Hansford R., Lakatta E.G. & Capogrossi M.C. (1989) *Cell Calcium* **10**, 527–534.

Tang S. & Beeler T. (1990) *Cell Calcium* **11**, 425–429.

Thomas M.V. (1982) *Techniques in Calcium Research.* Academic Press, London.

Timmers A.C.J., de Vries S.C. & Schel J.H.N. (1989) *Protoplasma* **153**, 24–29.

Timmers A.C.J., Reiss H.-D. & Schel J.H.N. (1991) *Cell Calcium* **12**, 515–521.

Tsien R.Y. & Harootunian A.T. (1990) *Cell Calcium* **11**, 93–109.

Uto A., Arai H. & Ogawa Y. (1991) *Cell Calcium* **12**, 29–37.

Wahl M., Lucherini M.J. & Grunstein E. (1990) *Cell Calcium* **11**, 487–500.

Wang M., van Duijin B. & Schram A.W. (1991) *FEBS Lett* **278**, 69.

Williams D.A., Cody S.H., Gehring C.A., Parish R.W. & Harris P.J. (1990) *Cell Calcium* **11**, 291–297.

Williamson R.E. & Ashley C.C. (1982) *Nature* **296**, 647–651.

Yin Z.-H., Neimanis S., Wagner U. & Heber U. (1990) *Planta* **182**, 244–252.

Zhang D.H., Callaham D.A. & Hepler P.K. (1990) *J. Cell Biol.* **111**, 171–182.

CHAPTER TWENTY-NINE

Assessment of Gap Junctional Intercellular Communication in Living Cells Using Fluorescence Techniques

ADRIAAN W. de FEIJTER,[1] JAMES E. TROSKO[2] & MARGARET H. WADE[1]

[1] Meridian Instruments Inc., Okemos, MI, USA
[2] Department of Pediatrics and Human Development, Michigan State University, MI, USA

29.1 INTRODUCTION

29.1.1 Intercellular communication: its role in maintaining homeostasis

The biological process by which cells influence control over each other in a multicellular organism is referred to as intercellular communication. One could classify intercellular communication as either systemic or local (Potter, 1983). However, with the discovery of a form of intercellular communication between contiguous cells via a membrane protein channel, the gap junction (Loewenstein, 1979, 1981), and the observation that the 'systemic' form of intercellular communication could modulate the gap junctional form of intercellular communication (Larsen & Risinger, 1985; Neyton & Trautman, 1986; Maldonado *et al.*, 1988; Madhukar *et al.*, 1989) via a variety of second messengers (Spray & Bennett, 1985), one needs to re-examine and integrate those different forms of cellular communication.

Three forms of interactive communication (extra-, intra- and inter-) could explain how the secretion of specific molecules by one cell could interact with another target cell, triggering a physiological response in that cell (e.g. stimulation of inhibition of mitogenesis; induction of differentiation; adaptive secretory products in a differentiated cell). The affected cell would, in turn, secrete other products which would signal the original signalling cell, thereby completing the feedback loop (Fig. 29.1).

Extracellular communication would refer to how cells communicate with distal cells via the secretion of a molecular signal (e.g. hormone, peptide, growth regulator, neurotransmitter). Intracellular communication would refer to those transmembrane-signalling mechanisms triggered by the extracellular signals. This would now modulate intercellular communication mediated by gap junctions.

One of the most important implications of this model of extra-, intra- and intercellular communication is that adaptive responses to changes in the environment can be made and that homeostasis of multiple functions of multiple systems can be restored after the challenge. This model provides a mechanistic integration of the various organ systems of the body, e.g. the neuroendocrine–immune system.

Disruption of any of these steps could lead to either adaptive or maladaptive consequences depending on circumstances (Trosko & Chang, 1984). Sustained or permanent blockage of any of the three communication pathways could, in principle, lead to disruption of homeostatic control of cell growth or differentiation.

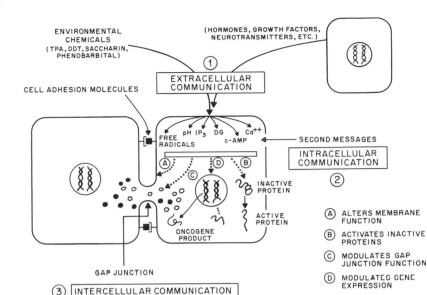

Figure 29.1 Diagram of the postulated link between extra-cellular communication and intercellular communication via various intracellular transmembrane signalling mechanisms. It provides an integrating view of how the neuroendocrine–immune system ('mind or brain/body connection') and other multisystem coordination could occur. While not shown here, activation or altered expression of various oncogenes (and 'anti-oncogenes') could also contribute to the regulation of gap junction function. (Reprinted from Trosko *et al.*, 1990b, with permission from Pergamon Press.)

This disruption could be the result of endogenous (e.g. genetic, developmental, sex) or exogenous (e.g. dietary, lifestyle, workplace, pollution, drug) factors (Trosko & Chang, 1988a,b, 1989).

Gap junctions are protein channels found in the membranes of most contiguous normal cells of metazoans (Hertzberg & Johnson, 1988). The subunit of the gap junction found in one membrane is referred to as a connexon and it joins the connexon in the adjoining membrane to form an aqueous channel which allows the passive transfer of ions and small molecules having molecular weights less than 1500 Da (Rammos *et al.*, 1990). Connexon subunits belong to a family of proteins, the connexins, which are related by their sequence but differ in their molecular weight (Nicholson *et al.*, 1987; Beyer *et al.*, 1987). Connexins 26, 30, 31, 32, 33, 36, 37, 38, 43 and 46 have been isolated from several organs in a variety of species (Willecke *et al.*, 1991). Several genes which code for these relatively highly conserved proteins have been cloned (Beyer *et al.*, 1987; Saez *et al.*, 1990).

Gap junction proteins are found in all metazoans and, while different gap junction genes/proteins can be found within one species and within one tissue in one species, the high degree of homology suggests their importance in the evolution of multicellular organisms (Saez *et al.*, 1990). It has been speculated that the variable regions of the gap junction protein might serve as the sites for specific regulation needed for gap junctions in different tissues (e.g. in excitable cells or non-excitable cells). In general, cells coupled by gap junctions are in equilibrium for ions and small regulatory and substrate molecules. The gap junctions can control episodic increases or decreases in these

ions or small molecules (Sheridan, 1987). As a result, cells in this communicating network can be synchronized for function (e.g. heart beats – DeMello, 1982; secretions of peptides – Meda *et al.*, 1987), or suppressed for cell division, i.e. contact inhibited (Azarnia *et al.*, 1988).

The biological roles of gap junctions seem to have been designed to be modulated by both endogenous factors (e.g. growth factors – Madhukar *et al.*, 1989; Maldonado *et al.*, 1988; hormones – Larsen & Risinger, 1985; neurotransmitters – Neyton & Trautman, 1986) and a wide variety of exogenous chemicals (pollutants, drugs, food additives, natural plant chemicals, nutrients, heavy metals, toxins, pesticides, herbicides, etc. – Trosko & Chang, 1988a). In other words, gap junctions have evolved to respond to external and internal signals in order to either increase or decrease communication (Trosko & Chang, 1984; Trosko *et al.*, 1990a). The biological consequences of the modulation of communication via gap junctions could be either adaptive or maladaptive (Trosko & Chang, 1984).

The regulation of gap junctional communication could, conceptually, occur by the modulation of cell–cell adhesion (Kanno *et al.*, 1984), by the production, assembly, and function of the gap junction proteins, at the transcriptional, translational or post-translational levels (Spray *et al.*, 1988), by the control of the amount of potential signals through the gap junctions (Loewenstein, 1966), or by the transduction of the gap junction signal passed via gap junctions (Trosko *et al.*, 1990a).

A wide number of intracellular communicating second messengers (intracellular calcium, pH, cAMP, free radicals, protein phosphorylation) have been reported to be factors in the regulation of gap junctions (Spray & Bennett, 1985; Saez *et al.*, 1987, 1990).

To date, a variety of approaches have been used to study functional gap junctional communication. Electrocoupling (Loewenstein, 1979), metabolic cooperation via the transfer of radioactive metabolites (Subak-Sharpe *et al.*, 1966), genetic mutants to metabolic cooperation (Yotti *et al.*, 1979), microinjection and transfer of fluorescent dyes (Enomoto & Yamasaki, 1984), scrape loading and dye transfer of fluorescent dyes (El-Fouly *et al.*, 1987), and fluorescence redistribution after photobleaching ('FRAP') (Wade *et al.*, 1986) have all been reported.

Glycyrrhetinic acid (GA), a folk remedy which has anti-inflammatory activity (Finney & Somers, 1958), was previously demonstrated to suppress 12-*O*-tetradecanoylphorbol-13-acetate (TPA)-induced ornithine decarboxylase activity (Okamoto *et al.*, 1983) and Epstein–Barr virus-associated early antigen (Okamoto *et al.*, 1983) *in vitro*. Furthermore, GA has been shown to possess anti-skin tumour-initiating (Wang *et al.*, 1991) and anti-skin tumour-promoting (Nishino *et al.*, 1984, 1986; Wang *et al.*, 1991) activities *in vivo*. On the other hand, GA and several derivatives have been shown to inhibit gap junctional communication *in vitro*, working by a mechanism other than activation of protein kinase C (PK-C) or mineralocorticoid or glucocorticoid receptors (Davidson *et al.*, 1986; Tsuda & Okamoto, 1986; Davidson & Baumgarten, 1988; Rosen *et al.*, 1988) and to enhance TPA-mediated inhibition of metabolic cooperation (Tsuda & Okamoto, 1986).

Several laboratories have reported that TPA induces a significant reduction in both the number and the area of gap junctions (Yancey *et al.*, 1982; Kalimi & Sirsat, 1984; van der Zandt *et al.*, 1990), and that the mechanism by which TPA exerts its inhibiting effect on gap junctional communication may involve PK-C (Nishizuka, 1984; van der Zandt *et al.*, 1990).

In this chapter, we discuss how fluorescence technology can be used to study the biological and physiological roles of gap junctions and the potential mechanisms of regulation of these gap junctions. In particular, we assess the effect of the number of contacting cells as well as the effect of two chemicals, TPA and 18β-GA, on intercellular gap junctional communication. In addition, an immunofluorescence technique to localize gap junction proteins is described.

29.2 MATERIALS AND METHODS

29.2.1 Instrument description

The ACAS 570 Interactive Laser Cytometer (Meridian Instruments Inc., Okemos, MI) was used in the studies

described to monitor fluorescence in living cells. The instrument consists of an inverted phase-contrast microscope, an *x*–*y* scanning stage, a 5 W argon ion laser for excitation, an acousto-optic modulator to control laser pulse intensity and duration during scanning, and photomultiplier tubes for fluorescence detection, all of which are under microprocessor and computer control. Unless otherwise noted, all data were acquired with a 40× long-working-distance objective. A schematic diagram of the ACAS optical path is depicted in Fig. 29.2. For the immunofluorescent localization of gap junction proteins, an ACAS 570 with confocal accessory was used. In addition to the components shown in Fig. 29.2, the confocal ACAS applies diffraction-limited optics, an adjustable pinhole in front of the detector, and an *x*–*y* scanning stage with *z*-axis control. Confocal optics reduce out-of-plane fluorescence, thus allowing an 'optical slice' to be scanned in the *z*-plane (see Chapter 17 in this volume for a description of confocal microscopy).

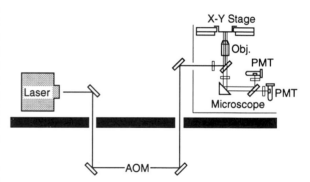

Figure 29.2 Optical path of the ACAS 570 showing the excitation pathway (laser, dichroic mirror, microscope objective) and emission pathway (microscope objective, dichroic mirror, photomultiplier tubes). Data are then digitized, collected and analysed using specific ACAS software.

29.2.2 Cell culture

The human teratocarcinoma cell line, PA-1 (ATTC CRL no. 1572) (HT) is near diploid and has functional gap junctional communication (Wade *et al.*, 1986; Schindler *et al.*, 1987). The human kidney epithelial cell line G401.2/6TG.1 (SB-3, a gift from E. Stanbridge, University of California) was originally derived from a Wilm's tumour patient and established as described previously (Weissman *et al.*, 1987). Fischer 344 rat liver epithelial cells (WB344), derived from normal liver (Tsao *et al.*, 1984), were used for the immunofluorescence study of the gap junction protein.

All cell cultures were maintained in D-medium, a modified Eagle's medium containing Earle's balanced salt solution with a 50% increase of vitamins and essential amino acids except glutamine, 100% increase of non-essential amino acids and 1 mM sodium pyruvate, 5.5 mM glucose, 14.3 mM NaCl, and 11.9 mM NaHCO$_3$ (pH 7.3). The medium was supplemented with 3% (SB-3), 5% (WB) or 10% (HT) fetal bovine serum (Gibco Laboratories) and 50 μg ml^{-1} gentamicin. Cells were grown at 37°C and 5% CO$_2$ in a humidified incubator. Low passage SB-3 cells were treated with the test chemicals 1 day after seeding. In both long- and short-term experiments, SB-3 cells were not trypsinized for the total duration of treatment. HT cells were plated and used within 48 h. All experiments were done at room temperature.

29.2.3 Chemicals

TPA (Sigma Chemical Co.) and 18β-glycyrrhetinic acid (18β-GA) (Aldrich Chemical Co.) were dissolved in ethanol and dimethylsulphoxide (DMSO), respectively. For both solvents, the final concentration in the medium was 0.1%. For the scrape loading experiments, Lucifer yellow CH (Sigma Chemical Co., molecular weight 457.2) and tetramethylrhodamine dextran (Molecular Probes Inc., molecular weight 10 000) were each dissolved in phosphate-buffered saline (PBS) at 0.5 mg ml^{-1}. For the FRAP assays, 5,6-carboxyfluorescein diacetate (molecular weight 376 as carboxyfluorescein, Molecular Probes Inc.) was dissolved in ethanol at 1 mg ml^{-1}.

29.2.4 Scrape loading

The scrape loading technique was performed as described previously by El-Fouly et al. (1987) and Oh et al. (1988). The cells were plated at high density and allowed to form a monolayer. They were then treated with non-cytotoxic concentrations of test chemicals and rinsed with PBS. Two millilitres of the dye (0.5 mg ml^{-1} Lucifer yellow in PBS) were added and several scrape lines were made in the monolayer with a surgical blade. After 3 min, to allow dye uptake and transfer, the cells were rinsed several times with PBS and examined under a Nikon epifluorescence phase-contrast microscope. Since the purpose of these experiments was to perform a quick assessment of various drug treatments on communication and not an exact quantitation, a fluorescence microscope was sufficient. Results were expressed as the average number of dye-coupled cells at a minimum of 15 random points along either side of the scrape line in each of two dishes. During the assay the cells were maintained at room temperature and care was taken to avoid exposure of the dye mixture to excessive room light. For the long-term incubations, media (with or without test chemicals) were changed every 2 days. In some experiments, tetramethyl-

rhodamine dextran (0.5 mg ml^{-1}) was included with Lucifer yellow as a control to distinguish the primary dye-loaded cells. Because of its large molecular weight, the dextran will neither diffuse through intact plasma membranes nor cross the junctional channels and can therefore serve to identify the primary loaded cells.

29.2.5 Fluorescence redistribution after photobleaching (FRAP)

For the FRAP assay, originally described by Wade et al. (1986), SB-3 or HT cells were plated at low density in 35 mm plates. After exposure to non-cytotoxic concentrations of test chemicals or solvent alone, the cells were rinsed with PBS containing calcium and magnesium (0.1 g litre^{-1} each) (PBS/Ca/Mg) and labelled with 7 μl ml^{-1} 5,6-carboxyfluorescein diacetate. After 15 min, the plates were rinsed several times with PBS/Ca/Mg and examined with the ACAS Interactive Laser Cytometer. An initial scan was made to determine the pre-bleach fluorescence levels. Then, selected cells were photobleached and monitored for the return of fluorescence at various time intervals (1–4 min, total of 5–10 scans).

For the HT experiments, care was taken to choose cells growing as single cells, in doublets, in triplets, or cells nested within clumps of four or more cells to monitor the rate of recovery after bleaching. In addition, the rate of decline in fluorescence of the neighbouring cells was assessed. The cells were examined within 40 min after labelling, as the recovery rates diminished when the cells were left out of culture for longer time periods. Some of the HT experiments were done with cells incubated in Hank's buffer (5.3 mM KCl, 0.34 mM KH$_2$PO$_4$, 0.5 mM MgCl$_2$, 1.3 mM CaCl$_2$, 0.8 mM MgSO$_4$, 138 mM NaCl, 4.2 mM NaHCO$_3$, 0.35 mM Na$_2$HPO$_4$, and 5.6 mM glucose).

For the experiments with SB-3 cells, 5–8 cells in each of five separate plates were monitored, resulting in at least 25 cells per treatment group. Cells growing in colonies and in direct contact with a minimum of three neighbouring cells were selected for photobleaching. In each plate a single cell was selected, but not photobleached to determine the background decline of fluorescence. Occasionally, a single cell was photobleached as a negative control for fluorescence redistribution.

Results of the FRAP experiments were expressed as the average percentage recovery of fluorescence with a standard error of the mean (SEM). The recovery is calculated as a percentage of the initial pre-bleach fluorescence value for each cell monitored. As in scrape loading, the complete assay was performed at room temperature, and media were changed every 2 days in long-term experiments.

382   A.W. de Feijter et al.

29.2.6 Immunofluorescence labelling for gap junction protein

WB rat liver epithelial cells were grown on coverslips until near confluency and fixed with 5% acetic acid in methanol and processed for immunofluorescence detection of gap junction proteins as described by Dupont et al. (1988) and El-Aoumari et al. (1990). In brief, fixed cells were incubated with 3% BSA in PBS to block non-specific binding, followed by incubation with a rabbit polyclonal antibody to connexin 43 (2 h). The antigen–antibody complex was localized using biotinylated anti-rabbit IgG and the streptavidin-fluorescein detection method. The coverslips were mounted with anti-fade solution (Aqua-Poly/Mount, Polysciences, Warrington, PA) and scanned with a confocal ACAS 570, using a 100× oil objective.

29.3 RESULTS

29.3.1 Morphological modulation of intercellular gap junction communication measured by FRAP analysis

To investigate a morphological aspect of gap junctional intercellular communication, we examined the effect of the number of contacting cells on the initial rate of fluorescence recovery after photobleaching in HT cells. Plate 29.1 depicts a triplet of cells where the cell in the centre was photobleached and fluorescence levels were monitored in that cell and both adjacent cells. As expected, while the bleached cell (area no. 3) recovered, the fluorescence levels in the neighbouring cells (areas nos 1 and 2) dropped as equilibrium between fluorescent and non-fluorescent carboxyfluorescein was attained. The fluorescence in the entire triplet was also monitored (area no. 4) and shows that the total fluorescence is not changing after photobleaching and that the recovery has nearly reached equilibrium. In some experiments, a single cell in the field was not bleached, but monitored, to correct for any photobleaching caused by repetitive scanning. Non-specific photobleaching was minimal (typically less than 1% per scan), due to the low excitation levels, the small area of the sample being illuminated, and the brief time of illumination. In some cases, the results were corrected for this non-specific bleaching effect.

Similar experiments were done with single cells, doublets, and cells with more than three cells in contact. The average rate over the first 3 min, expressed as the initial percentage increase per minute, for each of these combinations is shown in Table 29.1. As expected, the single cells did not

Table 29.1 FRAP analysis of gap junctional communication between HT cells with varying numbers of cell contacts.

	Recovery rate ± SEM	
Single cells	0.65 ± 0.16	($N = 23$)
Doublets	6.35 ± 0.47	($N = 21$)
Triplets	9.54 ± 0.95	($N = 10$)
Multiple contacts	11.59 ± 0.39	($N = 71$)

recover when bleached since there are no contacting cells. Recovery rates in cells growing in groups of two, three or more cells increase in a linear fashion as the number of contacting cells increases (Wade & Schindler, 1988).

29.3.2 Chemical modulation of gap junction communication in SB-3 cells measured by scrape loading and FRAP techniques

Another set of experiments was designed to examine the effects of various chemical treatments on gap junctional communication in SB-3 cells. These cells normally show a high level of intercellular communication as assessed by both scrape loading and FRAP (Madhukar et al., 1993). Figure 29.3 depicts several photomicrographs of scrape-loaded SB-3 cells after treatment with TPA. In control SB-3 cultures, Lucifer yellow migrated to 8–12 rows of adjacent cells on either side of the scrape line (Fig. 29.3(B)). When the cells were exposed to various concentrations of TPA for 1 h, an obvious inhibition of dye transfer was observed at 0.49 nM, resulting in 2–6 rows of dye transfer (Fig. 29.3(D)). At 4.9 nM, communication was completely blocked and the dye was retained only in the initially loaded cells (Fig. 29.3(E)). When the cells were treated with 18β-GA, dye transfer measured by scrape loading was partially inhibited at 25 μM, and virtually completely blocked at 50 μM (Table 29.2). As shown in Table 29.2, treatment with 30, 35, 40 and 45 μM 18β-GA resulted in significant, but not complete, inhibition of communication to 2–4 rows of cells without showing a clear, reproducible dose–response relationship. Similar results were obtained when the exposure was limited to a 10 min incubation with the compound (data not shown).

Intercellular communication was then quantitated in these cells after identical treatments using the FRAP technique. As expected, the control SB-3 cells show a high level of recovery (Plate 29.2 (A)); after 16 min, an average recovery of 61% was noted. Recovery was inhibited in a dose-dependent fashion by 0.16–1.6 nM TPA (Fig. 29.4). At concentrations over 1.6 nM communication was virtually completely blocked (Plate 29.2(B), Fig. 29.4). When the cells were treated

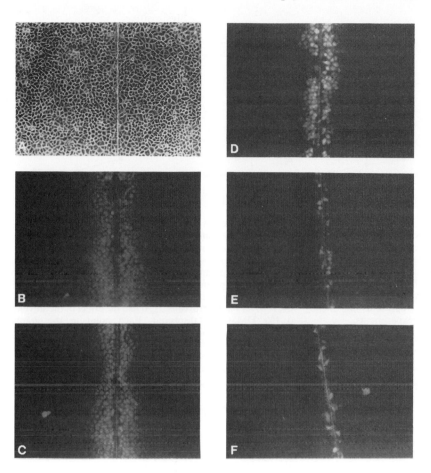

Figure 29.3 Dose–response relationship of TPA effect on gap junctional communication between SB-3 cells. Transfer of Lucifer yellow, 3 min after scrape loading of control cells is shown (B). Panel (A) shows the same field under phase contrast. Cells were treated for 1 h with 0.16 nM (C), 0.49 nM (D), 1.6 nM (E), and 4.9 nM (F) of TPA.

Table 29.2 Scrape loading analysis of dose-dependent inhibition of gap junctional communication in SB-3 cells after a 1 h exposure to 18β-GA.

Treatment	Transfer of Lucifer yellow (number of rows of cells)
0.1% DMSO	8–12
20 μM 18β-GA	8–12
25 μM 18β-GA	3–7
30 μM 18β-GA	2–4
35 μM 18β-GA	2–3
40 μM 18β-GA	2–4
45 μM 18β-GA	2–4
50 μM 18β-GA	1–3

with 18β-GA, only a moderate effect was seen, and concentrations needed to exceed 50 μM before significant reductions in recovery were noted (Fig. 29.5). The maximum inhibition observed under non-cytotoxic conditions (100 μM, 1 h) still resulted in 26% recovery.

The absence of a clear and reproducible dose–response relationship after treatment with 18β-GA in the scrape loading experiments, and the limited effect revealed by FRAP, led us to suppose that removal of 18β-GA immediately prior to the photobleaching experiment permits a fast recovery of communication after initial blockage. After ending exposure to the test chemical, the overall FRAP experiment takes approximately 30–40 min during which staining, rinsing, selection, photobleaching and scanning of the cells take place. Scrape loading, on the other hand, is completed within 4–6 min after drug treatment. To test this hypothesis, the FRAP protocol was slightly modified to allow continuation of 18β-GA treatment during the staining and scanning phases. As shown in Fig. 29.6, this modification had a drastic effect on the dose–response curve of the compound. No significant effect was observed at 1 μM, but at a concentration as low as 3 μM, dye transfer was virtually completely blocked. To establish the rate of return of communication after removal of 18β-GA, a FRAP experiment was performed in which the length of the 'recovery' period was varied. After incubation with the chemical for 1 h, cells were allowed to restore communication in the absence of 18β-GA, for 0, 20, 40, 60 or 80 min prior to measurement. In this experiment two concentrations of 18β-GA were tested to reveal any effect of

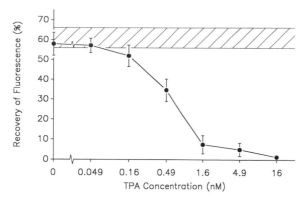

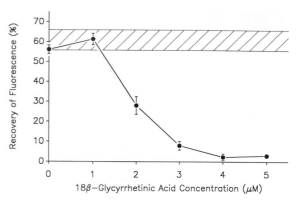

Figure 29.4 Dose–response relationship of TPA effect on gap junctional communication measured by FRAP. SB-3 cells were treated with different concentrations of TPA for 1 h and fluorescence recovery was determined 16 min after photobleaching. TPA concentration is plotted on a logarithmic scale. The shaded area shows the recovery in control cells. At 0 nM, the effect of the solvent (0.1% ethanol) is displayed. Each point in the graph represents the mean fluorescence recovery ± SEM of five experiments.

Figure 29.6 Dose–response relationship of 18β-GA effect on gap junctional communication measured by FRAP. SB-3 cells were treated with different concentrations of 18β-GA for 1 h and fluorescence recovery was determined 16 min after photobleaching. In contrast to the original protocol (Fig. 29.5), exposure to the chemical was continued during staining and scanning of the cells. The shaded area shows the recovery in control cells. At 0 μM, the effect of the solvent (0.1% DMSO) is displayed. Each point in the graph represents the mean fluorescence recovery ± SEM of five experiments.

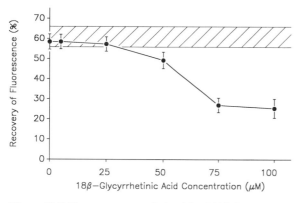

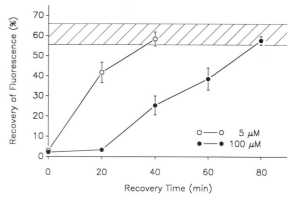

Figure 29.5 Dose–response relationship of 18β-GA effect on gap junctional communication measured by FRAP. SB-3 cells were treated with different concentrations of 18β-GA for 1 h and fluorescence recovery was determined 16 min after photobleaching. The shaded area shows the recovery in control cells. At 0 μM, the effect of the solvent (0.1% DMSO) is displayed. Each point in the graph represents the mean fluorescence recovery ± SEM of five experiments.

Figure 29.7 FRAP analysis of time-dependent recovery of gap junctional communication between SB-3 cells after complete blockage by a 1 h treatment with 5 or 100 μM 18β-GA. The shaded area shows the recovery in control cells. Each point in the graph represents the mean fluorescence recovery ± SEM of five experiments.

the dose on the rate of return of communication (Fig. 29.7). The results indicated that the return rate was very fast at the lower concentration of 5 μM 18β-GA; within 20 min communication returned to 69% of the control value while being completely blocked if the chemical was kept on the cells during analysis. After 40 min, no inhibitory effect remained. At 100 μM 18β-GA, more time was required for the recovery process; control levels were reached after 80 min. This fast recovery after 18β-GA removal was confirmed using the scrape loading technique: recovery was evident as soon as 5 min and complete within 40 min after

removal of 18β-GA (50 μM, 1 h) (data not shown).

Since inhibition of intercellular communication has been shown to be a transient effect in many *in vitro* studies using different chemicals and cell types (Enomoto & Yamasaki, 1985; Stedman & Welsch, 1985; Jongen *et al.*, 1987; Enomoto *et al.*, 1984; Madhukar *et al.*, 1993), the time–effect relationships of TPA and 18β-GA were examined using the FRAP assay during continuous exposure for up to 4 days. In contrast to the effect exhibited by TPA which was partially reversible, 18β-GA induced blockage of communication remained complete for at least 4 days

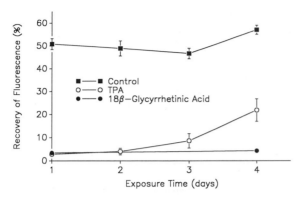

Figure 29.8 Time–response relationships of TPA (16 nM) and 18β-GA (5 μM) effects on gap junctional communication measured by FRAP. SB-3 cells were continuously exposed to the chemicals and their solvents for up to 4 days of culture, and fluorescence recovery was determined 16 min after photobleaching. Exposure to 18β-GA was continued during FRAP. Four days of exposure to 0.1% ethanol and 0.1% DMSO resulted in recoveries of respectively 49.6% ± 2.8 and 50.7 ± 3.1.

of incubation (Fig. 29.8). Removal of 18β-GA after a 4 day exposure resulted in a fast return of dye transfer to control levels within 40 min (data not shown). None of the experiments revealed any effects of the solvents, ethanol (for TPA) and DMSO (for 18β-GA), on gap junctional communication between SB-3 cells.

29.3.3   Immunofluorescent localization of gap junction protein

In addition to fluorescence assays to examine functional properties of gap junctions (i.e. intercellular communication), we used an immunofluorescence technique to directly visualize the gap junctions. Fixed WB rat liver epithelial cells were labelled with a rabbit polyclonal antibody against connexin 43, the major protein constituent of rat heart junctions (Dupont *et al.*, 1988) followed by a biotinylated anti-rabbit IgG, and streptavidin-fluorescein. The stained cells were examined with the ACAS 570, equipped with a confocal option. Figure 29.9 clearly demonstrates that the gap junction protein is located at areas of cellular contact. In areas where cells are not touching, there is no visible fluorescence.

29.4   DISCUSSION

FRAP and scrape loading have been shown to be very useful techniques for studies on gap junction mediated intercellular communication (Wade *et al.*, 1986; El-Fouly *et al.*, 1987; Schindler *et al.*, 1987; Oh *et al.*, 1988; Evans & Trosko, 1988; Evans *et al.*, 1988; de

Feijter *et al.* 1990; Madhukar *et al.*, 1991). Morphological and biochemical aspects of intercellular communication were examined with these techniques.

The studies with the human teratocarcinoma (HT) cells show that the number of cell contacts is directly related to the extent of intercellular communication. As the number of neighbouring cells increased, the initial rate of recovery after photobleaching increased in a linear fashion. It has been shown that the initial rate of recovery is a good measure of total recovery (Bombick & Doolittle, 1991). Similarly, reduced rates of communication have been noted in doublets when compared to cells with multiple contacts using rat glial cells, WB rat liver cells and normal human diploid fibroblasts (Bombick, pers. commun.). It has been shown that one can allow the bleached cell to recover, and bleach it again and obtain a similar rate of recovery (Wade *et al.*, 1986), and that the bleaching process is non-toxic as evidenced by continued cell growth (Wade, data not shown).

If the area of membrane contact is proportional to the number of gap junctions between cells, it should be possible, using antibodies to various connexins, to directly correlate the rate of gap junctional communication with the number of gap junctions between cells in doublets, triplets or groups with more than three contacting cells. Biegon *et al.* (1987) suggest that variations in size of the junction and dye permeance may also be related to cell cycle.

Intercellular communication between two different cell populations can be examined using the FRAP technique by culturing one of the cell types in the presence of fluorescent microspheres before co-culturing (Kalimi *et al.*, 1990, 1992). The microspheres serve to identify a particular cell type, but, unlike carboxyfluorescein molecules, are too large to move through gap junctions.

For the chemical studies on gap junction function, the techniques of scrape loading and FRAP were employed. The scrape loading experiments in this study primarily served to establish appropriate concentrations of test chemicals and confirm FRAP data; exact quantitation of the scrape loading data therefore was not attempted in this study. A more accurate quantitation of scrape loading data is possible using the ACAS (Oh *et al.*, 1988). Because of its low molecular weight, Lucifer yellow can be transferred between adjacent cells via gap junctions. The human kidney epithelial SB-3 cell line exerts a high level of intercellular communication as demonstrated by both the scrape loading and FRAP techniques.

Both TPA and GA proved to be very potent inhibitors of communication in SB-3 cells at non-cytotoxic concentrations. Many authors have observed that the TPA effect on gap junctional intercellular communication was partially or completely reversible

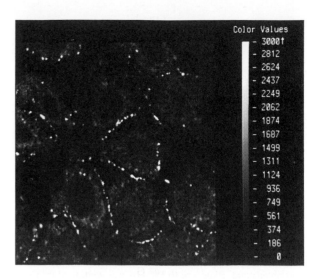

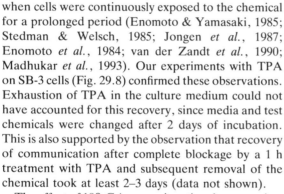

Figure 29.9 Optical section (approximately 0.7 μm thick) through WB rat liver epithelial cells, fluorescently labelled for the gap junction protein, connexin 43. The image (72 × 72 μm) was scanned with a confocal ACAS 570, using the 100× oil objective and displays fluorescence in areas of cell contact.

when cells were continuously exposed to the chemical for a prolonged period (Enomoto & Yamasaki, 1985; Stedman & Welsch, 1985; Jongen *et al.*, 1987; Enomoto *et al.*, 1984; van der Zandt *et al.*, 1990; Madhukar *et al.*, 1993). Our experiments with TPA on SB-3 cells (Fig. 29.8) confirmed these observations. Exhaustion of TPA in the culture medium could not have accounted for this recovery, since media and test chemicals were changed after 2 days of incubation. This is also supported by the observation that recovery of communication after complete blockage by a 1 h treatment with TPA and subsequent removal of the chemical took at least 2–3 days (data not shown).

The effect of 18β-GA on gap junctional communication has been described (Davidson *et al.*, 1986; Tsuda & Okamoto, 1986; Davidson & Baumgarten, 1988; Rosen *et al.*, 1988). In contrast to TPA, no recovery of communication was observed when cells were continuously exposed to 18β-GA for up to 4 days. Removal of 18β-GA after complete blockage of communication, however, led to full recovery within 40–80 min, depending on the concentration of the compound. This fast recovery was observed regardless of the exposure time (1 h or 4 days). Davidson *et al.* (1986) previously provided strong indications that the working mechanism of GA-induced inhibition of intercellular communication does not involve steroid receptors or the PK-C pathway. Data presented here are in accordance with this assumption. Rate of onset and reversal of the effect suggest that this compound does not exert its inhibitory effect on gap junctional communication via any biochemical route, but rather via a more direct mechanism. The fast recovery rate after chemical removal favours a mechanism in which direct and reversible binding of the molecule to existing gap junctions, as proposed by Davidson and Baumgarten (1988), might play a role. Other

mechanisms for inhibition of intercellular communication may involve intracellular calcium ions, pH, decrease in cAMP levels, or free radical production. Further studies may reveal the basis for the different mechanisms by which TPA and 18β-GA inhibit gap junctional communication.

Antibodies to connexin 43 were used to localize this gap junction protein in WB rat liver epithelial cells using immunofluorescence. This method has been used to show that the number of gap junctions in normal human keratinocytes, exposed to specific growth factors, is directly correlated to the extent of intercellular communication, measured by scrape loading (Dupont *et al.*, manuscript in preparation; Madhukar *et al.*, 1989). Jongen *et al.* (1991) demonstrated a correlation between the level of immunocytochemically stained connexin 43 and gap junctional communication, measured by microinjection/dye transfer, in cultured mouse epidermal cells at varying calcium concentrations.

The studies reported here have shown the use of three fluorescent tools, scrape loading, FRAP and immunofluorescent labelling, for studying gap junction function in cultured mammalian cells. Various mechanisms which affect intercellular communication, such as calcium perturbation, pH changes, peroxide formation (Bombick, 1990), lateral diffusion in the membrane, and translocation of PK-C (Rupp *et al.*, 1991) can also be assessed using fluorescent probes in combination with the ACAS 570.

ACKNOWLEDGEMENTS

The research on which this manuscript was based was supported, in part, by grants from the USAFOSR (no.

USAFOSR-89-0325) and the NIEHS (1P42ESOY91) to J.E.T.

The authors would like to thank Dr B.V. Madhukar and Dr Mel Schindler for helpful discussions, Dr Emmanuel Dupont for the antibody labelled specimens and Jeanne McHugh for secretarial assistance.

REFERENCES

Azarnia R., Reddy S., Kimiecki T.E., Shalloway D. & Loewenstein W.R. (1988) *Science* **239**, 398–400.

Beyer E., Paul D.L. & Goodenough D.A. (1987) *J. Cell Biol.* **195**, 2621–2623.

Biegon R.P., Atkinson M.M., Liu T-F., Kam E.Y. & Sheridan, J.D. (1987) *J. Membr. Biol.* **96**, 225–233.

Bombick D.W. (1990) *In Vitro Toxicol.* **3**, 27–39.

Bombick D.W. & Doolittle D.G. (1991) *The Toxicologist* **11**, 214.

Davidson J.S. & Baumgarten I.M. (1988) *J. Pharmac. Exp. Ther.* **246**, 1104–1107.

Davidson J.S., Baumgarten I.M. & Harley E.H. (1986) *Biochem. Biophys. Res. Commun.* **1**, 29–36.

de Feijter A.W., Ray J.S., Weghorst C.M., Klaunig J.E., Goodman J.I., Chang C.C., Ruch R.J. & Trosko J.E. (1990) *Molec. Carcin.* **3**, 54–67.

DeMello W.C. (1982) *Prog. Biophys. Molec. Biol.* **39**, 147–182.

Dupont E., El Aoumari A., Roustiau-Severe S., Briand J.P. & Gros D. (1988) *J. Membr. Biol.* **104**, 119–128.

El Aoumari A., Fromaget C., Dupont E., Reggio H., Durbec P., Brian J.-P., Boller K., Kreitman B. & Gros D. (1990) *J. Membr. Biol.* **115**, 229–240.

El-Fouly M.H., Trosko J.E. & Chang C.C. (1987) *Exp. Cell Res.* **168**, 422–430.

Enomoto T. & Yamasaki H. (1984) *Cancer Res.* **44**, 5200–5203.

Enomoto T. & Yamasaki H. (1985) *Cancer Res.* **45**, 2681–2688.

Enomoto T., Martel N., Kanno Y. & Yamasaki H. (1984) *J. Cell. Physiol.* **121**, 323–333.

Evans M.G. & Trosko J.E. (1988) *Cell Biol. Toxicol.* **4**, 163–171.

Evans M.G., El-Fouly M.H., Trosko J.E. & Sleight S.D. (1988) *J. Toxicol. Environ. Hlth* **24**, 261–271.

Finney R.S.H. & Somers G.J. (1958) *J. Pharmac. Pharmacol.* **10**, 613–620.

Hertzberg E.G. & Johnson R.G. (1988) *Gap Junctions.* Alan R. Liss, New York.

Jongen W.M.F., Sijtsma S.R., Zwijsen R.M.L. & Temmink J.H.M. (1987) *Carcinogenesis* **8**, 767–772.

Jongen W.M.F., Fitzgerald D.J., Asamoto M., Piccoli C., Slaga T.J., Gros D., Takeichi M. & Yamasaki H. (1991) *J. Cell Biol.* **114**, 545–555.

Kalimi G.H. & Sirsat S.M. (1984) *Cancer Lett.* **22**, 343–350.

Kalimi G.H., Chang C.C., Edwards P., Dupont E., Madhukar B.B., Stanbridge E. & Trosko J.E. (1990) *Am. Assoc. Cancer Res.* **31**, 319.

Kalimi G.H., Hampton L.L., Trosko J.E., Thorgeirsson S.S. & Huggett A.C. (1992) *Molec. Carcinogenesis* **5**, 301–310.

Kanno Y., Sasaki Y., Shiba Y., Yoshida-Noro C. & Takeichi M. (1984) *Exp. Cell Res.* **152**, 270–274.

Larsen W.J. & Risinger M.A. (1985) *Mod. Cell Biol.* **4**, 151–216.

Loewenstein W.R. (1966) *Ann. NY Acad. Sci.* **137**, 441–472.

Loewenstein W.R. (1979) *Biochim. Biophys. Acta* **560**, 1–65.

Loewenstein W.R. (1981) *Physiol. Rev.* **61**, 829–913.

Madhukar B.V., Oh S.Y., Chang C.C., Wade M.H. & Trosko J.E. (1989) *Carcinogenesis* **10**, 13–20.

Madhukar B.V., de Feijter A.W., Hasler C.M., Lockwood B., Oh S.Y., Chang C.C., Stanbridge E. & Trosko J.E. (1993) *In Vitro Toxicol.* (in press).

Maldonado P.E., Rose B. & Loewenstein W.R. (1988) *J. Membr. Biol.* **106**, 203–210.

Meda P., Bruzzone R., Chanson M., Bosco D. & Orci L. (1987) *Proc. Natl. Acad. Sci. USA* **84**, 4901–4904.

Murray A.W. & Fitzgerald D.J. (1979) *Biochem. Biophys Res. Commun.* **91**, 395–401.

Neyton J. & Trautman A. (1986) *J. Physiol.* **377**, 285–295.

Nicholson B.J., Dermietzel R., Teplov D.B., Traub O., Willecke K. & Revel J.P. (1987) *Nature* **329**, 732–734.

Nishino H., Kitagawa K. & Iwashima A. (1984) *Carcinogenesis* **5**, 1529–1530.

Nishino H., Yoshioka K., Iwashima A., Takizawa H., Konishi S., Okamoto H., Okabe H., Shibata S., Fujiki H. & Sugimura T. (1986) *Japan. J. Cancer Res. (GANN)* **77**, 33–38.

Nishizuka Y. (1984) *Nature* **308**, 693–698.

Oh S.Y., Madhukar B.M. & Trosko J.E. (1988) *Carcinogenesis* **9**, 135–139.

Okamoto H., Yosida D. & Mizusaki S. (1983) *Cancer Lett.* **21**, 29–35.

Pitts J.D. & Finbow M.E. (1986) *J. Cell Sci.* **4**, 239–266.

Potter V.R. (1983) *Prog. Nucleic Acid Res. Molec. Biol.* **29**, 161–173.

Rammos S., Gittenberger-de Groot A.C. & Oppenheimer-Dekker A. (1990) *Int. J. Cardiol.* **29**, 285–295.

Rosen A., Van der Merwe P.A. & Davidson J.S. (1988) *Cancer Res.* **48**, 3485–3489.

Rupp H.L., Trosko J.E. & Madhukar B.V. (1991) *The Toxicologist* **11**, 273 (abstract).

Saez J.D., Bennett M.V.L. & Spray D.C. (1987) *Science* **236**, 967–969.

Saez J.C., Spray D.C. & Hertzberg E.L. (1990) *In Vitro Toxicol.* **3**, 69–86.

Schindler M., Trosko J.E. & Wade M.H. (1987) *Methods Enzymol.* **141**, 439–447.

Sheridan J.D. (1987) In *Cell-to-Cell Communication*, W.C. DeMello (ed.). Plenum Press, New York, pp. 187–222.

Spray D.C. & Bennett M.V.L. (1985) *Ann. Rev. Physiol.* **47**, 281–303.

Spray D.C., Saeiz J.D. & Burt J.M. (1988) In *Gap Junctions*, E. Hertzberg & R. Johnson (eds). Alan R. Liss, New York, pp. 227–244.

Stedman D.B. & Welsch F. (1985) *Carcinogenesis* **6**, 1599–1605.

Subak-Sharpe H., Burk R.R. & Pitts J.D. (1966) *Hereditary* **21**, 342–343.

Trosko J.E. & Chang C.C. (1984) *Pharmacol. Rev.* **36**, 137–144.

Trosko J.E. & Chang C.C. (1988a) In *Tumor Promoters: Biological Approaches for Mechanistic Studies and Assay*

Systems, R. Langenbach, J.C. Barrett & E. Elmore (eds). Raven Press, New York, pp. 97–111.

Trosko J.E. & Chang C.C. (1988b) In *Banbury Report 31: Carcinogen Risk Assessment: New Directions in the Qualitative and Quantitative Aspects*, R.W. Hart & F.G. Hoerger (eds). Cold Spring Harbor Lab, Cold Spring Harbor, NY, pp. 139–170.

Trosko J.E. & Chang C.C. (1989) In *Biologically Based Methods for Cancer Risk Assessment*, C.C. Travis (ed.). Plenum Press, New York, pp. 165–179.

Trosko J.E., Chang C.C., Madhukar B.V. & Klaunig J.E. (1990a) *Pathobiology* **58**, 265–278.

Trosko J.E., Chang C.C., Madhukar B.V. & Oh S.Y. (1990b) *In Vitro Toxicol.* **3**, 9.

Tsao M.S., Smith J.D., Nelson K.G. & Grisham J.W. (1984) *Exp. Cell Res.* **154**, 38–52.

Tsuda H. & Okamoto H. (1986) *Carcinogenesis* **7**, 1805–1807.

van der Zandt P.T.J., de Feijter A.W., Homan E.C., Spaaij C., de Haan L.H.J., van Aelst A.C. & Jongen W.M.F. (1990) *Carcinogenesis* **11**, 883–888.

Wade M.H. & Schindler M. (1988) *FASEB* **2**, A320.

Wade M.H., Trosko J.E. & Schindler M. (1986) *Science* **232**, 525–528.

Wang Z.Y., Agarwal R., Zhou Z.C., Bickers D.R. & Mukhtar H. (1991) *Carcinogenesis* **12**, 187–192.

Weissman B.E., Saxon P.J., Pasquale S.R., Jones G.R., Geiser A.G. & Standbridge E.J. (1987) *Science* **236**, 175–180.

Willecke K., Jungbluth S., Dahl E., Hennemann H., Heynkes R. & Grzeschik K.-H. (1991) *Eur. J. Cell Biol.* **53**, 275–280.

Yamasaki H. (1988) In *Gap Junctions*, E.L. Hertzberg & R.G. Johnson. (eds). Alan R. Liss, New York, pp. 449–465.

Yancey S.B., Edens J.E., Trosko J.E., Chang C.C. & Revel J.P. (1982) *Exp. Cell Res.* **139**, 329–340.

Yotti L.P., Chang C.C. & Trosko J.E. (1979) *Science* **206**, 1089–1091.

CHAPTER THIRTY

Fast Multisite Optical Measurement of Membrane Potential

JIAN-YOUNG WU & LAWRENCE B. COHEN

Department of Physiology, Yale University School of Medicine, New Haven, CT, USA

30.1 INTRODUCTION

An optical measurement of membrane potential using a molecular probe can be beneficial in a variety of circumstances. 'Such a probe could, we believe, provide a powerful new technique for measuring membrane potential in systems where, for reasons of scale, topology, or complexity, the use of electrodes is inconvenient or impossible' (B.M. Salzberg, pers. commun.). The possibility of using optical methods was first suggested in 1968 by the discovery of potential-dependent changes in intrinsic optical properties of squid giant axons (Cohen et al., 1968). Shortly thereafter, Tasaki et al. (1968) found stimulus-dependent changes in fluorescence of stained axons, and in 1971 a search was begun for dyes that would give signals large enough to be useful for monitoring membrane potential. To date more than 1500 dyes have been tested for their ability to act as molecular transducers of membrane potential into three types of optical signals: absorption, birefringence, and fluorescence. This screening effort resulted in the discovery of dyes with a signal-to-noise ratio 100 times larger than any available in 1971. Several of these dyes (see, for example, Fig. 30.1) have been used to monitor changes in potential in a variety of preparations. For reviews, see Cohen and Salzberg (1978), Waggoner (1979), Salzberg (1983), Grinvald et al. (1988), and Dasheiff (1988). Earlier discussions of methods have been published (Cohen & Lesher, 1986; Grinvald et al., 1988).

We begin with the evidence showing that optical signals are potential-dependent. Then we discuss the selection of signal type, dye, light source, photodetectors, optics, and computer hardware and software. This concern about apparatus arises because the signal-to-noise ratios in optical measurements are often small and attention to detail is required to reduce noise. While some of the discussion is most relevant to our own apparatus, other aspects of the paper would apply to any optical measurement.

All of the optical signals described in this paper are 'fast' signals (Cohen & Salzberg, 1978). These signals are presumed to arise from membrane-bound dye; they follow changes in membrane potential with time-courses that are rapid compared to the rise time of an action potential.

30.2 SOME OPTICAL SIGNALS ARE POTENTIAL-DEPENDENT

The squid giant axon has provided a useful preparation for distinguishing among possible origins for optical

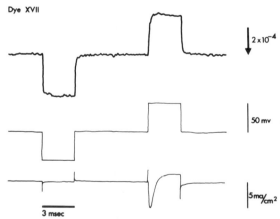

XVII, Merocyanine, Absorption,Birefringence

RH155, Oxonol, Absorption

RH414, Styryl, Fluorescence

XXV, Oxonol, Fluorescence, Absorption

Figure 30.1 Structures of four dyes that have been used to monitor membrane potential. The merocyanine (XVII) was the dye used in the experiments illustrated in Figs. 30.2 and 30.3. Dye XVII and the oxonol XXV are available from Dr A.S. Waggoner, Center for Fluorescence, Carnegie Mellon University, 4400 Fifth Ave., Pittsburgh, PA, as WW375 and WW781. Dye XVII is available commercially as NK 2495 from Nippon Kankoh-Shikiso Kenkyusho Co. Ltd. The oxonol, RH155, and styryl, RH414, are available from Amiram Grinvald, Department of Neurobiology, Weizmann Institute, Rehovot, Israel. RH414 is available commercially as dye 1112 from Molecular Probes, Junction City, OR. RH155 is available as NK3041.

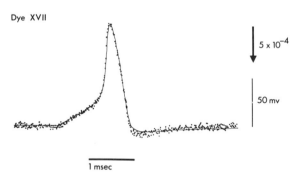

Dye XVII

5×10^{-4}

50 mv

1 msec

Figure 30.2 The change in absorption (dots) of a giant axon stained with dye XVII during a membrane action potential (smooth trace) simultaneously recorded with an internal electrode. The change in absorption and the action potential had the same time-course. In this figure and in Fig. 30.3 the direction of the arrow adjacent to the optical trace indicates the direction of an increase in absorption; the size of the arrow represents the stated value of a change in absorption, ΔA, in a single sweep divided by the resting absorption due to the dye, A_r. Incident light of 750 nm was used; 32 sweeps were averaged. The response time constant of the light measuring system was 5 µs. (Redrawn from Ross *et al.*, 1977.)

signals and for screening new dyes. A simple filter spectrofluorometer and spectrophotometer have been used to measure changes in dye-related optical properties. Figure 30.2 shows the results of a measurement of light absorption during an action potential in a squid axon stained with a merocyanine dye (XVII; Roman numerals refer to dyes in Fig. 30.1 or in table I of Gupta *et al.*, 1981). The dotted trace is the light intensity transmitted through the axon at 750 nm. The smooth curve is the potential measured between

internal and external electrodes. Because the two measurements have very similar time-courses, it seemed likely that the absorption signal was related to the changes in membrane potential not to the ionic currents or the increases in membrane permeability that occur during the action potential. More direct evidence for potential dependence comes from voltage clamp experiments such as the one illustrated in

Dye XVII

2×10^{-4}

50 mv

$5 \, ma/cm^2$

3 msec

Figure 30.3 Changes in absorption of a giant axon stained with dye XVII (top trace) during hyperpolarizing and depolarizing steps (middle trace). The bottom trace is the current density. The absorption changes had the same shape as the potential changes and were insensitive to the large currents and conductance changes during depolarization. The holding potential was the resting potential, and hyperpolarization is represented downward. Inward currents are downward. Incident light of 750 nm was used. 128 sweeps were averaged; the time constant of the light-measuring system was 20 µs. (Redrawn from Ross *et al.*, 1977.)

Fig. 30.3. The top trace is the absorption signal. Its time-course is similar to the membrane potential (middle trace) and distinctly different from the permeability changes or the ionic currents (bottom trace). This kind of result has been obtained for many dyes, including all four illustrated in Fig. 30.1.

Inspection of Fig. 30.3 might suggest that signals with a time-course similar to the currents or permeability are smaller than 5% of the total signal. In fact, a conclusion this strong is unwarranted because the result in Fig. 30.3 was obtained with a somewhat arbitrary amount of compensation for the resistance in series with the axon membrane. Although the compensation used implied a series resistance within the range of previously reported values, the series resistance was not measured independently in this experiment. This ambiguity has been resolved by Salzberg and Bezanilla (1983).

In a voltage clamp experiment with four potential steps, the absorption change was linearly related to membrane potential over the range ± 100 mV from the resting potential. A linear relationship between optical signal and membrane potential has been obtained measuring absorption, fluorescence and birefringence signals from many dyes. Thus, in many instances there was strong evidence that the signals obtained with millisecond potential steps in squid axons depended in some manner on changes in the transmembrane potential (Cohen et al., 1970; Conti & Tasaki, 1970; Patrick et al., 1971; Davila et al., 1974; Conti, 1975; Ross et al., 1977, Gupta et al., 1981), although there was initial disagreement about this conclusion (Conti et al., 1971; Tasaki et al., 1972).

However, it is certain that non-potential-dependent signals can also be found with the same dyes. Russell et al. (1979) reported slow ion-dependent optical changes from suspensions of sarcoplasmic reticulum. Recently Irena Klodos and Biff Forbush have found fluorescence changes related to the conformation of the Na–K ATPase with styryl dyes.

Several kinds of dye-related optical signals are quite slow (100–1000 ms) e.g. Orbach & Cohen, 1983; Orbach et al., 1985; Kauer et al., 1987; Lev-Ram & Grinvald, 1986; Blasdel & Salama, 1986). Clearly, it would be useful to have independent evidence from electrode recordings that these slow signals represent a change in membrane potential. But it is not easy to think of practical experiments to obtain quantitative evidence. Dye-related optical signals whose potential dependence cannot be confirmed with electrode measurements must be interpreted with caution.

Dye signals can be very rapid. The initial attempts to measure differences in time-course between the potential change and the optical signal showed that some signals lagged behind the change in membrane potential by less than 10 μs (Ross et al., 1974).

Measurements with a faster apparatus by Salzberg, Bezanilla and Obaid showed that the signals obtained with dyes XXVI (styryl), XVII and XXII (merocyanines) were still too rapid to measure and thus lagged behind the change in membrane potential by less than 2 μs (Loew et al., 1985; B.M. Salzberg, F. Bezanilla & A.L. Obaid, unpublished results). However, when relatively low concentrations of dye I were used, time constants as slow as 70 μs were obtained, and with high concentrations, very slow components may appear (Ross et al., 1974; Ross & Krauthamer, 1984).

A number of studies have been made to determine the molecular mechanisms that result in potential-dependent optical properties. This subject is discussed in Waggoner and Grinvald (1977), Huang et al. (1988), and Fromherz et al. (1991).

30.3 CHOOSING ABSORPTION, BIREFRINGENCE OR FLUORESCENCE

Sometimes it is possible to decide in advance which kind of optical signal will give the best signal-to-noise ratio, but in other situations an experimental comparison is necessary.

The choice of signal type may depend on the optical characteristics of the preparation. Birefringence signals are relatively large in preparations that, like giant axons, have a cylindrical shape and radial optical axis. However, in preparations with spherical symmetry (e.g. molluscan cell soma), the birefringence signals in adjacent quadrants will cancel (Boyle & Cohen, 1980). In these preparations, birefringence can only be measured using a detection system with high spatial resolution. Achieving this spatial resolution degrades the signal-to-noise ratio. Because birefringence can be measured at wavelengths outside the absorption band of the dye, eliminating photodynamic damage and dye bleaching, birefringence would be preferable to absorption or fluorescence in measurements of propagation along axons. In one experiment where birefringence should have been tested, it was not (Shrager et al., 1987).

An instance where the preparation dictated the choice of signal involved mammalian cortex. Here transmitted light measurements are not easy (subcortical implantation of a light guide would be necessary), and the small absorption signals that detected in reflected light (Ross et al., 1974; Orbach & Cohen, 1983) meant that fluorescence would be optimal (Orbach et al., 1985). Blasdel and Salama (1986) suggest that dye-related reflectance changes can be measured from cortex; this conclusion has been questioned (Grinvald et al., 1986).

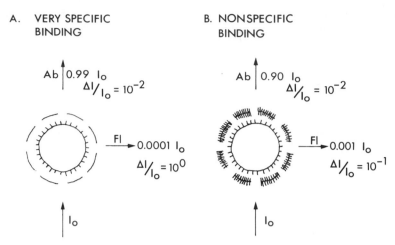

Figure 30.4 (A) The light transmission and fluorescence intensity when only a neuron binds dye and (B) when both the neuron and extraneous material binds dye. In (A), assuming that one dye molecule is bound per 2.5 phospholipid molecules, 0.99 of the incident light is transmitted. If a change in membrane potential causes the dye to disappear, the fractional change in transmission is 1%, but in fluorescence it is 100%. In (B), nine times as much dye is bound to extraneous material. Now the transmitted intensity is reduced to 0.9, but the fractional change is still 1%. The fluorescence intensity is increased 10-fold, and therefore the fractional change is reduced by the same factor. Thus extraneously bound dye degrades fluorescence fractional changes and signal-to-noise ratios more rapidly. (Redrawn from Cohen & Lesher, 1986.)

Another factor that affects the choice of absorption or fluorescence is that the signal-to-noise ratio in fluorescence is degraded by dye bound to extraneous material. Figure 30.4 illustrates a spherical cell surrounded by extraneous material. In Fig. 30.4(A), dye binds only to the cell; in Fig. 30.4(B), there is 10 times as much dye bound to extraneous material. To calculate the transmitted intensity we assume that there is one dye molecule for every 2.5 phospholipid molecules (this might be a large concentration if one would also expect to maintain physiological function) and a large extinction coefficient (10^5). The amount of light absorbed by the cell is still only 0.01 of the incident light and thus the transmitted light is $0.99 I_o$. Thus, even if this dye completely disappeared due to a change in potential, the fractional change in transmission, $\Delta I/I_o$, would be only 1% (10^{-2}). The amount of light reaching the photodetector in fluorescence will be much lower, say $0.0001 I_o$. Several factors account for the lower intensity. First, only 0.01 I_o is absorbed by the dye; second, we assume a fluorescence efficiency (photons emitted/photons absorbed) of 0.1; and third, if we assume a light-collecting system of 0.8 NA (numerical aperture), only 0.1 of the emitted light reaches the photodetector. But even though the light reaching the fluorescence detector is small, disappearance of dye would result in a 100% decrease in fluorescence – a fractional change of 10^0. Thus, in situations where dye is bound only to the cell membrane and there is only one cell

in the light path, the fractional change in fluorescence can be much larger than the fractional change in transmission.

However, the relative advantage of fluorescence is reduced if dye binds to extraneous material. When 10 times as much dye is bound to the extraneous material as was bound to the cell membrane (Fig. 30.4(B)), the transmitted intensity is reduced to approximately $0.9 I_o$. If a potential change again causes the cell-bound dye to disappear, the fractional change in transmission is nearly unaffected. In contrast, the resting fluorescence intensity is now higher by a factor of 10, so the fractional fluorescence change is reduced by the same factor. Thus the fluorescence fractional change is more severely affected. It doesn't matter whether the extraneous material happens to be connective tissue, glial membrane, or neighbouring neuronal membranes.

In Fig. 30.4(B), the fractional change in fluorescence was still larger than in transmission. However, the light intensity in fluorescence was about 10^3 smaller, and this reduces the signal-to-noise ratio in fluorescence (see below). Partly because of the signal degradation due to extraneous dye, fluorescence signals have most often been used in monitoring activity from tissue-cultured neurons, whereas absorption has been preferred in measurements from ganglia and brain slices. Recently very large signals have been obtained in the absorption mode from *Aplysia* neurons in tissue culture (Parsons *et al.*, 1989). In ganglia and brain slices the fractional changes in both transmission and fluorescence are small; they range between 10^{-4} and 10^{-2} for a 100 mV potential change.

30.4 DYES

Clearly, the choice of dye is important in maximizing the signal-to-noise ratio. Only in a few instances where there is a large density of synchronously active membrane, has it been possible to obtain large signals without testing a number of dyes. In addition, in some preparations, photodynamic damage due to illumination of the dye in the presence of oxygen may also affect the choice of dye. Pharmacological effects and dye bleaching are also considered in this section.

30.4.1 Screening

Using giant axons, more than 1500 dyes have been tested for signal size in response to changes in membrane potential. This screening was made possible by synthetic efforts of three laboratories: Alan Waggoner, Jeff Wang and Ravender Gupta of Amherst College; Rina Hildesheim and Amiram Grinvald at the Weizmann Institute; and Joe Wuskell and Leslie Loew at the University of Connecticut Health Center. Included in these syntheses were about 100 analogues of each of the four dyes illustrated in Fig. 30.1. In each of these four groups there were 10 or 20 dyes that gave similarly large signals (within a factor of 2 of the dye illustrated) on squid axons.

However, dyes that gave nearly identical signals on squid axons gave very different responses on other preparations, and thus many dyes had to be tested to maximize the signal. Examples of preparations where a number of dyes had to be screened are the *Navanax* and *Aplysia* ganglia (London *et al.*, 1987; Zecevic *et al.*, 1989; Morton *et al.*, 1991), rat cortex (Orbach *et al.*, 1985), and tissue-cultured neurons (Ross & Reichardt, 1979; Grinvald *et al.*, 1981a). Some of the dyes did not penetrate through connective tissue or along intercellular spaces to the membrane of interest. Others appeared to have a relatively low affinity for neuronal versus non-neuronal tissue. In some cases, the dye penetrated well and the staining appeared to be specific, but nonetheless the signals were small. Ross and Krauthamer (1984) have reported a case where neurons from different species of the same genus (*Balanus*) had qualitatively different signals.

30.4.2 Pharmacological effects

In most, if not all, preparations high concentrations of dye will have pharmacological effects. However, in many instances, the dye concentration needed to obtain the maximum signal size was lower than the concentration at which pharmacological effects were detected. These include the squid giant axon (Cohen *et al.*, 1974; Gupta *et al.*, 1981), neuroblastoma cells in tissue culture (Grinvald *et al.*, 1982), the barnacle supra-oesophageal ganglion (Salzberg *et al.*, 1977; Grinvald *et al.*, 1981b), embryonic semilunar ganglion (Sakai *et al.*, 1985), and the *Navanax* buccal ganglion (London *et al.*, 1987). In the *Navanax* experiments, the buccal ganglion was stained in a minimally dissected preparation and feeding behaviour was measured with and without staining. Complex synaptic interactions are probably required to generate the correct behaviour and thus, a strong statement can be made about the absence of pharmacological effects. However, in the optical experiments on salamander olfactory bulb, frog optic tectum, and mammalian cortex, the ability to detect pharmacological effects is limited since our understanding of the function(s) and outputs of these brain areas are far from complete. In these preparations one can only say that pharmacological effects were not disastrous.

In experiments on the *Aplysia* abdominal ganglion, a 0.5 mg ml^{-1} solution of the oxonol dye RH155 (Fig. 30.1) will suppress the spontaneous activity detected with extracellular electrode measurements from the siphon nerve. Nakashima *et al.* (1992) found this effect can be eliminated by reducing the staining concentration to 0.2 mg ml^{-1}.

30.4.3 Photodynamic damage and dye bleaching

In certain experiments – for example, on neuroblastoma neurons using the styryl dye RH414 (Fig. 30.1) – photodynamic damage (due to the interaction of light, dye, and oxygen) limited the duration of the experiments (Grinvald *et al.*, 1982). In others – for example, on *Navanax* buccal ganglia using the oxonol dye RH155 (Fig. 30.1) – it was difficult to detect photodynamic damage (London *et al.*, 1987). Similarly, dye bleaching has caused difficulties in some preparations but not in others. Grinvald *et al.* (1982) reported a 3% bleaching from 350 ms of illumination using a styryl dye, whereas we found bleaching difficult to detect after 10 min of illumination using the oxonol (London *et al.*, 1987). This difference in severity is, in part, due to the difference in dyes that were used; in addition, higher light intensities were used in the experiments where damage and bleaching were severe. Thus, advantages of increased intensities in terms of signal-to-noise ratio (see below) may be counterbalanced by increased damage and bleaching. Since both effects are dye-dependent (Cohen *et al.*, 1974; Ross *et al.*, 1977; Gupta *et al.*, 1981), additional dye screening may be necessary in preparations where they cause difficulty. Although bleaching and photo-dynamic damage are sometimes correlated (Gupta *et al.*, 1981), in other instances bleaching can occur without detectable damage (Ross & Krauthamer, 1984).

30.5 MEASURING TECHNOLOGY

The limit of accuracy with which light can be measured is currently set by the shot noise arising from the statistical nature of photon emission and detection. Fluctuations in the number of photons emitted per unit time will occur, and if an ideal light source (tungsten filament) emits an average of 10^{14} photons ms^{-1}, the root-mean-square (RMS) deviation in the number emitted is the square root of this number or 10^7 photons ms^{-1}. In the shot-noise-limited case, the signal-to-noise ratio is proportional to the square roots of the number of measured photons and the band-width of the photodetection system (Braddick, 1960; Malmstadt *et al.*, 1974). The shot-noise-dependence on the number of measured photons is illustrated in Fig. 30.5. Figure 30.5(A) shows the result of using a random number table to distribute 20 photons into 20 time windows. In Fig. 30.5(B) the same procedure was used to distribute 200 photons into the same 20 bins. Relative to the average light level there is more noise in the top trace with 20 photons than in the bottom

trace with 200 photons. On the right side of Fig. 30.5 the signal-to-noise ratios are measured and the improvement is similar to that expected from the square root relationship. This result, of signal-to-noise ratio proportional to the square root of measured intensity, is indicated by the dotted line in Fig. 30.6. In a shot-noise-limited measurement, improvements in the signal-to-noise ratio can only be obtained by (1) increasing the illumination inten-sity, (2) improving the light-gathering efficiency of the measuring system, or (3) reducing the system bandwidth.

Because only a small fraction of the 10^{14} photons ms^{-1} emitted by a 3300°F tungsten filament will be measured, a signal-to-noise ratio of 10^{-7} cannot be achieved. A 0.7 NA lamp collector lens would collect 0.06 of the emitted light. Only 0.2 of the emitted photons are in the visible wavelength range; the remainder are in the infrared (heat). An interference filter of 30 nm width at half-height might transmit only 0.05 of the visible light. Additional losses will occur at all air–glass interfaces. Thus, the light reaching the preparation might typically be reduced to 10^{10} photons ms^{-1}. If the light-collecting system has high efficiency, e.g. in an absorption measurement, about 10^{10} photons ms^{-1} will reach the photodetector, and if the photodetector has a quantum efficiency (photoelec-trons per photon) of 1.0, then 10^{10} electrons ms^{-1} will be measured. The RMS shot noise will be 10^5 electrons ms^{-1}; thus the relative noise is 10^{-5} (a signal-to-noise ratio of 100 db).

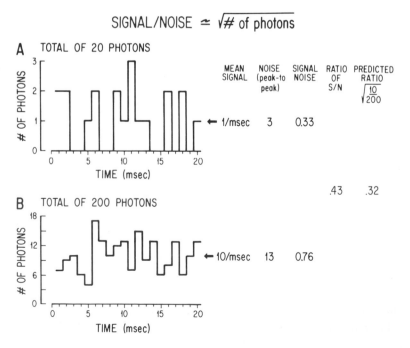

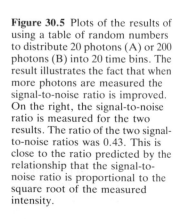

Figure 30.5 Plots of the results of using a table of random numbers to distribute 20 photons (A) or 200 photons (B) into 20 time bins. The result illustrates the fact that when more photons are measured the signal-to-noise ratio is improved. On the right, the signal-to-noise ratio is measured for the two results. The ratio of the two signal-to-noise ratios was 0.43. This is close to the ratio predicted by the relationship that the signal-to-noise ratio is proportional to the square root of the measured intensity.

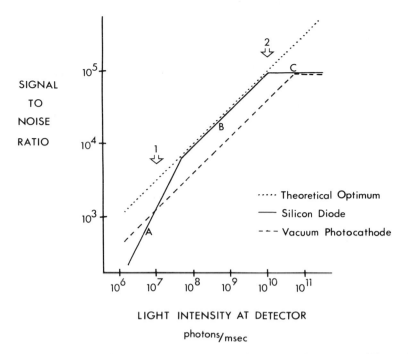

Figure 30.6 Signal-to-noise ratio as a function of light intensity in photons per millisecond. The approximate light intensity per detector in fluorescence measurements from ganglia or vertebrate cortex using a tungsten filament bulb is indicated by arrow 1. The approximate intensity in absorption measurements in ganglia or brain slices is indicated by arrow 2. The theoretical optimum signal-to-noise ratio (dotted line) is the shot-noise limit. The signal-to-noise ratio expected with a silicon diode detector is indicated by the solid line. The silicon diode signal-to-noise ratio approaches the theoretical maximum at intermediate light intensities (segment B) but falls off at low intensities (segment A) because of dark noise, and falls off at high intensities (segment C) because of extraneous noise. The expected signal-to-noise ratio for a vacuum photocathode detector is indicated by the dashed line. At low intensities the vacuum photocathode is better than a silicon diode because it has less dark noise. At intermediate intensities it is not as good because of its lower quantum efficiency. (Redrawn from Cohen and Lesher, 1986.)

30.5.1 Extraneous noise and dark noise

30.5.1.1 Extraneous noise

A second type of noise, termed extraneous or technical noise, is more apparent at higher light intensities where the sensitivity of the measurement is high because the fractional shot noise is low. One type of extraneous noise, caused by fluctuations in the output of the light source, will be discussed in Section 30.5.2. Two other sources of extraneous noise are vibrations in the light path and movement of the preparation. A number of precautions for reducing vibrational noise are described in Salzberg *et al.* (1977). Embedding ganglia in 1–3% agar reduced vibrational noise (London *et al.*, 1987). We recently found that the pneumatic isolation mounts on two vibration isolation tables which we used were providing only minimal isolation in the frequency range 20–60 Hz. By replacing the pneumatic mounts with air-filled soft rubber tubes we further reduced the vibration noise

(D. Zecevic, J.A. London & L.B. Cohen, unpublished results). Nevertheless, it has been difficult to reduce vibrational noise to less than 10^{-5} of the total light. With this amount of vibrational noise, increases in measured intensity beyond 10^{10} photons ms^{-1} would not improve the signal-to-noise ratio (segment C of Fig. 30.6).

Noise due to movement is a major problem in measurement from *in vivo* preparations. Methods for reducing the movements or the resulting artifacts in molluscan and mammalian experiments have been described (Orbach *et al.*, 1985; London *et al.*, 1987; Ts'o *et al.*, 1990).

30.5.1.2 Dark noise

Dark noise will degrade the signal-to-noise ratio at low light levels. In general silicon diode detectors will have a larger dark noise than vacuum photodetectors. This subject will be discussed in Section 30.5.4.

30.5.2 Light sources

Three kinds of sources have been used. Tungsten filament lamps are a stable source, but their intensity is relatively low, particularly at wavelengths less than 480 nm. Arc lamps and lasers are less stable but can provide more intense illumination.

30.5.2.1 Tungsten filament lamps

It is not difficult to provide a power supply stable enough that the output of the bulb fluctuates by less than 1 part in 10^5. In absorption measurements, where the fractional changes in intensity are relatively small, only tungsten filament sources have been used. On the other hand, fluorescence measurements often have larger fractional changes that will better tolerate light sources with systematic noise, and the measured intensities are low, making possible improvements in signal-to-noise ratio from more intense sources attractive. Hence arc lamps or laser sources have sometimes been used in fluorescence measurements.

30.5.2.2 Arc lamps

Both mercury and mercury–xenon arc lamps have been used to obtain higher intensities, but comparison of the excitation intensity from tungsten filament and arc sources is not simple. Grinvald et al. (1982) reported that the intensity from a mercury arc lamp ws 50–100 times higher than that from a tungsten filament lamp using a 540 nm filter with a width at half-height of 18 nm. However, the advantage implied by such a comparison may be misleading. Because the excitation 'action' spectrum of some dyes (i.e. the styryl, RH414) is quite broad, it is preferable to use a filter with a width at half-height of 90 nm to increase the incident intensity from a tungsten filament source. Because the mercury lamp has a distinct emission line at 546 nm, using a wider filter adds little intensity with this source. Furthermore, the tungsten filament bulb can be overrun to increase its colour temperature, increasing its output intensity by about 75%. Finally, the intensity will depend on the area of the object that is illuminated. Using critical illumination, the arc lamp, which approximates a point source, will be relatively preferred for smaller objects. Thus the increase in intensity obtained by using an arc lamp in lieu of tungsten will often not be as great as a factor of 50–100 (P. Saggau, L.B. Cohen & A. Grinvald, unpublished results).

The main problem with arc lamps is output intensity fluctuations (in part resulting from arc wander). Using a high-speed constant current power supply (KEPCO JQE 75–15(M) HS), the peak output fluctuations were 4×10^{-4} of the total intensity over the bandwidth of 0.5 Hz–1 kHz (Davila et al., 1974). Chien and Pine (1991) utilized a feedback circuit onto the lamp power supply to further reduce output fluctuations. With such a circuit the output noise was 10^{-4} over the bandwidth of 10 Hz–10 kHz. Ultimately the decision between a tungsten filament and an arc source for fluorescence measurements depends on the fractional change, the predominant source of noise, the level of difficulty associated with dye bleaching or photodynamic damage, and possible improvements in quieting an arc lamp.

30.5.2.3 Lasers

It has been possible to take advantage of two useful characteristics of laser sources. First, in preparations with no scattering, the laser output can be focused onto a small spot allowing measurement of membrane potential from small processes in tissue-cultured neurons (Grinvald & Farber, 1981). Second, the laser beam can be positioned flexibly and rapidly using acousto-optical deflectors (Dillon & Morad, 1981; Hill & Courtney, 1985). However, lasers have thus far only been used in situations where the fractional change in intensity was large. There appears to be excess noise that may be due to laser speckle (Dainty, 1984). Commercially available (Uniphase) helium–neon lasers with modified power supplies have intensity fluctuations of less than 2×10^{-5} of the total intensity, but the noise at the photodetector can be surprisingly large – even when the noise in the laser source appears to be small. In fluorescence measurements on invertebrate ganglia, we found that the fractional noise on each detector was substantially larger than the fractional noise in the incident light (B.M. Salzberg, D. Senseman, L.B. Cohen & A. Grinvald, unpublished results). This excess noise may be due to laser speckle and interference from reflected light in the fluorescence measurement. This noise can be reduced by introducing high-frequency mode scrambling into the laser beam (G. Ellis & B.M. Salzberg, pers. commun.).

30.5.3 Optics

The need to maximize the number of measured photons has also been a factor in the choice of optical components. The number of photons collected by an objective lens in forming the image is proportional to the square of the numerical aperture (NA) of the lens. In epifluorescence, both the excitation light and the emitted light pass through the objective, and the intensity reaching the photodiodes is proportional to the fourth power of its numerical aperture (A. Grinvald, pers. commun.). Accordingly, objectives (and condensers) with high numerical apertures have been employed. Normal microscope optics have very low numerical apertures at low magnifications. Ratzlaff and Grinvald (1991) have reported the design of a high numerical aperture microscope for use in monitoring activity from large fields.

Direct comparison of the intensity reaching the image plane has shown that objectives vary in their efficiency. With a Leitz 25× 0.4 NA lens we obtained 100% more intensity than with a Nikon 20 × 0.4 NA lens after correction for differences in magnification. Two Olympus lenses appeared to be intermediate in efficiency. (J-Y. Wu, H-P. Hopp, C.X. Falk & L.B. Cohen, unpublished results).

30.5.3.1 Depth of focus

Salzberg *et al.* (1977) determined the effective depth of focus for a 0.4 NA objective lens on an ordinary microscope by recording an optical signal from a neuron when it was in focus and then moving the neuron out of focus by various distances. They found that the neuron had to be moved 300 μm out of focus to cause a 50% reduction in signal size. (This result will be obtained only when the diameter of the neuron image and the diameter of the detector are similar.) Using 0.6 NA optics, the effective depth of focus was reduced by about 50% (A. Grinvald, unpublished results). A large effective depth of focus can be advantageous in some circumstances. If, for instance, one would like to record from all the neurons in a 500-μm-thick invertebrate ganglion, then, using 0.4 NA optics one could focus at the middle of the ganglion and the signals from neurons on the top and bottom of the ganglion would be reduced in size by less than 50%. This situation will, of course, result in the superposition of signals from two or more neurons on some detectors; subsequent sorting out will be required. On the other hand, the large effective depth of field makes it difficult to determine the depth in a preparation at which a signal originates.

30.5.3.2 Effects of light scattering

Light scattering can limit the spatial resolution of an optical measurement. London *et al.* (1987) measured the scattering of 705 nm light in *Navanax* buccal ganglia. They found that inserting a ganglion in the light path caused light from a 30 μm spot to spread so that the diameter of the circle of light that included intensities greater than 50% of the total was roughly 50 μm. The spread was greater, to about 100 μm, with light of 510 nm. Since the blurring is not large compared to the average cell diameter at the wavelengths used (705 nm), it does not lead to a large overestimate of cell size; but it does degrade the signal-to-noise ratio. Figure 30.7 illustrates the results of similar experiments carried out on the salamander

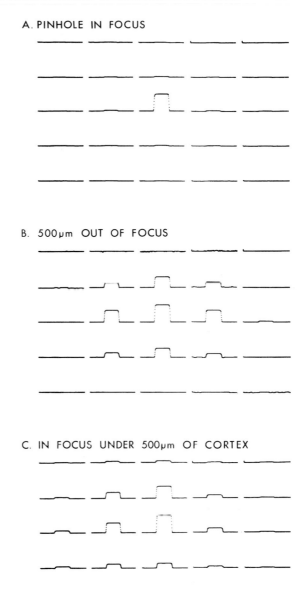

Figure 30.7 Effects of focus and scattering on the distribution of light from a point source onto the array. (A) A 40 μm pinhole in aluminium foil covered with saline was illuminated with light at 750 nm. The pinhole was in focus. More than 90% of the light fell on one detector. (B) The stage was moved downward by 500 μm. Light from the out-of-focus pinhole was now seen on several detectors. (C) The pinhole was in focus but covered by a 500 μm slice of salamander cortex. Again the light from the pinhole was spread over several detectors. A 10 × 0.4 NA objective was used. Kohler illumination was used before the pinhole was placed in the object plane. The recording gains were adjusted so the largest signal in each of the three trials would be approximately the same size in the figure. (Redrawn from Orbach & Cohen, 1983.)

Table 30.1 Detector comparison.

	Silicon diode	Vacuum photocathode
Quantum efficiency	0.9	0.15
Dark noise equivalent power	$\sim 10^7$ photons s^{-1}	$< 10^5$ photons s^{-1}
$1/f$ noise	Some diodes	No

olfactory bulb (Orbach & Cohen, 1983). The top section indicates that when no tissue is present, essentially all of the light (750 nm) from the small spot falls on one detector. The bottom section illustrates the result when a 500-μm-thick slice of olfactory bulb is present. The light from the small spot is spread to about 200 μm. Certainly mammalian cortex does not appear to scatter less than the olfactory bulb. Thus, light scattering will cause considerable blurring of signals in intact vertebrate preparations.

A second possible source of blurring is signal from regions that are out of focus. For example, if the active region is a cylinder (a column) perpendicular to the plane of focus, and the objective is focused at the middle of the cylinder, then the light from the middle will have the correct diameter at the image plane. Unfortunately, the light from the regions above and below are out of focus and will have a diameter that is too large. The middle section of Fig. 30.7 illustrates the effect of moving the small spot of light 500 μm out of focus. The light from the small spot is spread to about 200 μm. Thus, in preparations with considerable scattering or with out-of-focus signals the actual spatial resolution may be limited by the preparation and not by the number of pixels in the imaging device.

30.5.3.3   Confocal microscope

Petran and Hadravsky (1966) patented a modification of the microscope that substantially reduces both the scattered and out-of-focus light that contributes to the image. With a confocal microscope one might be able to obtain signals from intact vertebrate preparations with much better spatial resolution than was achieved with ordinary microscopy. P. Saggau (pers. commun.) is attempting to construct a laser-based confocal microscope with frame rates in the millisecond range.

30.5.4   Photodetectors

Since the signal-to-noise ratio in a shot-noise-limited measurement is proportional to the square root of the number of photons converted into photoelectrons (see above), quantum efficiency is critical. As indicated in Table 30.1, silicon photodiodes have quantum efficiencies approaching the ideal at wavelengths where most dyes absorb or emit light (500–900 nm). In contrast,

only specially chosen vacuum photocathode devices (phototubes, photomultipliers or image intensifiers) have a quantum efficiency as high as 0.15. Thus, in shot-noise-limited situations, a silicon diode will have a signal-to-noise ratio that is at least 2.5 times larger. This advantage is indicated in Fig. 30.6 by the fact that the diode curve (solid line) is higher than the vacuum photocathode curve (dashed line) over much of the intensity range (segment B).

There are three types of noise that can degrade the signal-to-noise ratio from the theoretical limit. The first is dark noise – the system's noise in the absence of light. Dark noise is generally far larger in a silicon diode system than in a vacuum photocathode system (Table 30.1). Thus, at low light levels ($< 10^7$ photons ms^{-1}), a vacuum photocathode device will provide a larger signal-to-noise ratio. When the light level is reduced so that the shot noise is less than the dark noise (about 10^8 photons ms^{-1} for a silicon photodiode), the signal-to-noise ratio decreases linearly with light intensity (segment A, Fig. 30.6). The crossover in signal-to-noise ratio between the silicon diode and vacuum photocathode device occurs at about 10^7 photons ms^{-1} (arrow 1, Fig. 30.6) which is near the intensities obtained in fluorescence measurements from ganglia and intact cortex. Thus the choice of photodetector for such preparations would benefit from a direct experimental comparison.

Some silicon diodes may have an additional light-dependent noise. David Kleinfeld (pers. commun.) found this second type of noise, called excess noise or $1/f$ noise, in PIN 6D and PIN 10D diodes (United Detector Technology) and in a Hewlett-Packard 5082–4203 diode. One indication of excess noise was that the noise current was directly proportional to the photocurrent rather than having the square root proportionality of shot noise. We measured the relationship between noise and intensity in three diodes – a United Detector Technology PIN 5D, and EG&G PV 444, and one element of a 12 × 12 Centronic array. For both the PV 444 and the Centronic diode the noise was proportional to intensity to the 0.55 power at the bandwidth we tested (10–100 Hz) (W.N. Ross, J.A. London, D. Zecevic & L.B. Cohen, unpublished results), close to the expected relationship for shot noise, However, with the PIN 5D the noise was proportional to intensity to the 0.65 power. This increased deviation from 0.5 may suggest the presence of $1/f$ noise in this kind of diode, in agreement with the results obtained by Kleinfeld. Optical measurements on vertebrate preparations are made using frequencies lower than those we tested. Since measurements at low frequencies are more likely to suffer from interference from $1/f$ noise, additional testing for $1/f$ noise might be useful.

30.5.5 Image recording devices

The major motivation for developing optical methods for monitoring membrane potential was the possibility of making simultaneous multisite measurements of potential changes. Many factors must be considered in choosing an imaging system. Perhaps the most important considerations are the requirements for spatial and temporal resolution. Since the signal-to-noise ratio in a shot-noise-limited measurement is proportional to the number of measured photons, increases in temporal resolution will reduce that number and reduce the signal-to-noise ratio. Similarly, increases in spatial resolution will reduce the number of photons per pixel and degrade the signal-to-noise ratio. For example, replacing a 10×10 array with a 100×100 array will reduce the photons per pixel by a factor of 100, reducing the signal-to-noise ratio by a factor of 10. While several types of imaging devices

will be considered, the emphasis will be on systems that allow frame rates near 1000 s^{-1}.

30.5.5.1 Film

One type of imager that has outstanding spatial and temporal resolution is movie film. But, because it is difficult to obtain quantum efficiencies of even 1% with film (Shaw, 1979), there would be an automatic factor of 10 degradation in signal-to-noise ratio in comparison with a silicon diode. This and other difficulties, including frame-to-frame and within-frame emulsion non-uniformity, has discouraged attempts to use film.

30.5.5.2 Silicon diode arrays (parallel read-out)

Arrays of silicon diodes are attractive because they have nearly ideal quantum efficiencies (close to 1.0). An array that was recently introduced in several laboratories is illustrated in Fig. 30.8. The dark squares (0.72×0.72 mm) in the array (a 24×24 array with the corners missing) are the individual detectors (pixels). Centronics also manufactures lower resolution 10×10 and 12×12 arrays. The 464 element array is similar to systems with 256 (Iijima *et al.*, 1989) and 444 elements (Nakashima *et al.*, 1989, 1992) that were developed earlier. A 464 element array can also be obtained from WuTech (New Haven, CT). In

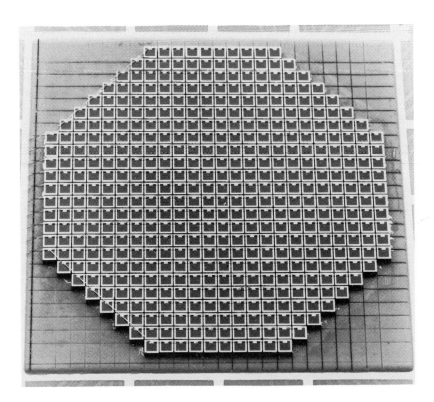

Figure 30.8 A 464 element array of silicon photodiodes manufactured by Centronic (Newbury Park, CA). The dark squares (0.72×0.72 mm) are the photodiodes. The insulating regions between the diodes are 0.04 mm thick. The development of this array by Centronics was supported by David Senseman. The output of each diode is carried from the array to a current-to-voltage converter (Fig. 30.9 top). We found that the 464 element array was less sensitive than the 12 × 12 array. Using the same illumination, the 464 element array generated a photocurrent per unit image plane area that was only 0.67 as large as that generated by the 12 × 12 array (A. Tang & L.B. Cohen, unpublished results). This decrease in photosensitivity will lead to a reduction in the signal-to-noise ratio by 18% in a shot-noise-limited measurement.

addition, the laboratory of Kohtaro Kamino at the Tokyo Medical and Dental University is presently constructing an array with 1024 elements, and Hammamatsu has constructed a system with 2500 elements. All of the above arrays are designed for parallel read-out and can be followed by data-acquisition systems that allow frame rates of 1000 s^{-1}. One factor which limits the number of pixels in parallel read-out systems is the need to provide a separate amplifier for each diode element (see below). A second factor is the need for a high-speed, multichannel multiplexer.

30.5.5.3 Silicon diode arrays (serial read-out)

By using serial read-out, the number of amplifiers can be reduced; however, the signal-to-noise ratio will be degraded in some measuring situations. Gen Matsumoto and his collaborators at the Electrotechnical Laboratory in Tsukuba City together with Fuji Film have implemented a 128×128 (16 384 pixel) array that can be read out at a frame rate of 2000 s^{-1} (Ichikawa et al., 1992). At 2 kHz, the data rate from this system is 64 Mbytes s^{-1} (requiring relatively specialized hardware for recording). This system has a signal-to-noise ratio of 70 db, considerably less than the 100 db that can be achieved with a parallel read-out scheme. The Grinvald laboratory has also developed a much faster CCD device (A. Grinvald, pers. commun.).

Several slower systems based on CCD cameras have been used to measure activity-dependent optical signals in neurobiological preparations. Lasser-Ross et al. (1991) have modified the software for a Photometrics CCD camera to allow several choices of spatial and temporal resolution. Table 30.2 illustrates these choices. In addition, CCD devices with very large numbers of pixels (but substantially slower frame rates) have been used (Connor, 1986; Gross et al., 1986; Kim et al., 1989; Ts'o et al., 1990.

The serial read-out devices presently in use will saturate at the light intensities that obtain in absorption measurements. Reducing the light intensity to levels below saturation will degrade the signal-to-noise ratio. It is probably this effect which accounts for the lower signal-to-noise ratio of these devices.

30.5.5.4 Vacuum photocathode cameras

While the lower quantum efficiencies of vacuum photocathodes is a disadvantage, these devices have high spatial resolution and may be preferable in low-light measurements (Fig. 30.6). One of these, a Radechon (Kazan & Knoll, 1968), has an output proportional to the changes in the input. Cameras have been used in recordings from mammalian cortex (Blasdel & Salama, 1986) and salamander olfactory

Table 30.2 Spatial and temporal resolution from a photometrics camera.

Frame rate (s^{-1})	Pixels
100	324
40	2 500
20	10 000

From Lasser-Ross et al. (1991).

bulb (Kauer, 1988). Kauer et al. have used their camera at 60 frames s^{-1}.

In all of the systems now in use, the image recorder has been placed in the objective image plane of a microscope. However, Tank and Ahmed (1985) suggested a scheme by which a hexagonal close-packed array of optical fibres is positioned in the image plane, and individual photodiodes are connected to the other end of the optical fibres. WuTech (New Haven, CT) sells a 464 pixel camera based on this scheme.

30.5.6 Amplifiers

30.5.6.1 Parallel versus serial amplification

The output of each detector in an array like that shown in Fig. 30.8 can be followed by its own amplifier. This scheme has two advantages. First, a lower digitizing accuracy is required. In a serial read-out, determining the changes in intensity, which are a very small fraction of the total intensity, can be done only by subtracting two relatively large numbers: the total intensity at time t_2 minus the total intensity at an earlier time, t_1. To measure the total intensity with the accuracy of one part in 10^5 necessary for some absorption signals, an analogue-to-digital conversion accurate to 17 bits would be needed. To achieve this kind of signal resolution in experimental situations, a 20 bit analogue-to-digital converter (ADC) would be needed. This kind of accuracy is not inexpensively achieved at the required data rates. However, in a parallel read-out, either capacity coupling or DC subtraction (see below) can be used to remove the DC component of the light so that the ADC needs only enough resolution to measure the change in intensity. In this situation, 12 bit resolution is adequate.

Ichikawa et al. (1992) have designed a serial read-out system to achieve the DC subtraction by first recording the DC intensity in one frame and then subtracting the first frame value from each subsequent frame before the analogue-to-digital conversion.

The second advantage of a parallel amplification scheme is that large intensities can be measured – between 100 and 10 000 times larger than serial read-out devices, which are limited by the amount

PHOTODIODE AMPLIFIER

First Stage

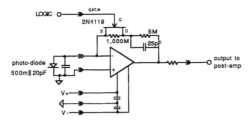

Second Stage

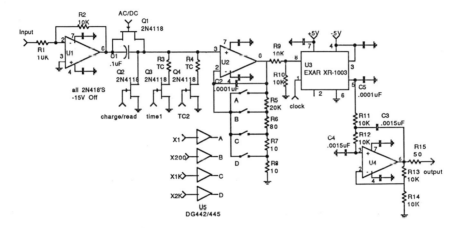

Figure 30.9 Schematic drawing of the amplifiers used to make parallel amplification recordings from photodiode arrays. This is the system presently in use. The output of each diode, used in a photovoltaic mode, is converted into a voltage signal by the amplifier in the first stage (Texas Instruments 118E). The photodiode signals are connected to the first-stage amplifiers by a printed circuit card that has solder points for the diode outputs and connectors for cards, each of which contains 32 of the first-stage amplifiers. The outputs of the first stage are cabled from the microscope to second-stage amplifiers. The second stage provides for several modes of coupling to the gain amplifier (see text). The output of the gain amplifier is filtered by a switched-capacity filter (Exar XR-1003) that implements a four-pole Bessel function whose corner frequency is 100 times smaller than the clock signal fed to the filter chip. In addition to allowing an easily adjusted low-pass filter frequency, the four-pole bessel filter substantially reduced the high-frequency components of the noise compared to a one-pole RC filter. The amplifiers can be purchased from the Department of Physiology Electronics Shop, Yale University School of Medicine, New Haven, CT (about $30 per channel).

of photoinduced charge that can be stored in each bucket.

30.5.6.2 One example

In our apparatus amplifier noise does not make a substantial contribution to either the dark noise or the total noise. The dark noise did not decrease when a National Semiconductor, LF 356N, was replaced with a Burr-Brown OPA 111AM, an amplifier with substantially lower noise specifications. We conclude that the electrical characteristics of the diode itself dominate

the dark noise. Figure 30.9 is a schematic drawing of the amplifiers we presently use. These amplifiers provide three kinds of coupling between the first stage and the gain amplifier. There is DC, AC (choice of two time constants) and a pseudo sample-and-hold scheme. This scheme uses a brief (30 ms) discharge of the coupling capacitor allowing current to flow through the FET at Q2 just after turning on the light. With all the FETs in their high resistance mode, there is a very long time-constant AC recording whose time constant is determined by the characteristics of the coupling capacitor and the various leakage currents.

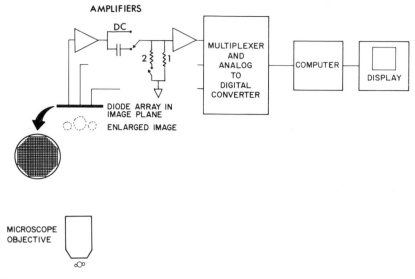

Figure 30.10 Schematic drawing of the apparatus used to make parallel-amplification recordings. This is the system presently in use. The output of each diode, used in a photovoltaic mode, is converted into a voltage signal which is then converted to a digital number by a multiplexer and analogue-to-digital converter (Abildgaard Input Module). The Input Module provides for 512 channels of input with 16 bit digitization and a 400 kHz throughput rate. For the VMEbus system, the module is designed to be used with a FORCE-OPIO1, optically isolated, parallel, DMA interface. The Input Module uses two ADCs (Analog Solutions AD2836) running in parallel with each ADC used to digitize from a selection of up to 256 inputs. The module has on-board RAM which is used to store the multiplexer selection list (loaded via software). The module has an on-board 4 MHz clock and a 24 bit count-down timer also loaded via software. The Input Module can be purchased from the Department of Physiology Electronics Shop, Yale University School of Medicine, New Haven, CT (about $7000).

The measured drift is about 0.001 I_o s^{-1}. In addition, when using ordinary AC coupling, allowing current to flow through the FET at Q2 very rapidly reduces the voltage presented to the gain amplifier, and thus dramatically reduces the settling time after turning on the light.

30.5.8 Computer

The discussion in this section is limited to the hardware and software currently in use in our laboratory. There are two sorts of problems. First, large amounts of data are generated by optical recordings. Recording from a 464 element array at 1000 frames s^{-1} generates 1 Mbyte of data per second. Second, while functions required for data acquisition are relatively circumscribed, there seems to be a never-ending need for new data-analysis software and, therefore, a convenient programming environment is important.

We are presently using a 32 bit Motorola VME 1000 system based on a 68000 series CPU and the VME bus (Fig. 30.10). We have used Pascal and assembler under Motorola's proprietary operating system, VERSADOS, which has both real-time and multi-user features. An optically isolated, parallel interface card (FORCE OPIO-1) with DMA controller is used to send

and receive information from a 512 channel multiplexer module with two 16 bit ADCs (Abildgaard Input Module). The maximum throughput rate of this system is 400 kHz when the data are stored in RAM memory. At present, a data rate of 300 kHz can be achieved when the data are deposited in RAM and simultaneously written to a SCSI disk. With the present programs this process can continue for 56 k samples per detector.

While the VERSADOS programming environment is dramatically improved over the previously used DEC RT11, analysis programs are much easier to write using high-level languages such as Mathematica or IDL running on a workstation. Richard Lombardi and Guy Salama (pers. commun.) have written data-acquisition programs using UNIX and a memory resident real-time kernel (RTUX from Emerge Systems Inc., Indialantic, FL).

30.6 TRADE-OFFS IN THE CHOICE OF RECORDING SYSTEMS

First it is important to determine the minimum acceptable spatial and temporal resolution because unneeded resolution will lead to a degraded signal-to-noise ratio. For example, in an optical recording from an *Aplysia* abdominal ganglion where one is attempting to monitor every action potential in every neuron, a temporal resolution of greater than 1000 frames s^{-1} leads to an unnecessarily degraded signal-to-noise ratio since the action potential in these neurons has a FWHM of more than 2 ms. Similarly, a spatial resolution of greater than 1000 pixels would also degrade the signal-to-noise ratio because there are only 1000 neuron cell bodies in the ganglion.

Second is the choice between a serial and parallel amplification scheme. At present the serial read-out systems have better spatial resolution but at the cost of a lower signal-to-noise ratio for relatively large light levels. CCD systems can store only a limited amount of electrical charge for each pixel. For most CCD cameras the charge storage well will be saturated in about 10 ns at the intensities reaching the image plane in an absorption measurement. Thus these systems cannot be used for absorption measurements without degrading the signal-to-noise ratio by reducing the light intensity.

30.7 FUTURE DEVELOPMENTS

Clearly improvements in the signal-to-noise ratios of optical measurements of membrane potential would be useful. A number of avenues remain partially or completely unexplored.

Only three optical properties of stained membranes have been examined for signals in response to changes in membrane potential. The possibilities of finding large changes in other optical properties – for example, energy transfer, circular dichroism (optical rotation), or absorption-enhanced Raman scattering – have been largely neglected. Ehrenberg and Berezin (1984) have used resonance Raman to study surface potential, and Ehrenberg and Loew (pers. commun.) are planning to investigate its use in measuring transmembrane potential. There was a report of holographic signals in leech neurons (Sharnoff *et al.*, 1978a), but signals were not found in subsequent experiments on squid axons (Sharnoff *et al.*, 1978b). There were also reports of changes in intrinsic infrared absorption (Sherebrin, 1972; Sherebrin *et al.*, 1972), but these have not been pursued further. Thus, one approach to looking for larger signals would be to investigate new types of optical phenomena.

A second approach involves improvement of the apparatus. One useful improvement would be further quieting of arc and laser light sources and investigation of new kinds of light sources. The successful implementation of a confocal microscope for preparations

with substantial scattering or thickness would greatly improve spatial resolution.

The third approach is to find or design better dyes. All of the dyes in Fig. 30.1 and the vast majority of those synthesized in recent years are of the general class named cyanines (Hamer, 1964), a class that is used to extend the wavelength response of photographic film. It is certainly possible that improvements in signal size can be obtained with new cyanine dyes (see Waggoner & Grinvald, 1977, and Fromherz *et al.*, 1991, for a discussion of maximum possible fractional changes in absorption and fluorescence). On the other hand, the fractional change on squid axons has not increased in recent years (Gupta *et al.*, 1981; L.B. Cohen, A. Grinvald, K. Kamino & B.M. Salzberg, unpublished results), and most improvements (Grinvald *et al.*, 1982) have involved synthesizing analogues that work well on new preparations. Radically different synthetic approaches or a modicum of cooperation from corporations like Eastman Kodak – might prove to be very useful.

The measurements that have been made with optical methods provide new and previously unobtainable information about cell and organ function. Clearly there has been dramatic progress since the first recording of an action potential in a leech neuron (Salzberg *et al.*, 1973). We hope that additional improvements will further increase the utility of these methods.

ACKNOWLEDGEMENTS

The authors are indebted to their collaborators Vicencio Davila, Chun Falk, Amiram Grinvald, Kohtaro Kamino, Jill London, Bill Ross, Brian Salzberg, Alan Waggoner and Dejan Zecevic for numerous discussions about optical methods. We also thank Pancho Bezanilla, David Kleinfeld, Richard Lombardi, Ana Lia Obaid, Jerome Pine, Guy Salama, Peter Saggau and David Senseman, who have allowed us to cite unpublished results. The experiments carried out in our laboratory were supported by NIH grant NS-08437.

REFERENCES

Blasdel G.G. & Salama G. (1986) *Nature* **321**, 579–585.
Boyle M.B. & Cohen L.B. (1980) *Fedn Proc.* **39**, 2130.
Braddick H.J.J. (1960) *Rep. Prog. Phys.* **23**, 154–175.
Chien C.B. & Pine J. (1991) *J. Neurosci. Methods* **38**, 93–105.
Cohen L.B. & Lesher S. (1986) *Soc. Gen. Physiol. Ser.* **40**, 71–99.
Cohen L.B. & Salzberg B.M. (1978) *Rev. Physiol. Biochem. Pharmacol.* **83**, 35–88.

Cohen L.B., Keynes R.D. & Hille B. (1968) *Nature* **218**, 438–441.

Cohen L., Landowne D., Shrivastav B.B. & Ritchie J.M. (1970) *Biol. Bull.* **139**, 418–419.

Cohen L.B., Salzberg B.M., Davila H.V., Ross W.N., Landowne D., Waggoner A.S. & Wang C.H. (1974) *J. Membr. Biol.* **19**, 1–36.

Connor J.A. (1986) *Proc. Natl. Acad. Sci. USA* **83**, 6179–6183.

Conti F. (1975) *Ann. Rev. Biophys. Bioengng* **4**, 287–310.

Conti F. & Tasaki I. (1970) *Science* **169**, 1322–1324.

Conti F., Tasaki I. & Wanke E. (1971) *Biophys. J.* **8**, 58–70.

Dainty J.C. (1984) *Laser Speckle and Related Phenomena.* Springer-Verlag, New York.

Dasheiff R.M. (1988) *J. Clin. Neurophysiol.* **5**, 211–235.

Davila H.V., Cohen L.B., Salzberg B.M. & Shrivastav B.B. (1974) *J. Membr. Biol.* **15**, 29–46.

Dillon S. & Morad M. (1981) *Science* **214**, 453–456.

Ehrenberg B. & Berezin Y. (1984) *Biophys. J.* **45**, 663–670.

Fromherz P., Dambacher K.H., Ephardt H., Lambacher A., Muller C.O., Neigl R., Schaden H., Schenk O. & Vetter T. (1991) *Ber. Bunsenges. Phys. Chem.* **95**, 1333–1345.

Grinvald A. & Farber I.C. (1981) *Science* **212**, 1164–1167.

Grinvald A., Ross W.N. & Farber I. (1981a) *Proc. Natl. Acad. Sci. USA* **78**, 3245–3249.

Grinvald A., Cohen L.B., Lesher S. & Boyle M.B. (1981b) *J. Neurophysiol.* **45**, 829–840.

Grinvald A., Hildesheim R., Farber I.C. & Anglister L. (1982) *Biophys. J.* **39**, 301–308.

Grinvald A., Lieke E., Frostig R.D., Gilbert C.D. & Wiesel T.N. (1986) *Nature* **324**, 361–364.

Grinvald A., Frostig R.D., Lieke E. & Hildesheim R. (1988) *Physiol. Rev.* **68**, 1285–1366.

Gross D., Loew L.M. & Webb W. (1986) *Biophys. J.* **50**, 339–348.

Gupta R.K., Salzberg B.M., Grinvald A., Cohen L.B., Kamino K., Lesher S., Boyle M.B., Waggoner A.S. & Wang C.H. (1981) *J. Membr. Biol.* **58**, 123–137.

Hamer F.M. (1964) *The Cyanine Dyes and Related Compounds.* Wiley, New York.

Hill B.C. & Courtney K.R. (1985) *Biophys. J.* **47**, 496a.

Huang J.Y., Lewis A. & Loew L.M. (1988) *Biophys. J.* **53**, 665–670.

Ichikawa M., Iijima T. & Matsumoto G. (1993) Simultaneous 16, 384-site optical recording of neural activities in the brain. In *Brain Mechanisms of Perception and Memory: From Neuron to Behavior.* T. Ono, L.R. Squire, M.E. Raichle, D.I. Perrett & M. Fukuda (eds). Oxford University Press, New York.

Iijima T., Ichikawa M. & Matsumoto G. (1989) *Abstr. Soc. Neurosci.* **15**, 398.

Kauer J.S. (1988) *Nature* **331**, 166–168.

Kauer J.S., Senseman D.M. & Cohen L.B. (1987) *Brain Res.* **418**, 255–261.

Kazan B. & Knoll M. (1968) *Electronic Image Storage.* Academic Press, New York.

Kim J.H., Dunn M.B., Hua Y., Rydberg J., Yae H., Elias S.A. & Ebner T.J. (1989) *Neuroscience* **31**, 613–623.

Lasser-Ross N., Miyakawa H., Lev-Ram V., Young S.R. & Ross W.N. (1991) *J. Neurosci. Methods* **36**, 253–261.

Lev-Ram V. & Grinvald A. (1986) *Proc. Natl. Acad. Sci. USA* **83**, 6651–6655.

Loew L.M., Cohen L.B., Salzberg B.M., Obaid A.L. & Bezanilla F. (1985) *Biophys. J.* **47**, 71–77.

London J.A., Zecevic D. & Cohen L.B. (1987) *J. Neurosci* **7**, 649–661.

Malmstadt H.V., Enke C.G., Crouch S.R. & Harlick G. (1974) *Electronic Measurements for Scientists.* Benjamin, Menlo Park, CA.

Morton D.W., Chiel H.J., Cohen, L.B. & Wu J.Y. (1991) *Brain Res.* **564**, 45–55.

Nakashima M., Yamada S., Shiono S. & Maeda M. (1989) *Abstr. Soc. Neurosci.* **15**, 1046.

Nakashima M., Yamada S., Shiono S., Maeda M. & Sato F. (1992) *IEEE Trans. Biomed. Engng* **39**, 26–36.

Orbach H.S. & Cohen L.B. (1983) *J. Neurosci.* **3**, 2251–2262.

Orbach H.S., Cohen L.B. & Grinvald A. (1985) *J. Neurosci.* **5**, 1886–1895.

Parsons T.D., Kleinfeld D., Raccuia-Behling F. & Salzberg B.M. (1989) *Biophys. J.* **56**, 213–212.

Patrick J., Valeur B., Monnerie L. & Changeux J.-P. (1971) *J. Membr. Biol.* **5**, 102–120.

Petran M. & Hadravsky M. (1966) Czechoslovakian patent 7720.

Ratzlaff E.H. & Grinvald A. (1991) *J. Neurosci. Methods* **36**, 127–137.

Ross W.N. & Krauthamer V. (1984) *J. Neurosci.* **4**, 659–672.

Ross W.N. & Reichardt L.F. (1979) *J. Membr. Biol.* **48**, 343–356.

Ross W.N., Salzberg B.M., Cohen L.B. & Davila H.V. (1974) *Biophys. J.* **14**, 983–986.

Ross W.N., Salzberg B.M., Cohen L.B., Grinvald A., Davila H.V., Waggoner A.S. & Wang C.H. (1977) *J. Membr. Biol.* **33**, 141–183.

Russell J.T., Beeler T. & Martonosi A. (1979) *J. Biol. Chem.* **254**, 2047–2052.

Sakai T., Hirota A., Komuro H., Fujii S. & Kamino K. (1985) *Brain Res.* **349**, 39–51.

Salzberg B.M. (1983) In *Current Methods in Cellular Neurobiology*, J.L. Barker & J.F. McKelvy (eds). Wiley, New York, pp. 139–187.

Salzberg B.M. & Bezanilla F. (1983) *J. Gen. Physiol* **82**, 807–817.

Salzberg B.M., Davila H.V. & Cohen L.B. (1973) *Nature* **246**, 508–509.

Salzberg B.M., Grinvald A., Cohen L.B., Davila H.V. & Ross W.N. (1977) *J. Neurophysiol.* **40**, 1281–1291.

Sharnoff M., Henry R.W. & Belleza D.M.J. (1978a) *Biophys. J.* **21**, 109A.

Sharnoff M., Romer N.J., Cohen L.B., Salzberg B.M., Boyle M.B. & Lesher S. (1978b) *Biol. Bull.* **155**, 465–466.

Shaw R. (1979) *Appl. Optics Optical Engng* **7**, 121–154.

Sherebrin M.H. (1972) *Nature* **235**, 122–124.

Sherebrin M.H., MacClement B.A.E. & Franko A.J. (1972) *Biophys. J.* **12**, 977–989.

Shrager P., Chiu S.Y., Ritchie J.M., Zecevic D. & Cohen L.B. (1987) *Biophys. J.* **51**, 351–355.

Tank D. & Ahmed Z. (1985) *Biophys. J.* **47**, 476A.

Tasaki I., Watanabe A., Sandlin R. & Carnay L. (1968) *Proc. Natl. Acad. Sci. USA* **61**, 883–888.

Tasaki I., Watanabe A. & Hallett A. (1972) *J. Membr. Biol.* **8**, 109–132.

Ts'o D.Y., Frostig R.D., Lieke E.E. & Grinvald A. (1990) *Science* **249**, 417–420.

Waggoner A.S. (1979) *Ann. Rev. Biophys. Bioengng* **8**, 47–68.

Waggoner A.S. & Grinvald A. (1977) *Ann. NY Acad. Sci.* **303**, 217–241.

Zecevic D., Wu J.Y., Cohen L.B., London J.A., Hopp H.P. & Falk X.C. (1989) *J. Neurosci.* **9**, 3681–3689.

CHAPTER THIRTY-ONE

Photoactivation of Fluorescence as a Probe for Cytoskeletal Dynamics in Mitosis and Cell Motility

K.E. SAWIN,[1] J.A. THERIOT[1] & T.J. MITCHISON[2]

[1] Department of Biochemistry and Biophysics, [2] Department of Pharmacology, University of California, San Francisco, CA, USA

31.1 INTRODUCTION

Microtubules (MTs) and actin microfilaments (AFs) are cytoskeletal polymers composed of subunits of the proteins tubulin and actin, respectively, with associated proteins. Tubulin and actin are among the most abundant proteins in cells, and MTs and AFs have important functions in nearly all aspects of the life of the cell – in cell structure and morphogenesis, motility, secretion, and mitosis, as well as in additional processes often specific to specialized cell types (for reviews, see Dustin, 1984; Bershadsky & Vasiliev, 1988). It is our opinion that understanding the role of the cytoskeleton in these and other phenomena depends upon a much deeper understanding of fundamental aspects of the biology of MTs and AFs, which are unusual as biopolymers, yet similar in many respects (for reviews, see Oosawa & Asakura, 1975; Inoue, 1982; Carlier, 1989).

MTs and AFs represent dynamic rather than static networks. This is primarily the result of two factors, both of which have been well-studied *in vitro*: (1) In biophysical terms, MTs and AFs are best described not as self-assembling equilibrium polymers but rather as reversible steady-state polymers. Their assembly and continued persistence involves a classical nucleation/condensation reaction (Oosawa & Asakura, 1975) but normally also requires the presence of monomer-associated nucleotide (ATP for actin, GTP for tubulin) which is hydrolysed at some point after subunit incorporation onto the ends of existing filaments, (for review see Carlier, 1989). Under normal steady-state conditions, MTs and AFs constantly turn over, exchanging with a monomer pool, with a continuous input of energy obtained through nucleotide hydrolysis. Interestingly, under certain conditions nucleotide hydrolysis *per se* is not required for MT or AF polymerization; assembly is essentially entropically driven (Oosawa & Asakura 1975). (2) Both MTs and AFs have a defined polarity that reflects the asymmetric structure of constituent subunits arranged in a head-to-tail fashion, and as a consequence the assembly properties of the two ends of the polymer can be functionally distinct. In conjunction with reversible, nucleotide-dependent assembly, this confers unusual and unexpected dynamic properties to both MTs and AFs. This was first recognized on theoretical grounds by Wegner (1976), who showed that a requirement for nucleotide hydrolysis could permit one end of an AF to undergo net assembly (the 'plus', or 'barbed' end) while the other (the 'minus', or 'pointed' end) undergoes a net disassembly, the result being a net flux, or treadmill, of subunits through the filament at steady state. Wegner demonstrated this treadmilling experimentally *in vitro*, and a similar flux was inferred for MT protein by Margolis and Wilson (1978, 1981). It remains controversial what function, if any, this *in vitro* behaviour may serve within cells (Margolis & Wilson, 1981; Kirschner, 1980; Hill & Kirschner, 1982).

At least in the case of MTs, treadmilling appears to depend on the presence of MT-associated proteins (Horio & Hotani, 1986; Hotani & Horio, 1988). By contrast, pure tubulin *in vitro* and most MTs *in vivo* display a different behaviour known as 'dynamic instability' (Mitchison & Kirschner, 1984), in which individual MTs in a steady-state population coexist in growing and shrinking phases and transit stochastically between these two states with characteristic frequencies in interphase and mitosis (Walker *et al.*, 1988; Belmont *et al.*, 1990). The mechanism of dynamic instability is still not well-understood but is probably best ultimately explained through conformational changes in the structure of the MT lattice, induced by the process of nucleotide hydrolysis.

In vivo, the dynamics of MTs and AFs have been studied indirectly for many years. In the 1950s and 1960s, Inoue and collaborators used polarization optics to observe the rapid loss of mitotic spindle birefringence under (what were later found to be) MT-depolymerizing conditions; when such conditions were withdrawn, spindle birefringence quickly recovered. These and other experiments indicated that, *in vivo*, MTs exist in a kind of dynamic equilibrium with the tubulin monomer pool, with rapid turnover rates (reviewed in Inoue & Sato, 1967; Sato *et al.*, 1975; Inoue, 1981). Similar results have also been obtained with inhibitors of the actin cytoskeleton, using phase-contrast and differential-interference contrast microscopy, most recently in the neuronal growth cone (Forscher & Smith, 1988).

The characteristic turnover and instability of AFs and MTs *in vivo* raises questions of fundamental biological importance: What is the function of these unusual dynamics? Why does the cell invest so much energy (in the form of nucleotide hydrolysis) in processes that essentially allow it only to 'run in place'? Are dynamics primarily designed for morphogenetic and organizational purposes, or can they also perform useful work within cells? How are the basic dynamic properties of actin and tubulin regulated *in vivo* to achieve a variety of morphogenetic and motile processes, and what are the components of such regulatory systems? How is the regulation of dynamics coupled to other cytoskeleton-based activities such as force generation and movement through motor proteins such as myosin, dynein, and kinesin?

To begin to answer these and related questions requires new technologies for the quantitative study of cytoskeletal polymer dynamics in a non-perturbing, physiological context. The introduction of vital fluorescent probes for the cytoskeleton has been singularly useful in this regard. Specifically, questions of polymer dynamics have been addressed by microinjection and incorporation of fluorescently labelled subunit proteins (e.g. actin or tubulin) to steady-state, followed by local photobleaching of incorporated fluorescence with a laser microbeam. The disposition and/or recovery of the bleached mark can then be tracked, in both time and space, by fluorescence microscopy. This technique, known as fluorescence redistribution after photobleaching (FRAP) has been used to study subunit turnover of both MTs and AFs in a number of systems and has been of particular value to students of mitosis (Salmon *et al.*, 1984; Saxton *et al.*, 1984; Gorbsky & Borisy, 1989) and cell motility (Wang *et al.*, 1982; Wang, 1985; Lee *et al.*, 1990).

For some years we have been interested in the role of polymer dynamics in the regulation of MT and AF function *in vivo*. The purpose of this paper is to summarize work from our laboratory concerning a novel fluorescence marking technique for studying cytoskeletal polymer dynamics, photoactivation of fluorescence (Mitchison, 1989). In principle, photoactivation is similar to FRAP, but whereas photobleaching involves the generation of a non-fluorescent mark on a bright background, photoactivation generates a fluorescent mark on a dark background. As described below, this is achieved by 'caging' a fluorochrome with a photolabile group that renders it non-fluorescent, whereupon the caged compound is covalently coupled to a specific protein and microinjected in cells. After incorporation of this tagged molecule into structures of interest, fluorescence is locally activated by a brief exposure to a UV microbeam, and followed by time-lapse fluorescence video microscopy. Here we describe the structure and properties of caged fluorochromes synthesized in the laboratory, the design of a computer-controlled microscope to study cytoskeletal dynamics, and some of the results obtained using these technologies. We expect that photoactivation of fluorescence will have wide applicability as a non-perturbing marking technique in a number of areas besides cytoskeletal polymer dynamics (Mitchison, 1989; Sawin & Mitchison, 1991; Theriot & Mitchison, 1991; Reinsch *et al.*, 1991) – for example, in studying membrane dynamics, embryonic fate mapping, intracellular membrane transport, and any other situations where it may be informative to locally mark structures of living cells or organisms under the microscope.

31.2   PHOTOACTIVATABLE FLUORESCENT PROBES

The use of photosensitive compounds in microscopy and biochemistry is well-established. A surprisingly wide variety of biologically active compounds have been made in photoactivatable or photoinactivatable forms, including nucleotides (Walker *et al.*, 1989), phosphatidylinositol metabolites (Walker *et al.*, 1987), neurotransmitters (Marque, 1989), amino acids (Patchornik *et al.*, 1970), calcium chelators (Tsien & Zucker 1986), proton donors (Janko & Reichert, 1987), and microtubule drugs (Hiramoto *et al.*, 1984).

31.2.1   Caged fluorescein

Fluorescein itself exists as two tautomeric forms; a fluorescent carboxylic acid tautomer favoured in neutral or basic aqueous solution, and a non-fluorescent lactone tautomer favoured in acidic or non-aqueous solvents. Fluorescein may be constrained in its lactone tautomer by alkylation of its two phenolic oxygens. Such a non-fluorescent derivative is photo-activatable if one or both of the ether groups are subject to photocleavage. The photolysis of 2-nitrobenzyl compounds to form 2-nitrosobenzaldehyde has been exploited in the caging of compounds

containing carboxylic acid, phosphate or hydroxyl functionalities, and this chemistry lends itself well to caging fluorescein (Krafft *et al.*, 1988; Mitchison, 1989; Fig. 31.1(a)). The resultant compound, bis-caged carboxyfluorescein, or C2CF, is colourless and non-fluorescent, and is photoactivated by 360 nm light with simple first-order kinetics to regenerate the fluorescent parent compound (Mitchison, 1989). The sidechain carboxyl group of carboxyfluorescein can be activated for protein labelling, for example, as the *N*-hydroxysulphosuccinimide ester (Mitchison, 1989) (Fig. 31.1(b)). The caged dye can be coupled to specific proteins or dextrans (e.g. as fluid-phase markers) using these and related chemistries.

C2CF has been used to label covalently tubulin purified from bovine brain. C2CF tubulin polymerizes to form fairly short microtubules which appear identical in structure to microtubules formed from unlabelled purified tubulin (Mitchison, 1989) and which can be rapidly photoactivated under the microscope (Mitchison, 1989; Fig. 31.2). C2CF-tubulin injected into cells incorporates into the endogenous microtubule structures including mitotic spindles (Mitchison, 1989) and neuronal axons (Reinsch *et al.*, 1991).

Although C2CF has only limited solubility in water, its solubility can be greatly improved by the addition of oxyacetyl sidechains to the caging nitrobenzyl rings

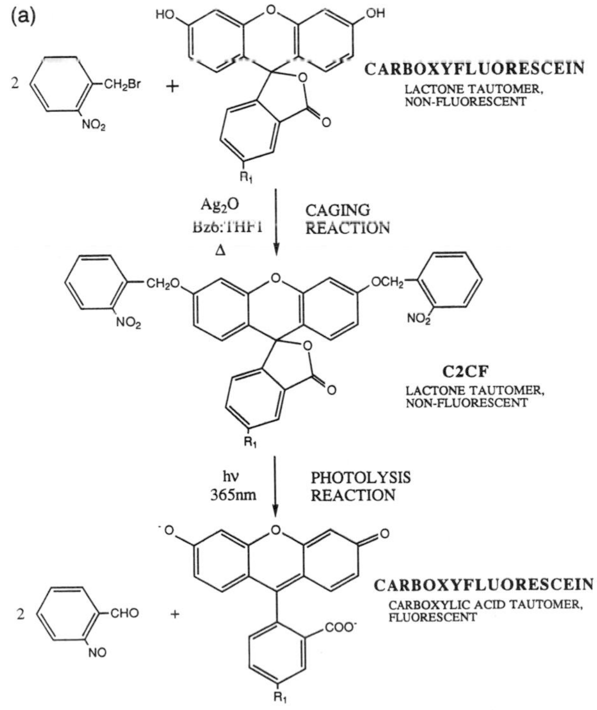

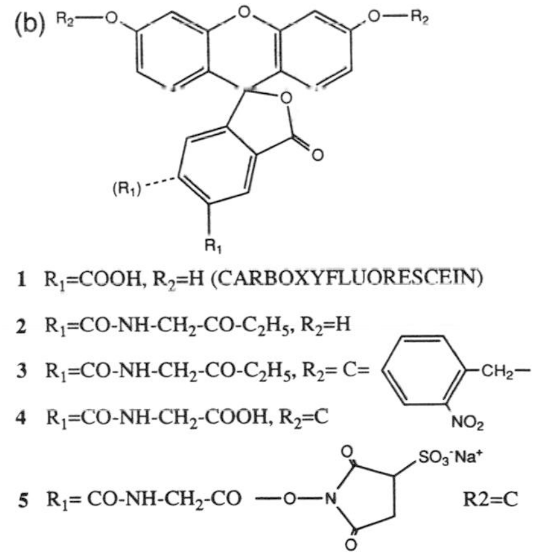

Figure 31.1 (a) Caging and photoactivation reactions of C2CF. Carboxyfluorescein is drawn as its non-fluorescent lactone tautomer, which reacts with *o*-nitrobenzylbromide to yield C2CF. Upon illumination at 365 nm the caging groups are cleaved off as *o*-nitrosobenzaldehyde, regenerating carboxyfluorescein, drawn as the fluorescent carboxylic acid tautomer. (b) Intermediates in the synthesis of C2CF (see Mitchison, 1989 for details). (Reproduced from the *Journal of Cell Biology*, 1989, **109**, 637–652, by copyright permission of the Rockefeller University Press.)

(a)   (b)   (c)

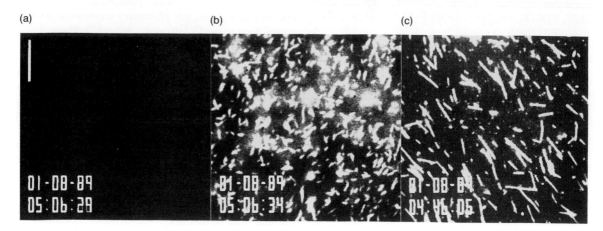

Figure 31.2 C2CF-labelled microtubules before and after photoactivation. (a) Microtubules polymerized from C2CF-tubulin in the dark, before photoactivation. (b) The same microscope field, after photoactivation (1 s exposure under the microscope, using a Hoechst filter set and a mercury arc lamp. (c) Microtubules polymerized from C2CF-labelled tubulin that was photoactivated before polymerization. Bar is 10 μm. (Micrographs from Mitchison, 1989. Reproduced from the *Journal of Cell Biology*, 1989, **109**, 637–652, by copyright permission of the Rockefeller University Press.)

(T.J. Mitchison, unpublished observations). The more water-soluble compound is superior to the original C2CF for labelling antibodies, but the efficiency of tubulin labelling is similar for the two compounds (T.J. Mitchison, unpublished observations).

31.2.2 Caged resorufin

The size and relative hydrophobicity of C2CF render it inefficient for labelling some proteins; specifically, we found it difficult to label actin or myosin to high stoichiometry with C2CF (M. Symons, pers. commun.; J.A. Theriot, unpublished observations). We therefore developed a smaller, more hydrophilic caged fluorescent compound based on the fluorochrome resorufin (Fig. 31.3). The non-fluorescent O-alkyl and O-acyl derivatives of resorufin are widely used as fluorogenic enzyme substrates (Haugland, 1989, and references therein), and resorufin modified with convenient protein-labelling groups (either iodoacetamide or *N*-hydroxysuccinimide ester) is available from Boehringer-Mannheim. Both compounds are readily caged on one phenolic oxygen in a single step using the standard protocol for caging nucleotides (Walker *et al.*, 1989), although we have obtained good yields only with the iodoacetamide derivative (Theriot & Mitchison, 1991).

Caged-resorufin (CR) iodoacetamide readily labels actin to high dye-to-protein stoichiometries and does not appreciably inhibit copolymerization of labelled with unlabelled monomer. CR-actin injected into cells incorporates into endogenous actin structures including stress fibres and lamellipodia (Theriot & Mitchison, 1991). Photoactivation with 360 nm light

is rapid and efficient, with a time-course somewhat faster than that of C2CF (probably because CR contains only one caging group, as opposed to two for C2CF). However, there are two major drawbacks to the use of CR. First, resorufin bleaches even more rapidly than fluorescein under full mercury arc illumination. Second, the caged compound retains some visible absorption at 450 nm (Fig. 31.3(b)), and can be photoactivated (albeit inefficiently) by visible light. To minimize bleaching, we usually attenuate the mercury arc with neutral-density filters and shutter the illumination (see below). To avoid uncaging with visible light, we use a narrow band-pass 577 nm excitation filter for fluorescence microscopy (see below). Within this band there is a strong mercury line corresponding to the peak of resorufin absorption (about 575 nm); this set-up provides optimal excitation with no undesired uncaging. The uncaging of CR by visible blue light limits its usefulness in double-label applications where fluorescein (or C2CF) is also observed.

Useful protein probes for photoactivation must fulfil the following six requirements (Mitchison, 1989):

(1) They must be stably non-fluorescent inside cells.
(2) They must be readily photoactivated to a stable, highly fluorescent form.
(3) They must be efficiently photoactivated by light at a wavelength and intensity which does not perturb biological structures.
(4) They must be easily coupled to proteins of interest.
(5) They must not perturb the function of the carrier protein.
(6) The side-products of activation must be non-toxic to the cell and non-perturbing to the carrier molecule.

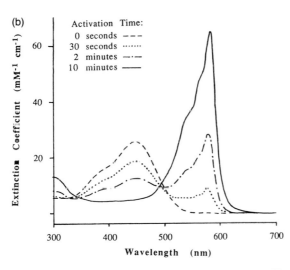

(a)

CAGING
REACTION

CR
Non-Fluorescent

hv
365 nm   PHOTOLYSIS
REACTION

RESORUFIN
Fluorescent

(b)

Figure 31.3 (a) Caging and photoactivation reactions of CR. (b) Optical absorption spectra after 0 s, 30 s, 2 min and 10 min of activation in a cuvette with a 360 nm hand-held lamp. (Adapted from Theriot & Mitchison, 1991, with permission.)

C2CF and CR fulfil these requirements to varying degrees: (1) Both C2CF and CR are stably non-fluorescent in the dark inside cells for hours or days (J.A. Theriot, unpublished observations; Reinsch *et al.*, 1991). As mentioned above, CR is slowly activated by visible light, and both compounds particularly must be guarded from the UV components of fluorescent overhead lights and sunlight. (2) C2CF and CR are readily photoactivated in cells, but both fluorescein and resorufin are easily photobleached. This must be kept in mind when determining the rate of turnover

of photoactivated subunits (see Fig. 31.9) and when considering possible photodynamic damage to the cell. (3) Both compounds are activated by 360 nm light, which is relatively benign to biological specimens (Hiramoto *et al.*, 1984). Photoactivation has not been observed to have adverse consequences for injected or uninjected cells, and photoactivation does not cause breakage of C2CF-labelled microtubules (Mitchison, 1989). (4) C2CF is capable of labelling tubulin, but only to stoichiometries substantially lower than those achieved with NHS-rhodamine or NHS-biotin (Hyman *et al.*, 1991). It does not efficiently label actin or myosin. CR does efficiently label actin and myosin, but its ability to label tubulin is similar to that of C2CF (J.A. Theriot, unpublished observations). (5) C2CF-labelled tubulin does not polymerize quite as efficiently as unlabelled tubulin, and tends to promote microtubule nucleation (Mitchison, 1989). CR-labelled actin also does not polymerize quite as efficiently as unlabelled actin (Theriot & Mitchison, 1991). However, in most *in vivo* applications, where labelled protein represents only a tiny fraction of the total protein present, the behaviour of labelled structures is presumably dominated by the characteristics of the more abundant unlabelled endogenous protein. (6) Both compounds generate potentially toxic photoactivation products, nitrosobenzaldehyde (from C2CF) or nitrosoaceto-phenone (from CR). Both are reactive toward sulphydryls and are presumably scavenged by gluta-thione and related compounds inside cells, but it is possible that they may interfere with the biological activity of some proteins. To date we have not observed any toxicity associated with photoactivation.

31.3 A COMPUTER-CONTROLLED, MULTIPLE-CHANNEL FLUORESCENCE MICROSCOPE FOR PHOTOACTIVATION

31.3.1 The photoactivation beam

A microbeam for local photoactivation of fluorescence requires only (1) a small aperture (typically a pinhole or slit) that can be demagnified by the microscope objective and focused onto a specimen of interest, and (2) a 360 nm light source for illuminating this aperture. An epifluorescence mercury arc lamp is ideal for this purpose, and in principle an existing epifluor-escence set-up could also be used for photoactivation. This method, however, suffers from a number of immediate disadvantages. First, under such conditions accurate positioning of a microbeam aperture will be difficult, because the aperture must be placed directly in the epi-illumination light path and therefore must also be easily removed and replaced for illumination

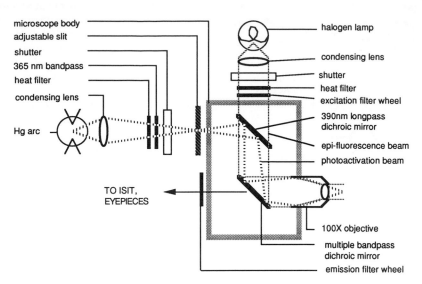

microscope body
adjustable slit
shutter
365 nm bandpass
heat filter
condensing lens

Hg arc

TO ISIT,
EYEPIECES

halogen lamp
condensing lens
shutter
heat filter
excitation filter wheel
390nm longpass
dichroic mirror
epi-fluorescence beam
photoactivation beam
100X objective
multiple bandpass
dichroic mirror
emission filter wheel

Figure 31.4 Schematic of the optical apparatus used for photoactivation microbeaming. Microscope, Zeiss IM35, with 100 × 1.3 NA Ph3 Neofluar; slit, homemade from Newport, Oriel, and Melles-Griot components; shutters, Vincent associates; 365 nm filter, Schott UG11; heat filter, Schott KG1; Hg arc HBO100, Oriel; halogen lamp, Zeiss 12V; filter wheels, AZI, with Omega filters; dichroic mirrors, Omega Optical. Further details are given in text. (Adapted from Mitchison, 1989.)

of the entire field during image acquisition. Moreover, the most efficient (i.e. intense) configuration for microbeaming requires critical illumination of the sample, while full-field epifluorescence nearly always involves Koehler illumination. For these reasons we use a separate, additional mercury arc lamp (HBO 100) exclusively for photoactivation in our microscope system. This lamp is oriented orthogonal to the epifluorescence light path and is brought into the microscope via a dichroic interference filter (390 nm long-pass) set at a 45° angle in the epifluorescence tube (Fig. 31.4). The photoactivation beam is focused with a secondary lens to critically illuminate the sample. This also focuses the beam as a second, conjugate plane that is close to but outside the body of the microscope. The microbeam aperture or slit is placed at this second plane, first mounted onto *x–y* positioners and then onto an optical rail bolted to a convenient part of the microscope. (Slits, rails, etc. can be purchased from companies such as Oriel or Newport and modified as needed by most machine shops.) The slit can then be adjusted along *x*, *y* and *z* axes and fixed into position. We usually adjust the slit by looking at a drop of coumarin in glycerol on a microscope slide. Coumarin is excited well by the 360 nm Hg line and can be seen easily using a filter set for fluorescein, yielding a sharp, fluorescent image of the slit through the oculars when the slit is in focus. Care should be taken to avoid exposing one's eyes to too much light when focusing the arc and the slit; in addition to reducing the spectrum of the microbeam at all times with a broad 360 nm band-pass filter (Schott UG11) and a UV/IR absorption filter (Schott KG1), we often add a neutral-density filter to the system when focusing the beam and slit under the microscope.

With nearly all possible geometries the image of the

arc at the slit plane is much larger than both the original arc and the slit itself, so a great deal of the available light does not pass through the aperture. (In our system, the mercury arc focuses to a spot of about 1 × 1.5 cm at a distance of about 30 cm from the focusing lens and is used to illuminate a slit of 0.01 × 0.3 cm.) Because the arc is not a true point source, there is nothing one can do about this loss. That is, any attempts to 'shrink' the beam through the use of a telescope (e.g. a laser beam expander placed in reverse orientation) or by repositioning the arc will be countered by an increase in beam divergence, and a corresponding loss of irradiance at the exit pupil of the objective. Because of the high intensity of the source, however, this 'masking effect' is typically not a problem for rapid photoactivation and is therefore of little concern.

We have also examined the possibility of using a UV laser for photoactivation. To date we have not found any suitable lasers operating in the range of 360 nm. Small tunable dye lasers pumped by a pulsed nitrogen laser (337 nm) do not offer much more total average power than a mercury arc, and also may cause photodamage at peak power. The only other potentially reasonable photoactivation source, a continuous-wave helium–cadmium laser (325 nm), requires expensive quartz objectives as well as additional modification to the microscope, since most microscope optics do not transmit below about 340 nm. It is also possible to double or triple a longer wavelength laser, but these lasers must be fairly high power and require considerably more care in operation.

31.3.2 Multiple-channel fluorescence microscopy

In most cases one would like to follow the fate of a photoactivated fluorescent mark with respect to some known landmark. In simple cases some fiduciary background fluorescence may be available. In our original work (Mitchison, 1989) we collected alternate fluorescence and phase-contrast images of cells before and after photoactivation by alternating between epi- and trans-illumination and manually sliding back and forth a two-place filter cube containing a fluorescein set in one place and no filters in the other. Where a more complete quantitation of fluorescence is desired, it is essential to be able to normalize locally photo-activated fluorescence to total fluorescence, observed in a separate channel (Sawin & Mitchison, 1991). To avoid the extreme tedium entailed in sliding a two-place fluorescence filter cube back and forth in long-term experiments (such as fate mapping of embryos; see below) we developed a simple computer-controlled multichannel fluorescence microscope that uses filter wheels for fluorescence excitation and emission and a multiple band-pass dichroic mirror in the fluorescence filter cube.

Based on the success of fluorescence ratio indicators, a number of motorized, computer-addressable filter wheels are currently commercially available (e.g. from AZI, Oriel, LEP, Sutter). They differ in many aspects, including: speed; 'intelligence' (e.g. the ability to find the shortest path to a new position, to know what is the current filter position, and to relay this information back to the computer); the number of wheels in a unit (e.g. for combining fluorescence and neutral-density filters); mode of address (serial port vs. TTL, via an additional digital input/output (I/O) board; see below); and price. Unless extreme precision or speed is required, wheels that are relatively simple and slow (0–2 s between positions) are usually sufficient. A low-

cost set-up, for example, could probably be put together using bottom-of-the-line filter wheels from Oriel in conjunction with a separately purchased low-cost I/O board (e.g. a Data Translation DT2817), that can also control any other electronics that respond to standard TTL signals.

If one is setting up a microscope for use exclusively with filter wheels, one should avoid purchasing any filters with the microscope itself altogether and instead obtain filters (typically 25 mm diameter for wheels, as opposed to 17–18 mm diameter for most microscopes) from a manufacturer specializing in high-quality interference filters (e.g. Omega Optical or Chroma Technologies). To avoid any signal cross-over one should use fairly narrow band-pass interference filters for both fluorescence excitation and emission, and filters should be anti-reflection (AR) coated to prevent any spurious reflection images. Interference-coated fluorescence excitation filters should be 'blocked' with coloured glass, which improves specificity, with an accompanying loss in total transmission. Emission filters are almost always similarly blocked, although in our experience this may not be completely essential.

To avoid sliding the fluorescence cube back and forth also requires having a dichroic mirror that can be used with more than one fluorochrome. Such interference filters ('multiple band-pass dichroic mirrors') can be custom-ordered from Omega Optical or Chroma Technologies. Specifications for such filters should take into consideration: (1) optimizing for the particular fluorochromes to be used, within limits of manufacturing technology; (2) matching the dichroic to excitation and emission filters, and at the same time avoiding any cross-over between fluorochromes; (3)

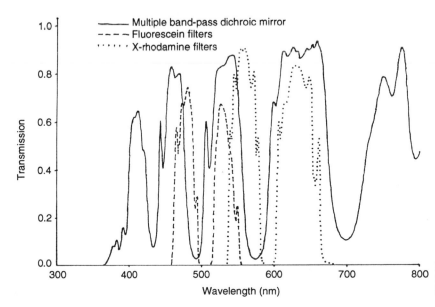

Figure 31.5 Filter sets used for double-label fluorescence imaging. See text for details.

making sure that the dichroic will reflect the 360 nm photoactivation beam onto the specimen. For good spectral separation we typically use C2CF and X-rhodamine or Texas Red as fluorochromes in double-label experiments, using a dichroic and filter set designed for fluorescein and Texas Red, made by Omega Optical (Fig. 31.5). We have also made a triple-label set, for imaging fluorescein, X-rhodamine and the far-red dye CY5 (Ernst *et al.*, 1989).

31.3.3 Putting it all together

Most fluorescence video microscopy requires that the total sample exposure to light be limited, either because of the photosensitivity of the sample and/or the rapid photobleaching rates characteristic of many fluorochromes, including fluorescein. This is achieved by shuttering illumination, usually under the control of an image-processing system (see below). At the same time, a low-light-level camera should be used to minimize the amount of incident light required to collect a decent image. Either a silicon-intensified target camera (SIT) or an intensified SIT (ISIT) may be appropriate, or, in some cases, a much more expensive cooled charge-coupled device (CCD). In our experience, SIT cameras can easily detect anything visible to the dark-adapted eye (which is quite sensitive). The ISIT is more sensitive, but has poor spatial resolution and is much noisier. For example, imaging single fluorescent MTs is quite easy with a SIT camera, but impossible with an ISIT. The ISIT may be preferred, however, where a need for sensitivity outweighs that for resolution. Both cameras are rather sensitive in the red, beyond the range of the human eye, so to avoid unnecessary degradation of signal-to-noise it is advisable to use band-pass rather than long-pass emission filters (as described above). As mentioned above, CCDs are very expensive and require additional hardware support, but they also have the benefit of an enormous linear dynamic range, which may be of great value in quantitating fluorescence. Under normal sampling frequency and duration in our experiments, a cooled CCD has approximately the sensitivity of a SIT.

For low-light, time-lapse experiments, video-frame summing or averaging a digital image processor is useful, if not essential, and in any long-term experiment (e.g. fate-mapping) it is easiest to perform as many operations as possible under computer control. Many types of image processors to perform these functions are now available, and most of these have fairly sophisticated I/O capabilities to control shutters, filter-wheels, and recording devices (see below) in addition to doing standard image acquisition, processing, and analysis. Image processors can vary widely in price, depending on colour capabilities, available

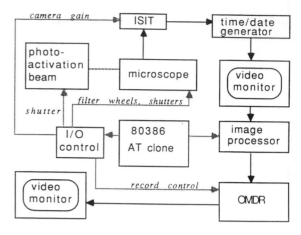

Figure 31.6 Schematic of the imaging system used in photoactivation experiments. The microscope and photoactivation beam are as in Fig. 31.4. Intensified silicon-intensified target camera (ISIT), Cohu; time-date generator, Panasonic; image processor, Datacube (AT-1), 80386-based AT computer, Samsung, optical memory disk recorder (OMDR), Panasonic; digital input/output (I/O) board, Data Translation, (no. 2817).

software, support, etc. Because of our high use of video microscopy in our laboratory we have found it useful to have a number of relatively low-cost image processors for data acquisition (typically the most time-consuming step in experiments), while more complicated analyses are done on a better-supported instrument. For data collection we use a Datacube MaxVision AT-1 system coupled to an inexpensive PC clone (Fig. 31.6). With this system it is particularly easy to write 'scripts' or macros that string together specific operations (e.g. rotate wheel, open shutter, average *n* frames, record image, wait *n* seconds), and to modify and extend existing scripts as source files. To make some operations run more smoothly and to connect the system to a number of I/O devices we found it necessary to modify the original source code (which is made available by the manufacturer at no extra cost, with some supporting documentation) in a few key places. These alterations, as well as additional tricks for using MaxVision for video microscopy, are available to interested investigators.

Most applications using photoactivation of fluorescence involve time-lapse recording and shuttered illumination rather than recording at 30 frames per second video-rates, so it is important to record images on an instrument that has good single-frame capability, i.e. either an optical memory disk recorder (OMDR) or direct digital storage. The drawback of the OMDR is its relatively high cost, while any digital storage system requires high-capacity disk storage, as well as memory buffers and a computer system capable of handling all of this. We currently do most recording with an OMDR; however, since most recording is done with an image processor that should be easily interfaced with direct digital image acquisition, and electronics prices are always dropping, it seems likely that in the future direct digital recording will be the method of choice.

31.4 EXPERIMENTS USING PHOTOACTIVATION OF FLUORESCENCE

31.4.1 Poleward MT flux in the mitotic spindle

C2CF was first synthesized in order to study the MT dynamics in the kinetochore fibres of the mitotic spindle (Mitchison, 1989). As described above, MTs are polar structures, and in each half of the spindle, MT 'minus' ends are proximal to the centrosome, while MT 'plus' ends are distal and interact with MTs from the opposing half-spindle and with kinetochores. While many photobleaching studies have shown that most MTs in the spindle are highly dynamic (Salmon *et al.*, 1984; Saxton *et al.*, 1984) MTs of the kinetochore fibre appear to be much more stable, both in terms of their turnover (Gorbsky & Borisy, 1989) and resistance to MT-depolymerizing conditions. We originally demonstrated *in vitro* that isolated kinetochores could 'capture' free MT plus ends and that MT ends thus captured could nevertheless incorporate exogenous, labelled tubulin without disrupting the attachment at the kinetochore (Mitchison & Kirschner,

1985). These studies were followed by *in vivo* micro-injection experiments, in which incorporation of a biotin-labelled tubulin into spindle MTs of mitotic tissue culture cells was followed by immunoelectron microscopy (Mitchison *et al.*, 1986). Whereas most spindle MTs incorporated the biotin-tubulin rather quickly, kinetochore MTs incorporated the biotin-tubulin signal more slowly, at the site of kinetochore attachment, and at increasing times after microinjection kinetochore MTs became progressively labelled from their plus ends polewards towards their minus ends at spindle poles. Because the metaphase spindle is at steady state, this result suggested that incorporation of labelled tubulin at kinetochore-attached MT plus ends might be balanced by a net disassembly at MT minus ends, resulting in the steady movement of tubulin subunits in the MT lattice towards spindle poles, as was suggested by Margolis and Wilson on the basis of experiments *in vitro* (Margolis & Wilson, 1981). The notion of a poleward MT flux is of great interest to students of mitosis because of its implications for MT organization and regulation of MT dynamics, and for modes of force generation and chromosome movement. However, pulse-labelling of spindles with biotin-tubulin could not show unequivocally that such a flux was occurring.

This inferred poleward movement was demonstrated directly in living tissue culture cells using C2CF covalently coupled to tubulin (Mitchison, 1989). After microinjection of unactivated C2CF-tubulin into metaphase tissue culture cells and incorporation into spindles, fluorescence was locally activated in the

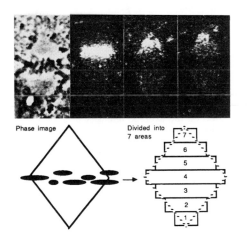

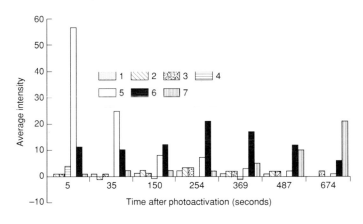

Figure 31.7 Fluorescence intensity in predefined zones of an LLC-PK1 spindle at different times after photoactivation. Left-hand panel shows (above) phase-contrast image of the spindle and three ISIT fluorescence images, at 5, 429, and 674 s after photoactivation, and (below) a diagram of the same spindle, arbitrarily divided into seven zones. Right-hand panel shows the average intensity in each of these zones at various times after photoactivation. Fluorescence intensity in zone 5, the site of photoactivation, decreases with time, while intensity in zone 6 rises, then falls. As intensity in zone 6 falls, intensity in zone 7 rises, indicating the vectorial movement of the microtubule lattice. The spindle remained in metaphase throughout the experiment. (From Mitchison, 1989. Reproduced from the *Journal of Cell Biology*, 1989, **109**, 637–652, by copyright permission of the Rockefeller University Press.)

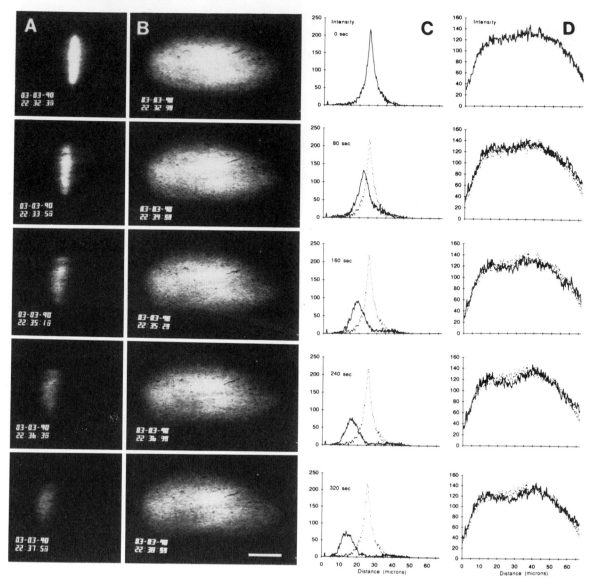

Figure 31.8 Photoactivation and double-label fluorescence imaging in a spindle assembled *in vitro* from a *Xenopus* egg extract. (A and B) ISIT images of photoactivated (fluorescein) and total (X-rhodamine) fluorescence at various times after photoactivation. (C and D) Digitized intensity-tracings of the same images (based on the average intensity of a 15-pixel-wide line, running from the left pole to the right pole), relative to the original photoactivated mark at time zero. Both the vectorial movement of photoactivated fluorescence and the essentially unchanged total fluorescence are apparent. (From Sawin & Mitchison, 1991. Reproduced from the *Journal of Cell Biology*, 1991, **112**, 941–954, by copyright permission of the Rockefeller University Press.)

spindle by UV microbeam and followed by low-light fluorescence video microscopy. As predicted, activated fluorescence moved polewards, and observed poleward flux in these experiments was apparent only in what appeared to be kinetochore MTs; other spindle MTs lost fluorescence rapidly, presumably because of their high rate of turnover (Fig. 31.7). This movement could be quantitated using digital image analysis. Photoactivated fluorescence increased near spindle poles at later times after photoactivation, while

decreasing at the initial site of activation, indicating true movement of fluorescence. Quantitation of fluorescence was particularly important in this instance because if MT dynamics were anisotropic in the spindle (so-called 'tempered instability'; Sammak *et al.*, 1987) fluorescence might falsely appear to the eye to move polewards over time, in the absence of flux (Sawin & Mitchison, 1991). By contrast, the measured increase of fluorescence in 'polar' regions of the spindle at increasing times after photoactivation indicates

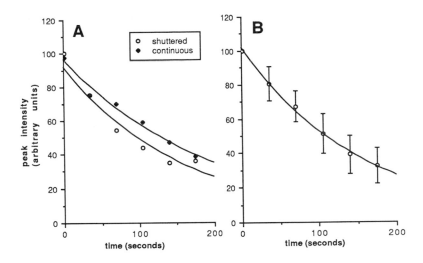

Figure 31.9 Turnover in fluxing microtubules in spindles assembled *in vitro*. (A) Decline of the peak of fluorescence intensity over time, from profiles such as those shown in the preceding figure. The same curve is obtained with continuous and carefully shuttered illumination, indicating that loss of fluorescence is not due to photobleaching. (B) Average rate of loss of fluorescence, from six different experiments. (From Sawin & Mitchison, 1991. Reproduced from the *Journal of Cell Biology*, 1991, **112**, 941–954, by copyright permission of the Rockefeller University Press.)

the directed movement of the microtubule lattice (Mitchison, 1989).

More recently we identified a poleward MT flux in spindles reconstituted *in vitro* from extracts of *Xenopus* eggs (Sawin & Mitchison, 1991). In these experiments we were interested in more accurately quantitating fluorescence, to determine rates of flux and of MT turnover. For these reasons we made spindles that contained both C2CF-tubulin and non-caged, X-rhodamine-tubulin, and by collecting alternate images in fluorescein and rhodamine channels using filter wheels we were able to normalize photoactivated fluorescence with respect to total fluorescence during image analysis. As shown in Figs 31.8 and 31.9, this allowed us to align intensity profiles over time to near-perfect register (Fig. 31.8) and to determine approximate rates of MT turnover, based on fluorescence decay after photoactivation (Fig. 31.9).

FRAP experiments have not yet detected any significant poleward flux of MTs in spindles (Gorbsky & Borisy, 1989; Wadsworth & Salmon, 1986, and references therein) with one possible exception (Hamaguchi *et al.*, 1987), even in the same cell types in which flux has been observed by photoactivation (Gorbsky & Borisy, 1989; Mitchison, 1989). The reasons for this apparent negative result are not yet clear. It is possible that photobleaching may not actually be non-perturbing. At very high incident illumination fluorescent MTs will experience significant damage and break up (Vigers *et al.*, 1988). However, when photobleaching experiments are done carefully, spindles remain morphologically normal and progress through mitosis at normal rates after photobleaching, so we consider it unlikely that photodamage is a serious concern. A second possibility is that signal-to-noise problems may make it difficult to detect small changes

in the position of a dark spot (representing a stable, bleached kinetochore MT) against a light background (representing dynamic, non-bleached and recovering spindle MTs) as opposed to the converse. One can imagine, for example, that as the more dynamic components of a multi-component system recover quickly after photobleaching, the dark spot of slowly recovering component will be 'obscured' by newly recovered fluorescence, making accurate measurements more difficult (Fig. 31.10(a)). From a theoretical point of view, it would seem immaterial whether a mark is bright against a dark background, or vice versa. However, we note that non-linearities in camera gain may indeed affect the ability to detect movement of bleached fluorescence as compared with photoactivated fluorescence, since identifying movement unambiguously often depends on distinguishing fairly subtle differences in intensity (Fig. 31.10(b,c)). Unfortunately, this explanation may be insufficient to account for the differences seen with photoactivation vs. FRAP, since recent FRAP experiments using CCDs with extremely broad linear dynamic range have also failed to detect poleward flux in the spindle. Further experiments using photoactivation and photobleaching in the same system will help to resolve these discrepancies.

31.4.2 Actin dynamics in the leading edge of cells

The leading edge, or lamellipodium, of motile cells is a dynamic cytoskeletal structure in some ways analogous to the mitotic spindle. Actin filaments in the lamellipodium are arranged in a loosely organized meshwork and are at least partially polarized with their barbed (dynamic) ends toward the membrane (Small, 1982; Yin & Hartwig, 1988). The most direct observation of actin behaviour in the lamellipodium was the

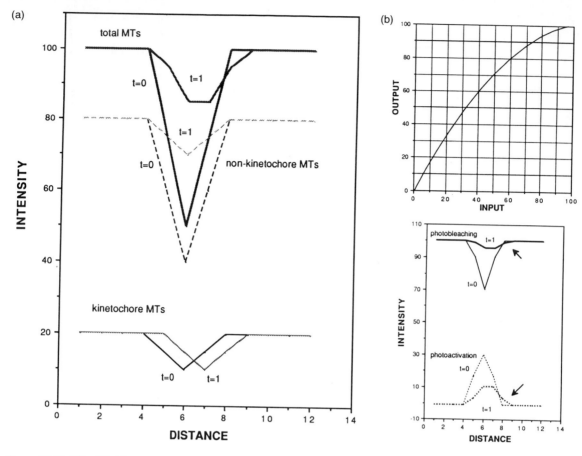

Figure 31.10 Why FRAP and photoactivation of fluorescence may yield different results. (a) Schematic drawing in which microtubules in the spindle are represented as two distinct populations: 80% non-kinetochore MTs, which turn over quickly but are stable in space; and 20% kinetochore MTs, which do not turn over but move from left to right. In this example fluorescence is bleached to 50% of its initial value at $t = 0$, and non-kinetochore MTs recover significantly by $t = 1$. The movement of kinetochore MTs at $t = 1$ is apparent in the profile of total MT fluorescence (kinetochore MTs plus non-kinetochore MTs), and theoretically should be equally obvious using photoactivation or photobleaching tehniques. (b) An example of a typical, slightly non-linear gain function for a video camera. 'Input' denotes actual light levels; 'output' is what the camera 'sees' (and what the investigator measures). (c) How non-linear gain may affect one's ability to judge movement of fluorescence. The original profiles of total MT fluorescence shown in (a) at $t = 0$ and $t = 1$were transformed using the function shown in (b) (inverting the profile in order to represent photoactivation), such that 'photobleaching' curves have intensities near the flat (top) portion of the gain function, while 'photoactivation' curves (i.e. the photobleaching profiles inverted) have intensities near the steeper (bottom) portion. While after this transformation, the contribution of the movement of the kinetochore MTs is still visible in both curves, it is more obvious in the 'photoactivation' curve, i.e. in the steeper portion of the gain function, where small differences in intensity are exaggerated. This is especially apparent where the profile at $t = 1$ has moved beyond the 'shoulder' of the profile at $t = 0$ (arrows).

classic photobleaching experiment of Wang (1985) which demonstrated that a spot bleached on the lamellipodium of a fibroblast injected with rhodamine-actin would move slowly backwards toward the cell body. This movement was interpreted as arising due to treadmilling within individual long actin filaments, that is, addition of monomers to the filament barbed end at the membrane coupled with concomitant loss of monomers at the pointed end at the rear of the lamellipodium. This rearward movement of actin polymer may be analogous to the polewards movement of tubulin in the spindle. The continual influx of

actin and actin-rich structures from the leading edge was supported by DIC observations of lamellipodia of neuronal growth cones (Forscher & Smith, 1988) and fibroblasts (Fisher *et al.*, 1988). However, the rate of actin filament transport observed by photobleaching was substantially slower than the rates of movement of structures observable by DIC. In addition, these observations of actin filament movements *in vivo* have generally been performed on cells which were immobile during the period of observation. Thus the relationships among rearward flux of actin and other lamellipodial structures, protrusion of the leading edge, and translocation of the bulk cytoplasm and nucleus were unclear.

We addressed these issues by injecting CR-actin into rapidly locomoting goldfish keratocytes (Euteneuer & Schliwa, 1984) and activating fluorescence in a narrow bar in the lamellipodium. We confirmed that actin filaments move backward with respect to the leading edge (Fig. 31.11). In this rapidly moving cell type, however, the rate of forward cell translocation in any given cell was approximately the same as the rate of relative rearward actin movement, so the activated filaments in the lamellipodium remained fixed in space as the cell moved forward over them, regardless of the rate of cell speed. We also found that the rate of actin filament turnover in the lamellipodium was remarkably rapid, about 23 s. This rapid turnover rate implies that the filaments must be much shorter than the width of the lamellipodium. This result is not consistent with the treadmilling model proposed by Wang (1985) for fibroblasts. Instead we think that

in the keratocyte, short filaments are nucleated continually at the tip of the leading edge, and then move backwards as a cross-linked meshwork (Theriot & Mitchison, 1991).

CR-actin will also be a useful probe for determining the rate of actin filament turnover in lamellipodia in fibroblasts, and for relating the rate of rearward filament movement to the rate of forward translocation in these cells, as in keratocytes. It will also provide a convenient way of observing actin filament movements directly in cells in which actin-based structures can be observed to move by DIC or phase microscopy, to determine whether these rates of movement are in fact different from one another in a given individual cell. In addition to studies of actin-based cell motility, CR-actin may prove itself useful in investigations of other actin-dependent phenomena, such as cytokinesis.

31.4.3 Fate mapping

A promising application for photoactivation of fluorescence involves marking not subcellular components but rather cells themselves. Over the years a number of different general, non-genetic techniques have been developed for fate mapping of embryos, but no single method is without problems. One early method, direct observation (see, for example, Poulson, 1950), is often difficult for a number of reasons, not least among them the opacity and light-scattering qualities of many embryos. A second, common approach involves labelling specific cells or groups of cells with a non-perturbing marker of some sort (often fluorescent or enzymatic), either externally or by microinjection (see, for example, Weisblat *et al.*, 1978). The major problem with this method is that in a multicellular embryo the number of cells accessible by this technique may be limited – for example, it would be impossible to inject cells of an inner germ layer without disrupting outer layers. Moreover, microinjection of single cells becomes increasingly difficult as embryonic cleavages progress. A third method, local killing of cells with a laser microbeam (e.g. Lohs-Schardin *et al.*, 1979), avoids the problems faced by microinjection approaches but only with the clear disadvantage that cell-killing may induce unforeseen abnormalities in development. As a marking technique, photoactivation of fluorescence combines the positive aspects of these latter two methods without their obvious drawbacks, as follows: embryos can be first 'pre-loaded' with caged fluorochromes coupled to soluble carrier proteins or dextrans at times early in development, when microinjection is fairly easy (for example, at the one- or two-cell stage), and then allowed to progress to some predefined point, when single cells or fields of cells can be marked by photoactivation. Preliminary experiments have suggested that this is a reasonable

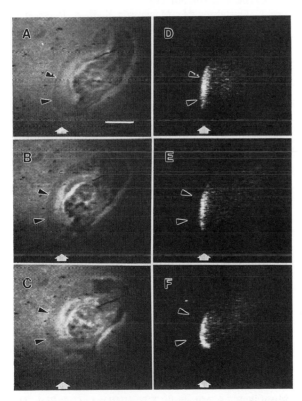

Figure 31.11 Behaviour of actin microfilaments in a moving goldfish keratocyte. Paired phase (A–C) and resorufin fluorescence (D–F) images of a keratocyte injected with CR-actin 4 s (A, D), 48 s (B, E) and 81 s (C, F) after activation of a bar-shaped region in the lamellipodium. Arrowheads indicated the position of the cell's leading edge in each frame; white arrows mark the same position in all panels. Note that the activated region remains approximately fixed in space as the cell moves from right to left. Bar is 10 μm. (Adapted from Theriot & Mitchison, 1991, with permission.)

approach to fate mapping (Goetz and T.J. Mitchison, unpublished observations); its success in real experiments appears to depend on coupling photoactivatable fluorochromes to appropriate carriers (J. Minden, pers. commun.; J.-P. Vincent, pers. commun.).

31.5 FUTURE PROSPECTS AND CONCLUSIONS

Photoactivation of fluorescence represents a novel technology for marking structures in living cells and organisms with the precision of a light microbeam. In addition to the experiments described above, future applications of photoactivation of fluorescence may include studying membrane and membrane skeleton dynamics during cell locomotion and adhesion, and perhaps also in following membrane transport and recycling pathways directly under the microscope. In addition to 'standard' embryonic fate mapping, one can imagine that photoactivation of fluorescence could also be used to mark cells or fields of cells in order to study morphogenetic movements in development, nerve growth, and cell–cell coupling with much greater precision than has previously been available *in situ*. More broadly speaking, photoactivation of fluorescence is one of a number of light-directed techniques – for example, photoactivation or photoablation of enzyme activities or enzyme substrates, or even the potential photoactivation of gene expression – which may be applied to complex biological problems. It is exciting to consider how these techniques may be combined in the future in order to uncover the inner workings of cells.

REFERENCES

Belmont L.D., Hyman, A.A., Sawin K.E. & Mitchison T.J. (1990) *Cell* **62**, 579–589.

Bershadsky A.D. & Vasiliev J.M. (1988) *Cytoskeleton.* New York, Plenum Press.

Carlier M.-F. (1989) *Int. Rev. Cytol.* **115**, 139–170.

Dustin P. (1984) *Microtubules.* Berlin, Springer-Verlag.

Ernst L.A., Gupta R.K., Mujumdar R.B. & Waggoner A.S. (1989) *Cytometry* **10**(1), 3–10.

Euteneuer U. & Schliwa M. (1984) *Nature (Lond.)* **310**, 58–61.

Fisher G.W., Conrad P.A., DeBiasio R.L. & Taylor D.L. (1988) *Cell Motil. Cytoskeleton* **11**, 235–247.

Forscher P. & Smith S.J. (1988) *J. Cell Biol.* **107**, 1505–1516.

Gorbsky G.J. & Borisy G.G. (1989) *J. Cell Biol.* **109**(2), 653–662.

Hamaguchi Y., Toriyama M., Sakai H. & Hiramoto Y. (1987) *Cell Struct. Funct.* **12**, 43–52.

Haugland R.P. (1989) *Handbook of Fluorescent Probes and Research Chemicals.* Molecular Probes Inc., Eugene, OR.

Hill T.L. & Kirschner M.W. (1982) *Proc. Natl. Acad. Sci. USA* **79**, 490–494.

Hiramoto Y., Hamaguchi M.S., Nakano Y. & Shoji Y. (1984) *Zool. Sci.* **1**, 29–34.

Horio T. & Hotani H. (1986) *Nature* **321**, 605–607.

Hotani H. & Horio T. (1988) *Cell Motil. Cytoskeleton* **10**, 229–236.

Hyman A., Drechsel D., Kellogg D., Salser S., Sawin K., Steffen P., Wordeman L. & Mitchison T. (1991) *Methods Enzymol.* **196**, 478–485.

Inoue S. (1981) *J. Cell Biol.* **91**, 131s–147s.

Inoue S. (1982) In *Developmental Order: Its Origin and Regulation (40th Symposium of the Society for Developmental Biology)*, S. Subtelny & P.B. Green (eds). Alan R. Liss, New York, pp. 35–76.

Inoue S. & Sato H. (1967) *J. Gen. Physiol.* **50**, 259–292.

Janko K. & Reichert J. (1987) *Biochim. Biophys. Acta* **905**, 409–416.

Kirschner M.W. (1980) *J. Cell Biol.* **86**, 330–334.

Krafft G.A., Sutton W.R. & Cummings R.T. (1988) *J. Am. Chem. Soc.* **110**, 301–303.

Lee J., Gustafsson M., Magnusson K.-E. & Jacobson K. (1990) *Science (Washington, D.C.)* **247**, 1229–1233.

Lohs-Schardin M., Sander K., Cremer C., Cremer T. & Zorn C. (1979) *Devel. Biol.* **68**, 533–545.

Margolis R.L. & Wilson L. (1978) *Cell* **13**, 1–8.

Margolis R.L. & Wilson L. (1981) *Nature (Lond.)* **293**, 705–711.

Marque J.J. (1989) *Nature* **337**, 583–584.

Mitchison T.J. (1989) *J. Cell Biol.* **109**, 637–652.

Mitchison T.J. & Kirschner M.W. (1984) *Nature* **312**, 237–242.

Mitchison T.J. & Kirschner M.W. (1985) *J. Cell Biol.* **101**, 767–777.

Mitchison T.J., Evans L., Schultze E. & Kirschner M.W. (1986) *Cell* **45**, 515–527.

Oosawa F. & Asakura S. (1975) *Thermodynamics of the Polymerization of Protein.* Academic Press, London.

Patchornik A., Amit B. & Woodward R.B. (1970) *J. Am. Chem. Soc.* **92**, 6333–6335.

Poulson D.F. (1950) In *The Biology of Drosophila.* Wiley-Interscience, New York.

Reinsch S.S., Mitchison T.J. & Kirschner M.W. (1991) *J. Cell Biol.* **115**, 365–379.

Salmon E.D., Leslie R.J., Karow W.M., McIntosh J.R. & Saxton R.J. (1984) *J. Cell Biol.* **99**, 2165–2174.

Sammak P.J., Gorbsky G.J. & Borisy G.G. (1987) *J. Cell Biol.* **104**(3), 395–405.

Sato H., Ellis G.W. & Inoue S. (1975) *J. Cell Biol.* **67**, 501–517.

Sawin K.E. & Mitchison T.J. (1991) *J. Cell Biol.* **112**, 941–954.

Saxton W.M., Stemple D.L., Leslie R.J., Salmon E.D., Zavortink M. & McIntosh J.R. (1984) *J. Cell Biol.* **99**, 2175–2186.

Small J.V. (1982) *Electron Microsc. Rev.* **1**, 155–174.

Theriot J.A. & Mitchison T.J. (1991) *Nature* **352**, 126–131.

Tsien F.Y. & Zucker R.S. (1986) *Biophys. J.* **50**, 843–853.

Wadsworth P. & Salmon E.D. (1986) *J. Cell Biol.* **102**, 1032–1038.

Walker J.W., Somlyo A.V., Goldman Y.E., Somlyo A.P. & Trentham D.R. (1987) *Nature* **327**, 249–252.

Walker J.W., Reid G.P. & Trentham D.R. (1989) *Methods Enzymol.* **172**, 288–301.

Walker R.A., O'Brien E.T., Pryer N.K., Sobeiro M.F., Voter W.A., Erickson H.P. & Salmon E.D. (1988) *J. Cell Biol.* **107**, 1437.

Wang Y.-L. (1985) *J. Cell Biol.* **101**, 597–602.

Wang Y.-L., Lanni F., McNeil P.L., Ware B.R. & Taylor D.L. (1982) *Proc. Natl. Acad. Sci. USA* **79**, 4660–4664.

Wegner A. (1976) *J. Molec. Biol.* **108**, 139–150.

Weisblat D.A., Sawyer R.T. & Stent G.S. (1978) *Science* **202**, 1295–1297.

Vigers G.P.A., Coue M. & McIntosh J.R. (1988) *J. Cell Biol.* **107**, 1011–1024.

Yin H.L. & Hartwig J.H. (1988) *J. Cell Sci. Suppl.* **9**, 169–184.

NOTE ADDED IN PROOF

Since the original submission of this manuscript, a number of papers on photoactivation of fluorescence have been published. They are listed below.

Mitchison T.J. & Salmon E.D. (1992) *J. Cell Biol.* **119**, 569–582.

Okabe S. & Hirokawa N. (1992) *J. Cell Biol.* **117**, 105–120.

Theriot J.A. & Mitchison T.J. (1992) *J. Cell Biol.* **119**, 367–377.

Theriot J.A., Mitchison T.J., Tilney L.G. & Portnoy D.A. (1992) *Nature* **357**, 257–260.

Vincent J.-P. & O'Farrell P.H. (1992) *Cell* **68**, 923–931.

CHAPTER THIRTY-TWO

Video Imaging of Lipid Order

KATHRYN FLORINE-CASTEEL,[1]
JOHN J. LEMASTERS[2] & BRIAN HERMAN[2]

[1] Department of Pathology, Duke University Medical Center, Durham, NC, USA [2] Department of Cell Biology and Anatomy, University of North Carolina at Chapel Hill, NC, USA

32.1 INTRODUCTION

Membrane lipid order may be an important parameter in the regulation of certain cellular functions, hence the interest in its measurement at the single cell level. A widely used method of measuring lipid order, usually performed on membrane suspensions, has been fluorescence polarization spectroscopy. In this technique, membranes are labelled with lipophilic fluorophores, and polarized fluorescence is measured in order to determine the average constraint of probe motion in the bilayer. In theory, this is a measure of the degree of lipid order. With the increasing sophistication of microscope/imaging systems as well as the availability of an increasingly diverse selection of suitable fluorescent probes, it is now possible to attempt these measurements microscopically on individual cells. This enables lipid order to be determined with subcellular spatial and temporal resolution.

Microscopic fluorescence polarization measurements on single cells present certain problems not associated with fluorometric measurements. The major difficulties are the depolarizing effect of the microscope optics and the orientation dependence of the observed fluorescence polarization. The first problem is due mainly to the use of a high aperture objective lens for collection of fluorescence, rather than a narrow slit of effectively zero aperture. A theoretical treatment of high aperture observation has been presented (Axelrod, 1979) and will not be discussed in detail here. The problem of oriented samples, such as surface-labelled single cells, requires a knowledge of the probe excited state orientation distribution in order to determine the degree of lipid order from the measured orientation-dependent fluorescence polarization.

In this chapter, we outline a method for obtaining spatially resolved measurements of lipid order using steady-state fluorescence polarization microscopy in combination with digital image processing. We treat in detail the case of spherical membrane surfaces labelled with rod-shaped, diphenylhexatriene-derivative probes. The general applicability of the technique is also discussed.

32.2 THEORY

32.2.1 Description of the experiment

Figure 32.1 depicts a spherical membrane surface on the microscope stage, focused at the X_2–X_3 plane. In

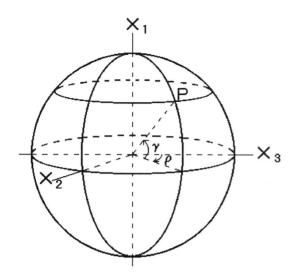

Figure 32.1 Definition of coordinates. The X_1-axis is the optical axis of the microscope, and the X_2- and X_3-axes define the focal plane. An arbitrary point P on the spherical membrane surface is specified by the angles ρ and γ.

the conventional polarization experiment employing an inverted microscope and epi-illumination, incident excitation light polarized in the X_3 direction is reflected from a dichroic mirror below the stage and propagates along the X_1 direction, through the condenser/objective, to the sample. Fluorescence is collected at 180° to the incident light propagation direction at each of two orthogonal orientations of a polarizer in the emission path, one in the X_2 direction and the other in the X_3 direction (i.e. perpendicular and parallel to the excitation polarization)

The polarized fluorescence observed in the microscope images will depend on four factors: (1) the orientation of probe absorption dipole moments relative to the excitation light polarization direction (X_3); (2) probe rotational diffusion and consequent emission dipole reorientation during the excited state lifetime; (3) objective lens numerical aperture, which determines the amount of mixing in of fluorescence components in the X_1 and X_3 (X_1 and X_2) directions in the recorded $F_\perp$ ($F_\parallel$) image; and (4) the range of probe orientations on the three-dimensional membrane surface corresponding to each pixel location in the two-dimensional image. The following section outlines how these factors are dealt with in order to infer lipid order from polarized fluorescence images.

32.2.2   Fluorescence polarization as a function of lipid order and membrane surface location

The general procedure for determining the functional dependence of fluorescence polarization on lipid order as well as probe location on the membrane surface is

as follows. First, a model of probe rotational diffusion in the membrane is adopted which includes a parameter that indicates the degree of lipid order. For diphenylhexatriene (DPH) probes, we have used the 'wobbling-in-a-cone' model (Kinosita $et\ al.$, 1977) in which lipid order is expressed in terms of a cone angle, θ_{max}, representing the fluorophore's maximum angular motional freedom in the bilayer. The cone angle θ_{max} is related to the lipid order parameter, S, by $S = 0.5 \cos\theta_{max} (1 + \cos\theta_{max})$ (Lipari & Szabo, 1980). One then solves the appropriate diffusion equation in terms of a Green function to obtain the probability $p(\Omega',t'|\Omega,t)$ that a probe with orientation Ω' (in the membrane reference frame) at time t' will have orientation Ω at time t, where t' and t are the time of absorption and emission, respectively, of a photon. The probe excited state distribution function f can then be found from the relation

$$f = \frac{1}{\tau} \iint (x_3')^2\, p(\Omega',t'|\Omega,t)\mathrm{e}^{-(t-t')/\tau}\, \mathrm{d}\Omega'\, \mathrm{d}(t - t') \quad [32.1]$$

where x_3' is the component of a unit magnitude absorption dipole moment along the X_3-axis (the excitation light polarization direction) at the time of absorption ($t - t' = 0$), and τ is the probe excited state lifetime. Finally, the distribution of excited fluorophores, f, is used to calculate the polarized fluorescence collected from the location (ρ,γ) on the membrane surface, for an arbitrary orientation ψ of the emission polarizer, according to

$$F_\psi(\theta_{max},\rho,\gamma) = N \int f\, [K_a x_1^2 + (K_b \cos^2\psi + K_c \sin^2\psi)x_2^2$$
$$+ (K_b\sin^2\psi + K_c\cos^2\psi)x_3^2 + 2(K_c - K_b)\sin\Psi\cos\psi x_2 x_3]\mathrm{d}\Omega \quad [32.2]$$

where x_1, x_2, and x_3 are the components of a unit magnitude emission dipole moment along the X_1-, X_2-, and X_3-axes, respectively, ψ is the angle in the X_2–X_3 plane between the X_3-axis and the transmission axis of the emission polarizer, and N represents the combined normalization constants. K_a, K_b and K_c are weighting factors which are functions of the numerical aperture of the objective and the index of refraction of the sample medium (Axelrod, 1979). For $\psi = 0°$ and 90°, which correspond to $F_\parallel$ and $F_\perp$, respectively, equation [32.2] reduces to

$$F_\parallel (\theta_{max},\rho,\gamma) = N \int f\, [K_a x_1^2 + K_b x_2^2 + K_c x_3^2]\mathrm{d}\Omega \quad [32.3]$$

$$F_\perp (\theta_{max},\rho,\gamma) = N \int f\, [K_a x_1^2 + K_c x_2^2 + K_b x_3^2]\mathrm{d}\Omega \quad [32.4]$$

The dependence of F_ψ on lipid order, in terms of θ_{max}, arises in the integration of equations [32.1] and [32.2] over Ω' and Ω, where in each case θ_{max} is an integration limit. The dependence on membrane surface location,

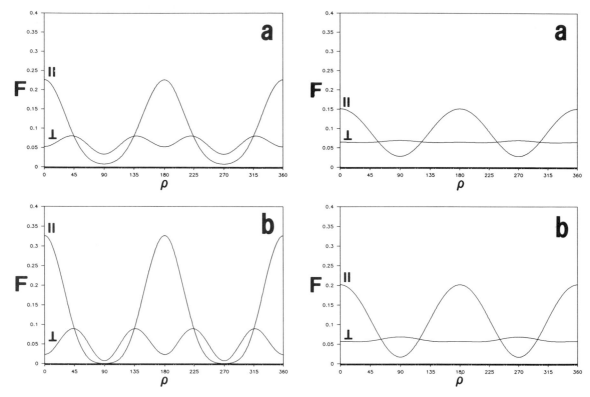

Figure 32.2 Theoretical relative polarized fluorescence, $F_{\parallel}$ and $F_{\perp}$, versus position on the membrane perimeter for cone angle $\theta_{max} = 40.9°$ (a) or $16.7°$ (b), with the cone axis normal to the membrane surface.

Figure 32.3 Theoretical relative polarized fluorescence, $F_{\parallel}$ and $F_{\perp}$, versus position on the membrane perimeter for cone angle $\theta_{max} = 40.9°$ (a) or $16.7°$ (b), with the cone axis tilted $30°$ from the bilayer normal.

i.e. the coordinates (ρ, γ), is contained in the absorption and emission dipole moment components x_i' and x_i, respectively, and results from the coordinate transformation from the membrane to the laboratory reference frame.

The fluorescence intensity recorded in each pixel in the two-dimensional fluorescence image is found by integration of F_ψ over the appropriate range of ρ and γ. For example, the spherical membrane surface shown in Fig. 32.1 is in focus at $\gamma = 0°$ (the X_2-X_3 plane). However, each image pixel around the in-focus membrane perimeter ($0 \le \rho \le 360°$, $\gamma = 0°$) actually contains fluorescence from the region $-\gamma_o$ to $+\gamma_o$, where the angle γ_o depends on pixel size and is typically about $10°$ (Florine-Casteel, 1990).

32.2.3   Examples

Some examples of theoretical polarized fluorescence patterns for the in-focus membrane perimeter region are illustrated in Figs 32.2–32.4. Fluorescence was calculated according to equations [32.3] and [32.4], with variations in the probe rotational diffusion model (and hence in the distribution function f). The explicit

form of these equations for each of the three diffusion models discussed is presented elsewhere (Florine-Casteel, 1990). The first model (Fig. 32.2), appropriate for DPH, is that in which the probe 'wobbles' with a maximum amplitude of angle θ_{max} about an axis normal to the plane of the membrane. In the second model (Fig. 32.3), the symmetry axis is tilted $30°$ from the bilayer normal, which is the case for probes such as TMA-DPH (a cationic derivative of DPH) that are located at or near the headgroup region of membrane phospholipids (Florine-Casteel, 1990). Finally, we consider the case of probe molecules trapped in the centre of the bilayer and oriented parallel to the plane of the membrane (Fig. 32.4), which has been reported in the literature for DPH and to a lesser extent, TMA-DPH.

As Figs 32.2 and 32.3 show, an increase in lipid order (i.e. a decrease in θ_{max}) results in a greater change in relative fluorescence between maxima and minima around the membrane perimeter. The presence of a cone axis tilt of $30°$ reduces the magnitude of this change. Probe molecules aligned parallel to the plane of the membrane (Fig. 32.4) display a different pattern of polarized fluorescence

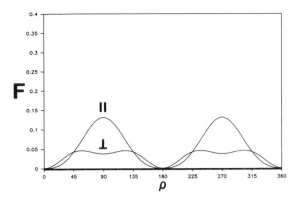

Figure 32.4 Theoretical relative polarized fluorescence, $F_\parallel$ and $F_\perp$, versus position on the membrane perimeter for the case of probe rotation in the plane of the membrane.

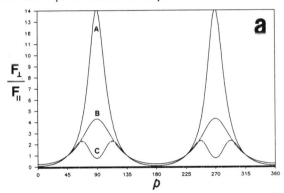

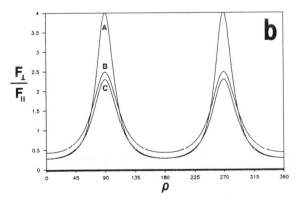

Figure 32.5 Theoretical polarization ratio, $F_\perp/F_\parallel$, versus position on the membrane perimeter in the absence (a) or presence (b) of a cone tilt of 30°, for $\theta_{max} = 16.7°$ (A), $\theta_{max} = 40.9°$ (B), or $\theta_{max} = 16.7°$ with 10% of probe molecules oriented parallel to the plane of the membrane (C).

that is not dependent on the degree of lipid order. In all cases, fluorescence is greatest in those regions where the average orientation of probe absorption and emission dipole moments (for DPH, the probe long axis) is parallel to the excitation and emission polarizer orientations, respectively. For example, in the first model, probes in the X_2–X_3 plane (see Fig. 32.1) are,

on average aligned radially around the membrane perimeter, along the X_3-axis (the excitation polarization direction) at $\rho = 0°$ and 180° and along the X_2-axis at $\rho = 90°$ and 270°. Therefore fluorescence will be maximum at $\rho = 0°$ and 180° with the emission polarizer oriented along the X_3-axis ($F_\parallel$).

Figure 32.5 illustrates how the polarization ratio, $F_\perp/F_\parallel$, changes with lipid order and with position on the membrane perimeter for the various models. The ratio is most sensitive to lipid order (θ_{max}) in the vicinity of $\rho = 0°$, 90°, 180°, and 270°. Accurate measurement of lipid order over the entire membrane perimeter would require an additional pair of emission polarizer orientations, $\psi = 45°$ and 135°. The polarization ratio is also very sensitive to probe location in the bilayer, with probe molecules located in the hydrophobic core (Fig. 32.5(a)) displaying a greater variation in $F_\perp/F_\parallel$ than those located in the headgroup region where the symmetry axis is tilted (Fig. 32.5(b)). The presence of a significant fraction of probe molecules aligned parallel to the plane of the membrane results in a dramatic decrease in the polarization ratio in the vicinity of $\rho = 90°$ and 270°, while leaving it virtually unchanged in the vicinity of $\rho = 0°$ and 180°, where these probes are oriented perpendicular to the excitation polarization direction and thus only weakly excited.

32.3 EXPERIMENT

32.3.1 Probe selection

As Fig. 32.5 illustrates, a knowledge of the probe orientation in the membrane is important in interpreting the fluorescence polarization images in terms of lipid order. For example, a significant fraction of probe molecules trapped between the inner and outer leaflets of the bilayer or adsorbed to the outer membrane surface such that the predominant orientation is not along the lipid acyl chains, could lead to ratio values that underestimate the degree of lipid order. Another important consideration is the probe's quantum yield, since light transmission through the emission polarizer can be reduced by as much as 90%. Because a minimum of two fluorescence images is required for a polarization measurement, the probe's susceptibility to photobleaching is also important.

32.3.2 Fluorescence polarization imaging

The optical arrangement for fluorescence polarization microscopy and imaging has been described in detail (Axelrod, 1989; Florine-Casteel, 1990). The essential feature is the addition to the conventional epifluorescence set-up of a pair of polarizers, one in the excitation light path and one in the emission light path. It is important to place the polarizers such that the number of optical elements, i.e. lenses and mirrors, 'downstream' from each polarizer is a minimum. On the excitation side, additional optical elements can affect the polarization purity of the light incident on the sample. Similarly, on the emission side, the transmittivity of fluorescent light to the detector will be polarization-dependent. It is therefore necessary to measure this effective birefringence of the microscope/imaging system before performing experiments and to correct the polarization images accordingly. Since a change in emission polarizer orientation can sometimes cause a slight optical shift of the image, it is also important to make sure that each pair of images is properly aligned before ratioing.

An example of polarization imaging is shown in Plate 32.1. Here, cell-size phosphatidylcholine liposomes have been labelled with TMA-DPH. The polarized fluorescence images were obtained using a 1.3 NA objective lens in a Zeiss IM-35 inverted microscope, with an ISIT video camera as the detector. Each image is an average of 64 frames (2 s illumination time). Fluorescence polarization ratios were computed from digitized image pairs on a pixel-by-pixel basis, after background subtraction and outlining (mapping) of the liposome perimeter, with all pixel intensities outside of the map set to zero for image clarity. Pseudo-colour ratio images were obtained by converting ratios to grey levels (0–255), using a multiplication factor of typically 100, and then assigning colour values to the grey levels. Based on the variation in image intensity between adjacent pixels in the fluorescence images, we estimate that polarization ratios can be accurately determined with a surface spatial resolution of about 1 μm^2.

The polarization ratio images in Plate 32.1 illustrate the difference in the degree of lipid order between fluid-phase POPC and gel-phase DPPC liposomes. Around the membrane perimeter, which is in the focal plane, the ratios are lower at $\rho = 0°$ and 180° and higher at $\rho = 90°$ and 270° for DPPC than for POPC. Moving inward toward the centre of the ratio image, fluorescence becomes increasingly out of focus and ratios contain contributions from both the upper and lower membrane surfaces. In the centre of the image, probe molecules are oriented predominantly along the microscope optical axis, where the probability of excitation is low and fluorescence polarization is only weakly dependent on lipid order. By comparing the polarization ratio around the membrane perimeter with theoretical curves such as those illustrated in Fig. 32.5, we can determine both the degree of lipid order, i.e. θ_{max}, and the fraction of probe molecules, if any, that are aligned parallel to the plane of the membrane.

In this example, we find $\theta_{max} = 32 \pm 4°$ for POPC and $17 \pm 2°$ for DPPC, with 10% and 5%, respectively, of probes aligned parallel to the plane of the bilayer (Florine-Casteel, 1990).

32.4 BIOLOGICAL APPLICATIONS

We have used the methodology outlined above to measure lipid order in the blebbing plasma membrane of single, cultured rat hepatocytes during ATP depletion (Florine-Casteel *et al.*, 1990, 1991). Cells on coverslips were incubated in medium containing the metabolic inhibitors cyanide and iodoacetic acid to induce the ATP depletion and bleb formation observed in hypoxic injury (Lemasters *et al.*, 1987). After subsequent transfer to the microscope stage, cells were labelled with the plasma membrane probe TMA-DPH by brief incubation in probe-containing buffer. We chose the membrane probe TMA-DPH because of its significantly slower rate of internalization compared to DPH (Kuhry *et al.*, 1983; Florine-Casteel *et al.*, 1990).

Plate 32.2 illustrates a polarization measurement on a pair of plasma membrane blebs. Experimental conditions were the same as described for Plate 32.1, except that an intensified CCD camera was used as the detector. The blebs are structurally similar to liposomes and give similar polarized fluorescence patterns (compare Plate 32.1 with the left-hand column of Plate 32.2). In this example, an additional pair of fluorescence images, $F_\perp$ and $F_\parallel$, was acquired at a second orientation of the excitation polarizer, orthogonal to the first (right-hand column of Plate 32.2), in order to more accurately determine lipid order over the entire bleb perimeter. Because of small amounts of probe internalization during the course of the measurements, some probe molecules will be transiently oriented parallel to the plane of the membrane. Lipid order is most accurately measured in those regions of the membrane perimeter where the in-plane component is perpendicular to the excitation polarization direction and thus only weakly excited, i.e. at $\rho = 0°$ and 180° (see Fig. 32.5). Therefore, by combining the information from two ratio images, obtained using orthogonal excitation polarizer orientations, we can measure lipid order over most of the bleb perimeter without regard to probe internalization. As in the case of the liposomes, the central regions of the bleb images contain weak, out-of-focus fluorescence from the upper and lower surfaces, making lipid order measurements difficult in these regions. However, the ratios at $\rho = 0°$ and 180° on the bleb perimeters range from 0.22 to 0.28, corresponding to $\theta_{max} = 10–20°$ (Florine-Casteel *et al.*, 1991; Fig.

32.5(b)), indicating a uniformly rigid membrane surface. The plasma membrane of the main cell body displays a fairly uniform polarization ratio of 0.6–0.7, with little variation around the in-focus cell perimeter, due to the randomizing effect of the microvillous structure on probe orientation.

The information on lipid order that is obtainable in a single cell by the method we have described depends on three factors: cell geometry, the properties of the membrane probe, and the particular combination of excitation/emission polarizer orientations used in image acquisition. In the above example, we looked at the simple case of a spherical membrane surface (a plasma membrane bleb) labelled with a rod-shaped probe of known orientation in the membrane (TMA-DPH). We were able to completely describe the plasma membrane lipid order profile around the bleb perimeter with a set of four fluorescence images. A different cell geometry, or a probe with different absorption/emission dipole orientations or diffusion properties, would require the appropriate modifications to equations [32.1]–[32.4] in order to determine the functional dependence of the polarization ratio on lipid order and membrane surface location. Image acquisition at other polarizer orientations might also be necessary.

ACKNOWLEDGEMENTS

This work was supported, in part, by grant AG07218 from the National Institutes of Health and grant J-1433 from the Office of Naval Research.

REFERENCES

Axelrod D. (1979) *Biophys. J.* **26**, 557–574.

Axelrod D. (1989) *Methods Cell Biol.* **30**, 333–352.

Florine-Casteel K. (1990) *Biophys. J.* **57**, 1199–1215.

Florine-Casteel K., Lemasters J.J. & Herman B. (1990) In *Optical Microscopy for Biology*, B. Herman & K. Jacobson (eds). Wiley-Liss, New York, pp. 559–573.

Florine-Casteel K., Lemasters J.J. & Herman B. (1991) *FASEB J.* **5**, 2078–2084.

Kinosita Jr. K., Kawato S. & Ikegami A. (1977) *Biophys. J.* **20**, 289–305.

Kuhry J.-G., Fonteneau P., Duportail G., Maechling C. & Laustriat G. (1983) *Cell Biophys.* **5**, 129–140.

Lemasters J.J., DiGuiseppi J., Niemien A.-L. & Herman B. (1987) *Nature* **325**, 78–81.

Lipari G. & Szabo A. (1980) *Biophys. J.* **30**, 489–506.

ACKNOWLEDGEMENTS

Index

Pluronic F-127   322
Polyclonal antibodies   165
Polytene chromosomes   247–9
Potassium indicators   40
Potential well concept   279
Potentiometric membrane dyes
   150–60
Prolactin hormone   165
Protein binding   38
Pyrene-labelled lipid probes   116
Pyridine nucleotides   48
Pyrophorus phagiophthalamus   71

Quantitative real-time imaging, optical
   probes in living cells   161–95
Quantum efficiency (QE)   265, 268
Quenching   2

Radiant sensitivity   268
Radioisotopes, detection of   197–8
Raman scattering   238, 40
Raman spectroscopy   238–42
   basophilic granulocytes   253–4
   instrumentation   240–5
   molecular biology   240–5
   single cells   241–2
   see also Confocal Raman
      microspectroscopy
Ratio dye   206
Ratio imaging
   maximum ratio level   225
   saturation artifacts   225
Ratiometric imaging system,
   calibration   225–6
Ratiometric measurements   224
Read-out noise   266
Real-time drug uptake analysis   316–17
Real-time fluorescence ratio imaging,
   cameras employed for   171
Real-time image acquisition   218–19
Real-time observation of biological
   activity   162
   Real-time processing of video
   signals   176
Real-time quantitative microscopy,
   biological applications   181–92
Real-time video imaging of ion-
   sensitive fluorescent dyes
   164–5
Receptor occupancy   307
Recording systems, choice of   402–3
Redox confocal imaging   44–57
   comparison with other
      techniques   55–6
   previous reviews   45–6
Redox fluorometry   44–5
   advantage of   44
   applications   45
   basis of   44
Redox imaging
   based on flavoprotein
      fluorescence   54
   based on NAD(P)H
      fluorescence   52–4

Redox ratio-scanning instrument   48
Regulatory light chain
   phosphorylation   355–6
RELACS III   314, 316
Renilla reniformis   71
Resolution enhancement using
   deconvolution   170
Resonance energy transfer   70
Resorufin   408–9
Responsive quantum efficiency
   (RQE)   265, 267
RGD sequence   185
RH155   153
RH421   152
Rhod-2   321
Rhodamines   154, 350
RNA   19, 23
Rubicon hypothesis   65

S. cerevisiae   75
Salmonella   70, 75
SBFI   40, 287, 304
Scanning optical microscopy   233–4
   advantages of   229–30
Scrambling with endogenous
   kinase   146
Scrape loading technique   381
Semiconductor detectors   265
Serum albumin   105
Short-pass (SP) filter   8
Signal maximizing   265
Signal-to-noise ratio   228, 265, 391,
   398, 399
   choice of dye   393
Silicon, characteristics of   278
Silicon detectors   234
Silicon diode arrays
   parallel read-out   399
   serial read-out   400
Silicon intensified target (SIT)   172,
   275, 278, 412
Single cells
   Raman spectroscopy   241–2
   study of living biological
      systems   162
SIT (silicon intensified target)   172,
   274–5, 278, 412
Small unilamellar vesicles (SUV)   105
SNAFL   36, 37, 41, 206, 321
SNARF   36, 37, 41, 206, 321
SNARF-1   329, 330, 332
SNARF-1-AM   332
Sodium indicators   40
Spectral response   268
Spectrophotometric analysis   125
Sphingo(lipid) trafficking   109
Spontaneous calcium changes in
   isolated cardiac myocytes
   324–6
Squid axons   389, 393
Stimulus-contraction coupling in
   non-muscle cells   350
Streptomyces aureofaciens   363
Super video graphics adapter (sVGA)
   graphics   196

Supravital fluorochroming   16
Surface markers   307
Surface proteins   307
Synchronization   271

Tandem scanning microscope   232
TARDIS software   176, 180
Television standards   272–3
Television technology   216
Thapsigargin (THG)   184
Thermal noise   265–6
Thresholding method   178, 217
Thyrotropin-releasing hormone
   (TRH)   165, 181, 184, 188–91,
   201
TIGA   221
TMA-DPH   422, 424
TMRE   154, 157
TMRM   154
TPA   380, 382, 383, 384, 385, 386
Trans-Golgi network (TGN)   108
Transgenic plants   78
Transmembrane potentials   304–5
Triton X-100   110, 117
Trypaflavin   15
Trypanosomes   15, 16
Tubercle bacilli   18
Tungsten filament lamps   396
Tungsten halogen lamps   6
Two-photon confocal NAD(P)H
   imaging   51

V. fischeri   75
Vaccinia   73
Vacuole, pH measurements   374
Vacuum photocathode cameras   400
Vargula hilgendorfii   71
Vargula luciferase   72
Vibrio   70
Video cameras   277
   intensified   277–8
Video-enhanced light microscopy   166
Video graphics adapter (VGA)
   220–1
Video image-acquisition products
   264
Video-rate cameras   272
Video technology, concepts of   272–3
Vital fluorochroming   16

Wavelength selection   173–4, 206,
   208–10
Wavelength switching   210, 211

Xenon lamps   6
Xenopus   415
Xenorhabdus   70
XGA   221

ZAPPER   314, 316
Ziehl–Neelsen procedure   15, 18